Lecture Notes in Computer Science 5162

Commenced Publication in 1973
Founding and Former Series Editors:
Gerhard Goos, Juris Hartmanis, and Jan van Leeuwen

T0190201

Edward Ochmański   Jerzy Tyszkiewicz (Eds.)

Mathematical Foundations of Computer Science 2008

33rd International Symposium, MFCS 2008
Toruń, Poland, August 25-29, 2008
Proceedings

 Springer

Volume Editors

Edward Ochmański
Nicolaus Copernicus University
87-100 Toruń, Poland
and
Institute of Computer Science PAS
01-237 Warsaw
E-mail: edoch@mat.uni.torun.pl

Jerzy Tyszkiewicz
Warsaw University
02-097 Warsaw Poland
E-mail: jty@mimuw.edu.pl

Library of Congress Control Number: 2008932590

CR Subject Classification (1998): F.1, F.2, F.3, F.4, G.2, E.1

LNCS Sublibrary: SL 1 – Theoretical Computer Science and General Issues

ISSN   0302-9743
ISBN-10   3-540-85237-9 Springer Berlin Heidelberg New York
ISBN-13   978-3-540-85237-7 Springer Berlin Heidelberg New York

Springer is a part of Springer Science+Business Media

springer.com

© Springer-Verlag Berlin Heidelberg 2008
Printed in Germany

Typesetting: Camera-ready by author, data conversion by Scientific Publishing Services, Chennai, India
Printed on acid-free paper   SPIN: 12463121   06/3180   5 4 3 2 1 0

Preface

The symposia on Mathematical Foundations of Computer Science, generally known under the acronym MFCS, have been organized alternately in Poland, the Czech Republic and Slovakia since 1972. They have a well-established tradition and international recognition as an event bringing together researchers in all branches of theoretical computer science. The previous meetings took place in Jabłonna 1972, Štrbské Pleso 1973; Jadwisin 1974, Mariánske Lázně, 1975, Gdańsk 1976, Tatranská Lomnica 1977, Zakopane 1978, Olomouc 1979, Rydzyna 1980, Štrbské Pleso 1981, Prague 1984, Bratislava 1986, Karlovy Vary 1988, Porąbka-Kozubnik 1989, Banská Bystrica 1990, Kazimierz Dolny 1991, Prague 1992, Gdańsk 1993, Košice 1994, Prague 1995, Kraków 1996, Bratislava 1997, Brno 1998, Szklarska Poręba 1999, Bratislava 2000, Mariánske Lázně 2001, Otwock 2002, Bratislava 2003, Prague 2004, Gdańsk 2005, Stará Lesná 2006 and Český Krumlov 2007.

The 33rd Symposium on Mathematical Foundations of Computer Science was organized during August 25–29, 2008 by the Faculty of Mathematics and Computer Science of Nicolaus Copernicus University (Uniwersytet Mikołaja Kopernika, UMK) in Toruń, a medieval Polish town, the birthplace of Nicolaus Copernicus. We gratefully acknowledge the support received from UMK.

This volume contains 5 invited and 45 contributed papers, which were presented at the Symposium. We would like to thank the authors of the invited papers; they accepted our invitations and delivered lectures, sharing with us their insights on their research areas. It should be stressed that all of the invited papers are extensive full versions. The 45 accepted papers were selected from 119 submissions. We thank the authors for their contributions to the scientific program of the symposium. The papers were reviewed by at least three members of the Program Committee (PC) each, with the assistance of external reviewers. We would like to thank the members of the PC for their excellent work, which contributed to the quality of the meeting, and for the friendly atmosphere of our discussions. We are also very grateful to all external reviewers that supported us with their expertise. The work of the PC was carried out using the EasyChair system, and we would like to take the opportunity to acknowledge this contribution. We thank Springer for the professional cooperation in printing this volume and all participants of the conference.

The greatest thanks are due to Barbara Klunder and her organizing team. Without them this meeting could not have taken place.

June 2008

Edward Ochmański
Jerzy Tyszkiewicz

Preface

The symposia on Mathematical Foundations of Computer Science, generally known under the acronym MFCS, have been organized alternately in Poland, the Czech Republic and Slovakia since 1972. They have a well-established tradition and international recognition as an event bringing together researchers in all branches of theoretical computer science. The previous meetings took place in Jabłonna 1972, Štrbské Pleso 1973; Jadwisin 1974, Mariánske Lázně, 1975, Gdańsk 1976, Tatranská Lomnica 1977, Zakopane 1978, Olomouc 1979, Rydzyna 1980, Štrbské Pleso 1981, Prague 1984, Bratislava 1986, Karlovy Vary 1988, Porąbka-Kozubnik 1989, Banská Bystrica 1990, Kazimierz Dolny 1991, Prague 1992, Gdańsk 1993, Košice 1994, Prague 1995, Kraków 1996, Bratislava 1997, Brno 1998, Szklarska Poręba 1999, Bratislava 2000, Mariánske Lázně 2001, Otwock 2002, Bratislava 2003, Prague 2004, Gdańsk 2005, Stará Lesná 2006 and Český Krumlov 2007.

The 33rd Symposium on Mathematical Foundations of Computer Science was organized during August 25–29, 2008 by the Faculty of Mathematics and Computer Science of Nicolaus Copernicus University (Uniwersytet Mikołaja Kopernika, UMK) in Toruń, a medieval Polish town, the birthplace of Nicolaus Copernicus. We gratefully acknowledge the support received from UMK.

This volume contains 5 invited and 45 contributed papers, which were presented at the Symposium. We would like to thank the authors of the invited papers; they accepted our invitations and delivered lectures, sharing with us their insights on their research areas. It should be stressed that all of the invited papers are extensive full versions. The 45 accepted papers were selected from 119 submissions. We thank the authors for their contributions to the scientific program of the symposium. The papers were reviewed by at least three members of the Program Committee (PC) each, with the assistance of external reviewers. We would like to thank the members of the PC for their excellent work, which contributed to the quality of the meeting, and for the friendly atmosphere of our discussions. We are also very grateful to all external reviewers that supported us with their expertise. The work of the PC was carried out using the EasyChair system, and we would like to take the opportunity to acknowledge this contribution. We thank Springer for the professional cooperation in printing this volume and all participants of the conference.

The greatest thanks are due to Barbara Klunder and her organizing team. Without them this meeting could not have taken place.

June 2008

Edward Ochmański
Jerzy Tyszkiewicz

Organization

Invited Speakers

Ursula Goltz (Braunschweig, Germany)
Yuri Gurevich (Redmond, USA)
Rastislav Královič (Bratislava, Slovakia)
Jean-Éric Pin (Paris, France)
Piotr Sankowski (Zurich, Switzerland and Warsaw, Poland)

Program Committee

Piotr Bała (Toruń, Poland)
Alberto Bertoni (Milan, Italy)
Viviana Bono (Turin, Italy)
Véronique Bruyère (Mons-Hainaut, Belgium)
Anuj Dawar (Cambridge, UK)
Manfred Droste (Leipzig, Germany)
Paul Gastin (Cachan, France)
Giovanna Guaiana (Rouen, France)
Maciej Koutny (Newcastle, UK)
Luděk Kučera (Prague, Czech Republic)
Kamal Lodaya (Chennai, India)
Klaus Meer (Cottbus, Germany)
Hung Son Nguyen (Warsaw, Poland)
Damian Niwiński (Warsaw, Poland)
Edward Ochmański (Toruń, Poland) – Co-chair
Wojciech Plandowski (Warsaw, Poland)
Yves Roos (Lille, France)
Branislav Rovan (Bratislava, Slovakia)
Marian Srebrny (Warsaw, Poland)
Colin Stirling (Edinburgh, UK)
Andrzej Tarlecki (Warsaw, Poland)
Wolfgang Thomas (Aachen, Germany)
Jerzy Tyszkiewicz (Warsaw, Poland) – Co-chair
Paul Vitanyi (Amsterdam, The Netherlands)
Wiesław Zielonka (Paris, France)

Organizing Committee

Barbara Klunder (Chair)
Kamila Barylska

Andrzej Kurpiel
Łukasz Mikulski
Robert Mroczkowski
Marcin Piątkowski

External Referees

Marcella Anselmo
Sarmad Abbasi
Juergen Albert
Julien Allali
Eric Allender
Jean-Paul Allouche
Yves Andre
Marcella Anselmo
N.R. Aravind
Pierre Arnoux
Albert Atserias
Franz Aurenhammer
Bruno Beaufils
Chantal Berline
Rana Barua
Marek Bednarczyk
Valerie Berthe
Dietmar Berwanger
Somenath Biswas
Johannes Blömer
Paolo Boldi
Benedikt Bollig
Andrzej M. Borzyszkowski
Thomas Brihaye
Danilo Bruschi
Peter Bürgisser
Thierry Cachat
Jin-Yi Cai
Anne-Cecile Caron
Balder ten Cate
Davide Cavagnino
Pavel Chalmoviansky
Jeremie Chalopin
Thomas Chatain
Krishnendu Chatterjee
Hubie Chen
Bogdan Chlebus
Christian Choffrut

Krzysztof Choromański
Giorgos Christodoulou
Thomas Colcombet
Matteo Comin
Nadia Creignou
Maxime Crochemore
Felipe Cucker
Jerzy Cytowski
Artur Czumaj
Robert Dąbrowski
Philippe Darondeau
Samir Datta
Jean-Charles Delvenne
Raymond Devillers
Mariangiola Dezani
Stefan Dobrev
Jean-Philippe Dubernard
Lech Duraj
Bruno Durand
Roman Durikovic
Christoph Dürr
Michael Elkin
Bruno Escoffier
Paul Ezhilchelvan
Ina Fichtner
Camillo Fiorentini
Diana Fischer
Tom Friedetzky
Christiane Frougny
Marco Gaboardi
Peter Gacs
Anna Gambin
Pietro Di Gianantonio
Hugo Gimbert
Elio Giovannetti
Mathieu Giraud
Andreas Goerdt
Massimiliano Goldwurm

Hans-Joachim Goltz
Serge Grigorieff
Dmitry Grigoryev
Giuliano Grossi
Dimitar Guelev
Michel Habib
Vesa Halava
Magnus Halldorsson
Michael Harrison
Denis Hirschfeldt
Christopher Homan
Peter Hoyer
Adam Jakubowski
Florent Jacquemard
Marek R. Janicki
Emmanuel Jeandel
Jan Jezabek
Joanna Jółkowska
Marcin Jurdzinski
Juhani Karhumäki
Jarkko Kari
Tomi Karki
Marek Karpinski
Tomasz Kazana
Ralf Koetter
Victor Khomenko
Daniel Kirsten
Leszek Kolodziejczyk
Eryk Kopczynski
Łukasz Kowalik
Rastislav Královič
Richard Královič
Pavel Krcal
Kamala Krithivasan
Andrei Krokhin
Petr Kurka
Laks V.S. Lakshmanan
Kim Skak Larsen

Sławomir Lasota
Michel Latteux
Aurélien Lemay
Stephane Lengrand
Giacomo Lenzi
Francesco Librizzi
Katrina Ligett
Christof Löding
Violetta Lonati
Krzysztof Loryś
Etienne Lozes
Alesssandro De Luca
Adam Malinowski
Edita Mácajová
Meena Mahajan
Jean Mairesse
Marcin Malawski
Andreas Maletti
Guillaume Malod
Amaldev Manuel
Victor Marek
Oliver Matz
Jacques Mazoyer
John McCabe-Dansted
Ingmar Meinecke
Nouredine Melab
Carlo Mereghetti
Wolfgang Merkle
Roland Meyer
Antoine Meyer
Martin Middendorf
Angelo Montanari
Ben Moszkowski
Marcin Mucha
Rahul Muthu
Slava Nesterov
Joachim Niehren
Anna Niewiarowska

Yahav Nussbaum
Kathryn Nyman
Jan Obdržálek
Michele Pagani
Beatrice Palano
Luca Paolini
Dana Pardubska
Paweł Parys
Jakub Pawlewicz
Marcin Pawłowski
Wiesław Pawłowski
Marcin Peczarski
Dominique Perrin
Giuseppe Persiano
Claudine Picaronny
Giovanni Pighizzini
Nir Piterman
Alexander Rabinovich
Roberto Radicioni
George Rahonis
Rajiv Raman
R. Ramanujam
Raghavendra Rao
ean-François Raskin
Pierre-Alain Reynier
Romeo Rizzi
Giovanna Rosone
Zdenek Ryjacek
Wojciech Rytter
Patrice Seebold
Piotr Sankowski
Miklos Santha
Ingo Schiermeyer
Philippe Schnoebelen
Tomasz Schreiber
Oded Schwartz
Stefan Schwoon
Marinella Sciortino

Martin Skoviera
Srikanth Srinivasan
Martin Stanek
Slawomir Staworko
Daniel Stefankovic
Michal Strojnowski
C.R. Subramanian
S.P. Suresh
Andrzej Szałas
Maciej Szreter
Alain Terlutte
Jean-Marc Talbot
German Tischler
Szymon Torunczyk
Angelo Troina
Marco Trubian
Jean-Stephane Varre
Ugo Vaccaro
Kasturi Varadarajan
Nikolai Vereshchagin
Mirko Viroli
Heribert Vollmer
Imrich Vrto
Gerhard Woeginger
Damian Wójtowicz
Bożena Woźna
Jean-Baptiste Yunes
Jeff Yan
Stanislav Zak
Marc Zeitoun
Alexander Zvonkin
Franciszek Seredynski
Olivier Serre
Pedro V. Silva
Sunil Simon
Luc Segoufin

Organizing Institutions

 Nicolaus Copernicus University, Toruń, Poland

 Faculty of Mathematics and Computer Science

 Department of Formal Languages and Concurrency

Table of Contents

One Useful Logic
That Defines Its Own Truth

Andreas Blass[1] and Yuri Gurevich[2]

[1] Math Dept, University of Michigan, Ann Arbor, MI 48109, USA
[2] Microsoft Research, One Microsoft Way, Redmond, WA 98052, USA

Abstract. Existential fixed point logic (EFPL) is a natural fit for some applications, and the purpose of this talk is to attract attention to EFPL. The logic is also interesting in its own right as it has attractive properties. One of those properties is rather unusual: truth of formulas can be defined (given appropriate syntactic apparatus) in the logic. We mentioned that property elsewhere, and we use this opportunity to provide the proof.

> *Believe those who are seeking the truth. Doubt those who find it.*
>
> *—André Gide*

1 Introduction

First-order logic lacks induction but first-order formulas can be used to define the steps of an induction. Consider a first-order (also called elementary) formula $\varphi(P, x_1, \ldots, x_j)$ where a j-ary relation P has only positive occurrences. The formula may contain additional individual variables, relation symbols, and function symbols. In every structure whose vocabulary is that of φ minus the symbol P and where the additional individual variables are assigned particular values, we have an operator

$$\Gamma(P) = \{\bar{x} : \ \varphi(P, \bar{x})\}.$$

A relation P is a *closed point* of Γ if $\Gamma(P) \subseteq P$, and P is a *fixed point* of Γ if $\Gamma(P) = P$. Since P has only positive occurrences in $\varphi(P, \bar{x})$, the operator is monotone: if $P \subseteq Q$ then $\Gamma(P) \subseteq \Gamma(Q)$. By the Knaster-Tarski Theorem, Γ has a least fixed point P^* which is also the least closed point of Γ [20].

There is a standard way to construct P^* from the empty set by iterating the operator Γ. Let $P^0 = \emptyset$, $P^{\alpha+1} = \Gamma(P^\alpha)$ and $P^\lambda = \bigcup_{\alpha < \lambda} P^\alpha$ if λ is a limit ordinal. There is an ordinal α such that $P^\alpha = P^{\alpha+1} = P^*$. The least such ordinal α is the *closure ordinal* of the iteration. Such elementary inductions have been extensively studied in logic [17,1].

Notice that we have not really used the fact that $\varphi(P, \bar{x})$ is first-order. One property of $\varphi(P, \bar{x})$ that we used was that $\varphi(P, \bar{x})$ is monotone in P, that is that, in every structure of the appropriate vocabulary with fixed values for the additional individual variables, Γ is a monotone operator. $\varphi(P, \bar{x})$ could be e.g. a second-order formula monotone in P.

E. Ochmański and J. Tyszkiewicz (Eds.): MFCS 2008, LNCS 5162, pp. 1–15, 2008.
© Springer-Verlag Berlin Heidelberg 2008

The least fixed point P^* can be denoted $\mathrm{LFP}_{P,\bar{x}}\varphi(P,\bar{x})$ and viewed as a j-ary relation, so that $[\mathrm{LFP}_{P,\bar{x}}\varphi(P,\bar{x})](y_1,\ldots,y_j)$ functions semantically as a formula. This observation gives rise to an idea to use LFP as a new formula constructor, in addition to propositional connectives and quantifiers. Aho and Ullman [2] indeed suggested to enrich first-order logic with the LFP constructor. The new logic became known as FOL+LFP.

Model checking is polynomial time for any FOL+LFP formula ψ. In other words, it can be checked in time polynomial in the size of a finite structure X of the vocabulary of ψ whether X with some values for the free individual variables of ψ is a model of ψ. Immerman [16] and Vardi [21] proved that, over ordered finite structures, the converse is true: every property that model checks in polynomial time is expressible in FOL+LFP. In that sense, FOL+LFP captures polynomial time.

Existential fixed point logic (EFPL) is essentially an extension of the existential fragment of first-order logic with the LFP construct. It does not have the universal quantifier and lacks means to simulate universal quantification; see the definition of EFPL in the next section. As far as we know, it was first introduced — in a different guise — by Chandra and Harel [10] in the context of database theory where vocabularies are relational, that is, consist of relation symbols and constants and do not have function symbols of positive arity. Chandra and Harel observed that relational EFPL is equi-expressive with Datalog, a popular database query language.

Existential fixed point logic (EFPL) was further developed by the present authors in [7]; see Section 3. The motivation came from program verification. We noticed that EFPL was appropriate for formulating pre- and post-conditions in Hoare's logic of asserted programs [15]. In particular, the heavy expressivity hypothesis needed for Cook's completeness theorem [12] in the context of first-order logic is automatically satisfied in the context of EFPL.

More recent developments include a deductive system for EFPL introduced by Compton [11] and a normal form for EFPL formulas discovered by Grohe [13], who also studied connections between EFPL and other logics. One of the present authors found connections with topos theory and showed that these connections imply some of the other, previously known, nice properties of EFPL [6,5]. The other of the present authors, together with Neeman, applied a logic equivalent to EFPL, called liberal Datalog, to develop a powerful authorization language [14]; the equivalence between liberal Datalog and EFPL is shown in detail in [9].

In this note, we recall the definition and known properties of EFPL, and then we prove that the truth definition of EFPL formulas can be given in EFPL.

Remark 1. Nikolaj Bjørner [4] observed that writing a truth definition for EFPL in EFPL is related to writing an interpreter for EFPL in EFPL. This is indeed the case when one views EFPL as a programming language rather than as a logic. But this interesting issue is outside the scope of this paper, and will have to be addressed elsewhere.

2   Existential Fixed-Point Logic: Definition

As indicated in the introduction, existential fixed-point logic differs from first-order logic in two respects, the absence of the universal quantifier and the presence of the least-fixed-point operator. Both of these deserve some clarification.

First we define existential logic EL. Notice that mere removal of the universal quantifier $\forall$ has no real effect on first-order logic, since $\forall x\, \varphi$ can be expressed as $\neg\exists x\, \neg\varphi$. To correctly define the existential fragment of first-order logic, one must prevent such surreptitious reintroduction of the universal quantifier. A traditional way to do that is to insist that all formulas have the prenex existential form $\exists x_1 \ldots \exists x_n \varphi(x_1, \ldots, x_n)$ where φ is quantifier-free.

But there is an alternative and more convenient form of the existential fragment proposed in [7]: Allow as propositional connectives only conjunction, disjunction, and negation; use only the existential quantifier; and apply negation only to atomic formulas. It is easy to see that every formula in this alternative fragment is equivalent to one in prenex existential form, and the other way round.

With an eye on the forthcoming introduction of recursion, we stipulate that all relation symbols are divided into two categories: *negatable* and *positive*. And we restrict further the use of negation in the alternative existential fragment of first-order logic: negation can be applied only to atomic formulas with negatable relation symbols. The resulting fragment of first-order logic will be called *existential logic* and denoted EL.

Now we extend existential logic by adding a new formula constructor. As usual, formulas are built by induction from atomic formulas by means of formula constructors. In the case of EFPL, the formula constructors are those of existential logic — the three propositional connectives and the existential quantifier — and one additional LET-THEN constructor that is used to construct induction assertions. We explain how the new constructor works.

Let $\mathcal{F}$ be the collection of formulas constructed so far. A *logic rule* has the form $P(x_1, \ldots, x_j) \leftarrow \delta(P, x_1, \ldots, x_j)$ where P is a positive relation symbol of arity j, the x_i's are distinct variables and δ is any formula in $\mathcal{F}$. We wrote δ as $\delta(P, x_1, \ldots, x_j)$ to emphasize that it is allowed to contain the relation symbol P and the individual variables $x_1, \ldots, x_j$, but it may also contain additional individual variables, relation symbols, and function symbols. P is the *head symbol* of the rule and δ is its *body*. Note that the arrow $\leftarrow$ in a logic rule is not the (reverse) implication connective but a special symbol whose only use, in our syntax, is in forming logic rules. A *logic program* is a finite collection of logic rules. (To write a program as text, one needs to order its rules, but the choice of ordering will never matter.) To be compatible with [7], we require that different rules have different head symbols; we could remove this restriction. If Π is a program and φ is a formula in $\mathcal{F}$ then

$$\text{LET } \Pi \text{ THEN } \varphi$$

is an EFPL formula, an *induction assertion*. If $P(x_1, \ldots, x_j) \leftarrow \delta$ is a rule in Π then all occurrences of the variables $x_1, \ldots, x_j$ in the rule are bound occurrences in the induction assertion. And P is a bound relation variable in the induction assertion.

In general, an occurrence of an individual variable v in a formula ψ is bound if it belongs to a subformula of the form $\exists v\, \alpha$ or to a rule of the form $P(\ldots, v, \ldots) \leftarrow \delta$; otherwise the occurrence is free. The free individual variables of ψ are those with free occurrences in ψ. An occurrence of relation symbol P in ψ is bound if it belongs to subformula LET Π THEN φ of ψ and P is a head symbol of Π; otherwise the occurrence is free. The vocabulary of ψ consists of all the function symbols in ψ and all relation symbols with free occurrences in ψ.

It remains to define the semantics of the induction assertion $\psi =$ LET Π THEN φ. To simplify the exposition, we presume that the program Π consists of two rules, $P(x_1, \ldots, x_j) \leftarrow \alpha$ and $Q(y_1, \ldots, y_k) \leftarrow \beta$. In every structure of the vocabulary of ψ with fixed values for the free individual variables of ψ, the program gives rise to an operator

$$\Gamma(P, Q) \leftarrow (\{\bar{x} : \ \alpha\}, \{\bar{y} : \ \beta\}).$$

Since P and Q are positive relation symbols, Γ is monotone and thus has a least fixed point (P^*, Q^*). To evaluate ψ, evaluate φ using P^* and Q^* as the values of relations P and Q.

3   EFPL: Some Properties

We describe some properties of EFPL. The default reference is [7].

Capturing Polynomial Time

EFPL captures polynomial time computability over structures of the form $\{0, 1, \ldots, n\}$ with (at least) the successor relation and names for the endpoints. In contrast to the corresponding result for FOL+LFP mentioned above, we use the successor relation here rather than the ordering relation $<$. In fact, both proofs depend on the successor relation rather than the order, but in FOL one can define successor in terms of order (but not vice versa), whereas in EFPL one can define order in terms of successor (but not vice versa).

Validity Is r.e. Complete

The set of logically valid EFPL formulas is recursively enumerable (in short r.e.). Furthermore, every r.e. set reduces, by means of a recursive function, to the set of valid EFPL formulas. Thus the set of valid EFPL formulas is a complete r.e. set.

Satisfiability Is r.e. Complete

The set of satisfiable EFPL formulas is a complete r.e. set.

Finite Validity Is co-r.e. Complete

The set of EFPL formulas that hold in all finite structures is a complete co-r.e. set. In other words, the set of EFPL formulas ψ such that ψ fails in some finite structure is a complete r.e. set.

Finite Model Property

When an EFPL formula ψ is satisfied in a structure X, this fact depends on only a finite part of the structure X. More precisely, there is a finite subset D of the elements of X such that ψ is satisfied in every structure X' of the vocabulary of X that coincides with X on D. Note that X' can be always chosen to be finite. If we allow basic functions of a structure to be partial, then the property in question can be formulated in a particularly simple way: If an EFPL formula is satisfied in a structure then it is satisfied in a finite substructure.

No Transfinite Induction Is Needed

The closure ordinal of any monotone induction

$$P \mapsto \{\bar{x} : \varphi(P, \bar{x})\},$$

where φ is in EFPL is at most ω, the first infinite ordinal. The definition of the closure ordinal generalizes in a straightforward way to simultaneous monotone induction. The closure ordinal of the induction given by any logic program is at most ω.

Truth Is Preserved by Homomorphisms

Truth of EFPL formulas is preserved by homomorphisms. Here a homomorphism is a function h from one structure to another such that

- h commutes with (the interpretations of) function symbols,
- $P(a_1, \ldots, a_j)$ implies $P(ha_1, \ldots, ha_j)$
  for every positive relation symbol P of any arity j, and
- $P(a_1, \ldots, a_j)$ if and only if $P(ha_1, \ldots, ha_j)$
  for every negatable relation symbol P of any arity j.

EFPL ∩ FOL ⊆ EL

If an EFPL formula φ is expressible in first-order logic then φ is equivalent to an existential formula. Only a limited form of this result survives in finite model theory. If an EFPL formula φ without function symbols and without negations is equivalent, on finite structures, to a first-order formula, then φ is equivalent, on finite structures, to an existential formula without negations [3,18]. This fails even if φ has no function symbols and only the equality relation is negatable [3, Section 10].

4 Prerequisites for Truth

Our objective in the rest of the article is to show that EFPL can formalize its own truth definition. That is, we shall define, in EFPL with suitable vocabulary, truth of EFPL sentences (that is formulas with no free variables) of the same vocabulary. We use the term predicate to mean a relation symbol or a relation depending on the context.

Since sentences are built from subformulas that may have free variables, we shall actually define the slightly more general concept of satisfaction of formulas by assignments of values to the free variables. The need to define the more general notion of satisfaction of formulas in order to obtain truth for sentences is familiar from first-order logic.[1] A new complication, of the same general nature, arises in EFPL. The bound predicates of a sentence φ are free in some subformulas of φ. We should define satisfaction of φ in a structure whose vocabulary does not include those predicates. But the definition will pass through subformulas of φ whose satisfaction will depend on the interpretations of those predicates. As a result, we need to define satisfaction of φ in a context that includes not only a structure (for the vocabulary of φ) and an assignment of values to the free variables of φ (as in FOL) but also the logic rules that provide the meaning of all other predicates that occur in φ — or that occur in the bodies of those rules.

Let Υ be a vocabulary and X a structure of vocabulary Υ. Any predicate that does not occur in Υ will be called an *extra predicate*. We shall define satisfaction in X for Υ-formulas. Requirements will be imposed shortly on Υ and X, but for now Υ is just some vocabulary and X some Υ-structure. We intend to define, in EFPL, a ternary predicate Sat such that, when

- the value of its first argument is a formula φ, of vocabulary Υ plus (possibly) some extra predicates,
- the value of its second argument is a logic program Π whose head predicates include all extra predicates that occur in φ or Π, and
- the value of its third argument is an assignment s of elements of X to (at least) all individual variables that are free in φ or in Π,

then the truth value of $\mathrm{Sat}(\varphi, \Pi, s)$ in X is the same as the truth value, in X, of φ with values for its variables given by s and with the extra predicates interpreted by the least fixed point of (the monotone operator defined by) Π.

Furthermore, we do not intend to use any clever tricks in our definition of Sat. It will be a formalization of the explanation given above (and in [7]) of the meaning of EFPL formulas. The point of this work is to show that this formalization can be carried out in EFPL itself.

For all this to make sense, the structure X must contain the formulas φ of EFPL, the logic programs Π, and the assignments s. Furthermore, the vocabulary must be adequate to express the basic syntactic properties of formulas and to allow basic constructions of assignments, rules, and programs. We do not, however, wish to specify the exact syntactic nature of formulas — for example, are they sequences of symbols, or are they parse trees, or are they Gödel numbers? Our work is independent of such details. So we shall merely assume that certain notions (e.g., the operation of forming the conjunction of two formulas)

---

[1] A few authors, notably Shoenfield [19], define truth directly. To do so, they expand the vocabulary by adding constants for all elements of the structure under consideration, and instead of assigning values to variables they substitute constants for variables. We could have used this approach for EFPL, but we chose to parallel the more widely used approach in FOL, via satisfaction.

are expressible; the details of how they are expressed (and which notions are primitive and which are derived) are irrelevant.[2]

In the rest of this section, we list what we require of our vocabulary Υ and structure X, occasionally adding some comments about the reasons for particular requirements.

Υ should be finite. The reason is that the definition of satisfaction must, in the clauses for atomic formulas, use all the relation and function symbols of Υ.

The equality predicate should be negatable. The reason is that the notion of EFPL formula requires some things to be distinct, for example the variables in the head of a rule and the head symbols of different rules in a program.

X should contain a copy $\mathbb{N}$ of the natural numbers, and Υ should have a constant symbol for 0 and a unary function symbol S for successor. $\mathbb{N}$ itself, as a unary relation, is definable:

$$\mathbb{N}(x) :\equiv \text{LET } N(z) \leftarrow z = 0 \ \lor \ \exists y\, (N(y) \land z = S(y)) \text{ THEN } N(x).$$

We could also define addition and multiplication as ternary relations, and the ordering, and similarly for other primitive recursive functions and relations.

We need $\mathbb{N}$ primarily to index elements of lists, for example the list of terms that serves as the argument of a relation or function symbol. Since Υ is finite, we could handle the argument lists of its own relation and function symbols in an ad hoc manner, without a general notion of natural number or of list. But EFPL imposes no bound on the arities of the head symbols of logic rules, so atomic formulas can involve arbitrarily long argument lists, and natural numbers are needed for treating these.

Although EFPL does not allow universal quantification in general, it can simulate universal quantification over finite initial segments of $\mathbb{N}$, as shown by the following lemma from [7].

Lemma 2. *For any EFPL formula $\varphi(x)$, there is an EFPL formula $\psi(y)$ equivalent, for all $y \in \mathbb{N}$, to $(\forall x < y)\, \varphi(x)$.*

Proof. The most natural choice of $\psi(y)$ describes a search from 0 up to y:

$$\text{LET } K(x) \leftarrow x = 0 \ \lor \ \exists w\, \big(x = S(w) \ \land \ K(w) \ \land \ \varphi(w)\big) \text{ THEN } K(y). \qquad \square$$

Convention 3. Consider the definition of $\mathbb{N}$ exhibited above, and notice that its essential content is contained in the rule

$$N(z) \leftarrow z = 0 \ \lor \ \exists y\, (N(y) \land z = S(y)),$$

which makes the bound predicate symbol N denote the set of natural numbers. The rest of the definition,

$$\mathbb{N}(x) :\equiv \text{LET } \ldots \text{ THEN } N(x),$$

---

[2] We shall occasionally indicate how certain notions can be defined from others in EFPL. Those indications can help to reduce the assumptions needed about Υ.

merely transfers this denotation to the defined notation $\mathbb{N}$. Instead of introducing a bound predicate variable N to, in effect, duplicate the desired predicate $\mathbb{N}$, we could convey the same information by writing

$$\mathbb{N}(z) :\leftarrow z = 0 \vee \exists y \, (\mathbb{N}(y) \wedge z = S(y)).$$

Although this is not an EFPL formula, we adopt the convention that it is to serve as an abbreviation of the definition of $\mathbb{N}$ displayed earlier. In general, when we write a rule with a colon before the $\leftarrow$, it is to be interpreted as defining a formula. Thus,

$$\mathbb{P}(\bar{x}) :\leftarrow \delta(\mathbb{P}, \bar{x})$$

means that $\mathbb{P}(\bar{x})$ is defined as the formula

$$\text{LET } Q(\bar{z}) \leftarrow \delta(Q, \bar{z}) \text{ THEN } Q(\bar{x}).$$

Convention 4. Later, we shall also need to deal with definitions of this sort in which the body δ is a disjunction of many subformulas. For example, our ultimate goal, the definition of Sat, will have several disjuncts, covering the different syntactic constructs of EFPL. In such cases, it is convenient to present one disjunct (or a small number of them) at a time. Thus, for a small example, the definition of $\mathbb{N}$ above could be broken into two parts:

$$\mathbb{N}(z); \leftarrow z = 0$$
$$\mathbb{N}(z); \leftarrow \exists y \, (\mathbb{N}(y) \wedge z = S(y)).$$

We use a semicolon before $\leftarrow$ (instead of a colon) to indicate that the full definition involves more disjuncts. (This use of a semicolon as a partial colon is suggested by the word "semicolon.") In general, if we write several semicolon definitions $\mathbb{P}(\bar{x}); \leftarrow \delta_i$ for the same $P(\bar{x})$, then they are to be understood as meaning $\mathbb{P}(\bar{x}) :\leftarrow \bigvee_i \delta_i$.

Returning to the requirements on X and Υ, we require X to contain the variables and the assignments. The latter are finite partial functions from the variables into (the universe of) X. Υ should define a predicate Vbl for the set of variables, a constant symbol $\varnothing$ for the empty assignment, and a ternary function symbol Modify for the function defined as follows: Given an assignment s, a variable v, and an element a of X, Modify(s, v, a) is the assignment t that sends v to a and otherwise agrees with s (whether or not a is in the domain of s).

Convention 5. Here and in what follows, we use the terminology "Υ should define a predicate for" some relation on X to mean that there should be an EFPL formula in vocabulary Υ whose truth set in X is the desired relation. Of course, the easiest way to arrange this would be for the given relation to be one of the basic relations of X, so that the required EFPL formula would be atomic. But it will never matter whether the formula is atomic or not.

Similarly, when we ask that Υ should have certain function symbols, we could weaken that to require only some terms, possibly involving nesting of function symbols, and our proofs would be unchanged.

We also need to express "s is an assignment," "v is in the domain of s," and "$s(v) = a$," but we need not assume these separately, as they are definable from $\varnothing$ and Modify. They are given, using our conventions above and the familiar convention of (existentially) quantifying several variables at once, by

$$\text{Assgt}(s); \leftarrow s = \varnothing$$
$$\text{Assgt}(s); \leftarrow \exists t, v, a \, (\text{Assgt}(t) \wedge \text{Vbl}(v) \wedge s = \text{Modify}(t, v, a))$$
$$v \text{ inDom } s :\leftarrow \exists t, a \, (s = \text{Modify}(t, v, a)).$$
$$s(v) = a :\leftarrow \exists t \, (s = \text{Modify}(t, v, a))$$

Note that here $s(v) = a$ is defined as a ternary relation, not as an instance of equality.

We shall also need to have, among the elements of X, the relation and function symbols of Υ as well as the extra predicates available as head symbols of rules. Each relation symbol P or function symbol f of Υ, should be denoted by a closed term $\dot{P}$ or $\dot{f}$ of Υ. (We remain flexible as to what the symbols of Υ should be. For example, they could be Gödel numbers, and then their names $\dot{P}$ and $\dot{f}$ could be terms of the form $SS\ldots S(0)$. But there are many other options, and all will work. Note, however, that we cannot take all the $\dot{f}$'s to be simple constant symbols, as they would then be among the f's, and there would not be enough room in a finite Υ for all of these names to have names.)

The extra predicates available as head symbols of rules should have specified numbers of arguments. That is, there should be an Υ-definable predicate Arity such that $\text{Arity}(a, n)$ holds in X (for elements $a, n \in X$) if and only if a is one of these head predicate symbols and $n \in \mathbb{N}$ is the number of its argument places.

As mentioned earlier, we shall need lists, so we require that X contain all lists (i.e., finite sequences) of elements of X. The vocabulary Υ should contain at least the constant Nil, denoting the empty list, and the binary function symbol Append, for the function that lengthens a list by adding one element at the end. Thus, for example,

$$\langle a, b, c \rangle = \text{Append}(\text{Append}(\text{Append}(\text{Nil}, a), b), c).$$

Other predicates and functions that we shall need for dealing with lists can be defined in terms of Nil and Append.

$$\text{List}(l); \leftarrow l = \text{Nil}$$
$$\text{List}(l); \leftarrow \exists x, a \, (\text{List}(x) \wedge l = \text{Append}(x, a))$$
$$l \text{ hasLength } n; \leftarrow l = \text{Nil} \wedge n = 0$$
$$l \text{ hasLength } n; \leftarrow \exists x, a, m \, (l = \text{Append}(x, a) \wedge x \text{ hasLength } m \wedge n = S(m))$$
$$(l)_i = a; \leftarrow \exists x \, (x \text{ hasLength } i \wedge l = \text{Append}(x, a))$$
$$(l)_i = a; \leftarrow \exists x, b \, ((x)_i = a \wedge l = \text{Append}(x, b))$$
$$\text{Cat}(a, b, l); \leftarrow b = \text{Nil} \wedge l = a$$
$$\text{Cat}(a, b, l); \leftarrow \exists c, x, m \, (\text{Cat}(a, c, m) \wedge$$
$$b = \text{Append}(c, x) \wedge (l = \text{Append}(m, x)).$$

Here $(l)_i = a$, though it looks like an equation, is really a defined ternary relation, whose meaning is that a is the ith component of the list l, where we start counting with 0, and where the length of l must be at least $i + 1$ so that there is an ith term. And "Cat" alludes to "concatenation". If a, b, l are lists and $\text{Cat}(a, b, l)$ holds, then l is the concatenation $a * b$ of a and b.

We note the following consequence of Lemma 2, allowing universal quantification over the elements of a list.

Corollary 6. *For any EFPL formula $\varphi(x)$, there is an EFPL formula $\psi(y)$ that holds, when the value of y is a list, if and only if φ holds of all elements of that list. That is, $\psi(y)$ is the result of universally quantifying $\varphi(x)$ over all elements x of the list y.*

Proof. Use Lemma 2 to express

$$\exists n \left(y \text{ hasLength } n \ \wedge \ (\forall i < n) \, \exists z \, ((y)_i = z \ \wedge \ \varphi(z)) \right). \qquad \square$$

It will be convenient to write $(\forall x \in y) \, \varphi(x)$ for the formula ψ given by this corollary.

Finally, X must contain the syntactic entities relevant to EFPL, such as terms, logic rules, logic programs, and formulas. The precise nature of these entities depends on arbitrary choices of how to represent syntax. We require merely that some representation be present and that Υ be able to describe fundamental syntactic relationships.

First, Υ should have a binary function symbol Apply, used to form a compound term $f(t_1, \ldots, t_n)$ from an n-ary function symbol f and a list $\langle t_1, \ldots, t_n \rangle$ of n terms, and also used similarly to form atomic formulas $P(t_1, \ldots, t_n)$. Depending on how syntax is represented, Apply could, for example, be simply a pairing function, or it could be the operation of prepending an element to a list, or it could produce a tree from a root and its immediate subtrees, or it could be an arithmetical operation on Gödel numbers.

There should also be a unary function symbol Neg and binary function symbols Conj, Disj, Quant, and IndAsrt for the operations of negating a formula, forming conjunctions, forming disjunctions, forming existential quantifications, and forming induction assertions LET Π THEN φ. The arguments of these operations are intended to be formulas, except that the first argument of Quant is the variable being quantified and the first argument of IndAsrt is the program that goes between LET and THEN.

There should also be a binary function symbol Rule for the operation building a logic rule from its head and its body. We shall take logic programs to be (certain) lists of rules, so we do not need additional capabilities in Υ to handle these. (We could have used sets of rules instead, but then Υ would need additional capabilities.) Finally, there is a ternary relation RenameAway such that, if Π is a program and φ is a formula and RenameAway(φ, Π, φ') holds, then φ' is a formula obtained from φ by renaming the bound predicates of φ away from the head predicates of Π, so that the formula φ' is equivalent to φ, and no head predicate of Π is bound in φ'.

This completes our requirements on Υ and X. They can be summarized thus: EFPL syntax and basic combinatorial ingredients for EFPL semantics (like assignments) are available in X and expressible in EFPL in vocabulary Υ.

5   Semantics of Terms

Terms are built, as in FOL, by starting with variables and iteratively applying function symbols. The definition is formalized as follows.

$\mathrm{Term}(t); \leftarrow \mathrm{Vbl}(t)$

$\mathrm{Term}(t); \leftarrow \exists l \big(t = \mathrm{Apply}(\dot{f}, l) \wedge \mathrm{List}(l) \wedge l \text{ hasLength } \hat{n} \wedge (\forall x \in l)\mathrm{Term}(x) \big).$

Here the second line is to be repeated for each function symbol f of Υ, n is the arity of f, and $\hat{n}$ is the numeral for n, namely $SS \ldots S(0)$ with n occurrences of S. Recall that the universal quantification $\forall x \in l$ was introduced after Corollary 6 as an abbreviation of an EFPL formula. Recall also that Υ is finite, so there is no difficulty writing the appropriate line for each f.

Semantically, a term gets a value (in the given structure X) once an assignment provides values for all the variables in t. So the values of terms are given by a binary function, whose arguments are a term and an assignment. To define it recursively, we regard this binary function as a ternary relation, and we define it as follows.

$$\mathrm{Val}(t, s, a); \leftarrow \mathrm{Vbl}(t) \wedge \mathrm{Assgt}(s) \wedge s(t) = a$$
$$\mathrm{Val}(t, s, a); \leftarrow \exists l, u_0, \ldots, u_{n-1}, b_0, \ldots, b_{n-1}$$
$$\big(t = \mathrm{Apply}(\dot{f}, l) \wedge \mathrm{List}(l) \wedge l \text{ hasLength } \hat{n} \wedge \mathrm{Assgt}(s)$$
$$\wedge \bigwedge_{i<n} ((l)_i = u_i \wedge \mathrm{Val}(u_i, s, b_i)) \wedge a = f(b_1, \ldots, b_n) \big).$$

The explanatory comments after the definition of Term apply here as well.

Remark 7. In principle, we could do without the definition of Term. The definition of Val assigns values only to terms in any case. But it would do no harm if Val were defined in some extraneous cases, as long as it worked correctly for terms.

6   Semantics of Formulas

As indicated earlier, the semantics of a formula involves not only the structure X and an assignment s but also a collection Π of logic rules to determine the meaning of any extra predicates used in the formula but not bound by LET-THEN constructions in the formula. Ultimately, when we deal with Υ-formulas, there will be no such extra predicates, so Π will be irrelevant, but in the recursive construction of an Υ-formula (and in the recursive definition of its satisfaction), subformulas can occur that do use extra predicates. So we shall define Sat as a ternary predicate, where the intended meaning of $\mathrm{Sat}(\varphi, \Pi, s)$ is that the formula φ is true, in our given structure X, when the extra predicates are interpreted by the least fixed point of Π and the free variables are assigned values by s.

The definition of Sat will have numerous clauses, according to the last constructor used in building φ, so we shall make much use of the ";←" convention. This way, we can present the clauses one (or a few) at a time and insert comments and even other definitions between them.

We begin with the case of atomic formulas whose predicates are from Υ. The definition is quite analogous to the earlier definition of the values of terms.

$$\mathrm{Sat}(\varphi, \Pi, s); \leftarrow \exists l, u_0, \ldots, u_{n-1}, b_0, \ldots, b_{n-1}$$
$$\left(\varphi = \mathrm{Apply}(\dot{P}, l) \wedge \mathrm{List}(l) \wedge l \text{ hasLength } \hat{n} \wedge \mathrm{Assgt}(s)\right. \tag{1}$$
$$\left. \wedge \bigwedge_{i<n}((l)_i = u_i \wedge \mathrm{Val}(u_i, s, b_i)) \wedge P(b_1, \ldots, b_n)\right).$$

This is to be repeated for all of the (finitely many) predicates P of Υ with n being the arity of P. As before, $\hat{n}$ is the numeral for n.

The case of negated atomic formulas is almost the same; of course it is to be repeated only for negatable P.

$$\mathrm{Sat}(\varphi, \Pi, s); \leftarrow \exists l, u_0, \ldots, u_{n-1}, b_0, \ldots, b_{n-1}$$
$$\left(\varphi = \mathrm{Neg}(\mathrm{Apply}(\dot{P}, l)) \wedge \mathrm{List}(l) \wedge l \text{ hasLength } \hat{n} \wedge \mathrm{Assgt}(s)\right. \tag{2}$$
$$\left. \wedge \bigwedge_{i<n}((l)_i = u_i \wedge \mathrm{Val}(u_i, s, b_i)) \wedge \neg P(b_1, \ldots, b_n)\right).$$

Rather than continuing with the remaining atomic formulas, those that use extra predicates, let us first dispose of the remaining "easy" clauses, those not involving Π.

$$\mathrm{Sat}(\varphi, \Pi, s); \leftarrow \exists \alpha, \beta \left(\varphi = \mathrm{Conj}(\alpha, \beta) \wedge \mathrm{Sat}(\alpha, \Pi, s) \wedge \mathrm{Sat}(\beta, \Pi, s)\right)$$
$$\mathrm{Sat}(\varphi, \Pi, s); \leftarrow \exists \alpha, \beta \left(\varphi = \mathrm{Disj}(\alpha, \beta) \wedge (\mathrm{Sat}(\alpha, \Pi, s) \vee \mathrm{Sat}(\beta, \Pi, s))\right) \tag{3}$$
$$\mathrm{Sat}(\varphi, \Pi, s); \leftarrow \exists \alpha, v, a \left(\varphi = \mathrm{Quant}(v, \alpha) \wedge \mathrm{Sat}(\alpha, \Pi, \mathrm{Modify}(s, v, a))\right)$$

This completes the easier part of the definition of Sat, the part concerning just EL. To complete the definition for EFPL, we must deal carefully with programs in both of their roles — as the second argument of Sat and as a constituent of induction assertions.

This will require some preliminaries. First, we need the notion of a list with no repetitions.

$$\text{1-1-List}(l) :\equiv \exists n \left(l \text{ hasLength } n \wedge \right.$$
$$\left. (\forall i, j < n) \exists x, y ((l)_i = x \wedge (l)_j = y \wedge (i = j \vee \neg(x = y)))\right).$$

We also need a construction that amounts to applying a unary function to each element of a list, producing a new list. The situation is complicated by the fact that our unary functions are often given as binary relations. We therefore adopt the following notation. If we have defined a binary relation R, then we write R^+ for the relation defined as follows.

$$R^+(l, m) :\equiv \exists n \, (l \text{ hasLength } n \wedge m \text{ hasLength } n \wedge$$
$$(\forall i < n) \exists u, v \, ((l)_i = u \wedge (m)_i = v \wedge R(u, v))).$$

For example, let us define HS (abbreviating "head symbol") by

$$HS(r, p) :\equiv \exists y, z \, (r = Rule(Apply(p, y), z)).$$

Then when Π is a list of rules, $HS^+(\Pi, m)$ means that m is the list of their head symbols. One of the requirements for a program is that this list m be one-to-one, so there will be a clause $\exists m \, (HS^+(\Pi, m) \wedge 1\text{-}1\text{-List}(m))$ in the definition of program.

We shall also use the plus-notation with a parameter. Specifically, we think of $\text{Val}(u, s, b)$ as the graph of a function $u \mapsto b$ with s fixed, so the plus-notation makes $\text{Val}^+(\bar{u}, s, \bar{b})$ the relation between a list of terms and their values, all for the same assignment s. We refrain from writing out the definition, since it's just like the definition of R^+ above, with the extra argument s inserted into both R and R^+.

We need an improved version of the function Modify, to modify an assignment by mapping all the variables in a list l to the corresponding values in another list q (of the same length).

$$\text{Change}(s, l, q, r); \leftarrow l = \text{Nil} \wedge q = \text{Nil} \wedge s = r$$
$$\text{Change}(s, l, q, r); \leftarrow \exists l', q', r', v, a \, (l = \text{Append}(l', v) \wedge q = \text{Append}(q', a)$$
$$\wedge \text{ Change}(s, l', q', r') \wedge r = \text{Modify}(r', v, a)).$$

With these preliminaries, we can write down the definition of satisfaction for atomic formulas that begin with one of the extra predicates. The idea is to find, in Π, the rule having that symbol as its head symbol, and to use the body of that rule as the criterion of truth for our atomic formula. It will be useful later to make sure that the Π in the second argument place of Sat has no repeated head symbols, so we include that in the definition.

$$\text{Sat}(\varphi, \Pi, s); \leftarrow \exists p, t, k, i, m, l, r, q, \delta$$
$$(\varphi = \text{Apply}(p, t) \wedge t \text{ hasLength } k \wedge \text{Arity}(p, k) \wedge$$
$$(\forall x \in t) \text{ Term}(x) \wedge HS^+(\Pi, m) \wedge 1\text{-}1\text{-List}(m) \wedge$$
$$(\Pi)_i = \text{Rule}(\text{Apply}(p, l), \delta) \wedge 1\text{-}1\text{-List}(l) \wedge \tag{4}$$
$$l \text{ hasLength } k \wedge (\forall x \in l) \text{ Vbl}(x) \wedge \text{Val}^+(t, s, q) \wedge$$
$$\text{Change}(s, l, q, r) \wedge \text{Sat}(\delta, \Pi, r)).$$

In prose, the essential part of this says that φ has the form $p(\bar{t})$ for an extra predicate of arity k, with $\bar{t}$ being a k-tuple of terms; that Π contains a rule

$p(\bar{l}) \leftarrow \delta$ with head p, $\bar{l}$ being a k-tuple of distinct variables; and that δ is satisfied by the assignment r obtained from s by replacing each of the variables in the list $\bar{l}$ by the value of the corresponding element of $\bar{t}$. This replacement amounts, intuitively, to taking the definition of $p(\bar{l})$ as $\delta(\bar{l})$ and applying it to $p(\bar{t})$, the terms $\bar{t}$ replacing the variables $\bar{l}$. Instead of doing a syntactic substitution of $\bar{t}$ for $\bar{l}$ in δ, we have made the corresponding semantic change, assigning to the variables in $\bar{l}$ the values of the terms in $\bar{t}$.

It may seem strange that this clause in the definition of Sat says nothing about iterating the operator defined by δ. After all, p should be interpreted as the least fixed point of that operator. But the desired iteration is automatically accomplished by the iteration involved in the definition of Sat. That is, if p occurs in δ, then the true instances of p can contribute to the true instances of δ and can thereby contribute to additional true instances of p.

We must still provide the clause for induction assertions in our definition of Sat. Fortunately, this is relatively easy, since iteration is already implicitly done in the preceding clause.

$$\begin{aligned}
\mathrm{Sat}(\varphi, \Pi, s) \;;\leftarrow\; &\exists \varphi', \Sigma, \alpha, \Theta \\
&\bigl(\mathrm{RenameAway}(\varphi, \Pi, \varphi') \wedge \varphi' = \mathrm{IndAsrt}(\Sigma, \alpha) \qquad (5) \\
&\wedge \mathrm{Cat}(\Pi, \Sigma, \Theta) \wedge \mathrm{Sat}(\alpha, \Theta, s)\bigr).
\end{aligned}$$

Here φ' is equivalent to φ and so $\mathrm{Sat}(\varphi, \Pi, s)$ should be equivalent to $\mathrm{Sat}(\varphi', \Pi, s)$. Further, $\varphi' = \mathrm{LET}\ \Sigma\ \mathrm{THEN}\ \alpha$, and no head predicate of Π is bound in φ'. It follows that the head predicates of Π are disjoint from the head predicates of Σ, so that the concatenation Θ of Π and Σ is a legitimate program. Accordingly $\mathrm{Sat}(\varphi', \Pi, s)$ should be equivalent to $\mathrm{Sat}(\alpha, \Theta, s)$.

That concludes the definition of $\mathrm{Sat}(\varphi, \Pi, s)$. It is easy to see that it works as intended. In the case when φ is a sentence and when both Π and s are empty, $\mathrm{Sat}(\varphi, \Pi, s)$ holds in the structure X if and only φ does.

References

1. Aczel, P.: An introduction to inductive definitions. In: Barwise, J. (ed.) Handbook of Mathematical Logic. North Holland, Amsterdam (1977)
2. Aho, A.V., Ullman, J.D.: Universality of data retrieval languages. In: 6th ACM Symp. on Principles of Programming Languages (POPL 1979), pp. 110–119 (1979)
3. Ajtai, M., Gurevich, Y.: Datalog vs first-order logic. J. Comput. System Sci. 49, 562–588 (1994)
4. Bjørner, N.: Private communication (June 2008)
5. Blass, A.: Topoi and computation. Bull. Eur. Assoc. Theor. Comput. Sci. 36, 57–65 (1988)
6. Blass, A.: Geometric invariance of existential fixed-point logic. In: Gray, J., Scedrov, A. (eds.) Categories in Computer Science and Logic, Contemp. Math. vol. 92, pp. 9–22 (1989)
7. Blass, A., Gurevich, Y.: Existential fixed-point logic. In: Börger, E. (ed.) Computation Theory and Logic. LNCS, vol. 270, pp. 20–36. Springer, Heidelberg (1987)

8. Blass, A., Gurevich, Y.: The underlying logic of Hoare Logic. Bull. Eur. Assoc. Theor. Comput. Sci. 70, 82–110 (2000); Reprinted In: Paun, G., Rozenberg, G., Salomaa, A. (eds.) Current Trends in Theoretical Computer Science: Entering the 21st Century, pp. 409–436. World Scientific (2001)

9. Blass, A., Gurevich, Y.: Two forms of one useful logic: Existential fixed point logic and liberal Datalog. Bull. Eur. Assoc. Theor. Comput. Sci. 95 (June 2008)

10. Chandra, A., Harel, D.: Horn clause queries and generalizations. J. Logic Prog. 1, 1–15 (1985)

11. Compton, K.J.: A deductive system for existential fixed point logic. J. Logic Comput. 3, 197–213 (1993)

12. Cook, S.: Soundness and completeness of an axiom system for program verification. SIAM J. Computing 7, 70–90 (1978)

13. Grohe, M.: Existential least fixed-point logic and its relatives. J. Logic Comput. 7, 205–228 (1997)

14. Gurevich, Y., Neeman, I.: DKAL: Distributed knowledge authorization language. In: 21st IEEE Computer Security Foundations Symp. (CSF 2008) (2008)

15. Hoare, C.A.R.: An axiomatic basis for computer programming. Comm. ACM 2, 576–580, 583 (1969)

16. Immerman, N.: Relational queries computable in polynomial time. In: 14th ACM Symp. on Theory of Computing (STOC 1982), pp. 147–152 (1982)

17. Moschovakis, Y.N.: Elementary Induction on Abstract Structures. In: Studies in Logic and the Foundations of Mathematics, vol. 77. North-Holland, Amsterdam (1974)

18. Rossman, B.: Existential positive types and preservation under homomorphisms. In: 20th Annual IEEE Symp. on Logic in Computer Science (LICS 2005), pp. 467–476 (2005)

19. Shoenfield, J.: Mathematical Logic. Addison-Wesley, Reading (1967); Reprinted by the Association for Symbolic Logic and A K Peters (2001)

20. Tarski, A.: A lattice theoretical fixpoint theorem and its applications. Pacific J. Math. 5(2), 285–309 (1955)

21. Vardi, M.: The complexity of relational query languages. In: Fourteenth ACM Symp. on Theory of Computing (STOC 1982), pp. 137–146 (1982)

On Synchronous and Asynchronous Interaction in Distributed Systems

Rob van Glabbeek[1,2], Ursula Goltz[3], and Jens-Wolfhard Schicke[3,*]

[1] NICTA, Sydney, Australia
[2] School of Computer Sc. and Engineering, University of New South Wales, Sydney, Australia
[3] Institute for Programming and Reactive Systems, TU Braunschweig, Germany
rvg@cs.stanford.edu, goltz@ips.cs.tu-bs.de, drahflow@gmx.de

Abstract. When considering distributed systems, it is a central issue how to deal with interactions between components. In this paper, we investigate the paradigms of synchronous and asynchronous interaction in the context of distributed systems. We investigate to what extent or under which conditions synchronous interaction is a valid concept for specification and implementation of such systems. We choose Petri nets as our system model and consider different notions of distribution by associating locations to elements of nets. First, we investigate the concept of simultaneity which is inherent in the semantics of Petri nets when transitions have multiple input places. We assume that tokens may only be taken instantaneously by transitions on the same location. We exhibit a hierarchy of 'asynchronous' Petri net classes by different assumptions on possible distributions. Alternatively, we assume that the synchronisations specified in a Petri net are crucial system properties. Hence transitions and their preplaces may no longer placed on separate locations. We then answer the question which systems may be implemented in a distributed way without restricting concurrency, assuming that locations are inherently sequential. It turns out that in both settings we find semi-structural properties of Petri nets describing exactly the problematic situations for interactions in distributed systems.

1 Introduction

In this paper, we address interaction patterns in distributed systems. By a distributed system we understand here a system which is executed on spatially distributed locations, which do not share a common clock (for performance reasons for example). We want to investigate to what extent or under which conditions synchronous interaction is a valid concept for specification and implementation of such systems. It is for example a well-known fact that synchronous communication can be simulated by asynchronous communication using suitable protocols. However, the question is whether and under which circumstances these protocols fully retain the original behaviour of a system. What we are interested in here are precise descriptions of what behaviours can possibly be preserved and which cannot.

The topic considered here is by no means a new one. We give a short overview on related approaches in the following.

---
* Supported by DAAD (Deutscher Akademischer Austauschdienst) while visiting NICTA.

E. Ochmański and J. Tyszkiewicz (Eds.): MFCS 2008, LNCS 5162, pp. 16–35, 2008.
© Springer-Verlag Berlin Heidelberg 2008

Already in the 80th, Luc Bougé considered a similar problem in the context of distributed algorithms. In [5] he considers the problem of implementing symmetric leader election in the sublanguages of CSP obtained by allowing different forms of communication, combining input and output guards in guarded choice in different ways. He finds that the possibility of implementing leader election depends heavily on the structure of the communication graphs. Truly symmetric schemes are only possible in CSP with arbitrary input and output guards in choices.

Synchronous interaction is a basic concept in many languages for system specification and design, e.g. in statechart-based approaches, in process algebras or the π-calculus. For process algebras and the π-calculus, language hierarchies have been established which exhibit the expressive power of different forms of synchronous and asynchronous interaction. In [4] Frank de Boer and Catuscia Palamidessi consider various dialects of CSP with differing degrees of asynchrony. Similar work is done for the π-calculus in [15] by Catuscia Palamidessi, in [13] by Uwe Nestmann and in [8] by Dianele Gorla. A rich hierarchy of asynchronous π-calculi has been mapped out in these papers. Again mixed-choice, i.e. the ability to combine input and output guards in a single choice, plays a central rôle in the implementation of truly synchronous behaviour.

In [17], Peter Selinger considers labelled transition systems whose visible actions are partitioned into input and output actions. He defines asynchronous implementations of such a system by composing it with in- and output queues, and then characterises the systems that are behaviourally equivalent to their asynchronous implementations. The main difference with our approach is that we focus on asynchrony within a system, whereas Selinger focusses on the asynchronous nature of the communications of a system with the outside world.

Also in hardware design it is an intriguing quest to use interaction mechanisms which do not rely on a global clock, in order to gain performance. Here the simulation of synchrony by asynchrony can be a crucial issue, see for instance [10] and [11].

In contrast to the approaches based on language constructs like the work on CSP or the π-calculus, we choose here a very basic system model for our investigations, namely Petri nets. The main reason for this choice is the detailed way in which a Petri net represents a concurrent system, including the interaction between the components it may consist of. In an interleaving based model of concurrency such as labelled transition systems modulo bisimulation semantics, a system representation as such cannot be said to contain synchronous or asynchronous interaction; at best these are properties of composition operators, or communication primitives, defined in terms of such a model. A Petri net on the other hand displays enough detail of a concurrent system to make the presence of synchronous communication discernible. This makes it possible to study synchronous and asynchronous interaction without digressing to the realm of composition operators.

Also in Petri net theory, the topic which concerns us here has already been tackled. It has been investigated in [9] and [18] whether and how a Petri net can be implemented in a distributed way. We will comment on these and other related papers in the area of Petri net theory in the conclusion.

In a Petri net, a transition interacts with its preplaces by consuming tokens. In Petri net semantics, taking a token is usually considered as an instantaneous action, hence a synchronous interaction between a transition and its preplace. In particular when a transition has several preplaces this becomes a crucial issue. In this paper we investigate what happens if we consider a Petri net as a specification of a system that is to be implemented in a distributed way. For this we introduce locations on which all elements of a Petri net have to be placed upon. The basic assumption is that interaction between remote components takes time. In our framework this means that the removal of a token will be considered instantaneous only if the removing transition and the place where the token is removed from are co-located. Our investigations are now twofold.

In Section 3 of this paper, we consider under which circumstances the synchronous interaction between a transition and its preplace may be mimicked asynchronously, thus allowing to put places and their posttransitions on different locations. Following [6], we model the asynchronous interaction between transitions and their preplaces by inserting silent (unobservable) transitions between them. We investigate the effect of this transformation by comparing the behaviours of nets before and after insertion of the silent transitions using a suitable equivalence notion. We believe that most of our results are independent of the precise choice of this equivalence. However, as explained in Section 5, it has to preserve causality, branching time and divergence to some small extent, and needs to abstract from silent transitions. Therefore we choose one such equivalence, based on its technical convenience in establishing our results. Our choice is *step readiness equivalence*. It is a variant of the *readiness equivalence* of [14], obtained by collecting the set of *steps* of multiple actions possible after a certain sequence of actions, instead of just the set of possible actions. We call a net *asynchronous* if, for a suitable placement of its places and transitions, the above-mentioned transformation replacing synchronous by asynchronous interaction preserves step readiness equivalence. Depending on the allowed placements, we obtain a hierarchy of classes of asynchronous nets: *fully asynchronous* nets, *symmetrically asynchronous* nets and *asymmetrically asynchronous* nets. We give semi-structural properties that characterise precisely when a net falls into one of these classes. This puts the results from [6] in a uniform framework and extends them by introducing a simpler notion of asymmetric asynchrony.

In Sections 4 and 5 we pursue an alternative approach. We assume that the synchronisations specified in a Petri net are crucial system properties. Hence we enforce co-locality between a transition and all its preplaces while at the same time assuming that concurrent activity is not possible at a single location. We call nets fulfilling these requirement *distributed* and investigate which behaviours can be implemented by distributed nets. Again we compare the behaviours up to step readiness equivalence. We call a net *distributable* iff its behaviour can be equivalently produced by a distributed net. We give a behavioural and a semi-structural characterisation of a class of non-distributable nets, thereby exhibiting behaviours which cannot be implemented in a distributed way at all. Finally, we give a lower bound of distributability by providing a concrete distributed implementation for a wide range of nets.

2   Basic Notions

We consider here 1-safe net systems, i.e. places never carry more than one token, but a transition can fire even if pre- and postset intersect.

Definition 1. Let Act be a set of *visible actions* and $\tau \notin$ Act be an *invisible action*. A *labelled net* (over Act) is a tuple $N = (S, T, F, M_0, \ell)$ where

- S is a set (of *places*),
- T is a set (of *transitions*),
- $F \subseteq S \times T \cup T \times S$ (the *flow relation*),
- $M_0 \subseteq S$ (the *initial marking*) and
- $\ell : T \to$ Act $\dot{\cup} \{\tau\}$ (the *labelling function*).

Petri nets are depicted by drawing the places as circles, the transitions as boxes containing the respective label, and the flow relation as arrows (*arcs*) between them. When a Petri net represents a concurrent system, a global state of such a system is given as a *marking*, a set of places, the initial state being M_0. A marking is depicted by placing a dot (*token*) in each of its places. The dynamic behaviour of the represented system is defined by describing the possible moves between markings. A marking M may evolve into a marking M' when a nonempty set of transitions G *fires*. In that case, for each arc $(s, t) \in F$ leading to a transition t in G, a token moves along that arc from s to t. Naturally, this can happen only if all these tokens are available in M in the first place. These tokens are consumed by the firing, but also new tokens are created, namely one for every outgoing arc of a transition in G. These end up in the places at the end of those arcs. A problem occurs when as a result of firing G multiple tokens end up in the same place. In that case M' would not be a marking as defined above. In this paper we restrict attention to nets in which this never happens. Such nets are called *1-safe*. Unfortunately, in order to formally define this class of nets, we first need to correctly define the firing rule without assuming 1-safety. Below we do this by forbidding the firing of sets of transitions when this might put multiple tokens in the same place.

Definition 2. Let $N = (S, T, F, M_0, \ell)$ be a labelled net. Let $M_1, M_2 \subseteq S$.

We denote the preset and postset of a net element $x \in S \cup T$ by ${}^{\bullet}x := \{y \mid (y, x) \in F\}$ and $x^{\bullet} := \{y \mid (x, y) \in F\}$ respectively. These functions are extended to sets in the usual manner, i.e. ${}^{\bullet}X := \{y \mid y \in {}^{\bullet}x, x \in X\}$.

A nonempty set of transitions $G \subseteq T, G \neq \emptyset$, is called a *step from M_1 to M_2*, notation $M_1 [G\rangle_N M_2$, iff

- all transitions contained in G are *enabled*, that is

$$\forall t \in G.\ {}^{\bullet}t \subseteq M_1 \wedge (M_1 \setminus {}^{\bullet}t) \cap t^{\bullet} = \emptyset,$$

- all transitions of G are *independent*, that is *not conflicting*:

$$\forall t, u \in G, t \neq u.\ {}^{\bullet}t \cap {}^{\bullet}u = \emptyset \wedge t^{\bullet} \cap u^{\bullet} = \emptyset,$$

- in M_2 all tokens have been removed from the *preplaces* of G and new tokens have been inserted at the *postplaces* of G:

$$M_2 = (M_1 \setminus {}^{\bullet}G) \cup G^{\bullet}.$$

To simplify statements about possible behaviours of nets, we use some abbreviations.

Definition 3. Let $N = (S, T, F, M_0, \ell)$ be a labelled net.
We extend the labelling function ℓ to (multi)sets elementwise.
$$\longrightarrow_N \subseteq \mathcal{P}(S) \times \mathbb{N}^{\text{Act}} \times \mathcal{P}(S) \text{ is given by } M_1 \xrightarrow{A}_N M_2 \Leftrightarrow \exists G \subseteq T.\, M_1 \, [G\rangle_N \, M_2 \wedge$$
$$A = \ell(G)$$
$$\xrightarrow{\tau}_N \subseteq \mathcal{P}(S) \times \mathcal{P}(S) \text{ is defined by } M_1 \xrightarrow{\tau}_N M_2 \Leftrightarrow \exists t \in T.\, \ell(t) = \tau \wedge M_1 \, [\{t\}\rangle_N \, M_2$$
$$\Longrightarrow_N \subseteq \mathcal{P}(S) \times \text{Act}^* \times \mathcal{P}(S) \text{ is defined by } M_1 \xRightarrow{a_1 a_2 \cdots a_n}_N M_2 \Leftrightarrow$$
$$M_1 \xrightarrow{\tau}{}^*_N \xrightarrow{\{a_1\}}_N \xrightarrow{\tau}{}^*_N \xrightarrow{\{a_2\}}_N \xrightarrow{\tau}{}^*_N \cdots \xrightarrow{\tau}{}^*_N \xrightarrow{\{a_n\}}_N \xrightarrow{\tau}{}^*_N M_2$$

where $\xrightarrow{\tau}{}^*_N$ denotes the reflexive and transitive closure of $\xrightarrow{\tau}_N$.
We write $M_1 \xrightarrow{A}_N$ for $\exists M_2.\, M_1 \xrightarrow{A}_N M_2$, $M_1 \xrightarrow{A}\!\!\!\!/\,_N$ for $\nexists M_2$. $M_1 \xrightarrow{A}_N M_2$ and
similar for the other two relations. Likewise $M_1[G\rangle_N$ abbreviates $\exists M_2.\, M_1[G\rangle_N M_2$.
A marking M_1 is said to be *reachable* iff there is a $\sigma \in \text{Act}^*$ such that $M_0 \xRightarrow{\sigma}_N M_1$.
The set of all reachable markings is denoted by $[M_0\rangle_N$.

We omit the subscript N if clear from context.

As said before, here we only want to consider 1-safe nets. Formally, we restrict
ourselves to *contact-free nets*, where in every reachable marking $M_1 \in [M_0\rangle$ for all
$t \in T$ with $^\bullet t \subseteq M_1$
$$(M_1 \setminus {}^\bullet t) \cap t^\bullet = \emptyset\,.$$

For such nets, in Definition 2 we can just as well consider a transition t to be enabled
in M iff $^\bullet t \subseteq M$, and two transitions to be independent when $^\bullet t \cap {}^\bullet u = \emptyset$.

In this paper we furthermore restrict attention to nets for which $^\bullet t \neq \emptyset$, and $^\bullet t$ and $t^\bullet$
are finite for all $t \in T$. We also require the initial marking M_0 to be finite. A consequence
of these restrictions is that all reachable markings are finite, and it can never happen
that infinitely many independent transitions are enabled. Henceforth, with *net* we mean
a labelled net obeying the above restrictions.

In our nets transitions are labelled with *actions* drawn from a set $\text{Act} \,\dot{\cup}\, \{\tau\}$. This
makes it possible to see these nets as models of *reactive systems*, that interact with their
environment. A transition t can be thought of as the occurrence of the action $\ell(t)$. If
$\ell(t) \in \text{Act}$, this occurrence can be observed and influenced by the environment, but if
$\ell(t) = \tau$, t is an *internal* or *silent* transition whose occurrence cannot be observed or
influenced by the environment. Two transitions whose occurrences cannot be distinguished by the environment are equipped with the same label. In particular, given that
the environment cannot observe the occurrence of internal transitions at all, all of them
have the same label, namely τ.

We use the term *plain nets* for nets where ℓ is injective and no transition has the
label τ, i.e. essentially unlabelled nets. Similarly, we speak of *plain τ-nets* to describe
nets where $\ell(t) = \ell(u) \neq \tau \Rightarrow t = u$, i.e. nets where every observable action is
produced by a unique transition. In this paper we focus on plain nets, and give semi-
structural characterisations of classes of plain nets only. However, in defining whether
a net belongs to one of those classes, we study its implementations, which typically are
plain τ-nets. When proving our impossibility result (Theorem 3 in Section 5) we even
allow arbitrary nets as implementations.

We use the following variation of readiness semantics [14] to compare the behaviour
of nets.

Definition 4. Let $N = (S, T, F, M_0, \ell)$ be a net, $\sigma \in \text{Act}^*$ and $X \subseteq \mathbb{N}^{\text{Act}}$.
$<\sigma, X>$ is a *step ready pair* of N iff

$$\exists M.\ M_0 \overset{\sigma}{\Longrightarrow} M \wedge M \overset{\tau}{\nrightarrow} \wedge X = \{A \in \mathbb{N}^{\text{Act}} \mid M \overset{A}{\longrightarrow}\}.$$

We write $\mathscr{R}(N)$ for the set of all step ready pairs of N.
Two nets N and N' are *step readiness equivalent*, $N \approx_{\mathscr{R}} N'$, iff $\mathscr{R}(N) = \mathscr{R}(N')$.

The elements of a set X as above are multisets of actions, but as in all such multisets that will be mentioned in this paper the multiplicity of each action occurrence is at most 1, we use set notation to denote them.

3   Asynchronous Petri Net Classes

In Petri nets, an inherent concept of simultaneity is built in, since when a transition has more than one preplace, it can be crucial that tokens are removed instantaneously. When using a Petri net to model a system which is intended to be implemented in a distributed way, this built-in concept of synchronous interaction may be problematic.

In this paper, a given net is regarded as a *specification* of how a system should behave, and this specification involves complete synchronisation of the firing of a transition and the removal of all tokens from its preplaces. In this section, we propose various definitions of an *asynchronous implementation* of a net N, in which such synchronous interaction is wholly or partially ruled out and replaced by asynchronous interaction. The question to be clarified is whether such an asynchronous implementation faithfully mimics the dynamic behaviour of N. If this is the case, we call the net N *asynchronous* with respect to the chosen interaction pattern.

The above programme, and thus the resulting concept of asynchrony, is parametrised by the answers to three questions:

1. Which synchronous interactions do we want to rule out exactly?
2. How do we replace synchronous by asynchronous interaction?
3. When does one net faithfully mimic the dynamic behaviour of another?

To answer the first question we associate a *location* to each place and each transition in a net. A transition may take a token instantaneously from a preplace (when firing) iff this preplace is co-located with the transition; if the preplace resides on a different location than the transition, we have to assume the collection of the token takes time, and thus the place looses its token *before* the transition fires.

We model the association of locations to the places and transitions in a net $N = (S, T, F, M_0, \ell)$ as a function $D : S \cup T \to \text{Loc}$, with Loc a set of possible locations. We refer to such a function as a *distribution* of N. Since the identity of the locations is irrelevant for our purposes, we can just as well abstract from Loc and represent D by the equivalence relation $\equiv_D$ on $S \cup T$ given by $x \equiv_D y$ iff $D(x) = D(y)$.

In this paper we do not deal with nets that have a distribution built in. We characterise the interaction patterns we are interested in by imposing particular restrictions on the allowed distributions. The implementor of a net can choose any distribution that satisfies the chosen requirements, and we call a net asynchronous for a certain interaction pattern if it has a correct asynchronous implementation based on any distribution satisfying the respective requirements.

The *fully asynchronous* interaction pattern is obtained by requiring that all places and all transitions reside on different locations. This makes it necessary to implement the removal of every token in a time-consuming way. However, this leads to a rather small class of asynchronous nets, that falls short for many applications. We therefore propose two ways to loosen this requirement, thereby building a hierarchy of classes of asynchronous nets. Both require that all places reside on different locations, but a transition may be co-located with one of its preplaces. The *symmetrically asynchronous* interaction pattern allows this only for transitions with a single preplace, whereas in the *asymmetrically asynchronous* interaction pattern any transition may be co-located with one of its preplaces. Since two preplaces can never be co-located, this breaks the symmetry between the preplaces of a transition; an implementor of a net has to choose at most one preplace for every transition, and co-locate the transition with it. The removal of tokens from all other preplaces needs to be implemented in a time-consuming way. Note that all three interaction patterns break the synchronisation of the token removal between the various preplaces.

Definition 5. Let D be a distribution on a net $N = (S, T, F, M_0, \ell)$, and let $\equiv_D$ be the induced equivalence relation on $S \cup T$. We say that D is

- *fully distributed*, $D \in \mathcal{Q}_{\mathrm{FD}}$, when $x \equiv_D y$ for $x, y \in S \cup T$ only if $x = y$,
- *symmetrically distributed*, $D \in \mathcal{Q}_{\mathrm{SD}}$, when

$$p \equiv_D q \text{ for } p, q \in S \quad \text{only if } p = q,$$
$$t \equiv_D p \text{ for } t \in T, \ p \in S \text{ only if } {}^\bullet t = \{p\} \text{ and}$$
$$t \equiv_D u \text{ for } t, u \in T \quad \text{only if } t = u \text{ or } \exists p \in S.\ t \equiv_D p \equiv_D u,$$

- *asymmetrically distributed*, $D \in \mathcal{Q}_{\mathrm{AD}}$, when

$$p \equiv_D q \text{ for } p, q \in S \quad \text{only if } p = q,$$
$$t \equiv_D p \text{ for } t \in T, \ p \in S \text{ only if } p \in {}^\bullet t \text{ and}$$
$$t \equiv_D u \text{ for } t, u \in T \quad \text{only if } t = u \text{ or } \exists p \in S.\ t \equiv_D p \equiv_D u.$$

The second question raised above was: How do we replace synchronous by asynchronous interaction? In this section we assume that if an arc goes from a place s to a transition t at a different location, a token takes time to move from s to t. Formally, we describe this by inserting silent (unobservable) transitions between transitions and their remote preplaces. This leads to the following notion of an asynchronous implementation of a net with respect to a chosen distribution.

Definition 6. Let $N = (S, T, F, M_0, \ell)$ be a net, and let $\equiv_D$ be an equivalence relation on $S \cup T$. The *D-based asynchronous implementation* of N is defined as the net $I_D(N) := (S \cup S^\tau, T \cup T^\tau, F', M_0, \ell')$ with

$$S^\tau := \{ s_t \mid t \in T, \ s \in {}^\bullet t, \ s \not\equiv_D t \},$$
$$T^\tau := \{ t_s \mid t \in T, \ s \in {}^\bullet t, \ s \not\equiv_D t \},$$
$$F' := \{ (t, s) \mid t \in T, \ s \in t^\bullet \} \cup \{ (s, t) \mid t \in T, \ s \in {}^\bullet t, \ s \equiv_D t \}$$
$$\cup \{ (s, t_s), (t_s, s_t), (s_t, t) \mid t \in T, \ s \in {}^\bullet t, \ s \not\equiv_D t \},$$
$$\ell' \restriction T = \ell \quad \text{and} \quad \ell'(t_s) = \tau \text{ for } t_s \in T^\tau.$$

Proposition 1. *For any (contact-free) net N, and any choice of $\equiv_D$, the net $I_D(N)$ is contact-free, and satisfies the other requirements imposed on nets, listed in Section 2.*

Proof. For $D \in \mathcal{Q}_{\mathrm{FD}}$ and $D \in \mathcal{Q}_{\mathrm{SD}}$, this is established in [6]. The proof of the general case goes likewise.   □

The above protocol for replacing synchronous by asynchronous interaction appears to be one of the simplest ones imaginable. More intricate protocols, involving many asynchronous messages between a transition and its preplaces, could be contemplated, but we will not study them here. Our protocol involves just one such message, namely from the preplace to its posttransition. It is illustrated in Fig. 1.

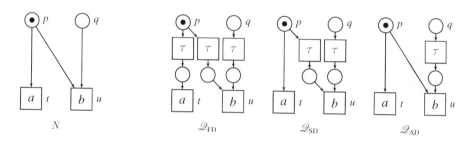

Fig. 1. Possible results for $I_D(N)$ given different requirements

The last question above was: When does one net faithfully mimic the dynamic behaviour of another? This asks for a *semantic equivalence* on Petri nets, telling when two nets display the same behaviour. Many such equivalences have been studied in the literature. We believe that most of our results are independent of the precise choice of a semantic equivalence, as long as it preserves causality and branching time to some degree, and abstracts from silent transitions. Therefore we choose one such equivalence, based on its technical convenience in establishing our results, and postpone questions on the effect of varying this equivalence for further research. Our choice is *step readiness equivalence*, as defined in Section 2. Using this equivalence, we define a notion of *behavioural asynchrony* by asking whether the asynchronous implementation of a net preserves its behaviour. This notion is parametrised by the chosen interaction pattern, characterised as a requirement on the allowed distributions.

Definition 7. Let $\mathcal{Q}$ be a requirement on distributions of nets.
A plain net N is *behaviourally $\mathcal{Q}$-asynchronous* iff there exists a distribution D of N meeting the requirement $\mathcal{Q}$ such that $I_D(N) \approx_{\mathscr{R}} N$.

Intuitively, the only behavioural difference between a net N and its asynchronous implementation $I_D(N)$ can occur when in N a place $s \in {}^\bullet u$ is marked, whereas in $I_D(N)$ this token is already on its way from s to its posttransition u. In that case, it may occur that a transition $t \neq u$ with $s \in {}^\bullet t$ is enabled in N, whereas t is not enabled in the described state of $I_D(N)$. We call the situation in N leading to this state of $I_D(N)$ a *distributed conflict*; it is in fact the only circumstance in which $I_D(N)$ fails to faithfully mimic the dynamic behaviour of N.

Definition 8. Let $N = (S, T, F, M_0, \ell)$ be a net and D a distribution of N. N has a *distributed conflict with respect to* D iff

$$\exists t, u \in T \; \exists p \in {}^{\bullet}t \cap {}^{\bullet}u. \, t \neq u \wedge p \not\equiv_D u \wedge \exists M \in [M_0\rangle_N. \, {}^{\bullet}t \subseteq M \, .$$

We wish to call a net N *(semi)structurally asynchronous* iff the situation outlined above never occurs, so that the asynchronous implementation does not change the behaviour of the net. As for behavioural asynchrony, this notion of asynchrony is parametrised by the set of allowed distributions.

Definition 9. Let $\mathcal{Q}$ be a requirement on distributions of nets.
A net N is *(semi)structurally $\mathcal{Q}$-asynchronous* iff there exists a distribution D of N meeting the requirement $\mathcal{Q}$ such that N has no distributed conflicts with respect to D.

The following theorem shows that distributed conflicts describe exactly the critical situations: For all plain nets the notions of structural and behavioural asynchrony coincide, regardless of the choice if $\mathcal{Q}$.

Theorem 1. Let N be a plain net, and $\mathcal{Q}$ a requirement on distributions of nets. Then N is behaviourally $\mathcal{Q}$-asynchronous iff it is structurally $\mathcal{Q}$-asynchronous.

Proof. In the full version of this paper [7]. □

Because of this theorem, we call a plain net $\mathcal{Q}$-asynchronous if it is behaviourally and/or structurally $\mathcal{Q}$-asynchronous. In this paper we study this concept for plain nets only. When taking $\mathcal{Q} = \mathcal{Q}_{\text{FD}}$ we speak of *fully asynchronous nets*, when taking $\mathcal{Q} = \mathcal{Q}_{\text{SD}}$ of *symmetrically asynchronous nets*, and when taking $\mathcal{Q} = \mathcal{Q}_{\text{AD}}$ of *asymmetrically asynchronous nets*.

Example 1. The net N of Fig. 1 is not fully asynchronous, for its unique D-based asynchronous implementation $I_D(N)$ with $D \in \mathcal{Q}_{\text{FD}}$ (also displayed in Fig. 1) is not step readiness equivalent to N. In fact $\langle \varepsilon, \emptyset \rangle \in \mathcal{R}(I_D(N)) \setminus \mathcal{R}(N)$. This inequivalence arises because in $I_D(N)$ the option to do an a-action can be disabled already before any visible action takes place; this is not possible in N.

The only way to avoid a distributed conflict in this net is by taking $t \equiv_D p \equiv_D u$. This is not allowed for any $D \in \mathcal{Q}_{\text{FD}}$ or $D \in \mathcal{Q}_{\text{SD}}$, but it is allowed for $D \in \mathcal{Q}_{\text{AD}}$ (cf. the last net in Fig. 1). Hence N is asymmetrically asynchronous, but not symmetrically asynchronous.

Since $\mathcal{Q}_{\text{FD}} \subseteq \mathcal{Q}_{\text{SD}} \subseteq \mathcal{Q}_{\text{AD}}$, any fully asynchronous net is symmetrically asynchronous, and any symmetrically asynchronous net is also asymmetrically asynchronous. Below we give semi-structural characterisations of these three classes of nets. The first two stem from [6], where the class of fully asynchronous nets is called $FA(B)$ and the class of symmetrically asynchronous nets is called $SA(B)$. The class $AA(B)$ in [6] is somewhat larger than our class of asymmetrically asynchronous nets, for it is based on a slightly more involved protocol for replacing synchronous by asynchronous interaction.

Definition 10. A plain net $N = (S, T, F, M_0, \ell)$ has a

– *partially reachable conflict* iff

$$\exists t, u \in T \; \exists p \in {}^\bullet t \cap {}^\bullet u. \; t \neq u \wedge \exists M \in [M_0\rangle_N. \; {}^\bullet t \subseteq M \;,$$

– *partially reachable* N iff

$$\exists t, u \in T \; \exists p \in {}^\bullet t \cap {}^\bullet u. \; t \neq u \wedge |{}^\bullet u| > 1 \wedge \exists M \in [M_0\rangle_N. \; {}^\bullet t \subseteq M \;,$$

– *left and right border reachable* M iff

$$\exists t, u, v \in T \; \exists p \in {}^\bullet t \cap {}^\bullet u \; \exists q \in {}^\bullet u \cap {}^\bullet v. \; \begin{array}{l} t \neq u \wedge u \neq v \wedge p \neq q \wedge \\ \exists M_1, M_2 \in [M_0\rangle_N. \; {}^\bullet t \subseteq M_1 \wedge {}^\bullet v \subseteq M_2 \;. \end{array}$$

Theorem 2. *Let N be a plain net.*
– *N is fully asynchronous iff it has no partially reachable conflict.*
– *N is symmetrically asynchronous iff it has no partially reachable* N*.*
– *N is asymmetrically asynchronous iff it has no left and right border reachable* M*.*

Proof. Straightforward with Theorem 1. □

In the theory of Petri nets, there have been extensive studies on classes of nets with certain structural properties like *free choice nets* [3,2] and *simple nets* [3], as well as extensions of theses classes. They are closely related to the net classes defined here, but they are defined without taking reachability into account. For a comprehensive overview and discussion of the relations between those purely structurally defined net classes and our net classes see [6]. Restricted to plain nets without dead transitions (meaning that every transition t satisfies the requirement $\exists M \in [M_0\rangle. \; {}^\bullet t \subseteq M$), Theorem 2 says that a net is fully synchronous iff it is conflict-free in the structural sense (no shared preplaces), symmetrically asynchronous iff it is a free choice net and asymmetrically asynchronous iff it is simple.

Our asynchronous net classes are defined for plain nets only. There are two approaches to lifting them to labelled nets. One is to postulate that whether a net is asynchronous or not has nothing to do with its labelling function, so that after replacing this labelling by the identity function one can apply the insights above. This way our structural characterisations (Theorems 1 and 2) apply to labelled nets as well. Another approach would be to apply the notion of behavioural asynchrony of Definition 7 directly to labelled nets. This way more nets will be asynchronous, because in some cases a net happens to be equivalent to its asynchronous implementation in spite of a failure of structural asynchrony. This happens for instance if all transitions in the original net are labelled τ. Unlike the situation for plain nets, the resulting notion of behavioural asynchrony will most likely be strongly dependent on the choice of the semantic equivalence relation between nets.

4   Distributed Systems

The approach of Section 3 makes a difference between a net regarded as a specification, and an asynchronous implementation of the same net. The latter could be thought of as a way to execute the net when a given distribution makes the synchronisations that are inherent in the specification impossible. In this and the following section, on the other hand, we drop the difference between a net and its asynchronous implementation. Instead of adapting our intuition about the firing rule when implementing a net in a distributed way, we insist that all synchronisations specified in the original net remain present as synchronisations in a distributed implementation. Yet, at the same time we stick to the point of view that it is simply not possible for a transition to synchronise its firing with the removal of tokens from preplaces at remote locations. Thus we only allow distributions in which each transition is co-located with all of its preplaces. We call such distributions *effectual*. For effectual distributions D, the implementation transformation I_D is the identity. As a consequence, if effectuality is part of a requirement $\mathscr{Q}$ imposed on distributions, the question whether a net is $\mathscr{Q}$-asynchronous is no longer dependent on whether an asynchronous implementation mimics the behaviour of the given net, but rather on whether the net allows a distribution satisfying $\mathscr{Q}$ at all.

The requirement of effectuality does not combine well will the requirements on distributions proposed in Definition 5. For if $\mathscr{Q}$ is the class of distributions that are effectual and asymmetrically distributed, then only nets without transitions with multiple preplaces would be $\mathscr{Q}$-asynchronous. This rules out most useful applications of Petri nets. The requirement of effectuality by itself, on the other hand, would make every net asynchronous, because we could assign the same location to all places and transitions.

We impose one more fundamental restriction on distributions, namely that when two visible transitions can occur in one step, they cannot be co-located. This is based on the assumption that at a given location visible actions can only occur sequentially, whereas we want to preserve as much concurrency as possible (in order not to loose performance). Recall that in Petri nets simultaneity of transitions cannot be enforced: if two transitions can fire in one step, they can also fire in any order. The standard interpretation of nets postulates that in such a case those transitions are causally independent, and this idea fits well with the idea that they reside at different locations.

Definition 11. Let $N = (S, T, F, M_0, \ell)$ be a net.
The *concurrency relation* $\smile \subseteq T^2$ is given by $t \smile u \Leftrightarrow t \neq u \wedge \exists M \in [M_0\rangle. \, M[\{t, u\}\rangle$.
N is *distributed* iff it has a distribution D such that

- $\forall s \in S, t \in T. \, s \in {}^\bullet t \Rightarrow t \equiv_D s$,
- $t \smile u \wedge l(t), l(u) \neq \tau \Rightarrow t \not\equiv_D u$.

It is straightforward to give a semi-structural characterisation of this class of nets:

Observation 1. *A net is distributed iff there is no sequence $t_0, \ldots, t_n$ of transitions with $t_0 \smile t_n$ and ${}^\bullet t_{i-1} \cap {}^\bullet t_i \neq \emptyset$ for $i = 1, \ldots, n$.*

A structure as in the above characterisation of distributed nets can be considered as a prolonged M containing two independent transitions that can be simultaneously enabled.

It is not hard to find a plain net that is fully asynchronous, yet not distributed. However, restricted to plain nets without dead transitions, the class of asymmetrically asynchronous nets is a strict subclass of the class of distributed nets. Namely, if a net is M-free (where an M is as in Definition 10, but without the reachability condition on the bottom line), then it surely has no sequence as described above.

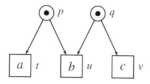

Fig. 2. A fully marked M

5   Distributable Systems

In this section, we will investigate the borderline for distributability of systems. It is a well known fact that sometimes a global protocol is necessary when concurrent activities in a system interfere. In particular, this may be necessary for deciding choices in a coherent way. Consider for example the simple net in Fig. 2. It contains an M-structure, which was already exhibited as a problematic one in Section 3. Transitions t and v are supposed to be concurrently executable (if we do not want to restrict performance of the system), and hence reside on different locations. Thus at least one of them, say t, cannot be co-located with transition u. However, both transitions are in conflict with u.

As we use nets as models of reactive systems, we allow the environment of a net to influence decisions at runtime by blocking one of the possibilities. Equivalently we can say it is the environment that fires transitions, and this can only happen for transitions that are currently enabled in the net. If the net decides between t and u before the actual execution of the chosen transition, the environment might change its mind in between, leading to a state of deadlock. Therefore we work in a branching time semantics, in which the option to perform t stays open until either t or u occurs. Hence the decision to fire u can only be taken at the location of u, namely by firing u, and similarly for t. Assuming that it takes time to propagate any message from one location to another, in no distributed implementation of this net can t and u be simultaneously enabled, because in that case we cannot exclude that both of them happen. Thus, the only possible implementation of the choice between t and u is to alternate the right to fire between t and u, by sending messages between them (cf. Fig. 3). But if the environment only sporadically tries to fire t or u it may repeatedly miss the opportunity to do so, leading to an infinite loop of control messages sent back and forth, without either transition ever firing.

In this section we will formalise this reasoning, and show that under a few mild assumptions this type of structures cannot be implemented in a distributed manner at all, i.e. even when we allow the implementation to be completely unrelated to the specification, except for its behaviour. For this, we apply the notion of a distributed net, as introduced in the previous section. Furthermore, we need an equivalence notion in order to specify in which way an implementation as a distributed net is required to preserve the behaviour of the original net. As in Section 3, we choose step readiness equivalence. We call a plain net *distributable* if it is step readiness equivalent to a distributed net. We speak of a *truly synchronous* net if it is not distributable, thus if it may not be transformed into any distributed net with the same behaviour up to step readiness equivalence, that is if no such net exists. We study the concept "distributable" for plain nets only, but in order to get the largest class possible we allow non-plain implementations, where a given transition may be split into multiple transitions carrying the same label.

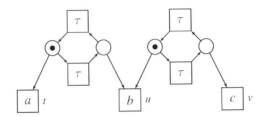

Fig. 3. A busy-wait implementation of the net in Fig. 2

Definition 12. A plain net N is *truly synchronous* iff there exists no distributed net N' which is step readiness equivalent to N.

We will show that nets like the one of Fig. 2 are truly synchronous.

Step readiness equivalence is one of the simplest and least discriminating equivalences imaginable that preserves branching time, causality and divergence to some small extend. Our impossibility result, formalised below as Theorem 3, depends crucially on all three properties, and thus needs to be reconsidered when giving up on any of them.When working in linear time semantics, every net is equivalent to an infinite net that starts with a choice between several τ-transitions, each followed by a conflict-free net modelling a single run. This net is N-free, and hence distributed. It can be argued that infinite implementations are not acceptable, but when searching for the theoretical limits to distributed implementability we don't want to rule them out dogmatically. When working in interleaving semantics, any net can be converted into an equivalent distributed net by removing all concurrency between transitions. This can be accomplished by adding a new, initially marked place, with an arc to and from every transition in the net. When fully abstracting from divergence, even when respecting causality and branching time, the net of Fig. 2 is equivalent to the distributed net of Fig. 3, and in fact it is not hard to see that this type of implementation is possibly for any given net. Yet, the implementation is suspect, as the implemented decision of a choice may fail to terminate. The clause $M \xrightarrow{\tau}\!\!\!\!\!/\,\,$ in Definition 4 is strong enough to rule out this type of implementation, even though our step readiness semantics abstracts from other forms of divergence.

We now characterise the class of nets which we will prove to be truly synchronous.

Definition 13. Let $N = (S, T, F, M_0, \ell)$ be a net. N has a *fully reachable visible pure* M iff $\exists t, u, v \in T$. ${}^\bullet t \cap {}^\bullet u \neq \emptyset \wedge {}^\bullet u \cap {}^\bullet v \neq \emptyset \wedge {}^\bullet t \cap {}^\bullet v = \emptyset \wedge \ell(t), \ell(u), \ell(v) \neq \tau \wedge \exists M \in [M_0\rangle. {}^\bullet t \cup {}^\bullet u \cup {}^\bullet v \subseteq M$.

Here a *pure* M is an M as in Definition 10 that moreover satisfies ${}^\bullet t \cap {}^\bullet v = \emptyset$, and hence $p \notin {}^\bullet v$, $q \notin {}^\bullet t$ and $t \neq v$. These requirements follow from the conditions above.

Proposition 2. *A net with a fully reachable visible pure* M *is not distributed.*

Proof. Let $N = (S, T, F, M_0, \ell)$ be a net that has a fully reachable visible pure M, so there exist $t, u, v \in T$ and $p, q \in S$ such that $p \in {}^\bullet t \cap {}^\bullet u \wedge q \in {}^\bullet u \cap {}^\bullet v \wedge {}^\bullet t \cap {}^\bullet v = \emptyset$ and $\exists M \in [M_0\rangle. {}^\bullet t \cup {}^\bullet u \cup {}^\bullet v \subseteq M$. Then $t \smile v$. Suppose N is distributed by the distribution D. Then $t \equiv_D p \equiv_D u \equiv_D q \equiv_D v$ but $t \smile v$ implies $t \not\equiv_D v$. ⨪   □

Now we show that fully reachable visible pure M's that are present in a plain net are preserved under step readiness equivalence.

Lemma 1. *Let $N = (S, T, F, M_0, \ell)$ be a plain net. If N has a fully reachable visible pure M, there exists $<\sigma, X> \in \mathcal{R}(N)$ such that $\exists a, b, c \in$ Act. $a \neq c \wedge \{b\} \in X \wedge \{a, c\} \in X \wedge \{a, b\} \notin X \wedge \{b, c\} \notin X$. (It is implied that $a \neq b \neq c$.)*

Proof. N has a fully reachable visible pure M, so there exist $t, u, v \in T$ and $M \in [M_0\rangle$ such that ${}^\bullet t \cap {}^\bullet u \neq \emptyset \wedge {}^\bullet u \cap {}^\bullet v \neq \emptyset \wedge {}^\bullet t \cap {}^\bullet v = \emptyset \wedge \ell(t), \ell(u), \ell(v) \neq \tau \wedge {}^\bullet t \cup {}^\bullet u \cup {}^\bullet v \subseteq M$. Let $\sigma \in$ Act* such that $M_0 \stackrel{\sigma}{\Longrightarrow} M$. Since N is a plain net, $M \stackrel{\tau}{\nrightarrow}$ and $\ell(t) \neq \ell(u) \neq \ell(v) \neq \ell(t)$. Hence there exists an $X \subseteq \mathbb{N}^{\text{Act}}$ such that $<\sigma, X> \in \mathcal{R}(N) \wedge \{\ell(u)\} \in X \wedge \{\ell(t), \ell(v)\} \in X \wedge \{\ell(t), \ell(u)\} \notin X \wedge \{\ell(u), \ell(v)\} \notin X$. $\qquad\square$

Lemma 2. *Let $N = (S, T, F, M_0, \ell)$ be a net. If there exists $<\sigma, X> \in \mathcal{R}(N)$ such that $\exists a, b, c \in$ Act. $a \neq c \wedge \{b\} \in X \wedge \{a, c\} \in X \wedge \{a, b\} \notin X \wedge \{b, c\} \notin X$, then N has a fully reachable visible pure M.*

Proof. Let $M \subseteq S$ be the marking which gave rise to the step ready pair $<\sigma, X>$, i.e. $M_0 \stackrel{\sigma}{\Longrightarrow} M$ and $M \stackrel{\{b\}}{\longrightarrow} \wedge M \stackrel{\{a,c\}}{\longrightarrow} \wedge M \stackrel{\{a,b\}}{\nrightarrow} \wedge M \stackrel{\{b,c\}}{\nrightarrow}$.

As $a \neq b \neq c \neq a$ there must exist three transitions $t, u, v \in T$ with $\ell(t) = a \wedge \ell(u) = b \wedge \ell(v) = c$ and $M[\{u\}\rangle \wedge M[\{t, v\}\rangle \wedge \neg(M[\{t, u\}\rangle) \wedge \neg(M[\{u, v\}\rangle)$. From $M[\{u\}\rangle \wedge M[\{t, v\}\rangle$ follows ${}^\bullet t \cup {}^\bullet u \cup {}^\bullet v \subseteq M$. From $M[\{t, v\}\rangle$ follows ${}^\bullet t \cap {}^\bullet v = \emptyset$. From $\neg(M[\{t, u\}\rangle)$ then follows ${}^\bullet t \cap {}^\bullet u \neq \emptyset$ and analogously for u and v. Hence N has a fully reachable visible pure M. $\qquad\square$

Note that the lemmas above give a behavioural property that for plain nets is equivalent to having a fully reachable visible pure M.

Theorem 3. *A plain net with a fully reachable visible pure M is truly synchronous.*

Proof. Let N be a plain net which has a fully reachable visible pure M. Let N' be a net which is step readiness equivalent to N. By Lemma 1 and Lemma 2, also N' has a fully reachable visible pure M. By Proposition 2, N' is not distributed. Thus N is truly synchronous. $\qquad\square$

Theorem 3 gives an upper bound of the class of distributable nets. We conjecture that this upper bound is tight, and a plain net is distributable iff it has no fully reachable visible pure M.

Conjecture 1. A plain net is truly synchronous iff it has a fully reachable visible pure M.

In the following, we give a lower bound of distributability by providing a protocol to implement certain kinds of plain nets distributedly. These implementations do not add additional labelled transitions, but only provide the existing ones with a communication protocol in the form of τ-transitions. Hence these implementations pertain to a notion of distributability in which we restrict implementations to be plain τ-nets. Note that this does not apply to the impossibility result above.

Definition 14. A plain net N is *plain-distributable* iff there exists a distributed plain τ-net N which is step readiness equivalent to N.

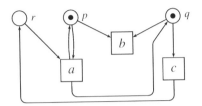

Fig. 4. An example net

Definition 15. Let $N = (S, T, F, M_0, \ell)$ be a net. We define the *enabled conflict relation* $\# \subseteq T^2$ as $t \# u \Leftrightarrow \exists M \in [M_0\rangle.\ M[\{t\}\rangle \wedge M[\{u\}\rangle \wedge \neg(M[\{t, u\}\rangle)$.

We now propose the following protocol for implementing nets. An example depicting it can be found in Fig. 5. As locations we take the places in a given net, and the equivalence classes of transitions that are related by the reflexive and transitive closure of the enabled conflict relation. We locate every transition t in its equivalence class, whereas every place s will have an embassy $s^{[t]}$ in every location $[t]$ where one of its posttransitions $t \in s^\bullet$ resides. As soon as s receives a token, it will distribute this information to its posttransitions by placing a token in each of these embassies. The arc from s to t is now replaced by an arc from $s^{[t]}$ to t, so if t could fire in the original net it can also fire in the implementation. So far the construction allows two transitions in different locations that shared the precondition s to fire concurrently, although they were in conflict in the original net. However, if this situation actually occurs, these transitions would have been in an enabled conflict, and thus assigned to the same location. The rest of the construction is a matter of garbage collection. If a transition t fires, for each of its preplaces s, all tokens that are still present in the various embassies of s in locations $[u]$ need to be removed from there. This is done by a special internal transition $t_s^{[u]}$. Once all these transitions (for the various choices of s and $[u]$) have fired, an internal transition t' occurs, which puts tokens in all the postplaces of t.

Definition 16. Let $N = (S, T, F, M_0, \ell)$ be a plain net. Let $[t] := \{u \in T \mid t \#^* u\}$. The transition-controlled-choice implementation of N is defined to be the plain τ-net $N' := (S \cup S^\tau, T \cup T^\tau, F', M_0, \ell')$ with

$$S^\tau := \{s^{[t]} \mid s \in S, t \in s^\bullet\} \cup \{\textcircled{t} \mid t \in T\} \cup$$
$$\{s_t^{[u]}, \overline{s}_t^{[u]} \mid s \in S,\ t, u \in s^\bullet, [u] \neq [t]\}$$
$$T^\tau := \{\boxed{s} \mid s \in S\} \cup \{t' \mid t \in T\} \cup$$
$$\{t_s^{[u]} \mid s \in S,\ t, u \in s^\bullet, [u] \neq [t]\}$$
$$F' := \{(s, \boxed{s}) \mid s \in S\} \cup$$
$$\{(\boxed{s}, s^{[t]}), (s^{[t]}, t) \mid s \in S, t \in s^\bullet\} \cup$$
$$\{(t, \textcircled{t}), (\textcircled{t}, t') \mid t \in T\} \cup$$
$$\{(t', s) \mid t \in T, s \in t^\bullet\} \cup$$
$$\{(t, s_t^{[u]}), (s_t^{[u]}, t_s^{[u]}), (t_s^{[u]}, \overline{s}_t^{[u]}), (\overline{s}_t^{[u]}, t'), (s^{[u]}, t_s^{[u]}) \mid s \in S,\ t, u \in s^\bullet, [u] \neq [t]\}$$

$\ell' \upharpoonright T = \ell$ and $\ell'(T^\tau) = \{\tau\}$.

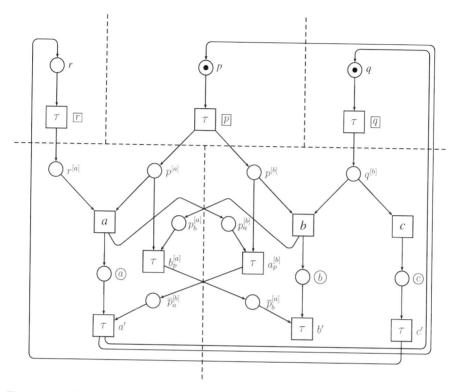

Fig. 5. A distributed implementation for the net in Fig. 4, partitioning into localities shown by dashed lines

Theorem 4. *A plain net N is plain distributable iff* $\#^* \cap \smile = \emptyset$.

Proof (sketch). "$\Rightarrow$": When implementing a plain net $N = (S, T, F, M_0, \ell)$ by a plain τ-net $N' = (S', T', F', M_0', \ell')$ that is step readiness equivalent to N, the $\#$ and $\smile$ relations between the transitions of N also exists between the corresponding visible transitions of N'. This is easiest to see when writing a_N, resp. $a_{N'}$, to denote a transition in N, resp. N', with label a, which must be unique since N is a plain net, resp. N' a plain τ-net. Namely if $a_N \# b_N$, then N has a step ready pair $<\sigma, X>$ with $\{a\}, \{b\} \in X$ but $\{a, b\} \notin X$. This must also be a step ready pair of N', and hence $a_{N'} \# b_{N'}$. Likewise, $a_N \smile b_N$ implies $a_{N'} \smile b_{N'}$.

Thus if $\#^* \cap \smile \neq \emptyset$ holds in N, then the same is the case for N', and hence N' is not distributed by Observation 1.

"$\Leftarrow$": If $\#^* \cap \smile = \emptyset$, N can be implemented as specified in Definition 16. In fact, the transition-controlled-choice implementation of any net N yields a net that is step readiness equivalent to N. See the full version of this paper [7] for a formal proof of this claim. Moreover, if $\#^* \cap \smile = \emptyset$ it never happens that concurrent visible transitions are co-located, and hence the implementation will be plain-distributed. $\qquad\square$

Our definition of distributed nets only enforces concurrent actions to be on different locations if they are visible, and our implementation in Definition 16 produces nets which actually contain concurrent unobservable activity at the same location. If this is undesired it can easily be amended by adding a single marked place to every location and connecting that place to every transition on that location by a self-loop. While this approach will introduce new causality relations, step readiness equivalence will not detect this.

6   Conclusion

In this paper, we have characterised different grades of asynchrony in Petri nets in terms of structural and behavioural properties of nets. Moreover, we have given both an upper and a lower bound of distributability of behaviours. In particular we have shown that some branching-time behaviours cannot be exhibited by a distributed system.

We did not consider connections from transitions to their postplaces as relevant to determine asynchrony and distributability. This is because we only discussed contact-free nets where no synchronisation by postplaces is necessary. In the spirit of Definition 6 we could insert τ-transitions on any or all arcs from transitions to their postplaces, and the resulting net would always be equivalent to the original.

We have already given a short overview on related work in the introduction of this paper. Most closely related to our approach are several lines of work using Petri nets as a model of reactive systems.

As mentioned in Section 3, classes of nets with certain structural properties like *free choice nets* [3,2] and *simple nets* [3], as well as extensions of theses classes, have been extensively studied in Petri net theory, and are closely related to the classes of nets defined here. In [3], Eike Best and Mike Shields introduce various transformations between free choice nets, simple nets and extended variants thereof. They use "essential equivalence" to compare the behaviour of different nets, which they only give informally. This equivalence is insensitive to divergence, which is relied upon in their transformations. It also does not preserve concurrency, which makes it possible to implement *behavioural free choice nets*, that may feature a fully reachable visible M, as free choice nets. They continue to show conditions under which liveness can be guaranteed for many of these classes.

In [1], Wil van der Aalst, Ekkart Kindler and Jörg Desel introduce two extensions to extended simple nets, by excluding self-loops from the requirements imposed on extended simple nets. This however assumes a kind of "atomicity" of self-loops, which we did not allow in this paper. In particular we do not implicitly assume that a transition will not change the state of a place it is connected to by a self-loop, since in case of deadlock, the temporary removal of a token from such a place might not be temporary indeed.

In [16], Wolfgang Reisig introduces a class of systems which communicate using buffers and where the relative speeds of different components are guaranteed to be irrelevant. The resulting nets are simple nets. He then proceeds introducing a decision procedure for the problem whether a marking exists which makes the complete system live.

Dirk Taubner has in [18] given various protocols by which to implement arbitrary Petri nets in the OCCAM programming language. Although this programming language offers synchronous communication he makes no substantial use of that feature in the protocols, thereby effectively providing an asynchronous implementation of Petri nets.

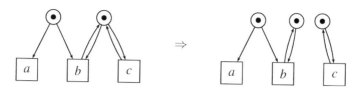

Fig. 6. A specification and its Hopkins-implementation which added concurrency

He does not indicate a specific equivalence relation, but is effectively using linear-time equivalences to compare implementations to the specification.

The work most similar to our approach we have found is the one by Hopkins, [9]. There he already classified nets by whether they are implementable by a net distributed among different locations. He uses an interleaving equivalence to compare an implementation to the original net, and while allowing a range of implementations, he does require them to inherit some of the structure of the original net. The net classes he describes in his paper are larger than those of Section 3 because he allows more general interaction patterns, but they are incomparable with those of Section 5. One direction of this inequality depends on his choice of interleaving semantics, which allows the implementation in Fig. 6. The step readiness equivalence we use does not tolerate the added concurrency and the depicted net is not distributable in our sense. The other direction of the inequality stems from the fact that we allow implementations which do not share structure with the specification but only emulate its behaviour. That way, the net in Fig. 7 can be implemented in our approach as depicted.

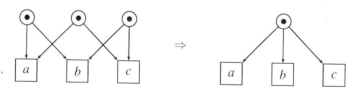

Fig. 7. A distributable net which is not considered distributable in [9], and its implementation

Still many open questions remain. While our impossibility result holds even when allowing labelled nets as implementations, our characterisation in Theorem 4 only considers unlabelled ones. This begs the question which class of nets can be implemented using labelled nets. We conjecture that a distributed implementation exists for every net which has no fully reachable visible pure **M**. We also conjecture that if we allow linear time correct implementations, all nets become distributable, even when only allowing finite implementations of finite nets. We are currently working on both problems.

Just as a distributable net is defined as a net that is behaviourally equivalent to, or implementable by, a distributed net, one could define an *asynchronously implementable* net as one that is implementable by an asynchronous net. This concept is again parametrised by the choice of an interaction pattern. It would be an interesting quest to characterise the various classes of asynchronously implementable plain nets.

Also, extending our work to nets that are not required to be 1-safe will probably generate interesting results, as conflict resolution protocols must keep track of which token they are currently resolving the conflict of.

In regard to practical applicability of our results, it would be very interesting to relate our Petri net based terminology to hardware descriptions in chip design. Especially in modern multi-core architectures performance reasons often prohibit using global clocks while a façade of synchrony must still be upheld in the abstract view of the system.

On a higher level of applications, we expect our results to be useful for language design. To start off, we would like to make a thorough comparison of our results to those on communication patterns in process algebras, versions of the π-calculus and I/O-automata [12]. Using a Petri net semantics of a suitable system description language, we could compare our net classes to the class of nets expressible in the language, especially when restricting the allowed communication patterns in the various ways considered in [4] or in [12]. Furthermore, we are interested in applying our results to graphical formalisms for system design like UML sequence diagrams or activity diagrams, also by applying their Petri net semantics. Our results become relevant when such formalisms are used for the design of distributed systems. Certain choice constructs become problematic then, as they rely on a global mechanism for consistent choice resolution; this could be made explicit in our framework.

References

1. van der Aalst, W.M.P., Kindler, E., Desel, J.: Beyond asymmetric choice: A note on some extensions. Petri Net Newsletter 55, 3–13 (1998)
2. Best, E.: Structure theory of Petri nets: The free choice hiatus. In: Brauer, W., Reisig, W., Rozenberg, G. (eds.) APN 1986. LNCS, vol. 254, pp. 168–206. Springer, Heidelberg (1987)
3. Best, E., Shields, M.W.: Some equivalence results for free choice nets and simple nets and on the periodicity of live free choice nets. In: Ausiello, G., Protasi, M. (eds.) CAAP 1983. LNCS, vol. 159, pp. 141–154. Springer, Heidelberg (1983)
4. de Boer, F.S., Palamidessi, C.: Embedding as a tool for language comparison: On the CSP hierarchy. In: Baeten, J.C.M., Groote, J.F. (eds.) CONCUR 1991. LNCS, vol. 527, pp. 127–141. Springer, Heidelberg (1991)
5. Bougé, L.: On the existence of symmetric algorithms to find leaders in networks of communicating sequential processes. Acta Informatica 25(2), 179–201 (1988)
6. van Glabbeek, R.J., Goltz, U., Schicke, J.-W.: Symmetric and asymmetric asynchronous interaction. Technical Report 2008-03, TU Braunschweig. Extended abstract in Proceedings 1st Interaction and Concurrency Experience (ICE 2008) on Synchronous and Asynchronous Interactions in Concurrent Distributed Systems, to appear in Electronic Notes in Theoretical Computer Science. Elsevier, Amsterdam (2008)
7. van Glabbeek, R.J., Goltz, U., Schicke, J.-W.: On synchronous and asynchronous interaction in distributed systems. Technical Report 2008-04, TU Braunschweig (2008)
8. Gorla, D.: On the Relative Expressive Power of Asynchronous Communication Primitives. In: Aceto, L., Ingólfsdóttir, A. (eds.) FOSSACS 2006. LNCS, vol. 3921, pp. 47–62. Springer, Heidelberg (2006)
9. Hopkins, R.P.: Distributable nets. In: Rozenberg, G. (ed.) APN 1991. LNCS, vol. 524, pp. 161–187. Springer, Heidelberg (1991)
10. Lamport, L.: Time, clocks, and the ordering of events in a distributed system. Communications of the ACM 21(7), 558–565 (1978)

11. Lamport, L.: Arbitration-free synchronization. Distributed Computing 16(2-3), 219–237 (2003)
12. Lynch, N.: Distributed Algorithms. Morgan Kaufmann Publishers, San Francisco (1996)
13. Nestmann, U.: What is a 'good' encoding of guarded choice? Information and Computation 156, 287–319 (2000)
14. Olderog, E.-R., Hoare, C.A.R.: Specification-oriented semantics for communicating processes. Acta Informatica 23, 9–66 (1986)
15. Palamidessi, C.: Comparing the expressive power of the synchronous and the asynchronous pi-calculus. In: Conference Record of the 24th ACM SIGPLAN-SIGACT Symposium on Principles of Programming Languages (POPL 1997), pp. 256–265. ACM Press, New York (1997)
16. Reisig, W.: Deterministic buffer synchronization of sequential processes. Acta Informatica 18, 115–134 (1982)
17. Selinger, P.: First-order axioms for asynchrony. In: Mazurkiewicz, A., Winkowski, J. (eds.) CONCUR 1997. LNCS, vol. 1243, pp. 376–390. Springer, Heidelberg (1997)
18. Taubner, D.: Zur verteilten Implementierung von Petrinetzen. Informationstechnik 30(5), 357–370 (1988); Technical report, TUM-I 8805, TU München

A Robust Class
of Regular Languages

Antonio Cano Gómez[1] and Jean-Éric Pin[2,*]

[1] Departamento de Sistemas Informáticos y Computación, Universidad Politécnica
de Valencia, Camino de Vera s/n, P.O. Box: 22012, E-46020 - Valencia
[2] LIAFA, Université Paris-Diderot and CNRS, Case 7014,
75205 Paris Cedex 13, France

Abstract. In this survey paper, we present known results and open
questions on a proper subclass of the class of regular languages. This
class, denoted by $\mathcal{W}$, is especially robust: it is closed under union, intersection, product, shuffle, left and right quotients, inverse of morphisms,
length preserving morphisms and commutative closure. It can be defined
as the largest positive variety of languages not containing the language
$(ab)^*$. It admits a nontrivial algebraic characterization in terms of finite
ordered monoids, which implies that $\mathcal{W}$ is decidable: given a regular language, one can effectively decide whether or not it belongs to $\mathcal{W}$. We
propose as a challenge to find a constructive description and a logical
characterization of $\mathcal{W}$.

Warning. In this paper, square brackets are used as a substitute to "respectively" to gather several definitions [properties] into a single one.

The search for robust classes of regular languages is an old problem of automata
theory, which occurs in particular in the study of regular model checking [3]. In
this survey paper, we present known results and open questions on a proper subclass of the class of regular languages, introduced a few years ago by the authors
in connection with the study of the shuffle product [6,7]. This class, denoted by
$\mathcal{W}$, is especially robust: it is closed under union, intersection, product, shuffle,
left and right quotients, inverse of morphisms, length preserving morphisms and
commutative closure. Furthermore, this class is decidable: there is an algorithm
to decide whether a given regular language belongs to $\mathcal{W}$ or not. As such, it
might offer an appropriate framework for modeling certain problems arising in
the verification of concurrent systems.

The class $\mathcal{W}$ is also interesting on its own and appears in the study of three
operations on languages: length preserving morphisms, inverse of substitutions
and shuffle product. More specifically, $\mathcal{W}$ is the largest proper positive variety
of languages closed under one of these operations. It is also the largest positive
variety of languages not containing the language $(ab)^*$.

---

* The authors acknowledge support from the AutoMathA programme of the European
Science Foundation.

E. Ochmański and J. Tyszkiewicz (Eds.): MFCS 2008, LNCS 5162, pp. 36–51, 2008.

All these results rely on an algebraic characterization of $\mathcal{W}$ in terms of ordered monoids (Theorem 5). It gives us the opportunity to review this algebraic approach and to apply it to a concrete example.

Our paper is organised as follows. In Section 1, we briefly introduce the definitions needed for this paper, including the notion of ordered automaton, which might be new to most readers. Section 2 presents the algebraic background. Again, the less familiar notions are probably those of ordered monoid and of profinite monoids. Section 3 is devoted to general results derived from the algebraic approach, including specific results on length preserving morphisms and the shuffle operation. The class $\mathcal{W}$, its algebraic characterization and its main properties are presented in Section 4. Closure under partial commutation is discussed in Section 5. Finally, we propose a few open problems on $\mathcal{W}$ in the final section. One of them is to find a logical characterization for $\mathcal{W}$, a problem which is widely open.

1   Languages and Automata

In this paper, an *alphabet* is a finite set whose elements are called *letters*. The free monoid A^* is the set of words on the alphabet A. The length of a word u is denoted by $|u|$. The *empty word*, denoted by 1, is the unique word of length 0.

1.1   Ordered Automata

An *ordered automaton* is a deterministic automaton equipped with a partial order on its set of states. This order is required to be compatible with the action of each letter. Formally, we are given an automaton $\mathcal{A} = (Q, A, \cdot\,, i, F)$ and a partial order $\leqslant$ on Q such that, for all $p, q \in Q$ and $a \in A$, $p \leqslant q$ implies $p \cdot a \leqslant q \cdot a$.

If $\mathcal{A}$ is a minimal deterministic automaton, there is a canonical way to define a partial order on Q, called the *syntactic order* on Q. Define a relation $\leqslant$ on Q by $p \leqslant q$ if and only if for each $u \in A^*$,

$$q \cdot u \in F \Rightarrow p \cdot u \in F$$

It is clear that $\leqslant$ is reflexive and transitive. To see it is a partial order, suppose that $p \leqslant q$ and $q \leqslant p$. Then, for all $u \in A^*$, one gets $q \cdot u \in F$ if and only if $p \cdot u \in F$, which gives $p = q$ since $\mathcal{A}$ is minimal.

Thus every language admits a *minimal ordered automaton*. In the remainder of this paper, we consider only regular languages and finite automata.

Example 1. For the minimal automaton of the language $(ab)^*$ represented in Figure 1, the order on the set of states is $1 < 0$ and $2 < 0$.

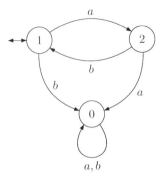

Fig. 1. The minimal automaton of $(ab)^*$

Example 2. For the minimal automaton of the language $(ab)^* \cup A^* aaA^*$ (with $A = \{a, b\}$) represented in Figure 2, the order on the set of states is $0 < 1 < 3$ and $0 < 2 < 4 < 3$.

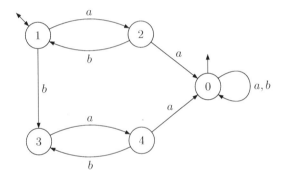

Fig. 2. The minimal automaton of $(ab)^* \cup A^* aaA^*$

1.2   Operations on Languages

A number of operations preserve regular languages: Boolean operations, product, star, shuffle, quotients, morphisms, inverse of morphisms, etc.

Boolean operations comprise union, intersection and complement. Let L_1 and L_2 be two languages of A^*. The *(concatenation) product* of L_1 and L_2 is the language

$$L_1 L_2 = \{x_1 x_2 \mid x_1 \in L_1, \ x_2 \in L_2\}$$

Their *shuffle* is the language

$$L_1 \ \mathrm{III} \ L_2 = \{w \in A^* \mid w = u_1 v_1 \cdots u_n v_n \text{ for some } n \geqslant 0 \text{ such that}$$
$$u_1 \cdots u_n \in L_1, v_1 \cdots v_n \in L_2\}$$

Given a language L and a word u, the *left and right quotients* of L by u are the languages

$$u^{-1} L = \{x \in A^* \mid ux \in L\}$$
$$L u^{-1} = \{x \in A^* \mid xu \in L\}$$

A *morphism* between two free monoids A^* and B^* is a map $\varphi : A^* \to B^*$ such that, for all $u, v \in A^*$, $\varphi(uv) = \varphi(u)\varphi(v)$. This condition implies in particular that $\varphi(1) = 1$. The morphism φ is *length preserving* if, for all $u \in A^*$, the condition $|\varphi(u)| = |u|$ is satisfied. This is equivalent to requiring that, for all $a \in A$, $\varphi(a) \in B$.

The languages of A^* form a monoid for the concatenation product, called the *monoid of languages* of A^*. A *substitution* from A^* into B^* is a monoid morphism σ from A^* into the monoid of languages on B^*. In particular, $\sigma(1) = \{1\}$, the language reduced to the empty word and if $u = a_1 \cdots a_n$, $\sigma(u) = \sigma(a_1) \cdots \sigma(a_n)$. Thus a substitution is completely determined by the languages $\sigma(a)$, for $a \in A$.

The *inverse substitution* σ^{-1} maps a language K of B^* onto the language $\sigma^{-1}(K)$ of A^*, defined by

$$\sigma^{-1}(K) = \{u \in A^* \mid \sigma(u) \cap K \neq \emptyset\}$$

1.3 Classes of Languages and Varieties of Languages

A *class of languages* is a correspondence $\mathcal{C}$ which associates with each alphabet A a set $\mathcal{C}(A^*)$ of regular languages of A^*. It is *closed under inverse of morphisms [substitutions]* if, for any morphism [substitution] $\varphi : A^* \to B^*$ and for any language $L \in \mathcal{C}(B^*)$, the language $\varphi^{-1}(L)$ belongs to $\mathcal{C}(A^*)$. Similarly, it is *closed under length-preserving morphism* if, for any length-preserving morphism $\varphi : A^* \to B^*$ and for any language $L \in \mathcal{C}(A^*)$, the language $\varphi(L)$ belongs to $\mathcal{C}(B^*)$. Finally, it is closed under union [intersection, complement, residuals, product, shuffle, etc.] if, for each alphabet A, the set $\mathcal{C}(A^*)$ is closed under union [intersection, complement, residuals, product, shuffle, etc.]

A class of regular languages is said to be *proper* if it is not the class of all regular languages.

A *positive variety of languages* is a class of regular languages closed under union, intersection, residuals and inverses of morphisms. A *variety of languages* is a positive variety closed under complement.

2 A Bit of Algebra

In this section, we gather the algebraic notions used in this paper: semigroups, monoids, ordered monoids, power monoids, profinite monoids and varieties.

2.1 Semigroups and Monoids

A *semigroup* is a set equipped with an associative operation, usually denoted multiplicatively. A *monoid* is a semigroup with an identity element, usually denoted by 1.

An element e of a monoid is *idempotent* if $e^2 = e$. Given an element s of a finite semigroup S, s^ω denotes the unique idempotent of the subsemigroup of S generated by s. Two elements s and t of a semigroup are *mutually inverse* if $sts = s$ and $tst = t$.

Let M be a finite monoid. The *exponent* of M is the least integer ω such that for all $x \in M$, x^ω is idempotent. Its *period* is the least integer p such that for all $x \in M$, $x^{\omega + p} = x^\omega$.

An *ideal* of a monoid M is a subset I of M such that $MIM \subseteq I$. An ideal I is called *minimal* if, for every J of M, the condition $J \subseteq I$ implies $J = \emptyset$ or $J = I$. Every finite monoid admits a unique minimal ideal. This minimal ideal I has a very constrained structure: in particular, if e is an idempotent of I and x is an element of M, then $(exe)^\omega = e$.

Let M and N be two monoids. A *morphism of monoids* from M into N is a function $\varphi : M \to N$ such that $\varphi(1) = 1$ and for all $x, y \in M$, $\varphi(xy) = \varphi(x)\varphi(y)$.

A *transformation* on a set Q is a map from Q to Q. We use the notation $q \cdot f$ to denote the image of an element $q \in Q$ by f, instead of the standard $f(q)$. The product of two transformations f and g is the transformation fg defined, for all $q \in Q$, by $q \cdot (fg) = (q \cdot f) \cdot g$. Note that, in traditional notation, the function fg would be denoted $g \circ f$. Equipped with this product, the set of transformations on Q form a monoid, denoted by $\mathcal{T}(Q)$.

Given a deterministic automaton $\mathcal{A} = (Q, A, \cdot, i, F)$, each word $u \in A^*$ defines a transformation on Q, which maps the state q onto the state $q \cdot u$. The set of all these transformations form a submonoid of $\mathcal{T}(Q)$, called the *transition monoid* of $\mathcal{A}$. One can also attach a finite monoid to a nondeterministic automaton. See [14,16] for more details.

The monoid attached to the minimal automaton of a language is called its *syntactic monoid*. It can be defined directly as follows. The *syntactic congruence* of a language L of A^* is the congruence $\sim_L$ defined on A^* by setting $u \sim_L v$ if and only if, for every $x, y \in A^*$,

$$xvy \in L \Leftrightarrow xuy \in L$$

The *syntactic monoid* is the quotient of A^* by $\sim_L$ and the natural morphism from A^* onto M is called the *syntactic morphism* of L.

2.2   Ordered Monoids

An *ordered monoid* is a monoid M equipped with a partial order $\leqslant$ compatible with the product on M: for all $x, y, z \in M$, if $x \leqslant y$ then $zx \leqslant zy$ and $xz \leqslant yz$.

Let $(M, \leqslant)$ be an ordered monoid. An *order ideal* of M is a subset I of M such that if $x \in I$ and $y \leqslant x$ then $y \in I$. A *filter* of M is a subset F of M such that if $x \in F$ and $x \leqslant y$ then $y \in F$. Note that a subset of M is a filter if and only if its complement is an order ideal.

Let M and N be two ordered monoids. A *morphism of ordered monoids* is an order-preserving monoid morphism from M into N. We say that N is a *quotient* of M if there exists a surjective morphism of ordered monoids from M onto N. An *ordered submonoid* of M is a submonoid of M, equipped with the restriction of the order on M.

The *product* of a family $(M_i)_{i \in I}$ of ordered monoids is the ordered monoid defined on the set $\prod_{i \in I} M_i$. The multiplication and the order relation are defined componentwise.

2.3   Monoids and Automata

There are two ways to make use of monoids to describe languages.

The first solution bypasses the notion of automata by defining directly languages recognized by an [ordered] monoid. We just recall this definition in the ordered case [14,16]. A language L of A^* is *recognized by an ordered monoid* $(M, \leqslant)$ if and only if there exist an order ideal I of M and a monoid morphism η from A^* into M such that $L = \eta^{-1}(I)$.

The second solution relies on the notion of transition monoid. If a deterministic automaton $\mathcal{A} = (Q, A, \cdot, i, F)$ is partially ordered, then its transition monoid M can be ordered in a natural way. It suffices to set $u \leqslant v$ if and only if, for every $x \in M$ and $q \in Q$,

$$q \cdot vx \in F \Rightarrow q \cdot ux \in F$$

If $\mathcal{A}$ is the ordered minimal automaton of a language L, we obtain the *syntactic ordered monoid* of L. The *syntactic order* $\leqslant$ on M can also be defined directly. Let $\eta : A^* \to M$ be the syntactic morphism of L. Then, given $u, v \in M$, one has $u \leqslant v$ if and only if, for all $x, y \in M$,

$$xvy \in \eta(L) \Rightarrow xuy \in \eta(L)$$

Example 3. The minimal automaton $\mathcal{A}$ of the language $(ab)^*$ is represented in Figure 1. The transition monoid of $\mathcal{A}$ contains six elements which correspond to the words 1, a, b, ab, ba and aa.

	a	b	aa	ab	ba
1	2	0	0	1	0
2	0	1	0	0	2

Furthermore aa is a zero of this monoid and thus can be denoted 0. Finally, the syntactic ordered monoid of $(ab)^*$ is the ordered monoid

$$B_2^{1-} = \{1, a, b, ab, ba, 0\}$$

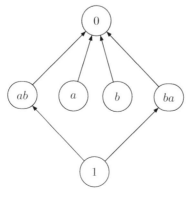

Fig. 3. The order on B_2^{1-}

presented by the relations $aba = a$, $bab = b$ and $aa = bb = 0$. The syntactic order is given by $1 \leqslant ab$, $1 \leqslant ba$ and $x \leqslant 0$ for every $x \in M$. This ordered monoid will play an important role in this paper.

2.4 Power Monoids

Let M be a monoid and let $\mathcal{P}(M)$ be the set of subsets of M. Define the *product of two subsets* X and Y of M as the set

$$XY = \{xy \mid x \in X \text{ and } y \in Y\}$$

This operation makes $\mathcal{P}(M)$ a monoid, called the *power monoid* of M.

It is possible to extend this notion to ordered monoids [7,19]. Let $(M, \leqslant)$ be an ordered monoid. Define the *product of two filters* F and G of M as the filter generated by the set FG:

$$\uparrow FG = \{z \in M \mid \text{there exist } x \in F \text{ and } y \in G \text{ such that } xy \leqslant z\}$$

This operation turns the set of filters on M into an ordered monoid, denoted by $\mathcal{P}^+(M, \leqslant)$, in which the order relation is the reverse inclusion $\supseteq$.

2.5 Profinite Monoids

We briefly recall the definition of a free profinite monoid. More details can be found in [1,2]. Let A be an alphabet. A monoid M *separates* two words u and v of the free monoid A^* if there exists a morphism φ from A^* onto M such that $\varphi(u) \neq \varphi(v)$. We set

$$r(u, v) = \min\{|M| \mid M \text{ is a monoid that separates } u \text{ and } v \}$$

and $d(u, v) = 2^{-r(u,v)}$, with the usual conventions $\min \emptyset = +\infty$ and $2^{-\infty} = 0$. Then d is an *ultrametric* on A^*, that is, satisfies the following properties, for all $u, v, w \in A^*$,

(1) $d(u, v) = 0$ if and only if $u = v$,

(2) $d(u, v) = d(v, u)$,

(3) $d(u, w) \leqslant \max\{d(u, v), d(v, w)\}$.

For the metric d, the closer are two words, the larger is the monoid needed to separate them.

As a metric space, A^* admits a completion, denoted by $\widehat{A^*}$. As A^* is dense in $\widehat{A^*}$ and since the product on A^* is uniformly continuous, it can be extended by continuity to $\widehat{A^*}$. The resulting monoid is called the *free profinite monoid* on A. This is a topological compact monoid which admits a unique minimal ideal. The elements of $\widehat{A^*}$ are called *profinite words*.

It can be shown that, for each profinite word x, the sequence $(x^{n!})_{n \geqslant 0}$ is a Cauchy sequence. It converges to an idempotent element of $\widehat{A^*}$, denoted by x^ω.

Every monoid morphism from A^* into a finite monoid M (considered as a discrete metric space), can be extended by continuity to a morphism from $\widehat{A^*}$

into M. In particular, the image of x^ω under any morphism $\varphi : A^* \to M$ into a finite monoid M is well defined: it is the unique idempotent of the subsemigroup of M generated by $\varphi(x)$. This fully justifies the natural formula $\varphi(x^\omega) = (\varphi(x))^\omega$, in which the ω on the right hand side denotes the exponent of M.

2.6   Varieties of Finite Monoids

A *variety of finite monoids* is a class of finite monoids closed under taking submonoids, quotients and finite direct products. *Varieties of finite ordered monoids* are defined analogously.

Given a variety of finite ordered monoids $\mathbf{V}$, the variety of finite ordered monoids $\mathbf{P}^+\mathbf{V}$ is generated by the monoids of the form $\mathcal{P}^+(M, \leqslant)$ where $(M, \leqslant) \in \mathbf{V}$.

In the same way as varieties in Birkhoff sense, varieties of finite monoids can be equationally defined, but this description involves *profinite equations*, which are formal equalities between two profinite words. More precisely, let u and v be two profinite words of $\widehat{A^*}$. A finite monoid M *satisfies the profinite equation* $u = v$ if and only if, for each morphism $\varphi : \widehat{A^*} \to M$, $\varphi(u) = \varphi(v)$. Similarly, a finite ordered monoid M *satisfies the profinite equation* $u \leqslant v$ if and only if, for each morphism $\varphi : \widehat{A^*} \to M$, $\varphi(u) \leqslant \varphi(v)$.

Given a set E of profinite equations, the class of finite [ordered] monoids satisfying all the equations of E form a variety of finite [ordered] monoids, denoted by $[\![E]\!]$. Reiterman's theorem [20] states that every variety of finite monoids can be defined by a set of profinite equations of the form $u = v$. As shown in [17], this result extends to varieties of finite ordered monoids, using equations of the form $u \leqslant v$.

For instance the variety $\mathbf{Com}$ of finite commutative monoids is defined by the single equation $xy = yx$. The variety of finite groups is defined by the single equation $x^\omega = 1$. The variety of finite ordered monoids $\mathbf{P}^+\mathbf{G}$ is defined by the single equation $x^\omega \leqslant 1$ [18].

3   The Algebraic Approach

The general idea of the algebraic approach is to classify regular languages through algebraic properties of their syntactic [ordered] monoid. We recall here two versions of the variety theorem. Extended versions were also obtained in [23,9] and a unified version is proposed in [11].

3.1   The Variety Theorem

Denote by $\mathbf{V} \to \mathcal{V}$ the correspondence which associates to a variety of finite [ordered] monoids the class $\mathcal{V}$ of all languages of A^* whose syntactic [ordered] monoid belongs to $\mathbf{V}$. One can show that $\mathcal{V}$ is a [positive] variety of languages.

Similarly, we denote by $\mathbf{V} \to \mathcal{V}$ the correspondence which associates to a [positive] variety of languages $\mathcal{V}$ the smallest variety of finite [ordered] monoids $\mathbf{V}$ containing the syntactic [ordered] monoids of the languages of $\mathcal{V}$.

The original variety theorem is due to Eilenberg [8]. Its ordered version was proved by the second author in [15].

Theorem 1 (Variety theorem). *The correspondences* $\mathbf{V} \to \mathcal{V}$ *and* $\mathcal{V} \to \mathbf{V}$ *are mutually inverse one to one correspondences between the varieties of finite [ordered] monoids and the [positive] varieties of languages.*

For instance, the variety of languages corresponding to **Com** is the variety $\mathcal{C}om$ of all commutative languages. Recall that a language L is *commutative* if $a_1 a_2 \cdots a_n \in L$ implies $a_{\sigma(1)} a_{\sigma(2)} \cdots a_{\sigma(n)} \in L$ for each permutation σ of $\{1, 2, \ldots, n\}$. Other descriptions of $\mathcal{C}om$ can be found in [8,13].

The variety of languages corresponding to **G** is the variety of group languages. Recall that a *group language* is a regular language whose syntactic monoid is a group, or, equivalently, is recognized by a finite deterministic automaton in which each letter defines a permutation of the set of states.

The languages of the positive variety corresponding to $\mathbf{P}^+\mathbf{G}$ are the polynomials of group languages. Recall that, given a class $\mathcal{C}$ of regular languages, the *polynomial languages* of $\mathcal{L}$ are the finite unions of languages of the form $L_0 a_1 L_1 \cdots a_k L_k$ where $a_1, \ldots, a_k$ are letters and $L_0, \ldots, L_k$ are languages of $\mathcal{C}$.

3.2   Length Preserving Morphisms and Inverse of Substitutions

Power monoids are the appropriate tool to study length preserving morphisms [12,21,22]. We recall here the ordered version of these results [7,19].

Given a positive variety of languages $\mathcal{V}$, the positive variety of languages $\Lambda^+\mathcal{V}$ is defined as follows. For each alphabet A, $\Lambda^+\mathcal{V}(A^*)$ is the lattice of languages generated by the languages of the form $\varphi(L)$, where $L \in \mathcal{V}(B^*)$ for some alphabet B and φ is a length preserving morphism from B^* into A^*.

Proposition 1. *Let* $\mathcal{V}$ *be a positive variety of languages and let* **V** *be the corresponding variety of finite ordered monoids. Then* $\Lambda^+\mathcal{V}$ *is a positive variety of languages and the corresponding variety of finite ordered monoids is* $\mathbf{P}^+\mathbf{V}$.

There is a similar result for inverse of substitutions. Given a positive variety of languages $\mathcal{V}$, the positive variety of languages $\Sigma^+\mathcal{V}$ is defined as follows. For every alphabet A, $\Sigma^+\mathcal{V}(A^*)$ is the lattice of languages generated by the languages of the form $\sigma^{-1}(L)$, where $L \in \mathcal{V}(B^*)$ for some alphabet B and σ is a substitution from A^* into B^*.

Proposition 2. *Let* **V** *be a variety of finite ordered monoids and* $\mathcal{V}$ *the corresponding positive variety of languages. Then* $\Sigma^+\mathcal{V}$ *is a positive variety of languages that corresponds to* $\mathbf{P}^+\mathbf{V}$. *In particular,* $\Sigma^+\mathcal{V} = \Lambda^+\mathcal{V}$.

3.3   The Shuffle Operation

Power monoids also make an important tool to study the shuffle product, due to the following result.

Proposition 3. *Let* L_1 *and* L_2 *be two languages of* A^*, *recognized respectively by the ordered monoids* M_1 *and* M_2. *Then* $L_1 \amalg L_2$ *is recognized by the ordered monoid* $\mathcal{P}^+(M_1 \times M_2)$.

A [positive] variety of languages $\mathcal{V}$ is *closed under shuffle* if the shuffle product of two languages of $\mathcal{V}$ is also in $\mathcal{V}$. It is closed under length preserving morphisms if $\Lambda^+\mathcal{V} = \mathcal{V}$ and it is closed under inverse of substitutions if $\Sigma^+\mathcal{V} = \mathcal{V}$. As a consequence of Propositions 1, 2 and 3, we get the following result.

Proposition 4

(1) *If a positive variety of languages is closed under length preserving morphisms, then it is closed under inverse of substitutions and under shuffle.*

(2) *If a positive variety of languages is closed under inverse of substitutions, then it is closed under length preserving morphisms and under shuffle.*

One may wonder whether a positive variety of languages is closed under length preserving morphisms if and only if it is closed under shuffle. This result holds for varieties of languages but depends on the classification of varieties closed under shuffle. It is still an open problem for positive varieties of languages.

It is easy to see that the variety of all commutative languages is closed under shuffle. Actually, the commutative varieties of languages closed under shuffle were characterised by Perrot [12]: they correspond to the varieties of commutative monoids whose groups belong to a given variety of commutative groups. Perrot also conjectured that the only non commutative variety of languages closed under shuffle was the variety of all regular languages, a result that was finally proved in 1998 by Esik and Simon [10]. Therefore the variety of commutative languages is the largest proper variety of languages closed under shuffle. This completes the classification of the varieties of languages closed under shuffle.

Classifying the positive varieties closed under shuffle seems to be a really challenging problem on which only partial results are known [4,7]. A first question is to know whether the result of Esik and Simon also holds for positive varieties: in other words, is there a largest proper positive variety closed under shuffle? This question was solved positively by the authors in [7].

Theorem 2. *There is a largest proper positive variety of languages closed under shuffle.*

This positive variety, denoted by $\mathcal{W}$ in the sequel, enjoys a number of interesting properties which are detailed in the next section.

4   A Robust Class of Languages

We start with a characterization of $\mathcal{W}$ in terms of languages, also given in [7]. The difficult part is to prove the existence of a largest positive variety of languages satisfying the condition of the theorem.

Theorem 3. *The positive variety $\mathcal{W}$ is the largest positive variety of languages such that, for $A = \{a, b\}$, the language $(ab)^*$ does not belong to $\mathcal{W}(A^*)$.*

Let us denote by **W** the variety of finite ordered monoids corresponding to $\mathcal{W}$. Theorem 3 can be translated immediately as follows:

Theorem 4. *The variety of finite ordered monoids* **W** *is the largest variety of finite ordered monoids not containing the ordered monoid B_2^{1-}.*

These characterizations are useful to prove that a language is not in $\mathcal{W}$. For instance, let $A = \{a, b\}$. We claim that the language $L = (aab)^* \cup A^* b(aa)^* ab A^*$ is not in $\mathcal{W}(A^*)$. Assume the contrary and let $\varphi : A^* \to A^*$ be the morphism defined by $\varphi(a) = aa$ and $\varphi(b) = b$. Then since a positive variety of languages is closed under inverse of morphisms, the language $\varphi^{-1}(L) = (ab)^*$ belongs to $\mathcal{W}(A^*)$, a contradiction with Theorem 3.

Theorems 3 and 4 are simple to state but they do not provide any algorithm to decide whether a given regular language belongs to $\mathcal{W}$ or, equivalently, whether a given finite ordered monoid belongs to **W**. A solution to this problem was given in [7].

Theorem 5. *A finite ordered monoid M belongs to* **W** *if and only if, for any pair (s, t) of mutually inverse elements of M, and any element z of the minimal ideal of the submonoid of M generated by s and t, $(stzst)^\omega \leqslant st$.*

Other equational descriptions are given in [7]. We now give a new formulation of Theorem 5 that is closer to automata theory. Before stating this result precisely, let us introduce some terminology.

Consider a deterministic automaton $\mathcal{A} = (Q, A, \cdot, i, F)$, a state p of Q and two words u and v of A^*. Let us say that u and v are *mutually inverse in* $\mathcal{A}$ if, for every state p, $p \cdot uvu = p \cdot u$ and $p \cdot vuv = p \cdot v$. This is clearly equivalent to saying that u and v define two mutually inverse transformations in the transformation monoid of $\mathcal{A}$.

We are interested in the graph $G(p, u, v)$ whose vertices are the states of the form $p \cdot z$, where $z \in \{u, v\}^*$ and the edges are of the form $q \to q \cdot u$ and $q \to q \cdot v$. As in any directed graph, the states of $G(p, u, v)$ are partially ordered by the reachability relation. To avoid any confusion with the syntactic order on Q, we will say that a state q_2 *is deeper that a state* q_1 if there is path from q_1 to q_2. Our new result can now be formulated as follows.

Theorem 6. *Let L be a regular language of A^*, let $\mathcal{A} = (Q, A, \cdot, i, F)$ be its minimal ordered automaton. Then L does not belong to $\mathcal{W}(A^*)$ if and only if there exist two states p and q of Q, two words u and v of A^*, mutually inverse in $\mathcal{A}$ such that $p \cdot u = q$, $q \cdot v = p$, a deepest state p' of the graph $G(p, u, v)$ and a word $r \in \{u, v\}^*$ such that $p \cdot r = p' \cdot r = p' \cdot uv = p'$ and $p' \not\leqslant p$ in the syntactic order on Q.*

Proof. Note that one may have $q = p$, but the condition $p' \not\leqslant p$ implies that p' is different from p. Let us denote by $(M, \leqslant)$ the syntactic ordered monoid of L and by $\eta : A^* \to M$ its syntactic morphism. Let ω be the exponent of M.

First assume that L does not belong to $\mathcal{W}(A^*)$. According to Theorem 5, there is a pair (s, t) of mutually inverse elements of M, generating a submonoid N of M and an element z of the minimal ideal of N such that $(stzst)^\omega \not\leqslant st$. Let us fix two words $u, v \in A^*$ such that $\eta(u) = s$ and $\eta(v) = t$. Then u and v

are by construction mutually inverse in $\mathcal{A}$. Since z belongs to N, there is also a word $w \in \{u, v\}^*$ such that $\eta(w) = z$. The condition $(stzst)^\omega \not\leqslant st$ implies that there exist two words $x, y \in A^*$ such that

$$xuvy \in L \quad \text{and} \quad x(uvwuv)^\omega y \notin L \tag{1}$$

Let us set $r = (uvwuv)^\omega$, $q = i \cdot xu$, $p = q \cdot v$ and $p' = p \cdot r$. Since s and t are mutually inverse, the words u and uvu define the same transformation on Q and in particular, $p \cdot u = i \cdot xuvu = i \cdot xu = q$. Further $p' \cdot r = p \cdot rr = p \cdot r = p'$ and $p' \cdot uv = p \cdot (uvwuv)^\omega uv = p \cdot (uvwuv)^\omega = p'$ since r and uv define idempotent transformations on Q.

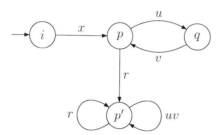

We claim that p' is a deepest state of the graph $G(p, u, v)$. Indeed, consider a state reachable from p', say $q' = p' \cdot f$ for some $f \in \{u, v\}^*$. Since $\eta(r)$ is an idempotent of the minimal ideal of N, one has $\eta((rfr)^\omega) = \eta(r)$. Since $p' \cdot r = p'$, it follows that $p' = p' \cdot r = p' \cdot (rfr)^\omega = q' \cdot r(rfr)^{\omega-1}$ and thus p' is reachable from q', which proves the claim.

Finally, it follows from (1) that $i \cdot xuvy \in F$ and $i \cdot x(uvwuv)^\omega y \notin F$, whence $p \cdot y \in F$ and $p' \cdot y \notin F$. Therefore $p' \not\leqslant p$ and the condition on $\mathcal{A}$ is satisfied.

Suppose now that the condition on $\mathcal{A}$ is satisfied. Since $\mathcal{A}$ is minimal, each state of Q is accessible and there exists a word x such that $i \cdot x = p$. Set $s = \eta(u)$ and $t = \eta(v)$. Then s and t are two mutually inverse elements of M which generate a submonoid N of M. Let I be the minimal ideal of N and let f be a word of $\{u, v\}^*$ such that $\eta(f)$ belongs to I. Since p' is a deepest state of $G(u, v)$, the state p' can be reached from $p' \cdot f$ and hence there is a word $g \in \{u, v\}^*$ such that $p' \cdot fg = p'$. Setting $w = rfg$, we get $p' \cdot w = p' \cdot rfg = p' \cdot fg = p'$. Therefore $p' = p' \cdot r = p' \cdot fg = p' \cdot uv$ and thus we obtain

$$i \cdot xuv = p \quad \text{and} \quad i \cdot x(uvwuv)^\omega = i \cdot x(uvrfguv)^\omega = p' \tag{2}$$

Now, since $p' \not\leqslant p$, there exists a word y such that $p \cdot y \in F$ but $p' \cdot y \notin F$. Consequently, it follows from (2) that $i \cdot xuvy \in F$ but $i \cdot x(uvwuv)^\omega y \notin F$, that is, $xuvy \in L$ but $x(uvwuv)^\omega y \notin L$. Setting $z = \eta(w)$, we get $(uvzuv)^\omega \not\leqslant uv$. Further, since $\eta(f) \in I$ and $z = \eta(r)\eta(f)\eta(g)$, z also belongs to I. Thus by Theorem 5, L does not belong to $\mathcal{W}(A^*)$. □

For instance, let us come back to Examples 1 and 2. If $\mathcal{A}$ is the minimal automaton of $(ab)^*$ represented in Figure 1, one can take $p = 1$, $q = 2$, $p' = 0$, $u = a$,

$v = b$ and $r = a$ to verify that $\mathcal{A}$ satisfies the conditions of Theorem 6. Thus $(ab)^*$ is not in $\mathcal{W}(A^*)$.

On the other hand, one can verify that the minimal automaton of $(ab)^* \cup A^*aaA^*$ represented in Figure 2 does not satisfy the conditions of Theorem 6. Thus $(ab)^* \cup A^*aaA^*$ belongs to $\mathcal{W}(A^*)$.

Note that the condition that u and v are mutually inverse in $\mathcal{A}$ is mandatory. Consider for instance the minimal automaton of the language $(ab)^* \cup (ab)^*bA^* \cup (ab)^*aabA^*$ on the alphabet $\{a, b\}$, represented in Figure 4. The order on the set of states is $3 < 1 < 4 < 0$ and $3 < 2 < 0$.

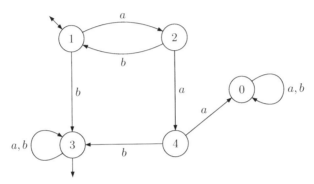

Fig. 4. The minimal automaton of $(ab)^* \cup (ab)^*bA^* \cup (ab)^*aabA^*$

Setting $p = 1$, $q = 2$, $p' = 0$ and $r = a^3$, one has $p \cdot a = q$, $q \cdot b = p$, $p \cdot r = p' \cdot r = p' \cdot uv = p'$ and $p' \not\leqslant p$. Further, $r \in \{a, b\}^*$ and p' is a deepest state in the graph $G(a, b)$. However, a and b are not mutually inverse and one can actually verify that $\mathcal{A}$ does not satisfy the conditions of Theorem 6. In particular, taking $u = aba$ and $v = b$ does not work, since there is no word $r \in \{u, v\}^*$ such that $p \cdot r' = p'$. Thus L belongs to $\mathcal{W}(A^*)$.

A key property of $\mathbf{W}$, also proved in [7], is stated in the next theorem

Theorem 7. *The equality* $\mathbf{W} = \mathbf{P}^+\mathbf{W}$ *holds.*

Propositions 1, 2 and 4 now give immediately:

Corollary 1. *The positive variety* $\mathcal{W}$ *is closed under length preserving morphisms, inverse of substitutions and shuffle.*

In fact, a stronger property holds [7].

Theorem 8

(1) *The positive variety* $\mathcal{W}$ *is the largest proper positive variety of languages closed under length preserving morphisms.*

(2) *The positive variety* $\mathcal{W}$ *is the largest proper positive variety of languages closed under inverse of substitutions.*

(3) *The positive variety* $\mathcal{W}$ *is the largest proper positive variety of languages closed under shuffle.*

Let us mention another important closure property of $\mathcal{W}$.

Proposition 5. *The positive variety $\mathcal{W}$ is closed under product.*

Proof. It suffices to use a standard trick to simulate a concatenation product by a shuffle product. Let L_1 and L_2 be two languages of $\mathcal{W}(A^*)$. Let $\bar{A} = \{\bar{a} \mid a \in A\}$ be a copy of A and let π_A and $\pi_{\bar{A}}$ be the morphisms from $(A \cup \bar{A})^*$ onto A^* defined by $\pi_A(a) = \pi_{\bar{A}}(\bar{a}) = a$ and $\pi_A(\bar{a}) = \pi_{\bar{A}}(a) = 1$ for all $\bar{a} \in A$. Consider the two languages of $(A \cup \bar{A})^*$

$$K_1 = \pi_A^{-1}(L_1) \cap A^* \qquad K_2 = \pi_{\bar{A}}^{-1}(L_2) \cap \bar{A}^*$$

Thus K_1 is just the same language as L_1, but on a larger alphabet, and K_2 is a copy of L_2. Finally, let

$$K = (K_1 \text{ III } K_2) \cap A^* \bar{A}^*$$

Since $\mathcal{W}$ is closed under intersection, inverse of morphisms and shuffle and since A^*, $\bar{A}^*$ and $A^* \bar{A}^*$ are languages of $\mathcal{W}((A \cup \bar{A})^*)$, one gets $K \in \mathcal{W}((A \cup \bar{A})^*)$. Finally let $\pi : (A \cup \bar{A})^* \to A^*$ be the morphism defined by $\pi(\bar{a}) = \pi(a) = a$. Now $\pi(K) = L_1 L_2$ and since $\mathcal{W}$ is closed under length preserving morphisms, $L_1 L_2 \in \mathcal{W}(A^*)$. $\qquad \square$

Note, however, that $\mathcal{W}$ is not the largest proper positive variety of languages closed under product. Indeed the variety of star-free languages is closed under product, but contains $(ab)^*$.

5 Closure under Commutation and Partial Commutation

The class $\mathcal{W}$ also occurs in the study of commutation relations.

Let A be an alphabet and let I be a symmetric and irreflexive relation on A (often called the *independence relation*). We denote by $\sim_I$ the congruence on A^* generated by the set $\{ab = ba \mid (a, b) \in I\}$. If L is a language on A^*, we also denote by $[L]_I$ the closure of L under $\sim_I$. When I is the relation $\{(a, b) \in A \times A \mid a \neq b\}$, we simplify the notation to $\sim$ and $[L]$, respectively. Thus $\sim$ is the commutation relation and $[L]$ is the *commutative closure* of L. A class $\mathcal{C}$ of languages is *closed under I-commutation* if $L \in \mathcal{C}$ implies $[L]_I \in \mathcal{C}$. It is *closed under commutation* if $L \in \mathcal{C}$ implies $[L] \in \mathcal{C}$.

Since the commutative closure of the language $(ab)^*$ is nonregular, a class of regular languages closed under commutation cannot contain $(ab)^*$. What happens for varieties and positive varieties of languages? One can show that there is a largest variety of languages $\mathcal{V}$ such that, for $A = \{a, b\}$, the language $(ab)^*$ does not belong to $\mathcal{V}(A^*)$. It is denoted by $\mathcal{DS}$ since the corresponding variety of finite monoids is the variety $\mathbf{DS}$. Recall that a finite monoid belongs to $\mathbf{DS}$ are if each of its regular $\mathcal{D}$-classes form a semigroup. In fact, one can show that $\mathcal{DS}$ is also the largest variety of languages closed under commutation. The corresponding result for positive varieties states that $\mathcal{W}$ is the largest positive variety of languages closed under commutation. Actually, a stronger result holds [5]. Define the *period* (respectively *exponent*) of a regular language as the period (respectively exponent) of its syntactic monoid.

Theorem 9. *Let L be a language of $\mathcal{W}(A^*)$. Then $[L]$ is regular and commutative (and hence belongs to $\mathcal{W}(A^*)$) and its period divides that of L.*

For partial commutations, a weaker result holds if I is transitive [5]. Recall that in this case, $A^*/\sim_I$ is a free product of free commutative monoids.

Theorem 10. *Let L be a language of $\mathcal{W}(A^*)$ and let I be a transitive independence relation. Then $[L]_I$ is a regular language.*

Although we know that $[L]_I$ is regular in this case, we don't know whether $[L]_I$ necessarily belongs to $\mathcal{W}(A^*)$.

6 Conclusion and Open Questions

We have seen that the class $\mathcal{W}$ is closed under the following operations: union, intersection, product, shuffle, left and right quotients, inverse of morphisms, length preserving morphisms and commutative closure. It is a decidable variety and the corresponding variety of finite ordered monoids **W** is precisely known. The positive variety $\mathcal{W}$ can be defined alternatively as the largest proper positive variety of languages satisfying (1) [(2), (3), (4)]:

(1) not containing the language $(ab)^*$;

(2) closed under shuffle;

(3) closed under length preserving morphisms;

(4) closed under inverse of substitutions.

Despite these numerous closure properties, we don't know of any constructive description of $\mathcal{W}$, similar to the definition of the star-free languages. For instance, the least positive variety of languages satisfying Conditions (1)-(4) is the variety of polynomials of group languages, which is strictly contained in $\mathcal{W}$. Is it possible to find more powerful operators to generate the languages of $\mathcal{W}$? We let this question as a research problem for the reader.

Another research problem is to find a logical description for $\mathcal{W}$. The fact that **W** contains all finite groups and even **DS** might be a problem, since no logical description is available for the corresponding subvarieties of $\mathcal{W}$.

Finally, it would be nice to have an evocative name for $\mathcal{W}$ and the authors would appreciate any motivated suggestion.

References

1. Almeida, J.: Finite semigroups and universal algebra. Series in Algebra, vol. 3. World Scientific, Singapore (1994)
2. Almeida, J., Weil, P.: Relatively free profinite monoids: an introduction and examples. In: Fountain, J. (ed.) NATO Advanced Study Institute Semigroups, Formal Languages and Groups, pp. 73–117. Kluwer Academic Publishers, Dordrecht (1995)
3. Bouajjani, A., Muscholl, A., Touili, T.: Permutation Rewriting and Algorithmic Verification. Information and Computation 205(2), 199–224 (2007)

4. Cano Gómez, A.: Semigroupes ordonnés et opérations sur les langages rationnels, PhD thesis, Université Paris 7 and Departamento de Sistemas Informáticos y Computación, Universidad Politécnica de Valencia (2003)
5. Cano Gomez, A., Guaiana, G., Pin, J.-E.: When does partial commutative closure preserve regularity? In: 35th ICALP. Springer, Heidelberg (2008)
6. Cano Gómez, A., Pin, J.-E.: On a conjecture of Schnoebelen. In: Ésik, Z., Fülöp, Z. (eds.) DLT 2003. LNCS, vol. 2710, pp. 35–54. Springer, Heidelberg (2003)
7. Cano Gómez, A., Pin, J.-E.: Shuffle on positive varieties of languages. Theoret. Comput. Sci. 312, 433–461 (2004)
8. Eilenberg, S.: Automata, Languages and Machines, vol. B. Academic Press, New York (1976)
9. Ésik, Z.: Extended temporal logic on finite words and wreath products of monoids with distinguished generators. In: DLT 2002. LNCS, vol. 2450, pp. 43–58. Springer, Heidelberg (2002)
10. Ésik, Z., Simon, I.: Modeling Literal Morphisms by Shuffle. Semigroup Forum 56, 225–227 (1998)
11. Gehrke, M., Grigorieff, S., Pin, J.-E.: Duality and equational theory of regular languages. In: 35th ICALP. Springer, Heidelberg (2008)
12. Perrot, J.-F.: Variétés de langages et operations. Theoret. Comput. Sci. 7, 197–210 (1978)
13. Pin, J.-E.: Varieties of formal languages. North Oxford, London and Plenum, New-York (1986) (Translation of Variétés de langages formels, Masson, 1984)
14. Pin, J.-E.: Finite semigroups and recognizable languages: an introduction. In: Fountain, J. (ed.) NATO Advanced Study Institute Semigroups, Formal Languages and Groups, pp. 1–32. Kluwer academic publishers, Dordrecht (1995)
15. Pin, J.-E.: A variety theorem without complementation. Russian Mathematics (Izvestija vuzov.Matematika) 39, 80–90 (1995)
16. Pin, J.-E.: Syntactic semigroups. In: Rozenberg, G., Salomaa, A. (eds.) Handbook of formal languages, ch. 10, vol. 1, pp. 679–746. Springer, Heidelberg (1997)
17. Pin, J.-E., Weil, P.: A Reiterman theorem for pseudovarieties of of finite first-order structures. Algebra Universalis 35, 577–595 (1996)
18. Pin, J.-E., Weil, P.: Semidirect products of ordered semigroups. Communications in Algebra 30, 149–169 (2002)
19. Polák, L.: Operators on classes of regular languages. In: Gomes, G., Pin, J.-E., Silva, P. (eds.) Semigroups, Algorithms, Automata and Languages (Coimbra, 2001), pp. 407–422. World Scientific Publisher, River Edge, NJ (2002)
20. Reiterman, J.: The Birkhoff theorem for finite algebras. Algebra Universalis 14, 1–10 (1982)
21. Reutenauer, C.: Sur les variétés de langages et de monoïdes. In: Theoretical computer science (Fourth GI Conf., Aachen), vol. 67, pp. 260–265. Springer, Berlin (1979)
22. Straubing, H.: Recognizable Sets and Power Sets of Finite Semigroups. Semigroup Forum 18, 331–340 (1979)
23. Straubing, H.: On logical descriptions of regular languages. In: Rajsbaum, S. (ed.) LATIN 2002. LNCS, vol. 2286, pp. 528–538. Springer, Heidelberg (2002)

Deterministic Models of Communication Faults[*]

Rastislav Královič[1] and Richard Královič[1,2]

[1] Department of Computer Science, Comenius University,
Mlynská dolina, 84248 Bratislava, Slovakia
[2] Department of Computer Science, ETH Zurich, Switzerland

Abstract. In this paper we survey some results concerning the impact of faulty environments on the solvability and complexity of communication tasks. In particular, we focus on deterministic models of faults in synchronous networks, and show how different variations of the model influence the performance bounds of broadcasting algorithms.

1 Introduction

One of the main challenges in the design of complex systems such as computers is the issue of robustness. Indeed, systems comprised of a large number, albeit highly resilient, components experience the malfunction of some of them with probability high enough to be taken into account. Computers use e.g. sophisticated error-correcting procedures transparent to the software environment that maintain integrity of data stored in memory so that it can be viewed as a reliable storage medium.

Large scale communication networks consist of very high numbers of computing elements (computers, routers, etc.) and due to this complexity the issue of fault tolerance plays an important role in their design. Moreover, communication networks are accessible to a number of people, and are often targets of misuse and hacker attacks. The aim of the design is to develop the communication protocols in such a way that the network can serve its purpose (communication, cooperative computation, etc.) even with a number of components being permanently or temporarily out of operation.

In order to analyze the impact of faulty environments on the performance of the distributed system, two things must be specified: the model of the distributed system (network), and the model of the faulty environment. For the former we use a standard model of point-to-point communication in which the system is composed of independent entities (processors, processes, etc.) which can pairwise communicate by exchanging messages. It is natural to model such system by a graph, where vertices represent the processors, and edges connect those pairs of processors that can communicate. Having said this, there is still a number of issues that have to be addressed: e.g. are the vertices synchronous or asynchronous? Can a vertex send a message to all its neighbors simultaneously?

---

[*] The research has been supported by grant APVV-0433-06.

E. Ochmański and J. Tyszkiewicz (Eds.): MFCS 2008, LNCS 5162, pp. 52–67, 2008.
© Springer-Verlag Berlin Heidelberg 2008

Do the vertices know the graph? It turns out that there is a number of parameters that significantly influence the results. Indeed, the situation in distributed computing is such that there is no simple and elegant model that would capture all the studied properties of communication. Thus, when speaking about the *network model*, a number of options must be specified. We briefly discuss these options in the next section.

Second thing that has to be formally modeled in order to obtain results about fault tolerance is the (faulty) environment. Standard approach of theoretical computer science uses the notion of an *adversary*: an entity that is responsible for introducing faults into the execution of the algorithm. What we call a *model of faults* is basically a set of restrictions that are imposed on the behavior of the adversary. It is always assumed that adversary produces the worst possible scenario within the limitations given by the model. Of course, without any restrictions on the adversary it is not possible to solve any non-trivial task. With some assumptions, certain tasks become solvable, and by restricting the adversary even further, more efficient solutions may be obtained.

There are two main approaches to the definition of the fault model: randomized, in which the behavior of the adversary is a random process (e.g. every message is lost with a given probability), and deterministic which states some worst-case bounds on the adversary's behavior (e.g. at most k messages are lost). Both approaches have advantages and weak-points, and we shall discuss them later in more detail.

Finally, one has also to specify the *task* performed by the network. In this paper we mostly consider the *broadcasting problem* (called also *one-to-all communication*) where the goal is to disseminate a piece of information known to only one vertex of the distributed system to every other vertex. This simply formulated task is an important communication primitive in many more complex distributed algorithms. Moreover, broadcasting can be used to solve other important communication tasks with reasonable efficiency. These tasks include e.g. the *gathering problem* (all-to-one communication), where each node knows a piece of information that has to be delivered to one dedicated *sink* node, and the *gossipping problem* (all-to-all communication), where each node needs to learn a piece of information from all other nodes. Other communication problems include, e.g. the *leader election* problem, where vertices start in identical states[1], and they all have to agree on one distinguished vertex, the *wakeup* problem etc.

As has just been mentioned, the setting of the research consists of three parts: the network model, the fault model, and the task. There are at least three approaches to the study of their relationship:

- Fix the network model (based on current technology), fault model (based on realistic assumptions), and task, and study the complexity (in terms of time or amount of communication) of the task.
- Fix the network model, and the task, and find the least restrictive fault model in which the task is still solvable.

---

[1] Up to some topology knowledge as described by the network model.

– Fix the fault model and the task and find the weakest possible network model in which the task is still solvable.

The rest of the paper is organized as follows: in the next two sections we give an overview of features that must be taken into account in the network model, and the model of faults, respectively. Then we present a case study to demonstrate the development of a particular fault model. We focus on the broadcasting problem and show how subsequent changes to the model influence the results. We finish by a brief mention of other considered problems, open questions, and possible future projects.

2   Network Models

In this section we briefly describe some of the properties of the networks that can fundamentally influence the results. The classification of the models of distributed systems presented in this section does not cover all models that have been considered in the literature so far; for information about some such models, see e.g. survey [27] and references therein. However, most of the models can be defined by imposing additional constraints on some model presented in this section.

We consider the *point-to-point* model of the communication, which is widely studied in the literature (see [1, 2, 3, 4, 5, 7, 8, 10, 11, 12, 14, 15, 18, 19, 20, 21, 25, 26, 28, 29, 30, 31, 32, 33, 34, 35, 36, 39, 45, 46, 48, 49, 50, 51, 53, 55, 56, 57, 58] or surveys [27, 37, 38, 52]).[2] In this model, the nodes of the distributed system can communicate only via a set of links, where each link connects exactly two nodes. Thus the distributed system can be viewed as a graph whose vertices represent nodes and edges represent links of the distributed system. We call this graph the *communication graph* of the distributed system. We assume that all links are symmetric, hence the communication graph is undirected.

An important aspect of the distributed system is its synchronicity. The cases most widely considered in the literature (see e.g. [60, 47]) are the extreme cases: the *synchronous* mode and the *asynchronous* mode. In the asynchronous mode, nothing is known about the timing of the computation (this mode is studied e.g. in [4, 14]). In the synchronous mode, the algorithm runs in time steps that are synchronized between the nodes. The algorithm can exploit this, since it can make decisions based on the global clock of the distributed system. For other modes of synchronicity see e.g. [58].

In this work we consider only the synchronous models. The reason for this is that it is impossible to design fault tolerant asynchronous algorithms without any restriction on the timing of the computation[3]. Indeed, in the asynchronous computation it is impossible to distinguish between a lost message and a message that will be delivered later (see [22]).

---

[2] Other considered models include e.g. the *radio network model* (see [54]), *ATM network model* (see [6] and references therein), etc.

[3] One way to overcome this is to enhance the system with additional devices as e.g. *failure detectors* from [9].

For the synchronous systems the notion of *time complexity* is naturally defined – it is the number of time steps needed for the algorithm to terminate. In the *constant* mode (which we shall consider in this survey) each message is delivered in one time step regardless of the length of the message. This mode is the most widely studied one, e.g. in [2,3,5,8,10,11,12,15,18,19,20,21,25,28,29,33,34,35,46,48,49,54,55,56]. On the other side, in the *linear* mode, the time of delivery depends on the size of the message: The message of length L is delivered in $\beta + L\tau$ time steps, where β (cost of the start-up) and τ (propagation time of data of unit length) are constants specified by the model. This mode is studied e.g. in [26], see also the survey [27].

There are several properties of the communication links that must be taken into consideration. First is the duplexity of the links: In the *one-way* mode (called also the *half-duplex* or the *telegraph* mode), during a given time step each link can be used to deliver a message only in one direction. In other words, if some node is active on some link, it can be active either as a sender or as a receiver, but not both at once. This is the mode used e.g. in [4,21,30,34]. In the *two-way* mode (called also the *full-duplex* or the *telephone* mode), each link can be used to deliver message in both directions in the same time. This means that one node can be active on one link both as a sender and as a receiver in one time step. This mode is considered e.g. in [15,25,33,50]. We also distinguish the communication modes based on the capability of the nodes to use multiple links at once: In the *one-port* mode (called also the *whispering* or the *processor-bound* mode), in one time step each node can be active on only one of its adjacent links. This model represents the situation where the bottleneck of the communication is the performance of the node. It is used e.g. in [2,3,5,11,12,21,26,29,33,34,35,50,56]. In the *multi-port* mode (called also the *shouting* or the *link-bound* mode), in one time step each node can be active on any number of its adjacent links. This mode represents the situation where the bottleneck is the bandwidth of the links. It is used e.g. in [10,18,19,20,25,26,28, 31,46,48,49,55,56]. In the *k-port* mode (called also the *DMA-bound* mode), in one time step each node can be active on at most k of its adjacent links where k is a constant specified by the model. This mode is used e.g. in [8]. We focus on the multi-port mode, unless stated otherwise.

When designing algorithms for communication networks it is important to know what information about the communication graph of the distributed system is available. This *topology knowledge* can be of two types: Some knowledge can be given *a-priori*, in the time when the algorithm is designed. For example, in some situations we are interested only in some particular class of graphs; the algorithm does not need to work correctly for other graphs (this is the situation e.g. in [8,14,18,19,20,26,31,46,48,49,50,55,56]). On the other side, some information may be available to the algorithm in the run-time and the algorithm can make decisions based on it. This information can be of any kind. We present some previously considered types of such information:

– *Full knowledge of the topology.* In this scenario each node knows the whole communication graph. It knows its own location and location of all its neighbors in this graph as well. This kind of topology knowledge is exploited e.g. in [53].

- *Blind map.* Each node knows the whole communication graph, but it does not know its location in the graph. The blind map has been considered (although not in connection with broadcasting) e.g. in [13].
- *Identifiers.* Each node knows its own unique identifier. Sometimes the knowledge of the identifiers of neighbors is assumed, too (e.g. in [53,40]). Note that full topology knowledge implies the knowledge of identifiers. Knowledge of the identifiers without full topology knowledge is used e.g. in [4,53,40].
- *Sense of direction.* Each node has associated some topological information (*label*) with each of its adjacent link. The sense of direction is usually defined separately for different topologies (see e.g. [60]). For example, it is the dimension of the link in case of hypercubes (used e.g. in [23]), orientation (up, down, left or right) in case of meshes, etc. A sense of direction has been defined for general graphs in [24].

3   Models of Faults

Similar to the model of the distributed system, the model of faults can be described by several independent aspects of the faults; any combination of these aspects yields a unique model.

There are two main approaches to the modelling of the faulty computations. In the *randomized model*, the faults occur with a certain probability, i.e. the fault model describes the expected behavior of the adversary. The goal is to design a distributed algorithm that is correct *with high probability*. In the randomized models studied in the literature there is usually a fixed probability of failure of certain node/link. This is the case in e.g. [5,11,12,15,51,54]. However, various variants of randomized models have been considered (e.g. both node and link failures, both permanent and transient failures, both Byzantine and omission failures, etc.). The main problem with these approaches is that in real life faults are usually not independent. Recently, a research direction aimed at analyzing dependency in randomized fault models has been introduced in [44]. The second approach is to analyze the worst-case scenario, i.e. the fault model sets rules for the adversary to obey. To do so, the model of faults usually imposes constraints on the number of faults that can occur. This approach has been used e.g. in [2,4,8,10,18,19,20,25,26,28,29,30,31,32,33,34,35,36,39,45,46,48,49,53,55,56]. The main goal is to design such a set of rules that are not too restrictive and are realistic in the sense that they do not create unwanted special cases the algorithm can rely upon (e.g. if there is a guarantee that at least one message in every time step is non-faulty, the algorithm may send important messages one at a time – a strategy that has little justification in reality).

Each distributed system consists of elements of two types: the nodes and the links. This implies that there are also two kinds of faults: the faults of the nodes (considered e.g. in [4,12,25,26,31,32,33,34,36,51,55,56]) and the faults of the links (considered e.g. in [2,5,7,8,10,11,12,15,18,19,20,25,26,28,29,31,32,35, 36,46,48,49,51,55,56]).

Another aspect is the type of faults that may occur. In the *Byzantine* faults model, the adversary can specify the behavior of the faulty entity arbitrarily. In case of the node failures this means that the faulty node does not need to follow the specified algorithm, but the adversary can maliciously choose the behavior most detrimental to the algorithm. In case of the link failures the adversary can discard or modify the message transmitted through the faulty link or create any fake message on it. Obviously, the Byzantine faults represent the most harmful type of faults, i.e. the type of faults that gives maximal power to the adversary. This type of faults is studied e.g. in [5, 46, 53, 54]. The *crash/omission* faults give less power to the adversary. In this model, the adversary can block the faulty entity, but it can not alter its behavior in any other way. In case of node failures this means that the node fails to send some messages or it can crash entirely. In case of link failures, some or all messages transmitted through the faulty link can be lost. However, the adversary can not modify the content of any message nor create any fake message. This type of faults is studied e.g. in [2,10,11,12,15,18,19,20,25,26,28,29,31,32,33,34,35,39,45,46,48,49,54,55,56]. The weakest variant of this aspect are *detectable* faults. In this model, the locations of faults are known to the nodes, hence it does not make sense to distinguish Byzantine and omission faults. This type of faults is considered e.g. in [4, 8, 31].

Yet another important aspect of the model is the duration of the faults. The models working with *static* (or *permanent*) faults assume that there is a set of faulty entities of the distributed system which are faulty through the whole computation. This means that the adversary must choose some fixed location of the faulty entities before the start of the algorithm. Static faults are considered e.g. in [2, 4, 5, 8, 25, 26, 31, 32, 33, 34, 36, 39, 45, 55]. In the models working with the *dynamic* (or *transient*) faults, entities can recover from faults. This means that the adversary can choose different location of the faults in every time step of the computation as long as it satisfies other constraints imposed by the model. Hence it is obvious that dynamic faults give more power to the adversary than static faults. Dynamic faults are studied e.g. in [10, 15, 18, 19, 20, 28, 29, 35, 41, 46, 48, 49, 54].

An important subclass of models with dynamic faults are models with *time-independent* dynamic faults. These models impose constraints on the number and type of failures independently on the history of the computation. The adversary can choose the location of the faults according to these fixed rules and does not need to consider its previous decisions. In this paper we focus on the time-independent faults. As opposed to time-independent dynamic faults, some models define constraints on the number and type of failures depending on the failures that occurred in the past part of the computation. We call such fault models as models with *time-dependent dynamic faults*. Time-dependent dynamic faults have been analyzed e.g. in [35] which deals with dynamic faults in the constant one-port communication model. In such model, however, even one fault per time step renders the broadcasting impossible. This has been the main reason to introduce the *linearly bounded* model of dynamic faults in [35]: The model of faults is with omission faults and faulty links. Given a constant $0 < \alpha < 1$, the adversary can block at most αi messages during the first i time steps of the computation, for every natural number i.

4   Case Study: Broadcasting

In this section we demonstrate an evolution of a particular deterministic time-independent model of omission faults on a problem of broadcasting. For the network model we shall consider synchronous, full-duplex, all-port network. If there are no faults, then obviously a simple flooding algorithm delivers the message to all vertices in $d(G)$ time steps, where $d(G)$ is the diameter of the graph. The type of faults we are focusing on are omission faults on links.

The simplest model of faults in this setting is when a fixed number of k static faults is allowed, i.e. at the beginning of the computation the adversary chooses k links that will remain broken during the whole execution; the algorithm is not a-priori aware of which links are functional. This model corresponds to a situation when the faults are long-lasting compared to the execution of the algorithm (e.g. broken cable). The worst-case broadcasting time on a graph G will be $T(k,G) = \max_{G' \subseteq G}\{d(G')\}$, where the maximum is taken over all subgraphs G' obtainable from G by deleting at most k edges. In [59], Schoone, Bodlaender and van Leeuwen studied the worst case value of $T(k,G)$ over all graphs G with given diameter D, and delivered upper bound of the form $(k+1)D$ and a lower bound of the form $(k+1)D-k$ for even D and $(k-1)D-2k+2$ for odd $D \geq 3$.

Sometimes, however, transient faults are more appropriate to model the situation in the network (e.g. packet loss in wireless networks due to background noise). A next step in the evolution of the model is thus to allow the faults to be dynamic: in the *constant k-bounded model*, the adversary can block at most k messages in each time step. Since the faults are dynamic, they may occur on different links during various time steps. This model is one of the oldest models of dynamic faults. It has been introduced in [57] and analyzed in numerous papers, e.g. [10, 18, 19, 20, 28, 46, 48, 49].

It is easy to check that in the considered model, the following greedy non-adaptive algorithm is optimal: each node v that has received the message, broadcasts the message to all its neighbors in every time step. Tight bounds on general graphs are due to Chlebusz, Diks, and Pelc:

Theorem 1. [10] *Let $\mathcal{G}_{d,k}$ be the class of graphs with diameter d and edge connectivity at least $k+1$. Let, for some class of graphs $\mathcal{G}$, denote $T(\mathcal{G})$ the worst case broadcasting time in the constant k-bounded model on graphs from $\mathcal{G}$.*

For a fixed constant d the following holds. $T(\mathcal{G}_{d,k}) = O(k^{\frac{d}{2}-1})$. Moreover, there exists a class of graphs $\mathcal{G}'_{d,k} \subseteq \mathcal{G}_{d,k}$ such that $T(\mathcal{G}'_{d,k}) = \Omega(k^{\frac{d}{2}-1})$.

For a fixed constant k the following holds. $T(\mathcal{G}_{d,k}) = O(d^{k+1})$. Moreover, there exists a class of graphs $\mathcal{G}''_{d,k} \subseteq \mathcal{G}_{d,k}$ such that $T(\mathcal{G}''_{d,k}) = \Omega(k^{\frac{d}{2}-1})$.

A number of special topologies has been investigated within this model. The broadcasting in complete graphs in constant k-bounded model is completely solved (together with the case of Byzantine faults) in [46]:

Theorem 2. *Let K_n be a complete graph with n vertices. The following holds for the broadcasting time in the constant k-bounded model in the graph K_n:*

- *The broadcasting time equals to 1 if and only if $k = 0$.*
- *The broadcasting time equals to 2 if and only if $1 \leq k < \frac{n}{2}$.*
- *The broadcasting time equals to 3 if and only if $\frac{n}{2} \leq k < n + \frac{1}{2} - \sqrt{n + \frac{1}{4}}$.*
- *The broadcasting time equals to 4 if and only if $n + \frac{1}{2} - \sqrt{n + \frac{1}{4}} \leq k < n - 1$.*
- *The broadcasting is impossible if $n - 1 \leq k$.*

The broadcasting in the constant k-bounded model in tori is discussed in [48] and [19]. The results from [48] are strengthened in [19] to the following form:

Theorem 3. (Theorem 1 in [19]) *Let d, k be integers such that k is even and one of the following holds:*

- *$k \geq 6$ and $d \geq k + 4$*
- *$k \geq d$ and $d \geq 10$*

The d-dimensional torus of order k, denoted as $C_{d,k}$, has the edge connectivity equal to $2d$ and diameter $\frac{dk}{2}$. The broadcasting time in the constant $(2d - 1)$-bounded model in $C_{d,k}$ is equal to $\frac{dk}{2} + 2$.

The broadcasting in the constant k-bounded model in hypercubes is considered e.g. in [18,20,28] and [49]. The results from these papers are summarized in the following theorem:

Theorem 4. *The d-dimensional hypercube, denoted as H_d, is a graph with both edge connectivity and diameter equal to d. The broadcasting time $T(d, k)$ in the constant k-bounded model in the graph H_d satisfies the following:*

1. *If $k = d - 1$, then $T(d, k) = \begin{cases} 1 & d = 1 \\ 3 & d = 2 \\ d + 2 & d > 2 \end{cases}$.*

2. *If $k = d - 2$, then $T(d, k) = \begin{cases} 1 & d = 2 \\ 3 & d = 3 \\ d + 1 & d > 3 \end{cases}$.*

3. *If $k \leq d - 3$, then $T(d, k) = d$.*

The above results are optimistic in what they claim: the adversary cannot delay the progress of the algorithm too much. However, it seems that these positive results are mainly due to the fact that the constant k has to be relatively small[4] compared to the number of edges that the algorithm can use during its computation. Indeed, the lower bounds are based on isoperimetric inequalities: once a small constant number of vertices are informed, it can be guaranteed that the edge-boundary of the set of informed vertices is large enough to allow progress. This is a somewhat unrealistic behavior of the model, since adding more messages to the network does not introduce new errors; indeed, flooding the networks with vast amount of traffic is the best solution in this model, although not so much in real-life networks.

---

[4] At most the connectivity of the graph, in order to grant non-trivial results.

Hence, a next step was to introduce a model in which the number of faults could vary during each time step, with the intuition borrowed from the probabilistic models: if every message has a fixed probability p of failure, and there are m messages sent in a particular step, the average number of delivered messages is pm. In the *fractional α-bounded* model, if there are m messages sent during a particular time step, at most $\lfloor \alpha m \rfloor$ of them can be lost.

In [41], the performance of the greedy algorithm is analyzed. Contrary to the constant k-bounded model, the greedy algorithm is not always optimal in the α-bounded model. However, this algorithm is appealing due to its simplicity and uniformity.

The main interest of [41] is the following question: In which graphs is the broadcasting time of the greedy algorithm[5] asymptotically equal to the time of the broadcasting without faults (i.e. to the diameter of the communication graph)? It is shown that in complete graphs and complete trees this is not the case, but in the d-dimensional tori for fixed d it is. More precisely, the time of broadcasting on complete graph K_n increases from $\Theta(1)$ to $\Theta(\log n)$. The following theorems were stated for $\alpha = 1/2$ in [41] and generalized to arbitrary α in [42]:

Theorem 5. *Let $T(n)$ be the broadcasting time of the greedy algorithm in the α-bounded model on the complete graph K_n. The following inequality holds:*

$$\lceil \log_{1/\alpha}(n(1 - \alpha) + \alpha) \rceil \leq T(n) \leq \lfloor \log_{1/\alpha}(n - 1) \rfloor + 1$$

The situation in complete trees is similar; e.g. the broadcasting time of the greedy algorithm on the complete d-ary tree of height h in presence of fractional $1/2$-bounded faults is $\Omega\left((d \log d - c)^{\frac{h}{d}}\right)$ for fixed d and some constant c as opposed to $\Theta(h)$ in a fault-free scenario. The result stated for arbitrary constant α is:

Theorem 6. *Consider a complete d-ary tree $T_n^{(d)}$ of height n. Let $A = \lceil 1/\alpha \rceil$ and a be integer constants such that $a \geq \frac{\log(A+1)}{\log d}$. Let $p = \lfloor \log_A\left(1 + (A-1)d^a\right) \rfloor$. The broadcasting time in fractional α-bounded model is at least $a\frac{p^{\lfloor n/a \rfloor} - 1}{(p-1)^2}$.*

Unlike complete graphs and complete trees, the broadcasting in fractional α-bounded model can not be slowed down asymptotically in multidimensional tori.

Theorem 7. *Let d be a fixed integer. The broadcasting time of the greedy algorithm in the fractional α-bounded model on the d-dimensional torus of order k for an even integer k and fixed d is $\Theta(k)$.*

Although the greedy flooding algorithm is not optimal in the fractional α-bounded model, the model exhibits another undesirable property: If there is only one message sent in one time step, this message is guaranteed to be delivered. In some cases, the broadcasting algorithm can gain advantage from this unrealistic

---

[5] In the greedy algorithm an informed vertex sends the message to all its uninformed neighbors in every step.

feature: the algorithm may identify some "critical" messages and send each of them in a time step of its own which would guarantee their delivery. This means that perfect reliability can be traded for time complexity which is not what was intended when designing the model.

Hence, a modification of the model has been proposed in [16] to avoid this property. The *α-fractional threshold* model uses another motivation from the randomized models: not only the number of faults is proportional to the number of messages in transit, but there must be enough messages for the statistics to work, as well. To model this, if the number of messages m sent by the algorithm in one step is less than the connectivity of the graph,[6] all of them may be lost. Otherwise, at most a fraction α of them can be lost. It is important to note that the model provides no fault-detection mechanism. If a node sends some messages, it can not detect which of them have been delivered. For sure, the destination node can send an acknowledgement in the next time step, but such acknowledgement is just an ordinary message and can be lost.

Before proceeding to the analysis of the α-fractional threshold model, let us first focus on its extreme setting when α is infinitely close to 1, i.e. the case where the adversary has maximum power. In this *simple threshold* model, at least $c(G)$ messages[7] have to be sent in one time step to guarantee at least[8] one delivery.

The motivation for the study of this model is twofold: first, since it is the extreme setting, any algorithm developed for the simple threshold model will work also for all other settings of fractional thresholds. Second, with an abundance of models and parameters, it is always useful to pay attention to the extreme cases in the hope of better understanding of the interplay among various parameters. Indeed, the simple threshold model can be viewed as an attempt to find the weakest restrictions to the adversary which still allow broadcasting to be performed.

The simple threshold model of faults is extremely harsh for any distributed algorithm. In fact, it is not easy to see if it is possible to perform the broadcasting in this model at all. However, several surprisingly positive results have been presented in [16]: Not only it is always possible to finish the broadcasting, but it is possible to do so fast (i.e. in polynomial time) for many topologies.

The complete overview of the results proven in [16] can be found in Table 1. The paper deals with rings, complete graphs, hypercubes and arbitrary communication graphs with various topology knowledge. Presented results prove that the broadcasting can be performed in polynomial time on any communication graph with edge-connectivity bounded by $O(\log n)$, where n is number of its vertices. Many interesting topologies, such as hypercubes, butterfly graphs,

---

[6] Or some other chosen threshold.

[7] Where $c(G)$ is the connectivity of the graph.

[8] Note that the "at least" part is important here. Indeed, if it is known that only one message is delivered, there are cases in which the algorithm may use this additional information about the adversary to its advantage (e.g. by using it to break symmetry).

Table 1. Results about broadcasting time in simple threshold model from [16]. The communication graph G has n vertices, m edges and is $c(G)$-edge-connected.

Topology	Condition	Time complexity
ring	n not necessarily known	$\Theta(n)$
complete graph	with sense of direction	$O(n^2)$ [9]
complete graph	unoriented	$\Omega(n^2)$, $O(n^3)$
hypercube	oriented	$O(n^2 \log n)$
hypercube	unoriented	$O(n^4 \log^2 n)$
arbitrary network	full topology knowledge	$O(2^{c(G)} nm)$
arbitrary network	no topology knowledge except $c(G)$, n, m	$O(2^{c(G)} m^2 n)$

multidimensional tori with fixed dimension, etc., satisfy this requirement. Furthermore, all presented algorithms provide explicit termination, i.e. when the algorithm terminates at an entity, it will not process any more messages (and, in fact, no messages should be arriving anyway).

In order to give a flavor of the techniques used, we present here a sketch of the algorithm for broadcasting in a ring. Even in this very limited scenario, a trivial algorithm would not work. The ring is a 2-connected network, i.e. $c(G) = 2$. Hence, at least two messages must be sent in a time step to ensure that at least one of them is delivered. Obviously the initiator has to start by sending 2 messages. At least one of then is delivered, but since there is no implicit acknowledgement the initiator does not know which one. Moreover it must avoid sending a message to an already informed neighbor since in this case the adversary would deliver this message, and no progress would be made.

For the ease of presentation let us suppose that the size of the ring, n, is known to the vertices. At any moment of time, the vertices can be either *informed* or *uninformed*. Since the information is spreading from the single initiator vertex s, informed vertices form a connected component. The initiator splits this component into the left part and the right part. Each informed vertex v can easily determine whether it is on the left part or on the right part of the informed component – this information is delivered in the message that informs the vertex v.

The computation is organized in phases where each phase takes four time steps. Each informed vertex can be either *active* or *passive*. A vertex is active if and only if it has received, in previous phases, messages from only one of its neighbors. A passive vertex has received a message from both neighbors. This implies that, as long as the broadcast has not yet finished, there is at least one active vertex in both left and right part of the informed component (the leftmost and the right-most informed vertices must be active; note, however, that also some intermediate vertices might be active).

The computation consists of $n-1$ phases. The goal of a phase is to ensure that at least one active vertex becomes passive. Each phase consists of the following four steps:

---

[9] This result can be improved to $O(n \log n)$.

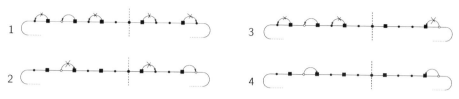

Fig. 1. A sample phase of the ring broadcasting algorithm. The initially active vertices are squares. In the first step, the only surviving message is delivered to the white vertex on the left, which then replies in the next step. At the end of the phase, two white vertices are sending acknowledgement, so at least one acknowledgement is delivered during that phase.

1. Each active vertex sends a message to its possibly uninformed neighbor.
2. Each active vertex in the right part sends a message to its possibly uninformed neighbor. Each vertex in the left part that received a message in step 1 replies to this message.[10]
3. Same as step 2, but left and right parts are reversed.
4. Each vertex that received a non-reply message in steps 1–3 replies to that message.

Initially, there are two active (virtual) vertices (the left- and right- part of the initiator). During each of the subsequent phases, at least one previously active vertex becomes passive. Since passive vertices never become active again, it follows that after at most $n-1$ phases, there are $n-1$ passive vertices. Once there are $n-1$ passive vertices, the remaining two must be informed (both are neighbors of a passive vertex), i.e. $n-1$ phases are sufficient to complete the broadcast.

The previous results were done in the simple threshold model, i.e. in the extreme case where α is infinitesimally close to 1. In [43], results summarized in Table 2 have been obtained for complete graphs and hypercubes for arbitrary constant α.

Table 2. Results for the broadcasting in the fractional model with threshold from [43]

Scenario	Broadcasting time
K_n, unoriented	$\Omega(\log n)$, $O(n^3)$
K_n, sense of direction	$\Theta(\log n)$
K_n, $\alpha \lesssim 0.55$	$\Theta(\log n)$
H_d $(n = 2^d)$	$\Omega(\log n)$, $O(n^4 \log^2 n)$

5   Other Problems in Fractional Threshold Model

While trying to solve the broadcasting problem, an interesting feature of the fractional threshold model was pointed out in [43]: Usually it is quite easy to inform vast majority of the vertices. But to inform the last remaining ones, all

---

[10] Note that passive vertices reply to such message, too.

informed vertices have to cooperate very tightly, which is often very hard to achieve. On the other hand, it is often vital to finish the broadcasting communication task fast, even subject to some small error. These facts give a motivation to study a natural relaxation of the broadcasting problem in which we allow a small constant number[11] c of vertices to stay uninformed in the end. We call such relaxation the *almost-complete broadcasting* problem. In [43] it was shown that the almost complete broadcasting can be solved in time $O(\log n)$ for complete graphs and $O(\log^2 n)$ for hypercubes.

In [17], the problem of leader election on rings has been investigated, and an $O(n \log n)$-time algorithm was developed in the case when all initiators start at the same time and there is no spontaneous wakeup, and an $O(n^2)$ algorithm in the general case.

6   Conclusion

This research leaves many open questions for further investigation. The line of the study is directed at specifying how various aspects of communication faults influence the performance of networks. One can thus consider a number of communication problems and study them under different models of faults. Also of interest is the relationship between the almost complete and complete broadcasts in various models.

For the fractional threshold model, non-constant values of α have not been considered. Results for more general classes of graphs would be beneficial. For the simple threshold model, one interesting question is whether the broadcasting can be done in polynomial time on all graphs. One can also try to compare the time complexity in the simple threshold model with the message complexity of a fault-free execution.

Another obvious question is to ask if it is possible to perform a complete broadcast in complete graphs also for large values of α in polylogarithmic time.

We finish by noting that there is a lack of any non-trivial lower bounds in the model of fractional faults with threshold.

References

1. Aguilera, M.K., Delporte-Gallet, C., Fauconnier, H., Toueg, S.: Thrifty generic broadcast. In: DISC 2000: Proceedings of the 14th International Conference on Distributed Computing, London, UK, pp. 268–282. Springer, Heidelberg (2000)
2. Ahlswede, R., Gargano, L., Haroutunian, H.S., Khachatrian, L.H.: Fault-tolerant minimum broadcast networks. Networks 27 (1996)
3. Ahlswede, R., Haroutunian, H., Khachatrian, L.H.: Messy broadcasting in networks. In: Blahut, Costello, Maurer, Mittelholzer (eds.) Communications and Cryptography: Two Sides of One Tapestry. Kluwer Academic Publishers, Dordrecht (1994)
4. Bagchi, A., Hakimi, S.L.: Information dissemination in distributed systems with faulty units. IEEE Transactions on Computers 43(6), 698–710 (1994)

---

[11] Independent on the network size.

5. Berman, P., Diks, K., Pelc, A.: Reliable broadcasting in logarithmic time with Byzantine link failures. Journal of Algorithms 22(2), 199–211 (1997)
6. Bermond, J.-C., Marlin, N., Peleg, D., Perennes, S.: Directed virtual path layouts in atm networks. Theoretical Computer Science 291(1), 3–28 (2003)
7. Bienstock, D.: Broadcasting with random faults. Discrete Applied Mathematics 20(1), 1–7 (1988)
8. Bruck, J.: On optimal broadcasting in faulty hypercubes. Discrete Applied Mathematics 53(1–3), 3–13 (1994)
9. Chandra, T.D., Toueg, S.: Unreliable failure detectors for reliable distributed systems. J. ACM 43(2), 225–267 (1996)
10. Chlebus, B., Diks, K., Pelc, A.: Broadcasting in synchronous networks with dynamic faults. Networks 27 (1996)
11. Chlebus, B.S., Diks, K., Pelc, A.: Optimal broadcasting in faulty hypercubes. In: FTCS, pp. 266–273 (1991)
12. Chlebus, B.S., Diks, K., Pelc, A.: Waking up an anonymous faulty network from a single source. In: HICSS (2), pp. 187–193 (1994)
13. Dessmark, A., Pelc, A.: Optimal graph exploration without good maps. Theoretical Computer Science 326(1–3), 343–362 (2004)
14. Diks, K., Kranakis, E., Pelc, A.: Broadcasting in unlabeled tori. Parallel Processing Letters 8(2), 177–188 (1998)
15. Diks, K., Pelc, A.: Almost safe gossiping in bounded degree networks. SIAM Journal on Discrete Mathematics 5(3), 338–344 (1992)
16. Dobrev, S., Královič, R., Královič, R., Santoro, N.: On fractional dynamic faults with threshold. In: Flocchini, P., Gasieniec, L. (eds.) SIROCCO 2006. LNCS, vol. 4056, pp. 197–211. Springer, Heidelberg (2006)
17. Dobrev, S., Královič, R., Pardubská, D.: Leader election in simple threshold model. Dept. of Computer Science, Comenius University, Bratislava, Slovakia (unpublished manuscript, 2008)
18. Dobrev, S., Vrťo, I.: Optimal broadcasting in hypercubes with dynamic faults. Information Processing Letters, 71(2), 81–85 (1999)
19. Dobrev, S., Vrťo.: Optimal broadcasting in tori with dynamic faults. Parallel Processing Letters 12(1), 17–22 (2002)
20. Dobrev, S., Vrťo, I.: Dynamic faults have small effect on broadcasting in hypercubes. Discrete Applied Mathematics 137(2), 155–158 (2004)
21. Even, S., Monien, B.: On the number of rounds necessary to disseminate information. In: SPAA 1989: Proceedings of the first annual ACM symposium on Parallel algorithms and architectures, pp. 318–327. ACM Press, New York (1989)
22. Fischer, M.J., Lynch, N.A., Paterson, M.S.: Impossibility of distributed consensus with one faulty process. Journal of the ACM 32(2), 374–382 (1985)
23. Flocchini, P., Mans, B.: Optimal elections in labeled hypercubes. Journal of Parallel and Distributed Computing 33(1), 76–83 (1996)
24. Flocchini, P., Mans, B., Santoro, N.: Sense of direction: definitions, properties, and classes. Networks 32(3), 165–180 (1998)
25. Fragopoulou, P., Akl, S.G.: Edge-disjoint spanning trees on the star network with applications to fault tolerance. IEEE Transactions on Computers 45(2), 174–185 (1996)
26. Fraigniaud, P.: Asymptotically optimal broadcasting and gossiping in faulty hypercube multicomputers. IEEE Transactions on Computers 41(11), 1410–1419 (1992)
27. Fraigniaud, P., Lazard, E.: Methods and problems of communication in usual networks. In: Proceedings of the international workshop on Broadcasting and gossiping 1990, pp. 79–133. Elsevier North-Holland, Inc., New York (1994)

28. Fraigniaud, P., Peyrat, C.: Broadcasting in a hypercube when some calls fail. Information Processing Letters 39(3), 115–119 (1991)

29. Gargano, L., Liestman, A.L., Peters, J.G., Richards, D.: Reliable broadcasting. Discrete Applied Mathematics 53(1-3), 135–148 (1994)

30. Gargano, L., Rescigno, A.A.: Communication complexity of fault-tolerant information diffusion. Theoretical Computer Science 209(1–2), 195–211 (1998)

31. Gargano, L., Rescigno, A.A., Vaccaro, U.: Minimum time broadcast in faulty star networks. Discrete Applied Mathematics 83(1-3), 97–119 (1998)

32. Gargano, L., Vaccaro, U., Vozella, A.: Fault tolerant routing in the star and pancake interconnection networks. Information Processing Letters 45(6), 315–320 (1993)

33. Gasieniec, L., Pelc, A.: Adaptive broadcasting with faulty nodes. Parallel Computing 22(6), 903–912 (1996)

34. Gasieniec, L., Pelc, A.: Broadcasting with a bounded fraction of faulty nodes. Journal of Parallel and Distributed Computing 42(1), 11–20 (1997)

35. Gasieniec, L., Pelc, A.: Broadcasting with linearly bounded transmission faults. Discrete Applied Mathematics 83(1–3), 121–133 (1998)

36. Han, Y.J., Igarashi, Y., Kanai, K., Miura, K.: Broadcasting in faulty binary jumping networks. Journal of Parallel and Distributed Computing 23(3), 462–467 (1994)

37. Hedetniemi, S., Hedetniemi, S., Liestman, A.: A survey of broadcasting and gossiping in communication networks. Networks 18, 319–349 (1988)

38. Hromkovic, J., Klasing, R., Monien, B., Peine, R.: Dissemination of information in interconnection networks (broadcasting and gossiping). In: Du, D.-Z., Hsu, D. (eds.) Combinatorial Network Theory, pp. 125–212. Kluwer Academic Publishers, Netherlands (1996)

39. Jovanovic, Z., Misic, J.: Fault tolerance of the star graph interconnection network. Information Processing Letters 49(3), 145–150 (1994)

40. Kowalski, D.R., Pelc, A.: Deterministic broadcasting time in radio networks of unknown topology. In: FOCS 2002: Proceedings of the 43rd Symposium on Foundations of Computer Science, Washington, DC, USA, pp. 63–72. IEEE Computer Society Press, Los Alamitos (2002)

41. Královič, R., Královič, R., Ružička, P.: Broadcasting with many faulty links. In: Sibeyn, J.F. (ed.) SIROCCO. Proceedings in Informatics, vol. 17, pp. 211–222. Carleton Scientific (2003)

42. Královič, R.: Broadcasting with dynamic faults. Phd. thesis, Dept. of Computer Science, Comenius University, Bratislava, Slovakia (2008)

43. Královič, R., Královič, R.: Rapid Almost-Complete Broadcasting in Faulty Networks. In: Prencipe, G., Zaks, S. (eds.) SIROCCO 2007. LNCS, vol. 4474, pp. 246–260. Springer, Heidelberg (2007)

44. Kranakis, E., Paquette, M., Pelc, A.: Communication in Networks with Random Dependent Faults. In: Kucera, L., Kucera, A. (eds.) MFCS 2007. LNCS, vol. 4708, pp. 418–429. Springer, Heidelberg (2007)

45. Latifi, S.: On the fault-diameter of the star graph. Information Processing Letters 46(3), 143–150 (1993)

46. Liptak, Z., Nickelsen, A.: Broadcasting in complete networks with dynamic edge faults. In: Butelle, F. (ed.) OPODIS, Studia Informatica Universalis, Suger, Saint-Denis, rue Catulienne, France, pp. 123–142 (2000)

47. Lynch, N.A.: Distributed Algorithms. Morgan Kaufmann Publishers Inc., San Francisco, CA, USA (1996)

48. Marco, G.D., Rescigno, A.A.: Tighter time bounds on broadcasting in torus networks in presence of dynamic faults. Parallel Processing Letters 10(1), 39–49 (2000)

49. Marco, G.D., Vaccaro, U.: Broadcasting in hypercubes and star graphs with dynamic faults. Information Processing Letters 66(6), 321–326 (1998)
50. Mendia, V.E., Sarkar, D.: Optimal broadcasting on the star graph. IEEE Transactions on Parallel and Distributed Systems 3(4), 389–396 (1992)
51. Pelc, A.: Broadcasting in complete networks with faulty nodes using unreliable calls. Information Processing Letters 40(3), 169–174 (1991)
52. Pelc, A.: Fault-tolerant broadcasting and gossiping in communication networks. networks 28(3), 143–156 (1996)
53. Pelc, A., Peleg, D.: Broadcasting with locally bounded Byzantine faults. Information Processing Letters 93(3), 109–115 (2005)
54. Pelc, A., Peleg, D.: Feasibility and complexity of broadcasting with random transmission failures. In: PODC 2005: Proceedings of the twenty-fourth annual ACM SIGACT-SIGOPS symposium on Principles of distributed computing, Las Vegas, NV, USA, pp. 334–341. ACM Press, New York, NY, USA (2005)
55. Peleg, D.: A note on optimal time broadcast in faulty hypercubes. Journal of Parallel and Distributed Computing 26(1), 132–135 (1995)
56. Ramanathan, P., Shin, K.G.: Reliable broadcast in hypercube multicomputers. IEEE Transactions on Computers 37(12), 1654–1657 (1988)
57. Santoro, N., Widmayer, P.: Distributed function evaluation in the presence of transmission faults. In: SIGAL 1990, pp. 358–367. Springer, Heidelberg (1990)
58. Schiper, A.: Group Communication: From Practice to Theory. In: Wiedermann, J., Tel, G., Pokorný, J., Bieliková, M., Štuller, J. (eds.) SOFSEM 2006. LNCS, vol. 3831, pp. 117–136. Springer, Heidelberg (2006)
59. Schoone, A., Bodlaender, H., van Leeuwen, J.: Diameter increase caused by edge deletion. J. Graph Theory 11, 409–427 (1987)
60. Tel, G.: Introduction to distributed algorithms. Cambridge University Press, New York, NY, USA (1994)

Algebraic Graph Algorithms

Piotr Sankowski

Warsaw University and ETH Zürich
piotr.sankowski@gmail.com

Abstract. The aim of this paper is to survey the results on dynamic algebraic algorithms, with main interest in matrix functions such as, determinant, inverse, rank and characteristic polynomial. First of all we summary the papers that in dynamic setup these problems can be solved faster than evaluating everything from scratch. The static complexity of these problem equals the matrix multiplication complexity, whereas the presented solutions work in subquadratic or quadratic (characteristic polynomial) time in the worst case. The dynamic matrix computations can be used to solve the following graphs problems in dynamic setup: computing transitive closure, computing shortest paths lengths, computing maximum matching size and computing vertex connectivity. For all of these problem the dynamic approach lead to the first known subquadratic algorithms. Astonishingly, the dynamic matrix algorithms can be used to obtain efficient static algorithms for the perfect matching problem as well. Using the $O(n^2)$ algorithms for the dynamic matrix inverse, one can obtain a very simple randomized algorithm for computing perfect matchings in $O(n^3)$ time. When the fast matrix multiplication is used, the complexity of this algorithm can be improved to $O(n^\omega)$ time, where ω is the exponent of the best known matrix multiplication algorithm. Since $\omega < 2.38$, this algorithm breaks through the $O(n^{2.5})$ barrier for the matching problem. The interplay between algebraic algorithms and graphs problems can be explored even further in order to obtain $O(Wn^\omega)$ time algorithms for single source shortest paths problem and weighted bipartite matching problem.

1 Dynamic Algebraic Algorithms

Let $\mathcal{R} = (S, +, \cdot, 0, 1)$ be a field with elements from set S and appropriately defined addition $+$ and multiplication $\cdot$. Let $f : S^n \to S^m$ be an algebraic function over this ring. A *dynamic algebraic algorithm* must handle the following types of requests:

- **initialize**$(x_1, x_2, \ldots, x_n)$: initialization with an input vector $(x_1, x_2, \ldots, x_n)$;
- **update**(k, x'_k): change input k to a new value x'_k;
- **query**(k): return the value of output k.

Our goal is to construct algorithms that support updates and queries as fast as possible. In particular, the updates must be faster than recomputing everything

E. Ochmański and J. Tyszkiewicz (Eds.): MFCS 2008, LNCS 5162, pp. 68–82, 2008.

from scratch. The initialization time is generally not very important. However, we also try to make the initialization step as fast as possible.

More specifically speaking, here, we consider the following types of dynamic matrix problems:

- **determinant** $\mathcal{R}^{n^2} \to \mathcal{R}$: The input is interpreted as an $n \times n$ matrix. The output is its determinant.
- **adjoint** $\mathcal{R}^{n^2} \to \mathcal{R}^{n^2}$: The input is interpreted as an $n \times n$ matrix A. The output is interpreted as $n \times n$ adjoint $\operatorname{adj}(A)$ of the input matrix.
- **inverse** $\mathcal{R}^{n^2} \to \mathcal{R}^{n^2}$, where $\mathcal{R}$ is a field: This is a function from non-singular $n \times n$ matrices that maps matrix A into the corresponding inverse matrix A^{-1}.
- **linear system of equations** $\mathcal{R}^{n^2+n} \to \mathcal{R}^n$, where $\mathcal{R}$ is a field: This is a function from non-singular $n \times n$ matrices and n dimensional vectors that maps matrix A and vector b into the solution x to the matrix equation $Ax = b$.
- **matrix rank** $\mathcal{R}^{n^2+n} \to \mathcal{N}$, where $\mathcal{R}$ is a field: This is a function from $n \times n$ matrices that maps matrix A into its rank.
- **characteristic polynomial** $\mathcal{R}^{n^2+n} \to \mathcal{R}[x]$, where $\mathcal{R}$ is a field: This is a function from $n \times n$ matrices that maps matrix A into its characteristic polynomial.

The first paper that studies dynamic algebraic problems is due to Reif and Tate [59]. They presented an $\Omega(n)$ lower bound for some simple dynamic algebraic problems such as multipoint polynomial evaluation, polynomial reciprocal, and extended polynomial GCD. They proved two time-space trade-off theorems applicable to dynamic algorithms for many algebraic functions. Moreover, they provided some general-purpose design techniques of dynamic algebraic algorithms. With the use of these techniques, they showed an $O(\sqrt{n})$ time per request algorithm for dynamic DFT and an $O(\sqrt{n \log n})$ time per request algorithm for polynomial multiplication. They also provided a technique for constructing parallel algorithms with optimal work.

Other lower bounds for dynamic algebraic problems were shown by Frandsen, Hansen and Miltersen [17]. They proved an almost tight $\Omega(\sqrt{n})$ lower bound for dynamic polynomial multiplication. They were also able to prove an $\Omega(n)$ lower bounds for the problems of computing determinant, adjoint, inverse, and solving linear system of equations. Till our paper [65] the best solution to these matrix problems was the so called Sherman-Morrison formula [71,72]. The Sherman-Morrison formula allows the recomputation of matrix inverse and determinant in $O(n^2)$ time, whereas the paper [65] presents first subquadratic algorithm for these problems. Table 1 summarizes some of the known results on dynamic algebraic problems.

In the paper [65] we assume that the matrix remains non-singular throughout the updates. The question whereas similar algorithm working with singular matrix exits was left open. However, algorithms supporting only non-singular updates are sufficient for many applications. Moreover, in order to break through the $O(n^2)$ barrier we need to restrict ourself only to *simple operations*. We show

Table 1. Upper and lower bounds for some dynamic algebraic problems. The bounds are with respect to both update and query operations.

Problem	Lower Bound	Upper Bound
matrix-vector multiplication matrix multiplication	$\Omega(n)$ [17]	$O(n)$ [simple multiplication]
polynomial multiplication	$\Omega(\sqrt{n})$ [17]	$O(\sqrt{n \log n})$ [59]
polynomial evaluation	$\Omega(n)$ [59,17]	$O(n)$ [Horner's algorithm]
DFT	$\Omega(\frac{\log^2 n}{\log \log n})$ [17]	$O(\sqrt{n})$ [59]
two dimensional DFT	$\Omega(\frac{\log^2 n}{\log \log n})$ [17]	$O(\sqrt{n \log n})$ [59]
polynomial reciprocal	$\Omega(n)$ [59]	$O(n \log n)$ [59]
extended polynomial GCD	$\Omega(n)$ [59]	$O(n \log^2 n)$ [59]
prefix sum		$O(\log n)$ [20]
matrix adjoint matrix inverse determinant linear system of equations	$\Omega(n)$ [59,17]	$O(n^{1.495})$ [65]
matrix rank	$\Omega(n)$ [18]	$O(n^{1.595})$ [18] $O(n^{1.495})$ [70]
characteristic polynomial – element operations – column operations	$\Omega(n)$ [19] $\Omega(n^2)$ [19]	$O(n^2 \log n)$ [19]

two trade-off algorithms for this problem. Both algorithms maintain the inverse in a lazy form, what allows to perform updates faster but increases the query time. The first algorithm uses $O(n^{\omega(\epsilon)-\epsilon} + n^{1+\epsilon})$ operations (worst-case) for updates, $O(1)$ operations to query the determinant and the solution of linear system of equations, and $O(n^\epsilon)$ operations to query the adjoint matrix and the inverse. Here, we denote by $O(n^{\omega(\epsilon)})$ the time needed to multiply an $n \times n^\epsilon$ matrix by $n^\epsilon \times n$ matrix. The second algorithm uses $O(n^{\omega(\epsilon)-\epsilon} + n^{2\epsilon})$ operations (worst-case) for updates, $O(1)$ operations to query the determinant and the solution of linear system of equations, and $O(n^{2\epsilon})$ operations to query the adjoint matrix and the inverse. Using the best known bounds on $\omega(\epsilon)$ [31] and minimizing the update cost, we get algorithms supporting updates in $O(n^{1.575})/O(n^{1.495})$ operations and queries in $O(n^{0.575})/O(n^{1.495})$ operations.

The problem of handling singular matrices was for the first time solved by Frandsen and Frandsen [18], who have presented an deterministic $O(n^{1.575})$ time algorithm. Latter on, we improved slightly this result by presenting an $O(n^{1.495})$ time algorithm [70]. The result is based on a black-box approach that can turn any dynamic algorithm that assumes non-singularity of the matrix into an algorithm working without this assumption. However, the reduction comes at the cost of randomization. By presenting, together with G.S.Frandsen, dynamic

algorithms for computing characteristic polynomial of a matrix we were able we
have generalized these even more, as the rank and determinant can be computed
via characteristic polynomial computation. Another result that extends the dynamic matrix problems in another direction is the paper with M.Mucha [51]
where we show that the update time can be improved to $O(n^{\omega-1}) = O(n^{1.38})$
when the algorithm is allowed some lookahead into the future operations. Such
results might be useful in ,e.g., batch processing.

2 Dynamic Graph Algorithms

A dynamic graph algorithm maintains actual information on a given property P
of a graph subject to dynamic change of the graph. Possible changes include edge
insertion, edge deletion or edge weight update. Any dynamic graph algorithm for
property P should process updates and after each update, it should be able to answer queries on P. In case of dynamic graph algorithms, we say that an algorithm is
incremental if it supports insertions only, and *decremental* if it supports deletions
only. We maintain a graph $G = (V, E)$ under intermixed sequence of **insert**(e) and
delete(e) operations, which insert or remove edge e from the graph. Updates of
this type are called *edge* updates throughout this survey. For the sake of concise-
ness we do not consider other types of updates here.

Additionally to the update operations we consider algorithms that support
one of the following queries:

- **test-matching**(e): test if an edge e is contained in any perfect matching,
  i.e., test if e it is *allowed*, assuming that G has a perfect matching;
- **matching-size:** compute the size of the maximum matching in G;
- **reachability**(v,w): test whether there is a path from v to w in G, i.e.,
  compute transitive closure;
- **shortest-path**(v,w): return the distance between v and w in G;
- k-**connected:** tests if the graph G is k-vertex connected.

2.1 Dynamic Transitive Closure

The problem of dynamic transitive closure has been studied by many researchers.
The first papers on this subject described only incremental and decremental algorithms [32,34,42,78,29]. In the case of incremental updates, the fastest algorithms were presented by Italiano [34], La Poutré and van Leeuwen [42]. Their
algorithms need $O(n)$ amortized time per insertion and $O(1)$ time per query. The
case of decremental algorithms seems to be inherently harder. An $O(n)$ amortized time algorithm is known only in the case of acyclic graphs. In the case of
general graphs only an $O(n^2)$ time algorithm [32] was known till the paper of
Henzinger and King who presented the randomized algorithm with query time
$O(n/\log n)$ and amortized update time $O(n \log^2 n)$.

The first fully dynamic transitive closure algorithm was given by Subramanian [75]. The algorithm was only limited to planar graphs, whereas Henzinger

and King [29] where the first to present algorithm for general graphs. They devised a Monte Carlo algorithm with one-sided error, $\tilde{O}(nm^{0.58})$ amortized time per update, and $\Theta(n/\log n)$ time per query. Khanna, Motwani and Wilson [38] showed that when a lookahead of $\Theta(n^{0.18})$ updates is permitted, there exists a deterministic algorithm with update time $\Theta(n^{2.18})$. Next, King and Sagert [40] devised an algorithm for general directed graphs supporting queries in $O(1)$ time and updates in $O(n^{2.26})$ time. They also showed an algorithm for acyclic graphs with $O(n^2)$ update time. These bounds were improved by King [39], who presented a deterministic algorithm with $O(n^2 \log n)$ amortized update time and $O(1)$ query time. In [8], Demetrescu and Italiano improved these bounds further by presenting an algorithm with $O(n^2)$ amortized update time and $O(1)$ query time for general digraphs. Slightly improved result was later presented in [61]. In [8], the authors also gave the first algorithm with subquadratic update time for directed acyclic graphs. This algorithm can answer queries in $O(n^\epsilon)$ time and perform updates in $O(n^{\omega(\epsilon)-\epsilon} + n^{1+\epsilon})$ time, for any $\epsilon \in [0, 1]$. The current best bounds on $\omega(\epsilon)$ [31] imply an $O(n^{0.575})$ query time and $O(n^{1.575})$ update time. This subquadratic algorithm is randomized with one-sided error. More recently, Roditty and Zwick [62] presented an algorithm with $O(m\sqrt{n})$ update time and $O(\sqrt{n})$ query time, and another [63] with $O(m + n \log n)$ update time and $O(n)$ query time. The history of the problem with algorithm complexities is presented in Table 2.

Table 2. The complexity of fully dynamic transitive closure algorithms

Author	Update	Query
Subramanian '93 [75] (planar graphs)	$\tilde{O}(n^{\frac{2}{3}})$	$\tilde{O}(n^{\frac{2}{3}})$
Henzinger and King '95 [29]	$\tilde{O}(nm^{0.58})$	$\Theta(n/\log n)$
Khanna, Motwani and Wilson '96 [38] (with lookahead of $\Theta(n^{0.18})$ operations)	$\Theta(n^{2.18})$	$O(1)$
King and Sagert '99 [40] (restricted to DAGs)	$O(n^{2.26})$ $O(n^2)$	$O(1)$ $O(1)$
King '99 [39]	$O(n^2 \log n)$	$O(1)$
Demetrescu and Italiano '00 [8] (restricted to DAGs)	$O(n^2)$ $O(n^{1.575})$	$O(1)$ $O(1)$
Roditty and Zwick '02 [62]	$O(m\sqrt{n})$	$O(\sqrt{n})$
Roditty '03 [61]	$O(n^2)$	$O(1)$
Roditty and Zwick '04 [63]	$O(m + n \log n)$	$O(n)$
Sankowski '04 [65]	$O(n^2)$ $O(n^{1.575})$ $O(n^{1.495})$	$O(1)$ $O(n^{0.575})$ $O(n^{1.495})$
Diks and Sankowski '07 [12] (plane graphs)	$\tilde{O}(\sqrt{n})$	$\tilde{O}(\sqrt{n})$

Notice that an update may change $\Omega(n^2)$ entries of the transitive closure matrix, and assuming that queries are answered in $O(1)$ time, the bound of $O(n^2)$ worst-case time seems to be the best we can hope for. However, in no such algorithm has been presented in the above mentioned papers. Moreover, the algorithms in [8,61] have only amortized time bounds. In dynamic setup, we are often interested in answering each query in a small worst-case time, therefore devising such algorithms might be very useful.

Most of the algorithms mentioned above contain (except [39,62,63]) a fast matrix multiplication as a subroutine. However, only the algorithm of Demetrescu and Italiano [8] explores the equivalence between matrix multiplication and transitive closure. The algorithm uses explicitly the formula for transitive closure

$$A^c = \sum_{i=0}^{i=n} A^i.$$

Using algorithms for dynamic evaluation of polynomials over matrices for this formula, the authors obtained dynamic algorithms for transitive closure. In the paper [65] we actually use a similar technique, that uses the informal observation

$$(I - A)^{-1} \simeq \sum_{i=0}^{\infty} A^i \simeq A^c.$$

Actually prove, via randomization, that A can be chosen in such a way that $(I - A)^{-1}$ encodes transitive closure with high probability. More general observations have been made in [55,6] for the case of vertex connectivity. Now, the transitive closure can be computed by inverting an appropriately defined adjacency matrix of the graph. Thus, algorithms for dynamic matrix inverse can be directly applied to the problem of dynamic transitive closure. Edge updates of the graph are translated into element updates of the matrix. The reduction is randomized with one-sided bounded error and gives the first algorithm with $O(n^2)$ worst-case update time and $O(1)$ query time. We also have shown two algorithms breaking the $O(n^2)$ barrier for general directed graphs. The first one achieves $O(n^{1.575})$ update time and $O(n^{0.575})$ query time, and generalizes the result from [8] for DAGs. The second one has even faster $O(n^{1.495})$ update time at the cost of much higher $O(n^{1.495})$ query time. Additionally, the results from [51] can be used in order to obtain faster algorithms when some information about the future operations is possible. In order to keep the story of the problem complete, one has to mention that, very recently, the complexity of the dynamic transitive closure problem in plane graphs has been improved to $\tilde{O}(\sqrt{n})$ [12].

2.2 Dynamic Shortest Paths

The study of algorithms for dynamic maintenance of shortest paths was started more than 35 year ago [44,54,60]. Since then many algorithms with times comparable to evaluating everything from scratch have been proposed [15,21,22,57,58,64]. The first decrease-only algorithm was proposed by Ausiello *et al.* [1]. The algorithm worked for graphs with integer weights less than some integer C and needed $O(Cn \log n)$ amortized time per edge insertion.

Next, two dynamic algorithms for planar graphs have been developed. Henziger *et al.* [30] developed an $O(n^{\frac{4}{3}} \log(nC))$ time algorithm in the case of integer weights. The second algorithm was proposed by Fakcharoemphol and Rao in [16] who showed how to support queries and updates in $O(n^{\frac{4}{5}} \log^{\frac{13}{5}} n)$ amortized time.

The first fully dynamic algorithm for general graphs in the case of integer weights was presented by King [39]. The running time of the algorithm is $O(n^{2.5} \sqrt{C \log n})$ per update. In the case of real edge weights similar results was obtained by Demetrescu and Italiano in [9,10]. They assumed that there are at most S different real edge weights and obtained an algorithm supporting updates in $O(n^{2.5} \sqrt{S \log^3 n})$ time and queries in constant time. They also presented two families of trade-off algorithms that have smaller update time but at the cost of bigger query time. In the case of unweighted graphs, Baswana, Hariharan and Sen [2] developed simpler deletion only algorithms. The final step in obtaining $\tilde{O}(n^2)$ update time was made by Demetrescu and Italiano in [11]. This result was generalized to graphs with negative weights by Thorup [76], who additionally one year latter presented algorithm working in $\tilde{O}(n^{2.75})$ worst-case time [77]. The summary of the results is presented in Table 3.

Table 3. The complexity of fully dynamic shortest paths algorithms

Author	Update	Query
Henzinger et al. '97 [30] (planar graphs)	$O(n^{\frac{4}{3}} \log(nC))$	$O(n^{\frac{4}{3}} \log(nC))$
Fakcharoemphol and Rao '02 [16] (plane graphs with positive weights) (plane graphs with negative weights)	$O(n^{\frac{2}{3}} \log^{\frac{7}{3}} n)$ $O(n^{\frac{4}{5}} \log^{\frac{13}{5}} n)$	$O(n^{\frac{2}{3}} \log^{\frac{7}{3}} n)$ $O(n^{\frac{4}{5}} \log^{\frac{13}{5}} n)$
King '99 [39]	$O(n^{2.5} \sqrt{C \log n})$	$O(1)$
Demetrescu and Italiano '00 [9]	$O(n^{2.5} \sqrt{S \log^3 n})$	$O(1)$
Demetrescu and Italiano '03 [10]	$\tilde{O}(n^2)$	$O(1)$
Thorup '04 [76] (with negative weights)	$\tilde{O}(n^2)$	$O(1)$
Thorup '05 [77]	$\tilde{O}(n^{2.75})$ *(worst case)*	$O(1)$
Sankowski '05 [68] (unweighted graphs)	$O(n^{1.932})$ *(worst case)*	$O(n^{1.288})$

The shortest paths problem is harder than the transitive closure problem. Similarly to the transitive closure problem, assuming $O(1)$ queries, the bound of $O(n^2)$ on the worst-case time for updates seems to be the best we can hope for.

Thus, it is interesting to know if allowing greater query time one can reduce the update time below $O(n^2)$.

In the paper [68], using extension of the algorithm for dynamic matrix inverse to rings, we have developed an algorithm for computing lengths of the shortest paths with at most k edges. This algorithm, combined with the standard technique of path decomposition, gives an algorithm supporting updates in $O(n^{1.932})$ time and queries in $O(n^{1.288})$ time. This result resolves the open question (see e.g. [9,10,11]) whether there exist algorithms with sub-quadratic update and query times. The problem of dynamic single source shortest distances seems inherently simpler than dynamic all pair shortest distances, but till that paper the best solution for this problem was evaluating everything again from the scratch, and this takes $O(n^2)$ time. Hence, we have also resolved the question whether more efficient algorithms for this problem exist. However, our result is only of theoretical importance, the $\tilde{O}(n^2)$ algorithm from [11] is surely practically more efficient.

3   Dynamic Graph Connectivity

In order to obtain the solution to the vertex connectivity, we have to extend the results from [65] to support so called rank one updates. In the *rank one updates* we are given two n dimensional vectors a and b and want to update the matrix A to $A + ab^T$. Using this extension we device new algorithm that allows to maintain information about determinants of submatrices of the matrix and then combine it with the ideas from [55] and [6] to obtain dynamic algorithms for k-vertex connectivity. The resulting algorithm supports edge updates in $O(n^{1.575} + nk^2)$ time. This means that for constant k the graph can be tested for directed vertex k-connectivity as fast as for strong connectivity (using dynamic transitive closure algorithms). The asymptotically fastest algorithms for static connectivity are summarized in Table 4. Our algorithm for testing connectivity improves in dynamic case substantially over these results even in the undirected case.

Table 4. The complexity results for the k-vertex connectivity

Author	Complexity
Linial, Lovász and Wigderson [55] (undirected)	$\tilde{O}(n^\omega + nk^\omega)$
Cheriyan and Reif [6]	$\tilde{O}(n^\omega + nk^\omega)$
Gabow [23] (undirected)	$O((k^{5/2} + n)m)$ $O((k + n^{1/4})n^{3/4}m)$ $O((k^{5/2} + n)kn)$ $O((k + n^{1/4})kn^{7/4})$

4   Matchings in Graphs

The algorithms for maintaining matrix inverse dynamically can be directly applied to the problems of testing if an edge is allowed in a graph having a perfect matching. We simply need to use the result of Rabin and Vazirani [56]. They showed that the inverse matrix of the adjacency matrix encodes allowed edges in the graph. This allows us to construct two algorithms:

- the first one supporting queries in $O(n^2)$ time and updates in $O(1)$ time,
- the second one supporting queries in $O(n^{1.575})$ time and updates in $O(n^{0.575})$ time.

The same idea can be used with the algorithms for computing matrix rank [70] in order to compute maximum matching size in dynamic graphs in $O(n^{1.575})$ time per update.

These algorithms can be used to construct a very simple algorithm for computing perfect matchings in graphs in $O(n^3)$ time, or a more complicated one that finds perfect matchings in $O(n^{2.575})$ time. This approach has been explored more deeply by the author in the papers with Marcin Mucha [49,48], where two algorithms were shown: an $O(n^\omega)$ time algorithm for finding maximum matchings in general graphs and an $O(n^{\frac{\omega}{2}})$ time algorithm for finding maximum matchings in planar graphs. In the case of dense graphs, these results improve over the fastest previously known algorithms of Micali and Vazirani [45], Blum [4], and Gabow and Tarjan [26], that worked in $O(\sqrt{n}m)$ time. In the case of planar graphs the situation is more complicated as the perfect matchings in planar graphs can be found in $\tilde{O}(n)$ time by combining the result of Miller and Naor [46] with the result of Fakcharoenphol and Rao [16]. However, in order to find maximum matchings in general planar graphs we need to use $O(n^{\frac{3}{2}})$ time algorithm of Liption and Tarjan [43] or solve the problem with the same complexity by applying the solutions for general graphs. Latter on, a parallel NC algorithm that does the some work as the algorithm from [49] was presented by the author in [66]. One has to note that in the case of general graphs the solution from [49] is rather complicated and a very interesting and simple algorithm with the same complexity was presented by Harvey [28].

5   Weighted Problems

The Single Source Shortest Paths (SSSP) problem is one of the most fundamental problems in combinatorial optimization. Let us focus on the case when negative weights are allowed but no negative weight directed cycles. In this case the first algorithm for SSSP problem was proposed by Shimbel in 1955 [73]. Some years later, the so called, Bellman-Ford method was developed in the papers [35,3,47]. The Bellman-Ford algorithm is strongly polynomial, i.e., its time complexity does not depend on the weights in the graph. Thirty years later three scaling algorithms were developed [24,25,27]. The fastest of them is the algorithm of Goldberg. It works only in the case of integer edge weights from the set $\{-W, \ldots, 0, \ldots, +\infty\}$, and its time complexity depends on $\log W$. This algorithm works faster than the Bellman-Ford method under the similarity assumption, i.e., when $W = O(poly(n))$. The complexity results for the SSSP problem with negative edge weights are summarized in Table 5.

Table 5. The complexity results for the SSSP problem with negative weights. The bold font indicates an asymptotically best bound in the table.

Complexity	Author
$O(n^4)$	Shimbel (1955) [73]
$O(n^2 m W)$	Ford (1956) [35]
$\mathbf{O(nm)}$	Bellman (1958) [3], Moore (1959) [47]
$O(n^{\frac{3}{4}} m \log W)$	Gabow (1983) [24]
$O(\sqrt{n} m \log(nW))$	Gabow and Tarjan (1989) [25]
$\mathbf{O(\sqrt{n} m \log(W))}$	Goldberg (1993) [27]
$\tilde{\mathbf{O}}(n^{\omega} W)$	Sankowski [67] and Yuster and Zwick [79]

In the paper [67] we have shown how the matrix multiplication can be used to obtain algorithm for the SSSP problem in the case of integer edge weights from the set $\{-W, \ldots, 0, \ldots, +W\}$. The algorithm works in time $\tilde{O}(Wn^{\omega})$, where ω is the matrix multiplication exponent, remember that the best known bound on omega is $\omega < 2.376$ given by Coppersmith and Winograd [7]. Hence, our result improves upon the previous fastest algorithms in the case of dense graphs with small integer weights. The same complexity result for the SSSP problem has been obtained independently by Yuster and Zwick [79]. Their result is based on a distance oracle that after $\tilde{O}(Wn^{\omega})$ preprocessing time can answer distance queries in $O(n)$ time.

Similar ideas that were used in the case of shortest paths problem can be applied to solve weighted bipartite maximum matching problem. The first algorithm for this problem was proposed in the fifties of the last century by Kuhn [41]. His result has been improved several times since then, the results are summarized in the Table 6.

In the above summary there are no algorithms that use matrix multiplication. However, in the papers studying the parallel complexity of the problem [37,52], such algorithms are implicitly constructed. These results might lead to $O(Wn^{\omega+2})$ sequential time algorithms. In the paper [69] we improve the complexity by factor of n^2. The improvement in the exponent by 1 is achieved with use of the very recent results of Storjohann [74], who had shown faster algorithms for computing polynomial matrix determinants. This result was used in the case of shortest paths as well. Further improvement is achieved by a novel reduction technique, that allows us to reduce the weighted version of the problem

Table 6. The complexity results for the bipartite weighted matching problem. The bold font indicates an asymptotically best bound in the table.

Complexity	Author
$O(n^4)$	Khun (1955) [41] and Munkers (1957) [53]
$O(n^2 m)$	Iri (1960) [33]
$O(n^3)$	Dinic and Kronrod (1969) [13]
$O(nm)$	Edmonds and Karp (1970) [14]
$O(n^{\frac{3}{4}} m \log W)$	Gabow (1983) [24]
$O(\sqrt{n} m \log(nW))$	Gabow and Tarjan (1989) [25]
$O(\sqrt{n} m W)$	Kao, Lam, Sung and Ting (1999) [36]
$O(n^\omega W)$	Sankowski (2006) [69]

to unweighted one. The unweighted problem is then solved with use of the $O(n^\omega)$ time algorithms developed last year by Mucha and Sankowski [50].

6 Conclusions

The summary of these results might seem as a story of success. It is indeed true that algebraic approach lead to some new solutions. However, still many open problems remain unsolved. Some of the more important include the following questions:

- Can one close the complexity gap between dynamic shortest paths and transitive closure?
- Can we find maximum weighted matchings in general graphs in matrix multiplication time?
- Can one dynamically compute characteristic polynomial in subquadratic time?
- Can one find perfect matchings in NC^2 on $O(n^\omega)$ processors?
- Is there a way to speed up maximum flow computations using matrix multiplication?

References

1. Ausiello, G., Italiano, G.F., Marchetti Spaccamela, A., Nanni, U.: Incremental algorithms for minimal length paths. J. Algorithms 12(4), 615–638 (1991)
2. Baswana, S., Hariharan, R., Sen, S.: Improved decremental algorithms for maintaining transitive closure and all-pairs shortest paths. In: Proceedings of the thiry-fourth annual ACM symposium on Theory of Computing, pp. 117–123. ACM Press, New York (2002)
3. Bellman, R.: On a Routing Problem. Quarterly of Applied Mathematics 16(1), 87–90 (1958)

4. Blum, N.: A new approach to maximum matching in general graphs. In: Proc.17th ICALP. LNCS, vol. 443, pp. 586–597. Springer, Heidelberg (1990)
5. Bugliesi, M., Preneel, B., Sassone, V., Wegener, I. (eds.): ICALP 2006. LNCS, vol. 4051. Springer, Heidelberg (2006)
6. Cheriyan, J., Reif, J.H.: Directed s-t numberings, rubber bands, and testing digraph k-vertex connectivity. In: SODA 1992: Proceedings of the third annual ACM-SIAM symposium on Discrete algorithms, pp. 335–344. Society for Industrial and Applied Mathematics, Philadelphia, PA, USA (1992)
7. Coppersmith, D., Winograd, S.: Matrix multiplication via arithmetic progressions. In: Proceedings of the nineteenth annual ACM conference on Theory of computing, pp. 1–6. ACM Press, New York (1987)
8. Demetrescu, C., Italiano, G.F.: Fully Dynamic Transitive Closure: Breaking Through the $O(n^2)$ Barrier. In: Proceedings of 41th annual IEEE Symposium on Foundations of Computer Science, pp. 381–389 (2000)
9. Demetrescu, C., Italiano, G.F.: Fully Dynamic All Pairs Shortest Paths with Real Edge Weights. In: Proceedings of 42th annual IEEE Symposium on Foundations of Computer Science, pp. 260–267 (2001)
10. Demetrescu, C., Italiano, G.F.: Improved Bounds and New Trade-Offs for Dynamic All Pairs Shortest Paths. In: Proceedings of the 29th International Colloquium on Automata, Languages and Programming, pp. 633–643. Springer, Heidelberg (2002)
11. Demetrescu, C., Italiano, G.F.: A new approach to dynamic all pairs shortest paths. In: Proceedings of the thirty-fifth annual ACM Symposium on Theory of Computing, pp. 159–166. ACM Press, New York (2003)
12. Diks, K., Sankowski, P.: Dynamic Plane Transitive Closure. In: Arge, L., Hoffmann, M., Welzl, E. (eds.) ESA 2007. LNCS, vol. 4698, pp. 594–604. Springer, Heidelberg (2007)
13. Dinic, E.A., Kronrod, M.A.: An Algorithm for the Solution of the Assignment Problem. Soviet Math. Dokl. 10, 1324–1326 (1969)
14. Edmonds, J., Karp, R.M.: Theoretical Improvements in Algorithmic Efficiency for Network Flow Problems. J. ACM 19(2), 248–264 (1972)
15. Even, S., Gazit, H.: Updating Distances in Dynamic Graphs. Methods of Operations Research 49, 371–387 (1985)
16. Fakcharoenphol, J.: Planar Graphs, Negative Weight Edges, Shortest Paths, and Near Linear Time. In: Proceedings of the 42nd IEEE symposium on Foundations of Computer Science, pp. 232–241. IEEE Computer Society Press, Los Alamitos (2001)
17. Frandsen, G.S., Hansen, J.P., Miltersen, P.B.: Lower Bounds for Dynamic Algebraic Problems. In: Meinel, C., Tison, S. (eds.) STACS 1999. LNCS, vol. 1563. Springer, Heidelberg (1999)
18. Frandsen, G.S., Frandsen, P.F.: Dynamic matrix rank. In: Bugliesi, et al. (eds.) [5], pp. 395–406
19. Frandsen, G.S., Sankowski, P.: Dynamic normal forms and dynamic characteristic polynomial. In: ICALP Proc. 35th ICALP. Springer, Heidelberg (2008)
20. Fredman, M.L.: The Complexity of Maintaining an Array and Computing Its Partial Sums. J. ACM 29(1), 250–260 (1982)
21. Frigioni, D., Marchetti-Spaccamela, A., Nanni, U.: Semi-dynamic Algorithms for Maintaining Single Source Shortest Paths Trees. Algorithmica 22(3), 250–274 (1998)
22. Frigioni, D., Marchetti-Spaccamela, A., Nanni, U.: Fully Dynamic Algorithms for Maintaining Shortest Paths Trees. J. Algorithms 34(2), 251–281 (2000)
23. Gabow, H.N.: Using expander graphs to find vertex connectivity. In: FOCS 2000: Proceedings of the 41st Annual Symposium on Foundations of Computer Science, p. 410. IEEE Computer Society, Washington, DC, USA (2000)

24. Gabow, H.N.: Scaling Algorithms for Network Problems. J. Comput. Syst. Sci. 31(2), 148–168 (1985)
25. Gabow, H.N., Tarjan, R.E.: Faster Scaling Algorithms for Network Problems. SIAM J. Comput. 18(5), 1013–1036 (1989)
26. Gabow, H.N., Tarjan, R.E.: Faster scaling algorithms for general graph matching problems. J. ACM 38(4), 815–853 (1991)
27. Andrew, V.: Scaling algorithms for the shortest paths problem. In: SODA 1993: Proceedings of the fourth annual ACM-SIAM Symposium on Discrete algorithms, pp. 222–231.Society for Industrial and Applied Mathematics (1993)
28. Nicholas, J.A.: Algebraic structures and algorithms for matching and matroid problems. In: FOCS 2006: Proceedings of the 47th Annual IEEE Symposium on Foundations of Computer Science, pp. 531–542. IEEE Computer Society, Washington, DC, USA (2006)
29. Henzinger, M.R., King, V.: Fully Dynamic Biconnectivity and Transitive Closure. In: Proceedings 36th annual IEEE Symposiumon Foundations of Computer Science, pp. 664–672 (1995)
30. Henzinger, M.R., Klein, P., Rao, S., Subramanian, S.: Faster Shortest-path Algorithms for Planar Graphs. J. Comput. Syst. Sci. 55(1), 3–23 (1997)
31. Huang, X., Pan, V.Y.: Fast Rectangular Matrix Multiplication and Applications. Journal of complexity 14(2), 257–299 (1998)
32. Ibaraki, T., Katoh, N.: On-line Computation of Transitive Closure for Graphs. Inform. Proc. Lett. 16, 95–97 (1983)
33. Iri, M.: A new method for solving transportation-network problems. Journal of the Operations Research Society of Japan 3, 27–87 (1960)
34. Italiano, G.F.: Amortized Efficiency of a Path Retrieval Data Structure. Theor. Comput. Sci. 48(2-3), 273–281 (1986)
35. Ford Jr., L.R.: Network Flow Theory. Paper P-923, The RAND Corperation, Santa Moncia, California (August 1956)
36. Kao, M.-Y., Lam, T.W., Sung, W.-K., Ting, H.-F.: A decomposition theorem for maximum weight bipartite matchings with applications to evolutionary trees. In: Proceedings of the 7th Annual European Symposium on Algorithms, pp. 438–449 (1999)
37. Karp, R.M., Upfal, E., Wigderson, A.: Constructing a perfect matching is in random nc. Combinatorica 6(1), 35–48 (1986)
38. Khanna, S., Motwani, R., Wilson, R.H.: On Certificates and Lookahead on Dynamic Graph Problems. In: Proceedings 7th annual ACM-SIAM Symposiumon on Discrete Algorithms, pp. 222–231 (1996)
39. King, V.: Fully Dynamic Algorithms for Maintaining All-Pairs Shortest Paths and Transitive Closure in Digraphs. In: Proceedings of 40th annual IEEE Symposium on Foundations of Computer Science, pp. 81–91 (1999)
40. King, V., Sagert, G.: A Fully Dynamic Algorithm for Maintaining the Transitive Closure. In: Proceedings of the thirty-first annual ACM Symposium on Theory of Computing, pp. 492–498. ACM Press, New York (1999)
41. Kuhn, H.W.: The Hungarian Method for the Assignment Problem. Naval Research Logistics Quarterly 2, 83–97 (1955)
42. La Poutré, J.A., van Leeuwen, J.: Maintenance of Transitive Closure and Transitive Reduction of Graphs. In: Proc. Workshop on Graph-Theoretic Concepts in Computer Science. LNCS, vol. 314, pp. 106–120. Springer, Berlin (1988)
43. Lipton, R.J., Tarjan, R.E.: Applications of a planar separator theorem. SIAM J. Comput. 9(3), 615–627 (1980)
44. Loubal, P.: A Network Evaluation Procedure. Highway Research Record 205, 96–109 (1967)

45. Micali, S., Vazirani, V.V.: An $O(\sqrt{|V|}|E|)$ algorithm for finding maximum matching in general graphs. In: Proceedings of the twenty first annual IEEE Symposium on Foundations of Computer Science, pp. 17–27 (1980)

46. Miller, G.L., Naor, J.: Flow in planar graphs with multiple sources and sinks. SIAM J. Comput. 24(5), 1002–1017 (1995)

47. Moore, E.F.: The Shortest Path Through a Maze. In: Proceedings of the International Symposium on the Theory of Switching, pp. 285–292. Harvard University Press (1959)

48. Mucha, M., Sankowski, P.: Maximum Matchings in Planar Graphs via Gaussian Elimination. In: Proceedings of 12th annual European Symposium on Algorithms, pp. 532–543 (2004)

49. Mucha, M., Sankowski, P.: Maximum Matchings via Gaussian Elimination. In: Proceedings of the 45th annual IEEE Symposium on Foundations of Computer Science, pp. 248–255 (2004)

50. Mucha, M., Sankowski, P.: Maximum matchings via gaussian elimination. In: Proceedings of the 45th annual IEEE Symposium on Foundations of Computer Science, pp. 248–255 (2004)

51. Mucha, M., Sankowski, P.: Fast dynamic transitive closure with lookahead. Algorithmica (2008)

52. Mulmuley, K., Vazirani, U.V., Vazirani, V.V.: Matching is as easy as matrix inversion. In: STOC 1987: Proceedings of the nineteenth annual ACM conference on Theory of computing, pp. 345–354. ACM Press, New York (1987)

53. Munkres, J.: Algorithms for the Assignment and Transportation Problems. Journal of SIAM 5(1), 32–38 (1957)

54. Murchland, J.: The Effect of Increasing or Decreasing the Length of a Single Arc on All Shortest Distances in a Graph. Technical report, LBS-TNT-26 (1967)

55. Lovász, L., Linial, N., Wigderson, A.: Rubber bands, convex embeddings and graph connectivity. Combinatorica 8, 91–102 (1988)

56. Rabin, M.O., Vazirani, V.V.: Maximum matchings in general graphs through randomization. Journal of Algorithms 10, 557–567 (1989)

57. Ramalingam, G., Reps, T.: An Incremental Algorithm for a Generalization of the Shortest-path Problem. J. Algorithms 21(2), 267–305 (1996)

58. Ramalingam, G., Reps, T.: On the Computational Complexity of Dynamic Graph Problems. Theor. Comput. Sci. 158(1-2), 233–277 (1996)

59. Reif, J.H., Tate, S.R.: On Dynamic Algorithms for Algebraic Problems. J. Algorithms 22(2), 347–371 (1997)

60. Rodionov, V.: The Parametric Problem of Shortest Distances. U.S.S.R. Computational Math. and Math. Phys. 8(5), 336–343 (1968)

61. Roditty, L.: A Faster and Simpler Fully Dynamic Transitive Closure. In: Proceedings of the fourteenth annual ACM-SIAM Symposium on Discrete Algorithms, pp. 404–412. Society for Industrial and Applied Mathematics (2003)

62. Roditty, L., Zwick, U.: Improved Dynamic Reachability Algorithms for Directed Graphs. In: Proceedings of the 43rd Symposium on Foundations of Computer Science, p. 679. IEEE Computer Society, Los Alamitos (2002)

63. Roditty, L., Zwick, U.: A Fully Dynamic Reachability Algorithm for Directed Graphs with an Almost Linear Update Time. In: Proceeding of the 36th annual ACM Symposium on Theory of Computing, pp. 184–191. ACM Press, New York (2004)

64. Rohnert, H.: A Dynamization of the All Pairs Least Cost Path Problem. In: Proceedings of the 2nd Symposium of Theoretical Aspects of Computer Science, pp. 279–286. Springer, Heidelberg (1985)

65. Sankowski, P.: Dynamic Transitive Closure via Dynamic Matrix Inverse. In: Proceedings of the 45th annual IEEE Symposium on Foundations of Computer Science, pp. 509–517 (2004)
66. Sankowski, P.: Processor efficient parallel matching. In: SPAA 2005: Proceedings of the seventeenth annual ACM symposium on Parallelism in algorithms and architectures. ACM Press, New York, NY, USA (2005)
67. Sankowski, P.: Shortest Paths in Matrix Multiplication Time. In: Brodal, G.S., Leonardi, S. (eds.) ESA 2005. LNCS, vol. 3669, pp. 770–778. Springer, Heidelberg (2005)
68. Sankowski, P.: Subquadratic Algorithm for Dynamic Shortest Distances. In: Wang, L. (ed.) COCOON 2005. LNCS, vol. 3595, pp. 461–470. Springer, Heidelberg (2005)
69. Sankowski, P.: Weighted bipartite matching in matrix multiplication time. In: Bugliesi, et al. (eds.) [5], pp. 274–285
70. Sankowski, P.: Faster dynamic matchings and vertex connectivity. In: SODA 2007: Proceedings of the eighteenth annual ACM-SIAM symposium on Discrete algorithms, pp. 118–126. Society for Industrial and Applied Mathematics, Philadelphia, PA, USA (2007)
71. Sherman, J., Morrison, W.J.: Adjustment of an inverse matrix corresponding to changes in the elements of a given column or a given row of the original matrix. Ann. Math. Statist. 20(5), 621 (1949)
72. Sherman, J., Morrison, W.J.: Adjustment of an inverse matrix corresponding to a change in one element of a given matrix. Ann. Math. Statist. 21(1), 124–127 (1950)
73. Shimbel, A.: Structure in Communication Nets. In: Proceedings of the Symposium on Information Networks, pp. 199–203. Polytechnic Press of the Polytechnic Institute of Brooklyn, Brooklyn (1955)
74. Storjohann, A.: High-order lifting and integrality certification. J. Symb. Comput. 36(3-4), 613–648 (2003)
75. Subramanian, S.: A fully dynamic data structure for reachability in planar digraphs. In: ESA 1993: Proceedings of the First Annual European Symposium on Algorithms, pp. 372–383. Springer, London, UK (1993)
76. Thorup, M.: Fully-dynamic all-pairs shortest paths: Faster and allowing negative cycles. In: Hagerup, T., Katajainen, J. (eds.) SWAT 2004. LNCS, vol. 3111, pp. 384–396. Springer, Heidelberg (2004)
77. Thorup, M.: Worst-case update times for fully-dynamic all-pairs shortest paths. In: STOC 2005: Proceedings of the thirty-seventh annual ACM symposium on Theory of computing, pp. 112–119. ACM, New York, NY, USA (2005)
78. Yellin, D.M.: Speeding up Dynamic Transitive Closure for Bounded Degree Graphs. Acta Informatica 30, 369–384 (1993)
79. Yuster, R., Zwick, U.: Answering distance queries in directed graphs using fast matrix multiplication. In: FOCS, pp. 389–396. IEEE Computer Society, Los Alamitos (2005)

Question/Answer Games on
Towers and Pyramids

Sarmad Abbasi[1] and Numan Sheikh[2]

[1] 117-BB, DHA, Lahore, Pakistan
sarmad_abbasi@yahoo.com
[2] Department of Computer Science,
Lahore University of Management Sciences,
Opposite Sector "U", DHA, Lahore, Pakistan
numan@lums.edu.pk

Abstract. Question/Answer games[3] (Q/A games) are a generalization of the game introduced in [1,2]. They are motivated by the classical game of twenty questions and are a generalization of Rényi-Ulam Game. A k-round Q/A game, $G = (D, (q_1, \ldots, q_k))$, is played on a rooted directed acyclic graph, $D = (V, E)$. In the i-th round, Paul selects a set, $Q_i \subseteq V$, of at most q_i non-terminal vertices. Carole responds by choosing an outgoing edge from each vertex in Q_i. At the end of k rounds, Paul wins if Carole's answers define a unique path from the root to one of the terminal vertices in D. Arbitrary Q/A games are known to be **PSPACE**-complete[3], and k-round games are known to be Σ_{2k-2}-complete[4]. In this paper we study Q/A games on two classes of graphs, towers and pyramids, respectively. We completely solve the problem of determining the winner for Q/A games on towers. We also solve an open problem on Q/A games on pyramids from [1,2]. Furthermore, we give some non-trivial lower and upper bounds for the rest of the cases for Q/A games on pyramids.

1 Introduction

A Question/Answer game[3] (or Q/A game for short) is a *perfect information game* that is played between two persons Paul and Carole. Q/A games can be considered as a model for resource-bound information extraction in parallel, where the amount of information that can be probed is limited.

A Q/A game is played on a rooted acyclic digraph. The game consists of k rounds, and in each round Paul inquires about some vertices; for each such vertex, v, Carole replies with an out-going edge from v. Paul wins the game if after k rounds, Carole's answers form a unique path, starting from the root and ending at a terminal vertex.

In the next subsections we give the definitions and provide the motivation and a brief history of the problem. In Section 2 we solve the Q/A games on towers. In Section 3 we define the notion of reducing one game to another. These reductions help us in proving some lower bound results. In Section 4 we study Q/A games on pyramids.

E. Ochmański and J. Tyszkiewicz (Eds.): MFCS 2008, LNCS 5162, pp. 83–95, 2008.
© Springer-Verlag Berlin Heidelberg 2008

1.1   Definitions

In this paper all the graphs mentioned will be rooted, directed and acyclic. A *terminal vertex* is defined as a vertex with out-degree zero. A vertex that is not a terminal vertex is called an *internal vertex*. Let $G = (V, E)$ be a directed graph, then the out-neighborhood of the vertex v in G is defined as $N_G^+(v) = \{w|\ (v, w) \in E\}$. Formal definitions of graph theoretic concepts can be found in any standard graph theory text; for example [5].

Formally, a Q/A game, $G = (D, (q_1, \ldots, q_k))$, is defined on a rooted directed acyclic graph, $D = (V, E)$. The game is played in k rounds and Paul is allowed to ask at most q_i questions in the i-th round. If the maximum number of questions in each round is the same; that is, $q = q_1 = q_2 = \cdots = q_k$, then we may denote the game by $G = (D, q, k)$.

In the i-th round, Paul chooses a subset, Q_i, of internal vertices such that $|Q_i| \leq q_i$. If $v \in Q_i$, we say that Paul *inquires* or *asks* about the vertex v. Carole chooses an outgoing edge $(v, f(v))$ for all $v \in Q_i$. We say that she *responds* by $f(v)$ or Carole *points* or *leads* v to $f(v)$. The goal in the game for Paul is to find a unique path, defined by Carole's answers, from the root to one of the terminal vertices. If he is able to identify such a unique path in k rounds, he wins; otherwise, Carole wins.

Consider a Q/A game in progress. We call a vertex v *reachable*, if there exists a path from the root to v that is "consistent" with Carole's answers. An internal vertex v is called *open*, if it is reachable and Paul has not inquired about v. A vertex v is called the *pseudo-root*, if v is open and all predecessors of v are not open. For simplicity, assume that Paul never repeats a question (or equivalently we can require that Carole, once having chosen an outgoing edge from a vertex v, consistently chooses the same edge when re-inquired about v).

An n-level, w-tower, T_n^w, consists of $n + 1$ levels where the vertices on the first n levels are the internal vertices. Level 0 consists of the root vertex which is labeled $(0, 0)$. For $1 \leq i \leq n$, level i has w vertices labeled $(i, 0), \ldots, (i, w - 1)$. Each vertex (i, j) has w outgoing edges to $(i + 1, 0), (i + 1, 1), \ldots, (i + 1, w - 1)$ for all $i < n$. Note that only the vertices on the n-th level are terminal vertices. A Q/A game on a tower can be described as $G_t = (T_n^w, (q_1, \ldots, q_k))$.

Similarly, an n-level pyramid, P_n, consists of $n + 1$ levels. Level 0 consists of the root vertex which is labeled $(0, 0)$. The i-th level, for $i \leq n$, consists of $i + 1$

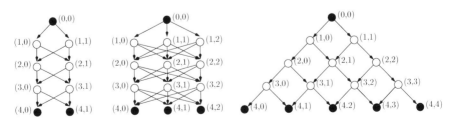

Fig. 1. Towers T_4^2 and T_4^3, and pyramid P_4

vertices. The vertices on the i-th level are labeled, $(i, 0), \ldots, (i, i)$. Each vertex (i, a) on the i-th level is connected to two vertices $(i + 1, a)$ and $(i + 1, a + 1)$ in $i + 1$-st level. Note again that only the vertices on the n-th level are terminal vertices. A Q/A game on pyramid can be described as $G_p = (P_n, (q_1, \ldots, q_k))$. In fact, the Q/A games studied in this paper are of the form $G_p = (P_n, q, k)$; that is, the number of questions are same in each round.

We denote the t-th triangular number by Δ_t; that is, $\Delta_t = \frac{t(t+1)}{2}$. We say that $q = \Delta_t + \delta$, where t is the largest number such that Δ_t is smaller than q; for example, if $q = 7$ then $t = 3$ and $\delta = 1$. We define the t-th *tree number*, T_t, as $2^t - 1$.

1.2 Motivation and Previous Work

It is interesting to compare Q/A games with the *Rényi-Ulam game*[7] and the classical *Twenty Questions*. The comparison is discussed in [3,4].

Originally, the Q/A game was defined as Chaudhuri's game on a mesh, or pyramid, to study how much information could be extracted in parallel on a grid. However, an easier version of the game played on complete binary trees was solved in [1,2]. Almost all the questions regarding the original Q/A game on the pyramid remained unsolved. By solving the game, we mean that given a description of a Q/A game, one can tell in polynomial time if Paul wins the game or not. In [1,2] it was shown that given a Q/A game on an n-level binary tree, T_n, Paul wins $(T_n, (q_1, \ldots, q_k))$ if and only if $\sum_{i=1}^{k} \lfloor \log_2(q_i + 1) \rfloor \geq n$.

Determining the winner of Q/A games on arbitrary graphs and arbitrary number of rounds, even with 2 questions in each round, is **PSPACE**-complete[3,4]. Q/A games with fixed number of rounds capture all the even levels of polynomial-time hierarchy, as it was shown in [4] that determining if Paul wins a k-round Q/A game is Σ_{2k-2}-complete. Odd levels of polynomial-time hierarchy can also be captured by an extended version of the game[6].

2 Q/A Games on the Towers

We have defined Q/A games on towers as $G_t = (T_n^w, (q_1, \ldots, q_k))$. In this section, we will assume that such a G_t is in progress with $w \geq 2$. The proof of the following theorem follows from Lemma 1 and Lemma 3.

Theorem 1. *Paul wins G_t if and only if $\sum_{i=1}^{k} (q_i + w - 1) \geq wn$.*   □

Lemma 1. *Let $\hat{T}_n^w$ denote a tower in which $z < w$ questions on the last level of T_n^w are already answered. If $\sum_{i=1}^{k} (q_i + w - 1) \geq wn - z$ then Paul wins $\hat{G}_t = (\hat{T}_n^w, (q_1, \ldots, q_k))$.*

Proof. We use induction on k. For $k = 1$, the lemma is easy to verify. For the induction step, consider the strategy in which Paul always asks the pseudo-root and asks the remaining questions on the open vertices as deep as possible (see Fig. 2). We can write $q_1 - 1 + z = wj + t$ where $t < w$. After the first round, the number of remaining questions satisfies $\sum_{i=2}^{k}(q_i + w - 1) \geq wn - z - q_1 - w + 1 = w(n - (j+1)) - t$. We note that in the remaining game $j+1$ levels have collapsed and t questions are answered on the last level. Hence the lemma follows by induction. $\square$

Consider a Q/A game, G_t, on T_n^w, after i rounds. We say that G_t is in a *standard position* if for all vertices, (a, b), that are not open, $f((a, b)) = (a + 1, b)$. In other words, in a standard position, all the answered vertices point directly downwards. For such vertices (a, b), define $\lambda((a, b))$ to be the maximum λ such that $(a, b), (a + 1, b), \ldots, (a + \lambda, b)$ are all answered. We define the *weight* of a standard position to be $wm - z$, where m is the number of levels in the graph, counted from the pseudo-root, and z is the number of answered questions.

Lemma 2. *Let G_t be in a standard position after i rounds, with weight W. Suppose Paul asks any q questions in the next round. Carole can answer the questions in such a way that G_t is in a standard position after the next round. Furthermore, the weight, W', of the new position satisfies $W' \geq W - q - w + 1$.*

Proof. We assume that the vertices of the tower are relabeled so that the pseudo-root is labeled $(0, 0)$. For $q = 1$, if Paul inquires about the pseudo-root, Carole finds a vertex, $(1, b)$, on level one that minimizes $\lambda((1, b))$ and points the pseudo-root to $(1, b)$. It is easy to see that in this case the next position is a standard position. Furthermore, the new position has weight at least $W - z$. On the other hand, if Paul inquires about any other vertex, (a, b), Carole simply points it to $(a + 1, b)$. In this case, we note that the new position has weight $W - 1$.

If $q > 1$, we observe that Paul must have asked at least $q - 1$ questions that are not about the pseudo-root. By ordering the questions in such a way that the question about the pseudo-root is the last question, Carole answers the vertices one by one using the strategy described above for answering a single question. Since at least $q - 1$ questions are not about the pseudo-root, the weight loss is at most one for these $q - 1$ questions. The last question can incur a loss of at most w. This shows that $W' \geq W - q - w + 1$. $\square$

Lemma 3. *If $\sum_{i=1}^{k}(q_i + w - 1) < wn$ then Carole wins G_t.*

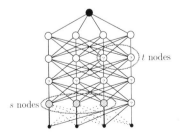

Fig. 2. A partially played T_5^4 and the strategy given in Lemma 1

Proof. The initial position has weight wn and is a standard position. If Carole uses the strategy given in Lemma 2, in each round, the weight of the game after k rounds is at least $nw - \sum_{i=1}^{k}(q_i + w + 1) \geq 1$. The Lemma follows from the observation that if the game ends in a standard position with positive weight then Carole wins. □

We apply these results on Q/A games on T_n^2 with q questions in each round to get the following useful theorem.

Corollary 1. *Paul wins on $G = (T_n^2, q, k)$ if and only if $k \geq \frac{2n}{q+1}$.* □

3   Reductions

Let $D_1 = (V_1, E_1)$ and $D_2 = (V_2, E_2)$ be two directed graphs with roots s_1 and s_2 respectively. In this section we describe how to reduce D_1 to D_2, such that, for a given question vector $(q_1, \ldots, q_k)$, Carole can use any strategy devised for a Q/A game on D_2 and use it to obtain a strategy for a game on D_1.

We say that (D_1, s_1) is *reducible* to (D_2, s_2) if there is a function $\lambda : V_1 \mapsto V_2$ such that $\lambda(s_1) = s_2$ and

$$N_{D_1}^+(v) \cap \lambda^{-1}(y) \neq \emptyset \text{ for all } y \in N_{D_2}^+(\lambda(v)). \tag{1}$$

To understand the condition (1) intuitively, consider D_1 and D_2 from Carole's point of view. Suppose Paul presents her with a question v. She has to choose an answer in $N^+(v)$. She maps the question v to $\lambda(v)$ via the reduction and pretends that the game is being played on D_2. Thus she pretends that Paul has inquired about $\lambda(v)$. Any strategy that she can use on D_2 will give her an answer $y \in N^+(\lambda(v))$. In order for her to "pull back" this answer to D_2 she must have a vertex in $N^+(v)$ that is mapped to y.

Theorem 2. *Let (D_1, s_1) be reducible to (D_2, s_2). Then for all question sequences $(q_1, \ldots, q_k)$, if Carole wins $G_2 = (D_2, (q_1, \ldots, q_k))$ then Carole wins $G_1 = (D_1, (q_1, \ldots, q_k))$.*

Proof. Let $\mathcal{S}$ be a winning strategy for Carole for G_2. We define a winning strategy $\mathcal{S}'$ for Carole for G_1. Every time Paul presents a set of questions to Carole, she maps the vertices of the D_1 to those of the D_2 via λ. Then she pretends that she is playing the game on D_2 and uses $\mathcal{S}$.

Formally, in each of the i-th round, when Carole is presented with a set of questions $Q_{i,1}$, she computes $Q_{i,2} = \{\lambda(v) : v \in Q_{i,1}\}$. Note that $|Q_{i,2}| \leq |Q_{i,1}|$ and she can find answers to all the questions in $Q_{i,2}$ using $\mathcal{S}$.

Fix any $v \in Q_{i,1}$ and let $w = \lambda(v)$. Let $y = f_2(w) \in N^+(w)$ be the answer to w obtained by using strategy $\mathcal{S}$. Since $N^+(v) \cap \lambda^{-1}(y) \neq \emptyset$ Carole sets $f_1(v)$ to some vertex in $N^+(v) \cap \lambda^{-1}(y)$ arbitrarily. This describes $\mathcal{S}'$. Now, we have to show that it is a winning strategy for Carole.

Consider the game after i rounds. We argue that if P is a path from the root in D_2 to any vertex that is consistent with Carole's answers then there is a path

P' that is consistent with Carole's answers in D_1 such that $\lambda(P') = P$. To see this, note that the result is trivial for paths of length 0 since $\lambda(s_1) = s_2$. For the induction step, let $s_2 = v_1, \ldots, v_k$ be a path of length k in D_2 that is consistent with Carole's answers. By induction on k, there exists a path $s_1 = w_1, \ldots, w_{k-1}$ that is mapped to $s_2 = v_1, \ldots, v_{k-1}$. Thus all we have to do is to find a $w_k \in V_1$ such that $\lambda(w_k) = v_k$ and the path $s_1 = w_1, \ldots, w_k$ is consistent in D_1. However, condition (1) and the strategy $\mathcal{S}'$ exactly allow us to do this. More precisely, if w_{k-1} is inquired about, then we let w_k be the vertex where Carole points w_{k-1} to when she uses $\mathcal{S}'$. In case w_{k-1} is not inquired about, then we can let w_k be any vertex in $N^+(w_{k-1}) \cap \lambda^{-1}(v_{k-1})$. In both cases it is clear that $s_1 = w_1, \ldots, w_k$ is consistent with Carole's answers.

The theorem follows from the observation that if P_1 and P_2 are two distinct consistent paths in D_2, then there exist paths P_1' and P_2' that are consistent in D_1. It is easy to see that P_1' and P_2' must also be distinct. □

4   Q/A Games on Pyramids

We now discuss the Q/A game on an n-level pyramid, $G_p = (P_n, q, k)$. The game consists of k rounds with q questions per round. G_p admits a "natural strategy" for Paul when q is a triangular number, Δ_t. In each round, Paul asks all the questions on the topmost t levels $(0, \ldots, t-1)$ of P_n. After getting Carole's answers he knows the first t vertices on the path. The game is now reduced to finding a path in P_{n-t}. This way he can finish the game in $\lceil \frac{n}{t} \rceil$ rounds.

Theorem 3 (Chaudhuri[1,2]). *Paul wins* (P_n, Δ_t, k) *if* $k \geq \frac{n}{t}$. □

Triangular numbers, Δ_t, seem to behave in similar way as tree numbers, $\mathcal{T}_t$, behave on a Q/A game on a complete binary tree. There is, however, a clear difference in the two games as pointed out in [1,2]. In a Q/A game on a complete binary tree, two questions per round are no better than one question per round, in both cases Paul gains only two levels in two rounds. However, Paul gains three levels in two rounds in a Q/A game on a pyramid, as shown in Fig.3. The optimality of this strategy was questioned in [1,2]. We show that not only is this strategy optimal for $q = 2$, natural strategy discussed above is also optimal for the first two triangular numbers; that is for $q = 1$ and $q = 3$.

We show that not only is this strategy optimal for $q = 2$, natural strategy discussed above is also optimal for the first two triangular numbers; that is for

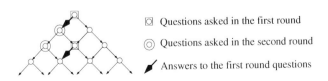

○ Questions asked in the first round

◎ Questions asked in the second round

✦ Answers to the first round questions

Fig. 3. Paul's strategy for $q = 2$

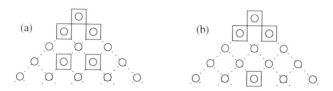

Fig. 4. Questions asked by Paul in the first round for strategy when $q = 5$

$q = 1$ and $q = 3$. If we play the game where Paul is allowed to ask four questions per round, the strategy is quite surprising. In the best strategy we are aware of for $q = 4$, Paul asks three questions on the first two rows of the pyramid, and asks the fourth question on the 11-th row. We can show using brute force analysis that Paul can gain 11 levels in five rounds. However, we are not aware of a simple description for such strategy. For $q = 5$, the strategy shown in Fig. 4(a) looks quite intuitive. If the number of levels, n, in the Pyramid is large, Paul requires $\frac{2}{5}$ rounds to gain a level, as no matter what vertex on the third level is the pseudo-root after the first round, Paul has enough questions to collapse three levels in the second round.

However, if in the first round Paul asks the questions shown in Fig. 4(b), he can ask the fifth question on the 316-th level. As we show later, he can not only gain five levels after every two rounds, he would gain an extra level in the 126-th round. In this case, Paul requires $\frac{2}{5 + \frac{1}{63}}$ rounds to gain a level, on the average.

Table 1 shows some best known bounds for Q/A games on pyramids. Some strategies to obtain these bounds are discussed in the following sections.

4.1 Lower Bounds

In this section we provide a lower bound on the number of rounds that Paul requires to win a Q/A game. Using the reductions defined in Section 3 we can show that:

Theorem 4. *If Carole wins* (T_n^2, q, k) *then Carole also wins* (P_n, q, k).

Proof. We can "wrap" the pyramid around the tower. Formally, let us define the reduction function $\lambda : V(P_n) \rightarrow V(T_n^2)$ as $\lambda(i, j) = (i, j \bmod 2)$. It is not hard to see that Carole can use her winning strategy for the game on tower to answers Paul's questions on the pyramid and win.   □

Theorem 4 and Corollary 1 immediately provide us with a lower bound on the number of rounds that Paul requires to win a game $G_p = (P_n, q, k)$.

Table 1. Paul's progress with the best known strategies for Q/A games on pyramids

No. of questions per round	1	2	3	4	5	6
No. of levels gained per round	1	3/2	2	11/5	316/126	3

Theorem 5. *Carole wins* (P_n, q, k) *if* $k < \frac{2n}{q+1}$. □

Applying these lower bound for the case $q = 1$, $q = 2$ and $q = 3$, we see that the strategies discussed earlier are optimal, resolving an open question that was posed in [1].

Theorem 6. *Given a game* $G_p = (P_n, q, k)$, *for* $q \in \{1, 2, 3\}$, *Paul wins* G_p *if and only if* $k \geq \frac{2n}{q+1}$. □

4.2 Some Strategies for Paul

When q is a triangular number; that is, $q = \Delta_t$ for some t, the natural strategy described above appears to be optimal. However the proof of its optimality is only known for $q = \Delta_1 = 1$ and $q = \Delta_2 = 3$. This leads to the following open problem.

Problem 1. Is the bound given in Theorem 3 tight for all $q = \Delta_t$?

On the other hand, one can ask what happens if k is not a triangular number. No optimal strategies are known in these cases except for when $q = 2$, as shown earlier. Lower bounds for Q/A games on pyramids, $G_p = (P_n, q, k)$, provided by Theorem 5 do not seem to be tight when $q > 3$. This leads to another interesting problem.

Problem 2. Devise optimal strategies for Paul when $q = \Delta_t + \delta$ where $0 < \delta \leq t$.

Taking Advantage of One Extra Question

Let $q = \Delta_t + 1$. In this section we devise a strategy (Algorithm 1) for Paul so that Paul can take advantage of this extra question in the long run.

At the beginning of the algorithm, we consider the B-th row where $B = t(2^{t+1} - 1) - 1$. We divide the B-th row into $2^{t+1} - 1$ blocks of t vertices each; formally, i-th block $= \{(p, B) : \lfloor \frac{p+1}{t} \rfloor = i\}$. The algorithm proceeds in $t + 1$ phases, $1, \ldots, t, t + 1$. The j-th phase consists of 2^{t-j+1} rounds. We now describe Paul's strategy for the first t phases. In each round, he asks Δ_t questions in the top t rows (just like in the natural strategy described earlier), and asks the extra question on the B-th level. After every round, t vertices become unreachable at the B-th level so number of reachable blocks is reduced by one. After every phase he makes sure that in each block, the number of questions asked is increased by one, so after the completion of t phases, all the reachable vertices on the B-th level are answered, and he is able to collapse the graph by one extra level. We will call a block *marked* for phase j, if Paul has already asked a question in that block, during the j-th phase.

Note that, after every round, the marked blocks are consecutive. Also, after every round, t vertices at the edges of B-th level become unreachable. These vertices are always in the unmarked block, except perhaps in the last round.

$B := t(2^{t+1} - 1) - 1;$
foreach $j = 1 \ldots t$ **do**
    divide the reachable vertices on the B-th row in blocks labeled
    $1 \ldots (2^{t-j+2} - 1)$ of t consecutive vertices each;
    unmark each block;
    foreach *round* $i = 1 \ldots (2^{t-j+1})$ **do**
        Ask Δ_t question in the top t rows;
        $x :=$ block below the pseudo-root. If the pseudo-root lies above the
        boundary of two blocks, pick the one on the right as a convention. If the
        one on the right is already marked, pick the one on the left;
        if x *already is marked* **then**
           | $x :=$ unmarked block that is nearest to x;
        end
        ask the right most question in the block x;
        mark x;
    end
end

Algorithm 1. Paul's strategy for the first t phases when $q = \Delta_t + 1$

Lemma 4. *After the j-th phase, in each reachable block there are j answered questions, in the same pattern.*

Proof. The proof is by induction on j. The base case is trivial, because before the first round, when $j = 0$, the blocks are identical with no question answered. Let us assume that after $j - 1$ phases, the claim holds and all the block have $j - 1$ questions asked in the same pattern. In each round of the j-th phase we ask the right most unasked vertex in the each block that falls either below the pseudo-root, or is the unmarked block nearest to the block below the pseudo-root. In this way all the blocks that are being asked in this phase will have the same pattern of asked vertices in them. Now let us assume that after the phase ends, the first reachable vertex in the B-th row is b vertices from the left edge of the block it is contained in. Let $a = t - b$. When we recreate the blocks before the next phase, the pattern is simply rotated by b vertices to the right (Fig. 5).

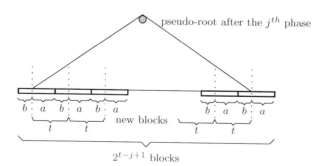

Fig. 5. Relabeling of the blocks after a phase

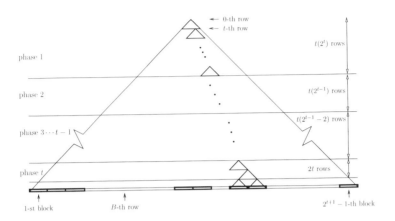

Fig. 6. Q/A game on Pyramid, where $q = \Delta_t + 1$

The new blocks would contain b vertices from one block followed by a vertices from the block on its right.   □

Note that after the j-th phase of 2^{t-j+1} rounds, the number of reachable vertices in B-th level are reduced by 2^{t-j+1} from $t(2^{t-j+2} - 1)$ to $t(2^{t-j+1} - 1)$. Hence, after $j = t$ phases, t vertices are reachable and they are all answered by Lemma 4.

Theorem 7. *Paul wins* $G_p = (P_n, \Delta_t + 1, k)$, *if* $k \geq \frac{n}{t + \frac{1}{2^{t+1} - 1}}$.

Proof. Applying the algorithm described above, all the vertices in the block below the pseudo-root are answered after the t-th phase. Paul now has enough questions to ask them on the $t - 1$ levels starting from the pseudo-root along with the other t reachable vertices on the $B + 1$-st level, to gain an extra level as compared to the natural strategy with Δ_t questions. We note that the number of rounds required by Paul to gain $B + 1$ levels is $k = \sum_{j=1}^{t+1} 2^{t-j+1} = 2^{t+1} - 1$. Paul gains $t(2^{t+1} - 1)$ levels in $2^{t+1} - 1$ rounds, therefore he requires $k \geq \frac{n}{t + \frac{1}{2^{t+1} - 1}}$ rounds to win an n level game.   □

It is important to remark here that although this algorithm works when $q = \Delta_t + 1$, for all values of t, this does not appear to be the most optimal, as mentioned earlier for the case when $q = 4$. Instead of asking the extra questions at the 14-th row, and gain 15 levels in seven rounds, Paul can asks the extra question on the 11-th row and gain 11 levels in five rounds, that would give him a better gain. However, the result of Theorem 7 can be easily generalized for other values of δ. By asking δ questions in a block per round and reducing the number of phases from $t + 1$ to $\lceil t/\delta \rceil + 1$.

Theorem 8. *Paul wins* $G = (P_n, \Delta_t + \delta, k)$, *if* $k \geq \frac{n}{t + \frac{1}{2^{\alpha+1} - 1}}$, *where* $\alpha = \lceil t/\delta \rceil$.   □

An m Round Strategy

We now describe a strategy for Paul that give better bounds than Theorem 8, when δ is large. First we show the strategy for the simplest case.

Theorem 9. *Paul wins* $(P_n, \Delta_t + \delta, k)$ *if* $k \geq \frac{n}{t+\frac{1}{2}}$ *and* $\delta \geq \frac{2t+1}{3}$ *in case δ is odd; and* $\delta \geq \frac{2t+2}{3}$ *if δ is even.*

Proof. Paul asks Δ_t questions on the first t levels, and uses the rest of the δ questions in such a way that they can combine with the $q = \Delta_t + \delta$ questions in the next round to gain an extra level. Paul asks these δ questions in the first round on the $2t$-th row, placing the questions at the vertices located at the center of the row. Note that, for any pseudo-root on the t-th level, after the first round, at least $\lfloor \frac{\delta+1}{2} \rfloor$ answered vertices will remain reachable on the $2t$-th level (Fig. 7). Paul gains $2t + 1$ levels in two rounds if $\lfloor \frac{\delta+1}{2} \rfloor + q \geq \Delta_{t+1}$, by asking the remaining δ open questions in the second round. □

Note that there is an odd number of vertices in the $2t$-th row. If δ is even, we can use one of the questions from the first round at the B-th level that is calculated using a recursive strategy of $\Delta_{2t+1} + 1$ questions according to Algorithm 1 and gain a further level after $2^{(2t+1)+1} - 1$ rounds.

We do not know of any strategy to gain $m(t+1)$ levels in m rounds as $\delta < t+1$. However, we can generalize the strategy above and gain $m(t+1) - 1$ levels in m rounds, if δ is large enough.

Theorem 10. *Paul wins* $(P_n, \Delta_t + \delta, k)$ *if* $k \geq \frac{n}{t+\left(1-\frac{1}{m}\right)}$
and $\delta \geq \frac{2t+1}{2}\left(1 - \frac{1}{m^2-m+1}\right)$.

Proof. In the first round, Paul asks the Δ_t questions on the first t levels, and uses the rest of the δ questions on rows $l_2, \ldots, l_m$, where $l_i = it + i - 2$, in such a way that they can combine with the $q = \Delta_t + \delta$ questions in the next $m - 1$ rounds to gain $t + 1$ levels in each round.

Paul splits these δ questions in batches of size p such that he asks $(i-1)$ batches on each of the l_i-th row, where $p = (2(t - \delta) + 1)$. Let us label the batches on

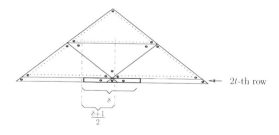

Fig. 7. Two round strategy where δ is odd

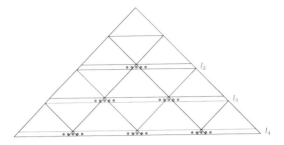

Fig. 8. Distribution of the δ questions for $m = 4$ round strategy

each row, l_i, as $p_1, \ldots, p_{i-1}$. The questions in a batch, p_j, are asked on vertices labeled $(l_i, j(t+1) - 1 - t + \delta), \ldots, (l_i, j(t+1) - 1), \ldots, (l_i, j(t+1) - 1 + t - \delta)$, for each j (see Fig. 8). Now at start of the i-th round, for $i = 2, \ldots, m$, any pseudo-root on row number $(i-1)(t+1) - 1$ will have at most δ open vertices that are reachable on the level l_i. In the i-th round, Paul asks Δ_t questions on the t levels starting at the pseudo-root, and δ questions on the row l_i to move down a total of $t+1$ levels. Note that $\delta \geq \sum_{i=2}^{m} (i-1)(2(t - \delta) + 1)$ in this case. Solving for δ and rearranging the terms give us our desired result.

5   Conclusions

We have seen that for Q/A games on the pyramid, number of rounds required for Paul to win Q/A games do not depict a clear pattern. Problems 1 and 2 are very intriguing questions. The following question can probably shed some more light on the nature of the game on pyramids.

Problem 3. Consider a Q/A game on a pyramid on n levels. Let L_q be the number of levels gained per round by Paul if he asks q questions in a round, playing with some optimal strategy. Can we show that $L_{q+1} > L_q$, for all values of $q > 0$ as $n \to \infty$?

There are many interesting questions about Q/A games on other classes of graphs also. The following, for example is another intriguing open problem.

Problem 4. Given an arbitrary tree, T, and a question vector, $\mathbf{q} = (q_1, \ldots, q_k)$, is it possible to determine in polynomial time, if Paul wins $(T, \mathbf{q})$?

References

1. Abbasi, S.: Do answers help in posing questions? Tech. Report 98-35, DIMACS (1998)
2. Abbasi, S.: Impact of parallelism on the efficiency of binary tree search, Ars Combinatoria (to appear)
3. Abbasi, S., Sheikh, N.: Some hardness results for Q/A games. INTEGERS: Electronic Journal of Combinatorial Number Theory 7, G08 (2007)

4. Abbasi, S., Sheikh, N.: Complexity of Question/Answer games. Theoretical Computer Science (conditionally accepted)
5. Bollobas, B.: Modern graph theory. Springer, NewYork (1998)
6. Sheikh, N.: Question/Answer Games, PhD thesis, Lahore University of Management Sciences (in preparation)
7. Spencer, J.: Ulam's searching game with fixed number of lies. Theoretical Computer Science 95 (1992)

The Maximum Independent Set Problem
in Planar Graphs

Vladimir E. Alekseev[1], Vadim Lozin[2,*],
Dmitriy Malyshev[3], and Martin Milanič[4,**]

[1] Department of Computational Mathematics, Nizhny Novgorod University, Russia
ave@uic.nnov.ru
[2] DIMAP and Mathematics Institute, University of Warwick, Coventry, UK
V.Lozin@warwick.ac.uk
[3] Department of Computational Mathematics, Nizhny Novgorod University, Russia
dsmalyshev@rambler.ru
[4] AG Genominformatik, Faculty of Technology, Bielefeld University, Germany
mmilanic@cebitec.uni-bielefeld.de

Abstract. We study the computational complexity of finding a maximum independent set of vertices in a planar graph. In general, this problem is known to be NP-hard. However, under certain restrictions it becomes polynomial-time solvable. We identify a graph parameter to which the complexity of the problem is sensible and produce a number of both negative (intractable) and positive (solvable in polynomial time) results, generalizing several known facts.

Keywords: maximum independent set problem, planar graphs, hereditary classes.

1 Introduction

Planar graphs form an important class both from a theoretical and practical point of view. The theoretical importance of this class is partly due to the fact that many algorithmic graph problems that are NP-hard in general remain intractable when restricted to the class of planar graphs. In particular, this is the case for the MAXIMUM INDEPENDENT SET (MIS) problem, i.e., the problem of finding in a graph a subset of pairwise non-adjacent vertices (an *independent set*) of maximum cardinality. Moreover, the problem is known to be NP-hard even for planar graphs of maximum vertex degree at most 3 [9] or planar graphs of large girth [15]. On the other hand, the problem can be solved in polynomial-time in some subclasses of planar graphs, such as outerplanar graphs [5] or planar graphs of bounded chordality [10].

* Vadim Lozin's research supported by DIMAP – Centre for Discrete Mathematics and its Applications at the University of Warwick.
** This author gratefully acknowledges the support by the group "Combinatorial Search Algorithms in Bioinformatics" funded by the Sofja Kovalevskaja Award 2004 of the Alexander von Humboldt Stiftung and the German Federal Ministry of Research and Education.

E. Ochmański and J. Tyszkiewicz (Eds.): MFCS 2008, LNCS 5162, pp. 96–107, 2008.

Which other graph properties are crucial for the complexity of the problem in the class of planar graphs? Trying to answer this question, we focus on graph properties that are *hereditary* in the sense that whenever a graph possesses a certain property the property is inherited by all induced subgraphs of the graph. Many important graph classes, such as bipartite graphs, perfect graphs, graphs of bounded vertex degree, graphs of bounded chordality are hereditary, including the class of planar graphs itself. Any hereditary property can be described by a unique set of minimal graphs that do not possess the property – the so-called *forbidden induced subgraphs*. We shall denote the class of graphs containing no induced subgraphs from a set M by $Free(M)$. Any graph in $Free(M)$ will be called M-free. All our results are expressed in terms of some restrictions on the set of forbidden induced subgraphs M. In particular, in Section 2 we will impose a condition on the set M that will imply NP-hardness of the maximum independent set problem in the class of M-free planar graphs. In Section 3, by violating this condition we will reveal new polynomially solvable cases of the problem that generalize some of the previously studied classes.

All graphs in this paper will be finite, undirected, without loops or multiple edges. For a vertex x, we denote by $N(x)$ the neighborhood of x. The *degree* of x, $deg(x)$, is the size of its neighborhood. The *independence number* of a graph G is the maximum cardinality of an independent set in G. The *girth* of a graph is the length of its shortest cycle, while the *chordality* is the length of its longest chordless cycle. The *subdivision* of an edge uv consists in replacing the edge with a new vertex adjacent to u and v. The *contraction* of an edge uv consists in replacing the two vertices u and v with a single vertex x adjacent to every vertex in $(N(u) \cup N(v)) \setminus \{u, v\}$. For two graphs G and H, we denote by $G + H$ the disjoint union of G and H. In particular, nG is the disjoint union of n copies of G. As usual, P_n, C_n and K_n denote the chordless path, the chordless cycle and the complete graph on n vertices, respectively. $K_{n,m}$ is the complete bipartite graph with parts of size n and m. By T_s we denote the graph obtained by subdividing each edge of $K_{1,s}$ exactly once. Also, A_k is the graph obtained by adding to a chordless cycle C_k a new vertex adjacent to exactly one vertex of the cycle. Following [6] we call this graph an *apple* of size k. $S_{i,j,k}$ and H_i are the two graphs shown in Figure 1.

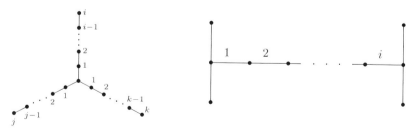

Fig. 1. Graphs $S_{i,j,k}$ (left) and H_i (right)

2   A Hardness Result

From [9] we know that the MIS problem is NP-hard for planar graphs of vertex degree at most 3. Murphy strengthened this result by showing that the problem is NP-hard for planar graphs of degree at most 3 and large girth [15]. This immediately follows from the fact that double subdivision of an edge increases the independence number of the graph by exactly one. The same argument can be used to prove the following lemma, which, for the case of general (not necessarily planar) graphs, was first shown in [1].

Lemma 1. *For any k, the* MAXIMUM INDEPENDENT SET *problem is NP-hard in the class of planar $(C_3, \ldots, C_k, H_1, \ldots, H_k)$-free graphs of vertex degree at most 3.*

We now generalize this lemma in the following way. Let $\mathcal{S}_k$ be the class of $(C_3, \ldots, C_k, H_1, \ldots, H_k)$-free planar graphs of vertex degree at most 3. To every graph G we associate the parameter $\kappa(G)$, which is the maximum k such that $G \in \mathcal{S}_k$. If G belongs to no class $\mathcal{S}_k$, we define $\kappa(G)$ to be 0, and if G belongs to all classes $\mathcal{S}_k$, then $\kappa(G)$ is defined to be ∞. Finally, for a set of graphs M, we define $\kappa(M) = \sup\{\kappa(G) \ : \ G \in M\}$.

Theorem 1. *Let M be a set of graphs and X the class of M-free planar graphs of degree at most 3. If $\kappa(M) < \infty$, then the* MAXIMUM INDEPENDENT SET *problem is NP-hard in the class X.*

Proof. To prove the theorem, we will show that there is a k such that $\mathcal{S}_k \subseteq X$. Denote $k := \kappa(M) + 1$ and let G belong to $\mathcal{S}_k$. Assume that G does not belong to X. Then G contains a graph $A \in M$ as an induced subgraph. From the choice of G we know that A belongs to $\mathcal{S}_k$, but then $k \leq \kappa(A) \leq \kappa(M) < k$, a contradiction. Therefore, $G \in X$ and hence, $\mathcal{S}_k \subseteq X$. By Lemma 1, this implies NP-hardness of the problem in the class X. □

This negative result significantly reduces the area for polynomial-time algorithms. But still this area contains a variety of unexplored classes. In the next section, we analyze some of them.

3   Polynomial Results

Unless $P = NP$, the result of the previous section suggests that the MIS problem is solvable in polynomial time for graphs in a class of M-free planar graphs only if $\kappa(M) = \infty$. Let us specify a few major ways to push $\kappa(M)$ to infinity.

One of the possible ways to unbind $\kappa(M)$ is to include in M a graph G with $\kappa(G) = \infty$. According to the definition, in order for $\kappa(G)$ to be infinite, G must belong to every class $\mathcal{S}_k$. It is not difficult to see that this is possible only if every connected component of G is of the form $S_{i,j,k}$ represented on the left of Figure 1. Let us denote the class of all such graphs by $\mathcal{S}$. More formally, $\mathcal{S} := \bigcap_{k \geq 3} \mathcal{S}_k$. Any other way to push $\kappa(M)$ to infinity requires the inclusion in M of infinitely many graphs. In particular, we will be interested in classes where the set of forbidden subgraphs M contains graphs with arbitrarily large chordless cycles.

The literature does not contain many results when M includes a graph from the class $\mathcal{S}$, that is, a graph G with $\kappa(G) = \infty$, and only a few classes of this type are defined by a single forbidden induced subgraph. Minty [14] and Sbihi [18] independently of each other found a solution for the problem in the class of claw-free (i.e., $S_{1,1,1}$-free) graphs. This result was then generalized to $S_{1,1,2}$-free graphs (see [3] for unweighted and [11] for weighted version of the problem) and to $(S_{1,1,1} + K_2)$-free graphs [13]. Another important example of this type is the class of mP_2-free graphs (where m is a constant). A solution to the problem in this class is obtained by combining an algorithm to generate all maximal independent sets in a graph [20] and a polynomial upper bound on the number of maximal independent sets in mP_2-free graphs [2,8].

Observe that all these results hold for general (not necessarily planar) graphs. In the case of planar graphs, the result for mP_2-free graphs can be further extended to the class of P_k-free graphs (for an arbitrary k) via the notion of tree-width (see e.g. [5] for several equivalent definitions of this notion). The importance of tree-width is due to the fact that many algorithmic graph problems, including the MIS problem, can be solved in polynomial time when restricted to graphs of bounded tree-width [4]. Clearly, the diameter of connected P_k-free planar graphs is bounded by a constant. Therefore, since the tree-width of planar graphs is bounded by a function of their diameter [7], the tree-width of P_k-free planar graphs is bounded by a constant. As a result, we conclude that the MIS problem is solvable in polynomial time for P_k-free planar graphs. An extension of this conclusion was recently proposed in [10] where the authors show that the tree-width is bounded in the class of $(C_k, C_{k+1}, \ldots)$-free planar graphs (for any fixed k).

Below we report further progress in this direction. In particular, we show that the MIS problem can be solved in polynomial time in the class of $(A_k, A_{k+1}, \ldots)$-free planar graphs. We thus generalize not only the analogous result for $(C_k, C_{k+1}, C_{k+2}, \ldots)$-free planar graphs, but also the results for $S_{1,1,1}$-free, $S_{1,1,2}$-free and $(S_{1,1,1} + K_2)$-free planar graphs. Observe that, in contrast to planar graphs of bounded chordality, the tree-width of $(A_k, A_{k+1}, \ldots)$-free planar graphs is not bounded, which makes it necessary to employ more techniques for the design of a polynomial-time algorithm. One of the techniques we use in our solution is known as *decomposition by clique separators* [19,21]. It reduces the problem to connected graphs without separating cliques, i.e., without cliques whose deletion disconnects the graph. We also use the notion of graph compression defined in the next section. In addition, in Section 3.2 we prove some auxiliary results related to the notion of tree-width. Finally, in Section 3.3 we describe the solution.

3.1   Graph Compressions and Planar T_s-Free Graphs

A *compression* of a graph $G = (V, E)$ is a mapping $\phi : V \to V$ which maps any two distinct non-adjacent vertices into non-adjacent vertices and which is not an automorphism. Thus, a compression maps a graph into its induced subgraph with the same independence number. Two particular compressions of interest will be denoted $\binom{a}{b}$ and $\binom{a\ b}{c\ d}$.

By $\binom{a}{b}$ we mean the compression which maps a to b and leaves all other vertices fixed. This map is a compression if and only if a is adjacent to b and $N(b) - \{a\} \subseteq N(a) - \{b\}$.

The compression $\binom{a\ b}{c\ d}$ is defined as follows: $\phi(a) = c$, $\phi(b) = d$ and the remaining vertices of the graph are fixed. This map is a compression if

- $c \neq d$,
- $ac, bd \in E$ and $ab, cd \notin E$,
- every vertex adjacent to c different from a and b is also adjacent to a,
- every vertex adjacent to d different from a and b is also adjacent to b.

A graph which admits neither $\binom{a}{b}$ nor $\binom{a\ b}{c\ d}$ will be called *incompressible*.

Lemma 2. *Let G be an incompressible T_s-free planar graph and a, b two vertices of distance 2 in G. Then $|N(a) \cap N(b)| \leq 4s + 1$.*

Proof. Let us call a vertex $x \in N(a) \cap N(b)$

- *specific* if every neighbor of x, other then a and b, belongs to $N(a) \cap N(b)$,
- *a-clear* (*b-clear*) if x has a neighbor non-adjacent to a (to b).

Notice that every vertex in $N(a) \cap N(b)$ is either specific or a-clear or b-clear. Let us estimate the number of vertices of each type in $N(a) \cap N(b)$.

First, suppose that $N(a) \cap N(b)$ contains 4 specific vertices. Then, due to planarity of G, two of these vertices are non-adjacent, say x and y. But then $\binom{a\ b}{x\ y}$ is a compression. Therefore, $N(a) \cap N(b)$ contains at most 3 specific vertices.

Now suppose $N(a) \cap N(b)$ contains $2s$ a-clear vertices. Consider a plane embedding of G. This embedding defines a cyclic order of the neighbors of each vertex. Let $x_1, x_2, \ldots, x_{2s}$ be the a-clear vertices listed in the cyclic order with respect to a. Also, for each $i = 1, 2, \ldots, 2s$, denote by y_i a vertex adjacent to x_i and non-adjacent to a. Some of the vertices in the set $\{y_1, y_2, \ldots, y_{2s}\}$ may coincide but the vertices $\{y_1, y_3, y_5 \ldots, y_{2s-1}\}$ must be pairwise distinct and non-adjacent. But then the set $\{a, x_1, x_3, x_5 \ldots, x_{2s-1}, y_1, y_3, y_5 \ldots, y_{2s-1}\}$ induces a T_s. This contradiction shows that there are at most $2s - 1$ a-clear vertices. Similarly, there are at most $2s - 1$ b-clear vertices.   □

Lemma 3. *Let G be an incompressible T_s-free planar graph. Then the degree of each vertex in G is at most $16s^2 - 1$.*

Proof. Let a be a vertex in G, A the set of neighbors of a and B the set of vertices at distance 2 from a. Consider the bipartite subgraph H of G formed by the sets A and B and all the edges connecting vertices of A to the vertices of B. Let the size of a maximum matching in H be π and the size of a minimum vertex cover in H be β. According to the theorem of König, $\pi = \beta$.

Observe that every vertex $x \in A$ has a neighbor in B, since otherwise $\binom{a}{x}$ is a compression. Thus, H contains a set D of $deg(a)$ edges no two of which share a vertex in A. By Lemma 2, the degree of each vertex of B in the graph H is at most $4s + 1$. Therefore, to cover the edges of D we need at least $deg(a)/(4s + 1)$ vertices, and hence $\pi \geq deg(a)/(4s + 1)$.

In the graph H, consider an arbitrary matching M with π edges. In the graph G, contract each edge of M into a single vertex obtaining in this way a planar graph G', and denote the subgraph of G' induced by the set of "contracted" vertices (i.e., those corresponding to the edges of M) by H'. If $deg(a) > (4s + 1)(4s - 1)$, then H' contains at least $4s$ vertices. By the Four Color Theorem [17], it follows that H' contains an independent set of size s. The vertices of this set correspond to s edges in the graph G that induce an sK_2. Together with vertex a these edges induce a T_s, a contradiction. □

3.2   Tree-Width and Planar Graphs

In this section, we derive several auxiliary results on the tree-width of planar graphs. More generally, our results are valid for any class of graphs excluding an apex graph as a minor. An *apex graph* is a graph that contains a vertex whose deletion leaves a planar graph. A graph H is said to be a *minor* of a graph G if H can be obtained from G by means of vertex deletions, edge deletions and edge contractions. We say that a class of graphs is *minor closed* if with every graph G it contains all minors of G. Both graphs of bounded tree-width and planar graphs are minor closed.

If H is not a minor of a graph G, we say that G is H-*minor-free* and call H a *forbidden minor* for G. Similarly, if X is a class of H-minor-free graphs, we call H a *forbidden minor* for X. By a result of Robertson and Seymour, every minor-closed graph class can be described by a unique *finite* set of minimal forbidden minors. For instance, the class of planar graphs is exactly the class of $(K_5, K_{3,3})$-minor-free graphs.

For brevity, let us call a family of graphs *apex-free*, if it is defined by a single forbidden minor H, which is an apex graph.

An $n \times n$ grid G_n is the graph with the vertex set $\{1, \dots, n\} \times \{1, \dots, n\}$ such that (i, j) and (k, l) are adjacent if and only if $|i - k| + |j - l| = 1$. In [16], Robertson and Seymour showed that every graph of large tree-width must contain a large grid as a minor. For apex-free graph families, even more is true. In the following lemma, an *augmented grid* is a grid G_n augmented with additional edges (and no additional vertices). Vertices (i, j) with $\{i, j\} \cap \{1, n\} \neq \emptyset$ are *boundary vertices* of the grid; the other ones are *nonboundary*.

Lemma 4. *[7] Let H be an apex graph. Let $r = 14|V(H)| - 22$. For every integer k there is an integer $g_H(k)$ such that every H-minor-free graph of tree-width at least $g_H(k)$ can be contracted into an $k' \times k'$ augmented grid R such that $k' \geq k$, and each vertex $v \in V(R)$ is adjacent to less than $(r + 1)^6$ nonboundary vertices of the grid.*

With extensive help of this lemma we shall derive the main result of this section, which we state now.

Lemma 5. *For any apex graph H and integers k, s and d, there is an integer $N = N(H, k, s, d)$ such that for every H-minor-free graph G of tree-width at least N and every nonempty subset $S \subseteq V(G)$ of at most s vertices, the graph G contains a chordless cycle C such that:*

– *every vertex $v \in V(G)$ is non-adjacent to at least k consecutive vertices of C.*
– *the distance between C and S is at least d.*

To prove Lemma 5, we will need a few auxiliary results. First, we recall that in apex-free graphs, large tree-width forces the presence of arbitrarily long chordless cycles [10]. More formally:

Lemma 6. *For every apex graph H and every integer k there is an integer $f_H(k)$ such that every H-minor-free graph of tree-width at least $f_H(k)$ contains a chordless cycle of order at least k.*

Next, we prove two additional lemmas that will be needed in the proof of Lemma 5.

Lemma 7. *For every apex graph H and every integer k there is an integer $f(H,k)$ such that every H-minor-free graph G of tree-width at least $f(H,k)$ contains a chordless cycle C such that every vertex $v \in V(G)$ is non-adjacent to at least k consecutive vertices of C.*

Proof. Let $r = 14|V(H)| - 22$, let f_H be the function given by Lemma 6, and let g_H be the function given by Lemma 4. Furthermore, let $f(H,k) = g_H\left(f_H\left((k+1)(r+1)^6\right)+2\right)$. We will show that the function $f(H,k)$ satisfies the claimed property.

Let G be an H-minor-free graph of tree-width at least $f(H,k)$. By Lemma 4, G can be contracted into an $k' \times k'$ augmented grid R where $k' \geq f_H((k+1)(r+1)^6)+2$ and such that each vertex $v \in V(R)$ is adjacent to less than $(r+1)^6$ non-boundary vertices of the grid. For $i,j \in \{1,\ldots,k'\}$, let $V(i,j)$ denote the subset of $V(G)$ that gets contracted to the vertex (i,j) of the grid. Furthermore, let R_0 denote the $(k'-2) \times (k'-2)$ augmented sub-grid, induced by the nonboundary vertices of R. Since the tree-width of an $n \times n$ grid is n, and the tree-width cannot decrease by adding edges, we conclude that the tree-width of R_0 is at least $k'-2 \geq f_H((k+1)(r+1)^6)$. Moreover, as R_0 is H-minor-free, Lemma 6 implies that R_0 contains a chordless cycle C_0 of length at least $(k+1)(r+1)^6$. By the above, every vertex $v \in V(R)$ is adjacent to less than $(r+1)^6$ vertices of R_0. Therefore, the neighbors of v on C_0 (if any) divide the cycle into less than $(r+1)^6$ disjoint paths whose total length is at least $|V(C_0)| - (r+1)^6$. In particular, this implies every vertex of $V(R)$ is non-adjacent to at least $\frac{|V(C_0)|-(r+1)^6}{(r+1)^6} \geq k$ consecutive vertices of C_0.

Let the cyclic order of vertices of R_0 on C_0 be given by $((i_1,j_1),(i_2,j_2),\ldots,(i_s,j_s))$. To complete the proof, we have to lift the cycle C_0 to a chordless cycle C in G. Informally, we will replace each pair of adjacent edges $(i_{p-1},j_{p-1})(i_p,j_p)$ and $(i_p,j_p)(i_{p+1},j_{p+1})$ in C_0 with a shortest path connecting vertex (i_{p-1},j_{p-1}) to vertex (i_{p+1},j_{p+1}) in the graph G whose internal vertices all belong to V_{i_p,j_p}. Implementation details of this "lifting" procedure are omitted due to the lack of space. □

Our second preliminary lemma states that the tree-width of apex-free graphs cannot be substantially decreased by contracting the set of vertices at constant distance from a set of constantly many vertices. We remark that this fails for minor-closed families that exclude no apex graph (to see this, take G in the statement of the lemma to be the graph obtained from an $n \times n$ grid by adding to it a dominating vertex).

Lemma 8. *Let H be an apex graph, and let s, d and m be integers. Then, there is an integer $t = t(H, s, d, m)$ such that the following holds:*

Let G be an H-minor-free graph of tree-width at least t, and let $S \subseteq V(G)$ be a set of at most s vertices of G. Furthermore, let U be the set of vertices in G that are at distance less than d from S, and let G' be the graph obtained from G by contracting the set U into a single vertex. Then, the tree-width of G' is at least m.

Proof. By an easy inductive argument on the number of connected components of $G[S]$, we may assume that S induces a connected subgraph of G. If $d = 0$, then $G' = G$, and we have $t = m$.

Let now $d \geq 1$. For $i = 1, \ldots, d$, let $G^{(i)}$ denote the graph obtained from G by contracting the set $V^{(i)}$ of vertices at distance less than i from S into a single vertex $v^{(i)}$. Furthermore, let $r = 14|V(H)| - 22$. Also, let g_H be the function given by Lemma 4.

Consider the following recursively defined function $h : \{1, \ldots, d\} \to \mathbb{N}$: $h(1) = m$, and $h(i+1) = g_H(2(r+1)^3 h(i))$, for all $i = 1, \ldots, d-1$. Let $t := t(H, s, d, m) := h(d) + s$.

With the above notation, we have $G' = G^{(d)}$. So, it suffices to show the following:

Claim. For all $i = 1, \ldots, d$, the tree-width of $G^{(i)}$ is at least $h(d + 1 - i)$.

We now prove the claim by induction on i. For $i = 1$, note that $G^{(1)}$ contains $G - S$ as an induced subgraph, and therefore $\mathrm{tw}(G^{(1)}) \geq \mathrm{tw}(G-S) \geq \mathrm{tw}(G) - s \geq t - s = h(d) = h(d - i + 1)$ (where $\mathrm{tw}(K)$ denotes the tree-width of a graph K).

For the induction hypothesis, assume that the statement holds for some $i \geq 1$: the tree-width of $G^{(i)}$ is at least $h(d + 1 - i) = g_H(2(r + 1)^3 h(d - i))$. By Lemma 4, $G^{(i)}$ can be contracted into an $k \times k$ augmented grid R such that $k \geq 2(r+1)^3 h(d-i)$, and each vertex $v \in V(R)$ is adjacent to less than $(r+1)^6$ nonboundary vertices of the grid.

Therefore, R must contain a large subgrid R' such that $v^{(i)} \in V(G^{(i)})$ does not belong to R', and has no neighbors in R'. More precisely, R' can be chosen to be of size $k' \times k'$, where $k' \geq \lfloor \frac{k-2}{\sqrt{(r+1)^6}} \rfloor \geq \frac{k}{2(r+1)^3} \geq h(d-i)$. By definition of $V^{(i+1)}$ and since $v^{(i)}$ has no neighbors in R', we conclude that the graph $G^{(i+1)}$ contains the grid R' as a minor. Thus, the tree-width of $G^{(i+1)}$ is at least the tree-width of R', which is at least $h(d - i) = h(d + 1 - (i + 1))$. The proof is complete. $\qquad\square$

We conclude this section with a short proof of Lemma 5, based on Lemmas 7 and 8.

Proof. (Lemma 5) Let $f(H, k)$ be given by Lemma 7. We let $N := N(H, k, s, d) := t(H, s, d + 1, f(H, k))$, where t is given by Lemma 8.

Let G' be the graph obtained from G by contracting the set of vertices at distance less than $d+1$ form S into a single vertex. Then, by Lemma 7, the tree-width of G' is at least $f(H,k)$. By Lemma 7, G' contains a chordless cycle C such that every vertex $v \in V(G')$ is non-adjacent to at least k consecutive vertices of C.

Using the same argument as in the proof of Lemma 7, C' can be lifted to a chordless cycle C of G such that every vertex $v \in V(G)$ is non-adjacent to at least k consecutive vertices of C. □

3.3 Solution to the Problem for Planar $(A_k, A_{k+1}, \ldots)$-Free Graphs

In this section, we prove polynomial-time solvability of the MIS problem in the class of planar $(A_k, A_{k+1}, \ldots)$-free graphs, for an arbitrary integer k.

Theorem 2. *For any k, the* MAXIMUM INDEPENDENT SET *problem can be solved in the class of planar $(A_k, A_{k+1}, \ldots)$-free graphs in polynomial time.*

Proof. Let k be an integer and G be a planar $(A_k, A_{k+1}, \ldots)$-free graph. If G is T_{11}-free, then we reapeteadly perform the two graph compressions until we obtain an incompressible graph G'. By Lemma 3 the degree of vertices in G' is bounded by a constant. It was recently shown in [12] that the MIS problem in the class of $(A_k, A_{k+1}, \ldots)$-free graphs of bounded vertex degree is polynomial-time solvable.

This enables us to assume that G contains a T_{11} as an induced subgraph. Moreover, we can assume that G has no clique separators. In the subgraph T_{11}, we will denote the vertex of degree 11 by a, the vertices of degree 2 by $b_1, \ldots, b_{11}$ and the respective vertices of degree 1 by $c_1, \ldots, c_{11}$.

Let $N = N(K_5, 6k+8, 23, k+2)$ be the constant defined in Lemma 5. We shall show that the tree-width of G is less than N. Assume by contradiction that the tree-width of G is at least N. Then by Lemma 5, with $S = V(T_{11})$, the graph G contains a chordless cycle C such that

- every vertex of G is non-adjacent to at least $6k+8$ consecutive vertices of C, and
- the distance between C and T_{11} is at least $k+2$.

Fact 1. No vertex of G can have more than 4 neighbors on C. Moreover, if a vertex v has 3 neighbors on C, then these neighbors appear in C consecutively. If v has 4 neighbors, they can be split into two pairs of consecutive vertices. If v has 2 neighbors, they are either adjacent or of distance 2 in C.

Indeed, if v has more then 4 neighbors on C, then a large portion of C containing at least $6k+8$ consecutive vertices together with v and one of its neighbors create a forbidden induced apple. The rest of Fact 1 also follows from $(A_k, A_{k+1}, \ldots)$-freeness of G, which can be verified by direct inspection.

Claim. G has a chordless cycle containing vertex a and some vertices of C.

Proof. Since G has no clique separators, it is 2-connected. Therefore, there exist two vertex-disjoint paths connecting a to C. Let $P = (x_1, \ldots, x_p)$ and $Q = (y_1, \ldots, y_q)$ be two such paths, where x_1 and y_1 are adjacent to a, while x_p and y_q have neighbors on C. Without loss of generality, we shall assume that the total length of P and Q is as small as possible. In particular, this assumption implies that x_1 and y_1 are the only neighbors of a on P, Q, and no vertex of P or Q different from x_p and y_q has a neighbor on C. Any edge connecting a vertex of P to a vertex of Q will be called a (P, Q)-*chord*.

Fact 2. *The neighborhood of x_p on C consists of two adjacent vertices and the neighborhood of y_q on C consists of two adjacent vertices.*

Obviously, to avoid a big induced apple, x_p must have at least two neighbors on C. Consider a longest sub-path P' of C such that x_p has no neighbors on P'. We know that P' has at least $6k + 8$ vertices. Moreover, by maximality of P', x_p is adjacent to the two (distinct!) vertices u, v on C outside P' each of which is adjacent to an endpoint of P'. Then, u and v must be adjacent, for otherwise G would contain a forbidden apple induced by the vertex set $P' \cup \{u, v, x_p, x_{p-1}\}$. The same reasoning shows that the neighborhood of y_q on C consists of two adjacent vertices.

Fact 3. *The neighborhood of x_p on C does not coincide with the neighborhood of y_q on C, and there are no (P, Q)-chords different from $x_1 y_1$.*

For the sake of contradiction, suppose that $N(x_p) \cap C = N(y_q) \cap C = \{x_{p+1}, y_{q+1}\}$. Denote by T^1 the triangle x_p, x_{p+1}, y_{q+1} and by T^2 the triangle y_q, x_{p+1}, y_{q+1}. To avoid a separating clique (one triangle inside the other), we must conclude that, without loss of generality, x_p is inside C while y_q is outside C in the planar embedding of G. If additionally a is inside C, then Q meets the cycle before it meets y_q. This contradiction completes the proof of the first part of Fact 2.

To prove the second part, suppose that G contains a (P, Q)-chord different from $x_1 y_1$. Let $x_i y_j$ be such a chord with maximum value of $i + j$. In order to prevent a large induced apple, x_i must be adjacent to y_{j-1}. By symmetry, y_j must be adjacent to x_{i-1}. This implies, in particular, that both $i > 1$ and $j > 1$. Denote by T^1 the triangle x_{i-1}, x_i, y_j, by T^2 the triangle y_{j-1}, x_i, y_j and by C' the cycle formed by vertices $x_p, x_{p-1}, \ldots, x_i, y_j, \ldots, y_q$ and a portion of C. The rest of the proof of Fact 3 is identical to the above arguments.

From Fact 3 we conclude that if $x_1 y_1$ is not a chord, then G has a desired cycle, i.e., a chordless cycle containing a and some vertices of C. From now on, assume x_1 is adjacent to y_1. Denote by C^* a big chordless cycle formed of P, Q and a portion of C containing at least half of its vertices, i.e., a portion containing at least $3k + 4$ consecutive vertices of C. We will denote this portion by P^*.

Observe that among vertices $b_1, \ldots, b_{11}$ there is a vertex, name it z_1, which is adjacent neither to x_1 nor to y_1, since otherwise G has a separating clique (a triangle with a vertex inside it). Let us show that z_1 has no neighbors on C^*. Indeed, z_1 cannot have neighbors on P^*, since the distance between z_1 and P^* is at least $k + 2$. If z_1 has both a neighbor on P and a neighbor on Q, then

G contains a big induced apple. If z_1 is adjacent to a vertex $x_i \in P$ and has no neighbors on Q, then either the pair of paths P, Q is not of minimum total length (if $i > 2$) or G has a big induced apple (if $i = 2$).

Since G has no clique separators, vertex z_1 must be connected to the cycle C^* by a path avoiding the clique $\{a, x_1, y_1\}$. Let $R = (z_1, \ldots, z_r)$ be a shortest path of this type. Since z_1 has no neighbors on C^*, r must be strictly greater than 1. According to Fact 1, z_r cannot have more than 4 neighbors on P^*. Moreover, these neighbors partition P^* into at most 3 portions (of consecutive non-neighbors of z_r) the largest of which has at least k vertices. Therefore, z_r has at least k consecutive non-neighbors on the cycle C^*. By analogy with Fact 2, we conclude that the neighborhood of z_r on C^* consists of two adjacent vertices. Also, by analogy with Fact 3, we conclude that the only possible chord between R and the other path connecting z_1 to C^* (i.e. (z_1, a)) is the edge az_2. Therefore, G has a chordless cycle containing vertex a and some vertices of C, and the proof of the claim is completed.   □

Denote by $C^a = (a, v_1, v_2, \ldots, v_s)$ a chordless cycle containing the vertex a and a part of C. The vertices of C^a belonging to C will be denoted $v_i, v_{i+1}, \ldots, v_j$. Since the distance between T_{11} and C is at least $k + 2$, none of the vertices $b_1, b_2, \ldots, b_{11}$ is adjacent to any of the vertices $v_{i-k}, v_{i-k+1}, \ldots, v_{k+j}$. Clearly among vertices $b_1, b_2, \ldots, b_{11}$ at least 9 do not belong to C^a. Among these 9, at least 5 vertices are adjacent neither to v_1 nor to v_s (since otherwise G contains a separating clique, i.e., a triangle with a vertex inside it). Without loss of generality, let the vertices $b_1, b_2, \ldots, b_5$ be not in C^a and non-adjacent to v_1, v_s. It is not difficult to see that none of these 5 vertices has a neighbor in the set $\{v_3, v_4, \ldots, v_{s-3}, v_{s-2}\}$, since otherwise a big induced apple arises (remember that none of these 5 vertices is adjacent to any of $v_{i-k}, v_{i-k+1}, \ldots, v_{k+j}$). For the same reason, none of $b_1, b_2, \ldots, b_5$ can be adjacent simultaneously to v_2 and v_{s-1} and none of them can be non-adjacent simultaneously to v_2 and v_{s-1}. Therefore, we may assume without loss of generality that in a fixed plane embedding of G, among these 5 vertices there are 2, say b_i, b_j, such that b_i is inside the 4-cycle a, b_j, v_2, v_1. Due to planarity, vertex c_i has no neighbors on the cycle C^a except possibly v_1 and v_2. However, regardless of the adjacency c_i to v_1 or v_2, the reader can easily find a big induced apple in G. This contradiction shows that if G contains a T_{11}, then the tree-width of G is bounded by a constant, which completes the proof of the theorem.   □

References

1. Alekseev, V.E.: The Effect of Local Constraints on the Complexity of Determination of the Graph Independence Number. In: Combinatorial-algebraic methods in applied mathematics, pp. 3–13. Gorkiy University Press, Gorky (1982) (in Russian)
2. Alekseev, V.E.: On the Number of Maximal Independent Sets in Graphs from Hereditary Classes. In: Combinatorial-algebraic methods in discrete optimization, pp. 5–8. University of Nizhny Novgorod (1991) (in Russian)
3. Alekseev, V.E.: A Polynomial Algorithm for Finding the Largest Independent Sets in Fork-free Graphs. Discrete Appl. Math. 135, 3–16 (2004)

4. Arnborg, S., Proskurowski, A.: Linear Time Algorithms for NP-hard Problems Restricted to Partial k-trees. Discrete Appl. Math. 23, 11–24 (1989)
5. Bodlaender, H.: A Partial k-Arboretum of Graphs with Bounded Treewidth. Theor. Comput. Sci. 209, 1–45 (1998)
6. De Simone, C.: On the Vertex Packing Problem. Graphs Combinator. 9, 19–30 (1993)
7. Demaine, E.D., Hajiaghayi, M.T.: Diameter and Treewidth in Minor-Closed Graph Families, Revisited. Algorithmica 40, 211–215 (2004)
8. Farber, M., Hujter, M., Tuza, Zs.: An Upper Bound on the Number of Cliques in a Graph. Networks 23, 207–210 (1993)
9. Garey, M.G., Johnson, D.S.: The Rectilinear Steiner Tree Problem is NP-Complete. SIAM J. Appl. Math. 32, 826–834 (1977)
10. Kamiński, M., Lozin, V.V., Milanič, M.: Recent Developments on Graphs of Bounded Clique-Width. Discrete Appl. Math. (accepted),
    http://rutcor.rutgers.edu/pub/rrr/reports2007/6_2007.pdf
11. Lozin, V., Milanič, M.: A Polynomial Algorithm to Find an Independent Set of Maximum Weight in a Fork-free Graph. In: Proceedings of the ACM-SIAM Symposium on Discrete Algorithms SODA 2006, pp. 26–30 (2006)
12. Lozin, V., Milanič, M.: Maximum Independent Sets in Graphs of Low Degree. In: Proceedings of the ACM-SIAM Symposium on Discrete Algorithms SODA 2007, pp. 874–880 (2007)
13. Lozin, V.V., Mosca, R.: Independent Sets in Extensions of $2K_2$-Free Graphs. Discrete Appl. Math. 146, 74–80 (2005)
14. Minty, G.J.: On Maximal Independent Sets of Vertices in Claw-free Graphs. J. Comb. Theory B 28, 284–304 (1980)
15. Murphy, O.J.: Computing Independent Sets in Graphs With Large Girth. Discrete Appl. Math. 35, 167–170 (1992)
16. Robertson, N., Seymour, P.D.: Graph Minors. V. Excluding a Planar Graph. J. Comb. Theory B 41, 92–114 (1986)
17. Robertson, N., Sanders, D., Seymour, P., Thomas, R.: The Four-Colour Theorem. J. Comb. Theory B 70, 2–44 (1997)
18. Sbihi, N.: Algorithme de Recherche d'un Stable de Cardinalité Maximum dans un Graphe Sans Étoile. Discrete Math. 29, 53–76 (1980)
19. Tarjan, R.E.: Decomposition by Clique Separators. Discrete Math. 55, 221–232 (1985)
20. Tsukiyama, S., Ide, M., Ariyoshi, H., Shirakawa, I.: A New Algorithm for Generating All the Maximal Independent Sets. SIAM J. Computing 6, 505–517 (1977)
21. Whitesides, S.H.: An Algorithm for Finding Clique Cut-sets. Inform. Process. Lett. 12, 31–32 (1981)

When Ignorance Helps:
Graphical Multicast Cost Sharing Games

Vittorio Bilò[1], Angelo Fanelli[2],
Michele Flammini[2], and Luca Moscardelli[2]

[1] Department of Mathematics, University of Salento
Provinciale Lecce-Arnesano, P.O. Box 193, 73100 Lecce - Italy
vittorio.bilo@unile.it
[2] Department of Computer Science, University of L'Aquila
Via Vetoio, Loc. Coppito, 67100 L'Aquila - Italy
{angelo.fanelli,flammini,moscardelli}@di.univaq.it

Abstract. In non-cooperative games played on highly decentralized networks the assumption that each player knows the strategy adopted by any other player may be too optimistic or even unfeasible. In such situations, the set of players of which each player knows the chosen strategy can be modeled by means of a social knowledge graph in which nodes represent players and there is an edge from i to j if i knows j. Following the framework introduced in [3], we study the impact of social knowledge graphs on the fundamental multicast cost sharing game in which all the players wants to receive the same communication from a given source. Such a game in the classical complete information case is known to be highly inefficient, since its price of anarchy can be as high as the total number of players ρ. We first show that, under our incomplete information setting, pure Nash equilibria always exist only if the social knowledge graph is directed acyclic (DAG). We then prove that the price of stability of any DAG is at least $\frac{1}{2}\log\rho$ and provide a DAG lowering the classical price of anarchy to a value between $\frac{1}{2}\log\rho$ and $\log^2\rho$. If specific instances of the game are concerned, that is if the social knowledge graph can be selected as a function of the instance, we show that the price of stability is at least $\frac{4\rho}{\rho+3}$, and that the same bound holds also for the price of anarchy of any social knowledge graph (not only DAGs). Moreover, we provide a nearly matching upper bound by proving that, for any fixed instance, there always exists a DAG yielding a price of anarchy less than 4. Our results open a new window on how the performances of non-cooperative systems may benefit from the lack of total knowledge among players and can be considered, in some sense, as another evidence of the famous Braess' paradox.

1 Introduction

The fast and striking affirmation of the Internet has quickly shifted researchers' attention from traditional centralized networks to unregulated non-cooperative ones. By introducing the notion of price of anarchy as a measure of the loss

E. Ochmański and J. Tyszkiewicz (Eds.): MFCS 2008, LNCS 5162, pp. 108–119, 2008.

of optimality in network performances due to the selfish behavior of its users, Koutsoupias and Papadimitriou [11] definitively started the topic of Algorithmic Game Theory. Since then there has been a flourishing of results on several different models of non-cooperative networks (see [7,13,14] for recent surveys).

As future networking scenarios are predicted to become more and more decentralized and pervasive, new and more stringent constraints need to be introduced in our models. For instance, the usual assumption that each player knows the strategy played by all the other ones may be too optimistic or even unfeasible. Thus, it becomes more realistic to assume that each player is aware only of the strategies played by a subset of players representing somehow her neighborhood. An interesting motivating discussion for this assumption in a selfish routing scenario can be found in [9] where oblivious players are introduced which can be seen as an extreme application of this concept: they cannot feel the consequences of other players' choices and hence do not participate actively to the game always choosing their best strategy regardless of what other players do. More in general, in [3] a new framework has been presented introducing the notion of social knowledge graphs, that is, graphs having as node set the set of players in the game and in which there is an edge from i to j if player i knows player j. According to a given social knowledge graph, the neighborhood of each node i models the set of players of which i is aware, that is, whose chosen strategies can influence i's payoff and hence her choices. Besides characterizing the convergence to equilibria with respect to the social graph topology (directed, undirected, directed acyclic), in [3] it has been shown that the performances in load balancing and congestion games decrease as a consequence of the stricter bounds on the players knowledge. More precisely, the price of anarchy and stability increase as the maximum outdegree of the social graph decreases.

The idea of modeling mutual influences among players by means of a graph was already used in graphical games [10]. However, in such a setting the graph is constructed in such a way that there is an edge from i to j if the choices of player i may influence j's payoff. Therefore, given a particular game, the graphical representation of [10] is completely induced by the underlying game, while in the framework of [3] the social knowledge graph is independent and causes a redefinition of the basic payoffs as a function of the induced mutual influences. Anyway, for analogy with [10], conventional games equipped with social graphs are also called *graphical*.

In this paper we analyze the consequences of the presence of social knowledge graphs in the fundamental multicast cost sharing game defined in [1], in which the players are network users interested in receiving the same communication from a given source and must share the incurred communication cost. The union of the strategies adopted by each player yields a particular solution graph and each player's payment or cost share is computed according to the well-known Shapley value method [15], which equally splits the cost of each edge in the solution among all the downstream players using it. The major drawback of this approach, as well as of the ones yielded by other reasonable methods considered in [8], is that the corresponding price of anarchy is equal to ρ, which makes all such methods excessively poorly performing in practice.

In order to partially cope with this problem, in [6] it is shown that the price of anarchy induced by the Shapley method in the case in which, starting from the empty configuration, the players join the network one at time by performing a best response, is upper bounded by $\log^3 \rho$, a best response corresponding to a strategy selection that minimizes the player's incurred cost share. Moreover, in [1] it is shown that the price of stability yielded by Shapley is upper bounded by H_ρ. However, both these approaches require a certain degree of centralized control on the network in order to let the players to enter the game starting from the empty configuration and to take their decisions in a perfectly sequential fashion, or to "suggest" them a best Nash equilibrium. Again, these assumptions may not be feasible in strongly decentralized systems. On this respect, if the presence of a social knowledge graph would be able to improve by chance the price of anarchy of a game (as we show indeed in our case), this might generate a useful instrument for limiting the bad effects due to the lack of cooperation among the users without directly interfering in their decisions. Consider for instance the design of P2P protocols which limit the visibility of the other peers, or simply, at a more foundational level, the possibility of using social graphs just as an intermediate methodological tool for defining cost shares and payoffs so as to induce good overall performances.

1.1   Model and Definitions

In the multicast cost sharing game we are given an undirected network $G = (V, E, c)$ with $c : E \rightarrow \mathbb{R}^+$, a source station $s \in V$, a set of ρ receivers $R \subseteq V$ and a cost sharing method distributing the cost of a solution among the receivers. Each receiver r_i wants to receive the same communication from the source s, so that her set of available strategies will be the set of all the $\langle s, r_i \rangle$-paths in G. Let π_i be the strategy played by r_i, $\pi = (\pi_1, \ldots, \pi_\rho)$ the strategy profile induced by the choices of all the receivers in the game, and $\Pi = \bigcup_{i=1}^{\rho} \pi_i$ be the subnetwork of G created by such choices. The cost sharing method divides the cost $c(\Pi) = \sum_{e \in \Pi} c(e)$ of Π among the receivers. We focus on the case in which $c(\Pi)$ is shared according to the Shapley value method. More precisely, if $n_e(\pi) = |\{r_i \in R : e \in \pi_i\}|$ is the number of receivers using edge e, the cost share of r_i in π is defined as $cost(\pi, r_i) = \sum_{e \in \pi_i} \frac{c(e)}{n_e(\pi)}$, that is, the cost of each edge in Π is equally shared among all its downstrem users. We denote with I a generic instance (G, R, s) of the multicast cost sharing game.

We associate with each game a social knowledge directed graph $K = (R, A)$ defining for each receiver r_i the set $R_i(K) = \{r_j \in R : (r_i, r_j) \in A\}$ of receivers of which r_i knows the chosen path. Let $n_e(\pi, K, r_i) = |\{r_j \in R_i(K) : e \in \pi_j\}| + 1$ be the number of users using known by r_i using edge e, r_i included. Then, the cost share of r_i in π becomes $cost(\pi, K, r_i) = \sum_{e \in \pi_i} \frac{c(e)}{n_e(\pi, K, r_i)}$. For the sake of simplicity, when clear from the context, we remove π and K from the notation by simply writing $cost(r_i)$.

Notice that, in this new social knowledge framework, in general $c(\Pi) \leq \sum_{r \in R} cost(r)$, that is the sum of all the cost shares can be strictly greater than the total cost of the induced network. This can be interpreted in at least two possible ways. First of all, given the incomplete information, one can assume that users are not able to exactly establish their cost shares, but just presumed ones estimated by observing only the strategies played by the known users; such cost shares provide suitable upper bounds on the actual costs that they will be finally asked to pay by the network provider. On the other hand, such presumed cost shares might coincide with the actual final ones, thus not yielding a cost sharing method in the strict sense, and the arising surplus $\sum_{r \in R} cost(r) - c(\Pi)$ translates into a revenue that the provider enjoys profiting from the users' incomplete information.

As usual in these settings, we assume that the possible solutions for the game are all (and only) its pure Nash equilibria, i.e. the solutions in which no agent can improve her own utility by unilaterally changing her strategy.

In order to measure the loss of optimality due to the selfish behavior of the receivers, we use the standard notions of price of anarchy [11] and price of stability [2]. The former is defined as the ratio between the total cost of the worst Nash equilibrium and the total cost of an optimal solution T^*, given by any optimal Steiner tree connecting the subset of vertices $R \cup \{s\}$. More precisely, let $N(I, K)$ be the set of pure Nash equilibria yielded by the social knowledge graph K on the instance I. We define $PoA(I, K) = \max_{\pi \in N(I,K)} \frac{c(\pi)}{c(T^*)}$ and $PoA(K) = \sup_{I=(G,R,s)} PoA(I, K)$ as the price of anarchy of K on the instance I and the *universal* price of anarchy or simply price of anarchy of K, respectively. Analogously, the latter measure is defined as the ratio between the best Nash equilibrium cost and the one of T^*. More formally, we define $PoS(I, K) = \min_{\pi \in N(I,K)} \frac{c(\pi)}{c(T^*)}$ and $PoS(K) = \sup_{I=(G,R,s)} PoS(I, K)$ as the price of stability of K on the instance I and the *universal* price of stability or simply price of stability of K, respectively.

Finally, given any numbering of the receivers, let $\hat{K}$ denote the complete DAG, i.e. the graph such that $A = \{(r_i, r_j) : 1 \leq i < j \leq \rho\}$.

1.2   Related Works

Multicast cost sharing games are instances of the well-known class of congestion games [12], which always converge to pure Nash equilibria. While a price of anarchy equal to ρ is folklore, in [1], by exploiting the properties of the exact potential function associated to congestion games, it is shown that the price of stability of multicast cost sharing games is equal to H_ρ when G is directed, while it is upper bounded by H_ρ when G is undirected.

Multicast cost sharing games are also investigated in [8] where, after noting that for several reasonable cost sharing methods the price of anarchy of the the game remains equal to ρ, the speed of convergence is estimated, that is the quality of the solutions reached after a limited number of selfish moves of the receivers. In [5] it is proved that the problem of computing a Nash equilibrium minimizing the potential function is NP-hard, while it is solvable in polynomial time if the receivers are allowed to arbitrarily split their requests among different paths.

It is also shown that the price of anarchy of a Nash equilibrium reached after any sequence of best responses in the case in which the receivers enter the game one at time starting from the empty configuration is between $\Omega(\frac{\log \rho}{\log \log \rho})$ and $O(\sqrt{\rho} \log^2 \rho)$. These bounds have been significantly improved in [6] to $\Omega(\log \rho)$ and $\log^3 \rho$, respectively. In [6] it is also shown that the price of anarchy of the solution obtained after a first round of best responses of the receivers entering the game one at time starting from the empty configuration is upper bounded by $\log^2 \rho$.

1.3 Our Contribution

When considering the presence of social knowledge graphs, since some receivers can be hidden to other ones, multicast cost sharing games are no longer a proper subclass of the congestion games. As a consequence, the existence and convergence to Nash equilibria is no longer guaranteed.

Therefore, on this respect we first provide a complete characterization of the game. In particular, while the convergence in the case in which K is a directed acyclic graph (DAG) can be directly inferred by the results in [3], which hold for every congestion game with a DAG social graph, we show that if K is undirected or directed cyclic, the existence of pure Nash equilibria is no longer guaranteed. This also closes the open question raised in [3] of whether in congestion games equilibria always exist for undirected social graphs. In fact, while such an existence was proven in case of linear latencies, we show that indeed this is not the case if the latency functions express the Shapley cost shares. Thus, this also completes also the general characterization picture of the congestion games with social knowledge.

As to the impact of social knowledge graphs on the performance of multicast cost sharing game, we show that the (universal) price of stability $PoS(K)$ of any DAG K is always at least equal to $\frac{1}{2} \log \rho$, that is $PoS(I, K) \geq \frac{1}{2} \log \rho$ for at least one instance I. Moreover, we show that the set of Nash equilibria induced by any complete DAG $\hat{K}$ on every instance I coincides with the set of solutions obtained after a first round of best responses in which, starting from the empty configuration, the receivers enter sequentially the game I according to the their topological ordering in $\hat{K}$. Therefore, by the results given in [6], an induced upper bound $PoA(\hat{K}) \leq \log^2 \rho$ on the price of anarchy (and thus of stability) for any complete DAG $\hat{K}$ holds.

Besides the above universal bounds, we show that there exist specific instances of the game I for which $PoS(I, K) \geq \frac{4\rho}{\rho+3}$ for any DAG K. The same bound holds also for the price of anarchy of every K (not only for DAGs, but also for all the other K inducing games admitting at least an equilibrium).

On the other side, we prove that, for any instance I, there always exists a DAG $K(I)$ such that $PoA(I, K(I)) \leq \frac{4\rho}{\rho+3}$ if $\rho = 2, 3$ and $PoA(I, K(I)) \leq \frac{4(\rho-1)}{\rho+1}$ if $\rho \geq 4$, hence obtaining an upper bound on the price of anarchy almost (surprisingly) matching the lower bound on the price of stability achievable with DAGs. Unfortunately, this is only an existential result and we do not know how to construct efficiently the graph $K(I)$. However, we can prove that given any r-approximation of T^* it is possible to compute in polynomial time, by using a simple depth first search, a DAG $K(I)$ such that $PoA(I, K(I)) \leq \frac{4\rho}{\rho+3}r$ if $\rho = 2, 3$ and $PoA(I, K(I)) \leq \frac{4(\rho-1)}{\rho+1}r$ if $\rho \geq 4$.

Our achievements are twofold: on one side, we shed some light on how the lack of knowledge among players can impact on the total cost of the self-emerging networks created by the interactions of selfish users; on the other side, we show that the idea of hiding some players to others is a powerful instrument that a designer of a decentralized application inducing the game can use in order to obtain solutions whose cost may be not too far from the optimal one without directly interfering on the choices performed by the players.

As a consequence of our study, we can conclude that the presence of social knowledge graphs can tremendously improve the performance of multicast cost sharing games. This situation can be seen as another evidence of the famous Braess' paradox [4]: there are cases in which adding fast links in a network results in a decrease of its performances or, symmetrically, hiding some fast links to the players can increase the network performances. This is exactly the final rationale of our analysis: hiding some of the players to other ones can yield better solutions, that is, the less players know the most they are "cooperative".

The paper is structured as follows. In the next section we completely characterize the existence of and convergence to Nash equilibria. In Sections 3 and 4 we present our results concerning the prices of anarchy and stability for given instances and the universal prices of anarchy and stability, respectively. Finally, in Section 5 we give some conclusive remark and discuss some open questions.

2   Existence of and Convergence to Pure Nash Equilibria

In this section we completely characterize the existence and convergence to pure Nash equilibria in graphical multicast cost sharing games.

As already proved in [3], if the social knowledge graph is acyclic, each congestion game (and thus each multicast cost sharing game) converges to a pure Nash equilibrium and one such an equilibrium can be efficiently computed.

Theorem 1 ([3]). *Each congestion game converges to a pure Nash equilibrium when the social knowledge graph is a DAG. Moreover, there always exists a sequence of at most n best responses which can be computed in polynomial time ending to a Nash equilibrium.*

We now show that if K is directed symmetric (or equivalently undirected) multicast cost sharing games may not admit pure Nash equilibria.

Theorem 2. *Multicast cost sharing games may not posses pure Nash Equilibria when the social knowledge graph is undirected.*

Proof. In order to prove the theorem we give an instance of the multicast cost sharing game in which each configuration is not a pure Nash equilibrium. Consider the instance presented in Figure 1(a), where ϵ represents any positive real number and the undirected social knowledge graph depicted in Figure 1(b).

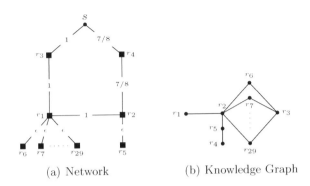

(a) Network   (b) Knowledge Graph

Fig. 1. Non-Existence of Nash Equilibrium

We say that a strategy is strictly dominant for a receiver r_i if, regardless of what any other receiver does, the cost of r_i is always strictly smaller than the one obtained by choosing any other strategy. It is not difficult to observe that the edges (s, r_3) and (s, r_4) are strictly dominant strategies for r_3 and r_4 respectively. This implies that the paths $\langle s, r_3, r_1, r_i \rangle$ with $6 \leq i \leq 29$, and $\langle s, r_4, r_2, r_5 \rangle$ are strictly dominant strategies for r_i and r_5 respectively. Specifically, in any configuration of the game only r_1 and r_2 may eventually perform a selfish move.

Table 1. A game not admitting a Nash equilibrium

r_1 \ r_2	$\langle s, r_3, r_1, r_2 \rangle$	$\langle s, r_4, r_2 \rangle$
$\langle s, r_3, r_1 \rangle$	$(1, \frac{28}{26})$	$(2, \frac{7}{8})$
$\langle s, r_4, r_2, r_1 \rangle$	$(\frac{9}{4}, \frac{29}{50})$	$(\frac{15}{8}, \frac{7}{12})$

Considering only the strategies of r_1 and r_2, the game can be represented in normal form as illustrated in Table 1. The theorem follows by observing that none of the resulting four configurations is a pure Nash equilibrium.   □

It is important to note that, since each undirected graph is also a cyclic directed one (by replacing each undirected edge $\{i, j\}$ with the pairwise opposite arcs (i, j) and (j, i)), we have that also for directed cyclic graphs the existence of pure Nash equilibria in the multicast cost sharing game is not guaranteed.

In the following sections we restrict to direct acyclic social graphs, since by Theorem 1 this is the only case of guaranteed convergence to equilibria.

3   Prices of Anarchy and Stability for Specific Instances

We first prove a lower bound holding for the price of stability of DAGs as well as for the price of anarchy of any social graph admitting equilibria.

Theorem 3. *There exists an instance I such that $PoS(I, K) \geq \frac{4\rho}{\rho+3}$ for any DAG K. Moreover, $PoA(I, K) \geq \frac{4\rho}{\rho+3}$ for any graph K (not only DAGs) admitting equilibria.*

Proof. Consider the instance depicted in Figure 2 where $c(T^*) = \rho(1 + \epsilon) + 3$. It is not difficult to see that for any possible social knowledge graph K the solution in which each receiver r_i uses the edge (s, r_i) is a Nash equilibrium. Hence, by the arbitrariness of ϵ, we have $PoA(I, K) \geq \frac{4\rho}{\rho+3}$. In order to prove the same bound for the price of stability of DAGs, we show that the solution in which each receiver r_i uses the edge (s, r_i) is the only Nash equilibrium. To this aim, fix a Nash equilibrium π and let r_i be the receiver with maximum index i according to the topological ordering induced by K among the ones not using the edge (s, r_i) in π. Since all the receivers $r_j \in R_i(K)$ are using edge (s, r_j), r_i's best strategy is to choose the edge (s, r_i) thus creating a contradiction. It follows, that if K is a DAG $PoS(I, K) \geq \frac{4\rho}{\rho+3}$, by the arbitrariness of ϵ. $\qquad \square$

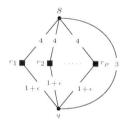

Fig. 2. An instance yielding $PoA(K) \geq \frac{4\rho}{\rho+3} \ \forall K$ and $PoS(K) \geq \frac{4\rho}{\rho+3} \ \forall K$ directed acyclic

We now show that for each instance I of the multicast cost sharing game there exists a complete DAG $\hat{K}(I)$ such that $PoA(I, \hat{K}(I))$ almost matches the above lower bound.

Theorem 4. *For any instance I of the multicast cost sharing game there exists a complete DAG $\hat{K}(I)$ such that $PoA(I, \hat{K}(I)) \leq \frac{4(\rho-1)}{\rho+1}$ if $\rho \geq 3$ and $PoA(I, \hat{K}(I)) \leq \frac{8}{5}$ if $\rho = 2$.*

Notice that the proven upper bounds on the price of anarchy exactly match the lower bounds holding for any social knowledge graph when $\rho = 2, 3$. Moreover, for $\rho > 3$ the upper bounds on the price of anarchy of $\hat{K}(I)$ almost match the lower bounds on the price of stability achievable by any DAG, which is quite a surprising result. Unfortunately, in order to compute $\hat{K}(I)$ we need do know an optimal solution T^* for instance I. Since it is NP-hard to compute T^*, we can resort to approximation algorithms for the Minimum Weighted Steiner Tree problem in order to compute in polynomial time a complete DAG $\hat{K}(I)$ yielding a slightly worse price of anarchy.

Theorem 5. *For any instance I of the multicast cost sharing game it is possible to compute in polynomial time a complete DAG $\hat{K}(I)$ such that $PoA(I, \hat{K}(I))$*

$\leq \frac{4(\rho-1)}{\rho+1} r$ if $\rho \geq 3$ and $PoA(I, \hat{K}(I)) \leq \frac{8}{5} r$ if $\rho = 2$, where r is the approximation ratio of the best algorithm for the Minimum Weighted Steiner Tree problem.

4   Universal Prices of Anarchy and Stability

From the analysis performed in the previous section it results that directed acyclic complete graphs are among the ones yielding the lowest possible price of anarchy for the multicast cost sharing game when the instance of the game is known in advance. We now show that such graphs achieve a good performance also when used as universal social knowledge graphs, that is without any assumption on the given instance. We obtain such a result by exploiting the strong similarity of the Nash equilibria induced by such graphs and the solutions obtained after a first round of best responses performed sequentially by the receivers starting from the empty space. For any instance I, let $FR(I)$ denote the set of all possible solutions which can be obtained, starting from the empty state, after a first round of best responses performed sequentially by the receivers (from r_ρ down to r_1).

Theorem 6. $N(I, \hat{K}) = FR(I)$ for any instance I of the multicast cost sharing game.

Proof. Consider a Nash equilibrium $\pi \in N(I, \hat{K})$. Since the strategy adopted by any receiver r_i is a best response strategy given the choices of all the receivers r_j such that $j > i$, we have that $\pi \in FR(I)$. On the other hand, consider a solution $\pi \in FR(I)$. Since each r_i enters the game by performing a best response given the strategies played by all the receivers r_j with $j > i$ and the choices performed by all the other receivers r_j with $j < i$ do not affect r_i's cost share, we have that $\pi \in N(I, \hat{K})$. □

By the above theorem and the results of [6] on the social performance achieved after one round of best response moves from empty state, the following corollary holds.

Corollary 1. $PoA(\hat{K}) \leq \log^2 \rho$.

We now provide a close lower bound on the price of stability achievable by any DAG.

Theorem 7. $PoS(K) \geq \frac{\lceil \log \rho \rceil}{2}$ for any DAG K.

Proof. Let $\rho = 2^\ell - 1$. Once fixed a DAG K number the ρ receivers according to the topological order induced by K and consider an instance $I = (G, R, s)$ defined as follows. Let s be the origin of a unit length segment, we locate r_i at position $\frac{1+2(i-2^{\lfloor \log i \rfloor})}{2^{\lceil \log(i+1) \rceil}}$ on the segment. We call the set $\{r_i \in R : \lceil \log(i+1) \rceil = j\}$ the set of receivers of level j. For any pair of consecutive receivers along the segment there exists a straight edge in G (the one ideally belonging to the segment) of cost $\frac{1}{2^\ell}$.

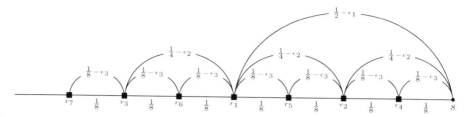

Fig. 3. An instance yielding $PoS(K) \geq \frac{1}{2} \log \rho$ for any DAG K. Only receivers of level ≤ 3 are shown.

Moreover, for each receiver r_i belonging to level j let $left(i,q)$ and $right(i,q)$ be the nearest receiver of level q lying respectively on the left and on the right of r_i along the segment. For each $q > j$ there exist two curve edges $(r_i, left(i,q))$ and $(r_i, right(i,q))$ both having cost $\frac{1}{2^q} - \epsilon_q$. Finally, for each level $j \in [\ell]$, letting $nearest(j)$ be the receiver of level j closest to s, there exists the curve edge $(s, nearest(j))$ of cost $\frac{1}{2^j} - \epsilon_j$. A picture of I is depicted in Figure 3 where $\ell = 3$. We define the values ϵ_j for $j \in [\ell]$ in such a way that $\epsilon_j > 2\epsilon_{j+1}$.

We claim that in any $\pi \in N(I,K)$ each receiver r_i of level j must use a curve edge starting at r_i and having cost equal to $\frac{1}{2^j} - \epsilon_j$ thus yielding a price of stability

$$PoS(K) \geq \sum_{j=1}^{\ell} \sum_{i=1}^{2^{j-1}} \left(\frac{1}{2^j} - \epsilon_j \right) \geq \sum_{j=1}^{\lceil \log \rho \rceil} \left(\frac{1}{2} - 2^{j-1}\epsilon_1 \right) \geq \left(\frac{1}{2} - \rho\epsilon_1 \right) \lceil \log \rho \rceil \approx \frac{\lceil \log \rho \rceil}{2}$$ for the arbitrariness of ϵ_1.

We show by induction on i that if r_i is of level j, r_i chooses only curve edges of level not greater than j. For $j = 1$ this reduces to show that the curve edge (s, r_1) is the unique shortest (s, r_1)-path. This easily follows by construction of the ϵ_js. Consider now a receiver r_i of level $j > 1$. Because of the non-intersecting structure of the curve edges, each (s, r_i)-path in G must pass through either $left(i, j-1)$ or $right(i, j-1)$. Since the only receivers of which r_i is aware are those belonging to levels which are not greater than j and they are only using curve edges of level not greater than j, the shortest path from r_i to one of the two receivers $left(i, j-1)$ and $right(i, j-1)$ is part of a best response for r_i. By construction, a curve edge starting at r_i and directly reaching $left(i, j-1)$ or $right(i, j-1)$ and having cost equal to $\frac{1}{2^j} - \epsilon_j$ is the unique desired shortest path. □

5   Conclusions and Open Problems

Following the framework introduced in [3], we have analyzed the impact of social knowledge graphs on the price of anarchy and stability of multicast cost sharing games. In particular, we have shown that any complete DAG $\hat{K}$ lowers the price of anarchy from ρ to $\log^2 \rho$ and that, when a particular instance of the game is fixed, there exists a complete DAG $\hat{K}$ (among the possible $\rho!$ ones which can be obtained by considering all permutations of the receivers) yielding a price of anarchy less than 4. Moreover, we can compute in polynomial time one complete DAG yielding a price of anarchy at most $4r$ (with $r \leq 1.55$). Interestingly, the presence of social knowledge graphs tremendously narrows the gap between the prices of anarchy and stability, as we have proved close or almost matching lower bounds, that is $PoS(K) \geq \frac{1}{2} \log \rho$ for every DAG K and there exist instances I such that $PoS(I, K) \geq \frac{4\rho}{\rho+3}$ for every DAG K.

We stress that complete DAGs reflects the situation in which players enter sequentially the game: when player r_i joins a non-cooperative system she can get knowledge of only the $i - 1$ players already involved in the game and will not be aware of those who may eventually join the system in the future. Thus, other than assuring good performances, such graphs also have a theoretical and technical motivation.

Possible applications of our results include the design of protocols and P2P systems which limit the visibility of the other peers, or simply, at a more foundational level, the possibility of using social graphs as an intermediate methodological tool for defining cost shares and payoffs so as to induce good overall performances without directly interfering in users decisions.

Several open problems arise from our approach.

Besides closing the gaps between upper and lower bounds on the prices of anarchy and stability, like the $\Omega(\log \rho) \div \log^3 \rho$ one on the price of anarchy in the universal case, it would be interesting to extend our study to other graphical games, like cost sharing congestion games.

Moreover, since $\sum_{i=1}^{\rho} cost(\pi, r_i)$ can be greater than $c(\Pi)$, it would be interesting to study the surplus (defined as $\sum_{i=1}^{\rho} cost(\pi, r_i) - c(\Pi)$) created by social knowledge graphs.

Finally, what about the effect of social graphs on the speed of convergence, that is on the number of selfish moves needed to reach equilibria or on the performances achieved after a limited number of steps?

References

1. Anshelevich, E., Dasgupta, A., Kleinberg, J., Tardos, E., Wexler, T., Roughgarden, T.: The Price of Stability for Network Design with Fair Cost Allocation. In: Proceedings of the 45th Annual IEEE Symposium on Foundations of Computer Science (FOCS), pp. 295–304. IEEE Computer Society, Los Alamitos (2004)
2. Anshelevich, E., Dasgupta, A., Tardos, E., Wexler, T.: Near-Optimal Network Design with Selfish Agents. In: Proceedings of the 35th Annual ACM Symposium on Theory of Computing (STOC), pp. 511–520. ACM, New York (2003)
3. Bilo, V., Fanelli, A., Flammini, M., Moscardelli, L.: Graphical Congestion Games with Linear Latencies. In: Proceedings of the 20th ACM Symposium on Parallelism in Algorithms and Architectures (SPAA). ACM, New York (to appear, 2008)
4. Braess, D.: Uber ein Paradoxon der Verkehrsplanung. Unternehmensforschung 12, 258–268 (1968)
5. Chekuri, C., Chuzhoy, J., Lewin-Eytan, L., Naor, J., Orda, A.: Non-Cooperative Multicast and Facility Location Games. In: Proceedings of the 7th ACM Conference on Electronic Commerce (EC), pp. 72–81. ACM, New York (2006)
6. Charikar, M., Mattieu, C., Karloff, H., Naor, J., Saks, M.: Best Response Dynamics in Multicast Cost Sharing. Personal communication

7. Czumaj, A.: Selfish Routing on the Internet. Handbook of Scheduling. CRC Press, Boca Raton (2004)
8. Fanelli, A., Flammini, M., Melideo, G., Moscardelli, L.: Multicast Transmissions in Non-cooperative Networks with a Limited Number of Selfish Moves. In: Královič, R., Urzyczyn, P. (eds.) MFCS 2006. LNCS, vol. 4162, pp. 363–374. Springer, Heidelberg (2006)
9. Karakostas, G., Kim, T., Viglas, A., Xia, H.: Selfish Routing with Oblivious Users. In: Prencipe, G., Zaks, S. (eds.) SIROCCO 2007. LNCS, vol. 4474, pp. 318–327. Springer, Heidelberg (2007)
10. Kearns, M.J., Littman, M.L., Singh, S.P.: Graphical Models for Game Theory. In: Proceedings of the 17th Conference in Uncertainty in Artificial Intelligence (UAI), pp. 253–260. Morgan Kaufmann, San Francisco (2001)
11. Koutsoupias, E., Papadimitriou, C.: Worst-Case Equilibria. In: Meinel, C., Tison, S. (eds.) STACS 1999. LNCS, vol. 1653, pp. 404–413. Springer, Heidelberg (1999)
12. Rosenthal, R.W.: A Class of Games Possessing Pure-Strategy Nash Equilibria. International Journal of Game Theory 2, 65–67 (1973)
13. Roughgarden, T.: Selfish Routing and the Price of Anarchy (Survey). OPTIMA 74 (2007)
14. Nisan, N., Roughgarden, T., Tardos, E., Vazirani, V.V.: Algorithmic Game Theory. Cambridge University Press, Cambridge (2007)
15. Shapley, L.S.: The value of n-person games. In: Contributions to the theory of games, pp. 31–40. Princeton University Press, Princeton (1953)

Shortest Synchronizing Strings
for Huffman Codes[*]
(Extended Abstract)

Marek Tomasz Biskup

Institute of Informatics, University of Warsaw,
Banacha 2, 02-097 Warszawa, Poland
mbiskup@mimuw.edu.pl
http://www.mimuw.edu.pl/~mbiskup

Abstract. Most complete binary prefix codes have a synchronizing string, that is a string that resynchronizes the decoder regardless of its previous state. This work presents an upper bound on the length of the shortest synchronizing string for such codes. Two classes of codes with a long shortest synchronizing string are presented. It is known that finding a synchronizing string for a code is equivalent to a finding a synchronizing string of some finite automaton. The Černý conjecture for this class of automata is discussed.

1 Introduction

Huffman codes are the most popular variable length codes. In the presence of channel errors a large part of an encoded message can be destroyed because of the loss of synchronization between the decoder and the coder. In case of some Huffman codes, under certain assumptions on the message source, the decoder will eventually resynchronize, and, from then on, symbols will be decoded correctly. These codes are called *synchronizing*. Capocelli et al. [1] proved that codes are synchronizing if and only if they have a *synchronizing string* — a string such that when received by the decoder always puts it into synchronization. Freiling et al. [2] proved that almost all Huffman codes have a synchronizing string. More precisely, they proved that the probability of drawing randomly a code without a synchronizing string decreases to zero with increasing code size.

Shützenberger [3] analyzed possible distribution of codewords' lengths in a synchronizing prefix codes. Rudner [4] gave an algorithm for the construction of a synchronizing Huffman code for a given distribution of codewords' lengths, that works under some assumptions on the distribution. His work was further extended in [5,6]. Capocelli et al. [7] showed how to modify a Huffman code by adding a little redundancy to create a synchronizing code. Ferguson and Rabinowitz [8] analyzed codes whose synchronizing string is a codeword.

---

[*] The research was partially supported by the grants of the Polish Ministry of Science and Higher Education N 206 004 32/0806 and N N206 376134.

E. Ochmański and J. Tyszkiewicz (Eds.): MFCS 2008, LNCS 5162, pp. 120–131, 2008.

The synchronization recovery of a Huffman code can be modeled with a finite automaton whose states are proper prefixes of codewords (or internal nodes of the code's tree). This automaton will be called a *Huffman automaton*. Such an automaton was used by Maxted and Robinson [9] to compute for a given code the average number of symbols lost before resynchronization.

A lot of research has been done in the area of automata synchronization. A synchronizing string for a finite automaton $\langle Q, \Sigma, \delta \rangle$ is a string s that brings all states to one particular state. That is $\delta(q_1, s) = \delta(q_2, s)$ for any states $q_1, q_2 \in Q$. An automaton will be called *synchronizing* if it has a synchronizing string. The famous Černý conjecture [10] states that a synchronizing finite automaton with N states has a synchronizing string of length $(N - 1)^2$.

Although there are proofs for certain classes of automata, for instance in [11,12], the problem remains open. There are some bounds on the length of the shortest synchronizing string. For instance Pin [13] proved that $\frac{1}{6}(N^3 - N)$ is an upper bound. Some research has also been done to find automata with long shortest synchronizing strings. Černý [10] constructed a series of automata with the shortest synchronizing string of length $(N - 1)^2$. Ananichev et al. [14] considered how long a synchronizing string can be if there is a letter that reduces the number of states by two. Trahtman [15] searched for worst-case automata.

Eppstein [16] gave an algorithm for testing whether an automaton is synchronizing and for the construction of a synchronizing string of length $O(N^3)$ for a synchronizing automaton. His algorithm requires $O(N^3)$ operations if the alphabet is of constant size. An overview of the area of automata synchronization is given in [17].

It is rather clear that a synchronizing string for a Huffman code is also a synchronizing string for the Huffman automaton of the code, and vice versa. Nevertheless, it seems that so far both areas of research have not been related. This paper fills this gap.

First we explain that Huffman code synchronization is equivalent to Huffman automaton synchronization. Then, we prove an upper bound on the length of the shortest *merging string* for a set of two states of a Huffman automaton: the root of the code's tree and another internal node of the tree. A merging string for a set of states is a string that brings all states of the set to the same state. The proof is constructive and an algorithm for the construction of the shortest merging string for such nodes is given. The execution of this algorithm also suffices for answering whether a code is synchronizing. Then we present an upper bound on the length of the shortest synchronizing string of a Huffman automaton. For most (but not all) codes the bound is better than the Černý conjecture. Also an algorithm for the construction of a synchronizing string for a Huffman automaton is presented. To the author's best knowledge, this class of automata has not been studied yet. The bounds presented here are better than the bounds $O(N^3)$ for general automata. Both algorithms are faster than the one of Eppstein [16].

Afterwards, results of experimental search for worst-case codes are shown. Three classes of Huffman codes are presented. The codes give a lower estimate on the possible upper bounds of the length of the shortest synchronizing or merging string. It is conjectured (but, unfortunately, not proved) that these classes of codes are the worst-case codes. It is interesting that the length of their synchronizing or merging strings is much lower than the bound proved.

Due to limited length of the paper, the most difficult proofs are omitted.

2   Definitions and Notation

A *word* is a string of letters, for instance $w = w_0 w_1 \ldots w_{k-1}$. The empty word is denoted by ϵ. The subword of a word w from the position p to $q - 1$ is denoted by $w[p..q)$. The length of a word w is denoted by $|w|$. A sequence of k letters a is denoted a^k. For instance, for $w=$'abc', $w[1..2)=$'b', $w[1..3)=$'bc', $w[0..1)=$'a', $|w| = 3$ and $0^4 =$ '0000'.

A *complete binary tree* is a tree with each node being either an *internal node* with two children, or a *leaf* with no children. Each left outgoing edge is labeled with 0 (0-edge). Each right outgoing edge is labeled with 1 (1-edge). The root of a tree is denoted by ε. Each node n has a unique binary string $\pi(n)$ that is formed of labels on the path from the root to n. We have $\pi(\varepsilon) = \epsilon$. The number of leaves in a tree is denoted by N. The height of a tree is denoted by h. In this paper, a code C such that $C = \{\pi(n) | n$ is a leaf of $T\}$ for some complete binary tree T, is called a *Huffman code*. The tree T is called a *Huffman tree*. We refer to a node n of T using the string $\pi(n)$. For instance, the node 10 is the left son of the right son of the root.

Let a *Huffman Automaton* (HA) $\mathcal{T}$ be an automaton whose states are internal nodes of the Huffman tree T. The transition function $\delta(n, b)$, $b \in \{0, 1\}$, brings an automaton from the node n to its b-edge child, if it is not a leaf, or to the root otherwise. The function δ^* is the extension of δ to strings: $\delta^*(q, b_0 \ldots b_{k-1}) = \delta(\delta^*(q, b_0 \ldots b_{k-2}), b_{k-1})$ and $\delta^*(q, \epsilon) = q$. For a subset S of states of a Huffman automaton we denote, $\delta(S, a) := \{\delta(q, a) | q \in S\}$. The same convention is used for δ^*.

We say that a word w *brings* a node n to a node n' if $n' = \delta^*(n, w)$. Then n' is the result of *applying* w to n. In addition, we say that w brings a node n to a leaf if $\delta^*(n, w) = \varepsilon$ and w is not empty. This is justified because the construction of the Huffman automaton $\mathcal{T}$ may be seen as merging the leaves of the tree T with the root of T. We say that w brings a node n to n' *without loops* if none of the nodes $\delta^*(n, w[0, 1)), \delta^*(n, w[0, 2)), \ldots, \delta^*(n, w[0, |w| - 2))$ is the root.

The values T, $\mathcal{T}$, δ, δ^*, N, h, ε, π depend on the code C. We assume that it is always clear from the context which code (or, equivalently, which Huffman tree) is being considered.

Definition 1. *A* synchronizing string *for a Huffman code is a string* w_s *such that* $w w_s$ *is a sequence of codewords for any binary word* w.

Equivalently, a synchronizing string is a string that brings any node of the Huffman automaton to the root.

Definition 2. *Let $\mathcal{A} = \langle Q, \Sigma, \delta \rangle$ be a finite automaton. A synchronizing string for $\mathcal{A}$ is a word w such that $|\delta^*(Q, w)| = 1$. A merging string for a set of states $R \subseteq Q$ of the automaton $\mathcal{A}$ is a word w such that $|\delta^*(R, w)| = 1$.*

Definition 3. *Let $\mathcal{A} = \langle Q, \Sigma, \delta \rangle$ be a finite automaton. The power automaton for $\mathcal{A}$ is the automaton $\mathcal{P}(\mathcal{A}) = (\mathcal{P}(Q), \Sigma, \delta_{\mathcal{P}})$, where $\mathcal{P}(Q)$ denotes the set of all subsets of Q and $\delta_{\mathcal{P}}(S, a) = \delta(S, a)$ for $S \in \mathcal{P}(Q)$.*

The operation of the power automaton $\mathcal{P}(\mathcal{A})$ can be seen as movements of coins that lie on some states of the automaton $\mathcal{A}$. If the power automaton is in a state $S \subseteq Q$, the coins lie on the states $q \in S$. Then, if the power automaton makes a transition by a letter a, the coins move according to the transition function δ of the automaton $\mathcal{A}$. If a coin is in the state p then it moves onto the state $\delta(p, a)$. If more than one coin goes to the same state only one of them is kept. It easy to see that after applying the letter a the set of states with coins is exactly $\delta_{\mathcal{P}}(S, a)$. This analogy helps to visualize the operation of the power automaton and gives some intuition. For instance, the string w is synchronizing if and only if applying w to the automaton $\mathcal{A}$ with a coin on each state results in just one coin left.

An automaton is *synchronizing* if it has a synchronizing string. A Huffman code is *synchronizing* if it has a synchronizing string.

Theorem 4. *A synchronizing string for a Huffman code C is a synchronizing string for the Huffman automaton $\mathcal{T}$ of the code. A synchronizing string s for the Huffman automaton $\mathcal{T}$, such that s brings all nodes to the root, is a synchronizing string for the Huffman code.*

Thus a Huffman code is synchronizing if and only if its Huffman automaton is synchronizing.

3   Merging String for a Pair of States

Theorem 5. *Let C be a synchronizing Huffman code of size N, let T be the Huffman tree for C, let $\mathcal{T}$ be the Huffman automaton of the code C. For any node n of $\mathcal{T}$ there is a merging string s_n for the set $\{n, \varepsilon\}$, with*

$$|s_n| \le \sum_{p \in Q(T) \setminus \{\varepsilon\}} h_p, \tag{1}$$

where $Q(T)$ is the set of the internal nodes of T and h_p is the height of the subtree of T rooted at p.

Proof. Let us consider a merging string s_n for $\{n, \varepsilon\}$ of minimal length (it exists because C is synchronizing, but it need not be unique). The string s_n brings both nodes to the root, because otherwise we could remove the last letter of s_n and the result would still merge n and ε.

Let $\{n_i, m_i\}$ be the unordered pairs of nodes that appear when consecutive prefixes of s_n are applied to the initial set $\{n, \varepsilon\}$, i.e.

$$\{n_i, m_i\} = \delta^* \left(\{n, \varepsilon\}, s_n[0..i)\right), \quad i = 0, \dots, |s_n|. \tag{2}$$

We have $\{n_0, m_0\} = \{n, \varepsilon\}$ and $\{n_{|s_n|}, m_{|s_n|}\} = \{\varepsilon\}$ (a singleton is also considered a pair).

Let us look at the subsequence $\{n_{i_k}, \varepsilon\}, k = 0, \dots, l$, of this sequence formed of pairs containing the root. Each node p, appears in this subsequence as the partner of ε at most once, because pairs do not repeat in $\{n_i, m_i\}$ (otherwise we could shorten the string s_n). The string $s_n[i_k, i_{k+1})$, that brings $\{n_{i_k}, \varepsilon\}$ to $\{n_{i_{k+1}}, \varepsilon\}$, is a string that either brings the node n_{i_k} to a leaf without loops or that brings ε to a leaf without loops. In either case the length of $s_n[i_k, i_{k+1})$ is at most $h_{n_{i_k}}$ (note that in the second case the node $n_{i_{k+1}}$ is in the subtree of n_{i_k}). We get

$$|s_n| = \sum_{k=0}^{l-1} |s_n[i_k, i_{k+1})| \leq \sum_{k=0}^{l-1} h_{n_{i_k}} \leq \sum_{p \in Q(T) \setminus \{\varepsilon\}} h_p. \tag{3}$$

The value of h_ε is not counted because the set $\{\varepsilon\}$ appears only as the last element of the sequence $\{n_{i_k}, \varepsilon\}$. □

Let H_T be the value of the bound in Theorem 5. H_T is the sum of heights of all the nontrivial subtrees of T apart from the whole tree. We will compare H_T with Π_T — the sum of depths of all the internal nodes, and with W_T — the sum of depths of all the leaves of T (that is the sum of codewords' lengths).

Lemma 6. *Let T be a complete binary tree, let $Q(T)$ be the set of internal nodes of T, let $L(T)$ be the set of leaves of T, let h_n and N_n be, respectively, the height and the number of leaves of the subtree rooted at the node n of T, let $|\pi(n)|$ be the distance from the root to n and let N_T be the number of leaves of T. Let us define*

$$H_T = \sum_{n \in Q(T) \setminus \{\varepsilon\}} h_n, \qquad\qquad \Pi_T = \sum_{n \in Q(T)} |\pi(n)|, \tag{4}$$

$$W_T = \sum_{n \in L(T)} |\pi(n)|, \qquad\qquad S_T = \sum_{n \in Q(T)} N_n. \tag{5}$$

Then the following holds:

$$H_T \leq \Pi_T = W_T - 2N_T + 2 \leq W_T = S_T - N_T. \tag{6}$$

Corollary 7. *Let w_i be codewords of a Huffman code. Then*

$$|s_n| \leq \sum_i |w_i| \qquad and \qquad |s_n| \leq (N-2)(h-1). \tag{7}$$

The result of Theorem 5 can be improved if we notice that the sequence $\{n_{i_k}, \varepsilon\}$, defined in the proof of Theorem 5, cannot contain two nodes n_{i_k} and $n_{i_{k'}}$ that are roots of identical subtrees of T. Indeed, otherwise we could shorten the string s_n in the same way as before. This gives the following result.

Corollary 8. *The bound of Theorem 5 can be improved to:*

$$|s_n| \leq \sum_{t \in \mathscr{T}(T) \setminus \{T\}} h_t \qquad (8)$$

where $\mathscr{T}(T)$ is the set all distinct subtrees of T.

The idea of identifying common subtrees can be formalized by introducing a *minimized Huffman automaton*. Although this does not give here a better estimate on the length of the shortest merging string for the set $\{n, \varepsilon\}$, it is interesting in itself.

Definition 9. *A* minimized Huffman automaton *for a Huffman code C is an automaton made of the Huffman automaton for C by merging the states that are roots of identical subtrees of the Huffman tree T for C.*

It is easy to see that minimized Huffman automata have exactly two edges, labeled with 0 and 1, going out of each node. An example of a minimized Huffman automaton is presented in Fig. 1.

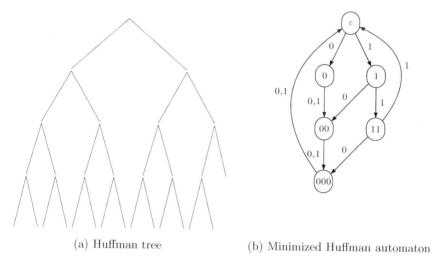

(a) Huffman tree   (b) Minimized Huffman automaton

Fig. 1. A Huffman tree and its minimized Huffman automaton

We will say that a set V of states of $\mathcal{T}$ corresponds to the set V_m of states of the minimized Huffman automaton $\mathcal{T}_m$ if V_m is the smallest set satisfying: if $q \in V$ and q is merged to a state q' of $\mathcal{T}_m$ then $q' \in V_m$.

Theorem 10. *Let C be a synchronizing Huffman code, let $\mathcal{T}$ be the Huffman automaton for C and let $\mathcal{T}_m$ be the minimized Huffman automaton for C. Let V be a set of states of $\mathcal{T}$ and V_m the corresponding set of states of $\mathcal{T}_m$. If s is a merging string for V then s is a merging string for V_m. If s' is a merging string for V_m that brings all nodes of V_m to the root then s' is a merging string for V.*

Note that the minimized Huffman automaton is implicitly used in Corollary 8, where we consider all non-identical subtrees of a Huffman tree. The roots of such subtrees are the states of the minimized Huffman automaton.

Theorem 5 leads to an algorithm for finding the shortest merging string for a set $\{n_0, \varepsilon\}$, where n_0 is any state of $\mathcal{T}$. First a graph $G = (V, E)$ is created. The vertices of G are unordered pairs $\{n, \varepsilon\}$, where n is a state of $\mathcal{T}$. The edges of G are weighted; $\{n_1, \varepsilon\} \rightarrow \{n_2, \varepsilon\}$ is an edge if there is a string w that brings $\{n_1, \varepsilon\}$ to $\{n_2, \varepsilon\}$ without passing through any other pair $\{n, \varepsilon\}$. The weight of the edge is the length of the shortest such string w (note that the string w need not be unique).

Such a string w will be the label of the edge $\{n_1, \varepsilon\} \rightarrow \{n_2, \varepsilon\}$, although it will not be stored explicitly. Instead, for retrieving the label w, we will store a mark M. The mark will depend on the target pair of the edge. If the target is a pair $\{n_2, \varepsilon\}$ with $n_2 \neq \varepsilon$, the mark is equal to either n_1 if $n_2 = \delta(n_1, w)$, or to ε if $n_2 = \delta(\varepsilon, w)$. We always have $n_2 = \delta(M, w)$. The node n_2 is in the subtree of the node M and w is formed of labels on the path from M to n_2. If the target of an edge is a singleton $\{\varepsilon\}$, that is $n_2 = \varepsilon$, the mark M is the leaf $\delta(\varepsilon, w)$. In this case the word w is formed of labels on the path from ε to M. In either case the label w can be recovered in $O(|w|)$.

The construction of the graph requires DFS-traversing the Huffman tree with a pair of nodes $\{n_1, n_2\}$, starting at $\{n, \varepsilon\}$ and applying transitions of the Huffman automaton to both nodes of the pair. The traversing goes forward until a set $\{n', m\}$ is reached, with m being a leaf. Then the edge $\{n, \varepsilon\} \rightarrow \{n', \varepsilon\}$ is added to the graph with the number of steps from $\{n, \varepsilon\}$ to $\{n', m\}$ as its weight. If such an edge has been added before, only the weight is updated to be the minimum of the previous weight and the new one. Finally, the mark M of the edge is set appropriately.

The cost of processing each pair $\{n, \varepsilon\}$ during the construction of the graph G is proportional to the size of the subtree rooted at n, because the DFS search is limited to the subtree of n. It follows that the construction of G uses the time proportional to the sum of sizes of the subtrees of T. By Lemma 6 this is $O(\sum |w_i|)$, where w_i are the codewords given by the tree T. The number of vertices in the graph is $|V| = N - 1$. The number of edges is bounded by the sum of sizes of all the subtrees of the tree, that is $|E| = O(\sum |w_i|)$.

The shortest merging string for a set $\{n, \varepsilon\}$ is given by the lightest path from $\{n, \varepsilon\}$ to $\{\varepsilon\}$. The tree of the lightest paths from any node to $\{\varepsilon\}$ can be constructed using the Dijkstra's algorithm in $O(|E| + |V| \log |V|)$. Since $|V| = O(N)$, $|E| = O(\sum |w_i|)$ and $\sum |w_i| \geq N \log N$, the lightest paths' tree can be computed in $O(\sum |w_i|)$.

Theorem 11. *Let $\mathcal{T}$ be a Huffman automaton. The algorithm for computing the shortest merging string for a set $\{n, \varepsilon\}$, where n is any state of $\mathcal{T}$, requires preprocessing time $O(\sum_i |w_i|)$. Then the shortest merging string for each pair $\{n, \varepsilon\}$ can be found in the time proportional to the length of the merging string.*

4   Length of a Synchronizing String

In this section we give an upper bound on the length of the shortest synchronizing string for any synchronizing Huffman code. We begin with a lemma that helps to prove the main theorem of this section (Theorem 13).

Lemma 12. *Let T be a complete binary tree with N leaves. There exists a string w of length at most $\lceil \log N \rceil$ such that for each node n of T some prefix of w labels a path from n to a leaf.*

Theorem 13. *For any synchronizing Huffman code of size N the length of the shortest synchronizing string s is at most*

$$|s| \leq \lceil \log N \rceil + (\lceil \log N \rceil - 1)X = O(Nh \log N) \tag{9}$$

where h is the length of the longest codeword,

$$X = \sum_{t \in \mathscr{T}(T) \setminus \{T\}} h_t, \tag{10}$$

$\mathscr{T}(T)$ is the set all different subtrees of T and h_t is the height of the subtree t.

Proof (sketch). We first find a string w given by Lemma 12 and apply it to coins on all states of $\mathcal{T}$. It reduces the number of coins to at most $\lceil \log N \rceil$, because the final position of each coin is determined by some proper suffix of w. We may assume that one of the coins is on the root of $\mathcal{T}$ (otherwise w could be shortened). Then, we pick a node n with a coin and we construct the shortest string s_n that merges n and the root (to get shorter synchronizing strings it is better to pick a node n with the shortest merging string for $\{n, \varepsilon\}$ among the nodes with a coin). By Corollary 8, $|s_n| \leq X$. This string applied the current set reduces the number of coins by at least one. Repeating this procedure additional $(\lceil \log N \rceil - 2)$ times leaves us with just one coin. The length of the merging string does not exceed $\lceil \log N \rceil + (\lceil \log N \rceil - 1)X$. Finally, $X < Nh$ gives the asymptotic bound $O(Nh \log N)$. □

The proof of Theorem 13 is constructive and gives an algorithm for the construction of a synchronizing string for a Huffman code. The algorithm works as follows.

First a string w from Lemma 12 is found. It is done by checking all the $O(N)$ strings of length less or equal $\lceil \log N \rceil$ in the following way. From each node n of T we traverse the subtree of n with DFS. Each time we are in a node m that is $l \leq \lceil \log N \rceil$ steps below n, we mark the string w on the path from n to m as *bad*. This means that no prefix of w brings n to a leaf.

After the traversal, the strings that have not been marked as bad bring any node through a leaf. By Lemma 12, there is at least one such a string of length $\lceil \log N \rceil$ or less. The cost of this algorithm is proportional to the sum of sizes of all subtrees of T, which is $O(\sum_i |w_i|)$.

After finding the string w we may apply it to the set of all internal nodes of T. This will take $O(N \log N)$ time. Then, at most $\log N$ merging strings for $\{n, \varepsilon\}$,

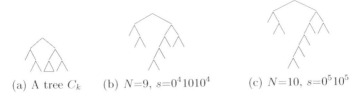

(a) A tree C_k (b) $N=9$, $s=0^4 1010^4$ (c) $N=10$, $s=0^5 10^5$

Fig. 2. The class of trees with the longest synchronizing string for a given number of nodes n. s denotes the synchronizing string for each tree. The triangle denotes a code $\{1, 01, 001, \ldots, 0^i 1, 0^i 0\}$, $i \geq 0$.

with some n, suffice to build a synchronizing string. Computing the merging strings require preprocessing time $O(\sum_i |w_i|)$ and then any string can be read in the time proportional to its length. The length of each merging string is bounded by X and there are at most $\log N$ vertices that have to be moved using each such string. Thus the total cost of the algorithm is $O(X \log^2 N + \sum_i |w_i|)$.

5   Experimental Tests

Tests were performed to find the worst-case trees for the length of the shortest synchronizing string and the worst-case trees for the length of the shortest merging strings for a pair $\{n, \varepsilon\}$, where n is an internal node of the tree. All trees of sizes, N, from 3 to 20 were analyzed first. Then the procedure was repeated for all trees of heights, h, from 2 to 5.

5.1   Long Synchronizing String

In most of the tested cases the worst-case trees for fixed size, N, were unique up to the reflection across the y axis (relabeling 0-edges to 1-edges and 1-edges to 0-edges). The exceptions were the trees with 7 nodes — three nonequivalent trees, 10 nodes — 5 trees, and 12 nodes — 2 trees. For trees with 9, 11 and 13-20 nodes the unique worst-case tree corresponds to one of the codes C_k, given below. The codes C_k also form one of the worst-case trees with 7, 10 and 12 nodes.

$$C_k = \{00, 010, 011, 110, 111\} \cup \{10^i 1 | i = 1, 2, \ldots, k-1\} \cup \{10^k\}, \quad k \geq 1. \quad (11)$$

The size of the code C_k is $k+5$. The structure of these trees is shown in Fig. 2(a) and examples can be found in Figs. 2(b) and 2(c).

Theorem 14. *The shortest synchronizing string for the tree C_k, $k \geq 1$, is $s_0 = 0^k 10^k$ for odd k (even number of codewords) and $s_1 = 0^k 1010^k$ or $s_2 = 0^k 1110^k$ for even k (odd number of codewords). The length of the shortest synchronizing string is $2N - 9$ for even code size, N, and $2N - 7$ for odd code size.*

The worst-case trees for fixed height h, with $h = 2, 3, 4$ and 5, are the trees given by the set of codewords

$$D_h = \big(\{0, 1\}^h \setminus \{1^{h-1}1, 1^{h-1}0\} \big) \cup \{1^{h-1}\}, \quad h \geq 2. \quad (12)$$

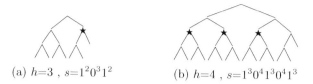

(a) $h{=}3$, $s{=}1^20^31^2$ (b) $h{=}4$, $s{=}1^30^41^30^41^3$

Fig. 3. Trees with the worst-case length of a synchronizing and merging string among trees of fixed height h, for $h = 2, 3, 4, 5$ and a scheme of these trees. The nodes n with the longest merging string for $\{n, \varepsilon\}$ are marked with a star.

These are full binary trees with two edges in the lower-right corner removed. The number of codewords in the code D_h is $2^h - 1$. They are unique worst-case trees up to the reflection across the y axis. The trees D_3 and D_4 are shown in Fig. 3(a) and Fig. 3(b).

Theorem 15. *The shortest synchronizing string for the tree D_h, $h \geq 2$, is $s = (1^{h-1}0^h)^{h-2}1^{h-1}$ with $|s| = 2h^2 - 4h + 1$ (however, the shortest synchronizing string is not unique).*

The minimized Huffman automaton for D_h has $K = 2(h-1)$ nodes. Even though it contains a letter that reduces the number of coins by $h - 2 = \frac{K}{2} - 1$ (a letter of deficiency $\frac{K}{2} - 1$), its shortest synchronizing string is of length $2h^2 - 4h + 1 = \frac{K^2}{2} - 1$, which is quadratic in K. This makes the automata D_h interesting in themselves.

The results of the search allow us to state the following conjecture.

Conjecture 16. The length of the shortest synchronizing string s for a code of N codewords, $N \geq 9$, with h being the length of the longest codeword, is at most:

$$|s| \leq \min(2N - a, 2h^2 - 4h + 1), \tag{13}$$

where a is 7 for odd N and 9 for even N.

5.2   Long Merging String

For trees of fixed size N the length of the shortest merging string in the worst case is equal $N - 2$, for $N = 3, \ldots, 20$, apart from $N = 6$. For $N = 6$ the worst-case length is equal $N - 1 = 5$. Two families of trees have the worst-case shortest merging strings. The first one corresponds to the code

$$G_k = \{0, 10^k\} \cup \{10^i1 | i < k\}, \quad k \geq 1, \tag{14}$$

and gives the worst-case trees for N from 3 to 20, apart from $N = 6$. The size of the tree G_k is $N = k + 2$. The merging string of the set $\{1, \varepsilon\}$ is of length $N - 2$. The structure of these trees is shown in Fig. 4(a) and the tree G_4 is shown in Fig. 4(b). The latter figure also shows the node whose merging string with ε is the longest.

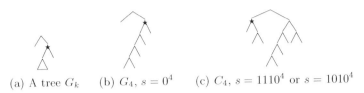

(a) A tree G_k (b) G_4, $s = 0^4$ (c) C_4, $s = 1110^4$ or $s = 1010^4$

Fig. 4. The nodes with the longest merging string for the two families C_k and G_k

The other family of trees is the family C_k (see (11) and Fig. 2(a)) with even k (odd number of codewords). The merging string for $\{0, \varepsilon\}$ is of length $N - 2$ and this is the worst case for $N = 7, 9, 11, \ldots 19$. The node with the longest merging string and the merging string itself for the tree C_4 are shown in Fig. 4(c).

There were also additional worst-case trees found for $N = 5, 6, 7, 9, 10, 12$. These do not correspond to neither the trees C_k nor G_k.

The worst-case trees among trees of fixed height are the trees D_h (Equation (12) and Fig. 3).

Theorem 17. *The upper bound on the length of the shortest merging string for any pair $\{n, \varepsilon\}$, where n is a state of D_h, is $\lceil h^2 - \frac{3}{2}h \rceil$. For odd h it is achieved by the pair $\{0^{(h-1)/2}, \varepsilon\}$. For even h it is achieved by pairs $\{x, \varepsilon\}$, where x is any binary string of length $\frac{h}{2}$ containing at least one 0.*

The results of the search allow us to state the following conjecture.

Conjecture 18. For any Huffman automaton $\mathcal{T}$ corresponding to a code with N codewords, with h being the length of the longest codeword, the length of the shortest merging string s_n for a set $\{n, \varepsilon\}$, where n is any state of $\mathcal{T}$ is at most:

$$|s| \leq \min(2N - 2, \lceil h^2 - \frac{3}{2}h \rceil), \tag{15}$$

if $N \neq 6$, and $|s| \leq 5$ for $N = 6$.

6   Summary

We presented a constructive upper bound on the length of the shortest merging string and the shortest synchronizing string for a Huffman code.

We tested the lengths of the shortest merging and synchronizing string on all codes of size from 3 to 20 and on all codes with the length of the longest codeword from 2 to 5. Three classes of worst-case codes were found. The length of the shortest synchronizing strings for these classes of codes is far from the bound proven before. This allowed us to formulate conjectures, which remain open.

Acknowledgement

The author wishes to thank dr Wojciech Plandowski of Institute of Informatics, University of Warsaw, Poland for useful discussions. Dr Plandowski improved the bounds in Theorems 5 and 13, originally found by the author, to the ones presented here.

References

1. Capocelli, R.M., Gargano, L., Vaccaro, U.: On the characterization of statistically synchronizable variable-length codes. IEEE Trans. Inform. Theory 34(4), 817–825 (1988)
2. Freiling, C.F., Jungreis, D.S., Theberge, F., Zeger, K.: Almost all complete binary prefix codes have a self-synchronizing string. IEEE Trans. Inform. Theory 49(9), 2219–2225 (2003)
3. Schützenberger, M.P.: On synchronizing prefix codes. Information and Control 11(4), 396–401 (1967)
4. Rudner, B.: Construction of minimum-redundancy codes with an optimum synchronizing property. IEEE Trans. Inform. Theory 17(4), 478–487 (1971)
5. Perkins, S., Escott, A.E.: Synchronizing codewords of q-ary Huffman codes. Discrete Math. 197-198, 637–655 (1999)
6. Huang, Y.M., Wu, S.C.: Shortest synchronizing codewords of a binary Huffman equivalent code. In: ITCC 2003: Proceedings of the International Conference on Information Technology: Computers and Communications, Washington, DC, USA, p. 226. IEEE Computer Society Press, Los Alamitos (2003)
7. Capocelli, R.M., Santis, A.D., Gargano, L., Vaccaro, U.: On the construction of statistically synchronizable codes. IEEE Trans. Inform. Theory 38(2), 407–414 (1992)
8. Ferguson, T.J., Rabinowitz, J.H.: Self-synchronizing Huffman codes. IEEE Trans. Inform. Theory 30(4), 687–693 (1984)
9. Maxted, J.C., Robinson, J.P.: Error recovery for variable length codes. IEEE Trans. Inform. Theory 31(6), 794–801 (1985)
10. Černý, J.: Poznámka k. homogénnym experimentom s konecnými automatmi. Mat. fyz.cas SAV 14, 208–215 (1964)
11. Ananichev, D.S., Volkov, M.V.: Synchronizing generalized monotonic automata. Theor. Comput. Sci. 330(1), 3–13 (2005)
12. Kari, J.: Synchronizing finite automata on eulerian digraphs. Theor. Comput. Sci. 295(1-3), 223–232 (2003)
13. Pin, J.E.: On two combinatorial problems arising from automata theory. Annals of Discrete Mathematics 17, 535–548 (1983)
14. Ananichev, D.S., Volkov, M.V., Zaks, Y.I.: Synchronizing automata with a letter of deficiency 2. Theor. Comput. Sci. 376(1-2), 30–41 (2007)
15. Trahtman, A.N.: An efficient algorithm finds noticeable trends and examples concerning the Černý conjecture. In: Kralovic, R., Urzyczyn, P. (eds.) MFCS 2006. LNCS, vol. 4162, pp. 789–800. Springer, Heidelberg (2006)
16. Eppstein, D.: Reset sequences for monotonic automata. SIAM J. Comput. 19(3), 500–510 (1990)
17. Sandberg, S.: Homing and Synchronizing Sequences. In: Broy, M., Jonsson, B., Katoen, J.P., Leucker, M., Pretschner, A. (eds.) Model-Based Testing of Reactive Systems. LNCS, vol. 3472, pp. 5–33. Springer, Heidelberg (2004)

Optimizing Conjunctive Queries over Trees Using Schema Information[*]

Henrik Björklund, Wim Martens[**], and Thomas Schwentick

Technical University of Dortmund

Abstract. We study the containment, satisfiability, and validity problems for conjunctive queries over trees with respect to a schema. We show that conjunctive query containment and validity are 2EXPTIME-complete w.r.t. a schema (DTD or Relax NG). Furthermore, we show that satisfiability for conjunctive queries w.r.t. a schema can be decided in NP. The problem is NP-hard already for queries using only one kind of axis. Finally, we consider conjunctive queries that can test for equalities and inequalities of data values. Here, satisfiability and validity are decidable, but containment is undecidable, even without schema information. On the other hand, containment w.r.t. a schema becomes decidable again if the "larger" query is not allowed to use both equalities and inequalities.

1 Introduction

In the context of relational databases, select-project-join queries are the ones most commonly used in practice. These queries are also known in database theory as *conjunctive queries*. The *containment problem* for conjunctive queries P and Q asks whether Q returns (at least) all answers of P. Ever since the seminal paper of Chandra and Merlin [5], conjunctive query containment has been a pivotal research topic; it is the most intensely researched form of query optimization in database theory. Moreover, the conjunctive query containment problem is essentially the same as the conjunctive query evaluation problem [5], and the Constraint Satisfaction Problem (CSP) in Artificial Intelligence [13].

The more recent rise of semi-structured data and XML initiated the investigation of conjunctive queries over trees [11]. As in the relational case, conjunctive queries over trees provide a very clean and natural querying formalism. XPath and (non-recursive) XQuery queries can both be naturally translated into conjunctive queries. However, as pointed out by Gottlob et al. [11], their applications are not at all limited to XML; they are also used for Web information extraction, as queries in computational linguistics, dominance constraints, and in higher-order unification.

As a matter of fact, containment for queries on tree-structured data was previously mainly studied for fragments of XPath 1.0. The investigations therefore concentrated on *acyclic* conjunctive queries (see, e.g., [16,17]).

---

[*] This work was supported by the DFG Grant SCHW678/3-1.
[**] Supported by a grant from the Nordrhein Westfälische Akademie der Wissenschaften.

In contrast to the relational setting, for conjunctive queries over trees, evaluation is not the same problem as containment. In relational databases, containment $P \subseteq Q$ holds if an only if there is a homomorphism from the canonical database of Q to the canonical database of P. Over trees, the existence of such a homomorphism is a sufficient, but not a necessary condition for containment [2].

Conjunctive query containment over trees is therefore investigated directly in [2], but was also treated more implicitly in the form of XPath 2.0 static analysis in, e.g., [12,14,19]. We elaborate on the relation with these papers below. The results in [2] were encouraging, as the complexities (compared with acyclic queries) did not increase too much: they remained inside Π_2^P.

The present paper extends our previous work [2] in the sense that we now take schema information into account, and that we consider queries that can test for equality and inequality of data values. In this framework, we study the complexities of the validity, satisfiability, and containment problems. Whereas our previous work outlined a quite complete picture of conjunctive query containment without schemas, one has to admit that, in practice, schema information is highly relevant. In XML, schema information is available for most documents, and the chances of being able to optimize queries are much better when it is taken into account. On the other hand, as we will see in this paper, there is also a tradeoff: the complexity of conjunctive query containment over trees is much higher with schema information than without.

Our work can be summarized as follows. First, we study conjunctive queries that cannot compare data values. Our main technical result here is that the practically most relevant problem, conjunctive query containment w.r.t. a DTD, is already 2EXPTIME-hard for queries using only the *Child* and *Child*$^+$ axes.[1] This result is quite surprising when one compares it to the known results for XPath 1.0 containment. For XPath 1.0, adding DTD information to the problem usually "only" increases the complexity from coNP [16] to (at most) EXPTIME [17,15]. Here, however, the complexity immediately jumps from Π_2^P to 2EXPTIME when DTDs are taken into consideration. In particular, the problem can provably not be solved in polynomial space in general. On the other hand, it remains in 2EXPTIME even when conjunctive queries can use all axes and the much more expressive Relax NG schemas are considered. In contrast, the satisfiability problem for even the most general conjunctive queries w.r.t. Relax NG schemas is in NP. Unfortunately the satisfiability problem is also already NP-hard for very simple cases using only DTD information.

Finally, we turn to the containment problem for queries that can compare data values for equality ($\sim$) and inequality ($\not\sim$). When data values are involved, static analysis problems are generally known to become undecidable very quickly. We show that conjunctive query containment is no exception: already without schema information, it is undecidable. However, the good news is that even very slight restrictions of this most general case become decidable, even without increasing the complexity over the setting without data values.

---

[1] Actually, we show hardness already for the validity problem.

Boolean versus n-ary queries. The conjunctive queries in our paper are *boolean* queries, i.e., they evaluate either to *true* or *false* on a tree. Our complexity results also carry over to containment for conjunctive queries that return an *n*-ary relation when evaluated on a tree.

Related work. We discuss the relation of our paper to some of the above mentioned work. Most relevant to us are the papers by ten Cate and Lutz [19], by David [8] (which evolved independently from ours), and by Lakshmanan et al. [14]. The connection with Hidders' work [12] is explained more elaborately in [2]. Hidders considers XPath 2.0 satisfiability, but does not take schema information into account. Ten Cate and Lutz study query containment for expressive fragments of XPath 2.0, which is closely related to our conjunctive queries. They also take schema information into account (at least for DTDs and XML Schema Definitions) and get 2EXPTIME-completeness, but their queries have negation, disjunction, and union while conjunctive queries do not.

The precise relation between our conjunctive queries and XPath 2.0 is not entirely obvious. Conjunctive queries are at least as expressive as the XPath 2.0 fragment that consists of Core XPath 1.0 without union, disjunction or negation, but augmented with the XPath 2.0 path intersection operator (see [19]). This implies that our upper bound proofs also apply to this XPath 2.0 fragment. On the other hand, such XPath expressions are syntactically constrained and cannot use path intersection arbitrarily. Our lower bound proofs can, however, also be adapted to these XPath 2.0 expressions. In this light, our results significantly strengthen the lower bound proof of Theorem 20 in [19] when DTD information is considered, since we do not make use of negation or disjunction.[2]

David studies the complexity of satisfiability for Boolean combinations of *data tree patterns* with respect to DTDs [8]. Different fragments are investigated, and the complexity results range from NP to undecidable. This formalism is on the surface quite similar to CQs with data value predicates, but there are some decisive differences. First, the data tree patterns are always tree-shaped, like XPath queries without path intersection. Second, the semantics used in [8] is injective, i.e., two variables cannot be assigned the same node, unlike the one for CQs. This means that boolean combinations of data tree patterns are in general more expressive but exponentially less succinct than CQs.

Lakshmanan et al. study satisfiability, with and without schema information, of tree pattern queries, where the tree patterns are also equipped with a node identity operator and can compare data values. In particular, they claim (Theorem 3.2 in [14]) that query satisfiability for queries with structural constraints, Value Based Constraints (VBCs) and no wildcards is in PTIME. However, it is NP-complete.[3] The results of the paper do not really overlap with our results on satisfiability, since they only consider a limited, non-recursive, form of DTDs.

---

[2] Without DTD information, ten Cate and Lutz still have 2EXPTIME-completeness due to the presence of negation, but conjunctive query containment is Π_2^P-complete.

[3] Here, structural constraints include node identities and VBCs allow comparison of data values to constants. One of our NP-hardness proofs can be easily adapted to this case. However, we do not conclude PTIME = NP.

Furthermore, there is a large amount of work on static analysis for XPath 1.0 (see, e.g., [1,10,15,16,17,20]). XPath 1.0 relates to our conjunctive queries in a similar way as XPath 2.0, except that XPath 1.0 does not have a path intersection operator. In other words, complexity lower bounds for XPath 1.0 sometimes carry over to conjunctive queries. We indicate this in the paper whenever relevant.

Due to space constraints, most proofs have been omitted and will appear in the full version of the paper.

2   Preliminaries

2.1   Trees

By Σ we always denote a finite alphabet. The trees we consider are rooted, ordered, finite, labeled, unranked trees, which are directed from the root downwards. That is, we consider finite trees in which nodes can have arbitrarily many children, which are ordered from left to right. We view a tree t as a relational structure over a finite number of unary labeling relations $a(\cdot)$, for $a \in \Sigma$, and binary relations $Child(\cdot, \cdot)$ and $NextSibling(\cdot, \cdot)$. Here, $a(u)$ expresses that u is a node with label a, and $Child(u, v)$ (respectively, $NextSibling(u, v)$) expresses that v is a child (respectively, the right sibling) of u.

The reason that we can restrict ourselves to a finite set of labels is that an XML schema defines the set of labels allowed in a tree. In the rare cases where we consider trees without schema information, we also consider the set of possible labels to be infinite.

In addition to $Child$ and $NextSibling$, we use their transitive closures (denoted $Child^+$ and $NextSibling^+$) and their transitive and reflexive closures (denoted $Child^*$ and $NextSibling^*$). We also use the $Following$-relation, which is inspired by XPath [6] and defined as

$$Following(u, v) = \exists x \exists y\, Child^*(x, u) \wedge NextSibling^+(x, y) \wedge Child^*(y, v).$$

We refer to the binary relations above as *axes*. We denote the set of nodes of a tree t by $Nodes(t)$. For a node u, we denote by $lab^t(u)$ the unique a such that $a(u)$ holds in t. We often omit t from this notation when t is clear from the context. By $root(t)$ we denote the root node of t.

2.2   Conjunctive Queries

Let $X = \{x, y, z, \dots\}$ be a set of variables. A *conjunctive query* (CQ) over alphabet Σ is a positive existential first-order formula without disjunction over a finite set of unary predicates $a(x)$ where each $a \in \Sigma$, and the binary predicates $Child, Child^+, Child^*, NextSibling, NextSibling^+, NextSibling^*$, and $Following$. In this paper, we will mainly focus on Boolean satisfaction of conjunctive queries. We will therefore consider conjunctive queries without free variables, and we also consider the constants *true* and *false* to be CQs. As our conjunctive queries do not contain free variables, we sometimes omit the existential quantifiers to simplify notation. For a conjunctive query Q, we denote the set of variables appearing in Q by $\mathrm{Var}(Q)$. We use $\mathrm{CQ}(R_1, \ldots, R_k)$ or $\mathrm{CQ}(\mathcal{R})$ (where $\mathcal{R} = \{R_1, \ldots, R_k\}$) to denote the fragment of CQs that uses only the unary alphabet predicates and the binary predicates $R_1, \ldots, R_k$. We use the terminology on valuations of a query from Gottlob et al. [11]. That is, let Q be a CQ, and t a tree. A *valuation* of Q on t is a total function $\theta : \mathrm{Var}(Q) \to \mathrm{Nodes}(t)$. A valuation is a *satisfaction* if it satisfies the query, that is, if every atom of Q is satisfied by the assignment. A tree t *models* Q ($t \models Q$) if there is a satisfaction of Q on t. The language $L(Q)$ of Q is the set of all trees that model Q.[4] We denote the complement of $L(Q)$ by $\overline{L(Q)}$.

We sometimes refer to a query as confluent. Intuitively, this means that the atoms of the query, interpreted as directed edges, merge at some point, i.e., the graph they form is not a directed forest. More formally, query Q is confluent if there are three distinct variables $x, y, z \in \mathrm{Var}(Q)$ and binary predicates R_1 and R_2 such that $R_1(x, z)$ and $R_2(y, z)$ are both atoms of Q.

2.3   Schemas

We abstract from Document Type Definitions (DTDs) as follows:

Definition 1. A *Document Type Definition (DTD)* over Σ is a triple $D = (\mathrm{Alpha}(D), \mathrm{Rules}(D), \mathrm{start}(D))$ where $\mathrm{Alpha}(D) = \Sigma$, $\mathrm{start}(D) \in \Sigma$ is the start symbol and $\mathrm{Rules}(D)$ is a set of rules of the form $a \to R$, where $a \in \Sigma$ and R is a regular expression over Σ. Here, no two rules have the same left-hand-side.

A tree t *satisfies* D if *(i)* $\mathrm{lab}^t(\mathrm{root}(t)) = \mathrm{start}(D)$ and, *(ii)* for every $u \in \mathrm{Nodes}(t)$ with label a and n children $u_1, \ldots, u_n$ from left to right, there is a rule $a \to R$ in $\mathrm{Rules}(D)$ such that $\mathrm{lab}^t(u_1) \cdots \mathrm{lab}^t(u_n) \in L(R)$. By $L(D)$ we denote the set of trees satisfying D.

We abstract from Relax NG schemas [7] by unranked tree automata, which are formally defined as follows:

Definition 2. A *nondeterministic (unranked) tree automaton (NTA)* over Σ is a quadruple $A = (\mathrm{States}(A), \mathrm{Alpha}(A), \mathrm{Rules}(A), \mathrm{Final}(A))$, where $\mathrm{Alpha}(A) = \Sigma$, $\mathrm{States}(A)$ is a finite set of states, $\mathrm{Final}(A) \subseteq \mathrm{States}(A)$ is the set of final states, and $\mathrm{Rules}(A)$ is a set of transition rules of the form $(q, a) \to L$, where $q \in \mathrm{States}(A)$, $a \in \mathrm{Alpha}(A)$, and L is a regular string language over $\mathrm{States}(A)$.

For simplicity, we denote the regular languages L in A's rules by regular expressions. For our complexity results, it doesn't matter whether the languages L are represented by regular expressions or nondeterministic string automata.

A *run* of A on a tree t is a labeling $r : \mathrm{Nodes}(t) \to \mathrm{States}(A)$ such that, for every $u \in \mathrm{Nodes}(t)$ with label a and children $u_1, \ldots, u_n$ from left to right, there

---

[4] Notice that, as stated in the introduction, we assume that trees only take labels from a finite alphabet Σ. Hence, for a conjunctive query Q, $L(Q)$ also consists of trees over alphabet Σ. In the rare cases where we consider trees without schema information, we state this explicitly.

exists a rule $(q, a) \to L$ such that $r(u) = q$ and $r(u_1) \cdots r(u_n) \in L$. Note that, when u has no children, the criterion reduces to $\varepsilon \in L$, where ε denotes the empty string. A run is *accepting* if the root is labeled with an accepting state, that is, $r(\text{root}(t)) \in \text{Final}(A)$. A tree t is accepted if there is an accepting run of A on t. The set of all accepted trees is denoted by $L(A)$ and is called a *regular tree language*. We denote the complement of $L(A)$ by $\overline{L(A)}$. In the remainder of the paper, we sometimes view the run r of an NTA on t as a tree over $\text{States}(A)$, obtained from t by relabeling each node u with the state $r(u)$.

From now on, we use the word "schema" to refer to DTDs or NTAs.

2.4   Our Problems of Interest

Definition 3.   – Containment w.r.t. a schema: Given two CQs P and Q, and a schema S, is $L(P) \cap L(S) \subseteq L(Q)$?
  – Validity w.r.t. a schema: Given a CQ Q and a schema S, is $L(S) \subseteq L(Q)$?
  – Satisfiability w.r.t. a schema: Given CQ Q and schema S, is $L(Q) \cap L(S) \neq \emptyset$?

All of the above problems are in a sense instances of the containment problem. That is, validity of Q is testing whether $L(\text{true}) \subseteq L(Q)$ w.r.t. S, and satisfiability for Q is testing whether $L(Q) \not\subseteq L(\text{false})$ w.r.t. S.

3   Validity and Containment

3.1   Complexity Upper Bounds

We start the technical part of the paper by settling the upper bound for the containment problem. This is achieved through a standard translation of CQs into NTAs.

Lemma 4. *Let Q be a CQ. There exists an NTA A such that $L(A) = L(Q)$ and A can be computed from Q in exponential time.*

It is now easy to derive the following theorem. We note that Theorem 5 is not new. The 2EXPTIME upper bound is obtained by composing the exponential translation of [11] from CQs to Core XPath and the polynomial time translation of [19] from Core XPath expressions to two-way alternating tree automata. The result now follows as emptiness testing of two-way alternating tree automata is in EXPTIME.

Theorem 5. *Containment of CQs w.r.t. an NTA is in 2EXPTIME.*

3.2   Complexity Lower Bounds

In this section, we prove the following result.

Theorem 6. *Validity   of   CQ(Child, Child$^+$)   w.r.t.   a   tree   automaton   is 2EXPTIME-complete.*

The proof of the above theorem is long and rather technical. We sketch it's most interesting parts below.

The upper bound in Theorem 6 follows from Theorem 5. We show the corresponding lower bound by reduction from the word problem for alternating exponential space bounded Turing machines, which is 2EXPTIME-hard [4].

An *alternating Turing machine (ATM)* [4] is a tuple $M = (Q, \Sigma, \Gamma, \delta, q_0)$ where $Q = Q_\forall \uplus Q_\exists \uplus \{q_a\} \uplus \{q_r\}$ is a finite set of states partitioned into *universal states* from $Q_\forall$, *existential states* from $Q_\exists$, an *accepting state* q_a, and a *rejecting state* q_r. The (finite) input and tape alphabets are Σ and Γ, respectively. We assume that the tape alphabet contains a special *blank* symbol "␣". The *initial state* of M is $q_0 \in Q$. The transition relation δ is a subset of $(Q \times \Gamma) \times (Q \times \Gamma \times \{L, R, S\})$. The letters L, R, and S denote the directions *left*, *right*, and *stay* in which the tape head is moved.

A *computation tree* for an ATM M is a tree labelled by configurations (tape content, reading head position, and internal state) of M such that (1) if node v is labelled by an existential configuration, then v has one child, labelled by one of the possible successor configurations, (2) if v is labelled by a universal configuration, then v has one child for each possible successor configuration, (3) the root is labelled by an initial configuration, and (4) all leaves are labelled by accepting or rejecting configurations. A computation tree is *accepting* if it is finite and all leaves are labelled by accepting configurations.

The overall idea of our proof is as follows. Given ATM M and a word w of length n we construct, in polynomial time, (1) an ATM M_w which accepts the empty word if and only if M accepts w and (2) an NTA A_{CT} that checks most important properties of (suitably encoded) computation trees of M_w, except their consistency w.r.t. the transition relation of M_w. The consistency is tested by the query Q_{CT} that we define. To be precise, Q_{CT} is satisfied by a tree t in $L(A_{CT})$ if and only if the transition relation of M_w is *not* respected by t. This means that Q_{CT} is valid w.r.t. A_{CT}, iff there does not exist a consistent, accepting computation tree for M_w. Since 2EXPTIME is closed under complementation, we conclude that validity of CQs with respect to NTAs is 2EXPTIME-hard.

Encoding Computation Trees. The encoding $enc(t)$ of a possible computation tree t of M_w is illustrated in Fig. 1(a) and obtained from t by replacing each node u of t with a tree t_u, where

- root(t_u) is labeled CT;
- the leftmost child of root(t_u) is labeled r (and is the root of the tree encoding the actual configuration at u); and
- for each child u_i of u, root(t_u) has a subtree $enc(t/u_i)$ where t/u_i denotes the subtree of t rooted at u_i.

Encoding configurations. The most crucial part of the reduction is to use the query to detect when the transition relation of M_w is violated. To be able to do this, the query must be able to navigate from a node representing tape cell i in one configuration tree to the node representing cell i in a successor configuration. To this end, we encode configurations as follows.

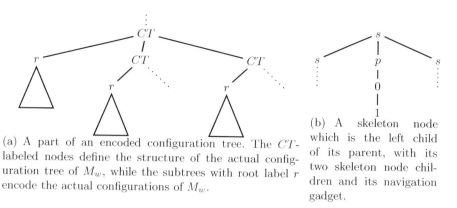

(a) A part of an encoded configuration tree. The CT-labeled nodes define the structure of the actual configuration tree of M_w, while the subtrees with root label r encode the actual configurations of M_w.

(b) A skeleton node which is the left child of its parent, with its two skeleton node children and its navigation gadget.

Fig. 1. The encoded computation tree

As we can assume w.l.o.g. that M_w never uses more than 2^n tape cells, we can encode configurations into the leaves of binary trees of height n, where each leaf represents a tape cell. A *configuration tree* is obtained from a full binary tree b of height n as follows. The root gets label r and the other nodes label s. The s-labeled nodes are called *skeleton nodes*. To each skeleton node v we attach a little gadget indicating whether v is a left or a right child in b. More precisely, we attach a path of length 3 labeled with $p, 0, 1$, respectively, to left children and a path labeled with $p, 1, 0$ to right children. Each leaf skeleton node (one that has no skeleton node children) is further provided with the relevant information about the tape cell it represents.

Thus, left and right children can be distinguished by the distance (1 or 2) of their 1-labelled gadget node from their p-labelled gadget node. More precisely, a skeleton node v at level i of a configuration tree and a skeleton node u at level i of a successor configuration tree are both left or both right children, if the nodes v^1 and u^1 with label 1 in their respective gadgets have a common ancestor which has distance $i + 4$ from v^1 and $i + 5$ from u^1.

Comparing configurations. Below, we construct a query $\text{SameCell}(t_1, t_t)$ which is true for two leaf skeleton nodes if and only if they belong to successive configuration trees and represent the same tape cell. In order to do this, we first define successively more complicated subqueries. The first one states that two nodes r_1 and r_2 are roots of two *successive* configuration trees, i.e., configuration trees such that the second encodes a successor configuration of the first.

$$\text{Succ}(r_1, r_2) \equiv \exists s_1, s_2 : r(r_1) \land r(r_2) \land CT(s_1) \land CT(s_2)$$
$$\land\ Child(s_1, r_1) \land Child(s_2, r_2) \land Child(s_1, s_2)$$

Next, we define a query $\Phi_i(x, y)$ to state that x and y belong to successive encoded configuration trees and are both at level $i > 0$ of their respective encoded configuration tree. Here, $Child^i(x, y)$ abbreviates the query stating that y can be reached from x by following the *Child*-axis i times.

$$\Phi_i(x,y) \equiv \exists r_1, r_2 : s(x) \wedge s(y) \wedge \mathrm{Succ}(r_1, r_2) \wedge \mathit{Child}^i(r_1, x) \wedge \mathit{Child}^i(r_2, y)$$

Now we can express that x and y fulfil Φ_i and, additionally, that they are either both left children of their parents, or both right children.

$$\begin{aligned}
\Psi_i(x,y) \equiv \exists p_x, p_y, t_x, t_y, z : {} & \Phi_i(x,y) \wedge p(p_x) \wedge p(p_y) \wedge 1(t_x) \wedge 1(t_y) \\
& \wedge \mathit{Child}(x, p_x) \wedge \mathit{Child}(y, p_y) \wedge \mathit{Child}^+(p_x, t_x) \wedge \mathit{Child}^+(p_y, t_y) \\
& \wedge \mathit{Child}^{i+4}(z, t_x) \wedge \mathit{Child}^{i+5}(z, t_y)
\end{aligned}$$

Using the above queries, we can now express that s_1 and s_2 are leaf skeleton nodes in successive configuration trees representing the same tape cell. Recall that n is the depth of the encoded configuration trees.

$$\begin{aligned}
\mathrm{SameCell}(s_1, s_2) \equiv {} & \\
\exists x_1, \ldots, x_{n-1}, y_1, \ldots, y_{n-1} : {} & \bigwedge_{1 \leq i < n-1} (\mathit{Child}(x_i, x_{i+1}) \wedge \mathit{Child}(y_i, y_{i+1})) \\
\wedge\ \mathit{Child}(x_{n-1}, s_1) \wedge \mathit{Child}(y_{n-1}, s_2) & \wedge \Psi_n(s_1, s_2) \wedge \bigwedge_{1 \leq i \leq n-1} \Psi_i(x_i, y_i)
\end{aligned}$$

DTDs. Actually, the 2EXPTIME lower bound from Theorem 6 can even be strengthened to the case where the schema is just a DTD instead of a tree automaton.

4   Satisfiability

4.1   Complexity Upper Bounds

In this section, we show that testing satisfiability for CQs with respect to a nondeterministic tree automaton is in NP. The idea is a kind of small model property for such queries. We start with the following lemma. The proof is by a standard pumping argument.

Lemma 7. *There is a polynomial p such that if a CQ Q is satisfiable with respect to an NTA A, then there is a tree $t \in L(Q) \cap L(A)$ and a satisfaction θ of Q on t such that for any variables $x, y \in \mathit{Var}(Q)$, the length of the path from $\theta(x)$ to $\theta(y)$ is at most $p(|A|, |Q|)$.*

Lemma 7 gives us the main machinery to prove the general NP upper bound on satisfiability:

Theorem 8. *Satisfiability of CQs with respect to an NTA is in NP.*

4.2   Complexity Lower Bounds

We show that our upper bound for satisfiability w.r.t. a schema is tight, in quite a strong sense. In particular, when considering a DTD as schema, satisfiability is NP-hard for queries using only a single axis, no matter which axis this is.

For the NP lower bounds, we will reduce from the SHORTEST COMMON SUPERSEQUENCE problem; or the SHORTEST COMMON SUPERSTRING problem, both of which are known to be NP-complete [18,9]. The SHORTEST COMMON SUPERSEQUENCE (respectively, SHORTEST COMMON SUPERSTRING) problem asks, given a set of strings S, and an integer k, whether there exists a string of length at most k which is a supersequence (respectively, superstring) of each string in S. Here, s is a supersequence of s_0 if s_0 can by obtained by deleting symbols from s, and s is a superstring of s_0 if s_0 can be obtained by deleting a prefix and a suffix of s.

Theorem 9. *Let Axis be an any element of { Child, Child$^+$, Child*, NextSibling, NextSibling$^+$, NextSibling*, Following}. Then Satisfiability of CQ(Axis) w.r.t. a DTD is NP-hard.*

5 Queries with Data Values

A *data tree* is a tree in which each node u, in addition to its label $\mathrm{lab}(u)$, carries a *data value* from a countably infinite data domain Δ (see also [3]).[5] We write $u \sim v$ if two nodes in a data tree have the same data value. Conjunctive queries over data trees can, in addition to the usual predicates, use the binary predicates $\sim$ and $\not\sim$ with the obvious interpretation. We adopt our notation to denote CQ fragments for data values as follows: CQ($\sim$), CQ($\not\sim$), and CQ($\sim, \not\sim$) denote the CQs that use only data equality, only data inequality, and both, respectively, and in which all axes are allowed. For $Q \in \mathrm{CQ}(\sim, \not\sim)$, $L(Q)$ is the set of all *data trees* t such that there exists a satisfaction of Q on t. Schemas do not constrain data values in any way, i.e., the set of data trees $L(A)$ defined by an NTA A is defined precisely as in Section 2.3, but with "tree" replaced by "data tree".

Our problems of interest for queries with data values are the same problems as defined in Section 2.4, but with the new definition of $L(Q)$. We first show that data values do not change the complexity of the satisfiability and validity problems.

Theorem 10. *Satisfiability of CQs($\sim, \not\sim$) w.r.t. an NTA is NP-complete.*

The proofs of Theorems 8 and 9 straightforwardly carry over to data trees.

The following result follows from Theorem 12, which is strictly stronger.

Theorem 11. *Validity of CQ($\sim, \not\sim$) w.r.t. an NTA is 2EXPTIME-complete.*

Next, we consider containment w.r.t. a schema. We write QC($X|Y$) for the problem of determining whether $L(P) \cap L(A) \subseteq L(Q)$ for a query $P \in \mathrm{CQ}(X)$, a query $Q \in \mathrm{CQ}(Y)$ and an NTA A. E.g., QC($\sim | \sim, \not\sim$) is about containment of queries with data equalities in queries with data equalities and inequalities.

It turns out that the consideration of data values does not change the complexity of the query containment problem for queries P, Q, unless P is allowed to use data inequalities and Q to use equalities and inequalities. In the latter

---

[5] We assume Δ to contain all the data values we use in our proofs and examples.

Table 1. Decidability for $QC(X|Y)$

$X \setminus Y$	$\sim$	$\not\sim$	$\sim, \not\sim$
$\sim$	2EXPTIME	2EXPTIME	2EXPTIME
$\not\sim$	2EXPTIME	2EXPTIME	undecidable
$\sim, \not\sim$	2EXPTIME	2EXPTIME	undecidable

case the problem is undecidable (Theorem 15). We summarize our results for $QC(X|Y)$ in Table 1.

Theorem 12. *Each of $QC(\sim, \not\sim \mid \sim)$, $QC(\sim, \not\sim \mid \not\sim)$, $QC(\sim \mid \sim, \not\sim)$, w.r.t. an NTA is 2EXPTIME-complete.*

Hence, $\sim$ and $\not\sim$ do no increase the complexity of query containment as long as they do not co-occur in Q. We show next, that the picture changes dramatically if they do co-occur and P uses $\not\sim$.

Theorem 13. *Validity of a disjunction of $CQ(\sim, \not\sim)$ w.r.t. an NTA is undecidable.*

With a little extra work, Theorem 13 can be extended to the following.

Theorem 14. *$QC(\not\sim \mid \sim, \not\sim)$ is undecidable.*

Actually, it turns out that if both queries can use $\sim$ and $\not\sim$, the schema automaton from Theorem 14 can be avoided.

Theorem 15. *$QC(\sim, \not\sim \mid \sim, \not\sim)$ is undecidable, even without a schema.*

6   Conclusion

We studied the query containment and the validity problem for conjunctive queries over trees (1) relative to a schema and (2) taking into account data values. It turned out that in the presence of a schema the complexity of the problem drastically increases. Thus, even though the query language does not have neither negation nor disjunction, it shares the bad complexity (2EXPTIME) of the language in [19].

Not surprisingly, with equalities and inequalities on data values the containment problem even becomes undecidable. Nevertheless, a slight restriction on the occurrence of inequalities yields a decidable problem.

Although conjunctive queries are a very natural query language, future research should identify tractable fragments, in particular with other restrictions than acyclicity. We found it interesting to observe that, from the lower bound proof of Theorem 6, we can conclude that there does *not* exist an exponential-size tree automaton recognizing the complement language of a conjunctive query.

Corollary 16. *In general, there does not exist an exponential-size nondeterministic tree automaton recognizing $\overline{L(Q)}$, where Q is a $CQ(Child, Child^+)$.*

References

1. Benedikt, M., Fan, W., Geerts, F.: XPath satisfiability in the presence of DTDs. J. ACM 55(2) (2007)
2. Bjorklund, H., Martens, W., Schwentick, T.: Conjunctive query containment over trees. In: DBPL, pp. 66–80 (2007)
3. Bojanczyk, M., David, C., Muscholl, A., Schwentick, T., Segoufin, L.: Two-variable logic on data trees and XML reasoning. In: PODS, pp. 10–19 (2006)
4. Chandra, A.K., Kozen, D.C., Stockmeyer, L.J.: Alternation. J. ACM 28(1), 114–133 (1981)
5. Chandra, A.K., Merlin, P.M.: Optimal implementation of conjunctive queries in relational data bases. In: STOC, pp. 77–90 (1977)
6. Clark, J., De Rose, S.: XML Path Language (XPath) version 1.0. Technical report, World Wide Web Consortium (1999), http://www.w3.org/TR/xpath/
7. Clark, J., Murata, M.: Relax NG specification (December 2001), http://www.relaxng.org/spec-20011203.html
8. David, C.: Complexity of data tree patterns over XML documents. In: MFCS (to appear, 2008)
9. Gallant, J., Maier, D., Storer, J.A.: On finding minimal length superstrings. JCSS 20(1), 50–58 (1980)
10. Geerts, F., Fan, W.: Satisfiability of XPath queries with sibling axes. In: DBPL, pp. 122–137 (2005)
11. Gottlob, G., Koch, C., Schulz, K.U.: Conjunctive queries over trees. J. ACM 53(2), 238–272 (2006)
12. Hidders, J.: Satisfiability of XPath expressions. In: DBPL, pp. 21–36 (2003)
13. Kolaitis, P.G., Vardi, M.Y.: Conjunctive query containment and constraint satisfaction. JCSS 61(2), 302–332 (2000)
14. Lakshmanan, L.V.S., Ramesh, G., Wang, H., Zhao, Z.: On testing satisfiability of tree pattern queries. In: VLDB, pp. 120–131 (2004)
15. Marx, M.: XPath with conditional axis relations. In: EDBT, pp. 477–494 (2004)
16. Miklau, G., Suciu, D.: Containment and equivalence for a fragment of XPath. J. ACM 51(1), 2–45 (2004)
17. Neven, F., Schwentick, T.: On the complexity of XPath containment in the presence of disjunction, DTDs, and variables. LMCS 2(3) (2006)
18. Räihä, K.J., Ukkonen, E.: The shortest common supersequence problem over binary alphabet is NP-complete. TCS 16(2), 187–198 (1981)
19. Cate, B.t., Lutz, C.: The complexity of query containment in expressive fragments of XPath 2.0. In: PODS, pp. 73–82 (2007)
20. Wood, P.T.: Containment for XPath fragments under DTD constraints. In: ICDT (2003); Full version, obtained through personal communication

Clustering with Partial Information[*]

Hans L. Bodlaender[1], Michael R. Fellows[2], Pinar Heggernes[3],
Federico Mancini[3], Charis Papadopoulos[3], and Frances Rosamond[2]

[1] Department of Information and Computing Sciences, Utrecht University,
The Netherlands
hansb@cs.uu.nl
[2] PCRU, Office of DVC (Research), University of Newcastle, Australia
{michael.fellows,frances.rosamond}@newcastle.edu.au
[3] Department of Informatics, University of Bergen, N-5020 Bergen, Norway
{pinar,federico,charis}@ii.uib.no

Abstract. The CORRELATION CLUSTERING problem, also known as the
CLUSTER EDITING problem, seeks to edit a given graph by adding and
deleting edges to obtain a collection of vertex-disjoint cliques, such that
the editing cost is minimized. The EDGE CLIQUE PARTITIONING problem
seeks to partition the edges of a given graph into edge-disjoint cliques,
such that the number of cliques is minimized. Both problems are known
to be NP-hard, and they have been previously studied with respect to
approximation and fixed parameter tractability. In this paper we study
these two problems in a more general setting that we term *fuzzy graphs*,
where the input graphs may have missing information, meaning that
whether or not there is an edge between some pairs of vertices of the
input graph can be *undecided*.

For fuzzy graphs the CORRELATION CLUSTERING and EDGE CLIQUE
PARTITIONING problems have previously been studied only with respect
to approximation. Here we give parameterized algorithms based on kernelization for both problems. We prove that the CORRELATION CLUS-
TERING problem is fixed-parameter tractable on fuzzy graphs when parameterized by (k, r), where k is the editing cost and r is the minimum
number of vertices required to cover the undecided edges. In particular
we show that it has a polynomial-time reduction to a problem kernel
on $O(k^2 + r)$ vertices. We provide an analogous result for the EDGE
CLIQUE PARTITIONING problem on fuzzy graphs. Using (k, r) as parameters, where k bounds the size of the partition, and r is the minimum
number of vertices required to cover the undecided edges, we describe a
polynomial-time kernelization to a problem kernel on $O(k^4 \cdot 3^r)$ vertices.
This implies fixed-parameter tractability for this parameterization. Furthermore we also show that parameterizing only by the number of cliques
k, is not enough to obtain fixed-parameter tractability. The problem remains, in fact, NP-hard for each fixed $k > 2$.

[*] This work is supported by The Research Council of Norway.

E. Ochmański and J. Tyszkiewicz (Eds.): MFCS 2008, LNCS 5162, pp. 144–155, 2008.
© Springer-Verlag Berlin Heidelberg 2008

1   Introduction

The CORRELATION CLUSTERING problem for general (ordinary) graphs was introduced and proved NP-hard by Bansal *et al.* [2,3]. Given a complete graph with labels $\langle + \rangle$ or $\langle - \rangle$ on each edge, the problem is to partition the vertices into clusters so that the number of $\langle - \rangle$ edges inside each cluster plus the number of $\langle + \rangle$ edges between the clusters, is minimized. Taking $\langle + \rangle$ edges as *edges* and $\langle - \rangle$ edges as *non-edges*, this problem is equivalent to the CLUSTER EDITING problem, where we are given an ordinary graph graph and asked to add and delete the total minimum number of edges so that the resulting graph is a collection of disconnected (i.e., vertex-disjoint) cliques. The CORRELATION CLUSTERING problem has been proven NP-hard several times, as it has been discovered and rediscovered in various applications areas, such as hierarchical tree clustering [23], computational biology [4,30], and phylogenetic trees [9]. General versions of the CORRELATION CLUSTERING problem have been defined and studied from the point of view of approximation [8,11,12,14]. The second problem that we study in this paper is the EDGE CLIQUE PARTITIONING problem, which asks to partition the edges of a given graph into the minimum number of edge-disjoint cliques. This problem is NP-hard [28] for general graphs, but also for K_4-free and even chordal graphs [24].

In a general way, one can view the problems we consider here, the CORRELATION CLUSTERING (equivalent to CLUSTER EDITING) problem and the EDGE CLIQUE PARTITIONING problem, as belonging to a loose class of problems, having to do with "clique-structuring" of graphs by means of editing or covering operations. For rhetorical convenience, we will refer to this loose class of problems as *GRAPH CLUSTERING PROBLEMS*. This class of problems, in which we would also include VERTEX COVER, CLIQUE COVER, and many others, has proved to be a highly productive source of practical applications for parameterized algorithms [1,10,20].

A key point of what we offer here is to expand the investigation of *GRAPH CLUSTERING PROBLEMS* to inputs consisting of *fuzzy graphs*, where some pairs of vertices of the input may have an *undetermined, unknown,* or *undecided* relation. For many applications, this clearly adds to the realism of the modeling in an important way. To mention one application area where a similar idea has been considered before, in bioinformatics the notion of "sandwich graph problems" has played a useful role [18].

NP-hard problems remain hard also on fuzzy graphs, as they are a generalization of ordinary graphs. Hence we investigate their tractability from a parameterized complexity point of view, and we try to understand which structural parameters are more suitable to attack problems on fuzzy graphs.

A problem is fixed parameter tractable (FPT) if its input can be partitioned into a main part of size n and a parameter (usually an integer) k so that there is an algorithm that solves the problem in time $O(n^c \cdot f(k))$, where f is a computable function and c is a fixed constant [13]. A *kernel* is an instance of the problem smaller than the input, such that the problem has a solution on the input if and only if it has a solution on the kernel. It is well known that a problem is FPT if and only if a kernel of size $g(k)$ can be computed from the input in polynomial time, for a computable function g [13,27].

The fixed parameter tractability of the CLUSTER EDITING problem (for ordinary non-fuzzy graphs) has been shown, with a series of improvements in [7,19,29], when using the editing cost k as parameter. The problem has also been shown to admit a linear kernelization [16,21]. On fuzzy graphs, it is not known whether using only k as parameter ensures fixed parameter tractability. Here we introduce a new parameter r, that represents the minimum number of vertices required to cover the undecided edges. By parameterizing the CLUSTER EDITING problem by (k, r), we show that the problem admits a quadratic kernel, specifically on $O(k^2 + r)$ vertices, and therefore FPT also for fuzzy graphs. Furthermore the results hold also when the fuzzy graph is weighted.

The EDGE CLIQUE PARTITIONING problem has been recently shown to be FPT in [26], when parameterized by the number k of cliques that the edges can be partitioned into. In their work the authors give a quadratic kernel for it. The corresponding parameterization on fuzzy graphs asks, given a fixed k, whether the fuzzy edges can be turned into edges and non-edges so that the resulting set of edges can be partitioned into at most k edge-disjoint cliques. We prove that the problem becomes hard when the input is a fuzzy graphs, namely NP-complete for any fixed $k \geq 3$. Parameterizing only by k is thus not enough to ensure fixed parameter tractability. However, if we parameterize by (k, r), where r is again the minimum number of vertices required to cover the undecided edges of the fuzzy graph, then the problem becomes FPT, and admits a polynomial time kernelization to a kernel on $O(k^4 \cdot 3^r)$ vertices.

Most proofs and figures have been removed due to page limitations, but they can be found in a full version of this paper [5].

2   Notation and Definitions

For an undirected graph $G = (V, E)$, we denote its vertex set by $V(G) = V$ and edge set by $E(G) = E$ with $n = |V|$. The set of *neighbors* of $v \in V$ is $N_G(v) = \{u \mid uv \in E\}$, and the *degree* of v is $d_G(v) = |N_G(v)|$. In addition, $N_G[v] = N_G(v) \cup \{v\}$. Analogously, for a set $S \subseteq V$, $N_G[S] = \cup_{x \in S} N_G[x]$ and $N_G(S) = N_G[S] \setminus S$. We omit subscripts when there is no ambiguity. An *induced subgraph* of G by $U \subseteq V$ is the graph $G[U] = (U, E_U)$, where $E_U = \{xv \in E \mid x, v \in U\}$. Given a vertex x of G, we denote the graph $G[V \setminus \{x\}]$ by $G - x$. In addition, for a set of edges $M \subset E$, we define $G(M) = (\{x \mid \exists u, xu \in M\}, M)$.

A graph is *complete* if every pair of vertices are adjacent. If a subgraph is complete then it is called a *clique*. If $G[K]$ is a clique for $K \subset V$, we also say that K is a clique. If $G(M)$ is a clique for $M \subset E$, we also say that M is a clique. A vertex subset $S \subseteq V$ is a *vertex cover* if every edge of G has at least one endpoint in S. A *connected component* is a maximal connected subgraph.

We define a *fuzzy graph* $G = (V, E, F)$ to be a graph with two types of edges: E is the set of *real edges*, and F is the set of *fuzzy* edges. Between all other pairs of vertices in the graph we say that we have *non-edges*. When we decide for each fuzzy edge whether it should become a real edge or a non-edge, we say that we *realize* the fuzzy edges. The resulting graph is called a *normalization* of the fuzzy graph. Formally we say that (R^+, R^-) with $F = R^+ \cup R^-$ is a *realization* of F into real edges R^+ and non-edges R^- such that $G' = (V, E \cup R^+)$ is the corresponding normalization of $G = (V, E, F)$. When we speak about the *connected components of a fuzzy graph*, we mean the connected components of the graph obtained by turning all fuzzy edges into non-edges. So a *connected fuzzy graph* is a fuzzy graph where between any two vertices there is a path of real edges.

3 Parameterized Cluster Editing with Partial Information

A *cluster graph* is a graph where each connected component is a clique. In this section we study the problem of editing a weighted fuzzy graph $G = (V, E, F)$ to obtain a cluster graph. *Editing* means turning some real edges into non-edges (*deleting*), turning some non-edges into real edges (*adding*), and turning all fuzzy edges into either real edges or non-edges. Each edge and non-edge is associated with a positive weight, whereas each fuzzy edge has weight 0. The *cost* of an edit is the sum of the weights of the deleted and added edges, and the goal is to minimize the cost. The problem is formally defined as follows.

WEIGHTED FUZZY CLUSTER EDITING (WFCE)
Instance: A fuzzy graph $G = (V, E, F)$, a weight function $w : V \times V \to \mathbb{N}$ such that $w(uv) = 0$ if $uv \in F$ and $w(uv) > 0$ if $uv \notin F$, and an integer $k \geq 0$.
Question: Is there a set $M \subseteq V \times V$ such that: $G' = (V, (E \setminus M) \cup (M \setminus E))$ is a cluster graph and $\sum_{uv \in M} w(uv) \leq k$?

First we characterize the fuzzy graphs that can be turned into a cluster graph just by realizing the fuzzy edges, that is, without any editing cost. We show that they can be defined by a family of forbidden induced (fuzzy) subgraphs. The result was already noted in [14], but we restate it in a form more suitable for our framework.

We define a *fuzzy path* $P_l^f = \{v_1, v_2, ..., v_l\}$ to be a fuzzy graph where for every $1 \leq i \leq l - 1$ we have that $v_i v_{i+1}$ is a real edge, $v_1 v_l$ is a non-edge, and all the other pairs of vertices are joined by fuzzy edges.

Theorem 1. *Let G be a fuzzy graph. Then there exists a realization of the fuzzy edges that results in a cluster graph without editing any real edge or non-edge if and only if G does not contain any induced subgraph isomorphic to P_l^f for $l \geq 3$.*

The k-WEIGHTED FUZZY CLUSTER EDITING problem (k-WFCE) is the WFCE problem where we choose k of the problem instance to be the parameter. The complexity of k-WFCE is open even for the unweighted case. The characterization given in Theorem 1 is through an infinite set of forbidden induced subgraphs, and hence an FPT algorithm for k-WFCE does not follow from the results of Cai [7].

In order to give an FPT algorithm, we introduce an additional parameter. We define a *fuzzy vertex cover* of a fuzzy graph to be a vertex subset S such that each fuzzy edge has an endpoint in S. The new parameter is $r = |S|$ where S is a smallest fuzzy vertex cover of G. We call the corresponding new problem the (k, r)-WEIGHTED FUZZY CLUSTER EDITING, or (k, r)-WFCE, problem. Observe that checking whether G has a fuzzy vertex cover of size at most r is FPT when parameterized by r. To do this we create a non-fuzzy graph G' from $G(F)$ by turning all real edges of $G(F)$ into non-edges and all fuzzy edges into real edges. It is easy to see that G has a fuzzy vertex cover with at most r vertices if and only if G' has a vertex cover of at most r vertices. Since the r-VERTEX COVER problem is well known to be FPT, our claim follows.

3.1   Kernel for the (k, r)-Weighted Fuzzy Cluster Editing Problem

We show fixed parameter tractability by giving a set of rules that either enable us to answer NO, or produce a kernel of size $O(k^2 + r)$ in polynomial time, for the (k, r)-WFCE problem. First we give a general result to simplify some later proofs.

Observation 1. *Let G be a weighted fuzzy graph with connected components $C_1, \ldots, C_l$. Then G can be made into a cluster graph with editing cost at most k if and only if each connected component C_i can be made into a cluster graph with editing cost at most k_i, such that $\sum_{1 \leq i \leq l} k_i \leq k$.*

Now we start presenting the rules, that are mostly self-explanatory. We will not give sharp bounds on the running time of each rule, but we will limit the explanation to why they can be executed in polynomial time.

Rule 1. *If there is a connected component C with no non-edges, remove C.*

Lemma 1. *Rule 1 is correct and can be applied in linear time.*

Rule 2. *If Rule 1 does not apply and there are more than $k + 1$ connected components, then answer NO.*

Lemma 2. *Rule 2 is correct and can be applied in linear time.*

For the following rule, note that a *minimum cut* between two vertices u and v is the minimum total weight of a collection of real edges that must be deleted so that u and v have no real paths between them. The idea is that, if two vertices cannot be disconnected deleting edges of total weight at most k, then they must belong to the same cluster in every solution, if any exists. For this rule we also need some new definitions. When we *contract* two vertices u and v into one new vertex x, then u and v are deleted from the graph, x is added to the graph, and each previous pair of real edge, fuzzy edge, or non-edge uz and vz, appears now as two *parallel* edges between x and z.

Rule 3. *If there are vertices u and v such that the value of a minimum cut between them is at least $k + 1$, then contract u and v into one vertex x, and do all of the following:*

1. *If uv was a non-edge then let $k = k - w(uv)$.*
2. *If there are parallel edges with endpoint x and at least one of them is fuzzy, remove the fuzzy edge.*
3. *If there are parallel real edges (resp. non-edges) with endpoint x, replace them with one real edge (resp. non-edge) with weight equal to the sum of the weights of the parallel real edges (resp. non-edges).*
4. *If there is a real edge $e = ax$ in parallel with a non-edge $f = ax$ then:*
   (a) *If $w(e) > w(f)$, then let $k = k - w(f)$ and replace e and f with a real edge e' such that $w(e') = w(e) - w(f)$.*
   (b) *If $w(e) < w(f)$, then let $k = k - w(e)$ and replace e and f with a non-edge f' such that $w(f') = w(f) - w(e)$.*
   (c) *If $w(e) = w(f)$, then let $k = k - w(e)$ and replace e and f with a fuzzy edge g.*

If now $k < 0$, answer NO.

Lemma 3. *Rule 3 is correct and can be applied in polynomial time.*

Theorem 2. *If Rules 1, 2 and 3 do not apply, and we have not answered* NO *yet, then either the current graph has at most $k^2 + 3k + r$ vertices, or the answer is* NO.

Proof. If Rule 1 and Rule 2 do not apply, it means that the graph has at most k connected components, and each of them must be edited. If Rule 3 does not apply, then there cannot be cliques of size greater than $k + 1$.

Let us now consider a connected fuzzy graph $G = (V, E, F)$ with no clique of size greater than $k + 1$ and that can be made into a cluster graph by editing a set of real edges and non-edges M of total weight at least 1 and at most k. Then we show that G cannot have more than $k^2 + 3k + r$ vertices. Let us define: $S \subseteq V$ the set of vertices that are incident to some edited edge or non-edge in M; R a minimum fuzzy vertex cover of G; and $X = V \setminus R$, so that $G[X]$ does not contain any fuzzy edge. We define also $X' = S \cap X$. It is easy to see that $|X'| \leq |S| \leq 2k$. Let us focus on the graph $G[X \setminus X']$. It does not contain fuzzy edges, and none of its vertices is incident to an edited real edge or non-edge. We can conclude that it must be a union of disjoint cliques. In particular we show that it must be the union of at most $k + 1$ disjoint cliques, and that each of them has specific neighbors in the rest of the graph. Since no vertex of these cliques is incident to an edited edge, and there are no fuzzy edges in between them, each of them must belong to a different cluster in the solution. However, to create more than $k + 1$ connected components from a connected graph, we need to remove at least $k + 1$ edges. Hence the first part of the claim is proved. From the previous argument, it also follows that all vertices in X' connected to a clique in $G[X \setminus X']$, must end up in the same cluster of the solution as the clique they are adjacent to. Assume they do not, then we should delete an edge incident to a vertex in $G[X \setminus X']$, getting a contradiction. Therefore every vertex in X' can be connected to at most one clique in $G[X \setminus X']$, and furthermore it must be adjacent to all vertices of this clique. This means that every clique in $G[X \setminus X']$ has either some neighbors in X' and size at most k, or it has neighbors only in R and size at most $k+1$. Going back to G, if we define N as the number of cliques in $G[X \setminus X']$ that have neighbors only in R, we can give the following bound: $|V| \le (k+1) \cdot N + k \cdot (k+1-N) + (2k-N) + |R| = k^2 + 3k + r$. The first two terms give a bound on the number of vertices in $G[X \setminus X']$ according to the previous discussion, while the term $(2k-N)$ represents a tighter bound on $|X'|$. In fact, for every clique with neighbors only in R, there must be at least one distinct vertex in R incident to an edited edge. This because we need to disconnect the clique from the rest of the graph, but we cannot touch edges incident to its vertices. Besides at least one endpoint of the edge we have to remove will belong to the same cluster as the clique.

Consider now a fuzzy graph G with l connected components $C_1, \ldots, C_l$, where $1 \le l \le k$. By Observation 1 we know that there is a solution for G that edits at most k edges if and only if there is a solution for each $G[C_i]$ that edits at most k_i edges, such that $\sum_{i=1}^{l} k_i \le k$. This means that, by what we just proved for connected fuzzy graphs, if there is a solution for G then $|V(G[C_i])| \le k_i^2 + 3k_i + r_i$ for $1 \le i \le l$, where r_i is the size of a minimum vertex cover of $G[C_i]$. Hence $|V(G)| = \sum_{i=1}^{l} k_i^2 + 3k_i + r_i$, that is $\sum_{i=1}^{l} k_i^2 + 3\sum_{i=1}^{l} k_i + \sum_{i=1}^{l} r_i \le (\sum_{i=1}^{l} k_i)^2 + 3k + r = k^2 + 3k + r$, completing the proof. $\qquad \square$

It is easy to construct examples where we have $(k+1) \cdot N + k \cdot (k+1-N) + (k+1-N) + r = k^2 + 2k + 1 + r$ vertices in a *yes* instance for any given k (see [5]). Hence Theorem 2 gives a quite tight bound on the size of the kernel, that is in any case $O(k^2 + r)$.

We have thus proved that the (k,r)-WFCE problem has a kernel of size $O(k^2 + r)$. We can now conclude the following.

Theorem 3. *The (k,r)-Weighted Fuzzy Cluster Editing problem can be solved in time $n^{O(1)} + O((k^2 + r)^{2k})$.*

4   Parameterized Edge Clique Partition with Partial Information

In this section, we study the problem of partitioning the edges of a fuzzy graph $G = (V, E, F)$ into edge-disjoint cliques. In this problem, no editing of the edges or non-edges of G is involved, but we have to decide for each fuzzy edge whether or not it should become a real edge or a non-edge. Below is a formal definition of the problem.

Fuzzy Edge Clique Partitioning (FECP)
Instance: A fuzzy graph $G = (V, E, F)$ and an integer $k \ge 0$.
Question: Is there a realization (R^+, R^-) of the fuzzy edges such that the edges of $G' = (V, E \cup R^+)$ can be partitioned into at most k edge-disjoint cliques?

Naturally, being a more general version of the problem on non-fuzzy graphs, the Fuzzy Edge Clique Partitioning problem is NP-hard as well. Interestingly, we show that it remains NP-hard also when k is a fixed constant and not a part of the input, for every $k \ge 3$. Recall that, in contrast, the Edge Clique

PARTITIONING problem is EPT when parameterized by k. We show the FECP problem parameterized by both k and r, where r is again the size of a minimum fuzzy vertex cover, is FPT. We call this version of the problem (k,r)-FECP.

4.1   The k-Fuzzy Edge Clique Partitioning Problem Is NP-Complete

Here we prove that is it NP-complete for every fixed $k \geq 3$ to decide whether the edges of a fuzzy graph can be partitioned into k edge-disjoint cliques. The problem we reduce from, is the classical k-Coloring problem. In this problem, the input is a graph $G = (V, E)$, and the problem is to decide whether the vertices of G can be colored with at most k colors, such that no two adjacent vertices have the same color. Since it is well known that this problem is NP-complete for every fixed $k \geq 3$, the result follows. We omit the proof of the reduction, but we give an example in Figure 1.

Theorem 4. *For every fixed $k \geq 3$ the k-FUZZY EDGE CLIQUE PARTITIONING problem is NP-complete.*

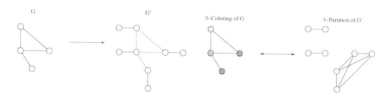

Fig. 1. An example of the reduction. The graph G' is obtained from a graph G by adding a pendant vertex to each original vertex of G, replacing the edges of G with non-edges and letting all remaining edges of G' be fuzzy edges. We show also the equivalence of a solution of the 3-Coloring problem on G with a solution of the 3-Clique Partition problem on G'. In G' the fuzzy edges are not drawn to keep the figure clean, while in the corresponding clique partition the non edges are not drawn.

For all values of k smaller than 3, it is possible to show that the problem is solvable in polynomial time, by careful but not too complicated arguments [5].

4.2   The (k,r)-Fuzzy Edge Clique Partitioning Problem Is FPT

To obtain a kernel for (k,r)-FECP, we first give some observations that apply to any valid solution of the problem on non-fuzzy graphs, i.e., the k-EDGE CLIQUE PARTITIONING problem.

For a non-fuzzy graph $G = (V, E)$, and a fixed $k \geq 0$, we call a *feasible solution* a partition $K = \{K_1, K_2, ..., K_l\}$ of E such that $G(K_i)$ is a clique for each i, and $l \leq k$. For $K_i \in K$, we define $V(K_i)$ as the union of the endpoints of the edges in K_i, i.e. $V(G[K_i])$. We call *gateways* the vertices that are in the intersection of some cliques defined by elements of K, while the vertices contained only in one clique are called *normal*. Two normal vertices in the same clique are said to be *co-normal.* We define a set $V' \subseteq V$ to be a *type* if there is at least one vertex v such that $N[v] = V'$. So we say that two vertices are of the same type if their closed neighborhood is identical, and that they are of different type otherwise. Finally notice that the intersection of two cliques in any solution cannot consist of more than one vertex, or there would be one edge covered by two cliques.

Theorem 5 ([6]). *Every edge clique partition of a complete graph on n vertices, except the trivial one of a single clique, contains at least n cliques.*

Lemma 4. *If the answer to the k-EDGE CLIQUE PARTITIONING problem on a graph $G = (V, E)$ is YES, then the answer is YES also on each induced subgraph of G.*

Lemma 4 implies that if there is even only one induced subgraph of G for which the answer is NO, then G itself is a NO instance. We will use this observation often.

Observation 2. *In any solution of k-EDGE CLIQUE PARTITIONING, there cannot be more than $\binom{k}{2}$ gateway vertices.*

Proof. Every two cliques can intersect in at most one vertex, or they would cover the same edge. Since there are at most $\binom{k}{2}$ possible intersection among k cliques, the result follows.   □

Examples exist to show that the bound given above is tight. For the next result, remember that the closed neighborhood of a vertex is the union of the cliques it belongs to in a feasible solution.

Observation 3. *If two vertices have the same type, then in any feasible solution either they are co-normal or they are both gateways.*

Observation 4. *If there are more than $k + \binom{k}{2}$ vertices of pairwise different type, then the answer to k-EDGE CLIQUE PARTITIONING is NO.*

Now we show that a simple generalization of the observations given until now, can be used as rules to give a polynomial time kernelization for the (k, r)-FECP problem. From now on we assume a fuzzy input graph $G = (V, E, F)$.

First we need to introduce a generalization of the type of a vertex for fuzzy graphs. The *fuzzy neighborhood* of a vertex v is the set of the vertices w such that $vw \in F$. We say that two vertices are of the same *absolute type* if their closed and fuzzy neighborhoods are equal.

Consider a fuzzy graph $G = (V, E, F)$, and let $S \subset V$ be a minimum fuzzy vertex cover of G, such that $|S| \leq r$. Then for each vertex in $X = V \setminus S$, there can be at most 3^r possible ways to have adjacencies in S. So we can classify the vertices of X into 3^r categories, so that the vertices in the same category have the same absolute type with respect to the vertices in S. Since $G[X]$ is a non-fuzzy graph, if there is no solution to k-EDGE CLIQUE PARTITIONING for $G[X]$, then there is no solution to (k, r)-FECP on G no matter how we realize the fuzzy edges, due to Lemma 4.

Rule 4. *If there are more than $(k + \binom{k}{2}) \cdot 3^r$ vertices of different absolute types in X, then the answer is* NO.

Lemma 5. *Rule 4 is correct and can be executed in polynomial time.*

Proof. If there are more than $(k + \binom{k}{2}) \cdot 3^r$ absolute types of vertices, then $G[X]$ must have more than $(k + \binom{k}{2})$ vertices of different types. Hence by Observation 4, there is no solution for $G[X]$. By Lemma 4, this implies that there is no solution for G as well, proving the first part of the statement.

The rule can be easily executed in polynomial time by listing the absolute closed neighborhoods of the vertices of G, and checking whether there are more than $(k + \binom{k}{2}) \cdot 3^r$ different ones. Since k and r are constants, the result follows.
$\square$

Rule 5. *If Rule 4 does not apply and there are more than $\binom{k}{2} + 1$ vertices of the same absolute type in X, then remove one.*

Lemma 6. *Rule 5 is correct and can be executed in polynomial time.*

Lemma 7. *If Rules 4 and 5 do not apply, then the graph has at most $(\binom{k}{2} + 1) \cdot ((k + \binom{k}{2}) \cdot 3^r) + r$ vertices.*

Theorem 6. (k, r)-FUZZY EDGE CLIQUE PARTITIONING *is FPT with a kernel of size* $O(k^4 \cdot 3^r)$.

5   Concluding Remarks

In this paper we have studied the parameterized complexity of two important examples of *GRAPH CLUSTERING PROBLEMS* on inputs consisting of fuzzy graphs: graphs that represent incomplete information about relationships. We believe that the investigation of "problems on fuzzy graphs" is extremely well-motivated by applications, particularly in areas such as machine learning and bioinformatics, where complete information about the graphs modeling various computational objectives is often not available. In this general context, much more remains to be done.

We have described two FPT algorithms, respectively, for the WEIGHTED FUZZY CLUSTER EDITING problem, and the FUZZY EDGE CLIQUE PARTITIONING problem, where both are parameterized by the compound parameter (k, r), and where k is a *cost parameter*: respectively, the total cost of the editing in the case of WEIGHTED FUZZY CLUSTER EDITING, and the number of cliques in the partition in the case of FUZZY EDGE CLIQUE PARTITIONING; and where r is a *structural parameter*: the minimum number of vertices required to cover the undecided edges of the fuzzy graph taken as input. This structural parameter could be well-motivated by applications where only a small number of "trouble-maker" vertices are the "cause" of the uncertain information about the input.

We have also shown that in the case of FUZZY EDGE CLIQUE PARTITIONING, it is not possible to extend the above positive outcome to a parameterization only by k.

In the case of the FUZZY CLUSTER EDITING problem, the analogous question remains open, and this is in fact a prominent concrete open problem in parameterized complexity. Apart from the important machine learning applications noted in [2,3], it has recently been shown that for the special case where all weights are 1, the FUZZY CLUSTER EDITING problem (parameterized only by k) is FPT-equivalent [17,14], to the MINIMUM TERMINAL EDGE SEPARATION problem left open by Marx in [25].

Another area of open problems concerning this work is that of *improving kernelization bounds*. Because FPT kernelization is of great practical significance due to the general connection to efficient pre-processing (see [15,22,27] for background and discussion of this point), it is an outstanding open problem as to whether FUZZY EDGE CLIQUE PARTITIONING admits a $poly(k, r)$ kernelization.

References

1. Abu-Khzam, F.N., Collins, R.L., Fellows, M.R., Langston, M.A., Suters, W.H., Symons, C.T.: Kernelization algorithms for the vertex cover problem: Theory and experiments. In: Proceedings of ALENEX/ANALC 2004 - 6th Workshop on Algorithm Engineering and Experiments and the First Workshop on Analytic Algorithmics and Combinatorics, pp. 62–69 (2004)

2. Bansal, N., Blum, A., Chawla, S.: Correlation clustering. In: Proceedings of FOCS 2002 - 43rd Symposium on Foundations of Computer Science, p. 238 (2002)

3. Bansal, N., Blum, A., Chawla, S.: Correlation clustering. Machine Learning 56(1-3), 89–113 (2004)

4. Ben-Dor, A., Shamir, R., Yakhini, Z.: Clustering gene expression patterns. Journal of Computational Biology 6(3-4), 281–297 (1999)

5. Bodlaender, H.L., Fellows, M.R., Heggernes, P., Mancini, F., Papadopoulos, C., Rosamond, F.: Clustering with partial information. Reports in Informatics 373, University of Bergen, Norway (2008)

6. De Bruijin, N.J., Erdos, P.: On a combinatorial problem. Ind. Math. 10, 421–423 (1948)

7. Cai, L.: Fixed-parameter tractability of graph modification problems for hereditary properties. Inf. Process. Lett. 58(4), 171–176 (1996)

8. Charikar, M., Guruswami, V., Wirth, A.: Clustering with qualitative information. J. Comput. Syst. Sci. 71(3), 360–383 (2005)

9. Chen, Z.-Z., Jiang, T., Lin, G.: Computing phylogenetic roots with bounded degrees and errors. SIAM J. Comput. 32(4), 864–879 (2003)

10. Dehne, F.K.H.A., Langston, M.A., Luo, X., Shaw, P., Zhang, Y., Pitre, S.: The Cluster Editing Problem: Implementations and Experiments. In: Bodlaender, H.L., Langston, M.A. (eds.) IWPEC 2006. LNCS, vol. 4169, pp. 13–24. Springer, Heidelberg (2006)

11. Demaine, E.D., Emanuel, D., Fiat, A., Immorlica, N.: Correlation clustering in general weighted graphs. Theor. Comput. Sci. 361(2-3), 172–187 (2006)

12. Demaine, E.D., Immorlica, N.: Correlation clustering with partial information. In: Proceedings of RANDOM-APPROX 2003 - 7th International Workshop on Randomization and Approximation Techniques in Computer Science, pp. 1–13 (2003)

13. Downey, R.G., Fellows, M.R.: Parameterized Complexity. Springer, Heidelberg (1999)

14. Emanuel, D., Fiat, A.: Correlation clustering - minimizing disagreements on arbitrary weighted graphs. In: Proceedings of ESA 2003 - 11th Annual European Symposium on Algorithms, pp. 208–220 (2003)

15. Fellows, M.R.: The Lost Continent of Polynomial Time: Preprocessing and Kernelization. In: Bodlaender, H.L., Langston, M.A. (eds.) IWPEC 2006. LNCS, vol. 4169, pp. 276–277. Springer, Heidelberg (2006)

16. Fellows, M.R., Langston, M.A., Rosamond, F.A., Shaw, P.: Efficient parameterized preprocessing for cluster editing. In: Proceedings of FCT 2007 - Fundamentals of Computation Theory, 16th International Symposium, pp. 312–321 (2007)

17. Fellows, M.R., Mnich, M., Rosamond, F., Saurabh, S.: Manuscript (2008)

18. Golumbic, M.C., Kaplan, H., Shamir, R.: Graph sandwich problems. J. Algorithms 19(3), 449–473 (1995)

19. Gramm, J., Guo, J., Huffner, F., Niedermeier, R.: Graph-modeled data clustering: Exact algorithms for clique generation. Theory Comput. Syst. 38(4), 373–392 (2005)

20. Gramm, J., Guo, J., Huffner, F., Niedermeier, R.: Data reduction, exact, and heuristic algorithms for clique cover. In: Proceedings of the 8th ACM-SIAM ALENEX, pp. 86–94. ACM-SIAM, New York (2006)

21. Guo, J.: A more effective linear kernelization for cluster editing. In: Proceedings of ESCAPE 2007 - Combinatorics, Algorithms, Probabilistic and Experimental Methodologies, First International Symposium, pp. 36–47 (2007)

22. Guo, J., Niedermeier, R.: Invitation to data reduction and problem kernelization. SIGACT News 38(1), 31–45 (2007)

23. Krivánek, M., Morávek, J.: NP-hard problems in hierarchical-tree clustering. Acta Inf. 23(3), 311–323 (1986)

24. Ma, S.H., Wallis, W.D., Wu, J.L.: The complexity of the clique partition number problem. Congr. Numer. 67, 59–66 (1988)

25. Marx, D.: Parameterized Graph Separation Problems. In: Downey, R.G., Fellows, M.R., Dehne, F. (eds.) IWPEC 2004. LNCS, vol. 3162, pp. 71–82. Springer, Heidelberg (2004)

26. Mujuni, E., Rosamond, F.A.: Parameterized complexity of the clique partition problem. In: Proceedings of ACiD 2007 - 2nd Workshop of Algorithms and Complexity in Durham (2007)

27. Niedermeier, R.: Invitation to Fixed-Parameter Algorithms. Oxford University Press, Oxford (2006)

28. Orlin, J.: Contentment in graph theory: Covering graphs with cliques. Indagationes Math. 39, 406–424 (1977)

29. Protti, F., da Silva, M.D., Szwarcfiter, J.L.: Applying modular decomposition to parameterized bicluster editing. In: Proceedings of IWPEC 2006 - Parameterized and Exact Computation, Second International Workshop, pp. 1–12 (2006)

30. Shamir, R., Sharan, R., Tsur, D.: Cluster graph modification problems. Discrete Applied Mathematics 144(1-2), 173–182 (2004)

Reoptimization of the Metric Deadline TSP*
(Extended Abstract)

Hans-Joachim Böckenhauer and Dennis Komm

Department of Computer Science, ETH Zurich, Switzerland
{hjb,dennis.komm}@inf.ethz.ch

Abstract. The reoptimization version of an optimization problem deals with the following scenario: Given an input instance together with an optimal solution for it, the objective is to find a high-quality solution for a locally modified instance.

In this paper, we investigate several reoptimization variants of the traveling salesman problem with deadlines in metric graphs (Δ-DlTSP). The objective in the Δ-DlTSP is to find a minimum-cost Hamiltonian cycle in a complete undirected graph with a metric edge cost function which visits some of its vertices before some prespecified deadlines. As types of local modifications, we consider insertions and deletions of a vertex as well as of a deadline.

We prove the hardness of all of these reoptimization variants and give lower and upper bounds on the achievable approximation ratio which are tight in most cases.

1   Introduction

The traditional approach for dealing with optimization problems is to find good feasible solutions to input instances nothing about which is known in advance. Unfortunately, most of the practically relevant problems are computationally hard, and so we use different approaches such as approximation algorithms or heuristics for computing good (but not necessarily optimal) solutions. In many applications, however, we might have some prior knowledge about our input instance at hand. For instance, if we want to maintain a timetable for a railway system or a routing scheme for a communications network, small changes to the railway system or the network require a new timetable or routing scheme, but we might be able to profit from the information about the old solution.

These considerations lead to the concept of reoptimization problems: Given an instance of an optimization problem together with an optimal solution for it, the objective is to compute an optimal solution for a locally modified input instance. For a graph problem, we might for instance consider the deletion or insertion of a vertex or an edge or the change of the cost of a single edge as a local modification. For an optimization problem U and a type of local modification LM,

---

* This work was partially supported by SBF grant C 06.0108 as part of the COST 293 (GRAAL) project funded by the European Union.

E. Ochmański and J. Tyszkiewicz (Eds.): MFCS 2008, LNCS 5162, pp. 156–167, 2008.

we denote the resulting reoptimization problem by LM-U. Obviously, LM-U may be easier than U because we have the optimal solution for the original problem instance as additional knowledge for free. But there also exist examples where the concept of reoptimization does not help since the reoptimization version is exactly as hard as the standard version of the problem [4].

The concept of reoptimization was already successfully applied to several variants of the TSP [1,2,4,5,13] and the Steiner tree problem [3,8,11]. A survey of reoptimization problems can also be found in [9].

A related question was also considered in operations research [12,14,15,16,17], where it was studied how much a given instance of an optimization problem may be changed without destroying the optimality of solutions. In contrast to this so called "postoptimality analysis", we are also interested in local modifications causing the loss of optimality for solutions to the old instance.

In this paper, we will apply the concept of reoptimization to the Deadline TSP. In the well-known traveling salesman problem (TSP), the objective is to find a minimum-cost Hamiltonian cycle in a complete graph with edge costs. The Deadline TSP (DLTSP) is a generalization of the TSP, where additionally a subset of the vertices is given which have deadlines imposed on them. Any feasible Hamiltonian tour, starting from a prespecified start vertex s, has to visit every deadline vertex v before v's deadline is expired, i.e., the partial tour from s to v has to have a total cost of at most the deadline value of v. The DLTSP is an important special case of the TSP with time windows which is one of the most prominent optimization problems in operations research occurring in a number of applications like for instance vehicle routing, for a survey, see [10].

We deal with the metric version of the problem only, that is, we assume that the cost function c satisfies the triangle inequality, i.e., $c(\{u,v\}) \leq c(\{u,w\}) + c(\{w,v\})$ for all vertices u, v and w. The approximation hardness of the metric Deadline TSP, or Δ-DLTSP for short, has been shown in [6,7]. Some reoptimization versions of DLTSP, where the local modifications consisted of changing the value of one deadline or the cost of one edge, have been investigated in [4]. As local modifications, we will here consider the insertion or deletion of a vertex with or without a deadline, and the insertion or deletion of a deadline (without changing the vertex set). For our results, we have to distinguish two cases depending on the number of deadline vertices. If the number of deadline vertices is bounded by a constant, the reoptimization problems are approximable within a constant, but APX-hard. If the number of deadline vertices is unbounded, most of the considered variants are approximable with a linear approximation ratio, and this bound is tight for adding or deleting deadlines or deadline vertices. For adding a vertex without a deadline, the problem is still approximable within a factor of 2. A complete overview of the results is shown in Table 1.

The paper is organized as follows: In Section 2, we formally define the reoptimization problems under consideration. In Section 3, we prove the lower bounds for the case of a constant number of deadline vertices; Section 4 is devoted to the lower bounds for an unbounded number of deadlines. In Section 5, we present the upper bounds on the approximation ratio.

Table 1. Lower and upper bounds on the approximability of different reoptimization variants of Δ-DLTSP

LOCAL MODIFICATION	BOUNDED # DEADLINES		UNBOUNDED # DEADLINES	
	lower bound	upper bound	lower bound	upper bound
add vertex without deadline	$2 - \varepsilon$	2	$2 - \varepsilon$	2
delete vertex without deadline	$2 - \varepsilon$	2.5	$2 - \varepsilon$	$0.5n$
add deadline to existing vertex	$2 - \varepsilon$	2.5	$(0.5 - \varepsilon)n$	$0.5n$
delete deadline from vertex	$2 - \varepsilon$	2.5	$(0.5 - \varepsilon)n$	$0.5n$
add vertex with deadline	$2 - \varepsilon$	2.5	$(0.5 - \varepsilon)n$	$0.5n$
delete vertex with deadline	$2 - \varepsilon$	2.5	$(0.5 - \varepsilon)n$	$0.5n$

2 Preliminaries

We start with formally defining the DLTSP. Consider a complete graph $G = (V, E)$ and a cost function $c \colon E \to \mathbb{Q}^+$. A *deadline triple* for (G, c) is a triple $\mathcal{D} = (s, D, d)$ where $s \in V$, $D \subseteq V \setminus \{s\}$, and $d \colon D \to \mathbb{Q}^+$. We call D the set of *deadline vertices* of G. A Hamiltonian path $P = (w_1, w_2, \ldots w_n)$ satisfies the deadlines according to $\mathcal{D}$, if $s = w_1$ and, for all $w_i \in D$, the following holds:

$$\sum_{j=2}^{i} c(\{w_{j-1}, w_j\}) \leq d(w_i).$$

A Hamiltonian cycle $C = (w_1, w_2, \ldots, w_n, w_1)$ *satisfies the deadlines* according to $\mathcal{D}$ if it contains a path $(w_1, w_2, \ldots, w_n)$ satisfying the deadlines according to $\mathcal{D}$.

Definition 1 (DLTSP). *The* TSP with deadlines (DLTSP) *is the following optimization problem. The input consists of a complete graph $G = (V, E)$, an edge cost function $c : E \to \mathbb{Q}^+$, a deadline triple $\mathcal{D} = (s, D, d)$, and a Hamiltonian cycle (of arbitrary cost) in G satisfying the deadlines according to $\mathcal{D}$. The objective is to find a minimum-cost Hamiltonian cycle of G satisfying the deadlines according to $\mathcal{D}$.*

By Δ-DLTSP we denote the restriction of DLTSP where the edge cost function c satisfies the triangle inequality, and k-Δ-DLTSP is the subproblem of Δ-DLTSP where the number of deadlines of any input instance is bounded by some constant k (i.e., $|D| = k$).

Note that, for a DLTSP instance, already finding a feasible solution might be a hard problem. Since we are not interested in this aspect of hardness, we have defined the problem as to contain an (arbitrarily bad) feasible solution as part of the input. In this way, it is easy to see that DLTSP is contained in $\mathcal{NPO}$.

Obviously, any instance of TSP can be regarded as an instance of DLTSP with $D = \emptyset$. Thus, all lower bounds for TSP directly carry over to DLTSP.

We are now ready to define the reoptimization variants of Δ-DLTSP.

Definition 2. *Let $G_O = (V_O, E_O)$ and $G_N = (V_N, E_N)$ be two complete undirected graphs with metric edge cost functions $c_O \colon E_O \to \mathbb{Q}^+$ and $c_N \colon E_N \to \mathbb{Q}^+$,* let $\mathcal{D}_O = (s, D_O, d_O)$ be a deadline triple for G_O and let $\mathcal{D}_N = (s, D_N, d_N)$ be a deadline triple for G_N such that $(G_N, c_N, \mathcal{D}_N)$ can be constructed from $(G_O, c_O, \mathcal{D}_O)$ by a local modification. We will consider the following six local modifications:

LM(D^-): *Deletion of a deadline:* In this case, we have $(G_O, c_O) = (G_N, c_N)$, $D_N = D_O \setminus \{x\}$ for some $x \in D_O$, and $d_N = d_O|_{D_N}$.

LM(D^+): *Addition of a deadline to an already existing vertex:* In this case, we have $(G_O, c_O) = (G_N, c_N)$, $D_O = D_N \setminus \{x\}$ for some $x \in D_N$, and $d_O = d_N|_{D_O}$.

LM(V^-): *Deletion of a vertex without deadline:* In this case, we have $V_N = V_O \setminus \{x\}$ for some $x \in V_O \setminus D_O$, E_N and c_N are the canonical restrictions of E_O and c_O to the vertices of V_N, and $\mathcal{D}_O = \mathcal{D}_N$.

LM(V^+): *Addition of a vertex without deadline:* In this case, we have $V_O = V_N \setminus \{x\}$ for some $x \in V_N \setminus D_N$, E_O and c_O are the canonical restrictions of E_N and c_N to the vertices of V_O, and $\mathcal{D}_O = \mathcal{D}_N$.

LM($(D \wedge V)^-$): *Deletion of a deadline vertex:* In this case, we have $D_N = D_O \setminus \{x\}$ for some $x \in D_O$, $d_N = d_O|_{D_N}$, $V_N = V_O \setminus \{x\}$, and E_N and c_N are the canonical restrictions of E_O and c_O to the vertices of V_N.

LM($(D \wedge V)^+$): *Addition of a deadline vertex:* In this case, we have $D_O = D_N \setminus \{x\}$ for some $x \in D_N$, $d_O = d_N|_{D_O}$, $V_O = V_N \setminus \{x\}$, and E_O and c_O are the canonical restrictions of E_N and c_N to the vertices of V_O.

For $X \in \{D^-, D^+, V^-, V^+, (D \wedge V)^-, (D \wedge V)^+\}$, we define the problem LM(X)-Δ-DLTSP *as to find an optimum solution for the Δ-DLTSP instance $(G_N, c_N, \mathcal{D}_N)$, given the Δ-DLTSP instance $(G_O, c_O, \mathcal{D}_O)$ together with an optimal solution $\overline{C}$ for it and an arbitrary feasible solution $\widetilde{C}$ for $(G_N, c_N, \mathcal{D}_N)$.*

Moreover, for any constant k, let LM(X)-k-Δ-DLTSP *denote the subproblem of* LM(X)-Δ-DLTSP *where $|D_N| = k$.*

3 Lower Bounds for a Bounded Number of Deadlines

In this section, we will give lower bounds of $(2 - \varepsilon)$ for any $\varepsilon > 0$ for all reoptimization variants of Δ-DLTSP as defined above. For the reductions we will employ the following decision problem RHP (*Restricted Hamiltonian Path Problem*).

Definition 3 (RHP). *Let $G = (V, E)$ be a graph where $|V| = n+1$. Let $s, t \in V$ be two distinct vertices and let $P' = (s, \ldots, t)$ be a Hamiltonian path in G from s to t. The objective is to decide whether there exists a second Hamiltonian path P in G from s to some vertex $v_i \neq t$.*

The problem RHP is known to be $\mathcal{NP}$-complete (for a proof, see [4]). We will now show that any approximation algorithm with a ratio better than $(2 - \varepsilon)$ for any of the reoptimization variants of Δ-DLTSP from Definition 2 could be used to decide RHP in polynomial time which contradicts $\mathcal{NP} \neq \mathcal{P}$.

We start with a proof for LM(D^+)-k-Δ-DLTSP.

Theorem 1. *Let $\varepsilon > 0$, let $k \geq 4$. There is no polynomial-time approximation algorithm for* LM(D^+)-k-Δ-DLTSP *with a ratio of* $(2 - \varepsilon)$, *unless* $\mathcal{P} = \mathcal{NP}$.

Proof. Given an instance (G', P') of RHP where $|V(G')| = n + 1$ and $P' = (s', \ldots, t')$, we will construct a complete graph $K_{G'}$ where $V(K_{G'}) = V(G')$, with an edge cost function c defined by

$$c(e) = \begin{cases} 1, \text{ if } e \in E(G') \\ 2, \text{ otherwise} \end{cases}$$

for all $e \in E(K_{G'})$. Note that there exists a Hamiltonian path P from s' to $v_i \neq t'$ of cost n in $K_{G'}$ if and only if P is a Hamiltonian path in G'.

As a second step, we extend $K_{G'}$ to a complete, weighted graph G which will be part of an instance of LM(D^+)-k-Δ-DLTSP. The deadline triples are $\mathcal{D}_O = (s, D_O, d_O)$ and $\mathcal{D}_N = (s, D_N, d_N)$ where $D_O = \{D_2, D_3, D_4\}$ and $D_N = \{D_1, D_2, D_3, D_4\}$. The deadline function d_O is shown in Figure 1 and d_N is shown in Figure 2. Observe that $D_N = D_O \cup \{D_1\}$. For any given $\varepsilon > 0$, let $\gamma := \gamma(\varepsilon) > \frac{9n}{2\varepsilon}$. All edge costs and deadlines are as depicted in Figure 1. Edges which are not shown have the largest cost possible (respecting the triangle inequality).

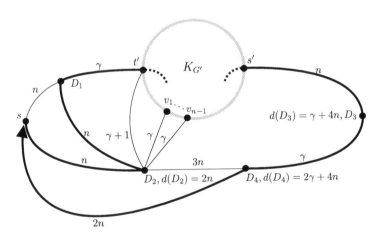

Fig. 1. $(G, \mathcal{D}_O)$ before the local modification is applied and the optimal tour $\overline{C}$

An optimal solution for $(G, \mathcal{D}_O)$, which will serve as part of the input, is $\overline{C} = (s, D_2, D_1, \overleftarrow{P'}, D_3, D_4, s)$ which uses the known Hamiltonian path P' backwards. The costs are $n + n + \gamma + n + n + \gamma + 2n = 2\gamma + 6n$.

To prove that $\overline{C}$ indeed is an optimal solution, we will show that any other solution will be at least as expensive as $\overline{C}$, independent of the existence of a second Hamiltonian path P of cost n in $K_{G'}$. Note that it is only possible to start with (s, D_2) or (s, D_1, D_2). Any other path would immediately violate D_2's deadline of $2n$. After that there are several options. If the path starts with (s, D_1, D_2), it cannot visit D_4 before D_3, because it would then violate the deadline at D_3. However, it is possible to visit all vertices in $K_{G'}$ before D_3 if there exists a Hamiltonian path P of cost n in $K_{G'}$ from s' to $v_i \neq t'$. (Note that $c(s', D_2) = \gamma + 1$.) The path will then visit D_3 just in time. After that, this tour will have to be completed by visiting D_4 and returning to s. Hence, this solution will be $(s, D_1, D_2, \overleftarrow{P}, D_3, D_4, s)$ with cost $n + n + \gamma + n + n + \gamma + 2n = 2\gamma + 6n$ which is as expensive as $\overline{C}$.

On the other hand, if the path starts with (s, D_2), it can go on to D_3 via D_4. It will arrive at D_3 at cost $\gamma + 4n$, and it will then have to visit D_1 and $K_{G'}$. No matter in which order it does so, this path will at least cost another $\gamma + 3n$. Obviously, visiting D_3 before D_4 will not improve this.

Observe, that the subgraph $K_{G'} \cup \{D_3\}$ is connected to the rest of G only with edges that cost at least γ. Therefore, any possible solution that does not visit the vertices in $K_{G'} \cup \{D_3\}$ consecutively, will cost more than 4γ. Thus, we have shown that $\overline{C}$ is indeed an optimal solution for $(G, \mathcal{D}_O)$.

Now we apply the local modification by adding a deadline of n to D_1. The new instance $(G, \mathcal{D}_N)$ is shown in Figure 2. If there exists a Hamiltonian path P from s' to $v_i \neq t'$ in G', and thus a second Hamiltonian path of cost n in $K_{G'}$, we claim that $(s, D_1, D_2, \overleftarrow{P}, D_3, D_4, s)$, which contains P backwards (w.l.o.g. let $P = (s', \ldots, v_1)$), is an optimal solution for $(G, \mathcal{D}_N)$. Again, the costs are $2\gamma + 6n$.

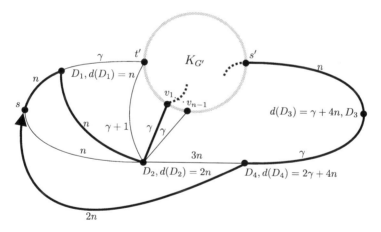

Fig. 2. $(G, \mathcal{D}_N)$ and an optimal solution using $P = (s', \ldots, v_1)$ backwards

To prove the claim, we note that any feasible solution has to start with the path (s, D_1, D_2) now. If there is no Hamiltonian path $P = (s', \ldots, v_i)$ with $v_i \neq t'$ of cost n in $K_{G'}$, it is impossible to visit all vertices in $K_{G'}$ between D_2 and D_3. After visiting D_3, the path must immediately visit D_4. Overall, this will cost more than $2\gamma + 3n$. (Note that $c(D_2, s') = c(D_2, t') = \gamma + 1$ and $c(t', D_3) = c(s', D_3) = n + 1$.) After that, all deadline vertices are visited and the path can visit those vertices it left out in $K_{G'}$. This will cost at least another 2γ, adding up to a total cost of at least $4\gamma + 3n$.

As an alternative, the path could start with (s, D_1, D_2, D_3, D_4) which costs more than $2\gamma + 2n$. Continuing by visiting the vertices in $K_{G'}$, the rest of the path costs at least another $2\gamma + n$. Thus, without a Hamiltonian path from s' to $v_i \neq t'$, no Hamiltonian cycle in G will be cheaper than $4\gamma + 3n$.

Since we chose $\gamma = \frac{9n}{2\varepsilon}$, an easy calculation shows $4\gamma + 3n > (2 - \varepsilon)(2\gamma + 6n)$. Thus, for any $\varepsilon > 0$, an approximation algorithm for $\text{LM}(D^+)\text{-}k\text{-}\Delta\text{-DLTSP}$ with a ratio of $(2 - \varepsilon)$ could be used to decide RHP which contradicts $\mathcal{NP} \neq \mathcal{P}$. $\square$

Theorem 2. *Let $\varepsilon > 0$, let $k \geq 4$. There is no polynomial-time approximation algorithm for any of the problems $\text{LM}(D^-)\text{-}k\text{-}\Delta\text{-DLTSP}$, $\text{LM}(V^+)\text{-}k\text{-}\Delta\text{-DLTSP}$, $\text{LM}(V^-)\text{-}k\text{-}\Delta\text{-DLTSP}$, $\text{LM}((D \wedge V)^+)\text{-}k\text{-}\Delta\text{-DLTSP}$, and $\text{LM}((D \wedge V)^-)\text{-}k\text{-}\Delta\text{-DLTSP}$ with a ratio of $(2 - \varepsilon)$, unless $\mathcal{P} = \mathcal{NP}$.*

Due to space constraints, the proof of Theorem 2 is omitted here.

4   Lower Bounds for an Unbounded Number of Deadlines

For proving lower bounds of $(0.5 - \varepsilon)n$ for an unbounded number of deadline vertices we need the following lemma which is a simplified version of the Zigzag Lemma from [4].

Lemma 1 (Zigzag Lemma). *Let $k, \gamma \in \mathbb{N}^+$ such that k is even and $\gamma \geq n$. Let $G^* = (V^*, E^*, c^*)$ be a complete, weighted graph with a deadline triple (s^*, D^*, d^*) such that any Hamiltonian path in G^* respecting the deadlines (which implies starting at s^*) ends in the same vertex t^*. Then, we can construct a complete, weighted graph $G \supset G^*$ and a deadline triple (s, D, d) such that $D \supset D^*, d|_{D^*} = d^*$, and any path that reaches t in time $7n$ can be extended to a Hamiltonian cycle which costs at most*

$$(k + 7)n + 2\gamma, \tag{1}$$

while any path that reaches t after $8n$, but before $9n$, can only be extended to a Hamiltonian cycle which costs at least

$$\left(9 + \frac{k - 3}{2}\right)n + k\gamma. \tag{2}$$

Proof sketch. Figure 3 shows the idea of the zigzag construction as presented in [4]. Note that, for clarity of exposition, only some edges of the complete graph G are shown, and only the expensive edges of cost γ are labeled in Figure 3. All shown edges without labels have cost of at most n and all edges not depicted have the maximum possible cost as to satisfy the triangle inequality. If a path arrives at t^* without having spent too much yet (i.e., at $7n$ in the special case we are looking at), it can directly go to E_{k-1} and traverse the zigzag construction by using the path $(E_{k-1}, E_{k-3}, \ldots, E_1, E_2, E_4, \ldots, E_k)$ and finally return to s^* avoiding the expensive γ-edges connecting consecutive vertices E_j and E_{j+1}. On

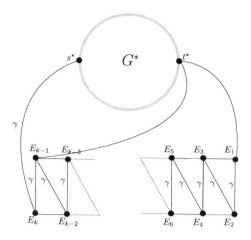

Fig. 3. The zigzag construction

the other hand, if a path arrives at t^* with too high cost (i.e., more than $8n$) it is forced to visit E_1 right after t^* and then to traverse $G \setminus G^*$ via the path $(E_1, E_2, \ldots, E_{k-1})$. Doing so, this path uses k expensive edges of cost γ.   □

For a formal proof of a generalized version of Lemma 1, see [4]. Furthermore, we need to prove the hardness of a modified version of RHP (*Modified Restricted Hamiltonian Path Problem*) for the reductions.

Definition 4 (MRHP). *Let $G = (V, E)$ be a graph where $|V| = n + 1$. Let $s, t \in V$ be two distinct vertices and let $P' = (s, v_1, \ldots, v_{n-1}, t)$ be a Hamiltonian path in G from s to t. The objective is to decide whether a Hamiltonian path P from v_i to t exists in $G \setminus \{s\}$ for some vertex $v_i \in V \setminus \{s, t, v_1\}$.*

Note that, if we would not require $v_i \neq v_1$, this problem would be trivial since, by knowing P', we also know a Hamiltonian path $(v_1, \ldots, v_{n-1}, t)$ in $G \setminus \{s\}$.

Lemma 2. MRHP *is $\mathcal{NP}$-hard.*

Proof. We give a reduction from RHP. Let (G', P'') be an instance of RHP where $P'' = (s', \ldots, t')$. We construct an instance (G, P') of MRHP where $G = G' \cup \{s\}, P' = (s, \overleftarrow{P''}), t = s'$. The given Hamiltonian path is therefore $P' = (s, v_1, \ldots, v_{n-1}, t)$ where $v_1 = t'$. Obviously, this construction can be done in polynomial time. Then, in both problems we want to decide the existence of the same path P (namely $(v_i, \ldots, t)$ in $G \setminus \{s\}$ and $(v_i, \ldots, s')$ in G').   □

Theorem 3. *Let $\varepsilon > 0$. There is no polynomial-time approximation algorithm for $\mathrm{LM}(D^-)$-Δ-DLTSP with a ratio of $(0.5 - \varepsilon)n$.*

Proof. In the following, we construct a graph G^* as described in the Zigzag Lemma (Lemma 1). We will then be able to apply this lemma to show a linear lower bound.

Consider an instance (G', P') of MRHP. Again, we construct a complete, weighted graph $K_{G'}$ as described before in the proof of Theorem 1. We then extend $K_{G'}$ to a complete, weighted graph G^* (which is a subgraph of G) as shown in Figure 4. Again, all costs not shown in the figure are chosen as large as possible, such that the triangle inequality is still satisfied. All vertices in $K_{G'}$ (except for s') get a deadline of $3n + 1$. The deadline vertices D_O and the deadlines d_O are also shown in Figure 4.

Every Hamiltonian path in G^* starting at s^* has to begin with the edge (s^*, s'). After that, all vertices in $K_{G'}$ have to be visited before continuing. One best way to traverse $K_{G'}$ is the known Hamiltonian path $P' = (s', \ldots, t')$. This path arrives at t' at cost $2n + 1$. Due to deadline constraints, there is no other way but ending this path by (D_1, D_2, D_3, D_4) where $D_4 = t^*$. An optimal Hamiltonian path in G^* is therefore $\overline{P} = (s^*, P', D_1, D_2, D_3, t^*)$. The costs are $8n + 1$.

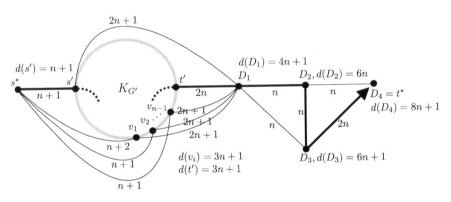

Fig. 4. The graph G^* before the local modification is applied and an optimal Hamiltonian path from s^* to t^*

We then apply the local modification by taking away the deadline from s'. Now things change in G^* as it is not necessary anymore to visit s' before reaching t^*. If there exists a Hamiltonian path P from $v_i \neq v_1$ to t' with cost $(n-1)$ in $K_{G'} \setminus \{s'\}$, an optimal Hamiltonian path in $G^* \setminus \{s'\}$ is the path $(s^*, P, D_1, D_3, D_2, D_4)$. The costs are $(n+1) + (n-1) + 2n + n + n + n = 7n$. This path is then able to "cheaply" traverse G's zigzag construction and visit s' before returning to s^*.

On the other hand, if there is no such path from $v_i \neq v_1$ to t' with cost $(n-1)$ in $K_{G'} \setminus \{s'\}$, $\overline{P}$ stays the best Hamiltonian path in G^* from s^* to t^* respecting the deadlines. Furthermore, due to $c(s^*, v_1) = n + 2$, a best solution cannot use any Hamiltonian path from v_1 to t' of costs $n - 1$ in $K_{G'}$. Similar arguments hold for any other path.

The path therefore arrives at t^* at time $7n$ if and only if the given MRHP-instance is a yes-instance, and not before time $8n + 1$ if it is a no-instance. When extending G^* by the zigzag construction (see Figure 3) we will set $c(E_k, s') = \gamma - (n + 1)$ and the path therefore needs to spend $(\gamma - (n+1)) + (n+1) = \gamma$

when going from E_k to s^* via s'. The Zigzag Lemma then allows us to choose γ in a way such that every $(0.5 - \varepsilon)n$-approximation algorithm could be used to decide MRHP which contradicts $\mathcal{NP} \neq \mathcal{P}$. □

Theorem 4. *Let $\varepsilon > 0$. There is no polynomial-time approximation algorithm for any of the problems* LM(D^+)-Δ-DLTSP, LM($(D \wedge V)^+$)-Δ-DLTSP, *and* LM($(D \wedge V)^-$)-Δ-DLTSP*with a ratio of $(0.5 - \varepsilon)n$, unless $\mathcal{P} = \mathcal{NP}$.*

Again, due to space contraints we have to omit the proof of Theorem 4.

5   Upper Bounds

Since we know a 2.5-approximation algorithm for Δ-DLTSP with a bounded number of deadline vertices (see [6,7]), we can directly apply this upper bound to any reoptimization version. Considering the results from Section 3, there is a gap between the lower and upper bounds.

In Section 4, we proved a lower bound of $(0.5 - \varepsilon)n$ for the general case (i.e., for an unbounded number of deadline vertices) for several local modifications. It is easy to see that this bound is tight: Remember that a feasible solution is part of the input for any of the considered reoptimization versions of Δ-DLTSP. No matter how bad this solution is, it is a $0.5n$-approximation, since any edge of this Hamiltonian cycle cannot be more expensive than 0.5 times the cost of an optimal solution (due to the triangle inequality) and there are exactly n edges in this cycle.

In one case, however, we are able to improve the upper bound by giving a 2-approximation for the reoptimization version in which we add a vertex to G, even for an unbounded number of deadline vertices.

Theorem 5. *There is a 2-approximation algorithm for* LM(V^+)-Δ-DLTSP.

Proof. Let $(G_O, c, \mathcal{D})$ be the given old instance. We will give a simple algorithm $\mathfrak{A}$ for LM(V^+)-Δ-DLTSP that has an approximation ratio of 2. Let $\overline{C} = (s, v_1, \dots, v_{n-1}, s)$ be the given optimal solution with cost Opt_O for the old instance. For the local modification, a vertex v is inserted into G (i.e., $G_N = G_O \cup \{v\}$). The algorithm $\mathfrak{A}$ will simply output $C_{\mathfrak{A}} = (s, v_1, \dots, v_{n-1}, v, s)$.

Since there is no deadline at v and $\overline{C}$ is feasible for the old instance, $C_{\mathfrak{A}}$ is also feasible for the new instance. For proving an approximation ratio of 2 we will need to look at some estimations: Let C_N be an optimal solution for the modified instance with cost Opt_N and let v_i and v_j be the neighbours of v in C_N. We claim that

$$Opt_N \geq Opt_O + c(v_i, v) + c(v, v_j) - c(v_i, v_j). \tag{3}$$

For the proof of Equation (3), suppose $Opt_N < Opt_O + c(v_i, v) + c(v, v_j) - c(v_i, v_j)$. Then $C_N - \{\{v_i, v\}, \{v, v_j\}\} + \{v_i, v_j\}$ is feasible for $(G_O, c, \mathcal{D})$ due to $c(v_i, v_j) \leq c(v_i, v) + c(v, v_j)$ and costs strictly less then Opt_O. This contradicts the optimality of Opt_O for $(G_O, c, \mathcal{D})$.

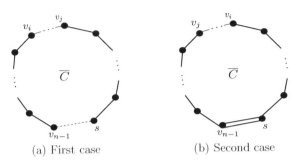

Fig. 5. The two cases of possible locations of v_i and v_j in $\overline{C}$

The costs of $\mathfrak{A}$'s output are $C_{\mathfrak{A}} = Opt_O + c(v_{n-1}, v) + c(v, s) - c(v_{n-1}, s)$. We have to distinguish two locations of v_i and v_j in $\overline{C}$ (see Figure 5). In the first case, we have $c(v_{n-1}, v_i) + c(v_j, s) \leq Opt_O - c(v_i, v_j) - c(v_{n-1}, s)$. In the second case, we have $c(v_{n-1}, v_i) + c(v_j, s) \leq Opt_O - c(v_i, v_j) + c(v_{n-1}, s)$.

It suffices to take the second (weaker) estimation into consideration. Together with Equation (3), we get

$$
\begin{aligned}
cost_{\mathfrak{A}} &= Opt_O + c(v_{n-1}, v) + c(v, s) - c(v_{n-1}, s) \\
&\leq Opt_O + c(v_{n-1}, v_i) + c(v_i, v) + c(v, v_j) + c(v_j, s) - c(v_{n-1}, s) \\
&\leq Opt_O + Opt_O + c(v_{n-1}, s) + c(v_i, v) + c(v, v_j) - c(v_{n-1}, s) - c(v_i, v_j) \\
&= Opt_O + Opt_O + c(v_i, v) + c(v, v_j) - c(v_i, v_j) \\
&\leq Opt_O + Opt_N \\
&\leq Opt_N + Opt_N = 2 \cdot Opt_N.
\end{aligned}
$$

The algorithm $\mathfrak{A}$ is therefore a 2-approximation for $\text{LM}(V^+)$-Δ-DLTSP. $\qquad\square$

References

1. Archetti, C., Bertazzi, L., Speranza, M.G.: Reoptimizing the traveling salesman problem. Networks 42, 154–159 (2003)
2. Ausiello, G., Escoffier, B., Monnot, J., Paschos, V.T.: Reoptimization of Minimum and Maximum Traveling Salesman's Tours. In: Arge, L., Freivalds, R. (eds.) SWAT 2006. LNCS, vol. 4059, pp. 196–207. Springer, Heidelberg (2006)
3. Bilò, D., Böckenhauer, H.-J., Hromkovič, J., Královič, R., Mömke, T., Widmayer, P., Zych, A.: Reoptimization of Steiner trees. In: Proc. of the 11th Scandinavian Workshop on Algorithm Theory (SWAT 2008) (to appear, 2008)
4. Böckenhauer, H.-J., Forlizzi, L., Hromkovic, J., Kneis, J., Kupke, J., Proietti, G., Widmayer, P.: Reusing optimal TSP solutions for locally modified input instances (extended abstract). In: Proc. of the 4th IFIP International Conference on Theoretical Computer Science (IFIP TCS 2006), pp. 251–270. Springer, Heidelberg (2006)
5. Böckenhauer, H.-J., Forlizzi, L., Hromkovič, J., Kneis, J., Kupke, J., Proietti, G., Widmayer, P.: On the approximability of TSP on local modifications of optimally solved instances. Algorithmic Operations Research 2(2), 83–93 (2007)

6. Böckenhauer, H.-J., Hromkovič, J., Kneis, J., Kupke, J.: On the Approximation Hardness of Some Generalizations of TSP. In: Arge, L., Freivalds, R. (eds.) SWAT 2006. LNCS, vol. 4059, pp. 184–195. Springer, Heidelberg (2006)
7. Böckenhauer, H.-J., Hromkovič, J., Kneis, J., Kupke, J.: The parameterized approximability of TSP with deadlines. Theory of Computing Systems 41(3), 431–444 (2007)
8. Böckenhauer, H.-J., Hromkovič, J., Královič, R., Mömke, T., Rossmanith, P.: Reoptimization of Steiner trees: changing the terminal set. Theoretical Computer Science (to appear)
9. Böckenhauer, H.-J., Hromkovič, J., Mömke, T., Widmayer, P.: On the Hardness of Reoptimization. In: Geffert, V., Karhumäki, J., Bertoni, A., Preneel, B., Návrat, P., Bieliková, M. (eds.) SOFSEM 2008. LNCS, vol. 4910, pp. 50–65. Springer, Heidelberg (2008)
10. Cordeau, J.-F., Desaulniers, G., Desrosiers, J., Solomon, M.M., Soumis, F.: VRP with time windows. In: Toth, P., Vigo, D. (eds.) The Vehicle Routing Problem, SIAM 2001, pp. 157–193 (2001)
11. Escoffier, B., Milanič, M., Paschos, V.T.: Simple and fast reoptimizations for the Steiner tree problem. DIMACS Technical Report 2007-01 (2007)
12. Greenberg, H.: An annotated bibliography for post-solution analysis in mixed integer and combinatorial optimization. In: Woodruff, D.L. (ed.) Advances in Computational and Stochastic Optimization, Logic Programming, and Heuristic Search, pp. 97–148. Kluwer Academic Publishers, Dordrecht (1998)
13. Královič, R., Mömke, T.: Approximation hardness of the traveling salesman reoptimization problem. In: Proc.of the 3rd Doctoral Workshop on Mathematical and Engineering Methods in Computer Science (MEMICS 2007), pp. 97–104 (2007)
14. Libura, M.: Sensitivity analysis for minimum Hamiltonian path and traveling salesman problems. Discrete Applied Mathematics 30, 197–211 (1991)
15. Libura, M., van der Poort, E.S., Sierksma, G., van der Veen, J.A.A.: Stability aspects of the traveling salesman problem based on k-best solutions. Discrete Applied Mathematics 87, 159–185 (1998)
16. Sotskov, Y.N., Leontev, V.K., Gordeev, E.N.: Some concepts of stability analysis in combinatorial optimization. Discrete Appl.Math. 58, 169–190 (1995)
17. Van Hoesel, S., Wagelmans, A.: On the complexity of postoptimality analysis of 0/1 programs. Discrete Applied Mathematics 91, 251–263 (1999)

On the Shortest Linear Straight-Line Program for Computing Linear Forms

Joan Boyar[1,*], Philip Matthews[**], and René Peralta[2]

[1] Dept. of Math. and Computer Science, University of Southern Denmark
joan@imada.sdu.dk
[2] Computer Security Division, Information Technology Laboratory, NIST
rene.peralta@nist.gov

Abstract. We study the complexity of the Shortest Linear Program (SLP) problem, which is to minimize the number of linear operations necessary to compute a set of linear forms. SLP is shown to be NP-hard. Furthermore, a special case of the corresponding decision problem is shown to be MAX SNP-Complete.

Algorithms producing cancellation-free straight-line programs, those in which there is never any cancellation of variables in GF(2), have been proposed for circuit minimization for various cryptographic applications. We show that such algorithms have approximation ratios of at least 3/2 and therefore cannot be expected to yield optimal solutions to non-trivial inputs.

1 Introduction

Let $\mathbb{F}$ be an arbitrary field and let

$$\alpha_{1,1}x_1 + \alpha_{1,2}x_2 + \ldots + \alpha_{1,n}x_n$$
$$\alpha_{2,1}x_1 + \alpha_{2,2}x_2 + \ldots + \alpha_{2,n}x_n$$
$$\ldots$$
$$\alpha_{m,1}x_1 + \alpha_{m,2}x_2 + \ldots + \alpha_{m,n}x_n$$

be linear forms where the $\alpha_{i,j}$'s are constants from $\mathbb{F}$ and the x_i's are variables over $\mathbb{F}$. We would like to design an algorithm for computing the linear forms given the x_i's as input. We consider this question in the model of computation known as *linear straight-line programs*. A linear straight-line program is a variation on a straight-line program which does not allow multiplication of variables. That is, every line of the program is of the form $u := \lambda v + \mu w$; where λ, μ are in $\mathbb{F}$ and v, w are variables. Some of the lines are output lines; these are the lines where the linear forms in the set are produced. For brevity, we will use the terms *linear*

---

* Partially supported by the Danish Natural Science Research Council. Some of this work was done while visiting the University of California, Irvine, and the Universtiy of Aarhus.
** This work was done while at the University of Aarhus.

E. Ochmański and J. Tyszkiewicz (Eds.): MFCS 2008, LNCS 5162, pp. 168–179, 2008.

programs or simply *programs* to refer to linear straight-line programs. The *length* of the program is the number of lines it contains. A program is *optimal* if it is of minimum length.

The linear straight-line program model (see [7] for a discussion of linear complexity) has the advantage of being very structured, but is nevertheless optimal to within a constant factor as compared with arbitrary straight-line programs when the computation is over an infinite field. Over finite fields the optimality of linear straight-line programs is unknown[1], but we restrict our attention to this form and consider minimizing the length of the program.

The standard algorithm for computing the linear forms Ax where A is an $m \times n$ matrix requires $m(n - 1)$ operations. Savage [9], however, showed that $O(mn/\log_r m)^2$ is sufficient in many cases, including computations over $GF(2)$ if $m \geq 4$. Williams [11] improved this to $O(n^2/\log^2 n)$ on a RAM with word length $\Theta(n)$ for n by n matrices over finite semirings. In contrast, Winograd [12] has shown that most sets of linear forms have a non-linear complexity in the straight-line program model; in fact, for a "random" $m \times n$ matrix A the probability is high that its complexity is $\Omega(mn)$ (for infinite fields). However, there are non-trivial matrices which can be computed considerably faster than this.

Over $GF(2)$, finding the shortest linear straight-line program is equivalent to finding a circuit with only XOR gates and minimizing the number used. Linear forms have many applications, especially to problems in scientific computation, and there has been considerable success in finding efficient algorithms for computing them in special cases. The best known example is the Fast Fourier Transform, an $O(n \log n)$ algorithm, discovered by Cooley and Tukey in 1965 [5].

In section 2 we show that finding the shortest linear straight-line program is NP-hard. This can be seen in relation to Håstad's result [6] showing that tensor rank is NP-hard and thus finding the minimum bilinear program for computing bilinear forms is NP-hard.

In section 2.2 the NP-hardness result is used to prove a special case of the problem MAX SNP-Complete [8] (and also APX-Complete), which means that there is no ϵ-approximation algorithm for the problem unless P=NP [1].

A linear straight-line program over GF(2) is said to be a *cancellation-free straight-line program* if, for every line of the program $u := v+w$, none of variables in the expression for v are also present in the expression for w, i.e., there is no cancellation of variables in the computation. A small example showing that the optimal linear program is not always cancellation-free over $GF(2)$ is:

$$x_1 + x_2;\ x_1 + x_2 + x_3;\ x_1 + x_2 + x_3 + x_4;\ x_2 + x_3 + x_4.$$

It is not hard to see that the optimum cancellation-free straight-line program has length 5. A solution of length 4 which allows cancellations is

$$v_1 = x_1 + x_2;\ v_2 = v_1 + x_3;\ v_3 = v_2 + x_4;\ v_4 = v_3 + x_1.$$

---

[1] It is not known if multiplication of variables can ever be used to reduce program length when the program outputs only linear functions.

[2] The base r is the size of the set containing the coefficients.

In section 3 we show that the approximation ratio for cancellation-free techniques is at least $3/2$.

2   Hardness of Finding Short Linear Straight-Line Programs

The problem SHORTEST LINEAR PROGRAM (SLP) is as follows: Given a set of linear forms E over a field F, find a shortest linear program to compute E.

2.1   NP-Hardness

In order to prove NP-hardness, we consider the corresponding decision problem, SLPd: Given a set of linear forms E over a field F and a positive integer k, determine if there exists a straight-line linear program with at most k lines which computes E.

We will prove SLPd NP–hard, even if the constants in the set of linear forms to be computed are only zeros and ones. Furthermore, if the field F is finite, then SLPd is easily seen to be in NP, so SLPd is NP–complete over finite fields.[3]

The interest of this section is not just in the final result that SLP is NP–hard, but also in the method used to prove it. In particular, most of this section is devoted to the proof of Lemma 1, which gives the exact complexity for sets of linear forms of a certain simple type. This proof is *algorithmic* in form, and its algorithmic nature can be exploited to prove a further result in subsection 2.2.

In order to show NP-hardness, we reduce from VERTEX COVER. A *vertex cover* of a graph $G = (V, E)$ is a subset V' of V such that every edge of E is incident with at least one vertex of V'. VERTEX COVER is defined as follows: Given a graph $G = (V, E)$ and an integer k, determine if there exists a vertex cover of size at most k.

The following polynomial-time reduction f transforms an arbitrary graph $G = (V, E)$ and a bound k to a set of linear forms with another bound $\bar{k}$. The input variables are $X = V \cup \{z\}$, where z is a distinguished variable not occurring in V. The linear forms are $\bar{E} = \{ z + a + b \mid (a, b) \in E \}$, and the program length we ask about is $\bar{k} = k + |\bar{E}|$. This is an instance of SLPd, and it is clear that $f(G, k) = (\bar{E}, X, \bar{k})$ can be produced in polynomial time. We call a set of linear expressions in this restricted form, $z + x_i + x_j$, a set of **z-expressions**.

Before we proceed, we illustrate with an example: The graph, G, in Figure 1 has a vertex cover of size $k = 3$: $\{a, c, e\}$. The corresponding instance of SLPd, $f(G, 3)$ is $\bar{E} = \{z + a + b, z + b + c, z + c + d, z + d + e, z + e + f, z + a + f, z + c + g, z + e + g\}$, $X = \{z, a, b, c, d, e, f, g\}$, and $\bar{k} = 3 + 8$. A linear program for this of size 11 is

---

[3] We avoid the discussion of models for dealing with infinite fields, such as in [10] or [2], by proving NP-hardness when the constants in the forms are only zeros and ones and showing that a shortest linear straight-line program for the forms considered can be created with only zeros and ones as constants.

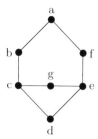

Fig. 1. Graph with 8 edges and cover size 3

$$v_1 := z + a; \quad v_2 := z + c; \quad v_3 := z + e; \quad v_4 := v_1 + b;$$
$$v_5 := v_2 + b; \quad v_6 := v_2 + d; \quad v_7 := v_3 + d; \quad v_8 := v_3 + f;$$
$$v_9 := v_1 + f; \quad v_{10} := v_2 + g; \quad v_{11} := v_3 + g;$$

where the computation of v_1, v_2, and v_3 correspond to the vertex cover in the graph G, and the remaining operations produce the eight forms in $\bar{E}$.

A *cover* for a set $\bar{E}$ of z-expressions is a subset W of $X - \{z\}$ such that every expression in $\bar{E}$ contains at least one variable in W. Note that if $(\bar{E}, X, \bar{k}) = f(G, k)$, a cover for $\bar{E}$ trivially defines a vertex cover for the graph G and vice versa.

Lemma 1. *Let $(\bar{E}, X)$ be a set of z-expressions without repetitions; that is, $\bar{E}$ is a set of expressions of the form $z + x_i + x_j$, where x_i, x_j are distinct variables in X, z is a distinguished variable in X, and no two of these z-expressions contain exactly the same variables. There is a cover of $\bar{E}$ of size k if and only if there is a linear straight-line program P for $\bar{E}$ of length $\bar{k} = k + |\bar{E}|$. In addition, given a linear straight-line program P for $\bar{E}$, a cover for $\bar{E}$ of size $|P| - |\bar{E}|$ can be computed in polynomial time.*

Proof. We will refer to the elements of $X - \{z\}$ as "the variables" and z as "the symbol", though as an element of a linear program, z is also an input variable.

Given a cover W of size k for $\bar{E}$, a (cancellation-free) linear straight-line program for $\bar{E}$ can be created consisting of $z + w_i$ for each $w_i \in W$, followed by linear expressions computing each output, created by adding a second variable to the appropriate $z + w_i$. This program has length $k + |\bar{E}|$.

It remains to be shown that, given a linear straight-line program P for $\bar{E}$, we can efficiently find a cover, W, for $\bar{E}$ of size no more than $|P| - |\bar{E}|$. This cover is computed by associating elements of $X - \{z\}$ with some non-output lines of the program—W will then be the union of all those variables so associated. Since we will assign at most one element of $X - \{z\}$ to each non-output line, the cover is of size at most $|P| - |\bar{E}|$.

Let $F^{(i)}$ be the linear function computed at line i. It will be convenient to use the notation $F^{(i)}$ to refer both to the function and to the minimal formal expression $\sum_j \beta_{i,j} x_j$ where the x_j's are distinct and the $\beta_{i,j}$'s are non-zero field elements. The association of variables with lines of the program will be denoted by a mapping $m : \mathbb{N} \to X \cup \{\emptyset\}$. Initially, we set $m(i) = \emptyset$ for all lines i. At any point in time, the current partial cover W is the set of all variables that are assigned to some $m(i)$.

The algorithm works as follows. Starting at the first line of P, the algorithm associates with each non-output line i a variable in $X - \{z\}$ which occurs in the formal expression computed at line i and which is currently unassociated (if there is no such variable, the line is assigned the null symbol, $\emptyset$). When an output is reached, the algorithm checks if the set W of all variables currently assigned covers that output, i.e. if there is some variable in W which occurs in the formal expression computed at that output line. If this is not the case, then a fix-up procedure is invoked. This fix-up procedure changes some of the associations until all the output expressions up to that point are covered. After the algorithm has terminated, all the output expressions will be covered, so W is the desired cover, and $|P| \geq |W| + |\bar{E}|$. Note that if the straight-line program P had been restricted to be cancellation-free, the fix-up procedure would never be necessary; it is only called if an output line was produced as a linear combination of two lines, where at one of those lines a cancelled variable was added to the cover, W.

We define the two properties that the algorithm seeks to establish for each line l of the program.

Property 1. If line l is not an output, either all variables in $F^{(l)}$ are in W, or some variable in both W and $F^{(l)}$ is associated uniquely with line l.

Property 2. There is at most one variable in $F^{(l)}$ which is not in W.

In terms of these two properties, the algorithm in Figure 2 can be described as follows. Given that Properties 1 and 2 hold for lines 1 to $i - 1$, establish Property 1 for line i and check if Property 2 holds for line i. If not, call the fix-up procedure.

Claim. If Properties 1 and 2 hold for lines 1 through $l-1$, after line l is processed, if $F^{(l)}$ is not an output, then Properties 1 and 2 hold for line l.

```
W ← ∅
for i = 1 to |P| do
    m(i) ← ∅
    if F^(i) is not an output then
        if ∃ a variable x in F^(i), but not in W then
            choose x
            m(i) ← x
    else { F^(i) is an output}
        if F^(i) contains more than one variable not in W then
            Fix-up(i, i)
```

Fig. 2. Computing the cover W

Proof. This holds by induction. Line 1 of P contains at most two variables, and one of them is assigned to $m(1)$ and W, so it holds initially. Suppose line i has the form $v_i := \lambda \cdot v_{i'} + \mu \cdot v_{i''}$.[4] By assumption, Property 2 holds for $F^{(i')}$ and $F^{(i'')}$, so there are at most two variables in $F^{(i)}$ but not in W before line i is processed. If there is at most one such variable, we are done. If there are exactly two such variables, then $m(i)$ is assigned some variable not in W. Thus, Property 1 holds. Since that variable is also set in W, $F^{(i)}$ has at most one variable that does not occur in W and Property 2 also holds.   □

Suppose the Fix-up procedure is called for line i:

$$v_i := \lambda \cdot v_{i'} + \mu \cdot v_{i''},$$

which produces the output expression $z + a + b$. If both a and b are present in the expression for line i' or both are in the expression for line i'', then at least one of a or b is in W. Since neither is there, we may assume, without loss of generality, that a is not present in line i' and b is not present in i''. All other variables present in line i' must also be present in line i''. In addition, at least one of those two lines is not an output, since exactly one contains the symbol z. Assume that line i' is not an output. Since it contains b, which is not in W, $m(i') \neq \emptyset$. Suppose $m(i') = c$.

Fix-up(i, l)
{ i is the current line being fixed; l is original line being fixed }
{ line i, $v_i := \lambda v_{i'} + \mu v_{i''}$, produces the expression $z + a + b$,
  a is not present in line i' and b is not present in i''}
{ line i' is not an output and $m(i') = c$}
set $m(i') \leftarrow b$
$W \leftarrow (W \cup \{b\}) \setminus \{c\}$
for $j \leftarrow 1$ **to** l **do**
    if $F^{(j)}$ is not an output **then**
        if $m(j) = \emptyset$ and $c \in F^{(j)}$ **then**
            $m(j) \leftarrow c$
            $W \leftarrow W \cup \{c\}$
            break { exit for loop }
for $j \leftarrow 1$ **to** l **do**
    if $F^{(j)}$ is an output **then**
        if $|F^{(j)} \setminus W| > 1$ **then** Fix-up(j, l)

Fig. 3. The Fix-up procedure

The fix-up procedure, as defined in Figure 3, backs up to line i', changes the mapping m to put b into W instead of c, and then scans to ensure, first, that Property 1 still holds, and then that Property 2 still holds. Since this change in

---

[4] So that all program lines can expressed in this way, we let $v_0 = z$ and $v_{-i} = x_i$ for each variable x_i.

the mapping may upset lines that were previously okay, this adjustment of the mapping may occur more than once.

Since $b \notin W$, it is okay to set $m(i')$ to b. The only way this can cause Property 1 to fail is that a line j might have $m(j) = \emptyset$, even though the expression there involved a c and c is no longer in W. The first "for" loop in "Fix-up" corrects this.

The removal of c from W may also cause Property 2 to fail. Note that this can only happen for an output line; the above claim still holds. Some of the failures at outputs may be rectified by the adjustment fixing Property 1. "Fix-up" is called recursively to fix the others.

We turn to the proof of termination.

Let $k_1, k_2, \ldots$ be the sequence of line numbers for output lines which require a call to the fix-up procedure, and let $W_1, W_2, \ldots$ be the corresponding values of W, the covers just before the fix-up procedure is called for the corresponding lines. Let $k_0 = i$ and define W_0 to be the value of the cover when the fix-up procedure is first called. Note that no two adjacent members of $k_0, k_1, k_2, \ldots$ are equal.

Let j be an index for which $k_j < k_{j+1}$. We claim that $|W_j| < |W_{j+1}|$. The size of the cover never decreases as the only operations done to change it are swaps and additions, so the claim follows if we show that a variable is added to the cover by the fix-up procedure when going from line k_j to k_{j+1}.

Consider how the fix-up procedure operates between the calls at lines k_j and k_{j+1}. Suppose that line k_j is

$$v_{k_j} := \lambda \cdot v_{k_j'} + \mu \cdot v_{k_j''}$$

We know that $k_j' < k_j'' < k_j < k_{j+1}$. Suppose the formal expressions computed at these lines are

$$F^{(k_j')} = (b - c - V)/\lambda; \quad F^{(k_j'')} = z + a + c + V;$$
$$F^{(k_j)} = z + a + b; \quad F^{(k_{j+1})} = \ldots,$$

where V is a sum of some variables (not including z, a, b, c). For line k_j to have caused a call to "Fix-up", neither a nor b could have been in the cover W_j. Thus the algorithm first visited line k_j' and changed the mapping $m(k_j')$ from c to b, then executed the first "for" loop, correcting lines not satisfying Property 1, and finally moved down the program, checking each line for Property 2, until reaching line k_{j+1}. But this means that Property 2 held at line k_j'' and this could only have happened if a or c was in the cover (if line k_j'' is not an output, it might have been added there). Since neither of them were in the cover immediately after the swap of b for c at line k_j', one of them must have been added by the fix-up procedure at one of the lines in between. Thus $|W_j| < |W_{j+1}|$.

Hence for each j where $k_j < k_{j+1}$, the size of the cover increases. Moreover, since k is always positive, there can be at most n^2 lines visited between these increases in the cover size (where n is the length of the program). And since $|W| < |X| < n$, it follows that the whole algorithm requires at most $O(n^3)$ time (The fact that the execution time is polynomial is irrelevant for the purposes of showing NP-hardness, but will be important later.). This completes the proof of Lemma 1.   □

The following theorem follows immediately, since we have given a polynomial time reduction from VERTEX COVER, which is NP-complete.

Theorem 1. *For any field* $\mathbb{F}$, *SHORTEST LINEAR PROGRAM is NP-hard.*

For finite fields, it is easy to see that SLPd $\in$ NP. Thus we have

Theorem 2. *For any finite field* $\mathbb{F}$, *the decision version of SHORTEST LINEAR PROGRAM is NP-Complete.*

Note that in the proof of Lemma 1, if the straight-line program P had been restricted to be cancellation-free, the proof would have been easier, because the fix-up procedure would never be necessary; it is only called if an output line was produced as a linear combination of two lines, where at one of those lines a cancelled variable was added to the cover, W. This immediately gives us the following:

Theorem 3. *For any finite field* $\mathbb{F}$, *SHORTEST LINEAR PROGRAM is NP-Complete even if the programs produced are restricted to being cancellation-free.*

2.2   Limits to Approximation

The major result of the previous section is that it is NP–hard to find an optimal linear program for computing a set of linear forms. Thus it is natural to turn our attention to approximation algorithms for this problem. In this section, we show that SHORTEST LINEAR PROGRAM has no ϵ–approximation scheme unless P=NP. Recall that these are families of algorithms, one for each $\epsilon > 0$, which are polynomial time and achieve an approximation ratio of $1 + \epsilon$. We use a concept called MAX SNP–completeness, which was introduced by Papadimitriou and Yannakakis [8]. Arora et.al. [1] have shown that no MAX SNP–complete problem has an ϵ–approximation scheme unless P=NP. We show that BOUNDED Z-EXPN (defined below), is MAX SNP–complete, showing that there is no ϵ–approximation scheme for the generalization, SHORTEST LINEAR PROGRAM, unless P=NP.

MAX SNP is a complexity class of optimization problems. It is contained within NP in the sense that the decision versions of the problems are all in NP. Papadimitriou and Yannakakis [8] proved that many problems are MAX SNP–complete, including the following: BOUNDED VERTEX COVER: Given a graph with maximum vertex degree bounded by a constant b, find the smallest vertex cover.

To talk about completeness for this class, we need a notion of reduction. The reductions Papadimitriou and Yannakakis defined, called **L-reductions**, preserve the existence of ϵ-approximation schemes. The following definitions and propositions are taken directly from the original paper.

Let Π and Π' be two optimization (maximization or minimization) problems, and let f be a polynomial-time transformation from problem Π to problem Π'. We say that f is an **L-reduction** if there are constants $\alpha, \beta > 0$ such that for each instance I of Π, the following two properties are satisfied:

(a) The optima of I and $f(I)$, written $\mathrm{OPT}(I)$ and $\mathrm{OPT}(f(I))$ respectively, satisfy the relation $\mathrm{OPT}(f(I)) \leq \alpha\mathrm{OPT}(I)$.

(b) For any solution of $f(I)$ with cost c', we can find in polynomial time a solution of I with cost c such that $|c - \mathrm{OPT}(I)| \leq \beta|c' - \mathrm{OPT}(f(I))|$.

The constant β will usually be 1. The following two propositions, stated in [8], follow easily from the definition.

Proposition 1. *L-reductions compose.*

Proposition 2. *If Π L-reduces to Π' and if there is a polynomial-time approximation algorithm for Π' with worst-case error ϵ, then there is a polynomial-time approximation algorithm for Π with worse-case error $\alpha\beta\epsilon$.*

BOUNDED Z-EXPN is the following problem: Given a set of z-expressions (as defined in Theorem 1) in which each non-z variable appears at most b times (b is a fixed constant), generate an optimal linear program for computing the expressions (over some fixed field F).

Theorem 4. *BOUNDED Z-EXPN is* Max *SNP–complete.*

Proof. First, we will show that BOUNDED Z-EXPN is in Max SNP. To show membership in Max SNP, we will exhibit an L-reduction of BOUNDED Z-EXPN to Bounded Vertex Cover, a problem in Max SNP.

For every non-z variable x_i, we associate a vertex $\bar{x}_i$. The L-reduction f maps z-expressions to edges as follows: $f(\text{``}z + x_i + x_j\text{''}) = \text{``edge } (i, j)\text{''}$. Since variable occurrences are bounded by b in BOUNDED Z-EXPN, the vertex degrees will by bounded by b in the graph.

We proved in the previous section that a set of z-expressions can be optimally computed by first computing $z + x_i$ for those x_i which are in the minimum vertex cover, and then using these intermediate results to compute the z-expressions. Thus $\mathrm{OPT}(f(I)) + |E| = \mathrm{OPT}(I)$ where $|E|$ is both the number of z-expressions and the number of edges in the graph.

We claim that this reduction is an L-reduction. Property (a) is satisfied because the equation above implies that $\mathrm{OPT}(f(I)) \leq \mathrm{OPT}(I)$. Property (b) is satisfied because, from a vertex cover, we can build a linear program which computes the z-expressions in the manner described above. This gives $c = \mathrm{OPT}(I) + [c' - \mathrm{OPT}(f(I))]$.

To show that the problem is Max SNP–hard we reverse the reduction so that it goes from Bounded Vertex Cover to Bounded Z-EXPN. The function f now maps "edge (i, j)" into "$z + x_i + x_j$".

Proof of Property (a): By Lemma 1 we have that $\mathrm{OPT}(I) + |E| = \mathrm{OPT}(f(I))$. Since the maximum degree in the graph is bounded by b and every edge must be adjacent to at least one vertex of the cover, there can be at most $b \cdot \mathrm{OPT}(I)$ edges, of the cover. Thus $\mathrm{OPT}(f(I)) \leq (b + 1)\mathrm{OPT}(I)$.

Proof of Property (b): The proof of Lemma 1 gave a polynomial-time procedure for converting any linear program computing a set of z-expressions into a vertex cover for the corresponding graph. By inspecting this procedure, one sees that $c = \mathrm{OPT}(I) + [c' - \mathrm{OPT}(f(I))]$.   □

The fact that BOUNDED Z-EXPN is complete for the class MAX SNP implies that there is no ϵ-approximation scheme for it unless P=NP. In fact, Clementi and Trevisan [4] have shown that BOUNDED VERTEX COVER is not approximable within $16/15 - \epsilon$ for sufficiently large maximum degree. By Proposition 2, this means that there is no $1+(1/15-\epsilon)/\alpha\beta = 1+(1/15-\epsilon)/(1+b)$-approximation algorithm for SLP unless P=NP. The fact that BOUNDED Z-EXPN is *in* the class MAX SNP means that there is an approximation algorithm for it with a constant approximation ratio. In fact, it is obvious that Z-EXPN, even without the boundedness constraint, has an approximation algorithm with a constant approximation. The straight-forward linear straight-line program for computing the $|E|$ forms only requires $2|E|$ lines, and every straight-line program for E must contain at least $|E|$ lines (assuming no repetitions within the set E). Thus, the straight-forward algorithm comes within a factor of 2 of optimal. Moreover, since there is an approximation algorithm for vertex cover which comes within a factor of two of optimal, we can do even better for Z-EXPN. Since the optimal linear program contains $|W| + |E|$ steps, where W is the minimum vertex cover, by Lemma 1, there is an algorithm which takes $2|W|+|E|$ steps. Since $|W| < |E|$, the ratio $(2|W|+|E|)/(|W|+|E|)$ is at most $3/2$, so there is a $(3/2)$-approximation algorithm for Z-EXPN. There are, however, no known approximation algorithms which obtain a constant ratio for the general SHORTEST LINEAR PROGRAM problem.

3   A Lower Bound on the Approximation Ratio for Cancellation-Free Techniques

As mentioned in the introduction, restricting the search for optimal straight-line programs for computing linear forms over GF(2) to cancellation-free programs can lead to sub-optimal solutions. In our counter-example, the optimal cancellation-free program has length $\frac{5}{4}$ that of the true shortest program. It is natural to ask how close to optimal cancellation-free programs can get as the number of variables increases. In this section we show that the best cancellation-free straight-line programs are not guaranteed to even have length within a factor $3/2$ that of the shortest straight-line linear program.

The following construction uses two integer parameters k and n, which can be made large to make the $3/2$ inapproximability result hold asymptotically. The parameter k is the number of variables in a *block*, and n is the number of distinct blocks. Blocks have disjoint sets of variables: Block i, where $0 \leq i \leq n - 1$, is the linear form $b_i = x_{ik+1} + x_{ik+2} + ... + x_{(i+1)k}$. The construction produces a linear straight-line program which is not cancellation-free. All intermediate linear forms (the linear forms produced at each line of the program) computed by this straight-line linear program will belong to the set of required outputs. The first part of the linear straight-line program will produce sums of consecutive pairs of blocks $s_i = b_i + b_{i+1}$, for $0 \leq i \leq n - 2$, mixing the variables in the two blocks in such a way that also producing a single block alone would require extra additions compared to the program here. Then, pairs of these consecutive sums are computed, $p_i = s_i + s_{i+1}$, for $0 \leq i \leq n - 3$. Each p_i is computed with only one further addition, but the two s_i's added share a common block which is cancelled, so $p_i = b_i + b_{i+2}$. We express this linear program, denoted P, using **for** loops in Figure 4, but for any fixed k and n it is a straight-line program of length $k(n - 1) + (k - 1)(n - 1) + n - 2 = 2kn - 2k - 1$.

$$
\begin{aligned}
&\textbf{for } i = 1 \text{ to } k(n - 1) \textbf{ do} \\
&\qquad u_i := x_i + x_{i+k} \\
&\textbf{for } i = 0 \text{ to } n - 2 \textbf{ do} \\
&\qquad s_i := u_{ik+1} + u_{ik+2} \\
&\qquad \textbf{for } j = 1 \text{ to } k - 2 \textbf{ do} \\
&\qquad\qquad s_i := s_i + u_{ik+j+2} \\
&\textbf{for } i = 0 \text{ to } n - 3 \textbf{ do} \\
&\qquad p_i := s_i + s_{i+1}
\end{aligned}
$$

Fig. 4. Straight-line program with cancellations

We claim that an optimal cancellation-free program (for computing all the linear forms which are the result of some line in this program) does at least enough additional operations to compute each of the blocks, and this would require at least $n(k - 1)$ additional lines. Let F denote the set consisting of the first $(2k - 1)(n - 1)$ lines of P, and let L denote the set of the last $n - 2$ lines. All of the $2kn - 2k - 1$ lines output by the above straight-line program are linear forms which must be output. The lines in L are the only ones with cancellations. None of the results from the lines in F can be used to compute the lines in P, because, for any two lines $f \in F$ and $l \in L$, f contains at least one variable which is not present in the form calculated by l. It is conceivable that some of the non-output results computed in the process of producing the outputs in L could be used in computing those in F, but, since they are all outputs, at least one extra operation is needed to produce each output from F. Thus, we can consider computing the outputs in L independently from those in F.

Blocks b_2 through b_{n-3} each appear in two of the outputs from L, but there is no other overlap between the outputs in L. Thus, the only reuse of forms computed which is possible is within the blocks. An optimal way to compute the forms in L is to first compute each of the n blocks, using $k - 1$ additions for each. After this, each form in L can be created by adding two blocks together, using one addition for each, as in P. The computation of the blocks gives $n(k - 1)$ extra additions, for a total of $3kn - 2k - n - 1$ additions. Asymptotically, the ratio $\frac{3kn-2k-n-1}{2kn-2k-1}$ is $3/2$ for large n and k.

Theorem 5. *Any algorithm for computing short straight-line linear programs, which only produces cancellation-free straight-line programs, has an approximation ratio of at least $3/2$.*

4   Conclusions and Open Problems

The result that SHORTEST LINEAR PROGRAM is NP-hard indicates that using heuristic techniques, is more realistic than expecting to find the smallest subcircuits for linear parts of a Boolean circuit. The result that a special case of SHORTEST LINEAR PROGRAM is MAX SNP-Complete indicates that there is a limit to how well these heuristic techniques can be guaranteed to do.

Since cancellation-free techniques cannot produce linear straight-line programs which are within a factor 3/2 of being optimal, new heuristics which are not restricted to cancellation-free operations are being developed [3]. It would still be interesting to determine how well cancellation-free techniques can do. We expect that the lower bound of 3/2 can be raised somewhat.

References

1. Arora, S., Lund, C., Motwani, R., Sudan, M., Szegedy, M.: Proof verification and the hardness of approximation problems. Journal of the Association for Computing Machinery 45, 501–555 (1998)
2. Blum, L., Shub, M., Smale, S.: On a theory of computation and complexity over the real numbers: NP-completeness, recursive functions and universal machines. Bull. Amer. Math. Soc. 21, 1–46 (1989)
3. Boyar, J., Peralta, R.: On building small circuits (2008) (manuscript in preparation)
4. Clementi, A.E.F., Trevisan, L.: Improved non-approximability results for vertex cover with density constraints. In: Computing and Combinatorics, pp. 333–342 (1996)
5. Cooley, J.W., Tukey, J.W.: An algorithm for the machine calculation of complex Fourier series. Math. Comp. 19, 297–301 (1965)
6. Håstad, J.: Tensor rank is NP-Complete. J. Algorithms 11(4), 644–654 (1990)
7. Shokrollahi, M.A., Bürgisser, P., Clausen, M.: Algebraic Complexity Theory, ch. 13. Springer, Heidelberg (1997)
8. Papadimitriou, C., Yannakakis, M.: Optimization, approximation, and complexity classes. Journal of Computer and System Sciences 43, 425–440 (1991)
9. Savage, J.E.: An algorithm for the computation of linear forms. SICOMP 3(2), 150–158 (1974)
10. Valiant, L.G.: Completeness classes in algebra. In: Proceedings of the 11th Annual ACM Symposium on the Theory of Computing, pp. 249–261 (1979)
11. Williams, R.: Matrix-vector multiplication in sub-quadratic time (some preprocessing required). In: Proceedings of the Eighteenth Annual ACM-SIAM Symposium on Discrete Algorithms, pp. 995–1001 (2007)
12. Winograd, S.: On the number of multiplications necessary to compute certain functions. Comm. Pure and Applied Math. 23, 165–179 (1970)

Flip Algorithm for Segment Triangulations

Mathieu Brévilliers, Nicolas Chevallier, and Dominique Schmitt

Laboratoire LMIA, Université de Haute-Alsace
4, rue des Frères Lumière, 68093 Mulhouse Cedex, France
{Mathieu.Brevilliers,Nicolas.Chevallier,Dominique.Schmitt}@uha.fr

Abstract. Given a set S of disjoint line segments in the plane, which we call sites, a segment triangulation of S is a partition of the convex hull of S into sites, edges, and faces. The set of faces is a maximal set of disjoint triangles such that the vertices of each triangle are on three distinct sites. The segment Delaunay triangulation of S is the segment triangulation of S whose faces are inscribable in circles whose interiors do not intersect S. It is dual to the segment Voronoi diagram. The aim of this paper is to show that any given segment triangulation can be transformed by a finite sequence of local improvements in a segment triangulation that has the same topology as the segment Delaunay triangulation. The main difference with the classical flip algorithm for point set triangulations is that local improvements have to be computed on non convex regions. We overcome this difficulty by using locally convex functions.

1 Introduction

In 1977, Lawson [16] has shown that any given triangulation of a planar point set can be transformed in a Delaunay triangulation (one whose triangles' circumcircles are empty of sites) by a sequence of local improvements. Every local improvement consisted in flipping a diagonal of a convex quadrilateral to the other diagonal. Since then, several extensions of flip algorithms have been proposed. On the one hand, they have been investigated in higher dimensions. The algorithm does not work as such in dimensions higher than two because flips should be applied to non convex polyhedrons, leading to geometrically unrealizable tetrahedrizations [13]. However, Joe [14] has shown that, once the Delaunay tetrahedrization of a point set in three dimensions is given, it can be updated by a sequence of flips, after the insertion of a new point. Cheng and Dey [7] have also proven that a surface triangulation that closely approximates a smooth surface with uniform density can be transformed to a Delaunay triangulation by a flip algorithm. On the other hand, flips have been studied for different types of triangulations such as constrained triangulations [9], weighted triangulations [10], pseudo-triangulations [2], pre-triangulations [1], ...

Independently of their efficiency when applied to a "not too bad" initial triangulation, flip algorithms have been implemented as subroutines for randomized algorithms [11]. They also enable to prove important properties of the manipulated triangulations. For example, they have been used for proving that, among

E. Ochmański and J. Tyszkiewicz (Eds.): MFCS 2008, LNCS 5162, pp. 180–192, 2008.

all triangulations of a point set in the plane, the Delaunay triangulation maximizes the minimum angle [16]. They also enable to structure and to enumerate triangulations as vertices of a graph in which two vertices are adjacent if they differ from each other by a flip [12], [5].

In this work we address the question of flip algorithm for the segment triangulations that have been introduced in [4]. Given a finite set S of disjoint line segments in the plane, a segment triangulation of S is a maximal set of disjoint triangles, each of them having its vertices on three distinct sites of S (see Figure 1). Segment triangulations form a very natural family of diagrams containing the dual of the segment Voronoi diagram. This dual diagram, called the segment Delaunay triangulation (or edge Delaunay triangulation), has been introduced much earlier by Chew and Kedem [8]. A topological dual of the segment Voronoi diagram has also been used to implement efficiently the construction of the segment Voronoi diagram in the CGAL Library [15]. In [4], we have given a local characterization of the segment Delaunay triangulation among the family of all segment triangulations of S as well as a local characterization of its topology.

An obstacle arises when trying to transform a segment triangulation into the segment Delaunay triangulation by a sequence of local improvements: As for three dimensional point sets, local transformations must be performed on non convex regions. We overcome this difficulty by allowing local improvements that not necessarily imply changes in the topology, as flips do. In order to characterize these local improvements and to prove that the constructed triangulations tend toward the segment Delaunay triangulation, we use a lifting on the three-dimensional paraboloid together with locally convex functions. The usefulness of locally convex functions in the context of flip algorithms has been already noticed by several authors (see [2], [3], ...).

Another difficulty comes out of segment triangulations: There are infinitely many segment triangulations of a given set of sites, while the number of triangulations usually handled by flip algorithms is finite. So, a flip algorithm that aims to construct a segment Delaunay triangulation explicitly, might need infinitely many steps. Fortunately, this drawback can be circumvented by stopping the algorithm when it reaches a segment triangulation that has the same topology as the segment Delaunay triangulation. We shall show that such a triangulation is obtained in finitely many steps, thanks to geometrical estimates about the angles of the triangles arising during the algorithm. The segment Delaunay triangulation can then be deduced from this triangulation in linear time.

2   Segment Triangulations

In this section, we recall the main results about segment triangulations given in [4]. They generalize the concept of triangulation to a set of disjoint segments in the plane. Afterwards, we slightly extend these results.

Throughout this paper, S is a finite set of $n \geq 2$ disjoint closed segments in the plane, which we call sites. A closed segment may possibly be reduced to a single point. We shall denote by $\mathbf{S}$ the set of points of the segments of S. We say that a circle is tangent to a site s if s meets the circle but not its interior. The sites of S are supposed to be in general position, that is, we suppose that no three segment endpoints are collinear and that no circle is tangent to four sites.

Definition 1. *A segment triangulation T of S is a partition of the convex hull $conv(\mathbf{S})$ of $\mathbf{S}$ in disjoint sites, edges, and faces such that:*
1. Every face of T is an open triangle whose vertices are in three distinct sites of S and whose open edges do not intersect $\mathbf{S}$,
2. No face can be added without intersecting another one,
3. The edges of T are the (possibly two-dimensional) connected components of $conv(\mathbf{S}) \setminus (F \cup \mathbf{S})$, where F is the union of faces of T.

In the following, the word "triangle" will only be used for faces and never for edges, even if they have the shape of a triangle.

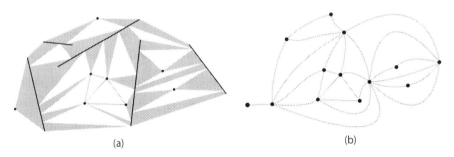

(a)   (b)

Fig. 1. A segment triangulation (a) (the sites appear in black, the faces in white, and the edges in gray) and its topology (b)

Using the fact that no triangle can be added to T, it has been shown that the closure of an edge of a segment triangulation meets exactly two sites (see Figure 1). Thus, a planar combinatorial map M can be associated with T in the following way:

- the vertices of M are the sites of S,
- the arcs connecting two sites s and t in M are the edges of T whose closures intersect s and t,
- for every vertex s of M, the cyclic ordering of the arcs out of s agrees with the counter-clockwise ordering of the associated edges around the site s in T.

M represents the topology of T. Using this topology, it has been shown that the number of faces of a segment triangulation of S depends only on S and is linear with the number of sites of S.

Definition 2. *A segment triangulation of S is Delaunay if the circumcircle of each face does not contain any point of $\mathbf{S}$ in its interior.*

If all the sites are points, a segment Delaunay triangulation is a usual point set Delaunay triangulation. It has been shown that the segment Delaunay triangulation exists for any set S, is unique if S is in general position, and is dual to the segment Voronoi diagram.

As for point sets, the legal edge property has been defined for segment triangulations in [4]. A more intuitive formulation is:

Definition 3. *Let e be an edge adjacent to two triangles T_1 and T_2 in a segment triangulation and let r, t, u, v be the sites adjacent to T_1 and T_2. The edge e is legal if there exists a segment triangulation $\mathcal{T}$ of $\{r, t, u, v\}$ with the same topology as the segment Delaunay triangulation of $\{r, t, u, v\}$ and such that T_1 and T_2 are two triangles of $\mathcal{T}$.*

An edge adjacent to zero or one triangle is legal.

This led to a local characterization of the segment Delaunay triangulation:

Theorem 1. *A segment triangulation of S whose all edges are legal has the same topology as the Delaunay one.*

Since the segment Delaunay triangulation of four sites contains at most four triangles, it can be checked in constant time whether an edge is legal or not.

Note that the segment Delaunay triangulation can be easily computed once its topology is known. It suffices to put each triangle in its tangency position, which means that the interior of its circumcircle does not meet the three sites that contain its vertices. Thus, computing the segment Delaunay triangulation comes down to compute its topology. Therefore, the goal of our flip algorithm is to lead in finitely many "local" steps to a segment triangulation whose edges are all legal. To this aim, we shall need to constrain the segment triangulations in some subsets of the convex hull of $\mathbf{S}$. So, we need to extend slightly the above results.

Definition 4. *A subset U of $conv(\mathbf{S})$ is S-polygonal if U is closed and if the boundary of U is a finite union of disjoint segments of two kinds:*

- *closed segments included in $\mathbf{S}$,*
- *open segments $]p, q[$ such that $\mathbf{S} \cap [p, q] = \{p, q\}$.*

Throughout this paper, U denotes an S-polygonal subset of $conv(\mathbf{S})$. Now, the definition of segment triangulations extends to U by replacing, in Definition 1, $conv(\mathbf{S})$ by U and $\mathbf{S}$ by $U \cap \mathbf{S}$. Here again we can show that:

Theorem 2. *The number of faces of a segment triangulation of U depends only on the couple (U, S).*

We say that a point q in U is *visible* from a point p in U if $]p, q[$ is included in $U \setminus \mathbf{S}$.

Definition 5. *1. A triangle t included in U with vertices in $\mathbf{S}$ is a Delaunay triangle of U if there exists a point p in the interior of t such that the interior of the circumcircle of t contains no point of $\mathbf{S}$ visible from p.*

2. A segment triangulation of U is Delaunay if all its triangles are Delaunay.

Theorem 3. *Every S-polygonal subset U admits a segment Delaunay triangulation.*

This result is a consequence of Theorem 7 of section 4. However, a segment Delaunay triangulation is not necessarily unique since four connected components of $U \cap \mathbf{S}$ may be cocircular even if S is in general position.

3   Description of the Flip Algorithm

The algorithm starts with a segment triangulation of S. The edges of the triangulation are stored in a queue.

One basic step of the algorithm goes as follows. The edge e at the head of the queue is popped. Let P be the closure of the union of e and of its at most two adjacent triangles: This region is called the input polygon of e (see Figure 2 (b) and (f)). Then, the algorithm computes a segment Delaunay triangulation of P. Since P meets at most four sites, the Delaunay triangles of P can be computed in constant time. The triangles adjacent to e are replaced with the Delaunay triangles of P. This gives rise to a new segment triangulation of S (it is a consequence of Theorems 2 and 3). Finally, the edge replacing e is pushed at the tail of the queue.

Beside this queue, the algorithm maintains the number of illegal edges in the current triangulation. The algorithm ends when all edges are legal.

If a basic step changes the topology of the current triangulation, we say that the processed edge is flipped.

In case of point set triangulations, when an illegal edge is processed by the flip algorithm, it is flipped, it becomes legal, and it will never reappear. Since there are finitely many edges, the flip algorithm reaches the Delaunay triangulation after a finite number of steps. Our flip algorithm looks very close to this classical flip algorithm, but we can not use the same idea to prove its convergence because of some important differences (see Figure 2):

- even if an edge is not flipped, its geometry may change,
- some illegal edges cannot be flipped,
- a new constructed edge is not necessarily legal.

For point set triangulations, another way to prove the convergence of the flip algorithm to the Delaunay triangulation, is to lift the point set on the three-dimensional paraboloid $z = x^2 + y^2$. It is well known that the downward projection of the lower convex hull of the lifting is the Delaunay triangulation of the point set. Conversely, every other triangulation lifts to a non convex polyhedral surface above the lower convex hull. Now, it is enough to notice that an edge flip brings down the polyhedral surface.

In the next two sections, we use the same approach to prove that our flip algorithm constructs a segment triangulation that has the same topology as the segment Delaunay triangulation. At first, for every S-polygonal subset U, the lower convex hull of the lifting of $U \cap \mathbf{S}$ on the paraboloid is defined with the help of locally convex functions and we show that it projects down to the segment Delaunay triangulation of U (Theorem 7). Then, we define the lifting of any segment triangulation that is not Delaunay (Definition 7) and we show that the lifting of the segment Delaunay triangulation is lower than or equal to the lifting of any other segment triangulation (Theorem 8). In order to show the correctness of the algorithm, we prove that, after a basic step, the lifting of the resulting segment triangulation is lower than or equal to the lifting of the segment triangulation before the basic step (Theorem 9). This leads to prove that the sequence of basic steps builds a sequence of segment triangulations that converges to the segment Delaunay triangulation (Theorem 9). It remains to see that, after a finite number of basic steps, the segment triangulation constructed by the flip algorithm has the same topology as the segment Delaunay triangulation (Corollary 2). From Theorem 1, there is no more illegal edge in this triangulation and the algorithm stops.

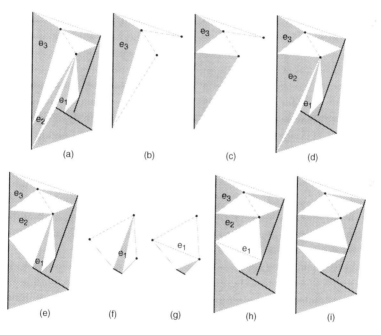

Fig. 2. The flip algorithm transforms the given segment triangulation (a) in a segment triangulation (h) that has the same topology as the segment Delaunay triangulation (i). The topology in (a) and the topology in (h) differ only by the flip of e_1, which is the only illegal edge of (a). However, the edge e_1 of (a) cannot be immediately flipped because its input polygon is not convex. So, the legal edges e_3 and e_2 have to be processed before e_1 becomes flippable. In (b), the algorithm considers the input polygon of the edge e_3. Then, in (c), it computes the segment Delaunay triangulation of the input polygon and this gives rise to a new segment triangulation in (d). In the same way, the processing of the edge e_2 leads to (e). Finally, the edge e_1 can be flipped (f, g), which leads to (h).

4   Locally Convex Functions and Segment Triangulations

Recall that, if V is a subset of $\mathbf{R}^2$, a function $\phi : V \to \mathbf{R}$ is locally convex if the restriction of ϕ to each segment included in V is convex.

We define now the lower convex hull of a function, which we shall use instead of the usual lower convex hull of a subset in $\mathbf{R}^3$. Note that it corresponds to this usual lower convex hull when the domain V is convex.

Definition 6. *Let $L(V)$ be the set of functions $\phi : V \to \mathbf{R}$ that are locally convex on V. Given a real-valued function f defined on $V \cap \mathbf{S}$, the lower convex hull of f on $(V, \mathbf{S})$ is the function $f_{V,\mathbf{S}}$ defined on V by*

$$f_{V,\mathbf{S}}(x) = \sup\{\phi(x) : \phi \in L(V), \ \forall y \in V \cap \mathbf{S}, \ \phi(y) \le f(y)\}.$$

In the following, the above definition will be used on an S-polygonal domain U with the function $f : \mathbf{R}^2 \to \mathbf{R}$ defined by $f(x, y) = x^2 + y^2$. The convexity of f implies that $f_{U,\mathbf{S}} = f$ on $U \cap \mathbf{S}$. Using the geometrical assumptions on U, it can also be proven that $f_{U,\mathbf{S}}$ is continuous.

The main aim of this section is to explain that the function $f_{U,\mathbf{S}}$ determines a segment Delaunay triangulation of U (see Figure 3). Next theorem gives information about the value of the function $f_{U,\mathbf{S}}$ at a point p. For every point p in $U \setminus \mathbf{S}$, denote S_p the closure of the set of points in $\mathbf{S}$ visible from p and V_p its convex hull (in general, V_p is not contained in U).

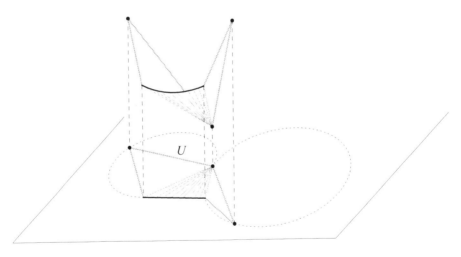

Fig. 3. An S-polygon U and the graph of $f_{U,\mathbf{S}}$. U is decomposed into two triangles and infinitely many line segments where $f_{U,\mathbf{S}}$ is affine. The triangles are Delaunay triangles of U and the union of the segments forms the five edges of the segment Delaunay triangulation of U.

Theorem 4. *Every point p of U belongs to a closed convex subset C of U whose extreme points are in $\mathbf{S}$ and such that the function $f_{U,\mathbf{S}}$ is affine on C. Moreover $f_{U,\mathbf{S}}(p) = f_{V_p, S_p}(p)$.*

Proof. (1) We begin with the case $U = conv(\mathbf{S})$. In this case, the graph of $f_{U,\mathbf{S}}$ is the classical lower convex hull E of the lifting of S on the paraboloid. Every (0-, 1-, or 2-dimensional) face of E is contained in a non vertical supporting hyperplane of E, which implies that $f_{U,\mathbf{S}}$ is affine on the downward projection C of each face of E in the plane $z = 0$. Every point p of the plane belongs to such a set C. Moreover the extreme points of C are the projections of the extreme points of a face and thus they belong to S.

(2) We consider now a general S-polygonal subset U. The Theorem is more difficult to prove and we only give the steps of its proof.

If p is in $\mathbf{S}$ or in the boundary of U, the Theorem is very easy to prove. So we may suppose that p is in the interior of U and not in $\mathbf{S}$. In this case, since U is S-polygonal, it can be shown that p is also in the interior of V_p. We use the the result of (1) with V_p and S_p instead of $conv(\mathbf{S})$ and $\mathbf{S}$: There exists a convex set C containing p, included in V_p whose extreme points are in S_p and such that f_{V_p,S_p} is affine on the convex set C. Since U is an S-polygonal subset, we can see that for every point in $U \setminus \mathbf{S}$ there is a ball centered at this point whose intersection with U is convex. This property enables to prove that the convex C is entirely contained in U.

We know that the function f_{V_p,S_p} is affine on the convex set C. Since p is in the interior of V_p and since f_{V_p,S_p} is convex there exists an affine function $h : \mathbf{R}^2 \to \mathbf{R}$ equal to f_{V_p,S_p} on the convex set C and lower than or equal to f_{V_p,S_p} on V_p. This implies $h \leq f$ on S_p.

The main idea of the proof is to construct a locally convex function $g : U \to \mathbf{R}$ which is equal to h on the convex set C and such that $g \leq f$ on $\mathbf{S}$. Indeed, $f_{U,\mathbf{S}} \geq g$ by definition of the lower convex hull of a function and since $f_{U,\mathbf{S}}$ is convex on C, the function $f_{U,\mathbf{S}}$ must be lower than or equal to the function h on C. Therefore $f_{U,\mathbf{S}} = h$ on C.

The function g is defined as follows. Consider the open disk $A = \{q \in \mathbf{R}^2 : f(q) < h(q)\}$ and let W be the connected component of $A \cap U$ containing p. The function $g : U \to \mathbf{R}$ is defined by $g(q) = h(q)$ if $q \in W$ and $g(q) = f(q)$ otherwise. One can check that g is convex on each segment included in U, hence g is locally convex. Moreover, it is not difficult to see that C is included in $W \cup \{q \in \mathbf{R}^2 : h(q) = f(q)\}$. The last and most difficult thing to check is that $g \leq f$ on $\mathbf{S}$. It is enough to prove that W contains no point of $\mathbf{S}$. Since the function h is lower than or equal to f on all the points of $\mathbf{S}$ visible from p and since $h < f$ on W, we know that W contains no point of $\mathbf{S}$ visible from p. Suppose that there exists a point q in $W \cap \mathbf{S}$. Since W is connected, we can join the point p to the point q by a path γ in W and we can suppose that the lenght of γ is finite. The distance $\delta = d(\gamma, \partial A)$ from γ to the boundary of A is positive. It is not difficult to show that $W_\delta = W \cap \{x \in A : d(x, \partial A) \geq \delta\}$ is a closed set. It follows that there exists a shortest path from p to q in W_δ. Now, we know that for every point in $U \setminus \mathbf{S}$ there is a ball centered at this point whose intersection with U is convex. This shows that this shortest path from p to q is straight unless it meets some point in $\mathbf{S}$. Therefore, either q is visible from p or the shortest path contains a point of $\mathbf{S}$ visible from p, which is impossible. □

Corollary 1. *Let t be a triangle included in U with vertices in $\mathbf{S}$. $f_{U,\mathbf{S}}$ is affine on t if and only if t is a Delaunay triangle of U.*

Proof. Suppose that $f_{U,\mathbf{S}}$ is affine on t. Let p be any point in the interior of t and $q \in S_p$. Denote $h : \mathbf{R}^2 \to \mathbf{R}$ the affine function equal to $f_{U,\mathbf{S}}$ on t. The function $f_{U,\mathbf{S}}$ is convex on $[p,q]$ and is equal to h on a neighborhood of p. Therefore $f_{U,\mathbf{S}} \geq h$ on $[p,q]$. Since $f_{U,\mathbf{S}} = f$ on $\mathbf{S}$, $f(q) = f_{U,\mathbf{S}}(q) \geq h(q)$. Hence q is not in the region of $\mathbf{R}^2$ where $f < h$, which is precisely the interior of the circumcircle of the triangle t.

Conversely, suppose that t is a Delaunay triangle. We begin by the case $U = conv(\mathbf{S})$. There exists a point p in the interior of t such that the interior of the circumcircle of t contains no point of $\mathbf{S}$ visible from p. Consider the affine function $h_t : \mathbf{R}^2 \to \mathbf{R}$ which is equal to f on the vertices of the triangle t. Since U is convex, the interior of the circumcircle contains no point of $\mathbf{S}$. Therefore $h_t \leq f$ on $\mathbf{S}$. It follows that $f_{U,\mathbf{S}} \geq h_t$ on the entire set U. On the other hand, $f_{U,\mathbf{S}} = f = h_t$ on the vertices of t. Thus, by convexity, $f_{U,\mathbf{S}} \leq h_t$ on t. It follows that $f_{U,\mathbf{S}} = h_t$ on t.

In the general case, if t is a Delaunay triangle of U then, by definition, it is also a Delaunay triangle of (V_p, S_p). Hence, by the convex case, f_{V_p,S_p} is affine on t. By the previous Theorem, we have $f_{U,\mathbf{S}}(p) = f_{V_p,S_p}(p)$. Since $f_{U,\mathbf{S}}$ is locally convex, we have $f_{U,\mathbf{S}} \leq f_{V_p,S_p}$ on t. Now, p is in the interior of t, therefore $f_{U,\mathbf{S}} = f_{V_p,S_p}$ on t.   □

The next step consists in showing that U can be partitioned into maximal convex subsets where the function $f_{U,\mathbf{S}}$ is affine. We are not going to prove this result, nevertheless we can explain why it is natural. On the one hand, the relative interiors of the faces of a closed convex set form a partition of this convex set (see [6]). In the case $U = conv(\mathbf{S})$, it follows that U is partionned by the downward projections of the relative interiors of the lower faces of the convex hull of the lifting of $\mathbf{S}$ on the paraboloid $\{z = x^2 + y^2\}$. On the other hand, in the case of an S-polygonal subset U, the maximal subsets of U where $f_{U,\mathbf{S}}$ is affine are intended to replace these downward projections. This leads to the Theorem:

Theorem 5. *For each point p in $U \setminus \mathbf{S}$ consider the set of all relatively open convex subsets of U containing p where $f_{U,\mathbf{S}}$ is affine. This set of convex subsets contains a maximal element C_p (maximal for the inclusion). Moreover, the extreme points of $\overline{C_p}$ are in $\mathbf{S}$ and, if q is another point of $U \setminus \mathbf{S}$, either $C_p \cap C_q = \emptyset$, or $C_p = C_q$.*

The last statement of Theorem 5 means that the subsets C_p form a partition of $U \setminus \mathbf{S}$. Now we have to establish that the two-dimensional convex subsets among the C_p induce the triangles of a segment triangulation.

Theorem 6. *By decomposing the two-dimensional $(C_p)_{p \in U \setminus \mathbf{S}}$ into triangles we get the triangles of a segment triangulation of U, which we call a triangulation induced by $f_{U,\mathbf{S}}$.*

Proof. As before, by lack of place, we can only give some hints about the proof of this Theorem. Denote by C_{S_1,S_2} the set of all one-dimensional C_p with one endpoint in the site S_1 and the other in the site S_2. Denote by U_{S_1,S_2} the union of all the segments of C_{S_1,S_2}. The continuity of $f_{U,\mathbf{S}}$ and the strict convexity of f allow to show that a point in $\mathbf{S} \setminus (S_1 \cup S_2)$ cannot be too close to a point in a segment $C_p \in C_{S_1,S_2}$. It follows that U_{S_1,S_2} is open in $V = U \setminus (\mathbf{S} \cup \{\text{the two-dimensional } C_p\})$. Now, by definition, a connected component A of V cannot meet more than one set U_{S_1,S_2}, $S_1 \in S$, $S_2 \in S$. Then, it is not difficult to prove that the closure of A meets exactly two sites. Therefore, it is impossible to add a triangle to the two-dimensional C_p without intersecting them. Thus we have a segment triangulation of U. □

From Corollary 1 and Theorem 6 we deduce:

Theorem 7. *For any S-polygonal subset U, a segment triangulation of U is induced by $f_{U,\mathbf{S}}$ if and only if it is Delaunay.*

Using locally convex functions, we are able to lift any segment triangulation in the following way:

Definition 7. *Let $\mathcal{T}$ be a segment triangulation of U. The function $f_{U,\mathbf{S},\mathcal{T}}$: $U \to \mathbf{R}$ is equal to f on $\mathbf{S}$, to $f_{\bar{e},\mathbf{S}}$ on every edge e of $\mathcal{T}$, and to $f_{\bar{t},\mathbf{S}}$ on the interior of every triangle t of $\mathcal{T}$.*

The lifting of $\mathcal{T}$ to $\mathbf{R}^3$ is the graph of the function $f_{U,\mathbf{S},\mathcal{T}}$.
    Since $f_{\bar{e},\mathbf{S}} \geq f_{U,\mathbf{S}}$ on $\bar{e}$ and $f_{\bar{t},\mathbf{S}} \geq f_{U,\mathbf{S}}$ on t, we get:

Theorem 8. *If $\mathcal{T}$ is a segment triangulation of U, then $f_{U,\mathbf{S}} \leq f_{U,\mathbf{S},\mathcal{T}}$.*

5   Convergence of the Flip Algorithm

In case of point set triangulations, it is well known that a flip increases the smallest angle of the triangles. A weaker result holds for segment triangulations.
    Given a segment triangulation $\mathcal{T}$ of U, let the slope of $\mathcal{T}$ be:

$$\sigma(\mathcal{T}) = \sup\{\frac{f_{U,\mathbf{S},\mathcal{T}}(p) - f_{U,\mathbf{S},\mathcal{T}}(q)}{|p - q|} : p \in U \setminus \mathbf{S},\ q \in U \cap \mathbf{S},\ [p,q] \subset U\}$$

Denoting by $\theta(\mathcal{T})$ the minimum angle of the triangles of $\mathcal{T}$, we can prove:

Proposition 1. *There exists a positive constant c depending only on f, S, and U such that, for every segment triangulation $\mathcal{T}$ of U, $\theta(\mathcal{T}) \geq c/(\max(1, \sigma(\mathcal{T})))$.*

It is not difficult to prove that $\sigma(\mathcal{T}) < +\infty$. Moreover, it is obvious that if $\mathcal{T}'$ is a segment triangulation of U such that $f_{U,\mathbf{S},\mathcal{T}} \leq f_{U,\mathbf{S},\mathcal{T}'}$, then $\sigma(\mathcal{T}) \leq \sigma(\mathcal{T}')$.
    Consider now our algorithm: It starts with a segment triangulation $\mathcal{T}_0$ of $conv(\mathbf{S})$ and computes a sequence $\mathcal{T}_1, \mathcal{T}_2, ..., \mathcal{T}_n, ...$ of triangulations.

Theorem 9. *The sequence $(f_n = f_{conv(\mathbf{S}),\mathbf{S},\mathcal{T}_n})_{n \in \mathbf{N}}$ decreases to $f_{conv(\mathbf{S}),\mathbf{S}}$ as n goes to infinity.*

Proof. At every stage n, we compute a Delaunay triangulation of the input polygon P_n of the edge at the head of the queue. Applying Theorem 8 to the S-polygonal subset U composed of P_n and of all the edges of $\mathcal{T}_n$ adjacent to P_n, we get that $f_{n+1} \leq f_n$ on U, which implies that $f_{n+1} \leq f_n$ on $conv(\mathbf{S})$.

It follows that the sequence of functions $(f_n)_{n \in \mathbf{N}}$ decreases to a function $g : conv(\mathbf{S}) \to \mathbf{R}$. The only thing to show is that g is (locally) convex. Since $g = f$ on $\mathbf{S}$ and $g \geq f$ on $conv(\mathbf{S})$, we are reduced to show that g is convex on any open segment $]p_0, p_1[$ included in the interior of $conv(\mathbf{S}) \setminus \mathbf{S}$. By Proposition 1 and since the sequence f_n decreases, the angles of the triangles generated by the algorithm are not too sharp. It allows to show that, for every point p in $]p_0, p_1[$, there exists $\varepsilon > 0$ such that the neighborhood $I_{p,\varepsilon}$ of p of radius ε in $]p_0, p_1[$ is included either in a triangle of $\mathcal{T}_n$ or in the input polygon P_n treated at stage n, for infinitely many integers n. Thus, for these integers n, either f_n or $f_{P_n,\mathbf{S}}$ is convex on $I_{p,\varepsilon}$, and since $f_{n+1} \leq f_{P_n,\mathbf{S}} \leq f_n$ on P_n, the function g is a limit of a sequence of convex functions on $I_{p,\varepsilon}$, hence g is convex. Finally, since $f_n \geq f_{U,\mathbf{S}}$ for all n, we get $g \geq f_{U,\mathbf{S}}$. Moreover, $g = f_n = f$ on $\mathbf{S}$ and g is convex, therefore $g \leq f_{U,\mathbf{S}}$. $\qquad\square$

Corollary 2. *There exists an integer N such that, for all integers $n \geq N$, the triangulation $\mathcal{T}_n$ has the same topology as the segment Delaunay triangulation of $conv(\mathbf{S})$.*

Proof. Since the set of topologies of all the segment triangulations of S is finite, if the corollary were false, then a non Delaunay topology would appear infinitely many times. Therefore, it is enough to prove that, if for an increasing sequence of integers $(k_n)_{n \in \mathbf{N}}$, the triangulations $\mathcal{T}_{k_n}$ have the same topology, then it is the topology of the segment Delaunay triangulation.

We can always suppose that, given a topological triangle t, its geometrical representation t_{k_n} in $\mathcal{T}_{k_n}$ converges to a triangle t_∞ when n goes to infinity (take subsequences of $\mathcal{T}_{k_n}$). Therefore, the function $f_{conv(\mathbf{S}),\mathbf{S}} = \lim_{n \to \infty} f_{conv(\mathbf{S}),\mathbf{S},\mathcal{T}_{k_n}}$ must be affine on each of these triangles t_∞. Together with Theorem 7, this shows that the set of all triangles t_∞ defines the segment Delaunay triangulation of S and that all the segment triangulations $\mathcal{T}_{k_n}$ have the same topology as the Delaunay one. $\qquad\square$

6   Conclusion

In this paper, we have shown that the segment Delaunay triangulation can be constructed by a flip algorithm in a finite number of steps.

The complexity of the flip algorithm seems difficult to estimate since the same edge is processed several times. Nevertheless, as for point set triangulations, we can expect that the practical complexity of the algorithm will be efficient if the input segment triangulation is not too bad. This practical complexity may be improved by a better handling of the queue: It is not necessary to systematically insert all the edges in the queue and we could establish an ordering of these edges.

The proof of the convergence of the flip algorithm uses the control of the angles of the triangles during the algorithm. Moreover, as for point set triangulations, the segment Delaunay triangulation is the only segment triangulation whose three-dimensional lifting is convex. These are two strong hints that make us believe that the segment Delaunay triangulations should have some optimality properties.

At last, possible extensions of segment triangulations should be mentioned: Extension to three-dimensional space, to more general sites, to more general distance functions, ... The three-dimensional extension is certainly a difficult problem; it will be easier to consider first more general convex sites in the plane. We believe that some of the results given in this paper can be extended to this more general setting.

References

1. Aichholzer, O., Aurenhammer, F., Hackl, T.: Pre-triangulations and liftable complexes. In: Proc. 22th Annu. ACM Sympos. Comput. Geom., pp. 282–291 (2006)
2. Aichholzer, O., Aurenhammer, F., Krasser, H., Brass, P.: Pseudotriangulations from surfaces and a novel type of edge flip. SIAM J. Comput. 32(6), 1621–1653 (2003)
3. Aurenhammer, F., Krasser, H.: Pseudo-simplicial complexes from maximal locally convex functions. Disc. Comput. Geom. 35(2), 201–221 (2006)
4. Brévilliers, M., Chevallier, N., Schmit, D.: Triangulations of Line Segment Sets in the Plane. In: Arvind, V., Prasad, S. (eds.) FSTTCS 2007. LNCS, vol. 4855, pp. 388–399. Springer, Heidelberg (2007)
5. Brönnimann, H., Kettner, L., Pocchiola, M., Snoeyink, J.: Counting and enumerating pointed pseudotriangulations with the greedy flip algorithm. SIAM J. Comput. 36(3), 721–739 (2006)
6. Bronsted, A.: An Introduction to Convex Polytopes. Graduate Texts in Mathematics. Springer, New York (1983)
7. Cheng, S., Dey, T.K.: Delaunay edge flips in dense surface triangulations. In: Proceedings of the 24th European Workshop on Computational Geometry, pp. 1–4 (2008)
8. Chew, L.P., Kedem, K.: Placing the largest similar copy of a convex polygon among polygonal obstacles. In: Proc. 5th Annu. ACM Sympos. Comput. Geom., pp. 167–174 (1989)
9. Edelsbrunner, H.: Triangulations and meshes in computational geometry. In: Acta Numerica, pp. 133–213 (2000)
10. Edelsbrunner, H., Shah, N.R.: Incremental topological flipping works for regular triangulations. Algorithmica 15(3), 223–241 (1996)
11. Guibas, L.J., Knuth, D.E., Sharir, M.: Randomized incremental construction of delaunay and voronoi diagrams. Algorithmica 7(4), 381–413 (1992)
12. Hurtado, F., Noy, M.: Graph of triangulations of a convex polygon and tree of triangulations. Comput. Geom. Theory Appl. 13(3), 179–188 (1999)

13. Joe, B.: Three-dimensional triangulations from local transformations. SIAM J. Sci. Stat. Comput. 10(4), 718–741 (1989)
14. Joe, B.: Construction of three-dimensional delaunay triangulations from local transformations. Comp. Aided Geom. Design. 8(2), 123–142 (1991)
15. Karavelas, M.I.: A robust and efficient implementation for the segment Voronoi diagram. In: Proceedings of the International Symposium on Voronoi Diagrams in Science and Engineering, pp. 51–62 (2004)
16. Lawson, C.L.: Software for C^1 surface interpolation. In: Rice, J.R. (ed.) Math. Software III, pp. 161–194. Academic Press, New York, NY (1977)

Computing Sharp 2-Factors in Claw-Free Graphs

Hajo Broersma and Daniël Paulusma

Department of Computer Science, Durham University, DH1 3LE Durham, United Kingdom
{hajo.broersma,daniel.paulusma}@durham.ac.uk

Abstract. In a recently submitted paper we obtained an upper bound for the minimum number of components of a 2-factor in a claw-free graph. This bound is sharp in the sense that there exist infinitely many claw-free graphs for which the bound is tight. In this paper we extend these results by presenting a polynomial algorithm that constructs a 2-factor of a claw-free graph with minimum degree at least four whose number of components meets this bound. As a byproduct we show that the problem of obtaining a minimum 2-factor (if it exists) is polynomially solvable for a subclass of claw-free graphs. As another byproduct we give a short constructive proof for a result of Ryjáček, Saito & Schelp.

1 Introduction

In this paper we consider 2-factors of claw-free graphs. Graph factors are well-studied. See [16] for a survey. Our motivation to study 2-factors goes back to the well-known NP-complete decision problem H-CYCLE (cf. [9]) in which the problem is to decide whether a given graph has a hamiltonian cycle, i.e., a connected 2-regular spanning subgraph. In the related problem 2-FACTOR the connectivity condition is dropped, hence the problem is to decide whether a given graph admits a 2-factor, i.e., a 2-regular spanning subgraph. This makes the problem considerably easier in the algorithmic sense: it is well-known that 2-FACTOR can be solved in polynomial time by matching techniques, and a 2-factor can be constructed in polynomial time if the answer is YES (cf [14]). Clearly, a hamiltonian cycle is a 2-factor consisting of one component, and the minimum number of components of a 2-factor can be seen as a measure for how far a graph is from being hamiltonian. So, from an algorithmic viewpoint a natural question is to consider the problem of determining a 2-factor of a given graph with a minimum number of components. Obviously, this is an NP-hard problem. Hence it makes sense to search for 2-factors with a reasonably small number of components if we aim for polynomial time algorithms. For this research we have restricted ourselves to the class of claw-free graphs. This is a rich class containing, e.g., the class of line graphs and the class of complements of triangle-free graphs. It is also a very well-studied graph class, both within structural graph theory and within algorithmic graph theory; see [7] for a survey. Furthermore, computing a 2-factor with a minimum number of components remains NP-hard for the class of claw-free graphs.

In a recently submitted paper [1] we already obtained an upper bound for the minimum number of components of a 2-factor in a claw-free graph. This bound is sharp in the sense that there exist infinitely many claw-free graphs for which the bound is tight; we will specify this later. When considering the related complexity problems, we

E. Ochmański and J. Tyszkiewicz (Eds.): MFCS 2008, LNCS 5162, pp. 193–204, 2008.

soon realized that the proof methods used in [1] need to be extended in order to obtain a polynomial algorithm that constructs a corresponding 2-factor, e.g., a 2-factor whose number of components is at most our upper bound. In the present paper we present this polynomial time algorithm.

2   Terminology and Background

We consider graphs that are finite, undirected and simple, i.e., without multiple edges and loops. For notation and terminology not defined in this paper we refer to [4].

Let $G = (V_G, E_G)$ be a graph of order $|G| = |V_G| = n$ and of size $e_G = |E_G|$. The neighbor set of a vertex x in G is denoted by $N_G(x) = \{y \in V_G \mid xy \in E_G\}$, and its cardinality by $d_G(x)$. We denote the minimum (vertex) degree of G by $\delta_G = \min\{d_G(x) \mid x \in V_G\}$. If no confusion can arise we often omit the subscripts.

Let K_n denote the complete graph on n vertices. A graph F is called a *2-factor* of a graph G if F is a 2-regular spanning subgraph of G, i.e., if F is a subgraph of G with $V_F = V_G$ and $d_F(x) = 2$ for all $x \in V_F$. A *claw-free* graph is a graph that does not contain an induced subgraph isomorphic to the four-vertex star $K_{1,3} = (\{u, a, b, c\}, \{ua, ub, uc\})$.

2.1   Known Results

Several interesting problems are still open for claw-free graphs such as the conjecture of Matthews and Sumner [15] that every 4-connected claw-free graph is hamiltonian. However, there is quite a lot known on 2-factors in claw-free graphs, including some very recent results. Results of both Choudum & Paulraj [3] and Egawa & Ota [5] imply that every claw-free graph with $\delta \geq 4$ contains a 2-factor.

Theorem 1 ([3,5]). *A claw-free graph with $\delta \geq 4$ has a 2-factor.*

We observe that every 4-connected claw-free graph has minimum degree at least four, and hence has a 2-factor. A 2-connected claw-free graph already has a 2-factor if $\delta = 3$ [20]. However, in general a claw-free graph with $\delta \leq 3$ does not have to contain a 2-factor. Examples are easily obtained.

Faudree et al. [6] showed that every claw-free graph with $\delta \geq 4$ has a 2-factor with at most $6n/(\delta + 2) - 1$ components. Gould & Jacobson [11] proved that, for every integer $k \geq 2$, every claw-free graph of order $n \geq 16k^3$ with $\delta \geq n/k$ has a 2-factor with at most k components. Fronček, Ryjáček & Skupień [8] showed that, for every integer $k \geq 4$, every claw-free graph G of order $n \geq 3k^2 - 3$ with $\delta \geq 3k - 4$ and $\sigma_k > n + k^2 - 4k + 7$ has a 2-factor with at most $k - 1$ components. Here σ_k denotes the minimum degree sum of any k mutually nonadjacent vertices.

If a graph G is claw-free, 2-connected and has $\delta \geq 4$, then G has a 2-factor with at most $(n+1)/4$ components [13]. If a graph G is claw-free, 3-connected and has $\delta \geq 4$, then G has a 2-factor with at most $2n/15$ components [13].

In [1] we considered claw-free graphs with $\delta \geq 4$. Our motivation for this is as follows. We first note that the number of components of a 2-factor in any graph on n vertices is obviously at most $n/3$. For claw-free graphs with $\delta = 2$ that have a 2-factor we cannot do better than this trivial upper bound. This is clear from considering a disjoint set of triangles (cycles on three vertices). For claw-free graphs with $\delta = 3$ that have a 2-factor, the upper bound $n/3$ on its number of components is also tight, as shown in [1]. Hence, in order to get a nontrivial result it is natural to consider claw-free graphs with $\delta \geq 4$.

Our two main results in [1] provide answers to two open questions posed in [20].

Theorem 2 ([1]). *A claw-free graph G on n vertices with $\delta \geq 5$ has a 2-factor with at most $(n-3)/(\delta-1)$ components unless G is isomorphic to K_n.*

Theorem 3 ([1]). *A claw-free graph G on n vertices with $\delta = 4$ has a 2-factor with at most $(5n-14)/18$ components, unless G belongs to a finite class of exceptional graphs.*

Both results are tight in the following sense. Let $f_2(G)$ denote the minimum number of components in a 2-factor of G. Then in [20], for every integer $d \geq 4$, an infinite family $\{F_i^d\}$ of claw-free graphs with $\delta(F_i^d) \geq d$ is given such that $f_2(F_i^d) > |F_i^d|/d \geq |F_i^d|/\delta(F_i^d)$. This shows we cannot replace $\delta - 1$ by δ in Theorem 2. The bound in Theorem 3 is tight in the following sense. There exists an infinite family $\{H_i\}$ of claw-free graphs with $\delta(H_i) = 4$ such that

$$\lim_{|H_i| \to \infty} \frac{f_2(H_i)}{|H_i|} = \frac{5}{18}.$$

This family can be found in [20] as well.

The exceptional graphs of Theorem 3 have at most seventeen vertices. They are described in [1], and we will not specify them here. In [1] we also explain that Theorem 2 and 3 together improve the previously mentioned result of Faudree et al. [6] and that Theorem 2 also improves the previously mentioned result of Gould & Jacobson [11].

2.2   Results of This Paper

The proofs in [1] do not yield algorithms for constructing 2-factors that satisfy the upper bounds in Theorems 2 and 3. In the remainder of this paper we will develop a new approach to these problems in order to establish polynomial algorithms that construct 2-factors of claw-free graphs with minimum degree at least four. Using our results in [1] we show that the number of components in these 2-factors are guaranteed to satisfy the upper bounds of Theorems 2 and 3. We will illustrate our approach by concentrating on Theorem 2, but the same approach works for Theorem 3 in exactly the same way. As a byproduct we show that the problem of obtaining a minimum 2-factor (if it exists) is polynomially solvable for a subclass of claw-free graphs which we describe later on. As another byproduct we give a short constructive proof for a result of Ryjáček, Saito & Schelp [19].

3   The Algorithm for Constructing 2-Factors of Claw-Free Graphs

We split the proof into six different parts. For the first two parts we do not have to develop any new theory or algorithms, but can rely on the beautiful existing machinery from the literature. The first part of this says that claw-free graphs behave the same with respect to our problem as line graphs obtained from them by performing some closure operation which will be explained shortly. The second part then describes the known equivalence of our problem with an analogous problem based on concepts and results in the preimage graphs of line graphs. Our new contributions are described and explained in the third, fourth, fifth and sixth part. In the third part we consider preimage graphs that are trees and in the fourth part we consider preimage graphs that are triangle-free. Finally, in the fifth and sixth part we translate the results back to the original domain of claw-free graphs and mention some special class for which our algorithm finds a 2-factor with a minimum number of components.

Step 1: Restrict to Line Graphs of Triangle-Free Graphs

The *line graph* of a graph H with edges $e_1, \ldots, e_p$ is the graph $L(H)$ with vertices $u_1, \ldots, u_p$ such that there is an edge between any two vertices u_i and u_j if and only if e_i and e_j share one end vertex in H. It is easy to verify and well-known (see e.g. [12]) that line graphs are claw-free graphs, but that the class of claw-free graph is much richer (in fact, line graphs have been characterized by a set of nine forbidden induced subgraphs). We show that we can restrict ourselves to an even smaller subclass of claw-free graphs, namely the class of line graphs of triangle-free graphs. For this purpose we use the *closure* concept as defined in [18].

The closure of a claw-free graph is defined as follows. Let G be a claw-free graph. Then, for each vertex x of G, the set of neighbors of x in G induces a subgraph with at most two components. If this subgraph has two components, both of them must be cliques. If the subgraph induced by $N(x)$ is connected, we add edges joining all pairs of nonadjacent vertices in $N(x)$. This operation is called the *local completion of G at x*. The *closure* $cl(G)$ of G is a graph we can obtain by recursively repeating the local completion operation, as long as this is possible. Ryjáček [18] showed that the closure of G is uniquely determined, i.e., that the ordering in which one performs the local completions does not matter. Ryjáček [18] also showed that G is hamiltonian if and only if $cl(G)$ is hamiltonian. This result was later extended to 2-factors [19].

Theorem 4 ([19]). *Let G be a claw-free graph. Then G has a 2-factor with at most k components if and only if $cl(G)$ has a 2-factor with at most k components.*

The following relationship between claw-free graphs and triangle-free graphs exists.

Theorem 5 ([18]). *If G is a claw-free graph, then there is a triangle-free graph H such that $L(H) = cl(G)$.*

It is well-known that apart from K_3 which is $L(K_3)$ and $L(K_{1,3})$, every connected line graph F has a unique H with $F = L(H)$ (see e.g. [12]). We call H the *preimage graph* of F. For K_3 we let $K_{1,3}$ be its preimage graph. For disconnected graphs we define the preimage graphs according to their components.

Recall that $f_2(G)$ denotes the minimum number of components in a 2-factor of a graph G. By Theorem 4 and Theorem 5, we deduce that for a claw-free graph G, $f_2(G) = f_2(cl(G)) = f_2(L(H))$, where H is the (triangle-free) preimage graph of $cl(G)$. Recall that the closure of a claw-free graph can be obtained in polynomial time. Since it is known that the preimage graph of a line graph can be obtained in polynomial (linear) time (see e.g. [17]) we can efficiently compute H.

Step 2: Translate the Problem into Finding Dominating Systems

An *even* graph is a graph in which every vertex has a nonzero even degree. A connected even graph is called a *circuit*. For $q \geq 2$, a *star* $K_{1,q}$ is a complete bipartite graph with independent sets $A = \{c\}$ and B with $|B| = q$; the vertex c is called the *center* and the vertices in B are called the *leaves* of $K_{1,q}$.

Let H be a graph that contains a set S consisting of stars with at least three edges and circuits, all (stars and circuits) mutually edge-disjoint. We call S a *system that dominates* H or simply a *dominating system* if for every edge e of H the following holds:

- e is contained in one of the stars of S, or
- e is contained in one of the circuits of S, or
- e shares an end vertex with an edge of at least one of the circuits in S.

Gould & Hynds [10] proved the following result.

Theorem 6 ([10]). *The line graph $L(H)$ of a graph H has a 2-factor with k components if and only if H has a dominating system with k elements.*

Combining Theorem 4 and Theorem 5 with Theorem 6 yields the following result.

Theorem 7. *Let G be a claw-free graph. Then G has a 2-factor with k components if and only if the (triangle-free) preimage graph of G has a dominating system with k elements.*

The *edge degree* of an edge xy in a graph H is defined as $d_H(x) + d_H(y) - 2$. We denote the minimum edge degree of H by $\delta_e = \delta_e(H)$. Due to the previous discussions it is clear that Theorem 2 is equivalent to the following theorem.

Theorem 8. *A triangle-free graph H with $\delta_e(H) \geq 5$ has a dominating system with at most $(e(H) - 3)/(\delta_e(H) - 1)$ elements unless H is isomorphic to $K_{1,e(H)}$.*

We will now concentrate on determining (in polynomial time) a *sharp dominating system*, i.e., one that satisfies the upper bound of Theorem 8. We first deal with the case that H is a tree. In this case we can even determine a minimum dominating system in polynomial time.

Step 3: Compute Minimum Dominating Systems for Trees

Here we present a polynomial time algorithm for computing the number of elements in a minimum dominating system of any given tree. We use the following new terminology. A *minimum dominating system*, or shortly, an *m-system* of a graph H is a dominating system of H with the smallest number of elements. We denote such a system by $\mathcal{M}(H)$, and its number of elements by $m(H)$. If H does not allow a dominating system we write $m(H) = \infty$.

A vertex with degree 1 in a graph F is called an *end vertex* or *leaf* of F. An edge which is incident with a leaf is called a *pendant* edge. We say that we *add a pendant edge* to F if we add a new vertex to F and join it to precisely one of the vertices of F. Two edges are called *independent* if they do not share any end vertices. A *matching* is a set of mutually independent edges.

We write $H^q(w)$ to denote a tree H that contains a vertex w to which we added q new pendant edges. Note that $H^0(w) = H$. Let $H_1, \ldots, H_p$ be a set of p mutually vertex-disjoint trees, where each H_i contains a vertex w_i. We say that we have *joined* trees $H_1, \ldots, H_p$ in $w_1, \ldots, w_p$ by u if we add a new vertex u with edges uw_i for $i = 1, \ldots, p$. If $p = 0$, then the resulting tree $H(u)$ is the single vertex u, which has a dominating system of 0 elements by definition. Before we present our algorithm we first deduce a number of equations. Note that $m(H^1(w)) = \infty$ if $H = (\{w\}, \emptyset)$.

Lemma 1. *Let $w_1, \ldots, w_p$ be a set of p vertices belonging to mutually disjoint trees $H_1, \ldots, H_p$, respectively. Let $H(u)$ be the tree obtained after joining $H_1, \ldots, H_p$ in $w_1, \ldots, w_p$ by u. Then $m(H(u)) =*

$$
\begin{cases}
0 & \text{if } p = 0 \\
\displaystyle\sum_{i=1}^{p} m(H^1(w_i)) & \text{if } p \in \{1, 2\} \\
\min\Big\{ \displaystyle\sum_{i=1}^{p} m(H^1(w_i)), \\
\quad 1 + \min_{i_1 < i_2 < i_3} \Big\{ \displaystyle\sum_{j=1}^{3} m(H_{i_j}) + \sum_{i \notin \{i_1, i_2, i_3\}} \min\{m(H_i), m(H^1(w_i))\} \Big\} \Big\} & \text{if } p \geq 3.
\end{cases}
$$

Proof. We prove each case separately.

- Let $p = 0$. Then $H(u) = (\{u\}, \emptyset)$. So, $m(H(u)) = 0$ by definition of a dominating system.
- Let $1 \leq p \leq 2$. Then, in any dominating system of $H(u)$, u is not a star center, and consequently, each w_i is the center of a star containing the edge uw_i. Note that in each tree $H^1(w_i)$, w_i is a star center (because the new pendant edge to w_i needs to be covered by a star). Hence, we can combine any m-systems $\mathcal{M}^1(w_i)$ of each $H^1(w_i)$ to obtain an m-system $\mathcal{M}(H(u))$ with $\sum_{i=1}^{p} m(H^1(w_i))$ elements.
- Let $p \geq 3$. First consider the set of dominating systems of $H(u)$ in which u is not a star center. In all these dominating systems, each w_i is the center of a star containing the edge uw_i. Similar to the previous case, we can combine any m-systems of each $H^1(w_i)$ to obtain a dominating system $\mathcal{S}$ of $H(u)$ with $\sum_{i=1}^{p} m(H^1(w_i))$ elements. We note that $\mathcal{S}$ has the minimum number of elements over all dominating systems of $H(u)$ in which u is not a star center.

  Secondly, consider the set of dominating systems of $H(u)$ in which u is a star center. In all these dominating systems, the star with center u contains at least three edges, say uw_{i_1}, uw_{i_2}, and uw_{i_3}, by definition of a dominating system. For the remaining edges uw_i we act as follows. In each dominating system of $H(u)$ that has a star with center u, such an edge uw_i either belongs to the star with center u, or else to the star with center w_i. We compute an m-system $\mathcal{M}(H_i)$ and an m-system $\mathcal{M}(H^1(w_i))$. Then we choose the one with the smallest number of elements, which we denote by $\mathcal{M}_i^*$. So, $|\mathcal{M}_i^*| = \min\{m(H_i), m(H^1(w_1))\}$. We now combine the m-systems of H_{i_j} for $j = 1, 2, 3$, together with the dominating systems $\mathcal{M}_i^*$ and a star that contains the edges uw_{i_j} for $j = 1, 2, 3$ plus possibly some more edges depending on our choice for each $\mathcal{M}_i^*$. We try all possible triples (i_1, i_2, i_3), and choose the combination with the smallest total number of elements. This way we obtain a dominating system $\mathcal{S}'$ of $H(u)$ that has

$$1 + \min_{i_1 < i_2 < i_3} \left\{ \sum_{j=1}^{3} m(H_{i_j}) + \sum_{i \notin \{i_1, i_2, i_3\}} \min\{m(H_i), m(H^1(w_i))\} \right\}$$

elements. We note that $\mathcal{S}'$ has the minimum number of elements over all dominating systems of $H(u)$ in which u is a star center.

Finally, we compare the numbers of elements of $\mathcal{S}$ and $\mathcal{S}'$, and we choose (the) one with the smallest number of elements. This yields an m-system $\mathcal{M}(H(u))$.   □

Using Lemma 1 we can prove the following theorem.

Theorem 9. *The problem of finding a minimum dominating system is polynomially solvable for the class of trees.*

Proof. Let H be a tree with a designated vertex v^0. We partition $V(H)$ into $L_0 \cup L_1 \cup \ldots \cup L_r$ such that for $i = 0, \ldots, r$, L_i is the set of vertices at distance i from v^0. Note that $L_0 = \{v^0\}$. For $v \in V(H) \setminus \{v^0\}$, we let $v^+ \in N(v)$ be the first vertex on the (unique) path from v to v^0 in H, and we let the subtree H_v be the component of $H - vv^+$ that contains v.

Now let $v \in V(H)$. Suppose v has neighbors $w_1, \ldots, w_p$ in H_v. Then H_v is obtained after joining the p mutually disjoint trees $H_{w_1}, \ldots, H_{w_p}$ in $w_1, \ldots, w_p$ by v. Suppose we have already computed the values $m(H_{w_i})$ and $m(H_{w_i}^1(w_i))$. Then, using Lemma 1, we can easily compute $m(H_v)$. We observe that the tree $H_v^1(v)$ is obtained after joining the trees $H_{w_i}, \ldots, H_{w_p}$ together with a new single vertex tree $(\{w_{p+1}\}, \emptyset)$ in $w_1, \ldots, w_{p+1}$ by v. Hence, we can use Lemma 1 to compute $m(H_v^1(v))$ as well. So, our strategy is to recursively compute the values $m(H_v)$ and $m(H_v^1(v))$: for $i = 1, \ldots, r$, we first compute the values $m(H_{v^i})$ and $m(H_{v^i}^1(v^i))$ for all $v^i \in L_i$, and use them to compute $m(H_{v^{i-1}})$ and $m(H_{v^{i-1}}^1(v^{i-1}))$ for all $v^{i-1} \in L_{i-1}$ according to Lemma 1. Clearly, computing $m(H)$ this way can be done in polynomial time.

In order to find an m-system $\mathcal{M}(H)$, we keep track of the stars as follows. Firstly, for each $v \in V(H)$, we remember whether v is a star center in an m-system of H_v when we compute $m(H_v)$. In case v is the center of a star S_v, we keep track of the edges in S_v as well. Secondly, we check whether v becomes the center of a star S_v (and which edges belong to S_v if S_v exists) both when we compute $m(H_{v^+})$ and when we compute $m(H_{v^+}^1(v^+))$. Note that we can do this in polynomial time when we use the formula in Lemma 1. With the above information we can efficiently compute an m-system $\mathcal{M}(H)$, as the following claim shows.

Claim. For all v in each L_i we can compute in polynomial time whether v is the center of a star S_v of an m-system $\mathcal{M}(H)$ and, if so, which edges of H belong to S_v.

We prove this claim by induction on i. Let $i = 0$. When we computed the value for $m(H_{v^0}) = m(H)$ by using Lemma 1, we checked whether v^0 is the center of a star in an m-system of H. In case v^0 is the center of such a star S_{v^0}, we also remembered which edges belong to S_{v^0}.

Now suppose $i \geq 1$. Let $v \in L_i$. By the induction hypothesis, we know whether v^+ is the center of a star in an m-system $\mathcal{M}(H)$ or not. First suppose v^+ is not the center of a star in an m-system $\mathcal{M}(H)$. Then v is the center of a star S_v in $\mathcal{M}(H)$, and S_v is a star in an m-system $\mathcal{M}(H_v^1(v))$ as well. So, we kept track of S_v. Now suppose v^+ is the center of a star S_{v^+} in an m-system $\mathcal{M}(H)$. By the induction hypothesis, we know which edges S_{v^+} has. Then there are two cases: either vv^+ belongs to S_{v^+}, or it does not. If vv^+ belongs to S_{v^+}, then v is the center of a star S_v in $\mathcal{M}(H)$ if and only if S_v is a star in $\mathcal{M}(H_v)$. If vv^+ does not belong to S_{v^+}, then v is the center of a star S_v in $\mathcal{M}(H)$, and S_v is a star in an m-system $\mathcal{M}(H_v^1(v))$. In both cases we kept track of all the edges of S_v.   □

Step 4: Compute Sharp Dominating Systems for General Triangle-Free Graphs

Suppose G is a claw-free graph. Let H be the preimage of $cl(G)$, i.e., the triangle-free graph with $L(H) = cl(G)$. We now assume that H is not a tree.

The key idea behind our approach in this case is to start with an even subgraph X of H, then to "break" the circuits in X by removing a number of edges, such that we obtain a new graph H^* that is a forest. Then we can apply our approach from the previous section to each component of H^* if we first add sufficiently many pendant edges to ensure that each component has minimum edge-degree at least $\delta_e(H)$. In this procedure we have to add more edges than we remove. However, we will have the following advantage. The added pendant edges have to be dominated by (extra) stars in any dominating system of H^*, and these stars can be merged together into fewer elements of a dominating system in the original graph H. In other words, the larger number of stars we get by applying the upper bounds to H^* will provide the necessary compensation for the larger number of edges that we created. This way we are able to establish our upper bound for H. In [1] we gave a nonconstructive proof to show that this approach works. This proof in [1] was based on a number of assumptions on the choice of the even subgraph X of H. Here we follow an alternative approach which enables a constructive proof.

Let X be an even subgraph of H with set of components $\mathcal{C}$. Let $\mathcal{C}_4 \subset \mathcal{C}$ be the set of components of order 4. For each C in $\mathcal{C}$ we choose an edge e_C of C and for each C in $\mathcal{C}_4$ we choose two independent edges e_C, e_C' of C. We call the set of all chosen edges the X-set and denote it by M. Note that M is a matching of H. Let $H^* = (H - E(X)) \cup M$. We call H^* the X-graph.

Lemma 2. *We can compute an even subgraph X of H such that H^* is a forest in polynomial time.*

Proof. We use an algorithm based on the following arguments:

Phase 1. We first construct an even subgraph X' of H. We can do this in polynomial time by adding mutually edge-disjoint cycles to X' until this is not possible anymore.

Phase 2. We choose an X'-set M' and check (in polynomial time) whether its X'-graph H' is a forest. If it is a forest, then we are done.

Suppose H' is not a forest. Let D be a cycle in H'. Consider the graph $X' \cup D$. For each e in $E(X' \cap D)$ we do as follows. Let e belong to a circuit C of X. Then C shares at most two edges with D. If C only shares e with D then we remove e from $X' \cup D$. In the other case C belongs to $\mathcal{C}_4$ and we remove the two edges of C that are not on D. This way we obtain an even subgraph Y' of H in polynomial time. We go to Phase 2 with Y' instead of X'.

We show that either $e(Y') > e(X')$ or else, if $e(Y') = e(X')$, then Y' contains fewer components than X'. This means that the algorithm will terminate at a certain moment with our desired graph X.

The above can be verified as follows. Note that we removed exactly $e(X' \cap D)$ edges and we added $e(D \setminus X')$ edges. So we are done if $e(D \setminus X') > e(X' \cap D)$. Since M' is a matching, we find that $e(X' \cap D) \leq e(D)/2$, so $e(D) = e(X' \cap D) + e(D \setminus X') \leq e(D)/2 + e(D \setminus X')$, so $e(D \setminus X') \geq e(D)/2 \geq e(X' \cap D)$. We are done unless $e(D \setminus X') = e(X' \cap D) = e(D)/2$.

Suppose the latter is the case. Then we are done if Y' contains fewer components than X'. Suppose Y' and X' have the same number of components. Then $X' \cap D$ belongs to exactly one circuit of X'. We already deduced that $e(X' \cap D) = e(D)/2 \geq 2$. Hence D is a four-cycle, but then the triangle-free graph H contains an induced K_4. With this contradiction we have completed the proof of this claim.   □

The remainder of the constructive proof is exactly the same as the corresponding parts in our nonconstructive proof in [1]. We do not include it here due to the page restriction.

Step 5: Translate the Dominating Systems Back into 2-Factors

Once we have obtained a dominating system S for the preimage graph H with $cl(G) = L(H)$, it is easy to translate this back into a 2-factor of $cl(G)$ in polynomial time:

- the stars in S correspond to complete graphs in $cl(G)$ on at least three vertices; a hamiltonian cycle can clearly be constructed in polynomial time;
- the circuits in S and the edges they dominate correspond to hamiltonian subgraphs in $cl(G)$; one can construct a hamiltonian cycle by traversing the circuit, picking up the edges (vertices in $cl(G)$) one by one and inserting dominated edges at the first instance an end vertex of a dominated edge is encountered. For traversing the circuits we use the polynomial algorithm that finds a eulerian tour in an even connected graph (cf. [4]).

Step 6: Translate 2-Factors in $cl(G)$ to 2-Factors in G

We first introduce some notations. Let $C = v_1 v_2 \ldots v_p v_1$ be a cycle with a fixed orientation. The successor v_{i+1} of v_i is denoted by $v_i^{+C} = v_i^+$ and its predecessor v_{i-1} by $v_i^{-C} = v_i^-$. The segment $v_i v_{i+1} \ldots v_j$ is denoted by $v_i \overrightarrow{C} v_j$, where the subscripts are to be taken modulo $|C|$. The converse segment $v_j v_{j-1} \ldots v_i$ is denoted by $v_j \overleftarrow{C} v_i$. We use similar notations for paths.

We assume we are given a 2-factor F' of $cl(G)$ of a claw-free graph G. Let k be the number of components of F'. Here, we show how to obtain in polynomial time a 2-factor F of G such that F has *at most* k components. We base our translation of the following new theorem, which generalizes a similar result for hamiltonicity [2] in algorithmic sense.

Theorem 10. *Let G be a graph and let $\{u, v, x, y\}$ be a subset of four vertices of V_G such that $uv \notin E_G$ and $\{x, y\} \subseteq N(u) \cap N(v)$. Let $N(x) \subseteq N(u) \cup N(v) \cup \{u, v\}$ and let $N(y) \setminus (N(x) \cup \{x\})$ induce a complete graph (or be empty). Then we can find a 2-factor of G with at most k components in polynomial time, if $G + uv$ has a 2-factor with k components.*

Proof. Suppose $G + uv$ has a 2-factor F' with at most k components. Below we give a number of polynomial time transformations of F' such that we obtain a 2-factor F of G with at most k components. If $uv \notin E_{F'}$ then F' is a 2-factor of G. Suppose $uv \in E_D$ for some (cycle) component D of F', say $v = u^-$. Let $P = u\overrightarrow{D}v$. We distinguish the following three cases.

First suppose $x \notin V_D$. Let $x \in V_{D'}$ for some (cycle) component D' of F. By our assumptions, we may without loss of generality assume that $x^{+D'}u \in E_G$. Then we replace D and D' by a new cycle $ux^{+D'}\overrightarrow{D'}xv\overleftarrow{P}u$, and we are done.

Second suppose $x \in V_D$ but $y \notin V_D$. Let $y \in V_{D^*}$ for some (cycle) component D^* of F. Let $y' = y^{+D^*}$ and $y'' = y^{-D^*}$ be the neighbors of y on D^*. Suppose $y'y'' \in E_G$. Then we replace D^* by $y'\overrightarrow{D^*}y''y'$ and D by $uyv\overleftarrow{P}u$, and we are done. Suppose $y'y'' \notin E_G$. Since $N(y) \setminus (N(x) \cup \{x\})$ induces a complete graph, we find that one of the edges xy', xy'', say xy', must exist in G. By our assumptions, we then find that $y'u$ or $y'v$ belongs to E_G, and we are done by the same argument as in the previous case.

Third suppose $\{x, y\} \subset V_D$. Say x is on $u\overrightarrow{P}y$. First suppose $xy \in E_D$. We replace D by $u\overrightarrow{P}xv\overleftarrow{P}yu$, and we are done. Now suppose $xy \notin E_D$. Then $x^+ \neq y$. By our assumptions, $x^+ \in N(u) \cup N(v)$. Suppose $ux^+ \in E_G$. We replace D by $ux^+\overrightarrow{P}vx\overleftarrow{P}u$. Hence we may assume $vx^+ \in E_G$. Suppose $y^- = x^+$. Then we replace D by $uy\overrightarrow{P}vy^-\overleftarrow{P}u$. Hence we may assume $y^- \neq x^+$. Suppose $y^-x \in E_G$. Then we replace D by $u\overrightarrow{P}xy\overleftarrow{P}x^+v\overrightarrow{P}yu$. Hence we may assume $y^-x \notin E_G$. Suppose $y^+ = v$. Then we replace D by $u\overrightarrow{P}xvx^+\overrightarrow{P}yu$. Hence we may assume $y^+ \neq v$. Suppose $y^+x \in E_G$. Then we replace D by $u\overrightarrow{P}xy^+\overrightarrow{P}vx^+\overrightarrow{P}yu$. Hence we may assume $y^+x \notin E_G$. As $y^-x \notin E_G$, we then find $y^-y^+ \in E_G$ due to our assumptions. Then we replace D by $u\overrightarrow{P}y^-y^+\overrightarrow{P}vyu$. This proves Theorem 10. $\qquad\square$

Note that in Theorem 10, x and y can be nonadjacent, and G does not have to be claw-free. However, the following observation is easy to see.

Observation 1 ([2]). *If G is claw-free, then the conditions of Theorem 10 are satisfied if x and y are adjacent.*

Then, by the following observation, we can indeed transform a 2-factor of $cl(G)$ that has k components to a 2-factor of G that has at most k components. This means we have proven our main result. For convenience we include the proof of the next observation.

Observation 2 ([2]). *Let x be a vertex of a claw-free graph G with $G[N(x)]$ connected and non-complete. Then the local completion of G at x can be obtained by iteratively joining pairs $\{u, v\} \subseteq N(x)$ that satisfy the conditions in Theorem 10 for some $y \in N(u) \cap N(v)$.*

Proof. Consider the subgraph H_x of G induced by $N(x) \cup \{a \in V_G \mid ab \in E_G$ for some $b \in N(x)\}$. Note that x is a vertex of H_x and that H_x is claw-free. Hence, by Observation 1, x and y satisfy the conditions of Theorem 10 (in H_x) for every $y \in N(x)$. Since we only join nonadjacent pairs in $N(x)$, $N(x)$ and $N(y)$ will keep these properties for all $y \in N(x)$.   □

Note that the above approach gives a short constructive proof for Theorem 4 (the result of Ryjáček, Saito & Schelp in [19]).

We note that Theorem 9 has the following consequence as a byproduct. We need a few definitions before we can state the result. A *cut vertex* of a graph G is a vertex whose removal increases the number of components. A *block* of G is a maximal subgraph of G without cut vertices (of itself). Hence if G has no isolated vertices, its blocks are either K_2s or (maximal) 2-connected subgraphs. For the purpose of our next result we call a block B of a claw-free graph G a *semiclique* if B becomes a complete subgraph of $cl(G)$. Since a claw-free graph in which every block is a semiclique has a forest as its preimage, we obtain the following consequence of Theorem 9.

Corollary 1. *Let G be a claw-free graph in which all blocks are semicliques. If G has a 2-factor, then we can construct a minimum 2-factor of G in polynomial time.*

4 Conclusions

In a recently submitted paper we obtained sharp upper bounds for the minimum number of components of a 2-factor in a claw-free graph. Here we extended these results by presenting a polynomial algorithm that constructs a 2-factor of a claw-free graph with minimum degree at least four whose number of components meets this bound. As a byproduct we showed that the problem of obtaining a minimum 2-factor (if it exists) is polynomially solvable for a subclass of claw-free graphs in which all blocks are semicliques. As another byproduct we gave a short constructive proof for a result of Ryjáček, Saito & Schelp.

Our polynomial time algorithm yields a 2-factor with a number of components below a guaranteed upper bound. This upper bound is completely determined by an upper bound we find for the number of elements of a dominating system of a certain tree (that is obtained from he corresponding triangle-free graph in Theorem 8). As this upper bound is sharp (cf. [20]), our next goal will be to determine the extremal tree cases and try to exclude these from happening. This refined analysis may lead to a better upper bound for claw-free graphs for which the current upper bound is not sharp. Another way to improve our algorithm is trying to refine the algorithm that constructs the tree H^* in Lemma 2 such that we have more information on the number of circuits in the even subgraph X of H.

Finally, Corollary 1 shows that our algorithm yields a 2-factor with a minimum number of components for claw-free graphs with arbitrary minimum degree in which all blocks are semicliques. In future research we aim to generalize this result, i.e, to find a larger class of claw-free graphs for which our (possibly modified) algorithm solves the problem of finding a minimum 2-factor. We will also analyze the class of claw-free graphs with minimum degree 3 that have a 2-factor more carefully.

References

1. Broersma, H.J., Paulusma, D., Yoshimoto, K.: Sharp upper bounds for the minimum number of components of 2-factors in claw-free graphs (submitted), http://www.dur.ac.uk/daniel.paulusma/Papers/Submitted/claw.pdf
2. Broersma, H.J., Trommel, H.: Closure concepts for claw-free graphs. Discrete Math. 185, 231–238 (1998)
3. Choudum, S.A., Paulraj, M.S.: Regular factors in $K_{1,3}$-free graphs. J. Graph Theory 15, 259–265 (1991)
4. Diestel, R.: Graph Theory, 2nd edn. Graduate Texts in Mathematics, vol. 173. Springer, Heidelberg (2000)
5. Egawa, Y., Ota, K.: Regular factors in $K_{1,n}$-free graphs. J. Graph Theory 15, 337–344 (1991)
6. Faudree, R.J., Favaron, O., Flandrin, E., Li, H., Liu, Z.: On 2-factors in claw-free graphs. Discrete Math. 206, 131–137 (1999)
7. Faudree, R., Flandrin, E., Ryjáček, Z.: Claw-free graphs—a survey. Disc. Math. 164, 87–147 (1997)
8. Fronček, D., Ryjáček, Z., Skupień, Z.: On traceability and 2-factors in claw-free graphs. Discussiones Mathematicae Graph Theory 24, 55–71 (2004)
9. Garey, M.R., Johnson, D.S.: Computers and Intractability. W.H. Freeman and Co, New York (1979)
10. Gould, R., Hynds, E.: A note on cycles in 2-factors of line graphs. Bull. of ICA. 26, 46–48 (1999)
11. Gould, R.J., Jacobson, M.S.: Two-factors with few cycles in claw-free graphs. Discrete Math. 231, 191–197 (2001)
12. Harary, F.: Graph Theory. Addison-Wesley, Reading MA (1969)
13. Jackson, B., Yoshimoto, K.: Even subgraphs of bridgeless graphs and 2-factors of line graphs. Discrete Math. 307, 2775–2785 (2007)
14. Lovasz, L., Plummer, M.D.: Matching Theory, North-Holland Mathematics Studies 121. North-Holland, Amsterdam
15. Matthews, M.M., Sumner, D.P.: Hamiltonian results in $K_{1,3}$-free graphs. J. Graph Theory 8, 139–146 (1984)
16. Plummer, M.D.: Graph factors and factorization: 1985-2003: A survey. Discrete Math. 307, 791–821 (2007)
17. Roussopoulos, N.D.: A max{m,n} algorithm for determining the graph H from its line graph G. Information Processing Letters 2, 108–112 (1973)
18. Ryjacek, Z.: On a closure concept in claw-free graphs. J. Combin. Theory Ser. B 70, 217–224 (1997)
19. Ryjacek, Z., Saito, A., Schelp, R.H.: Closure, 2-factor, and cycle coverings in claw-free graphs. J. Graph Theory 32, 109–117 (1999)
20. Yoshimoto, K.: On the number of components in a 2-factor of a claw-free graph. Discrete Math. 307, 2808–2819 (2007)

A 6/5-Approximation Algorithm for the Maximum 3-Cover Problem[*]

Ioannis Caragiannis[1] and Gianpiero Monaco[2]

[1] Research Academic Computer Technology Institute and
Department of Computer Engineering and Informatics
University of Patras, 26500 Rio, Greece
[2] Department of Computer Science, University of L'Aquila
Via Vetoio, Coppito 67100, L'Aquila, Italy

Abstract. In the maximum cover problem, we are given a collection of sets over a ground set of elements and a positive integer w, and we are asked to compute a collection of at most w sets whose union contains the maximum number of elements from the ground set. This is a fundamental combinatorial optimization problem with applications to resource allocation. We study the simplest APX-hard variant of the problem where all sets are of size at most 3 and we present a 6/5-approximation algorithm, improving the previously best known approximation guarantee. Our algorithm is based on the idea of first computing a large packing of disjoint sets of size 3 and then augmenting it by performing simple local improvements.

1 Introduction

In the maximum cover problem, we are given a collection of sets over a ground set of elements V and a positive integer w, and we are asked to compute a collection of at most w sets of maximum benefit, i.e., so that their union contains the maximum number of elements from the ground set. This is a fundamental combinatorial optimization problem with applications to the resource allocation scenario with w available resources, users wishing to access one of the resources, and compatibility constraints defined as sets of users that can simultaneously access the same resource. The problem of computing an assignment of the maximum number of users to the resources so that the compatibility constraints are satisfied is equivalent to the maximum cover problem.

The natural greedy algorithm which starts with an empty solution and iteratively includes in the solution a set of maximum size that consists of elements that have not been covered before until w sets are selected has approximation ratio $\frac{e}{e-1}$. In general, this result is tight due to an inapproximability result due

[*] This work was partially supported by the EU COST Action 293 "Graphs and Algorithms in Communication Networks" (GRAAL), by the EU IST FET Integrated Project 015964 AEOLUS, and by a "Caratheodory" research grant from the University of Patras.

E. Ochmański and J. Tyszkiewicz (Eds.): MFCS 2008, LNCS 5162, pp. 205–216, 2008.
© Springer-Verlag Berlin Heidelberg 2008

to Feige [4]. An interesting special case of the problem where this inapproximability result does not hold is when the size of the sets in $\mathcal{S}$ is small. In maximum
k-cover, k denotes the maximum size of each set in $\mathcal{S}$. Without loss of generality,
we assume that $\mathcal{S}$ is closed under subsets.

When applied to maximum k-cover, the greedy algorithm essentially belongs
to the following class of iterative algorithms. An iterative algorithm for the
maximum k-cover problem starts with an empty solution and works in phases,
one phase for each $i = k, k-1, ...$ In the phase associated with i, the algorithm
includes a maximal collection of disjoint sets each consisting of exactly i elements
that have not been included in previous phases. Maximality implies that any set
of size i (henceforth called i-set) intersects at least one of the selected sets. The
algorithm terminates when w sets have been selected in total. Intuitively, a large
number of disjoint sets in the early phases is desirable in order to obtain good
solutions. So, the problem that has to be solved in each phase is known as k-set
packing. In k-set packing, we are given a collection of sets of size exactly k over
a ground set of elements V and the objective is to select the maximum number
of disjoint sets among the collection. This problem is known to be APX-hard
for $k \geq 3$ [10] (see also [8]) while it is equivalent to maximum matching for
$k = 2$ and trivial for $k = 1$. For $k \geq 3$, a well-known local search heuristic has
an approximation ratio very close to $k/2$. An analysis technique for the class
of iterative algorithms is presented in [2]. In that paper, an upper bound of the
benefit of an iterative algorithm follows by the solution of a linear program whose
constraints capture the approximation guarantee of the i-set packing algorithm
used in phase i. Algorithms for k-set packing have also been used recently in [1]
in order to improve the known approximation guarantees for the related k-set
cover problem [3,11].

In this work, we study the maximum 3-cover problem. This is the simplest
variant of maximum cover which is still APX-hard while the maximum 2-cover
problem can be solved in polynomial time. Both statements follow by similar
statements for 3-set packing [10] and 2-set packing, respectively. The analysis
of [2] yields an approximation ratio of 18/13-approximation for the greedy algorithm while the best result that can be obtained using the techniques of [2] is a
9/7-approximation algorithm that first computes a maximal 3-set packing of any
size and then completes the solution by including the maximum possible number of elements outside the selected 3-sets into 2-sets and (if necessary) 1-sets so
that exactly w sets are used. In general, the upper bounds on the approximation
ratio of these algorithms cannot be improved as we observe in the next section,
and new algorithmic ideas are required in order to obtain better results.

Local search seems to be a promising approach since it has been proved to
be efficient for the related 3-set cover and 3-set packing problems [3,5,6,7,9,11].
For example, such an algorithm could start with any solution consisting of w
sets and repeatedly follow a better solution that is produced by the previous
one by changing (inserting and/or removing) a constant number of 3-sets and
updating the 2-sets and 1-sets. Unfortunately, such pure local search algorithms
seem to be complicated to analyze. In this paper, we present a more structured

algorithm which combines the idea of starting with a large packing of disjoint 3-sets and then augments this solution by 2-sets and 1-sets by performing appropriate simple local improvements while this is possible. The analysis of our algorithm distinguishes between several cases depending on the structure of the covering obtained. The algorithm is proved to have an approximation ratio of 6/5 by a series of combinatorial arguments applied on the covering produced.

The rest of the paper is structured as follows. In Section 2, we discuss the limitations of iterative algorithms and outline the main ideas behind our algorithm which is presented in detail in Section 3. The statement of our main result and its proof then follow in Section 4. Due to lack of space, some proofs of intermediate lemmas have been omitted.

2   Limitations of Iterative Algorithms

In this section, we briefly discuss the limitations of iterative algorithms when applied to maximum 3-cover and give the intuition behind our algorithm in four different examples presented in Figure 1. In the first (upper left) example, a solution produced by the greedy algorithm is depicted. The vertical 3-sets form the optimal solution with $w = 12$. The greedy algorithm first selects 4 3-sets that intersect each optimal 3-set in exactly one element. Then, it selects 6 2-sets again intersecting each optimal 3-set in exactly one element. So, in order to complete the solution, the algorithm can select only 2 additional elements (1-sets) for a total benefit of 26. The optimal benefit is 36 and this is an example where the greedy algorithm has approximation ratio 18/13. Of course, the algorithm could

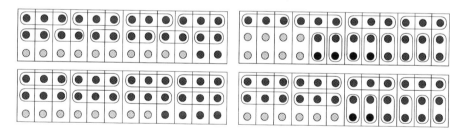

Fig. 1. Four examples of coverings with approximation ratios 18/13, 9/7, 9/7, and 6/5

do better when selecting 2-sets since the maximum number of disjoint 2-sets corresponds to a maximum matching computation in the graph whose nodes correspond to the elements not covered by 3-sets and whose edges correspond to 2-sets among these elements. Since exactly one element from each optimal 3-set has been included in the 3-sets selected by the algorithm, there are 12 disjoint 2-sets among the uncovered elements, one 2-set including the two uncovered elements of each optimal 3-set. Even in this case (see the second example at the upper right part of Figure 1), at most 8 additional 2-sets can be selected for a total benefit of 28. The algorithm of this example has approximation ratio 9/7.

One attempt to improve the benefit could be to use a better 3-set packing algorithm when selecting the disjoint 3-sets. The best known such algorithm is based on local search [9] and works as follows. It uses a constant parameter t. Starting with an empty solution, it repeatedly searches for a neighbor solution which results from the current one by removing $p-1$ 3-sets and inserting p new 3-sets for some $p \leq t$ (so that the 3-sets are disjoint). If such a neighbor solution is found, this is set as the current solution. The algorithm terminates when a locally maximum solution is reached, i.e., a solution with no neighbor solutions. The following result has been proved in [9].

Theorem 1 (Hurkens and Schrijver [9]). *The local search algorithm that computes a 3-set packing by performing at most t local improvements at each step has approximation ratio at most $\frac{3}{2} + \frac{3}{2(2 \cdot 2^r - 3)}$ when $t = 2r - 1$, and at most $\frac{3}{2} + \frac{1}{2 \cdot 2^r - 2}$ when $t = 2r$.*

For any constant value of t, this algorithm may result to a locally maximum solution with 2 of the elements of each optimal 3-set covered. So, at most 4 additional elements (1-sets) can be used to complete the solution. The approximation ratio is again 9/7. See the third example in Figure 1 where the 3-sets form a locally maximum solution for the local search 3-set packing algorithm that performs at most 2 local improvements per step.

We point out that another approach could be to consider the problem as an instance of 3-set cover, attempt to minimize the number of sets required in order to cover all elements, and then, simply keep the elements in the first w sets. Using the semi-local optimization algorithm of Duh and Furer [3], we would have ended up with a solution similar to the one in the second example above. Similar attempts through semi-local optimization algorithms for the partial set cover problem with sets of maximum size 3 [5] would have led to the same situation with approximation ratio 9/7.

Instead, if we could force the 3-sets to intersect half of the 3-sets in two elements and half of them in one element, the selection of 2-sets would result in at least 6 additional sets; this would give a 6/5-approximate solution (see the fourth example in Figure 1). This is what our algorithm is trying to achieve. Starting with a good 3-set packing completed with optimally selected 2-sets, it further considers improving the solution according to simple local improvement steps in order to achieve some balance between the number of 3-sets and 2-sets and avoid the situations in the second and third examples of Figure 1. Of course, the instances of the problem may be much different than those in Figure 1 and this should be taken into account in our analysis.

3   The Algorithm

Our algorithm receives as input an integer w, a set of elements V, a collection S of subsets of V each containing at most 3 elements (S is closed under subsets), and it produces as output w sets of S covering some of the elements of V. It works in three steps. Step A computes a covering that consists of a large 3-set packing and optimally selected 2-sets (i.e., using the maximum matching computation mentioned in the second example of the previous section). Phase B performs two simple local improvement steps while this is possible. Step C simply completes the solution with 1-sets if this is necessary. A detailed description of the algorithm follows.

A. Compute a maximal set of disjoint 3-sets using the local search heuristic for 3-set packing with parameter $t = 2$ and complete the solution optimally with 2-sets. Denote by α_2 and α_3 the number of 2-sets and 3-sets computed, respectively.
B. Set $Stuck = FALSE$
While $\alpha_2 + \alpha_3 < w$ and $Stuck = FALSE$ do:
Try to perform a *good local change*, otherwise set $Stuck = TRUE$.
C. If $\alpha_2 + \alpha_3 < w$ then
Include $w - \alpha_2 - \alpha_3$ elements of V that have not been included in 2-sets and 3-sets yet into 1-sets.

A *good local change* can be performed in the following two cases:

- A *replacement* means to remove one 3-set from the current solution and include another one that does not intersect with the remaining 3-sets and to complete the solution optimally with 2-sets. A replacement is performed only if it is good. A replacement is good if $\alpha_2 + \alpha_3 \leq w - 1$ at its beginning and the number of 2-sets increases after it. If the new number of 2-sets exceeds $w - \alpha_3$, the algorithm arbitrarily keeps only $w - \alpha_3$ of them and discards the rest. After performing a good replacement, the algorithm updates the numbers α_2 and α_3 of 2-sets and 3-sets in the current solution.
- A *removal* means to remove a 3-set from the current solution and complete optimally with 2-sets. A removal is performed only if it is good. A removal is good if $\alpha_2 + \alpha_3 \leq w - 2$ at its beginning and the number of 2-sets increases by 3. After performing a good removal, the algorithm updates the numbers α_2 and α_3 of 2-sets and 3-sets in the current solution.

4   Analysis of the Algorithm

In the analysis of the algorithm, we denote by b_3, b_2 and b_1 the number of 3-sets, 2-sets and 1-sets in the optimal solution, respectively. We will show that the algorithm computes a solution of benefit at least $\frac{5}{2}b_3 + \frac{7}{4}b_2 + b_1 - \frac{1}{2}$ while the optimal benefit is $3b_3 + 2b_2 + b_1$. Due to the integrality of b_3, b_2, and b_1, the only case in which the algorithm above may fail to produce a 6/5-approximate solution is when b_3 is odd and $b_2 = b_1 = 0$. In order to avoid this situation, we may run the algorithm $O(|V|^3)$ times and pick the best solution, each time guessing one of the optimal 3-sets and run our algorithm to the remaining instance. The total number of elements covered will then be at least $3 + \frac{5}{2}(b_3 - 1) + \frac{7}{4}b_2 + b_1 - \frac{1}{2} = \frac{5}{2}b_3 + \frac{7}{4}b_2 + b_1 \geq \frac{5}{6}(3b_3 + 2b_2 + b_1)$. In this way, we obtain our main result.

Theorem 2. *There exists a polynomial-time 6/5-approximation algorithm for the maximum 3-cover problem.*

The proof of the lower bound on the benefit of our algorithm has been divided into four parts. In Section 4.1, we consider the two simple cases where the algorithm finishes step B having used w sets (Lemma 2) or $w-1$ sets (Lemma 3). In Section 4.2, in order to handle the remaining cases where the algorithm finishes step B by having produced a covering on which no good local change is possible, we define a graph based on the structure of this covering compared to the optimal solution. Then, we distinguish between two cases which are studied in Sections 4.3 and 4.4, respectively. The results for the benefit of the algorithm in these cases are stated in Lemmas 6 and 9, respectively.

4.1 Some Easy Cases

We start with an important property maintained after step A of the algorithm.

Lemma 1. *At each step after step A of the algorithm, it always holds*

$$\alpha_2 + 3\alpha_3 \geq b_2 + 2b_3.$$

Proof. We first consider the case where the algorithm uses at least $b_2 + b_3$ 2-sets and 3-sets at step A, i.e., $\alpha_2 + \alpha_3 \geq b_2 + b_3$. Note that the maximum number of disjoint 3-sets in the instance of 3-set packing solved at step A is at least b_3. The local search 3-set packing algorithm with parameter $t = 2$ guarantees a 2-approximate solution and, hence, $2\alpha_3 \geq b_3$. The lemma then follows by summing the two inequalities.

If after step A it is $\alpha_2 + \alpha_3 < b_2 + b_3$, denote by T_i for $i = 1, 2, 3$, the number of optimal 3-sets from which the algorithm has included i elements in 3-sets. Denote by D_i for $i = 1, 2$, the number of optimal 2-sets from which the algorithm has included i elements in 3-sets. Since there are T_1 optimal 3-sets from which the algorithm does not include two elements in 3-sets and $b_2 - D_2 - D_1$ optimal 2-sets from which the algorithm has not included any elements in 3-sets, it will also use $\alpha_2 \geq T_1 + b_2 - D_2 - D_1$ 2-sets, otherwise the selection of 2-sets would not be optimal. Also, $3\alpha_3 \geq 3T_3 + 2T_2 + T_1 + 2D_2 + D_1$ since three elements in T_3 optimal 3-sets, two elements in T_2 optimal 3-sets and D_2 optimal 2-sets, and one element in T_1 optimal 3-sets and D_1 optimal 2-sets have been included in 3-sets by the algorithm. So, we have that

$$\begin{aligned}
\alpha_2 + 3\alpha_3 &\geq T_1 + b_2 - D_2 - D_1 + 3T_3 + 2T_2 + T_1 + 2D_2 + D_1 \\
&= 2T_1 + 2T_2 + 3T_3 + D_2 + b_2 \\
&\geq 2(T_1 + T_2 + T_3) + b_2 \\
&= b_2 + 2b_3,
\end{aligned}$$

where the second equality holds due to the maximality of the collection of 3-sets computed during step A.

So far, we have proved that the inequality holds just after step A. In order to prove that it holds at all steps of the algorithm we observe that neither removals nor replacements decrease $\alpha_2 + 3\alpha_3$. $\square$

Next, we consider two easy cases.

Lemma 2. *If the algorithm finishes step B with Stuck = FALSE then its benefit is at least $\frac{5}{2}b_3 + 2b_2 + \frac{3}{2}b_1$.*

Proof. Since the algorithm exits at step B with $Stuck = FALSE$ it must be that $\alpha_2 + \alpha_3 = w \geq b_1 + b_2 + b_3$. Using Lemma 1, we have that the benefit of the algorithm after exiting step B is

$$
\begin{aligned}
2\alpha_2 + 3\alpha_3 &= \frac{3}{2}(\alpha_2 + \alpha_3) + \frac{1}{2}(\alpha_2 + 3\alpha_3) \\
&\geq \frac{3}{2}(b_1 + b_2 + b_3) + \frac{1}{2}(b_2 + 2b_3) \\
&\geq \frac{5}{2}b_3 + 2b_2 + \frac{3}{2}b_1.
\end{aligned}
$$
□

Lemma 3. *If the algorithm finishes step B with Stuck = TRUE and $\alpha_2 + \alpha_3 = w - 1$, then its benefit is at least $\frac{5}{2}b_3 + 2b_2 + \frac{3}{2}b_1 - \frac{1}{2}$.*

Proof. The proof is the same as the proof of lemma 2 by considering the extra 1-set the algorithm will include at step C. □

4.2   Non-improving Coverings

In the following we consider the case when the algorithm finishes step B with $Stuck = TRUE$ and $\alpha_2 + \alpha_3 \leq w - 2$. In this case, we say that the algorithm has computed a *non-improving covering* i.e., a collection of 3-sets and optimally selected 2-sets in which no good removal or good replacement is possible. The algorithm will then enter step C and include $w - \alpha_2 - \alpha_3$ additional single elements in the solution for a total benefit of

$$
benefit = 2\alpha_2 + 3\alpha_3 + w - \alpha_2 - \alpha_3 \geq \alpha_2 + 2\alpha_3 + b_1 + b_2 + b_3. \qquad (1)
$$

We introduce the following notation. Given a non-improving covering, we say that an optimal 3-set is of type:

- 333 if all its elements are included in 3-sets by the algorithm.
- 332 if two of its elements are included in 3-sets and one element in a 2-set.
- 33* if two of its elements are included in 3-sets and one is not covered.
- 322 if one of its elements is included in a 3-set and two elements in 2-sets.
- 32* if one of its elements is included in a 3-set, one element in a 2-set and one element is not covered.
- 222 if all its elements are included in 2-sets.
- 22* if two of its elements are included in 2-sets and one is not covered.

We say that an optimal 2-set is of type:

- 33 if all its elements are included in 3-sets.
- 32 if one of its elements is included in a 3-set and one element in a 2-set.
- 3* if one of its elements is included in a 3-set and one is not covered.

- 22 if all its elements are included in 2-sets.
- 2∗ if one of its elements is included in a 2-set, and one element is not covered.

We say that an element that does not belong in optimal 2-sets or 3-sets is of type:

- 3 if the element is included in a 3-set.
- 2 if the element is included in a 2-set.
- ∗ if the element is not covered.

We remark that we use the term optimal sets to refer to these elements as well.

Now, we will define a graph associated with a non-improving covering as follows. Given an optimal solution and the 3-sets and 2-sets in a non improving covering, we define the graph G that has one node for each 3-set and 2-set of the optimal solution, one node for each element not included in 3-sets and 2-sets in the optimal solution, and an edge connecting two nodes if the algorithm has included an element from each corresponding set in the same 2-set. The graph may have self-loops (for example, when two of the elements of an optimal set of type 322 have been included in the same 2-set by the algorithm), or parallel edges. The degree of a node is the number of 2's in its type. We call nodes corresponding to optimal sets of type 33∗, 32∗, 3∗, 22∗, 2∗, or ∗ *faulty* nodes. We also call *faulty* a connected component of G if it contains at least one faulty node. The next lemma states an important property of faulty connected components of G.

Lemma 4. *Each faulty connected component of G contains exactly one faulty node.*

We distinguish between two cases depending on whether the graph of the non-improving covering has a faulty connected component that contains a cycle or not. In the former, we say that the non-improving covering is *cyclic*; in the latter, we say that the non-improving covering is *acyclic*.

4.3   The Case of Cyclic Non-improving Covering

We first consider the case where the non-improving covering computed by the algorithm after step B is cyclic. In this case, we can show the following property of the cyclic covering. The proof is omitted.

Lemma 5. *If the algorithm finishes step B by computing a cyclic non-improving covering, then (a) there are no optimal sets of type 33∗ or 3∗ and (b) the number of optimal sets of type 32∗ is at most the number of optimal sets of type 222.*

We are now ready to prove the next statement using Lemmas 1 and 5 and the lower bound on the benefit of the algorithm from inequality (1).

Lemma 6. *If the algorithm finishes step B by computing a cyclic non-improving covering, then its benefit at the end is at least $\frac{5}{2}b_3 + \frac{7}{4}b_2 + b_1$.*

Proof. We use the notation T_X to denote the number of optimal sets or elements out of optimal sets of type X. Taking into account that $T_{33∗} = T_{3∗} = 0$ (by Lemma 5), this definition immediately yields

$$b_2 = T_{33} + T_{32} + T_{22} + T_{2*} \tag{2}$$

$$b_3 = T_{333} + T_{332} + T_{322} + T_{32*} + T_{222} + T_{22*} \tag{3}$$

Since the three elements in optimal sets of type 222, two of the elements in optimal sets of type 322, 22*, and 22, and one of the elements in optimal sets of type 332, 32*, 32, 2*, and 2 are included in 2-sets by the algorithm, we have that

$$\alpha_2 \geq \frac{1}{2}T_{332} + T_{322} + \frac{1}{2}T_{32*} + \frac{3}{2}T_{222} + T_{22*} + \frac{1}{2}T_{32} + T_{22} + \frac{1}{2}T_{2*} + \frac{1}{2}T_2 \tag{4}$$

Similarly, since the three elements in optimal sets of type 333, two of the elements in optimal sets of type 332 and 33, and one of the elements in optimal sets of type 322, 32*, 32, and 3 are included in 3-sets by the algorithm we have that

$$\alpha_3 \geq T_{333} + \frac{2}{3}T_{332} + \frac{1}{3}T_{322} + \frac{1}{3}T_{32*} + \frac{2}{3}T_{33} + \frac{1}{3}T_{32} + \frac{1}{3}T_3 \tag{5}$$

We use inequality (1) and inequalities (2)-(5) to obtain

$$benefit \geq 2\alpha_3 + \alpha_2 + b_3 + b_2 + b_1$$
$$\geq 3T_{333} + \frac{17}{6}T_{332} + \frac{8}{3}T_{322} + \frac{13}{6}T_{32*} + \frac{5}{2}T_{222} + 2T_{22*}$$
$$+ \frac{7}{3}T_{33} + \frac{13}{6}T_{32} + 2T_{22} + \frac{3}{2}T_{2*} + \frac{2}{3}T_3 + \frac{1}{2}T_2 + b_1 \tag{6}$$

By Lemma 5, we have $T_{32*} \leq T_{222}$ which we express as

$$0 \geq \frac{1}{12}(T_{32*} - T_{222}) \tag{7}$$

By Lemma 1 and using inequalities (2)-(5), we obtain

$$T_{333} + \frac{1}{2}T_{332} + T_{33} + \frac{1}{2}T_{32} + T_3 + \frac{1}{2}T_2 \geq \frac{1}{2}T_{32*} + \frac{1}{2}T_{222} + T_{22*} + \frac{1}{2}T_{2*}$$

and, equivalently,

$$0 \geq \frac{1}{2}(-T_{333} - \frac{1}{2}T_{332} - T_{33} - \frac{1}{2}T_{32} - T_3 - \frac{1}{2}T_2 + \frac{1}{2}T_{32*} + \frac{1}{2}T_{222} + T_{22*} + \frac{1}{2}T_{2*}) \tag{8}$$

Summing (6), (7), (8) we obtain that

$$benefit \geq \frac{5}{2}T_{333} + \frac{31}{12}T_{332} + \frac{8}{3}T_{322} + \frac{5}{2}T_{32*} + \frac{8}{3}T_{222} + \frac{5}{2}T_{22*}$$
$$+ \frac{11}{6}T_{33} + \frac{23}{12}T_{32} + 2T_{22} + \frac{7}{4}T_{2*} + \frac{1}{6}T_3 + \frac{1}{4}T_2 + b_1$$
$$\geq \frac{5}{2}(T_{333} + T_{332} + T_{322} + T_{32*} + T_{222} + T_{22*})$$
$$+ \frac{7}{4}(T_{33} + T_{32} + T_{22} + T_{2*}) + b_1$$
$$= \frac{5}{2}b_3 + \frac{7}{4}b_2 + b_1$$

which completes the proof of the lemma. □

4.4   The Case of Acyclic Non-improving Covering

The only case that remains to be considered is when the algorithm finishes step B by computing an acyclic non-improving covering. In the analysis of this case, we will use non-improving coverings of a particular type.

Definition 1. *An acyclic non-improving covering is called* canonical *if each faulty connected component is singleton (i.e. of type* $*$, $3*$ *or* $33*$*).*

Lemma 7. *For each acyclic non-improving covering there is a connected non-improving covering with the same 3-sets and the same number of 2-sets.*

Proof. We will construct a new covering by replacing the 2-sets corresponding to some of the edges in a faulty connected component F of the corresponding graph containing a faulty node v_0 of type $32*$, $22*$, or $2*$ with an equal number of 2-sets so that the connected components induced by the nodes of F in the graph corresponding to the new covering belong either to non-faulty connected components or to a connected component of type $33*$, $3*$, or $*$.

Since F is acyclic, there is a non-faulty node of degree 1 in F; this node corresponds to an optimal set of type 332, 32, or 2. Let $p = \langle v_0, v_1, ..., v_t \rangle$ be the path from v_0 to such a node v_t. Let s_0 be the element of the optimal set corresponding to node v_0 which has not been included in 3-sets and 2-sets by the algorithm, and for $i = 0, 1, ..., t - 1$ let s_{2i+1} and $s_{2(i+1)}$ be the elements of the optimal sets v_i and v_{i+1}, respectively, which have been included in the same 2-set by the algorithm that corresponds to the edge (v_i, v_{i+1}) of p. By replacing the t 2-sets $\{s_{2i+1}, s_{2(i+1)}\}$ for $i = 0, 1, ..., t - 1$ with the t 2-sets $\{s_{2i}, s_{2i+1}\}$ for $i = 0, 1, ..., t - 1$ (and leaving element s_{2t} uncovered), we obtain a new covering whose subgraph replaces the edges of p with self-loops in each of the nodes v_0, v_1, ..., v_{t-1}, while node v_t now corresponds to an optimal set of type $33*$, $3*$, or $*$ (depending on whether it was originally of type 332, 32, or 2, respectively) and defines a singleton connected component. $\qquad\square$

An important property of a canonical non-improving covering is given by the next lemma.

Lemma 8. *Any 3-set in a canonical non-improving covering produced by applying Lemma 7 to the acyclic non-improving covering computed when algorithm finishes step B intersects at least one optimal set that corresponds to a non-faulty node.*

We are now ready to prove the following lemma.

Lemma 9. *If the algorithm finishes step B by computing an acyclic non-improving covering, then its benefit at the end is at least* $\frac{5}{2}b_3 + \frac{7}{4}b_2 + b_1$.

Proof. Consider the acyclic non-improving covering computed by the algorithm and the canonical non-improving covering obtained by it with the same 3-sets and the same number of 2-sets. By Lemma 7, in order to account for the number of elements covered in the non-improving covering, it suffices to account for the number of elements covered in the canonical non-improving covering. We use the notation T_X to denote the number of optimal 3-sets and 2-sets or nodes out of optimal 3-sets and 2-sets of type X in the canonical non-improving covering. Taking into account that $T_{32*} = T_{22*} = T_{2*} = 0$ (by the definition of the canonical non-improving covering), this definition immediately yields

$$b_2 = T_{33} + T_{32} + T_{3*} + T_{22} \tag{9}$$
$$b_3 = T_{333} + T_{332} + T_{33*} + T_{322} + T_{222} \tag{10}$$

Since the three elements in optimal sets of type 222, two of the elements in optimal sets of type 322 and 22, and one element in optimal sets of type 332, 32, and 2 are included in 2-sets by the algorithm, we have that

$$\alpha_2 \geq \frac{1}{2}T_{332} + T_{322} + \frac{3}{2}T_{222} + \frac{1}{2}T_{32} + T_{22} + \frac{1}{2}T_2 \tag{11}$$

Since the three elements in optimal sets of type 333, two of the elements in optimal sets of type 332, 33*, and 33, and one element in optimal sets of type 322, 32, 3*, and 3 are included in 3-sets by the algorithm, we have that

$$\alpha_3 \geq T_{333} + \frac{2}{3}T_{332} + \frac{1}{3}T_{322} + \frac{2}{3}T_{33*} + \frac{2}{3}T_{33} + \frac{1}{3}T_{32} + \frac{1}{3}T_{3*} + \frac{1}{3}T_3 \tag{12}$$

We use inequality (1) and inequalities (9)-(12) to obtain

$$benefit \geq 2\alpha_3 + \alpha_2 + b_3 + b_2 + b_1$$
$$\geq 3T_{333} + \frac{17}{6}T_{332} + \frac{7}{3}T_{33*} + \frac{8}{3}T_{322} + \frac{5}{2}T_{222} + \frac{7}{3}T_{33}$$
$$+ \frac{13}{6}T_{32} + \frac{5}{3}T_{3*} + 2T_{22} + \frac{2}{3}T_3 + \frac{1}{2}T_2 + b_1 \tag{13}$$

By Lemma 8, each 3-set of the canonical non-improving covering has at least one of its elements in optimal sets of type 333, 332, 322, 33, 32, 3, and at most two of its elements in optimal sets of type 33* and 3*. So,

$$\alpha_3 \leq 3T_{333} + 2T_{332} + 2T_{33} + T_{322} + T_{32} + T_3 \tag{14}$$

and

$$2\alpha_3 \geq T_{3*} + 2T_{33*}. \tag{15}$$

By multiplying all the terms in inequalities (14) and (15) with $-\frac{1}{6}$ and $\frac{1}{12}$, respectively, and summing them, we obtain

$$0 \geq \frac{1}{12}(T_{3*} + 2T_{33*} - 6T_{333} - 4T_{332} - 4T_{33} - 2T_{322} - 2T_{32} - 2T_3) \tag{16}$$

Summing (13) and (16), we obtain

$$benefit \geq \frac{5}{2}T_{333} + \frac{5}{2}T_{332} + \frac{5}{2}T_{33*} + \frac{5}{2}T_{322} + \frac{5}{2}T_{222} + 2T_{33} + 2T_{32} + \frac{7}{4}T_{3*}$$
$$+2T_{22} + \frac{1}{2}T_3 + \frac{1}{2}T_2 + b_1$$
$$\geq \frac{5}{2}(T_{333} + T_{332} + T_{33*} + T_{322} + T_{222}) + \frac{7}{4}(T_{33} + T_{32} + T_{3*} + T_{22}) + b_1$$
$$= \frac{5}{2}b_3 + \frac{7}{4}b_2 + b_1$$

which completes the proof of the lemma.   □

References

1. Athanassopoulos, S., Caragiannis, I., Kaklamanis, C.: Analysis of Approximation Algorithms for k-Set Cover Using Factor-Revealing Linear Programs. In: Csuhaj-Varjú, E., Ésik, Z. (eds.) FCT 2007. LNCS, vol. 4639, pp. 52–63. Springer, Heidelberg (2007)
2. Caragiannis, I.: Wavelength Management in WDM Rings to Maximize the Number of Connections. In: Thomas, W., Weil, P. (eds.) STACS 2007. LNCS, vol. 4393, pp. 61–72. Springer, Heidelberg (2007)
3. Duh, R., Furer, M.: Approximation of k-set cover by semi local optimization. In: Proceedings of the 29th Annual ACM Symposium on Theory of Computing (STOC 1997), pp. 256–264 (1997)
4. Feige, U.: A threshold of ln n for approximating set cover. Journal of the ACM 45(4), 634–652 (1998)
5. Gandhi, R., Khuller, S., Srinivasan, A.: Approximation algorithms for partial covering problems. Journal of Algorithms 53(1), 55–84 (2004)
6. Halldórsson, M.M.: Approximating discrete collections via local improvements. In: Proceedings of the 6th Annual ACM/SIAM Symposium on Discrete Algorithms (SODA 1995), pp. 160–169 (1995)
7. Halldórsson, M.M.: Approximating k-set cover and complementary graph coloring. In: Proceedings of the 5th Conference on Integer Programming and Combinatorial Optimization (IPCO 1996). LNCS, vol. 1084, pp. 118–131. Springer, Heidelberg (1996)
8. Hazan, E., Safra, S., Schwartz, O.: On the complexity of approximating k-set packing. Computational Complexity 15(1), 20–39 (2006)
9. Hurkens, C.A.J., Schrijver, A.: On the size of systems of sets every t of which have an SDR, with an application to the worst-case ratio of heuristics for packing problems. SIAM Journal on Discrete Mathematics 2(1), 68–72 (1989)
10. Kann, V.: Maximum bounded 3-dimensional matching is MAX SNP-complete. Information Processing Letters 37, 27–35 (1991)
11. Levin, A.: Approximating the Unweighted k-Set Cover Problem: Greedy Meets Local Search. In: Erlebach, T., Kaklamanis, C. (eds.) WAOA 2006. LNCS, vol. 4368, pp. 290–301. Springer, Heidelberg (2007)

Positional Strategies for Higher-Order Pushdown Parity Games

Arnaud Carayol[1] and Michaela Slaats[2]

[1] IGM–LabInfo, Université Paris-Est & CNRS
arnaud.carayol@univ-mlv.fr
[2] RWTH Aachen, Informatik 7, 52056 Aachen, Germany
slaats@automata.rwth-aachen.de

Abstract. Higher-order pushdown systems generalize pushdown systems by using higher-order stacks, which are nested stacks of stacks. In this article, we consider parity games defined by higher-order pushdown systems and provide a k-EXPTIME algorithm to compute finite representations of positional winning strategies for both players for games defined by level-k higher-order pushdown automata. Our result is based on automata theoretic techniques exploiting the tree structure corresponding to higher-order stacks and their associated operations.

1 Introduction

Two player games of infinite duration over possibly infinite game graphs (also called arenas) play an important role in computer science and in particular in the domain of automatic verification of infinite-state systems (see [14,19] for surveys). In such games, the vertices of the game graph are partitioned into two sets, one for each player. A play consists in moving a token following the edges of the game graph. The player owning the vertex where the token lies, moves the token. If at some point a player cannot move the token he loses, otherwise the play is infinite. The winning condition of the game describes the set of winning plays for one of the players.

We consider the parity winning condition which plays an important role in the context of verification. In a parity game, each vertex is assigned an integer from a finite range and the winning condition is based on the parity of the smallest integer appearing infinitely often during the play. These games are determined (i.e. from any node, one of the players has a winning strategy) and can be won using positional strategies (i.e. strategies that only depend on the current vertex and not on the whole history of the play) [20]. For these games to be accessible to automatic treatment, we assume that the arena, though infinite, has a finite representation. In this article, the arenas will be given by transition graphs of an extension of pushdown automata. The main algorithmic problems, given the finite description of such an arena, are to determine who is winning from a given vertex and to give finite descriptions of the winning regions as well as of the winning strategies for each player. In the context of automatic verification,

E. Ochmański and J. Tyszkiewicz (Eds.): MFCS 2008, LNCS 5162, pp. 217–228, 2008.
© Springer-Verlag Berlin Heidelberg 2008

these problems correspond respectively to deciding if the behavior of a system satisfies a property expressed in modal μ-calculus, to giving a finite description of the set of states of the system satisfying the property and to synthesizing a controller for the system against a modal μ-calculus formula [16].

The first class of infinite arenas for which parity games have been studied are the ones defined by pushdown automata. In [18], Walukiewicz gives an EXPTIME algorithm to compute the winner from a given configuration as well as a finite description of a winning strategy for one player. In [3,12] the winning region is shown to be regular when a configuration (q, w) is represented by the word qw. Furthermore, a finite representation of a positional winning strategy for one player can easily be derived from [17].

In this article, we consider parity games defined by an extention of pushdown automata called higher-order pushdown automata. Whereas an ordinary (*i.e.* level-1) pushdown automaton works with a stack of symbols (*i.e.* a level-1 stack), a pushdown automaton of level 2 works with a stack of (level-1) stacks. In addition to pushing a symbol onto and popping a symbol from the top-most level-1 stack, a level-2 pushdown automaton can duplicate or remove the entire top-most (level-1) stack. Pushdown automata of higher levels are defined in a similar way. Recently, the infinite structures defined by these automata have received a lot of attention. In [11], the families of infinite terms defined by higher-order pushdown automata were shown to correspond to the solutions of safe higher-order recursion schemes. Subsequently, in [9,8], the ε-closure of their configuration graphs were shown to be exactly those constructible from finite graphs using natural graph transformations (see [15] for a survey).

We consider these infinite structures as arenas for parity games. In [4], Cachat showed that the winner of a parity game defined by a level-k pushdown automaton starting from a given node can be decided in k-EXPTIME. We provide for each player a finite description of the winning region and of a positional winning strategy. These finite descriptions are based on a notion of regularity for higher-order stacks introduced independently in [6] and in [10]. A set of level-k stacks is said to be regular (for operations) if it can be constructed by applying a regular set of sequences of level-k operations to the empty level-k stack. For usual (level-1) stacks, this notion corresponds to the regularity for words. For higher levels it enjoys most of the good properties of the regular sets of words. In particular these sets form a Boolean algebra and are accepted by a natural model of finite automata. The finite description obtained in this article is expressed in terms of this model of finite acceptors. Our construction is based on tree automata techniques introduced in [17] and already used in [4] to solve these games.

The fact that the notion of regularity by operations can be used to describe the winning regions and the positional winning strategies was already known from [6] and [10] respectively. These results are based on the definability in monadic second-order logic which, though effective, only provide a ck-EXPTIME complexity for some constant $c \geq 2$.

Outline. Section 2 introduces the necessary notions. The main theorem is stated in Section 3 with an outline of its proof (developed in Section 4 and 5).

2   Preliminaries

Higher-order pushdown systems. A *level-1 stack* over a finite alphabet Γ can be seen as a word of Γ^*. The empty stack (corresponding to ε) is written $[\ ]_1$. We write $Stacks_1(\Gamma) := \Gamma^*$ for the set of all level-1 stacks over Γ. A *level-$(k+1)$ stack* for $k \geq 1$ is a non-empty sequence of level-k stacks. The empty stack of level $k+1$ (written $[\ ]_{k+1}$) is the level-$(k+1)$ stack containing only the empty stack of level k. The set of all level-$(k+1)$ stacks is defined by $Stacks_{k+1}(\Gamma) := (Stacks_k(\Gamma))^+$. A level-$(k+1)$ stack s corresponding to the sequence $s_1, \ldots, s_n$ of level-k stacks will be written $[s_1, \ldots, s_n]_{k+1}$ and by convention s_n is the top-most level-k stack of s. We write $top_k(s) = s_n$.

We define the following partial functions on higher-order stacks called *operations*. The level-1 operations are for each symbol $x \in \Gamma$ the operations $push_x$ and pop_x which are respectively defined on level-1 stacks by $push_x([s_0, \ldots, s_n]_1) = [s_0, \ldots, s_n, x]_1)$ and $pop_x([s_0, \ldots, s_n, x]_1) = [s_0, \ldots, s_n]_1$.

For each level $k+1 \geq 2$, we consider the level-$(k+1)$ operations $copy_k$ which copies the top-most level-k stack and its symmetric operation $\overline{copy}_k$ which removes the top-most level-k stack if it is equal to its predecessor. Formally, these operations are respectively defined on level-$(k+1)$ stacks by $copy_k([s_0, \ldots, s_n]_{k+1}) = [s_0, \ldots, s_n, s_n]_{k+1}$ and $\overline{copy}_k([s_0, \ldots, s_n, s_n]_{k+1}) = [s_0, \ldots, s_n]_{k+1}$. In addition, for each level k, we define a level-k operation written $T_{[\ ]_k}$ allowing to test emptiness at level k. Formally $T_{[\ ]_k}(s)$ is equal to s if $s = [\ ]_k$ and is undefined otherwise.

An operation ψ of level k is extended to stacks of level $\ell > k$ using the equation $\psi([s_0, \ldots, s_n]_\ell) = [s_0, \ldots, \psi(s_n)]_\ell$.

The set of *symmetric operations*[1] of level k over Γ is defined inductively by $Ops_1 = \{push_x, pop_x \mid x \in \Gamma\} \cup \{T_{[\ ]_1}\}$ and $Ops_{k+1} = Ops_k \cup \{copy_k, \overline{copy}_k, T_{[\ ]_{k+1}}\}$. Moreover, we denote by Ops_k^* the monoid for the composition of partial functions generated by Ops_k.

To obtain a symbolic representation of the operations, we associate to each operation a symbol called an *instruction*. At level 1, we define the set of instructions as $\Gamma_1 = \Gamma \cup \overline{\Gamma} \cup \{\perp_1\}$ where $\overline{\Gamma}$ is a set disjoint from Γ but in bijection with Γ and at level $k+1$, we take $\Gamma_{k+1} = \Gamma_k \cup \{k, \overline{k}, \perp_{k+1}\}$. Furthermore, we define $\Gamma_k^T = \{\perp_\ell \mid \ell \in [1, k]\}$ and $\Gamma_k^O = \Gamma_k \setminus \Gamma_k^T$. We extend the bar notation to all symbols in Γ_k^O by taking $\overline{\overline{x}} = x$. Consider the mapping φ from Γ_k to Ops_k defined by $x \to push_x$, $\overline{x} \to pop_x$, $k \to copy_k$, $\overline{k} \to \overline{copy}_k$ and $\perp_k \to T_{[\ ]_k}$ for all $x \in \Gamma, k \in \mathbb{N}$. The mapping φ induces a monoid morphism from Γ_k^* to Ops_k^*. In the following, we will not distinguish between the two monoids and omit φ.

Definition 1. *A* higher-order pushdown system *P of level k (k-HOPDS for short) is defined as a tuple $(Q, \Sigma, \Gamma, \Delta)$ where Q is the finite set of states, Σ is the input alphabet, Γ is the stack symbol alphabet and $\Delta \subseteq Q \times \Sigma \times \Gamma_k \times Q$ is the transition relation.*

---

[1] The usual definition of higher-order pushdown automata [11] considers the unconditional destruction of level-k stacks written pop_k. The choice of the symmetric operations and its consequences are discussed in the conclusion.

A configuration is a tuple $(p, s) \in Q \times Stacks_k(\Gamma)$. We write $(p, s) \xrightarrow{\alpha} (q, s')$ if there exists a transition $(p, \alpha, \rho, q) \in \Delta$ such that $s' = \rho(s)$. A k-HOPDS is *deterministic* if for all $\alpha \in \Sigma$ and all configurations c, c' and c'', $c \xrightarrow{\alpha} c'$ and $c \xrightarrow{\alpha} c''$ implies $c' = c''$.

Regular sets of higher-order pushdown stacks. The natural notion of regularity for sets of level-1 stacks is the regularity for words. Indeed the set of reachable stack contents of a pushdown automaton is regular [2].

Starting from level 2, two notions of regularity have been introduced: regularity for words and regularity for (symmetric) operations. We will use the second notion and discuss this choice in the conclusion.

The notion of regularity for words was introduced in [1]. A level-k stack is represented by a well-bracketed word of depth k (e.g. the level-2 stack $[[aa]_1[abb]_1]_2$ is represented by the word $[[aa][aab]]$). A set of level-k stacks is *regular for words* if the set of words representing it is a regular set of words. For example the set of level-2 stacks $\{[[a^n]_1[b^m]_1]_2 \mid n, m \geq 0\}$ is regular for words.

The notion of *regularity for (symmetric) operations* was introduced independently in [6] and [10]. A set of level-k stacks is *regular for operations* if it can be obtained by applying a regular subset of Ops_k^* to the empty level-k stack $[\ ]_k$. Formally, we define the set of all level-k stacks which are *regular for operations* as follows: $OReg_k(\Gamma) = Reg(Ops_k^*(\Gamma))([\ ]_k) = Reg(\Gamma_k^*)([\ ]_k)$ (i.e. $S \in OReg_k(\Gamma)$ if there exists $R \in Reg(\Gamma_k^*)$ and $S = \{\rho([\ ]_k) \mid \rho \in R\}$). At level 1, the notion of regularity for operations coincides with notion of regularity for words. For level $k > 2$, every set regular for words is also regular for operations but the converse does not hold. For instance, the set of level-2 stacks $S = \{[[a^n][a^n]]_2 \mid n \geq 0\}$ is regular for operations as $S = push_a^* copy_1([\ ]_2) = a^*1([\ ]_2)$ but it is not regular for words.

For every level $k \geq 1$, the set $OReg_k(\Gamma)$ is a Boolean algebra. These closure properties are due to the fact that a level-k stack s can be uniquely represented by the smallest sequence of instructions $\rho \in \Gamma_k^*$ such that $s = \rho([\ ]_k)$. This unique sequence, called the *reduced sequence* of s, will be written ρ_s. For instance the reduce sequence of the level-2 stack $[[aab][ab]]_2$ is $aab1\bar{b}ab$. Note that the reduced sequence of a level-1 stack is simply the stack itself. For a stack of level $k + 1 \geq 2$, its reduced sequence cannot contain $x\bar{x}$, $\perp_\ell$ nor $\bar{k}$ for any $x \in \Gamma_k^O$ and $\ell \in [1, k+1]$. In fact, the reduced sequence of a level-k stack s is the unique sequence $\rho \in \Gamma_k^*$ such that $s = \rho([\ ]_k)$ which does not contain such factors.

The sets in $OReg_k(\Gamma)$ can be characterized by a model of finite automata tightly linked to the notion of reduced sequences. A reduced automaton of level 1 is simply a finite automaton over Γ. A reduced automaton $\mathcal{A}$ of level $k+1 \geq 2$ is given by a tuple (Q, I, F, Δ) together with a finite set of tests $\mathcal{R} \subset OReg_k(\Gamma)$ where Q is the finite set of states, $I \subseteq Q$ and $F \subseteq Q$ are respectively the set of initial and final states, and Δ is the finite set of transitions of the form $p \xrightarrow{\gamma} q, T$ with $\gamma \in \Gamma_k^O \setminus \{\bar{k}\}$ and $T \subseteq \mathcal{R}$. Intuitively, the automaton in state p on a stack s can apply the transition to go to state q on the stack $\gamma(s)$ if the top-most level-k

stack of $\gamma(s)$ belongs to T. Furthermore, we impose that the automaton only follows reduced sequences: if $p \xrightarrow{\gamma} p', T \in \Delta$ and $p' \xrightarrow{\gamma'} p'', T \in \Delta$ then $\gamma' \neq \bar{\gamma}$.

A run of $\mathcal{A}$ is a sequence $(q_0, s_0), \ldots, (q_n, s_n) \in (Q \times Stacks_{k+1}(\Gamma))^+$ where $q_0 \in I$, $s_0 = [\ ]_{k+1}$ and for all $i \in [0, n-1]$, there exists a transition $q_i \xrightarrow{\gamma_{i+1}} q_{i+1}, T_{i+1}$ with $s_{i+1} = \gamma_{i+1}(s_i)$ and the top-most level k stack of s_{i+1} belongs to T for all $T \in T_{i+1}$. A run of $\mathcal{A}$ accepts a stack s if $q_n \in F$ and $s = s_n$. Note that in this case, the reduce sequence of s is $\gamma_1 \ldots \gamma_n$.

We will always consider reduced automata whose tests are given by reduced automata of one level below. The size of the automaton is the size of the transition relation together with the sum of the reduced automata accepting the set of tests.

Theorem 1 ([6,10]). *For all $k \geq 1$, the sets of level-k stacks regular for operations are exactly those sets accepted by reduced level-k automata. Moreover $OReg_k(\Gamma)$ forms a Boolean algebra.*

Parity games defined by higher-order pushdown systems. A *parity game* $\mathcal{G}$ played between Player 0 and Player 1 is given by a tuple (V_0, V_1, E), where V_i is the set of nodes of Player i for $i \in \{0, 1\}$ and $E \subseteq (V_0 \cup V_1) \times (V_0 \cup V_1)$ is the edge relation, and $\Omega : (V_0 \cup V_1) \to [1, n]$ is the coloring mapping for some fixed $n \in \mathbb{N}$.

Player 0 and Player 1 play in $\mathcal{G}$ by moving a token between vertices. A *play* from some initial vertex v_0 proceeds as follows: the player owning v_0 moves the token to a vertex v_1 such that $(v_0, v_1) \in E$. Then the player owning v_1 chooses a successor v_2 and so on. If at some point one of the players cannot move, he loses the play. Otherwise, the play is an infinite word $\pi \in (V_0 \cup V_1)^\omega$ and is won by Player 0 if the smallest color that is seen infinitely often in π is even.

A strategy for Player i is a partial function φ_i assigning to a partial play ending in some vertex $v \in V_i$ a vertex v' such that $(v, v') \in E$. Player i *respects a strategy* φ_i during some play $\pi = v_0 v_1 v_2 \cdots$ if $v_{i+1} = \varphi_i(v_0 \cdots v_i)$, for all $i \geq 0$ such that $v_i \in V_i$. A strategy φ_i for Player i is *winning* from some position $v \in V_0 \cup V_1$ if every play starting from v where Player i respects φ_i is won by him. A *positional* strategy for Player i is a strategy that only depends on the last vertex of the partial play (i.e. it is a partial function from V_i to $V_0 \cup V_1$). Finally, a vertex $v \in V_0 \cup V_1$ is *winning* for Player i if he has a winning strategy from v, and the winning region W_i consists of all winning vertices for Player i.

The positional determinacy theorem for parity games [20] states that from every vertex either Player 0 or Player 1 has a positional winning strategy. This assertion can be strengthened by saying that Player i has a *global* positional winning strategy φ_i such that φ_i is winning for Player i from all vertices in $Dom(\varphi_i)$ and $Dom(\varphi_i) = W_i \cap V_i$ (see [13]).

Definition 2. *A higher-order pushdown parity game $\mathcal{G}$ of level k is given by a deterministic k-HOPDS $P = (Q, \Sigma, \Gamma, \delta)$ together with a partition of the states $Q_0 \uplus Q_1$ and a coloring mapping $\Omega_P : Q \to \mathbb{N}$ and is the game (V_0, V_1, E, Ω) where: $V_0 = Q_0 \times Stacks_k(\Gamma)$, $V_1 = Q_1 \times Stacks_k(\Gamma)$, E is the Σ-labeled transition relation of P and Ω is defined for $(p, s) \in Q \times Stacks_k(\Gamma)$ by $\Omega(p, s) := \Omega_P(p)$.*

The labels in Σ on the transitions do not play any role in the game but together with the hypothesis of determinism of P, they permit to give a simpler
description of positional strategies. In fact a positional strategy φ for Player i
can be described by a partial function from V_i to Σ, or equivalently by a family
$(F_\alpha)_{\alpha \in \Sigma}$ of subsets of V_i (i.e. $F_\alpha := \{(p, s) \in V_i \mid \varphi((p, s)) = \alpha\}$). We say that
a positional strategy defined by a family $(F_\alpha)_{\alpha \in \Sigma}$ is regular if all these sets are
regular[2] for operations.

Tree automata models. Let Σ and W be two finite alphabets which are
respectively a labeling alphabet and a set of directions. A Σ-labeled W-tree t is
a partial function from W^* to Σ such that $Dom(t)$ is a non-empty prefix-closed
subset of W^*. An element of $Dom(t)$ is a *node* and ε is called the *root* of t. For
$d \in W$, a node $wd \in Dom(t)$ is a *d-son* of $w \in Dom(t)$ and w is the *parent* of
wd. Let Ξ be a finite alphabet, a Ξ-labeling of a Σ-labeled W-tree t is a $\Sigma \times \Xi$-
labeled W-tree t' such that $Dom(t) = Dom(t')$ and such that for $w \in Dom(t')$,
$t'(w) = (t(w), \sigma)$ for some $\sigma \in \Xi$.

We consider two-way alternating parity tree automata which can from a node
of an input tree send several copies to sons of this node but also to its parent. To
navigate through the tree, we consider an extended set of directions $ext(W) :=
W \uplus \{\varepsilon, \uparrow\}$. The symbol $\uparrow$ means "go to the parent node" and ε means "stay on
the present node". We take $\forall u \in W^*, d \in W, u.\varepsilon = u$ and $ud \uparrow = u$. The node $\varepsilon \uparrow$
is not defined. As we consider non-complete W-trees (i.e. $Dom(t) \neq W^*$), we
assume that the labeling of a tree provides the directions to all sons of a node
in the tree: the automaton runs on $\mathcal{P}(W) \times \Sigma$-labeled W-trees t where for all
$w \in Dom(t), t(w) = (\theta_w, \sigma_w)$ where $\theta_w = \{d \in W \mid wd \in Dom(t)\}$.

Definition 3. *A two-way alternating parity tree automaton (2-PTA for short)
running over $\mathcal{P}(W) \times \Sigma$-labeled W-trees is a tuple $\mathcal{A} = (Q, \Delta, I, \Omega)$ where Q is
the finite set of states, $\Delta \subseteq Q \times (\mathcal{P}(W) \times \Sigma) \times \mathcal{P}(ext(W) \times Q)$ is the transition
relation, I is set of initial states and $\Omega : Q \to \mathbb{N}$ the coloring mapping.*

A transition $(q, (\theta, \sigma), \{(d_1, q_1), \ldots, (d_n, q_n)\}) \in \Delta$ will be written $q, (\theta, \sigma) \to
(d_1, q_1) \wedge \ldots \wedge (d_n, q_n)$. We will always assume that $\{d_1, \ldots, d_n\}$ is a subset of
$\theta \cup \{\uparrow, \varepsilon\}$. The behavior of a 2-PTA $\mathcal{A} = (Q, \Delta, I, \Omega_{\mathcal{A}})$ over a $\mathcal{P}(W) \times \Sigma$-labeled
W-tree t is given by the parity game $G_{\mathcal{A},t} = (V_0, V_1, E)$ played between two
players called Automaton and Pathfinder. The set V_0 of vertices of Automaton
is $Dom(t) \times Q$ and the set V_1 of vertices of Pathfinder is $Dom(t) \times \Delta$. For all
$w \in Dom(t)$ and $q \in Q$, there is an edge $((w, q), (w, \delta)) \in E$ for all transitions
$\delta \in \Delta$ of the form $q, t(w) \to P$. Conversely for every transition $\delta = q, t(w) \to
P \in \Delta$, there is an edge $((w, \delta), (wd_i, q_i))$ for all $(d_i, q_i) \in P$. The automaton $\mathcal{A}$
accepts the tree t if Automaton wins $G_{\mathcal{A},t}$ from some vertex in $\{\varepsilon\} \times I$.

The classical notion of (one-way) non-deterministic parity tree automata (PTA)
coincide with 2-PTAs with transitions of the form $q, (\theta, \sigma) \to (d_1, q_1) \wedge \ldots \wedge (d_n, q_n)$
where for all $i, j \in [1, n], d_i \in W$ and $d_i = d_j$ implies $i = j$.

---

[2] We represent a configuration (p, s) by the stack $push_p(s)$.

3   Main Theorem and Outline of the Proof

Theorem 2. *Given a pushdown parity game of level k, we can construct in k-EXPTIME reduced level-k automata describing the winning region and a global positional winning strategy for each player.*

In Section 4, we define for every level k, a tree $\mathbf{t}_k$ (see e.g. Fig. 1) associated to the stacks of level k. The branches of $\mathbf{t}_k$ correspond to the reduced sequences of the level-k stacks. Starting with a parity game $\mathcal{G}$ described by a level-k pushdown automaton P, we construct a 2-PTA $\mathcal{A}_P$ running on $\mathbf{t}_k$ which captures the game $\mathcal{G}$ and whose size is polynomial in the size of P. More precisely, we can reduce the computation of regular representations of a global positional winning strategy and of the winning region for each player to the computation of a regular representation of a global positional winning strategy for the automaton $\mathcal{A}_P$ running on $\mathbf{t}_k$. Intuitively such a strategy consists for every node u of the tree and every state q of the automaton to either provide a transition of the automaton starting with state q which can be applied at node u or a set of directions and states which refute any transitions of the automaton that can be applied at node u in state q.

   In Section 5, we show how to compute regular global positional winning strategies for a 2-PTA running on the trees $\mathbf{t}_k$. The proof proceeds by induction on the level of the tree. First based on a construction from [17], we construct for any two-way alternating parity tree automaton $\mathcal{A}$ a non-deterministic one-way parity tree automaton $\mathcal{B}$ accepting the labelings of $\mathbf{t}_k$ corresponding to global positional winning strategies of $\mathcal{A}$ (see Proposition 3). Second for any non-deterministic one-way parity tree automaton $\mathcal{B}$ running on $\mathbf{t}_k$, we construct a two-way alternating parity tree automaton $\mathcal{C}$ running on $\mathbf{t}_{k-1}$ (see Proposition 4) such that from a global positional strategy of $\mathcal{C}$ over $\mathbf{t}_{k-1}$ defined by regular sets of level-$(k-1)$ stacks, we can construct a strategy (for $\mathcal{A}$) accepted by $\mathcal{B}$ defined by regular sets of level-k stacks.

4   From Games to Trees

In this section, we introduce the infinite tree $\mathbf{t}_k$ associated to the stacks of level k and based on the reduced sequences of these stacks. We show that the problem of computing a global positional winning strategy of a level-k pushdown parity game can be reduced in polynomial time to the problem of computing a global positional winning strategy for an alternating two-way parity tree automaton running on the tree $\mathbf{t}_k$.

   As we have seen in Section 2, a level-k stack s is uniquely characterized by its reduced sequence ρ_s. Hence the set of reduced sequences of all level-k stacks is a Γ_k^O-tree in which each node corresponds to one and only one level-k stack. In order to increase the expressivity of tree automata running on these trees, we label each node by a finite information about the surrounding of the stack corresponding to this node. The surrounding of a stack $s \in \mathrm{Stacks}_k(\Gamma)$ is a triple $\ell(s) = (d, D, e)$ where:

- $d \in \Gamma_k^O \cup \{\diamond\}$ is the last symbol of ρ_s if $\rho_s \neq \varepsilon$ and is equal to $\diamond$ otherwise,
- D is the set $\{\gamma \in \Gamma_k^O \mid \exists s' \in \text{Stacks}_k(\Gamma), \rho_{s'} = \rho_s\gamma\}$,
- $e \in [0,k]$ is the maximum of $\{n \in [1,k] \mid \perp_n(s) = s\} \cup \{0\}$.

Formally, the tree $\mathbf{t}_k$ is defined for all $s \in \text{Stacks}_k(\Gamma)$ by $\mathbf{t}_k(\rho_s) = \ell(s)$. When referring to the nodes of $\mathbf{t}_k$, we do not distinguish between the stack and its reduced sequence. In particular, we said that a Ξ-labeling t of $\mathbf{t}_k$ is regular if for all $x \in \Xi$, the set of level-k stacks $\mathcal{S}_x = \{s \in \text{Stacks}_k(\Gamma) \mid t(\rho_s) = (\mathbf{t}_k(\rho_s), x)\}$ is regular for operations. A finite representation of t is then given by a family of reduced level-k automata $(\mathcal{A}_x)_{x \in \Xi}$. For $\Gamma = \{a, b\}$, the tree $\mathbf{t}_1$ is essentially the full binary tree. The tree $\mathbf{t}_2$ (depicted in Figure 1) is not complete nor regular.

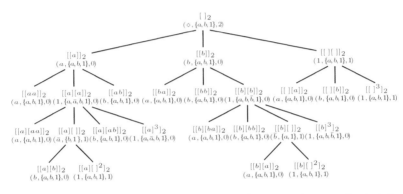

Fig. 1. The tree $\mathbf{t}_2$ for $\Gamma = \{a, b\}$ where the labels appear in parenthesis below the corresponding node

As it was done for (level-1) pushdown parity games in [17], we reduce the decision problem for level-k parity games to the acceptance problem for alternating two-way parity tree automata running over $\mathbf{t}_k$. Intuitively, the non-determinism of the automaton is used to reflect the choices of Player 0 and the alternation is used to reflect the choices of Player 1.

Proposition 1. *Given a pushdown parity game $\mathcal{G}$ of level k, we can construct a 2-PTA $\mathcal{A}$ running on $\mathbf{t}_k$ such that Player 0 wins $\mathcal{G}$ from $(q, [\ ]_k)$ if and only if $\mathcal{A}$ has an accepting run on $\mathbf{t}_k$ starting from state q. Furthermore the size of $\mathcal{A}$ is polynomial in the size of the level-k pushdown automaton defining $\mathcal{G}$.*

Proof (Sketch). Let $\mathcal{G}$ be a parity game defined by a level-k pushdown automaton $P = (Q_P, \Sigma, \Gamma, \Delta_P)$ with $Q_P = Q_0 \uplus Q_1$ and a coloring mapping Ω_P. We construct the 2-PTA $\mathcal{A} = (Q_A, \Delta_A, \Omega_A)$ with $Q_A = Q_P$, $\Omega_A = \Omega_P$. To define Δ_A, we introduce for a surrounding $\tau = (d, D, e)$ and an instruction $\gamma \in \Gamma_k$ the direction $[\![\gamma]\!]_\tau$ on $\mathbf{t}_k$. It is defined by $[\![\gamma]\!]_\tau = \gamma$ if $\gamma \in D$, $[\![\gamma]\!]_\tau = \uparrow$ if $\gamma = \bar{d}$, $[\![\gamma]\!]_\tau = \varepsilon$ if $\gamma = \perp_j$ and $e \geq j$ and $[\![\gamma]\!]_\tau$ is undefined otherwise.

For p $\in$ Q$_0$, $(p, \alpha, \gamma, q) \in \Delta_P$ and surrounding $\tau = (d, D, e)$ such that $[\![\gamma]\!]_\tau$ is defined, we have: $(p, \tau) \to ([\![\gamma]\!]_\tau, q) \in \Delta_A$.

For $p \in Q_1$, let $\delta_1, \ldots, \delta_n$ be the set of all transitions in Δ_P starting in p, i.e. having the form $\delta_i = (p, \alpha_i, \gamma_i, q_i)$ for all $i \in [1, n]$. For all labelings $\tau = (d, D, e)$, we take: $(p, \tau) \rightarrow \bigwedge_{i \in [1,n] \wedge [\![\gamma_i]\!]_\tau \text{ def.}} ([\![\gamma_i]\!]_\tau, q_i) \in \Delta_A$.

Note that the size of $\mathcal{A}$ is exponential in the size of Γ_k.   □

The relation between the game $\mathcal{G}$ and the 2-PTA $\mathcal{A}$ constructed in Proposition 1 can be lifted to strategies. Before stating this correspondence, we show how to represent a pair of global positional winning strategies φ_{aut} and φ_{path} in $\mathcal{G}(\mathcal{A}, \mathbf{t}_k)$ for Automaton and Pathfinder respectively as a labeling of $\mathbf{t}_k$ by a finite amount of information. The labeling set is $\mathcal{F}_0 \times \mathcal{F}_1$ where $\mathcal{F}_0$ is the set of all partial functions from Q to Δ and $\mathcal{F}_1$ is the set of all partial functions from Q to $\mathcal{P}(ext(W) \times Q)$.

The strategy φ_{aut} of Automaton at a node $w \in Dom(t)$ is given by a partial function ν_0^w from Q to Δ which when defined on a state q gives the transition to apply in the configuration (w, q). Formally, for all $q \in Q$, $\nu_0^w(q) = \delta$ iff $\varphi_{\text{aut}}(w, q) = (w, \delta)$.

The strategy φ_{path} of Pathfinder can be given by a partial function ν_1^w from Q to $\mathcal{P}(ext(W) \times Q)$. For all $q \in Q$, we have two cases depending on who wins the game from (w, q). If Automaton wins from (w, q) (i.e. there exists a transition $\delta = (q, t(w)) \rightarrow P$ such that $\varphi_{\text{path}}(w, \delta)$ is undefined) then $\nu_1^w(q)$ is undefined. If Pathfinder wins from (w, q) then for all transitions $\delta_1, \ldots, \delta_n$ starting with $(q, t(w))$ (i.e. $\delta_j = q, t(w) \rightarrow P_j$), we have $\varphi_{\text{path}}(w, \delta_j) = (wd_{j_i}, q_{j_i})$ for some $(d_{j_i}, p_{j_i}) \in P_j$ for all $j \in [1, n]$. In this case, $\nu_1^w(q)$ is equal to $\{(d_{j_i}, q_{j_i}) \mid j \in [1, n]\}$. Intuitively $\nu_1^w(q)$ is defined if Pathfinder wins from (w, q) and in this case it corresponds to a set of directions and states that can refute any transitions of the automaton that can be applied in this configuration. Note that for any node $w \in Dom(t)$, $Dom(\nu_0^w)$ and $Dom(\nu_1^w)$ form a partition of Q.

Conversely we say that a $\mathcal{F}_0 \times \mathcal{F}_1$-labeling of $\mathbf{t}_k$ is a global winning strategy for $\mathcal{A}$ on $\mathbf{t}_k$ if it induces a pair of global winning strategies for each player.

Proposition 2. *Let $\mathcal{G}$ be a pushdown parity game of level k and $\mathcal{A}$ the alternating two-way parity tree automaton given by Proposition 1. Given a regular global positional winning strategy Φ for $\mathcal{A}$ on $\mathbf{t}_k$, we can compute, for each player, a regular global positional winning strategy and a regular representation of the winning region.*

5 Computing Strategies over $\mathbf{t}_k$

In this section, we show how to compute a regular global positional strategy for a 2-PTA $\mathcal{A}$ running on $\mathbf{t}_k$. For this we proceed by induction on the level k.

The first step is based on [17] and consists in showing that for any 2-PTA $\mathcal{A}$ running on $\mathbf{t}_k$, one can construct a PTA $\mathcal{B}$ accepting the $\mathcal{F}_0 \times \mathcal{F}_1$-labelling of $\mathbf{t}_k$ representing a global winning strategy for $\mathcal{A}$ on $\mathbf{t}_k$. In [17], a PTA $\mathcal{B}$ is constructed that accepts the trees representing a positional strategy for Automaton winning from a given state of $\mathcal{A}$ (see [3] for a detailed presentation of the construction). The following proposition simply adapts the construction to make it symmetric between Automaton and Pathfinder.

Proposition 3. *Given a 2-PTA $\mathcal{A}$ running on $\mathbf{t}_k$, we can construct a PTA $\mathcal{B}$ accepting the trees representing global positional winning strategies of $\mathcal{A}$. Furthermore, the size of $\mathcal{B}$ is exponential in the size of $\mathcal{A}$ but the number of colors of $\mathcal{A}$ is linear in the number of colors of $\mathcal{B}$.*

Using Proposition 3, we have reduced our initial problem to computing a regular Ξ-labeling of $\mathbf{t}_{k+1}$ accepted by a given PTA $\mathcal{B}$. In the next step, we reduce this problem to the computation of a global winning strategy for a 2-PTA $\mathcal{C}$ running on $\mathbf{t}_k$.

The construction is based on the fact that $\mathcal{B}$ does not use the $\uparrow$ direction and hence when running on $\mathbf{t}_{k+1}$ only take directions in $\Gamma_k^O \cup \{k\}$. From the point of view of the operations on the stacks, $\mathcal{B}$ does not perform the $\overline{copy}_k$ operation and hence can only access the top-most level-k stack. Intuitively, we can simulate the same behavior at one level below by replacing the direction k by ε: we use alternation instead of performing the $copy_k$ operation. The construction is a bit more technical as we need to relate the surroundings in $\mathbf{t}_{k+1}$ which belong to a node corresponding to a level-$(k+1)$ stack s to the surroundings in $\mathbf{t}_k$ of the node corresponding to the level-k stack $top_k(s)$.

Proposition 4. *Given a PTA $\mathcal{B}$ accepting at least one Ξ-labeling of $\mathbf{t}_{k+1}$, we can construct a 2-PTA $\mathcal{C}$ running on $\mathbf{t}_k$ such that given a regular global positional strategy for $\mathcal{C}$ on $\mathbf{t}_k$, we can construct a regular Ξ-labeling of $\mathbf{t}_{k+1}$ accepted by $\mathcal{B}$. Furthermore, the size of $\mathcal{C}$ is polynomial in the size of $\mathcal{B}$.*

Proof (Sketch). Let $\mathcal{B} = (Q_B, \Delta_B, I_B, \Omega_B)$ be a PTA accepting Ξ-labelings of $\mathbf{t}_{k+1}$. We can assume w.l.o.g that $\mathcal{B}$ runs on $\mathbf{t}_{k+1}$ and guesses the Ξ-labeling (i.e. $Q_B = Q_B' \times \Xi$). Furthermore, we can assume that the surroundings appearing in transitions of $\mathcal{B}$ really occur in $\mathbf{t}_{k+1}$.

We define the 2-PTA $\mathcal{C} = (Q_C, \Delta_C, I_C, \Omega_C)$ with $Q_C = Q_B \times \Gamma_k^O \cup \{k, \diamond\}$, $I_C = I_B \times \{\diamond\}$ and $\Omega_C(p, d) = \Omega_B(p)$ for all $d \in \Gamma_k^O \cup \{k, \diamond\}$. For each transition $\delta := q, (d, D, e) \rightarrow (\gamma_1, q_1) \wedge \ldots \wedge (\gamma_n, q_n) \in \Delta_B$ and for each $d' \in \Gamma_k^O \cup \{\diamond\}$, we add the following transition to Δ_C:

$$\delta_{\downarrow_{d'}} := (q, d), (d', D', e') \rightarrow (\gamma_1', (q_1, \gamma_1)) \wedge \ldots \wedge (\gamma_n', (q_n, \gamma_n))$$

when
1. $d = \diamond \Rightarrow d' = \diamond$ and $d' \neq \diamond \wedge d' \neq d \Rightarrow \overline{d'} \in D$ $D' = (D \cup \{\overline{d}\}) \setminus \{k, \overline{d'}, \overline{k}\}$ and $e' = \min\{e, k\}$
2. for all $i \in [1, n]$, γ_i' is equal to ε if $\gamma_i = k$, to $\uparrow$ if $\gamma_i = \overline{d'}$ and to γ_i otherwise.

Intuitively, the automaton $\mathcal{C}$ simulates the actions of $\mathcal{B}$ on the top-most level-k stack. This is enough to capture the whole behavior of $\mathcal{B}$ as $\mathcal{B}$ never performs the $\overline{copy}_k$ operation. Condition 1 relates the surroundings (d, D, e) in $\mathbf{t}_{k+1}$ of a node correspond to a level-$(k+1)$ stack s to the surroundings (d', D', e') in $\mathbf{t}_k$ of the node corresponding to the level-k stack $top_k(s)$. Condition 2 reflects the fact that the $copy_k$ operation does not modify the top-most level-k stack (i.e the direction k is replace by ε).

An important property for the transitions of the 2-PTA $\mathcal{C}$ is that for every $\delta \in \Delta_C$, there exists a unique transition $\delta^\uparrow \in \Delta_B$ and a unique $d' \in \Gamma_k^O \cup \{\diamond\}$ such that $\delta = (\delta^\uparrow)_{\downarrow_{d'}}$. Moreover, if the labels of δ and $\delta^\uparrow$ are respectively (d', D', e') and (d, D, e), we have $D = (D' \cup \{\overline{d'}, k\}) \setminus \{\overline{d}\}$. This property allows us to lift a regular positional strategy for $\mathcal{C}$ on $\mathbf{t}_k$ to a regular positional strategy of $\mathcal{B}$ on $\mathbf{t}_{k+1}$. Consequently, we can compute a regular Ξ-labeling accepted by $\mathcal{B}$ following the regular global positional strategy for $\mathcal{B}$.   □

By an induction on the levels combining Proposition 3 and Proposition 4, we reduce the initial problem to the problem of computing a winning strategy in a finite parity game which is k-fold exponential in the size of original automaton. Indeed only the first step provides an exponential blow-up in the number of states but the number of colors remains linear. The computation of a winning strategy on finite parity games is only exponential in the number of colors [20]. Hence, we obtain a k-EXPTIME procedure.

Theorem 3. *Given a 2-PTA $\mathcal{A}$ running on $\mathbf{t}_{k+1}$, we can compute in k-EXPTIME a regular global positional strategy for $\mathcal{A}$ on $\mathbf{t}_{k+1}$.*

6   Conclusion

We have presented a k-EXPTIME algorithm to provide a finite representation of the winning regions and global positional winning strategies in higher-order pushdown parity games of level k. Note that deciding the winner of these games is already k-EXPTIME hard [5]. Our results can be extended to richer graph structures based on higher-order pushdown automata such as for example rooted higher-order pushdown parity games (where the game graph is restricted to the configurations reachable from a given configuration) and their ε-closure (see [8]).

One of the key feature of our approach is the use of the symmetric destruction of level-k stacks $\overline{copy}_k$ instead of the usual unconditional destruction pop_k. This choice is motivated by the closure properties of the notion of regularity induced by the set of symmetric operations as well as the tree structure it induces. The game graphs obtained when considering pop_k instead of $\overline{copy}_k$ can be obtained as ε-closure of rooted higher-order pushdown parity games (see [6]) and hence can be treated in our framework.

In [7], it was shown that when considering higher-order pushdown parity games defined using the unconditional destruction pop_k instead of the symmetric destruction $\overline{copy}_k$ considered in this article, the winning region is regular by words. This result is stronger than the one obtainable by our approach as we can only prove that the winning region is regular for operations. It is important to note that when considering the symmetric version this result no longer holds. The proof of [7] also provides a finite description of a winning strategy from a given vertex for one of the players based on a higher-order pushdown automaton reading the moves of the play and outputting the next move. It is unknown if the notion of regularity by words can be used to describe positional winning strategies for higher-order pushdown parity games.

Acknowledgments. The authors thank Olivier Serre, Wolfgang Thomas and an anonymous referee for their numerous helpfull comments.

References

1. Bouajjani, A., Meyer, A.: Symbolic Reachability Analysis of Higher-Order Context-Free Processes. In: Lodaya, K., Mahajan, M. (eds.) FSTTCS 2004. LNCS, vol. 3328, pp. 135–147. Springer, Heidelberg (2004)
2. Büchi, J.: Regular canonical systems. Arch. Math. Logik Grundlag. 6, 91–111 (1964)
3. Cachat, T.: Symbolic Strategy Synthesis for Games on Pushdown Graphs. In: Widmayer, P., Triguero, F., Morales, R., Hennessy, M., Eidenbenz, S., Conejo, R. (eds.) ICALP 2002. LNCS, vol. 2380, pp. 704–715. Springer, Heidelberg (2002)
4. Cachat, T.: Higher-order-pushdown automata, the Caucal hierarchy of graphs and parity games. In: Baeten, J.C.M., Lenstra, J.K., Parrow, J., Woeginger, G.J. (eds.) ICALP 2003. LNCS, vol. 2719, pp. 556–569. Springer, Heidelberg (2003)
5. Cachat, T., Walukiewicz, I.: The complexity of games on higher order pushdown automata. In: Electronic version (May 2007)
6. Carayol, A.: Regular Sets of Higher-Order Pushdown Stacks. In: Jedrzejowicz, J., Szepietowski, A. (eds.) MFCS 2005. LNCS, vol. 3618, pp. 168–179. Springer, Heidelberg (2005)
7. Carayol, A., Hague, M., Meyer, A., Ong, C.-H.L., Serre, O.: Winning regions of higher-order pushdown games. In: Proc. LICS 2008 (2008)
8. Carayol, A., Wöhrle, S.: The Caucal hierarchy of infinite graphs in terms of logic and higher-order pushdown automata. In: Pandya, P.K., Radhakrishnan, J. (eds.) FSTTCS 2003. LNCS, vol. 2914, pp. 112–123. Springer, Heidelberg (2003)
9. Caucal, D.: On Infinite Terms Having a Decidable Monadic Theory. In: Diks, K., Rytter, W. (eds.) MFCS 2002. LNCS, vol. 2420. Springer, Heidelberg (2002)
10. Fratani, S.: Automates á piles de piles.. de piles. PhD thesis, Universit Bordeaux 1 (2005)
11. Knapik, T., Niwiński, D., Urzyczyn, P.: Higher-Order Pushdown Trees Are Easy. In: Nielsen, M., Engberg, U. (eds.) FOSSACS 2002. LNCS, vol. 2303. Springer, Heidelberg (2002)
12. Serre, O.: Note on winning positions on pushdown games with omega-regular conditions. IPL 85(6), 285–291 (2003)
13. Thomas, W.: Languages, automata, and logic. In: Handbook of Formal Languages, vol. 3, pp. 389–455. Springer, Heidelberg (1997)
14. Thomas, W.: Infinite Games and Verification. In: Brinksma, E., Larsen, K.G. (eds.) CAV 2002. LNCS, vol. 2404, Springer, Heidelberg (2002)
15. Thomas, W.: Constructing Infinite Graphs with a Decidable MSO-Theory. In: Rovan, B., Vojtas, P. (eds.) MFCS 2003. LNCS, vol. 2747, pp. 113–124. Springer, Heidelberg (2003)
16. Thomas, W.: On the synthesis of strategies in infinite games. In: Mayr, E.W., Puech, C. (eds.) STACS 1995. LNCS, vol. 900, pp. 1–13. Springer, Heidelberg (2005)
17. Vardi, M.Y.: Reasoning about the Past with Two-Way Automata. In: Larsen, K.G., Skyum, S., Winskel, G. (eds.) ICALP 1998. LNCS, vol. 1443, pp. 628–641. Springer, Heidelberg (1998)
18. Walukiewicz, I.: Pushdown processes: Games and model checking. In: Alur, R., Henzinger, T.A. (eds.) CAV 1996. LNCS, vol. 1102, pp. 62–74. Springer, Heidelberg (1996)
19. Walukiewicz, I.: A landscape with games in the background. In: Proc. LICS 2004, pp. 356–366. IEE (2004)
20. Zielonka, W.: Infinite games on finitely coloured graphs with applications to automata on infinite trees. TCS 200, 135–183 (1998)

Arthur and Merlin as Oracles

Venkatesan T. Chakaravarthy and Sambuddha Roy

IBM India Research Lab, New Delhi, India
{vechakra,sambuddha}@in.ibm.com

Abstract. We study some problems solvable in deterministic polynomial time given oracle access to the promise version of the Arthur-Merlin class AM. The main result is that $BPP_{||}^{NP} \subseteq P_{||}^{prAM}$. An important property of the class $P_{||}^{prAM}$ is that it can be derandomized as $P_{||}^{prAM} = P_{||}^{NP}$, under a natural hardness hypothesis used for derandomizing the class AM; this directly follows from a result due to Miltersen and Vinodchandran [10]. As a consequence, we get that $BPP_{||}^{NP} = P_{||}^{NP}$, under the above hypothesis. This gives an alternative (and perhaps, a simpler) proof of the same result obtained by Shaltiel and Umans [16], using different techniques.

Next, we present an FP^{prAM} algorithm for finding near-optimal strategies of a succinctly presented zero-sum game. For the same problem, Fortnow et al. [7] described a ZPP^{NP} algorithm. As a by product of our algorithm, we also get an alternative proof of the result by Fortnow et. al. One advantage with an FP^{prAM} algorithm is that it can be directly derandomized using the Miltersen-Vinodchandran construction [10]. As a consequence, we get an FP^{NP} algorithm for the above problem, under the hardness hypothesis used for derandomizing AM.

1 Introduction

We study some problems solvable in deterministic polynomial time given oracle access to the (promise version of) Arthur-Merlin classes, namely the class P^{prAM} and its variants, such as $P_{||}^{prAM}$. These classes are interesting from a derandomization perspective.

Starting with the classical hardness-randomness tradeoff due to Nisan and Wigderson [13], several complexity classes have been derandomized, under various hardness hypotheses. Working under this framework, Miltersen and Vinodchandran [10] derandomized the class AM as AM = NP, under a suitable hardness hypothesis. As a direct consequence of their work, we get that $P^{prAM} = P^{NP}$ and $P_{||}^{prAM} = P_{||}^{NP}$, under similar hardness hypotheses. Thus, the advantage with a result of the form $\mathcal{C} \subseteq P^{prAM}$ (resp. $\mathcal{C} \subseteq P_{||}^{prAM}$) is that we get $\mathcal{C} \subseteq P^{NP}$ (resp. $\mathcal{C} \subseteq P_{||}^{NP}$), under a hypothesis similar to the one used for derandomizing AM. Motivated by such considerations, we prove two results involving P^{prAM} and its variants. The two results are concerned with $BPP_{||}^{NP}$ and succinct zero-sum games, discussed in detail in the following sections.

E. Ochmański and J. Tyszkiewicz (Eds.): MFCS 2008, LNCS 5162, pp. 229–240, 2008.

1.1 $\mathrm{BPP}_{||}^{\mathrm{NP}}$

Working under the above derandomization framework, Klivans and van Melke-beek [8] showed that $\mathrm{BPP}_{||}^{\mathrm{NP}} = \mathrm{P}_{||}^{\mathrm{NP}}$, under a hardness hypothesis naturally associated with the class $\mathrm{BPP}_{||}^{\mathrm{NP}}$, which we shall refer to as $\mathrm{BPP}_{||}^{\mathrm{NP}}$-hypothesis[1]. Subsequently, Shaltiel and Umans [16] obtained an improvement by showing $\mathrm{BPP}_{||}^{\mathrm{NP}} = \mathrm{P}_{||}^{\mathrm{NP}}$, under the hypothesis naturally associated with the class AM; we shall refer to this hypothesis as the AM-hypothesis[2]. This is a surprising re-sult, since they derandomize $\mathrm{BPP}_{||}^{\mathrm{NP}}$, under a weaker hypothesis associated with the smaller class $\mathrm{AM} \subseteq \mathrm{BPP}_{||}^{\mathrm{NP}}$.

Our main result here is that $\mathrm{BPP}_{||}^{\mathrm{NP}} \subseteq \mathrm{P}_{||}^{\mathrm{prAM}}$. As discussed before, it follows from the construction of Miltersen and Vinodchandran [10] that $\mathrm{P}_{||}^{\mathrm{prAM}} = \mathrm{P}_{||}^{\mathrm{NP}}$, under the AM-hypothesis; this immediately implies that $\mathrm{BPP}_{||}^{\mathrm{NP}} = \mathrm{P}_{||}^{\mathrm{NP}}$, under the same hypothesis. Thus our main result yields an alternative proof and a different explanation for why the surprising result by Shaltiel and Umans is true. The Shaltiel-Umans approach involved proving a more fundamental result that the AM-hypothesis and the $\mathrm{BPP}_{||}^{\mathrm{NP}}$-hypothesis are, in fact, equivalent. For this they utilize sophisticated arithmetization tools and derandomization constructs. In contrast, our alternative proof shows how to derandomize $\mathrm{BPP}_{||}^{\mathrm{NP}}$ using the Miltersen-Vinodchandran construction as a black-box.

The main result, in fact, yields the equivalence $\mathrm{BPP}_{||}^{\mathrm{NP}} = \mathrm{P}_{||}^{\mathrm{prAM}}$, since a simple argument shows that $\mathrm{P}_{||}^{\mathrm{prAM}} \subseteq \mathrm{BPP}_{||}^{\mathrm{NP}}$. Furthermore, it can easily be shown that $\mathrm{BPP}_{||}^{\mathrm{prAM}} = \mathrm{BPP}_{||}^{\mathrm{NP}}$. Thus, we get that $\mathrm{BPP}_{||}^{\mathrm{prAM}} = \mathrm{P}_{||}^{\mathrm{prAM}}$; this is an unconditional derandomization of the class $\mathrm{BPP}_{||}^{\mathrm{prAM}}$.

By extending the proof of $\mathrm{AM} \subseteq \Pi_2^p$ [2], it is easy to show that $\mathrm{P}_{||}^{\mathrm{prAM}} \subseteq \mathrm{P}_{||}^{\Sigma_2^p}$. It follows that $\mathrm{BPP}_{||}^{\mathrm{NP}} \subseteq \mathrm{P}_{||}^{\Sigma_2^p}$, or equivalently $\mathrm{BPP}^{\mathrm{NP}[\log]} \subseteq \mathrm{P}^{\Sigma_2^p[\log]}$. This can be seen as a baby-step towards resolving the much larger open problem of whether $\mathrm{BPP}^{\mathrm{NP}}$ is contained in $\mathrm{P}^{\Sigma_2^p}$.

We now present a brief outline of the proof of the main result. Following the work of Sipser [17] and Stockmeyer [18], it is known that the cardinality of a set $X \subseteq \{0,1\}^m$ can be approximately estimated in $\mathrm{P}_{||}^{\mathrm{prAM}}$, when the membership testing for X can be performed via a nondeterministic polynomial time compu-tation. More precisely, given a non-deterministic circuit accepting X, we can ap-proximately estimate the cardinality of X in $\mathrm{P}_{||}^{\mathrm{prAM}}$. Fix a language $L \in \mathrm{BPP}_{||}^{\mathrm{NP}}$ and consider an input string x. Denote by W the set of all random strings that lead to acceptance on input x. So, testing whether $x \in L$ or not, reduces to a question of distinguishing between the two cases of W being "large" (say, W has measure $\geq 3/4$) and W being "small" (say, W has measure $\leq 1/4$). However, notice that the membership testing for W requires a $\mathrm{P}_{||}^{\mathrm{NP}}$ computation. Nevertheless, we show how

---

[1] $\mathrm{BPP}_{||}^{\mathrm{NP}}$-hypothesis: $\mathrm{E}_{||}^{\mathrm{NP}}$ requires exponential size non-adaptive SAT-oracle circuits.
[2] AM-hypothesis: $\mathrm{E}_{||}^{\mathrm{NP}}$ requires exponential size SV-nondeterministic circuits.

to test whether W is large or small , using a black-box for approximate-counting of non-deterministic circuits.

1.2   Succinct Zero-Sum Games

A two-player zero-sum 0-1 game is specified by a $2^n \times 2^m$ Boolean payoff matrix M. The rows and the columns correspond to the *pure strategies* of the row-player and the column-player, respectively. The row-player chooses a row $y \in \{0,1\}^n$ and the column-player chooses a column $z \in \{0,1\}^m$, simultaneously. The row-player then pays $M[y, z]$ to the column-player. The goal of the row-player is to minimize its loss, while the goal of the column-player is to maximize its gain.

A *mixed strategy* (or simply, a strategy) for the row-player is a probability distribution P over the rows; similarly, a strategy for the column-player is a probability distribution Q over the columns. The expected payoff is defined as:

$$M(P, Q) = \sum_{y,z} P(y) M[y, z] Q(z).$$

The classical Minmax theorem of von Neumann [11] says that even if the strategies are chosen sequentially, who plays first does not matter:

$$\min_P \max_Q M(P, Q) = \max_Q \min_P M(P, Q) = v^*,$$

where v^* is called the *value* of the game. This means that there exist strategies P^* and Q^* such that

$$\max_Q M(P^*, Q) \leq v^* \text{ and } \min_P M(P, Q^*) \geq v^*.$$

Such strategies P^* and Q^* are called *optimal strategies*.

In some scenarios, it is sufficient to compute approximately optimal strategies. We shall be mainly concerned with additive errors. A row-player strategy $\widetilde{P}$ is said to be ϵ-optimal, if

$$\max_Q M(\widetilde{P}, Q) \leq v^* + \epsilon,$$

and similarly, a column-player strategy $\widetilde{Q}$ is said to be ϵ-optimal, if

$$\min_P M(P, \widetilde{Q}) \geq v^* - \epsilon,$$

Zero-sum games have been well explored, due to their diverse applications. The problem of finding the value, as well as optimal and near-optimal strategies of a given game have been well-studied. In particular, it is known that the value and optimal strategies can be computed in polynomial time (see [14]). Moreover, efficient parallel algorithms have been devised for finding near-optimal strategies. We refer to [7] for a brief account of these results and the applications of zero-sum games in computational complexity and learning theory.

This paper deals with computing near-optimal strategies when the payoff matrix M is presented *succinctly* in the form of a circuit C. We say that a boolean circuit $C : \{0,1\}^n \times \{0,1\}^m \to \{0,1\}$ *succinctly encodes* the matrix M, if for every y and z, $C(y,z) = M[y,z]$. Notice that the circuit C can be much smaller than the matrix M (for instance, while M is of size $2^n \times 2^m$, the size of C can be polynomial in n and m). In the succinct zero-sum game problem, the payoff matrix M would be presented implicitly by the circuit C and the running time of an algorithm is computed with respect to $|C|$.

It is known that computing the exact value of a succinctly presented zero-sum game is EXP-complete (see [6]) and that approximating the value within multiplicative factors is Π_2^p-hard [7]. We study the problem of approximating the value and finding near-optimal strategies, within additive errors. The above problem generalizes the problem of learning circuits for SAT (assuming such circuits exist) and the problems in the symmetric alternation class S_2^p. We refer the reader to the paper by Fortnow et. al [7] for a detailed account on these aspects.

Lipton and Young [9] showed that ϵ-optimal strategies of a succinctly presented zero-sum game can be computed in Σ_2^p. Fortnow et. al [7] presented a ZPP^{NP} algorithm for the same problem; their algorithm also finds an estimate of the value of the game within additive errors. Our main result is an FP^{prAM} algorithm for the same problems (i.e., the algorithm runs in deterministic polynomial time, given oracle access to a promise language in the promise version of the Arthur-Merlin class).

In the proof, we show that the problem of finding ϵ-optimal strategies reduces to the problem of finding "strong collectively irrefutable certificates (strong CIC)" of a given S_2-type matrix. Cai [3] presented a ZPP^{NP} algorithm for the latter problem of finding strong CICs. In [4], an FP^{prAM} algorithm for finding a "weak" CIC was described. By borrowing and extending components from these two algorithms, we devise an FP^{prAM} algorithm for computing strong CICs. Put together, we get the required FP^{prAM} algorithm for computing ϵ-optimal strategies.

The proof of the main result yields the following as by-products. We notice that by composing the above-mentioned reduction and Cai's algorithm, we get a ZPP^{NP} algorithm for finding ϵ-optimal strategies. This gives an alternative proof of the result by Fortnow et al. The details of this proof will be included in the full version of the paper. We also obtain a $\text{FP}^{\text{prS}_2^p}$ algorithm for approximating the value of a succinctly presented zero-sum game. However, we do not know whether ϵ-optimal strategies can be computed in $\text{FP}^{\text{prS}_2^p}$; we consider this to be an interesting open problem.

As discussed earlier, the Miltersen-Vinodchandran construction can be directly used to derandomize the FP^{prAM} algorithm. Thus, under the AM- hypothesis, we get an FP^{NP} algorithm for finding ϵ-optimal strategies. It is not clear whether such a derandomization can be achieved for the ZPP^{NP} algorithm by Fortnow et al. [7].

2 Preliminaries

In this section, we present relevant definitions and notation.

Complexity classes. We use standard definitions for complexity classes such as P, NP, AM, ZPP^{NP} and BPP^{NP} [5,15]. Below, we present definitions for promise and function classes, that are central to our paper.

Promise languages. A promise language Π is a pair (Π_1, Π_2), where $\Pi_1, \Pi_2 \subseteq \Sigma^*$, such that $\Pi_1 \cap \Pi_2 = \emptyset$. The elements of Π_1 are called the *positive instances* and those of Π_2 are called the *negative instances*. If $\Pi_1 \cup \Pi_2 = \Sigma^*$, then Π is a language in the usual sense.

Symmetric Alternation. A promise language $\Pi = (\Pi_1, \Pi_2)$ is said to be in the promise class prS_2^p, if there exists a polynomial time computable Boolean predicate $V(\cdot, \cdot, \cdot)$ and polynomials $p(\cdot)$ and $q(\cdot)$ such that for any x, we have

$$x \in \Pi_1 \Longrightarrow (\exists y \in \{0,1\}^n)(\forall z \in \{0,1\}^m)[V(x,y,z) = 1], \text{ and}$$
$$x \in \Pi_2 \Longrightarrow (\exists z \in \{0,1\}^m)(\forall y \in \{0,1\}^n)[V(x,y,z) = 0],$$

where $n = p(|x|)$ and $m = q(|x|)$. We refer to the y's and z's above as *certificates*. The predicate V is called the *verifier*. The complexity class S_2^p consists of languages in prS_2^p.

Promise AM (prAM). A promise language $\Pi = (\Pi_1, \Pi_2)$ is said to be in the promise class prAM, if there exists a polynomial time computable Boolean predicate $A(\cdot, \cdot, \cdot)$ and polynomials $p(\cdot)$ and $q(\cdot)$ such that, for all x, we have

$$x \in \Pi_1 \Longrightarrow (\forall y \in \{0,1\}^n)(\exists z \in \{0,1\}^m)[A(x,y,z) = 1], \text{ and}$$
$$x \in \Pi_2 \Longrightarrow \Pr_{y \in \{0,1\}^n}[(\exists z \in \{0,1\}^m) A(x,y,z) = 1] \leq \frac{1}{2},$$

where $n = p(|x|)$ and $m = q(|x|)$. The predicate A is called *Arthur's predicate.*

$\mathrm{BPP}_{||}^{\mathrm{NP}}$: A language L is said to be in the class $\mathrm{BPP}_{||}^{\mathrm{NP}}$, if there exists a $\mathrm{P}_{||}^{\mathrm{NP}}$ machine M and a polynomial $p(\cdot)$ such that for any input x,

$$x \in L \Longrightarrow \Pr_{y \in \{0,1\}^m}[M(x,y) = 1] \geq \frac{3}{4}, \text{ and}$$
$$x \notin L \Longrightarrow \Pr_{y \in \{0,1\}^m}[M(x,y) = 1] \leq \frac{1}{4},$$

where $m = p(|x|)$.

Succinct encoding of matrices and sets. A Boolean circuit $C : \{0,1\}^n \times \{0,1\}^m \to \{0,1\}$ is said to *succinctly encode* a Boolean $2^n \times 2^m$ matrix M, if for all $y \in \{0,1\}^n$ and $z \in \{0,1\}^m$, we have $C(y,z) = M[y,z]$. A Boolean circuit $C : \{0,1\}^m \to \{0,1\}$ is said to *succinctly encode* a set $X \subseteq \{0,1\}^m$, if for all $x \in \{0,1\}^m$, $x \in X \iff C(x) = 1$.

Oracle access to promise languages. Let A be an algorithm and $\Pi = (\Pi_1, \Pi_2)$ be a promise language. When the algorithm A asks a query q, the oracle behaves as follows: if $q \in \Pi_1$, the oracle replies "yes"; if $q \in \Pi_2$, the oracle replies "no"; if q is neither in Π_1 nor in Π_2, the oracle may reply "yes" or "no". We allow the algorithm to ask queries of the third type. The requirement is that the algorithm should be able to produce the correct answer, regardless of the answers given by the oracle to the queries of the third type.

Function classes. For a promise language Π, the notation FP^Π refers to the class that are computable by a polynomial time machine, given oracle access to Π. For a promise class $\mathcal{C}$, we denote by $\mathrm{FP}^{\mathcal{C}}$, the union of FP^{Π}, for all $\Pi \in \mathcal{C}$. Regarding $\mathrm{ZPP}^{\mathrm{NP}}$, we slightly abuse the notation and use this to mean both the standard complexity class and the function class. The function class $\mathrm{ZPP}^{\mathrm{NP}}$ contains functions computable by a zero-error probabilistic polynomial time algorithm given oracle to NP; the algorithm either outputs a correct value of the function or "?", the latter with a small probability.

3 $\mathrm{BPP}_{||}^{\mathrm{NP}} \subseteq \mathrm{P}_{||}^{\mathrm{prAM}}$

In this section, we prove that $\mathrm{BPP}_{||}^{\mathrm{NP}} \subseteq \mathrm{P}_{||}^{\mathrm{prAM}}$. The proof goes via approximate counting for non-deterministic circuits, a problem known to be solvable in $\mathrm{P}_{||}^{\mathrm{prAM}}$.

A *non-deterministic* circuit C is a boolean circuit that takes an m-bit string as input and an s-bit string z as auxiliary input and outputs 1 or 0, i.e., $C : \{0,1\}^m \times \{0,1\}^s \to \{0,1\}$. For a string $y \in \{0,1\}^m$, C is said to *accept* y, if there exists a $z \in \{0,1\}^s$ such that $C(y,z) = 1$; C is said to *reject* y, otherwise. Let $\mathrm{Count}(C)$ denote the number of strings from $\{0,1\}^m$ accepted by C.

The following result is implicit in the work of Sipser [17] and Stockmeyer [18] and it deals with the problem of approximately counting the number of strings accepted by a given non-deterministic circuit.

Theorem 1. *[17][18] There exists an* $\mathrm{FP}_{||}^{\mathrm{prAM}}$ *algorithm that takes as input a non-deterministic circuit C, and a parameter $\delta > 0$ and outputs a number e such that*

$$(1 - \delta)\mathrm{Count}(C) \le e \le (1 + \delta)\mathrm{Count}(C).$$

The running time is polynomial in $|C|$ and $1/\delta$.

The first step of the proof involves reducing the error probability of a given $\mathrm{BPP}_{||}^{\mathrm{NP}}$ machine. This is achieved via the standard method of repeated trials and taking majority.

Proposition 1. *Let L be a language in* $\mathrm{BPP}_{||}^{\mathrm{NP}}$. *Then there exists a* $\mathrm{P}_{||}^{\mathrm{NP}}$ *machine M and a polynomial $p(\cdot)$ such that for any input x,*

$$x \in L \implies \Pr_{y \in \{0,1\}^m}[M(x,y) = 1] \ge 1 - \frac{1}{8K}, \ and$$

$$x \notin L \implies \Pr_{y \in \{0,1\}^m}[M(x,y) = 1] \le \frac{1}{8K},$$

where $m = p(|x|)$ and $K \le p(|x|)$ is the maximum number of queries asked by M on input x for any string $y \in \{0,1\}^m$.

Theorem 2. $\mathrm{BPP}_{||}^{\mathrm{NP}} \subseteq \mathrm{P}_{||}^{\mathrm{prAM}}$.

Proof. Let L be a language in $\mathrm{BPP}_{||}^{\mathrm{NP}}$ and let M be a $\mathrm{BPP}_{||}^{\mathrm{NP}}$ machine for deciding L given by Proposition 1. Fix an input string x and let m be the number of random bits used by M. Without loss of generality, assume that M uses SAT as its oracle and that the number of queries is exactly K on all random strings $y \in \{0,1\}^m$, with $K \geq 1$.

Partition the set $\{0,1\}^m$ into the set of *good* strings G and the set of *bad* strings B, where $G = \{y : M(x,y) = \chi_L(x)\}$ and $B = \{0,1\}^m - G$, where $\chi_L(\cdot)$ is the characteristic function[3]. For a set $X \subseteq \{0,1\}^m$, let $\mu(X) = |X|/2^m$ denote its measure. Thus, $\mu(G) \geq 1 - 1/(8K)$ and $\mu(B) \leq 1/(8K)$.

Consider a string $y \in \{0,1\}^m$. Let $\Phi(y) = \langle \varphi_1^y, \varphi_2^y, \ldots, \varphi_K^y \rangle$, be the K SAT queries asked by M on the string y. Let $\underline{\mathbf{a}}^y = \langle a_1^y, a_2^y, \ldots, a_K^y \rangle$ be the correct answers to these queries, namely the bit $a_j^y = 1$, if $\varphi_j^y \in$ SAT and $a_j^y = 0$, otherwise. We shall consider simulating the machine M with arbitrary answer strings. For a bit string $\underline{\mathbf{b}} = \langle b_1 b_2 \ldots b_K \rangle$, let $M(x,y,\underline{\mathbf{b}})$ denote the outcome of the machine, if $\underline{\mathbf{b}}$ is provided as the answer to its K queries. Thus, $M(x,y) = M(x,y,\underline{\mathbf{a}}^y)$. When an arbitrary bit-string $\underline{\mathbf{b}}$ is provided as the answer, the outcome $M(x,y,\underline{\mathbf{b}})$ can be different from $M(x,y)$.

Let $N(y)$ denote the number of satisfiable formulas in $\Phi(y)$, i.e., $N(y) = |\{1 \leq i \leq K : \varphi_i^y \in \text{SAT}\}|$. Partition the set G into $K+1$ parts based the value $N(\cdot)$: for $0 \leq r \leq K$, let $S_r = \{y : y \in G \text{ and } N(y) = r\}$.

For each $0 \leq r \leq K$, define two sets C_r^1 and C_r^0 as below. The set C_r^1 consists of all strings $y \in \{0,1\}^m$ satisfying the following property: there exists an answer string $\underline{\mathbf{b}} = \langle b_1 b_2 \ldots b_K \rangle$ such that (i) if $b_j = 1$ then $\varphi_j^y \in$ SAT; (ii) $\underline{\mathbf{b}}$ has at least r ones in it; and (iii) $M(x,y,\underline{\mathbf{b}}) = 1$. The set C_r^0 is defined similarly. The set C_r^0 consists of all strings $y \in \{0,1\}^m$ satisfying the following property: there exists an answer string $\underline{\mathbf{b}} = \langle b_1 b_2 \ldots b_K \rangle$ such that (i) if $b_j = 1$ then $\varphi_j^y \in$ SAT; (ii) $\underline{\mathbf{b}}$ has at least r ones in it; and (iii) $M(x,y,\underline{\mathbf{b}}) = 0$. Notice that the sets C_r^1 and C_r^0 need not be disjoint and that there may be strings $y \in \{0,1\}^m$ that belong to neither C_r^1 nor C_r^0.

Claim 1

(i) If $x \in L$, then there exits an $0 \leq \ell \leq K$ such that $\mu(C_\ell^1) - \mu(C_\ell^0) \geq \frac{1}{4K}$.

(ii) If $x \notin L$, then for all $0 \leq r \leq K$, $\mu(C_r^1) - \mu(C_r^0) \leq \frac{1}{8K}$.

Proof of claim: We first make an observation regarding the concerned sets C_r^1 and C_r^0. Fix any $0 \leq r \leq K$. Notice that for any $j < r$, $S_j \cap C_r^1 = \emptyset$ and $S_j \cap C_r^0 = \emptyset$. This is becuase, for any $y \in S_j$, with $j < r$, $\Phi(y)$ has only j satisfying formulas and so, it does not meet the requirement for either C_r^1 or C_r^0.

Suppose $x \in L$. Consider any $0 \leq r \leq K$. We first derive a lowerbound on $|C_r^1|$. Consider any string $y \in S_j$, with $r \leq j \leq K$. Notice that $\underline{\mathbf{a}}^y$ has $j \geq r$ satisfying formulas and $M(x,y,\underline{\mathbf{a}}^y) = 1$. So, $y \in C_r^1$. Thus, for $r \leq j \leq K$, $S_j \subseteq C_r^1$. It follows that $|C_r^1| \geq \sum_{j=r}^{K} |S_j|$. We next derive an upperbound on C_r^0. We observed that $S_j \cap C_r^0 = \emptyset$, for all $j < r$. Let us now consider the set S_r. Pick any string $y \in S_r$. Notice that $\Phi(y)$ has exactly r satisfiable formulas. So, the only answer string satisfying the first two requirements of C_r^0 is $\underline{\mathbf{a}}^y$. Since $M(x,y,\underline{\mathbf{a}}^y) = 1$, we have that $y \notin C_r^0$. Thus, $S_r \cap C_r^0 = \emptyset$. Hence, for $j \leq r$,

---

[3] $\chi_L(x) = 1$, if $x \in L$ and $\chi_L(x) = 0$, if $x \notin L$.

$S_j \cap C_r^0 = \emptyset$. It follows that $|C_r^0| \leq |B| + \sum_{j=r+1}^{K} |S_j|$. As a consequence, we get that $|C_r^1| - |C_r^0| \geq |S_r| - |B|$. Since $\mu(G) \geq 1 - 1/(8K)$, by an averaging argument, there exists an $0 \leq \ell \leq K$ such that

$$\mu(S_\ell) \geq \frac{\left(1 - \frac{1}{8K}\right)}{K+1} \geq \frac{3}{8K}.$$

Such an ℓ satisfies $\mu(C_\ell^1) - \mu(C_\ell^0) \geq 1/(4K)$. We have proved the first part of the claim.

Suppose $x \notin L$. The argument is similar to the first part. Fix any $0 \leq r \leq K$. Observe that for $j \leq r$, $S_j \cap C_r^1 = \emptyset$. Hence, $|C_r^1| \leq |B| + \sum_{j=r+1}^{K} |S_j|$. On the other hand, for any $j \geq r$, $S_j \subseteq C_r^0$. So, $|C_r^0| \geq \sum_{j=r}^{K} |S_j|$. It follows that $|C_r^1| - |C_r^0| \leq |B|$. Since $\mu(B) \leq 1/(8K)$, we get the second part of the claim. This completes the proof of Claim 1.

Notice that the membership testing for the sets C_j^1 and C_j^0 can be performed in non-deterministic polynomial time. Thus, for $0 \leq j \leq K$, we can construct a non-deterministic circuit accepting C_j^1 (similarly, C_j^0) such that the size of the circuit is polynomial in $|x|$. Setting $\delta = 1/(80K)$, we invoke the algorithm given by Theorem 1 to get estimates e_j^1 and e_j^0 such that $(1-\delta)\mu(C_j^1) \leq e_j^1 \leq (1+\delta)\mu(C_j^1)$ and $(1-\delta)\mu(C_j^0) \leq e_j^0 \leq (1+\delta)\mu(C_j^0)$, for all $0 \leq j \leq K$. We output "$x \in L$", if there exists an ℓ such that $e_\ell^1 - e_\ell^0 \geq 9/(40K)$. If no such ℓ exists, then we output "$x \notin L$". The correctness follows from Claim 1 and the simple fact that $\mu(X) \leq 1$, for any $X \subseteq \{0,1\}^m$.   □

4   Succinct Zero-Sum Games

In this section, we present an FP^{prAM} algorithm for finding the near-optimal strategies of a zero-sum game presented succinctly in the form of a circuit.

The first task is to approximately find the value of the given game. This is treated in the next section.

4.1   Approximately Finding the Value

We shall first discuss an $\text{FP}^{\text{prS}_2^p}$ algorithm for approximating the value. The algorithm uses the following promise language as the oracle. Fortnow et. al showed that this promise language lies in prS_2^p.

Succinct Game Value (SGV): The input consists of a circuit C succinctly encoding a 0-1 zero-sum game and parameters v and ϵ.
Positive instances: $v^* \geq v + \epsilon$.
Negative instances: $v^* \leq v - \epsilon$.
Here, v^* refers to the value of the given game.

Theorem 3. [7] *The promise language SGV belongs to* prS_2^p.

Using SGV as an oracle, we can perform a linear search in the interval $[0, 1]$ and approximately find the value of a given game.

Theorem 4. *There exists an* $\text{FP}^{\text{prS}_2^p}$ *algorithm that takes as input a circuit C succinctly encoding a 0-1 zero-sum game and a parameter ϵ and outputs a number v such that $v - \epsilon \leq v^* \leq v + \epsilon$, where v^* is the value of the given game. The running time of the algorithm is polynomial in $|C|$ and $1/\epsilon$.*

Proof. We shall use SGV as the prS_2^p oracle. Set $\epsilon' = \epsilon/2$. For $j = 0,1,2,\ldots,\lceil n/\epsilon' \rceil$, ask the query $\langle C, j\epsilon', \epsilon' \rangle$. Let $\hat{j}$ be the first index such that the oracle answers "no". Set $v' = \hat{j}\epsilon'$. Return(v'). Notice that v' has the required property.   □

Fortnow et. al [7] presented a ZPP^{NP} algorithm for the problem of approximating the value of a succinct zero-sum game. By extending Cai's result [3] that $S_2^p \subseteq \text{ZPP}^{\text{NP}}$, it can be shown that $\text{FP}^{\text{prS}_2^p} \subseteq \text{ZPP}^{\text{NP}}$. Thus, Theorem 4 provides a mild improvement over the previously best known result for the problem under consideration.

It is known that $S_2^p \subseteq \text{P}^{\text{prAM}}$ [4]. An easy extension of this result shows that $\text{FP}^{\text{prS}_2^p} \subseteq \text{FP}^{\text{prAM}}$. Combining with Theorem 4, we get an FP^{prAM} algorithm for approximating the value of succinct zero-sum game.

Theorem 5. *There exists an* FP^{prAM} *algorithm that takes as input a circuit C succinctly encoding a 0-1 zero-sum game and a parameter ϵ and outputs a number v such that $v - \epsilon \leq v^* \leq v + \epsilon$, where v^* is the value of the given game. The running time of the algorithm is polynomial in $|C|$ and $1/\epsilon$.*

4.2 Finding Near-Optimal Strategies

In this section, we present an FP^{prAM} algorithm for finding near-optimal strategies of a succinctly presented zero-sum game. The following small support lemma is useful in designing our algorithms.

A mixed strategy of a player is said to be *k-uniform*, if it chooses uniformly from a multi-set of k pure strategies. Such a strategy can simply be specified by the multi-set of size k. The following lemma asserts the existence of k-uniform ϵ-optimal, for a small value of k.

Lemma 1 ([1][12][9]). *Let M be a $0-1$ $2^n \times 2^m$ payoff matrix. Then there are k-uniform ϵ-optimal strategies for both the row and the column-players, where $k = O(\frac{n+m}{\epsilon^2})$.*

The algorithm for finding near-optimal strategies goes via a reduction to the problem of finding strong collectively irrefutable certificates of a given S_2-type matrix. Such matrices and associated concepts are discussed next.

Definition 1 (S_2-type matrices and CIC's). *Let M be a $2^n \times 2^m$ boolean matrix. We view the matrix as specifying a bipartite tournament. For a row $y \in \{0,1\}^n$ and a column $z \in \{0,1\}^m$, we say that y beats z, if $M[y,z] = 1$; z is said to beat y, if $M[y,z] = 0$. A row y is said to be a row-side irrefutable certificate (shortened, IC), if y beats every column $z \in \{0,1\}^m$. Similarly, a column z is said to be a column-side IC, if z beats every row $y \in \{0,1\}^n$. The matrix M is called a*

S_2-*type matrix, if it has either a row-side* IC *or a column-side* IC. *In the former case,* M *is called a row-side* S_2-*type matrix and in the latter case, it is called a column-side* S_2-*type matrix. A set of rows* $Y \subseteq \{0,1\}^n$ *is called a row-side weak* collectively irrefutable certificate *(shortened,* CIC*), if for any column* $z \in \{0,1\}^m$, *at least one row in* Y *beats* z. *For* $0 \leq \alpha \leq 1$, *a set of rows* $Y \subseteq \{0,1\}^n$ *is called a row-side* α-*strong* CIC, *if for any column* z, *at least* α *fraction of rows from* Y *beat* z, *i.e.,* $|\{y \in Y : y \text{ beats } z\}| \geq \alpha|Y|$. *The notion of column-side weak* CIC *and column-side* α-*strong* CIC *are defined analogously.*

A $\mathrm{ZPP}^{\mathrm{NP}}$ algorithm for finding α-strong CIC of a given S_2-type matrix is implicit in the work of Cai ([3], Section 5). In [4], a $\mathrm{FP}^{\mathrm{prAM}}$ algorithm is presented for finding a weak CIC of a given S_2-type matrix. By combining and extending arguments from these two algorithms, we obtain an $\mathrm{FP}^{\mathrm{prAM}}$ algorithm for finding α-strong CIC for S_2-type matrices. The claim is encapsulated in the following theorem. Its proof will be included in the full version of the paper.

Theorem 6. *There exists an* $\mathrm{FP}^{\mathrm{prAM}}$ *algorithm that takes as input a circuit* C *succinctly encoding a row-side (resp. column side)* S_2-*type matrix* M *of dimension* $2^n \times 2^m$ *and a parameter* $\alpha < 1$, *and outputs an* α-*strong row-side* CIC *(resp. a column-side* α-*strong* CIC*) of size polynomial in* $1/\alpha$, n *and* m. *The running time is polynomial in* $|C|$ *and* $1/\alpha$.

We now present the $\mathrm{FP}^{\mathrm{prAM}}$ algorithm for finding near-optimal strategies of a succinctly presented zero-sum game.

Theorem 7. *There exists an* $\mathrm{FP}^{\mathrm{prAM}}$ *algorithm that takes as input a circuit* C *succinctly encoding a* $2^n \times 2^m$ *payoff matrix* M *of a 0-1 zero-sum game and a parameter* ϵ *and outputs a pair of* ϵ-*optimal mixed strategies* $(\widetilde{P}, \widetilde{Q})$. *The running time of the algorithm is polynomial in* $|C|$ *and* $1/\epsilon$.

Proof. Let v^* be the value of the game given by M. Our algorithm finds the required strategies $\widetilde{P}$ and $\widetilde{Q}$ in two phases. Here, we discuss the first phase of the algorithm that finds $\widetilde{P}$. The second phase for finding $\widetilde{Q}$ works in a similar manner.

The algorithm uses a parameter $\epsilon' = \epsilon/2$. Invoke the algorithm given in Theorem 5 with error parameter $\epsilon'/2$ and obtain an estimate v. Set $v^+ = v + \epsilon'/2$. Notice that v^+ is an upperbound on v^* satisfying $v^* \leq v^+ \leq v^* + \epsilon'$. Invoking Lemma 1 with error parameter ϵ' we get a number $k = k(\epsilon')$ such that M has k-uniform ϵ'-optimal strategies $(P_{\epsilon'}, Q_{\epsilon'})$.

Construct a matrix $\overline{M}$ as follows. Each row $\overline{y}$ of $\overline{M}$ corresponds to a sequence $\langle y_1, y_2, \ldots, y_k \rangle$, where y_i is a row in M; each column of $\overline{M}$ corresponds to a single column of M. Thus, $\overline{M}$ is a $2^{\overline{n}} \times 2^m$ matrix, where $\overline{n} = nk$. Its entries are defined as follows. Consider a row $\overline{y} = \langle y_1, y_2, \ldots, y_k \rangle$ and a column z. The entry $\overline{M}[\overline{y}, z]$ is defined as:

$$\overline{M}[\overline{y}, z] = \begin{cases} 1 \text{ if } \frac{1}{k}\left(\sum_{i=1}^{k} M[y_i, z]\right) \leq v^+ \\ 0 \text{ otherwise} \end{cases}$$

Let $\overline{y}$ be a row corresponding to $P_{\epsilon'}$. For any strategy Q of the column-player the expected payoff $M(P_{\epsilon'}, Q) \le v^+$. In particular, this is true for all pure strategies z of the column-player. Therefore, we see that $\overline{y}$ is a row full of 1's. In other words, $\overline{M}$ is a row-side S_2-type matrix.

Our next task is to find an α-strong row-side CIC of the matrix $\overline{M}$, where α is a parameter suitably fixed as $\alpha = (1 - \epsilon/2)$. For this, we invoke the algorithm given in Theorem 6 on $\overline{M}$ and obtain a row-side α-strong CIC $\overline{Y}$. Let the size of $\overline{Y}$ be t. Notice that each element $\overline{y}$ of $\overline{Y}$ is a sequence of k pure row-player strategies. Consider the collection S obtained by including all the pure strategies found in each $\overline{y} \in \overline{Y}$; thus, S is a multiset of size kt. Let $\widetilde{P}$ be the (kt)-uniform strategy over the multiset S. We next prove that $\widetilde{P}$ is an ϵ-optimal row-player strategy. The following easy claim is useful for this purpose.

Claim 1: Let P be a k-uniform row-player strategy. Let $v \le 1$ be such that for any pure column-player strategy z, $\Sigma_{y \in P} M[y, z] \le v$. Then, $\max_Q M(P, Q) \le v$, where Q ranges over all mixed strategies of the column-player.

Claim 2: $\widetilde{P}$ is an ϵ-optimal row-player strategy.

Proof. Consider any pure strategy of the column-player $z \in \{0, 1\}^m$. Since $\overline{Y}$ is an α-strong CIC, at least an α fraction of the rows in $\overline{Y}$ beat z. A row $\overline{y} \in \overline{Y}$ beating z means that $\overline{y} = \{y_1, y_2, \ldots, y_k\}$ satisfies $\sum_{y_i \in \overline{y}} M[y_i, z] \le kv^+$. Recall that $v^+ \le v^* + \epsilon'$. We now want to estimate the sum $\sum_{y \in S} M[y, z]$, which can be written as:

$$\sum_{y \in S} M[y, z] = \sum_{\overline{y} \in \overline{Y}} \sum_{y_i \in \overline{y}} M[y_i, z].$$

To estimate the sum on the RHS, we partition $\overline{Y}$ into two disjoint sets $\overline{Y}_{good}$ and $\overline{Y}_{bad}$: place all the $\overline{y} \in \overline{Y}$ that beat z in $\overline{Y}_{good}$ and the rest in $\overline{Y}_{bad}$. Notice that $|\overline{Y}_{good}| \ge \alpha t$.

$$
\begin{aligned}
\sum_{y \in S} M[y, z] &= \sum_{\overline{y} \in \overline{Y}} \sum_{y_i \in \overline{y}} M[y_i, z] \\
&= \sum_{\overline{y} \in \overline{Y}_{good}} \sum_{y_i \in \overline{y}} M[y_i, z] + \sum_{\overline{y} \in \overline{Y}_{bad}} \sum_{y_i \in \overline{y}} M[y_i, z] \\
&\le \sum_{\overline{y} \in \overline{Y}_{good}} kv^+ + \sum_{\overline{y} \in \overline{Y}_{bad}} k \\
&\le |\overline{Y}_{good}| kv^+ + |\overline{Y}_{bad}| k \\
&\le tk(v^* + \epsilon') + (1 - \alpha)tk \le tk(v^* + \epsilon)
\end{aligned}
$$

The last inequality follows from the choice of α and ϵ'. Now, Claim 2 follows from Claim 1.

The second phase of the algorithm that finds $\widetilde{Q}$ works in a similar manner. The details will be included in the full version of the paper. □

5   Conclusions and Open Problems

We showed that $BPP_{||}^{NP} \subseteq P_{||}^{prAM}$. A challenging open problem asks whether BPP^{NP} is contained in P^{prAM}. An affirmative answer would have two interesting implications. First, this would show that $BPP^{NP} \subseteq P^{\Sigma_2^p}$. The second implication is that BPP^{NP} can be derandomized under the hardness hypothesis used for derandomizing AM. We presented a $FP^{prS_2^p}$ algorithm for approximating the value of a succinct zero-sum game. It is open whether near-optimal strategies can be found in $FP^{prS_2^p}$.

Acknowledgments. We thank Eric Allender for his useful comments and Dieter van Melkebeek for his insightful suggestions and for sharing with us his alternative proof of Theorem 2.

References

1. Althöfer, I.: On sparse approximations to randomized strategies and convex combinations. Linear Algebra and its Applications 199 (1994)
2. Babai, L., Moran, S.: Arthur-Merlin games: A randomized proof system, and a hierarchy of complexity classes. Journal of Computer and System Sciences 36(2), 254–276 (1988)
3. Cai, J.: $S_2^p \subseteq ZPP^{NP}$. Journal of Computer and System Sciences 73(1) (2007)
4. Chakaravarthy, V., Roy, S.: Finding irrefutable certificates for S_2^p via Arthur and Merlin. In: STACS (2008)
5. Du, D., Ko, K.: Computational Complexity. John Wiley, Chichester (2000)
6. Feigenbaum, J., Koller, D., Shor, P.: A game-theoretic classification of interactive complexity classes. In: CCC (1995)
7. Fortnow, L., Impagliazzo, R., Kabanets, V., Umans, C.: On the complexity of succinct zero-sum games. In: CCC (2005)
8. Klivans, A., van Melkebeek, D.: Graph nonisomorphism has subexponential size proofs unless the polynomial hierarchy collapses. SIAM Journal on Computing 31(5), 1501–1526 (2002)
9. Lipton, R., Young, N.: Simple strategies for large zero-sum games with applications to complexity theory. In: STOC (1994)
10. Miltersen, P., Vinodchandran, N.: Derandomizing Arthur-Merlin games using hitting sets. In: FOCS (1999)
11. Neumann, J.: Zur theorie der gesellschaftspiel. Mathematische Annalen 100 (1928)
12. Newman, J.: Private vs. common random bits in communication complexity. Information Processing Letters 39, 67–71 (1991)
13. Nisan, N., Wigderson, A.: Hardness vs randomness. Journal of Computer and System Sciences 49(2), 149–167 (1994)
14. Owen, G.: Game Theory. Academic Press, London (1982)
15. Papadimitriou, C.: Computational Complexity. Addison-Wesley, Reading (1994)
16. Shaltiel, R., Umans, C.: Pseudorandomness for approximate counting and sampling. Computational Complexity 15(4), 298–341 (2007)
17. Sipser, M.: A complexity theoretic approach to randomness. In: STOC (1983)
18. Stockmeyer, L.: The complexity of approximate counting. In: STOC (1983)

A Decision Problem for Ultimately Periodic Sets in Non-standard Numeration Systems

Emilie Charlier and Michel Rigo

Institute of Mathematics, University of Liège, Grande Traverse 12 (B 37),
B-4000 Liège, Belgium
{echarlier,M.Rigo}@ulg.ac.be

Abstract. Consider a non-standard numeration system like the one built over the Fibonacci sequence where nonnegative integers are represented by words over $\{0, 1\}$ without two consecutive 1. Given a set X of integers such that the language of their greedy representations in this system is accepted by a finite automaton, we consider the problem of deciding whether or not X is a finite union of arithmetic progressions. We obtain a decision procedure under some hypothesis about the considered numeration system. In a second part, we obtain an analogous decision result for a particular class of abstract numeration systems built on an infinite regular language.

1 Introduction

Definition 1. *A* positional numeration system *is given by a (strictly) increasing sequence* $U = (U_i)_{i \geq 0}$ *of integers such that* $U_0 = 1$ *and* $C_U := \sup_{i \geq 0} \lceil U_{i+1}/U_i \rceil$ *is finite. Let* $A_U = \{0, \ldots, C_U - 1\}$. *The* greedy U-representation *of a positive integer* n *is the unique finite word* $\mathrm{rep}_U(n) = w_\ell \cdots w_0$ *over* A_U *satisfying*

$$n = \sum_{i=0}^{\ell} w_i U_i, \ w_\ell \neq 0 \ and \ \sum_{i=0}^{t} w_i U_i < U_{t+1}, \ \forall t = 0, \ldots, \ell.$$

We set $\mathrm{rep}_U(0)$ *to be the empty word* ε. *A set* $X \subseteq \mathbb{N}$ *of integers is* U-recognizable *if the language* $\mathrm{rep}_U(X)$ *over* A_U *is regular (i.e., accepted by a deterministic finite automaton, DFA). If* $x = x_\ell \cdots x_0$ *is a word over a finite alphabet of integers, then the* U-numerical value *of* x *is*

$$\mathrm{val}_U(x) = \sum_{i=0}^{\ell} x_i U_i.$$

Remark 1. Let x, y be two words over A_U. As a consequence of the greediness of the representation, if xy is a greedy U-representation and if the first letter of y is not 0, then y is also a greedy U-representation. Notice that for $m, n \in \mathbb{N}$, we have $m < n$ if and only if $\mathrm{rep}_U(m) <_{gen} \mathrm{rep}_U(n)$ where $<_{gen}$ is the genealogical ordering over A_U^*: words are ordered by increasing length and for words of same

E. Ochmański and J. Tyszkiewicz (Eds.): MFCS 2008, LNCS 5162, pp. 241–252, 2008.
© Springer-Verlag Berlin Heidelberg 2008

length, one uses the lexicographical ordering induced by the natural ordering of the digits in the alphabet A_U. Recall that for two words $x, y \in A_U^*$ of same length, x is lexicographically smaller than y if there exist $w, x', y' \in A_U^*$ and $a, b \in A_U$ such that $x = wax'$, $y = wby'$ and $a < b$.

For a positional numeration system U, it is natural to expect that $\mathbb{N}$ is U-recognizable. A necessary condition is that the sequence U satisfies a linear recurrence relation [12].

Definition 2. *A positional numeration system $U = (U_i)_{i \geq 0}$ is said to be* linear, *if the sequence U satisfies a homogenous linear recurrence relation. For all $i \geq 0$, we have*

$$U_{i+k} = a_1 U_{i+k-1} + \cdots + a_k U_i \tag{1}$$

for some $k \geq 1$, $a_1, \ldots, a_k \in \mathbb{Z}$ and $a_k \neq 0$.

Example 1. Consider the sequence defined by $F_0 = 1$, $F_1 = 2$ and for all $n \geq 0$, $F_{n+2} = F_{n+1} + F_n$. The *Fibonacci (linear numeration) system* is given by $F = (F_i)_{i \geq 0} = (1, 2, 3, 5, 8, 13, \ldots)$. For instance, $\text{rep}_F(15) = 100010$ and $\text{val}_F(101001) = 13 + 5 + 1 = 19$.

In this paper, we address the following decidability question.

Problem 1. Given a linear numeration system U and a set $X \subseteq \mathbb{N}$ such that $\text{rep}_U(X)$ is recognized by a (deterministic) finite automaton. Is it decidable whether or not X is ultimately periodic, i.e., whether or not X is a finite union of arithmetic progressions ?

Ultimately periodic sets of integers play a special role. On the one hand such infinite sets are coded thanks to a finite amount of information. On the other hand the celebrated Cobham's theorem asserts that these sets are the only sets that are recognizable in all integer base systems [3]. It is the reason why they are also referred in the literature as *recognizable* sets of integers (the recognizability being in that case independent of the base). Moreover, Cobham's theorem has been extended to various situations and in particular, to numeration systems given by substitutions [4].

 J. Honkala showed in [8] that Problem 1 turns out to be decidable for the usual integer base $b \geq 2$ numeration system defined by $U_n = bU_{n-1}$ for $n \geq 1$. Let us also mention [1] where the number of states of the minimal automaton accepting numbers written in base b and divisible by d is given explicitly.

 The question under inspection in this paper was raised by J. Sakarovitch during the "*Journées de Numération*" in Graz, May 2007. The question was initially asked for a larger class of systems that the one treated here, namely for any abstract numeration systems defined on an infinite regular language [9].

 The structure of this paper is the same as [8]. First we give an upper bound on the admissible periods of a U-recognizable set X when it is assumed to be ultimately periodic, then an upper bound on the admissible preperiods is obtained. These bounds depend essentially on the number of states of the (minimal) automaton recognizing $\text{rep}_U(X)$. Finally, finitely many such periods and preperiods have to be checked. Even if the structure is the same, our arguments and techniques are quite different from [8]. Actually they cannot be applied to integer base systems (see Remark 5).

In the next section, Theorem 1 gives a decision procedure for Problem 1 whenever U is a linear numeration system such that $\mathbb{N}$ is U-recognizable and satisfying a relation like (1) with $a_k = \pm 1$ (the main reason for this assumption is that 1 and -1 are the only two integers invertible modulo n for all $n \geq 2$). In the last section, we consider the same decision problem but restated in the framework of abstract numeration systems [9]. We apply successfully the same kind of techniques to a large class of abstract numeration systems (for instance, an example consisting of two copies of the Fibonacci system is considered). The corresponding decision procedure is given by Theorem 2. All along the paper, we try whenever it is possible to state results in their most general form, even if later on we have to restrict ourselves to particular cases. For instance, results about the admissible preperiods do not require any extra assumption.

2   Decision Procedure for Linear Systems with $a_k = \pm 1$

We will often consider positional numeration systems $U = (U_i)_{i \geq 0}$ satisfying the following condition:

$$\lim_{i \to +\infty} U_{i+1} - U_i = +\infty. \tag{2}$$

Lemma 1. *Let $U = (U_i)_{i \geq 0}$ be a positional numeration system satisfying (2). Then for all j, there exists L such that for all $\ell \geq L$,*

$$10^{\ell - |\operatorname{rep}_U(t)|} \operatorname{rep}_U(t), \ t = 0, \dots, U_j - 1$$

are greedy U-representations. Otherwise stated, if w is a greedy U-representation, then for r large enough, $10^r w$ is also a greedy U-representation.

Proof. Notice that $\operatorname{rep}_U(U_j - 1)$ is the greatest word of length j in $\operatorname{rep}_U(\mathbb{N})$, since $\operatorname{rep}_U(U_j) = 10^j$. By hypothesis, there exists L such that for all $\ell \geq L$, $U_{\ell+1} - U_\ell > U_j - 1$. Therefore, for all $\ell \geq L$,

$$10^{\ell - j} \operatorname{rep}_U(U_j - 1)$$

is the greedy U-representation of $U_\ell + U_j - 1 < U_{\ell+1}$ and the conclusion follows.

Remark 2. Bertrand numeration systems associated with a real number $\beta > 1$ are defined as follows. Let $A_\beta = \{0, \dots, \lceil \beta \rceil - 1\}$. Any $x \in [0, 1]$ can be written as

$$x = \sum_{i=1}^{+\infty} c_i \beta^{-i}, \ \text{with } c_i \in A_\beta$$

and the sequence $(c_i)_{i \geq 1}$ is said to be a β-representation of x. The maximal β-representation of x for the lexicographical order is denoted $d_\beta(x)$ and is called the

β-development of x (for details see [10, Chap. 8]). We say that a β-development $(c_i)_{i\geq 1}$ is *finite* if there exists N such that $c_i = 0$ for all $i \geq N$. If there exists $m \geq 1$ such that $d_\beta(1) = t_1 \cdots t_m$ with $t_m \neq 0$, we set $d_\beta^*(1) := (t_1 \cdots t_{m-1}(t_m - 1))^\omega$, otherwise $d_\beta(1)$ is infinite and we set $d_\beta^*(1) := d_\beta(1)$.

We can now define a positional numeration system $U_\beta = (U_i)_{i\geq 0}$ associated with β (see [2]). If $d_\beta^*(1) = (t_i)_{i\geq 1}$, then

$$U_0 = 1 \text{ and } \forall i \geq 1, \ U_i = t_1 U_{i-1} + \cdots + t_i U_0 + 1. \tag{3}$$

If β is a Parry number (i.e., $d_\beta(1)$ is finite or ultimately periodic) then $d_\beta^*(1)$ is ultimately periodic and one can derive from (3) that the sequence U_β satisfies a linear recurrence relation and as a consequence of Bertrand's theorem [2] linking greedy U_β-representations and finite factors occurring in β-developments, the language $\mathrm{rep}_{U_\beta}(\mathbb{N})$ of the greedy U_β-representations is regular. The automaton accepting these representations is well-known [6] and has a special form (all states — except for a sink — are final and from all these states, an edge of label 0 goes back to the initial state). We therefore have the following property being much stronger than the previous lemma. If x and y are greedy U_β-representations then $x0y$ is also a greedy U_β-representation.

Example 2. The Fibonacci system is the Bertrand system associated with the golden ratio $(1 + \sqrt{5})/2$. Since greedy representations in the Fibonacci system are the words not containing two consecutive ones [13], then for $x, y \in \mathrm{rep}_F(\mathbb{N})$, we have $x0y \in \mathrm{rep}_F(\mathbb{N})$.

Definition 3. *Let $X \subseteq \mathbb{N}$ be a set of integers. The* characteristic word *of X is an infinite word $x_0 x_1 x_2 \cdots$ over $\{0, 1\}$ defined by $x_i = 1$ if and only if $i \in X$.*

Consider for now $X \subseteq \mathbb{N}$ to be an ultimately periodic set. The characteristic word of X is therefore an infinite word over $\{0, 1\}$ of the form

$$x_0 x_1 x_2 \cdots = uv^\omega$$

where u and v are chosen of minimal length. We say that the length $|u|$ of u (resp. the length $|v|$ of v) is the preperiod *(resp. period) of X. Hence, for all $n \geq |u|$, $n \in X$ if and only if $n + |v| \in X$.*

The following lemma is a simple consequence of the minimality of the period chosen to represent an ultimately periodic set.

Lemma 2. *Let $X \subseteq \mathbb{N}$ be an ultimately periodic set of period p_X and preperiod a_X. Let $i, j \geq a_X$. If $i \not\equiv j$ mod p_X then there exists $t < p_X$ such that either $i + t \in X$ and $j + t \notin X$ or $i + t \notin X$ and $j + t \in X$.*

We assume that the reader is familiar with automata theory (see for instance [11]) but let us recall some classical results. Let $L \subseteq \Sigma^*$ be a language over a finite alphabet Σ and x be a finite word over Σ. We set

$$x^{-1}L = \{z \in \Sigma^* \mid xz \in L\}.$$

We can now define the Myhill-Nerode congruence. Let $x, y \in \Sigma^*$. We have $x \sim_L y$ if and only if $x^{-1}L = y^{-1}L$. Moreover L is regular if and only if $\sim_L$ has a finite index being the number of states of the minimal automaton of L.

For a sequence $(U_i)_{i \geq 0}$ of integers, $N_U(m) \in \{1, \ldots, m\}$ denotes the number of values that are taken infinitely often by the sequence $(U_i \bmod m)_{i \geq 0}$.

Proposition 1. *Let* $U = (U_i)_{i \geq 0}$ *be a positional numeration system satisfying* (2). *If* $X \subseteq \mathbb{N}$ *is an ultimately periodic* U-*recognizable set of period* p_X, *then any deterministic finite automaton accepting* $\mathrm{rep}_U(X)$ *has at least* $N_U(p_X)$ *states.*

Proof. Let a_X be the preperiod of X. By Lemma 1, there exists L such that for any $h \geq L$, the words

$$10^{h - |\mathrm{rep}_U(t)|} \, \mathrm{rep}_U(t), \; t = 0, \ldots, p_X - 1$$

are greedy U-representations. The sequence $(U_i \bmod p_X)_{i \geq 0}$ takes infinitely often $N_U(p_X) =: N$ different values. Let $h_1, \ldots, h_N \geq L$ be such that

$$i \neq j \Rightarrow U_{h_i} \not\equiv U_{h_j} \bmod p_X$$

and $h_1, \ldots, h_N$ can be chosen such that $U_{h_i} > a_X$ for all $i \in \{1, \ldots, N\}$.

By Lemma 2, for all $i, j \in \{1, \ldots, N\}$ such that $i \neq j$, there exists $t_{i,j} < p_X$ such that either $U_{h_i} + t_{i,j} \in X$ and $U_{h_j} + t_{i,j} \notin X$, or $U_{h_i} + t_{i,j} \notin X$ and $U_{h_j} + t_{i,j} \in X$. Therefore,

$$w_{i,j} = 0^{|\mathrm{rep}_U(p_X - 1)| - |\mathrm{rep}_U(t_{i,j})|} \, \mathrm{rep}_U(t_{i,j})$$

is a word such that either

$$10^{h_i - |\mathrm{rep}_U(p_X - 1)|} w_{i,j} \in \mathrm{rep}_U(X) \text{ and } 10^{h_j - |\mathrm{rep}_U(p_X - 1)|} w_{i,j} \notin \mathrm{rep}_U(X),$$

or

$$10^{h_i - |\mathrm{rep}_U(p_X - 1)|} w_{i,j} \notin \mathrm{rep}_U(X) \text{ and } 10^{h_j - |\mathrm{rep}_U(p_X - 1)|} w_{i,j} \in \mathrm{rep}_U(X).$$

Therefore the words $10^{h_1 - |\mathrm{rep}_U(p_X - 1)|}, \ldots, 10^{h_N - |\mathrm{rep}_U(p_X - 1)|}$ are pairwise non-equivalent for the relation $\sim_{\mathrm{rep}_U(X)}$ and the minimal automaton of $\mathrm{rep}_U(X)$ has at least $N = N_U(p_X)$ states.

The previous proposition has an immediate consequence.

Corollary 1. *Let* $U = (U_i)_{i \geq 0}$ *be a positional numeration system satisfying* (2). *Assume that*

$$\lim_{m \to +\infty} N_U(m) = +\infty.$$

Then the period of an ultimately periodic set $X \subseteq \mathbb{N}$ *such that* $\mathrm{rep}_U(X)$ *is accepted by a DFA with* d *states is bounded by the smallest integer* s_0 *such that for all* $m \geq s_0$, $N_U(m) > d$.

For a sequence $(U_i)_{i \geq 0}$ of integers, if $(U_i \mod m)_{i \geq 0}$ is ultimately periodic, we denote its (minimal) preperiod by $\iota_U(m)$ (we choose notation ι to remind the word index which is equally used as preperiod) and its (minimal) period by $\pi_U(m)$. The next lemma provides a special case where assumption about $N_U(m)$ in Corollary 1 is satisfied.

Lemma 3. *If $U = (U_i)_{i \geq 0}$ is a linear numeration system satisfying a recurrence relation of order k of the kind (1) with $a_k = \pm 1$, then $\lim_{m \to +\infty} N_U(m) = +\infty$.*

Proof. For all $m \geq 2$, since U is a linear numeration system, the sequence $(U_i \mod m)_{i \geq 0}$ is ultimately periodic but here it is even purely periodic. Indeed, for all $i \geq 0$, U_{i+k} is determined by the k previous terms $U_{i+k-1}, \ldots, U_i$. But since $a_k = \pm 1$, for all $i \geq 0$, U_i is also determined by the k following terms $U_{i+1}, \ldots, U_{i+k}$. So, by definition of $N_U(m)$, the sequence $(U_i \mod m)_{i \geq 0}$ takes exactly $N_U(m)$ different values because any term appears infinitely often.

Since U is increasing, the function α mapping m onto the smallest index $\alpha(m)$ such that $U_{\alpha(m)} \geq m$ is nondecreasing and $\lim_{m \to +\infty} \alpha(m) = +\infty$. The conclusion follows, as $N_U(m) \geq \alpha(m)$. Indeed, $U_0, \ldots, U_{\alpha(m)-1}$ are distinct. So $(U_i \mod m)_{i \geq 0}$ takes infinitely often at least $\alpha(m)$ values.

Remark 3. Let $U = (U_i)_{i \geq 0}$ be a positional numeration system satisfying hypothesis of Lemma 3 and let X be a U-recognizable set of integers. If $\mathrm{rep}_U(X)$ is accepted by a DFA with d states, then the constant s_0 (depending on d) given in the statement of Corollary 1 can be estimated as follows.

By Lemma 3, $\lim_{m \to +\infty} N_U(m) = +\infty$. Define t_0 to be the smallest integer such that $\alpha(t_0) > d$, where α is defined as in the proof of Lemma 3. This integer can be effectively computed by considering the first terms of the linear sequence $(U_i)_{i \geq 0}$. Notice that $N_U(t_0) \geq \alpha(t_0) > d$. Consequently $s_0 \leq t_0$.

Moreover, if U satisfies (2) and if X is an ultimately periodic set, then, by Corollary 1, the period of X is bounded by t_0. So t_0 can be used as an upper bound for the period and it can be effectively computed.

A result similar to the previous corollary (in the sense that it permits to give an upper bound on the period) can be stated as follows. One has to notice that $a_k = \pm 1$ implies that 1 occurs infinitely often in $(U_i \mod m)_{i \geq 0}$ for all $m \geq 2$.

Proposition 2. *Let $U = (U_i)_{i \geq 0}$ be a positional numeration system satisfying (2) and $X \subseteq \mathbb{N}$ be an ultimately periodic U-recognizable set of period p_X. If 1 occurs infinitely many times in $(U_i \mod p_X)_{i \geq 0}$ then any deterministic finite automaton accepting $\mathrm{rep}_U(X)$ has at least p_X states.*

Proof. Let a_X be the preperiod of X. Applying several times Lemma 1, there exist $n_1, \ldots, n_{p_X}$ such that

$$10^{n_{p_X}} 10^{n_{p_X}-1} \cdots 10^{n_1} 0^{|\mathrm{rep}_U(p_X-1)| - |\mathrm{rep}_U(t)|} \mathrm{rep}_U(t), \quad t = 0, \ldots, p_X - 1$$

are greedy U-representations. Moreover, since 1 occurs infinitely many times in the sequence $(U_i \mod p_X)_{i \geq 0}$, $n_1, \ldots, n_{p_X}$ can be chosen such that, for all $j = 1, \ldots, p_X$,

$$\mathrm{val}_U(10^{n_j}\cdots 10^{n_1+|\,\mathrm{rep}_U(p_X-1)|}) \equiv j \mod p_X$$

and

$$\mathrm{val}_U(10^{n_1+|\,\mathrm{rep}_U(p_X-1)|}) > a_X.$$

For $i, j \in \{1, \ldots, p_X\}$, $i \neq j$, by Lemma 2 the words

$$10^{n_i}\cdots 10^{n_1} \text{ and } 10^{n_j}\cdots 10^{n_1}$$

are nonequivalent for $\sim_{\mathrm{rep}_U(X)}$. This can be shown by concatenating some word of the kind $0^{|\,\mathrm{rep}_U(p_X-1)|-|\,\mathrm{rep}_U(t)|}\,\mathrm{rep}_U(t)$ with $t < p_X$, as in the proof of Proposition 1. This concludes the proof.

Now we want to obtain an upper bound on the preperiod of any ultimately periodic U-recognizable set.

Proposition 3. *Let $U = (U_i)_{i\geq 0}$ be a linear numeration system. Let $X \subseteq \mathbb{N}$ be an ultimately periodic U-recognizable set of period p_X and preperiod a_X. Then any deterministic finite automaton accepting $\mathrm{rep}_U(X)$ has at least $|\,\mathrm{rep}_U(a_X - 1)| - \iota_U(p_X)$ states.*

The arguments of the following proof are similar to the one found in [8].

Proof. W.l.o.g. we can assume that $|\,\mathrm{rep}_U(a_X - 1)| - \iota_U(p_X) > 0$. The sequence $(U_i \mod p_X)_{i\geq 0}$ is ultimately periodic with preperiod $\iota_U(p_X)$ and period $\pi_U(p_X)$. Proceed by contradiction and assume that $\mathcal{A}$ is a deterministic finite automaton with less than $|\,\mathrm{rep}_U(a_X - 1)| - \iota_U(p_X)$ states accepting $\mathrm{rep}_U(X)$. There exist words w, w_4 such that the greedy U-representation of $a_X - 1$ can be factorized as

$$\mathrm{rep}_U(a_X - 1) = ww_4$$

with $|w| = |\,\mathrm{rep}_U(a_X - 1)| - \iota_U(p_X)$. By the pumping lemma, w can be written $w_1 w_2 w_3$ with $w_2 \neq \varepsilon$ and for all $i \geq 0$,

$$w_1 w_2^i w_3 w_4 \in \mathrm{rep}_U(X) \Leftrightarrow w_1 w_2 w_3 w_4 \in \mathrm{rep}_U(X).$$

By minimality of a_X and p_X, either $a_X - 1 \in X$ and for all $n \geq 1$, $a_X + np_X - 1 \notin X$, or $a_X - 1 \notin X$ and for all $n \geq 1$, $a_X + np_X - 1 \in X$. Using the ultimate periodicity of $(U_i \mod p_X)_{i\geq 0}$, we observe that repeating a factor of length multiple of $\pi_U(p_X)$ exactly p_X times does not change the value mod p_X and we get

$$\mathrm{val}_U(w_1 w_2^{p_X \pi_U(p_X)} w_2 w_3 w_4) \equiv \mathrm{val}_U(w_1 w_2 w_3 w_4) \mod p_X,$$

leading to a contradiction.

For the sake of completeness, we restate some well-known property of ultimately periodic sets (see for instance [11] for a prologue on the Pascal's machine for integer base systems).

Lemma 4. *Let a, b be nonnegative integers and $U = (U_i)_{i \geq 0}$ be a linear numeration system. The language*

$$\mathrm{val}_U^{-1}(a\mathbb{N} + b) = \{w \in A_U^* \mid \mathrm{val}_U(w) \in a\mathbb{N} + b\} \subset A_U^*$$

is regular. In particular, if $\mathbb{N}$ is U-recognizable then a DFA accepting $\mathrm{rep}_U(a\mathbb{N}+b)$ can be obtained efficiently and any ultimately periodic set is U-recognizable.

Remark 4. In the previous statement, the assumption about the U-recognizability of $\mathbb{N}$ is of particular interest. Indeed, it is well-known that for an arbitrary linear numeration system, $\mathbb{N}$ is in general *not* U-recognizable. If $\mathbb{N}$ is U-recognizable, then U satisfies a linear recurrence relation [12], but the converse does not hold. Sufficient conditions on the recurrence relation that U satisfies for $\mathbb{N}$ to be U-recognizable are given in [7].

Theorem 1. *Let $U = (U_i)_{i \geq 0}$ be a linear numeration system such that $\mathbb{N}$ is U-recognizable and satisfying a recurrence relation of order k of the kind (1) with $a_k = \pm 1$ and condition (2). It is decidable whether or not a U-recognizable set is ultimately periodic.*

Proof. Let X be a U-recognizable set and d be the number of states of the minimal automaton of $\mathrm{rep}_U(X)$.

As discussed in Remark 3, if X is ultimately periodic, then the admissible periods are bounded by the constant t_0, which is effectively computable (an alternative and easier argument is provided by Proposition 2). Then, using Proposition 3, the admissible preperiods are also bounded by a constant. Indeed, assume that X is ultimately periodic with period $p_X \leq t_0$ and preperiod a_X. We have $\iota_U(p_X) = 0$ and any DFA accepting $\mathrm{rep}_U(X)$ must have at least $|\mathrm{rep}_U(a_X - 1)|$ states. Therefore, the only values that a_X can take satisfy $|\mathrm{rep}_U(a_X - 1)| \leq d$.

Consequently the sets of admissible preperiods and periods that we have to check are finite. For each pair (a, p) of admissible preperiods and periods, there are at most $2^a 2^p$ distinct ultimately periodic sets. Thanks to Lemma 4, one can build an automaton for each of them and then compare the language L accepted by this automaton with $\mathrm{rep}_U(X)$. (Recall that testing whether $L \setminus \mathrm{rep}_U(X) = \emptyset$ and $\mathrm{rep}_U(X) \setminus L = \emptyset$ is decidable algorithmically). $\square$

Remark 5. We have thus obtained a decision procedure for our Problem 1 when the coefficient a_k occurring in (1) is equal to ± 1. On the other hand, whenever $\gcd(a_1, \ldots, a_k) = g \geq 2$, for all $n \geq 1$ and for all i large enough, we have $U_i \equiv 0 \bmod g^n$ and assumption about $N_U(m)$ in Corollary 1 does not hold [5]. Indeed, the only value taken infinitely often by the sequence $(U_i \bmod g^n)_{i \geq 0}$ is 0, so $N_U(m)$ equals 1 for infinitely many values of m. Notice in particular, that the same observation can be made for the usual integer base $b \geq 2$ numeration system where the only value taken infinitely often by the sequence $(b^i \bmod b^n)_{i \geq 0}$ is 0, for all $n \geq 1$.

3   A Decision Procedure for a Class of Abstract Numeration Systems

An *abstract numeration system* $S = (L, \Sigma, <)$ is given by an infinite regular language L over a totally ordered alphabet $(\Sigma, <)$ [9]. By enumerating the words of L in genealogical order, we get a one-to-one correspondence denoted rep_S between $\mathbb{N}$ and L. In particular, 0 is represented by the first word in L. The reciprocal map associating a word $w \in L$ to its index in the genealogically ordered language L is denoted val_S. A set $X \subseteq \mathbb{N}$ of integers is S-*recognizable* if the language $\text{rep}_S(X)$ over Σ is regular (i.e., accepted by a finite automaton).

Let $S = (L, \Sigma, <)$ be an abstract numeration system built over an infinite regular language L having $\mathcal{M}_L = (Q_L, q_{0,L}, \Sigma, \delta_L, F_L)$ as minimal automaton. The transition function $\delta_L : Q_L \times \Sigma \to Q_L$ is extended on $Q_L \times \Sigma^*$. We denote by $\mathbf{u}_j(q)$ (resp. $\mathbf{v}_j(q)$) the number of words of length j (resp. $\leq j$) accepted from $q \in Q_L$ in $\mathcal{M}_L$. By classical arguments, the sequences $(\mathbf{u}_j(q))_{j \geq 0}$ (resp. $(\mathbf{v}_j(q))_{j \geq 0}$) satisfy the same homogenous linear recurrence relation for all $q \in Q_L$ (for details, see Remark 6).

In this section, we consider, with some extra hypothesis on the abstract numeration system, the following decidability question analogous to Problem 1.

Problem 2. Given an abstract numeration system S and a set $X \subseteq \mathbb{N}$ such that $\text{rep}_S(X)$ is recognized by a (deterministic) finite automaton, is it decidable whether or not X is ultimately periodic, i.e., whether or not X is a finite union of arithmetic progressions ?

Abstract numeration systems are a generalization of positional numeration systems $U = (U_i)_{i \geq 0}$ for which $\mathbb{N}$ is U-recognizable.

Example 3. Take the language $L = \{\varepsilon\} \cup 1\{0, 01\}^*$ and assume $0 < 1$. Ordering the words of L in genealogical order: $\varepsilon, 1, 10, 100, 101, 1000, 1001, \ldots$ gives back the Fibonacci system.

Example 4. Consider the language $L = \{\varepsilon\} \cup \{a, ab\}^* \cup \{c, cd\}^*$ and the ordering $a < b < c < d$ of the alphabet. If we order the first words in L we get

0	ε	5	cc	10	ccc	15	$aaba$	20	$ccdc$
1	a	6	cd	11	ccd	16	$abaa$	21	$cdcc$
2	c	7	aaa	12	cdc	17	$abab$	22	$cdcd$
3	aa	8	aab	13	$aaaa$	18	$cccc$	23	$aaaaa$
4	ab	9	aba	14	$aaab$	19	$cccd$	24	$aaaab$

Notice that there is no bijection between $\{a, b, c, d\}$ and a set of integers leading to a positional linear numeration system. Otherwise stated, a, b, c, d cannot be identified with usual "digits". For all $n \geq 1$, we have $\mathbf{u}_n(q_{0,L}) = 2F_n$ and $\mathbf{u}_0(q_{0,L}) = 1$. Consequently, for $n \geq 1$,

$$\mathbf{v}_n(q_{0,L}) = 1 + \sum_{i=1}^{n} \mathbf{u}_i(q_{0,L}) = 1 + 2\sum_{i=1}^{n} F_i.$$

Fig. 1. A DFA accepting L

Notice that for $n \geq 1$, $\mathbf{v}_n(q_{0,L}) - \mathbf{v}_{n-1}(q_{0,L}) = \mathbf{u}_n(q_{0,L}) = 2F_n$. Consequently, by definition of the Fibonacci sequence, we get for all $n \geq 3$,

$$\mathbf{v}_n(q_{0,L}) - \mathbf{v}_{n-1}(q_{0,L}) = (\mathbf{v}_{n-1}(q_{0,L}) - \mathbf{v}_{n-2}(q_{0,L})) + (\mathbf{v}_{n-2}(q_{0,L}) - \mathbf{v}_{n-3}(q_{0,L}))$$

and $\mathbf{v}_n(q_{0,L}) = 2\mathbf{v}_{n-1}(q_{0,L}) - \mathbf{v}_{n-3}(q_{0,L})$, with $\mathbf{v}_0(q_{0,L}) = 1$, $\mathbf{v}_1(q_{0,L}) = 3$, $\mathbf{v}_2(q_{0,L}) = 7$.

Remark 6. The computation given in the previous example to obtain a homogenous linear recurrence relation for the sequence $(\mathbf{v}_j(q_{0,L}))_{j\geq 0}$ can be carried on in general. Let $q \in Q_L$. The sequence $(\mathbf{u}_j(q))_{j\geq 0}$ satisfies a homogenous linear recurrence relation of order t whose characteristic polynomial is the characteristic polynomial of the adjacency matrix of $\mathcal{M}_L$. There exist $a_1, \ldots, a_t \in \mathbb{Z}$ such that for all $j \geq 0$, $\mathbf{u}_{j+t}(q) = a_1\mathbf{u}_{j+t-1}(q) + \cdots + a_t\mathbf{u}_j(q)$. Consequently, we have for all $j \geq 0$, $\mathbf{v}_{j+t+1}(q) - \mathbf{v}_{j+t}(q) = \mathbf{u}_{j+t+1}(q) = a_1(\mathbf{v}_{j+t}(q) - \mathbf{v}_{j+t-1}(q)) + \cdots + a_t(\mathbf{v}_{j+1}(q) - \mathbf{v}_j(q))$. Therefore the sequence $(\mathbf{v}_j(q))_{j\geq 0}$ satisfies a homogenous linear recurrence relation of order $t+1$.

As shown by the following lemma, in an abstract numeration system, the different sequences $(\mathbf{u}_j(q))_{j\geq 0}$, for $q \in Q_L$, are replacing the single sequence $(U_j)_{j\geq 0}$ defining a positional numeration system as in Definition 1.

Lemma 5. *[9] Let $w = \sigma_1 \cdots \sigma_n \in L$. We have*

$$\mathrm{val}_S(w) = \sum_{q \in Q_L} \sum_{i=1}^{|w|} \beta_{q,i}(w)\, \mathbf{u}_{|w|-i}(q) \tag{4}$$

where
$$\beta_{q,i}(w) := \#\{\sigma < \sigma_i \mid \delta_L(q_{0,L}, \sigma_1 \cdots \sigma_{i-1}\sigma) = q\} + \mathbf{1}_{q,q_{0,L}} \tag{5}$$

for $i = 1, \ldots, |w|$.

Recall that $\mathbf{1}_{q,q'}$ is equal to 1 if $q = q'$ and it is equal to 0 otherwise.

Proposition 4. *[9] Let $S = (L, \Sigma, <)$ be an abstract numeration system built over an infinite regular language L over Σ. Any ultimately periodic set X is S-recognizable and a DFA accepting $\mathrm{rep}_S(X)$ can be effectively obtained.*

Recall that an automaton is *trim* if it is accessible and coaccessible (each state can be reached from the initial state and from each state, one can reach a final state).

Proposition 5. *Let $S = (L, \Sigma, <)$ be an abstract numeration system such that for all states q of the trim minimal automaton $\mathcal{M}_L = (Q_L, q_{0,L}, \Sigma, \delta_L, F_L)$ of L,*

$$\lim_{j \to +\infty} \mathbf{u}_j(q) = +\infty$$

and $\mathbf{u}_j(q_{0,L}) > 0$ for all $j \geq 0$. If $X \subseteq \mathbb{N}$ is an ultimately periodic set of period p_X, then any deterministic finite automaton accepting $\mathrm{rep}_S(X)$ has at least $\lceil N_{\mathbf{v}}(p_X)/\#Q_L \rceil$ states where $\mathbf{v} = (\mathbf{v}_j(q_{0,L}))_{j \geq 0}$.

To prove this result one has to adapt the arguments given in the proof of Proposition 1 to the framework of abstract numeration systems.

Corollary 2. *Let $S = (L, \Sigma, <)$ be an abstract numeration system having the same properties as in Proposition 5. Assume that the sequence $\mathbf{v} = (\mathbf{v}_j(q_{0,L}))_{j \geq 0}$ is such that*

$$\lim_{m \to +\infty} N_{\mathbf{v}}(m) = +\infty.$$

Then the period of an ultimately periodic set $X \subseteq \mathbb{N}$ such that $\mathrm{rep}_S(X)$ is accepted by a DFA with d states is bounded by the smallest integer s_0 such that for all $m \geq s_0$, $N_{\mathbf{v}}(m) > d \#Q_L$, where Q_L is the set of states of the (trim) minimal automaton of L.

Proposition 6. *Let $S = (L, \Sigma, <)$ be an abstract numeration system. If $X \subseteq \mathbb{N}$ is an ultimately periodic set of period p_X such that $\mathrm{rep}_S(X)$ is accepted by a DFA with d states, then the preperiod a_X of X is bounded by a constant C depending only on d and p_X.*

To prove this result one has to adapt the arguments given in the proof of Proposition 3 to the framework of abstract numeration systems.

Remark 7. The constant C of the previous result can be effectively computed. Using notation of the previous proof, one has to choose a constant C such that $a_X > C$ implies $|\mathrm{rep}_S(a_X - 1)| - d \#Q_L > I(p_X)$. Since the abstract numeration system S, the period p_X and the number d of states are given, $I(p_X)$ and $\mathrm{rep}_S(n)$ for all $n \geq 0$ can be effectively computed.

Theorem 2. *Let $S = (L, \Sigma, <)$ be an abstract numeration system such that for all states q of the trim minimal automaton $\mathcal{M}_L = (Q_L, q_{0,L}, \Sigma, \delta_L, F_L)$ of L*

$$\lim_{j \to \infty} \mathbf{u}_j(q) = +\infty$$

and $\mathbf{u}_j(q_{0,L}) > 0$ for all $j \geq 0$. Assume moreover that $\mathbf{v} = (\mathbf{v}_i(q_{0,L}))_{i \geq 0}$ satisfies a linear recurrence relation of the form (1) with $a_k = \pm 1$. It is decidable whether or not a S-recognizable set is ultimately periodic.

Proof. The proof is essentially the same as the one of Theorem 1. Let X be a S-recognizable set and d be the number of states of the minimal automaton of $\mathrm{rep}_S(X)$. With the same reasoning as in the proof of Lemma 3, $\lim_{m \to +\infty}$

$N_\mathbf{v}(m) = +\infty$. If X is ultimately periodic, then its period is bounded by a constant t_0 that can be effectively estimated.

If X is ultimately periodic with period $p_X \leq t_0$, then using Proposition 6, its preperiod is bounded by a constant (which can also be computed effectively thanks to Remark 7).

Consequently, the sets of admissible periods and preperiods we have to check are finite. Thanks to Proposition 4, one has to build an automaton for each ultimately periodic set corresponding to a pair of admissible preperiods and periods and then compare the accepted language with $\text{rep}_S(X)$.

Example 5. The abstract numeration system given in Example 4 satisfies all the assumptions of the previous theorem.

Acknowledgments

We warmly thank Jacques Sakarovitch for casting a new light on Juha Honkala's paper and also Aviezri Fraenkel for fruitful electronic discussions about linear numeration systems that allowed us to reconsider reasonable hypothesis to tackle this problem.

References

1. Alexeev, B.: Minimal DFAS for testing divisibility. J. Comput. Syst. Sci. 69, 235–243 (2004)
2. Bertrand-Mathis, A.: Comment écrire les nombres entiers dans une base qui n'est pas entière. Acta Math. Acad. Sci. Hungar. 54, 237–241 (1989)
3. Cobham, A.: On the base-dependence of sets of numbers recognizable by finite automata. Math. Systems Theory 3, 186–192 (1969)
4. Durand, F.: A theorem of Cobham for non primitive substitution. Acta Arithmetica 104, 225–241 (2002)
5. Fraenkel, A.: Personal communication (February 2008)
6. Frougny, C., Solomyak, B.: On the representation of integers in linear numeration systems. In: Ergodic theory of Z_d actions (Warwick, 1993–1994). London Math. Soc. Lecture Note Ser, vol. 228, pp. 345–368. Cambridge Univ. Press, Cambridge (1996)
7. Hollander, M.: Greedy numeration systems and regularity. Theory Comput. Syst. 31, 111–133 (1998)
8. Honkala, J.: A decision method for the recognizability of sets defined by number systems. Theoret. Inform. Appl. 20, 395–403 (1986)
9. Lecomte, P.B.A., Rigo, M.: Numeration systems on a regular language. Theory Comput. Syst. 34, 27–44 (2001)
10. Lothaire, M.: Algebraic Combinatorics on Words. Encyclopedia of Mathematics and its Applications, vol. 90. Cambridge University Press, Cambridge (2002)
11. Sakarovitch, J.: Éléments de théorie des automates, Vuibert. English translation: Elements of Automata Theory. Cambridge University Press, Cambridge, Vuibert (2003)
12. Shallit, J.: Numeration systems, linear recurrences and regular sets. Inform. and Comput. 113, 331–347 (1994)
13. Zeckendorf, E.: Representation des nombres naturels par une somme des nombres de Fibonacci ou de nombres de Lucas. Bull. Soc. Roy. Sci. Liege 41, 179–182 (1972)

Regional Languages and Tiling: A Unifying Approach to Picture Grammars*

Alessandra Cherubini[1], Stefano Crespi Reghizzi[1], and Matteo Pradella[2]

[1] Politecnico di Milano
[2] CNR IEIIT-MI
P.zza L. da Vinci, 32, 20133 Milano, Italy
{alessandra.cherubini,stefano.crespireghizzi,
matteo.pradella}@polimi.it

Abstract. Several classical models of picture grammars based on array rewriting rules can be unified and extended by a tiling based approach. The right part of a rewriting rule is formalized by a finite set of permitted tiles. We focus on a simple type of tiling, named *regional*, and define the corresponding regional tile grammars. They include both Siromoney's (or Matz's) Kolam grammars, and their generalization by Průša. Regionally defined pictures can be recognized with polynomial time complexity by an algorithm extending the CKY one for strings. Regional tile grammars and languages are strictly included into the tile grammars and languages, and are incomparable with Giammarresi-Restivo tiling systems (or Wang's tilings).

Keywords: Picture language, tiling, picture grammar, 2D language, CKY algorithm, syntactic pattern recognition.

1 Introduction

Several classical models of picture grammars based on array rewriting rules can be unified by a tiling based approach. The right part of a rewriting rule can be specified by a finite set of permitted two by two tiles. We focus on a simple type of tiling, named regional and define the corresponding regional tile grammars. The new class generalizes some classical models, yet it permits efficient, polynomial-time recognition of pictures by an approach extending the classical CKY algorithm [12] of context-free (CF) string languages.

Regional tile grammars can be viewed from the standpoint of less, or of more powerful models. First, regional tile grammars are a generalization of the classical Kolam grammars of Siromoney [11] (which are equivalent to the grammars of Matz [7]), where the right parts of grammar rules are tiled in ways than cannot be obtained by 2D regular expressions.

From the standpoint of more powerful grammar models, regional tile grammars correspond to a natural restriction of the recently introduced tile (rewriting) grammars (TG). Such grammars have rewriting rules that replace a homogeneous non-terminal rectangular area with a picture belonging to a local language defined by tiles. It is known

---

* Work Partially Supported by ESF *Automata: from Mathematics to Applications (AutoMathA).*

E. Ochmański and J. Tyszkiewicz (Eds.): MFCS 2008, LNCS 5162, pp. 253–264, 2008.

that the TG family dominates the family of languages defined by the Tiling Systems of Giammarresi and Restivo [4] (which are equivalent to Wang's tilings [1]), and that the latter are NP-complete with respect to picture recognition time complexity.

The new model can be conveniently defined starting from TG grammars, by imposing the constraint that the local language used in a rule is made by assembling a finite number of homogeneous rectangular pictures. Such tiling is related to Simplot's [9] interesting closure operation on pictures.

The presentation continues in Sect. 2 with preliminary definitions, and in Sect. 3 with the definition of regional tile grammars and relevant examples. In Sect. 4 we present the parsing algorithm and prove its correctness and complexity. In Sect. 5 we compare regional tile grammars and languages with other picture language families.

2   Basic Definitions and Regional Local Languages

The following notation and definitions are mostly from [5] and [2].

Definition 1. *Let Σ be a finite alphabet. A two-dimensional array of elements of Σ is a picture over Σ. The set of all pictures over Σ is Σ^{++}. A picture language is a subset of Σ^{++}.*

For $h, k \geq 1$, $\Sigma^{(h,k)}$ denotes the set of pictures of size (h, k) (we will use the notation $|p| = (h, k), |p|_{row} = h, |p|_{col} = k$). $\# \notin \Sigma$ is used when needed as a boundary symbol; $\hat{p}$ refers to the bordered version of picture p. That is, for $p \in \Sigma^{(h,k)}$, it is

$$p = \begin{matrix} p(1,1) & \dots & p(1,k) \\ \vdots & \ddots & \vdots \\ p(h,1) & \dots & p(h,k) \end{matrix} \qquad \hat{p} = \begin{matrix} \# & \# & \dots & \# & \# \\ \# & p(1,1) & \dots & p(1,k) & \# \\ \vdots & \vdots & \ddots & \vdots & \vdots \\ \# & p(h,1) & \dots & p(h,k) & \# \\ \# & \# & \dots & \# & \# \end{matrix}$$

A pixel is an element $p(i, j)$ of p. If all pixels are identical to $C \in \Sigma$ the picture is called C-homogeneous or C-picture.

The domain of a picture p is the set $dom(p) = \{1, \dots, |p|_{row}\} \times \{1, \dots, |p|_{col}\}$.

Row and column concatenations are denoted $\ominus$ and $\oplus$, respectively. $p \ominus q$ is defined iff p and q have the same number of columns; the resulting picture is the vertical juxtaposition of p over q; $p^{k\ominus}$ is the vertical juxtaposition of k copies of p; $p^{*\ominus}$ is the corresponding closure. $\oplus, {}^{k\oplus}, {}^{*\oplus}$ are the column analogous.

Definition 2. *Let p be a picture over Σ. A subdomain of $dom(p)$ is a set d of the form $\{x, \dots, x'\} \times \{y, \dots, y'\}$ where $1 \leq x \leq x' \leq |p|_{row}, 1 \leq y \leq y' \leq |p|_{col}$. We will often denote a subdomain by using its top-left and bottom-right coordinates, in the previous case the quadruple $(x, y; x', y')$.*

The set of subdomains of p is denoted $D(p)$. Let $d = \{x, \dots, x'\} \times \{y, \dots, y'\} \in D(p)$, the subpicture $spic(p, d)$ associated to d is the picture of size $(x' - x + 1, y' - y + 1)$ such that $\forall i \in \{1, \dots, x' - x + 1\}$ and $\forall j \in \{1, \dots, y' - y + 1\}$ $spic(p, d)(i, j) = p(x + i - 1, y + j - 1)$.

A subdomain is called C-homogeneous when its associated subpicture is a C-picture. C is called the label *of the subdomain.*

Two subdomains $d_a = (i_a, j_a; k_a, l_a)$ and $d_b = (i_b, j_b; k_b, l_b)$ are horizontally adjacent *(resp.* vertically adjacent*) iff $j_b = l_a + 1$, and $k_b \geq i_a, k_a \geq i_b$ (resp. $i_b = k_a + 1$, and $l_b \geq j_a, l_a \geq j_b$).*

The translation *of a subdomain $d = (x, y; x', y')$ by displacement $(a, b) \in \mathbb{Z}^2$ is the subdomain $d = (x + a, y + b; x' + a, y' + b)$.*

We now introduce the central concepts of *regional language, tile,* and *local language.* The adjective "regional" is a metaphor of geographical political maps, such that different regions are filled with different colors.

Definition 3. *A* homogeneous partition *of a picture p is any partition π of* dom(p) *into homogeneous subdomains such that adjacent subdomains have different labels.*

A homogeneous partition is regional *(HR) iff distinct subdomains have distinct labels. We will call a picture p* regional *if it admits a HR partition.*

A language is regional *if all its pictures are so.*

We observe that if a picture p admits a homogeneous partition of dom(p) into subdomains, then the partition is unique and will be denoted by $\Pi(p)$.

Definition 4. *We call* tile *a square picture of size (2,2). We denote by $[\![p]\!]$ the set of all tiles contained in a picture p.*

Let Σ be a finite alphabet. A (two-dimensional) language $L \subseteq \Sigma^{++}$ is local *if there exists a finite set θ of tiles over the alphabet $\Sigma \cup \{\#\}$ such that $L = \{p \in \Sigma^{++} \mid [\![\hat{p}]\!] \subseteq \theta\}$. We will refer to such language as $LOC(\theta)$.*

The right parts of the rules presented in Sect. 3.1 are examples of regional local languages. Next, we characterize the form of tiles occurring in a regional local language.

Consider a tile set θ over the alphabet $\Sigma \cup \{\#\}$. For a tile $t = \begin{pmatrix} x & y \\ z & w \end{pmatrix}$ we define the *horizontal and vertical adjacency relations $\mathcal{H}_t, \mathcal{V}_t \subseteq (\Sigma \cup \{\#\})^2$ as*

$$x\mathcal{H}_t y, \quad z\mathcal{H}_t w, \quad x\mathcal{V}_t z, \quad y\mathcal{V}_t w$$

The adjacency relation is $\mathcal{A}_t = \mathcal{H}_t \cup \mathcal{V}_t$.

The relations can be extended to a tile set θ: $x\mathcal{H}_\theta y$ iff $\exists t \in \theta : x\mathcal{H}_t y$; and similarly for $\mathcal{V}_\theta$ and $\mathcal{A}_\theta$.

Proposition 1. *The local language defined by a tile set θ is regional if*

1. *the (finite) language $\theta \cap \Sigma^{(2,2)}$ is regional, and*
2. *the incidence graph of $(\mathcal{A}_\theta \cap \Sigma^2) \setminus \mathcal{I}$, where $\mathcal{I}$ is the identity relation, is acyclic.*

3   Regional Tile Grammars

We are going to introduce and study a grammar model specified by a set of rewriting rules. A typical rule has a left and a right part, both pictures of unspecified but equal size (isometric). The left part is an A-homogeneous picture, where A is a nonterminal symbol. The right part is a picture of a regional local language over nonterminal symbols. Thus a rule is a scheme defining a possibly unbounded number of isometric pairs: left picture, right picture. In addition there are rules whose right part is a single terminal.

Notice that regional tile grammars may be viewed as extending CF grammars from one to two dimensions: see [2] for the argument that such grammars in one dimension are essentially CF grammars allowing a local regular expression in right parts of rules.

The derivation process of a picture starts from a S(axiom)-homogeneous picture. At each step, an A-homogeneous subpicture is replaced with an isometric picture of the regional language, defined by the right part of a rule $A \to \dots$. The process terminates when all nonterminals have been eliminated from the current picture.[1]

Definition 5. *A regional tile grammar (RTG) is a tuple (Σ, N, S, R), where Σ is the terminal alphabet, N is a set of nonterminal symbols, $S \in N$ is the starting symbol, R is a set of rules.*
Let $A \in N$. There are two kinds of rules:

$$Fixed\ size: \quad A \to t, \quad where\ t \in \Sigma; \tag{1}$$
$$Variable\ size: \quad A \to \omega, \quad \omega\ is\ a\ set\ of\ tiles\ over\ N \cup \{\#\}, \tag{2}$$
$$LOC(\omega)\ is\ a\ regional\ language. \tag{3}$$

Picture derivation is defined as a relation between partitioned pictures.

Definition 6. *Consider a grammar $G = (\Sigma, N, S, R)$, let $p, p' \in (\Sigma \cup N)^{(h,k)}$ be pictures of identical size. Let π, π' be homogeneous partitions of $\mathrm{dom}(p)$, with $\pi = \{d_1, \dots, d_n\}$. We say that (p', π') derives in one step from (p, π), written*

$$(p, \pi) \Rightarrow_G (p', \pi')$$

iff, for some $A \in N$ and for some rule $\rho \in R$ with left part A, there exists in π an A-homogeneous subdomain $d_i = (x, y; x', y')$, called application area, such that:

- *p' is obtained substituting $\mathrm{spic}(p, d_i)$ in p with a picture s, defined as follows:*

    - *if ρ is of type (1), then $s = t$;*
    - *if ρ is of type (2), then $s \in LOC(\omega)$.*

- *π' is a homogeneous partition of $\mathrm{dom}(p)$ into the subdomains*

$$(\pi \setminus \{d_i\}) \cup \mathrm{transl}_{d_i}(\Pi(s))$$

where $\mathrm{transl}_{d_i}(\Pi(s))$ is the translation by displacement $(x-1, y-1)$ (intuitively the position of d_i in p) of the subdomains of $\Pi(s)$, the homogeneous partition of s.

---

[1] For brevity, this presentation focuses on nonterminal rules, thus excluding for instance that both terminal and nonterminal symbols are in the same right part. More concise and readable forms of rules should be used in applications.

We say that (q, π') derives from (p, π) in n steps, written $(p, \pi) \stackrel{n}{\Rightarrow}_G (q, \pi')$, iff $p = q$ and $\pi = \pi'$, when $n = 0$, or there are a picture r and a homogeneous partition π'' such that $(p, \pi) \stackrel{n-1}{\Longrightarrow}_G (r, \pi'')$ and $(r, \pi'') \Rightarrow_G (q, \pi')$. We use the abbreviation $(p, \pi) \stackrel{*}{\Rightarrow}_G (q, \pi')$ for a derivation with a finite number of steps.

Definition 7. The picture language *defined by a grammar* G *(written* $L(G)$*) is the set of* $p \in \Sigma^{++}$ *such that*

$$\left(S^{|p|}, \mathrm{dom}(p) \right) \stackrel{*}{\Rightarrow}_G (p, \mathcal{I})$$

where $\mathcal{I}$ *denotes the partition of* $\mathrm{dom}(p)$ *defined by single pixels. For short we also write* $S \stackrel{*}{\Rightarrow}_G p$.

If we drop the constraint (3), we obtain the more general model of *tile grammars* [2].

Definition 8. A tile grammar *(TG) is a tuple* (Σ, N, S, R) *as in Definition 5, but condition (3) is omitted.*

Clearly, the picture derivation process for TG and RTG is the same. Notice that a derivation is defined iff the picture admits a homogeneous partition (see [2] for details). What makes the difference between Definition 5 and Definition 8 is that in the former the homogeneous partition is regional.

To illustrate, we now list some examples that will be reconsidered in Sect. 5 to separate the family RTG from other ones.

3.1 Regional Tile Grammars Examples

Example 1. One row and one column of b's.

The set of all pictures such that there is one row and one column (both not at the border) that hold b's, and the remainder of the picture is filled with a's.

$$
S \rightarrow \begin{bmatrix} \# & \# & \# & \# & \# & \# & \# \\ \# & A_1 & A_1 & V_1 & A_2 & A_2 & \# \\ \# & A_1 & A_1 & V_1 & A_2 & A_2 & \# \\ \# & H_1 & H_1 & V_1 & H_2 & H_2 & \# \\ \# & A_3 & A_3 & V_2 & A_4 & A_4 & \# \\ \# & A_3 & A_3 & V_2 & A_4 & A_4 & \# \\ \# & \# & \# & \# & \# & \# & \# \end{bmatrix} ; \; A_i \rightarrow \begin{bmatrix} \# & \# & \# & \# \\ \# & X & X & \# \\ \# & A_i & A_i & \# \\ \# & A_i & A_i & \# \\ \# & \# & \# & \# \end{bmatrix} \Bigg| \begin{bmatrix} \# & \# & \# & \# \\ \# & X & X & \# \\ \# & \# & \# & \# \end{bmatrix}, \text{ for } 1 \le i \le 4
$$

$$
X \rightarrow \begin{bmatrix} \# & \# & \# & \# & \# \\ \# & A & X & X & \# \\ \# & \# & \# & \# & \# \end{bmatrix} \Big| \; a; \; H_i \rightarrow \begin{bmatrix} \# & \# & \# & \# & \# \\ \# & B & H_i & H_i & \# \\ \# & \# & \# & \# & \# \end{bmatrix} \Big| \; b, \text{ for } 1 \le i \le 2
$$

$$
A \rightarrow a; \; B \rightarrow b; \; V_i \rightarrow \begin{bmatrix} \# & \# & \# \\ \# & B & \# \\ \# & V_i & \# \\ \# & V_i & \# \\ \# & \# & \# \end{bmatrix} \Big| \; b, \text{ for } 1 \le i \le 2.
$$

We recall that $[\![\,]\!]$ denotes the set of tiles contained in the argument picture. This notation is more readable and concise than listing every tile:

$$S \to \left\{ \frac{\#\ \#}{\#\ A_1}\ \frac{\#\ \#}{A_1\ A_1}\ \cdots,\ \frac{A_1\ V_1}{H_1\ V_1}\ \frac{V_1\ A_2}{V_1\ H_2}\ \cdots,\ \frac{A_4\ A_4}{\#\ \#}\ \frac{A_4\ \#}{\#\ \#} \right\}.$$

Here is an example of derivation, with partitions outlined for better readability:

$$
\begin{array}{|cccc|}
\hline
S & S & S & S & S \\
S & S & S & S & S \\
S & S & S & S & S \\
S & S & S & S & S \\
\hline
\end{array}
\Rightarrow
\begin{array}{|cc|c|cc|}
A_1 & A_1 & V_1 & A_2 & A_2 \\
H_1 & H_1 & V_1 & H_2 & H_2 \\
A_3 & A_3 & V_2 & A_4 & A_4 \\
A_3 & A_3 & V_2 & A_4 & A_4 \\
\end{array}
\Rightarrow
\begin{array}{|cc|c|cc|}
A_1 & A_1 & V_1 & A_2 & A_2 \\
H_1 & H_1 & V_1 & H_2 & H_2 \\
X & X & V_2 & A_4 & A_4 \\
A_3 & A_3 & V_2 & A_4 & A_4 \\
\end{array}
\Rightarrow
\begin{array}{|cc|c|cc|}
A_1 & A_1 & V_1 & A_2 & A_2 \\
H_1 & H_1 & V_1 & H_2 & H_2 \\
A & X & V_2 & A_4 & A_4 \\
A_3 & A_3 & V_2 & A_4 & A_4 \\
\end{array}
\Rightarrow
$$

$$
\Rightarrow
\begin{array}{cc|c|cc}
A_1 & A_1 & V_1 & A_2 & A_2 \\
H_1 & H_1 & V_1 & H_2 & H_2 \\
A & A & V_2 & A_4 & A_4 \\
A_3 & A_3 & V_2 & A_4 & A_4 \\
\end{array}
\overset{2}{\Rightarrow}
\begin{array}{cc|c|cc}
A_1 & A_1 & V_1 & A_2 & A_2 \\
H_1 & H_1 & V_1 & H_2 & H_2 \\
a & a & V_2 & A_4 & A_4 \\
A_3 & A_3 & V_2 & A_4 & A_4 \\
\end{array}
\overset{+}{\Rightarrow}
\begin{array}{ccccc}
a & a & b & a & a \\
b & b & b & b & b \\
a & a & b & a & a \\
a & a & b & a & a \\
\end{array}
$$

Example 2. Picture with palindromic rows. Each row is an even palindrome over $\{a, b\}$.

$$
S_P \to \left[\!\!\left[
\begin{array}{cccc}
\# & \# & \# & \# \\
\# & R & R & \# \\
\# & S_P & S_P & \# \\
\# & S_P & S_P & \# \\
\# & \# & \# & \#
\end{array}
\right]\!\!\right]
\ \Big|\
\left[\!\!\left[
\begin{array}{cccc}
\# & \# & \# & \# \\
\# & R & R & \# \\
\# & \# & \# & \#
\end{array}
\right]\!\!\right]
$$

$$
R \to \left[\!\!\left[
\begin{array}{cccccc}
\# & \# & \# & \# & \# & \# \\
\# & A & R & R & A' & \# \\
\# & \# & \# & \# & \# & \#
\end{array}
\right]\!\!\right]
\ \Big|\
\left[\!\!\left[
\begin{array}{cccccc}
\# & \# & \# & \# & \# & \# \\
\# & B & R & R & B' & \# \\
\# & \# & \# & \# & \# & \#
\end{array}
\right]\!\!\right]
\ \Big|\
\left[\!\!\left[
\begin{array}{cccc}
\# & \# & \# & \# \\
\# & A & A' & \# \\
\# & \# & \# & \#
\end{array}
\right]\!\!\right]
\ \Big|\
\left[\!\!\left[
\begin{array}{cccc}
\# & \# & \# & \# \\
\# & B & B' & \# \\
\# & \# & \# & \#
\end{array}
\right]\!\!\right]
$$

$$A \to a; \quad B \to b; \quad A' \to a; \quad B' \to b.$$

Example 3. Misaligned palindromes.

A picture is a "ribbon" of two rows, divided into four fields: at the top-left and at the bottom right of the picture are palindromes as in Example 2 (where S_p is defined). The other two fields are filled with c's and must not be adjacent.

$$
S \to \left[\!\!\left[
\begin{array}{cccccccc}
\# & \# & \# & \# & \# & \# & \# & \# \\
\# & P_1 & P_1 & P_1 & P_1 & C_1 & C_1 & \# \\
\# & C_2 & C_2 & P_2 & P_2 & P_2 & P_2 & \# \\
\# & \# & \# & \# & \# & \# & \# & \#
\end{array}
\right]\!\!\right];
\quad P_1 \to S_P;\ P_2 \to S_P
$$

$$
C_i \to \left[\!\!\left[
\begin{array}{ccccc}
\# & \# & \# & \# & \# \\
\# & C & C_i & C_i & \# \\
\# & \# & \# & \# & \#
\end{array}
\right]\!\!\right] \Big|\ c,\ \text{for } 1 \le i \le 2;\ C \to c.
$$

Procedure *Compute*$\mathfrak{D}(i, j; k, l)$:
for each size $(v, h) \in \{1, \ldots, k - i + 1\} \times \{1, \ldots, l - j + 1\}$:

    for each coordinate $(i', j') \in \{i, \ldots, k\} \times \{j, \ldots, l\}$:
        for each nonterminal $A \in \mathfrak{M}(i', j'; i' + v - 1, j' + h - 1)$:
          put $(i', j'; i' + v - 1, j' + h - 1)$ into the set $\mathfrak{D}(i, j; k, l)|_A$;

for each nonterminal $A \in N$:

    if $\mathfrak{D}(i, j; k, l)|_A = \emptyset$ **then** put $(0, 0; 0, 0)$ into the set $\mathfrak{D}(i, j; k, l)|_A$.

Fig. 1. Compute$\mathfrak{D}$

4 Parsing for Regional Tile Grammars

To present our version of the CKY algorithm, we have to generalize from substrings to subpictures. As a substring is identified by the positions of its first and last characters, a subpicture is conveniently identified by its subdomain.

Let p be a picture, of size (m, n), to be parsed with a grammar $G = (\Sigma, N, R, S)$.

Definition 9. *A recognition matrix $\mathfrak{M}$ is a 4-dimensional $m \times n \times m \times n$ matrix, whose generic element $\mathfrak{M}(i, j; h, k)$ is a set of non-terminals. The meaning of $A \in \mathfrak{M}(i, j; h, k)$ is that A can derive the subpicture* $\text{spic}(p, (i, j; h, k))$ *of p.*

In fact, only cells $(i, j; h, k)$, with $h \geq i, k \geq j$, are used: these cells are the four-dimensional counterpart of the upper triangular matrix used in classical CKY.

Definition 10. *Consider a recognition matrix $\mathfrak{M}$, and a subdomain $d = (i, j; k, l)$. Let us order the nonterminal set N: $A_1, A_2, \ldots, A_{|N|}$. The subdomains vector $\mathfrak{D}(d, \mathfrak{M})$ is the cartesian product $D_1 \times D_2 \times \ldots \times D_{|N|}$, where every D_t is the set of subdomains d' such that $N_t \in \mathfrak{M}(d')$ and d' is a subdomain of d; if such set is empty, then D_t contains the rectangle $(0, 0; 0, 0)$.*

For any nonterminal A, we will use the notation $\mathfrak{D}(d, \mathfrak{M})|_A$ to denote the component of the vector corresponding to A.

To simplify the notation, we shall write $\mathfrak{D}(d)$ instead of $\mathfrak{D}(d, \mathfrak{M})$ at no risk of confusion, because the algorithm refers to a unique recognition matrix $\mathfrak{M}$. Figure 1 shows the procedure used to compute $\mathfrak{D}$.

Figure 2 shows the procedure to check if a rule ρ of the grammar can be applied to a given rectangle $(i, j; k, l)$.

The *Main* procedure, presented in Figure 3, is structured as a straightforward generalization to two dimensions of the CKY parsing algorithm. The input picture p is in $L(G)$ iff $S \in \mathfrak{M}(1, 1; m, n)$.

4.1 Correctness and Complexity of Parsing

We start with a technical lemma, used to prove the correctness of the CheckRule procedure.

Procedure *CheckRule* $(\omega, (i, j; k, l))$:

Compute $\mathfrak{D}(i, j; k, l)$;

for each $(d_1, d_2, \ldots, d_{|N|}) \in \mathfrak{D}(i, j; k, l)$:

> $f := True$;
>
> **for each** $(N_a, N_b) \in \mathcal{H}_\omega$:
>> **if** $d_a = (i_a, j_a; k_a, l_a)$ and $d_b = (i_b, j_b; k_b, l_b)$ are not such that
>> $j_b = l_a + 1$, and $k_b \geq i_a, k_a \geq i_b$,
>> **then** $f := False$;
>
> **for each** $(N_a, N_b) \in \mathcal{V}_\omega$:
>> **if** $d_a = (i_a, j_a; k_a, l_a)$ and $d_b = (i_b, j_b; k_b, l_b)$ are not such that
>> $i_b = k_a + 1$, and $l_b \geq j_a, l_a \geq j_b$,
>> **then** $f := False$;
>
> **for each** $(\#, N_a) \in \mathcal{H}_\omega$:
>> **if** $d_a = (i_a, j_a; k_a, l_a)$ and $j_a \neq j$ **then** $f := False$;
>
> **for each** $(N_a, \#) \in \mathcal{H}_\omega$:
>> **if** $d_a = (i_a, j_a; k_a, l_a)$ and $l_a \neq l$ **then** $f := False$;
>
> **for each** $(\#, N_a) \in \mathcal{V}_\omega$:
>> **if** $d_a = (i_a, j_a; k_a, l_a)$ and $i_a \neq i$ **then** $f := False$;
>
> **for each** $(N_a, \#) \in \mathcal{V}_\omega$:
>> **if** $d_a = (i_a, j_a; k_a, l_a)$ and $k_a \neq k$ **then** $f := False$;
>
> **if** $f = True$ **then return** $True$;

return $False$.

Fig. 2. CheckRule

Procedure *Main*:

Every set in $\mathfrak{M}$ is empty;

for each pixel $p(i, j) = t$,

> **if** there exists a fixed size rule $A \to t \in R$,
> **then** put A into the set $\mathfrak{M}(i, j; i, j)$;

for each size $(v, h) \in \{1, \ldots, m\} \times \{1, \ldots, n\}$:

> **for each** coordinate $(i, j) \in \{1, \ldots, m\} \times \{1, \ldots, n\}$:
>> **for each** variable size rule rule $(A \to \omega) \in R$:
>>> **if** $CheckRule(\omega, (i, j; i + v - 1, j + h - 1))$,
>>> **then** put A into the set $\mathfrak{M}(i, j; i + v - 1, j + h - 1)$.

Fig. 3. Main

Lemma 1. *Let ω be a regional set of tiles and d a subdomain. CheckRule(ω, d) returns true iff there exists a rule $C \to \omega$, such that $(p_0, \pi_0) \Rightarrow_G (p_1, \pi_1)$, where $d \in \pi_0$, and spic(p_0, d) is a C-picture.*

Proof. By construction, a true output of CheckRule(ω, d) is equivalent to the fact that there exists a partition of d in the subdomains $d_1, d_2, \ldots, d_r$, and $q \in LOC(\omega)$, such that:

1. every $\mathrm{spic}(q, d_j)$ is an A-picture, for some nonterminal $A \in \mathfrak{M}(d_j)$;
2. if $\mathrm{spic}(q, d_j)$ is an A-picture, then for no $d_k \neq d_j$ the subpicture $\mathrm{spic}(q, d_k)$ is an A-picture.

This means that $\mathrm{transl}_d(\Pi(q))$ is the HR partition $\{d_1, d_2, \ldots, d_r\}$. Moreover, starting from (p_0, π_0), where $\mathrm{spic}(p_0, d)$ is a C-picture, it is possible to apply a rule $C \rightarrow \omega$ in a derivation step $(p_0, \pi_0) \Rightarrow_G (p_1, \pi_1)$, where $\pi_0 = \{d, d'_1, d'_2, \ldots, d'_n\}$, $\pi_1 = \{d'_1, d'_2, \ldots, d'_n\} \cup \{d_1, d_2, \ldots, d_r\}$, and $q = \mathrm{spic}(p_1, d) \in LOC(\omega)$. □

After this, the correctness is easy to prove, analogously to the 1D case [12].

Theorem 1. $\mathfrak{M}(d) = \{A \in N \mid A \overset{*}{\Rightarrow}_G \mathrm{spic}(p, d)\}$.

Proof. The proof is by induction over derivation steps.

Base: $d = (i, j; i, j)$. This means that $|\mathrm{spic}(p, d)| = (1, 1)$. Hence, $A \overset{*}{\Rightarrow}_G \mathrm{spic}(p, d)$ iff $A \rightarrow \mathrm{spic}(p, d) \in R$. This case is handled by the first loop of procedure Main, the one over each pixel $p(i, j)$. If $\mathrm{spic}(p, d) = t$, and there exists a rule $A \rightarrow t$, then the algorithm puts A into $\mathfrak{M}(d)$. Vice versa, $A \in \mathfrak{M}(d)$ means that the algorithm has put A in the set, therefore there must exist a rule $A \rightarrow \mathrm{spic}(p, d)$.

Induction: let us consider $d = (i, j; i+v-1, j+h-1)$, $v > 1$, or $h > 1$, or both. We prove that $A \overset{*}{\Rightarrow}_G \mathrm{spic}(p, d)$ implies $A \in \mathfrak{M}(d)$. In this case, the size of the subpicture is not $(1, 1)$, therefore the first rule used in the derivation $A \overset{*}{\Rightarrow}_G \mathrm{spic}(p, d)$ is a variable size rule $A \rightarrow \omega$. Thanks to the two nested loops with control variables u and v, when the algorithm considers d, it has already considered all its subdomains $d_1, d_2, \ldots, d_k$. By the induction hypothesis, for every $1 \leq j \leq k$, $B \overset{*}{\Rightarrow}_G \mathrm{spic}(p, d_j)$ implies $B \in \mathfrak{M}(d_j)$. Hence (Lemma 1), CheckRule(ω, d) must be true, and the algorithm puts A in $\mathfrak{M}(d)$.

Next, we prove that $A \in \mathfrak{M}(d)$ implies $A \overset{*}{\Rightarrow}_G \mathrm{spic}(p, d)$. $A \in \mathfrak{M}(d)$ means that procedure Main has put A in the set. Therefore, CheckRule(ω, d) must be true. Thanks to Lemma 1, this is equivalent to the existence of an applicable variable size rule $A \rightarrow \omega$ for the first step of the derivation $A \overset{*}{\Rightarrow}_G \mathrm{spic}(p, d)$. The rest of the derivation holds by induction hypotesis. □

Theorem 2. *The parsing problem for RTG has polynomial time complexity.*

Proof. First, it is straightforward to see the time complexity of procedure *Compute$\mathfrak{D}$*: $T_{\mathrm{Compute}\mathfrak{D}} = O(|N| \cdot m^2 n^2)$. Let us now consider the *CheckRule* procedure. After computing $\mathfrak{D}$, the procedure performs a loop for each element of the subdomains vector, and nested loops on $\mathcal{H}_\omega$ and $\mathcal{V}_\omega$. Therefore, $T_{\mathrm{CheckRule}}(m, n) = O(|N| \cdot m^2 n^2 \cdot \max_{A \rightarrow \omega \in R} \{|\mathcal{H}_\omega|, |\mathcal{V}_\omega|\})$.

Coming finally to the *Main* procedure, we note that its core part consists of five nested loops, two on sets of m elements, two on sets of n elements, and the last one on $|R|$ elements. The body is a call to CheckRule. Therefore, $T_{\mathrm{Main}}(m, n) = O(|R| \cdot m^2 n^2 \cdot T_{\mathrm{CheckRule}}(m, n))$, i.e.

$$T_{\mathrm{Main}}(m, n) = O\left(|R||N| \cdot \max_{A \rightarrow \omega \in R} \{|\mathcal{H}_\omega|, |\mathcal{V}_\omega|\} \cdot m^4 n^4\right).$$ □

For the special case of CF Kolam grammars in Chomsky Normal form (CNF), we note that the parsing time complexity is $O(m^2 n^2 (m + n))$ [3]. Some of the reasons of this significant difference are the following. Kolam grammars in CNF are much simpler, because in the right part of a rule there are at most two distinct nonterminals (see [7] for details). So, checking if a rule is applicable has complexity which is linear with respect to the picture width or height. Moreover, we think that there is room for improvement e.g. in the CheckRule procedure, by using more complex data structures.

The next section will show that CF Kolam grammars are less expressive than RTG.

5   Comparison with Other Language Families

The property of having polynomial time complexity for picture recognition, united with the rather simple and intuitively pleasing form of RTG rules, should make them a worth addition to the series of array rewriting grammar models conceived in past years. In this section we prove or recall some inclusion relations between grammar models and corresponding language families. To this end we rely on the examples of Sect. 3, and on the separation of complexity classes.

Starting with the family of highest generative capacity, we focus on tile grammars.

Proposition 2. *The family of RTG languages is a proper subset of the family of TG languages.*

Proof. We have seen in Sect. 3 that RTG rules are a restricted form of TG rules, characterized by the constraint of regional tiling. To show that inclusion is strict, we observe that the picture recognition problem for tile grammars is NP-complete. This follows from the (strict) inclusion [2] of the tiling systems (or Wang's tiling) [4] family within the TG language family, and the fact that the recognition problem is NP-complete in time for the former [6].   □

We proceed by comparing RTG's and tiling systems.

Proposition 3. *The family of tiling system languages (i.e. Wang's tiling) and the family of RTG languages are incomparable.*

Proof. On one hand, it is easy to see that the language of palindromic columns, used in [2] to prove that tiling systems are strictly included in tile grammars, is also a RTG language, obtained by a 90° rotation of Example 2. On the other hand, we know that parsing tiling systems is NP-complete, and parsing RTG's has polynomial time complexity.   □

The remaining models are weaker than RTG, and will be taken in historical order.

Proposition 4. *The family of CF Kolam array grammar [11] (i.e. also [7]) languages is strictly included in the family of RTG languages.*

Proof. In [2] a construction is given to prove that a CF Kolam grammar (in the form defined by Matz [7]) can be transformed into a TG. It is easy to see that the construction used in the proof actually produces rules which satisfy the restriction of RTG's.

More directly, CF Kolam grammars in CNF can be seen as RTG's such that the tile-sets used in the right parts of rules must have one of the following forms:

$$
\text{Either}\quad
\begin{bmatrix}
\# \, \# \, \# \, \# \, \# \, \# \\
\# \, A \, A \, B \, B \, \# \\
\# \, A \, A \, B \, B \, \# \\
\# \, \# \, \# \, \# \, \# \, \#
\end{bmatrix},\ \text{or}\quad
\begin{bmatrix}
\# \, \# \, \# \, \# \\
\# \, A \, A \, \# \\
\# \, A \, A \, \# \\
\# \, B \, B \, \# \\
\# \, B \, B \, \# \\
\# \, \# \, \# \, \#
\end{bmatrix},\ \text{or}\quad
\begin{bmatrix}
\# \, \# \, \# \, \# \\
\# \, A \, A \, \# \\
\# \, A \, A \, \# \\
\# \, \# \, \# \, \#
\end{bmatrix},\ \text{with } A \neq B.
$$

The inclusion is strict, because the language of Example 1 was shown by Matz [7] to trespass the generative capacity of his grammars.   □

The fact that the picture recognition problem for CF Kolam grammars has been recently proved [3] to be polynomial in time of course follows from the above inclusion property and from Theorem 2.

In the quest for generality, D. Průša [8] has recently defined a grammar model that extends CF Kolam rules, gaining some generative capacity. The model is for instance able to generate the language of Example 1.

Essentially, this kind of grammars can be seen as RTG's with the additional constraint that tiles used in the right parts of rules must not have one of these forms: $\begin{pmatrix} A \, B \\ C \, C \end{pmatrix}, \begin{pmatrix} A \, C \\ B \, C \end{pmatrix}, \begin{pmatrix} C \, C \\ A \, B \end{pmatrix}, \begin{pmatrix} C \, A \\ C \, B \end{pmatrix}$, with A, B, C all different. Therefore the following inclusion holds.

Proposition 5. *A Průša's grammar "with productions in CF form" (PG) [8] is a restricted kind of RTG. The corresponding family of languages is strictly included in the family of RTG languages.*

The inclusion of the language families is strict, because the language of Example 3 cannot be defined by PG's. This fact can be sketchily proved as follows.

Proof. First, an obvious application of CF pumping lemma for strings (over the alphabet $\{a \ominus a, a \ominus b, a \ominus c, b \ominus a, \dots\}$) excludes that the language can be obtained by horizontal concatenation only. Therefore, it is necessary to generate the pictures either as vertical concatenations of strings of equal length, or using a grid-like rule, such as $\begin{bmatrix} \# \, \# \, \# \, \# \, \# \, \# \\ \# \, A \, A \, B \, B \, \# \\ \# \, C \, C \, D \, D \, \# \\ \# \, \# \, \# \, \# \, \# \, \# \end{bmatrix}$. By definition of the language, the two palindromes must span at least one common column, therefore we cannot use a simple vertical concatenation. The fact that $\{uu^R \mid u \in \{a,b\}^+\}$ cannot be factorized as a concatenation of CF languages is another simple corollary of the pumping lemma for CF languages. This means that each misaligned palindrome must be generated starting from a single nonterminal. But it is impossible for PG's to define a grid-like rule with single nonterminals partially overlapping on common columns.   □

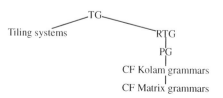

We finish with a synopsis of the previous language family inclusions. The early model of CF Matrix grammars [10] is a very limited kind of CF Kolam grammars.

6   Conclusions

The generalization of array rewriting grammars offered by the new and simple regional tile model is a convenient accomplishment of a series of generalizations, stemming from the early models of Rosenfeld up to the models of Siromoney, Matz and Průša. To our knowledge (but we may be missing something because the literature of picture grammars is rather fragmented), this is the most general family of polynomial time recognizable picture languages based on rewriting rules, which in one dimension collapse to CF string grammars.

Acknowledgement. We thank the anonymous referees for helpful suggestions.

References

1. Allauzen, C., Durand, B.: Tiling problems. In: Borger, E., Gradel, E. (eds.) The classical decision problem. Springer, Heidelberg (1997)
2. Crespi Reghizzi, S., Pradella, M.: Tile Rewriting Grammars and Picture Languages. Theoretical Computer Science 340(2), 257–272 (2005)
3. Crespi Reghizzi, S., Pradella, M.: A CKY parser for picture grammars. Information Processing Letters 105(6), 213–217 (2008)
4. Giammarresi, D., Restivo, A.: Recognizable picture languages. International Journal Pattern Recognition and Artificial Intelligence 6(2-3), 241–256; Special Issue on Parallel Image Processing (1992)
5. Giammarresi, D., Restivo, A.: Two-dimensional languages. In: Salomaa, A., Rozenberg, G. (eds.) Handbook of Formal Languages, vol. 3, pp. 215–267. Springer, Berlin (1997)
6. Lewis, H.: Complexity of solvable cases of the decision problem for predicate calculus. In: Proc. 19th Symposium on Foundations of Computer Science, pp. 35–47 (1978)
7. Matz, O.: Regular expressions and context-free grammars for picture languages. In: 14th Annual Symposium on Theoretical Aspects of Computer Science. LNCS, vol. 1200, pp. 283–294 (1997)
8. Průša, D.: Two-dimensional languages (PhD Thesis) (2004)
9. Simplot, D.: A characterization of recognizable picture languages by tilings by finite sets. TCS: Theoretical Computer Science 218 (1999)
10. Siromoney, G., Siromoney, R., Krithivasan, K.: Abstract families of matrices and picture languages. Computer Graphics and Image Processing 1 (1972)
11. Siromoney, G., Siromoney, R., Krithivasan, K.: Picture languages with array rewriting rules. Information and Control 23(5), 447–470 (1973)
12. Younger, D.H.: Recognition of context-free languages in time n^3. Information and Control 10(2), 189–208 (1967)

On a Special Class of Primitive Words

Elena Czeizler, Lila Kari, and Shinnosuke Seki

Department of Computer Science, University of Western Ontario,
London, Ontario, Canada, N6A 5B7
{elenac,lila,sseki}@csd.uwo.ca

Abstract. When representing DNA molecules as words, it is necessary to take into account the fact that a word u encodes basically the same information as its Watson-Crick complement $\theta(u)$, where θ denotes the Watson-Crick complementarity function. Thus, an expression which involves only a word u and its complement can be still considered as a repeating sequence. In this context, we define and investigate the properties of a special class of primitive words, called θ-primitive, which cannot be expressed as such repeating sequences. For instance, we prove the existence of a unique θ-primitive root of a given word, and we give some constraints forcing two distinct words to share their θ-primitive root. Also, we present an extension of the well-known Fine and Wilf Theorem, for which we give an optimal bound.

1 Introduction

Encoding information as DNA strands as in, e.g., DNA Computing, brings up for investigation new features based on the specific biochemical properties of DNA molecules. Recall that single-stranded DNA molecules can be viewed as words over the quaternary alphabet of bases $\{A, T, C, G\}$. Moreover, one of the main properties of DNA molecules is the Watson-Crick complementarity of the bases A and T and respectively G and C. Because of this property two Watson-Crick complementary single DNA strands with opposite orientation bind together to form a DNA double strand, in a process called base-pairing. Recently, there were several approaches to generalize notions from classical combinatorics on words in order to incorporate this major characteristic of DNA molecules, see, e.g., [6], [7], and [9]. Following these lines, in this paper, we generalize the concept of *primitivity* and define *θ-primitive words*.

The notion of primitivity plays an important role in various fields of theoretical computer science, such as algebraic coding theory, [11], and combinatorics on words, [8]. A word is called *primitive* if it cannot be decomposed as a power of another word. Thus, investigating the primitivity of a word is often the first step when analyzing its properties. Moreover, how a word can be decomposed and whether two words are powers of a common word are two questions which were widely investigated in language theory, see, e.g., [2], [8], and [12]. While, in classical combinatorics on words we look for repetitions of the form u^i for some word u and some $i \geq 2$, when dealing with DNA molecules (i.e., their abstract

E. Ochmański and J. Tyszkiewicz (Eds.): MFCS 2008, LNCS 5162, pp. 265–277, 2008.

representation as words) we should exploit the fact that a word u encodes the same information as its complement $\theta(u)$, where θ denotes the Watson-Crick complementarity function, or its mathematical formalization as an arbitrary antimorphic involution. In other words, we should look for expressions involving a word u and its complement $\theta(u)$. In this context, we define θ-primitive words as strings which cannot be decomposed using only some word u and its complement. Also, we define the θ-primitive root of a word w as the shortest word u such that w can be decomposed using only u and its complement. In classical combinatorics on words, there exist two equivalent definitions for the *primitive root* of a word w: the shortest word u such that $w = u^i$ for some $i \geq 1$, or the unique primitive word u such that $w = u^i$ for some $i \geq 1$. In our search for such equivalent definitions for the θ-primitive root of a word, we succeed to prove an extension of the well-known Fine and Wilf Theorem, one of the most widely used results on words. Although it was initially proved in connection with real functions, [5], the Fine and Wilf Theorem can be naturally interpreted also as a result on words, see, e.g., [2] and [8]. Moreover, several extensions of this theorem were proved so far, see, e.g., [1], [3], [4], and [10]. In this paper, we look at the case when a word w has two decompositions: one using a word u and its complement $\theta(u)$, and the other using some other word v and its complement $\theta(v)$. If w is longer than a given bound, then we prove that u and v share their θ-primitive root t and, thus w will have a refined decomposition depending on t and its complement. Moreover, we show that our bound is optimal, i.e., twice the length of the longer word (u or v) plus the length of the other word minus the greatest common divisor of the lengths of u and v.

The paper is organized as follows. In Section 2, we fix our terminology and recall some basic results. In Section 3 we investigate some basic properties of θ-primitive words. In particular, we give an extension of the Fine and Wilf Theorem which implies immediately that we can define the θ-primitive root of a word in two equivalent ways. In Section 4, we present several constraints forcing two words to share their θ-primitive root. In Section 5, we investigate some connections between the θ-primitive words that we introduced here and the θ-palindrome words, which were proposed and investigated in [7] and [9]. In Section 6, we present the optimal bound for our extension of the Fine and Wilf Theorem.

2   Preliminaries

Let Σ be a finite alphabet. We denote by Σ^* the set of all finite words over Σ, by ϵ the empty word, and $\Sigma^+ = \Sigma^* \setminus \{\epsilon\}$. The *length* of a word w, denoted by $|w|$, is the number of letters occurring in it, i.e., if $w = a_1 \ldots a_n$ with $a_i \in \Sigma$, $1 \leq i \leq n$, then $|w| = n$. We say that u is a *prefix* (resp. a *suffix*) of v if $v = ut$ (resp. $v = tu$) for some $t \in \Sigma^*$. For any $0 \leq k \leq |v|$, we use the notation $pref_k(v)$ (resp. $suf_k(v)$) for the prefix (resp. suffix) of length k of a word v and $\mathrm{Pref}(v)$ (resp. $\mathrm{Suff}(v)$) for the set of all prefixes (resp. all suffixes) of v. In particular $pref_0(v) = \epsilon$ for any word $v \in \Sigma^*$. An integer $p \geq 1$ is a *period* of a word $w = a_1 \ldots a_n$, with $a_i \in \Sigma$ for all $1 \leq i \leq n$, if $a_i = a_{i+p}$ for all $1 \leq i \leq n - p$.

A word $w \in \Sigma^+$ is called *primitive* if it cannot be written as a power of another word; that is, $w = u^n$ implies $n = 1$ and $w = u$. For a word $w \in \Sigma^+$, the shortest $u \in \Sigma^+$ such that $w = u^n$ for some $n \geq 1$ is called the *primitive root* of the word w and is denoted by $\rho(w)$. The following result gives an alternative, equivalent way for defining the primitive root of a word.

Theorem 1. *For a word $w \in \Sigma^*$, there exists a unique primitive word $t \in \Sigma^+$ such that $\rho(w) = t$, i.e., $w = t^n$ for some $n \geq 1$.*

The next result illustrates another useful property of primitive words.

Proposition 1. *Let $u \in \Sigma^+$ be a primitive word. Then, u cannot be a factor of u^2 in a nontrivial way, i.e., if $u^2 = xuy$, then necessarily either $x = \epsilon$ or $y = \epsilon$.*

We say that two words u and v *commute* if $uv = vu$. The following result characterizes the commutation of two words in terms of primitive roots.

Theorem 2. *For $u, v \in \Sigma^*$, the following conditions are equivalent: i) u and v commute; ii) u and v satisfy a non-trivial relation, i.e., an equation where the two sides are not graphically identical; iii) u and v have the same primitive root.*

Two words u and v are said to be *conjugate* if there exist words x and y such that $u = xy$ and $v = yx$. In other words, v can be obtained via a cyclic permutation of u. The next result characterizes the conjugacy of two words.

Theorem 3. *Let $u, v \in \Sigma^+$. Then, the following conditions are equivalent: i) u and v are conjugate; ii) there exists a word z such that $uz = zv$; moreover, this holds if and only if $u = pq$, $v = qp$, and $z = (pq)^i p$, for some $p, q \in \Sigma^*$ and $i \geq 0$; iii) the primitive roots of u and v are conjugate.*

Note that conjugacy is an equivalence relation, the *conjugacy class* of a word w consisting of all conjugates of w. The following is a well-known result.

Proposition 2. *If w is a primitive word, then its conjugacy class contains $|w|$ distinct primitive words.*

The following result, known as the Fine and Wilf theorem in its form for words, cf. [2] and [8], illustrates a fundamental periodicity property of words. As usual, $gcd(n, m)$ denotes the *greatest common divisor of n and m*.

Theorem 4. *Let $u, v \in \Sigma^*$, $n = |u|$, $m = |v|$, and $d = gcd(n, m)$. If two powers u^i and v^j of u and v have a common prefix of length at least $n + m - d$, then u and v are powers of a common word. Moreover, the bound $n + m - d$ is optimal.*

A mapping $\theta : \Sigma^* \to \Sigma^*$ is called a *morphism* (resp. an *antimorphism*) if for any words $u, v \in \Sigma^*$, $\theta(uv) = \theta(u)\theta(v)$ (resp. $\theta(uv) = \theta(v)\theta(u)$). Moreover, a mapping $\theta : \Sigma^* \to \Sigma^*$ is called an *involution* if, for all words $u \in \Sigma^*$, $\theta(\theta(u)) = u$.

For a mapping $\theta : \Sigma^* \to \Sigma^*$, a word $w \in \Sigma^*$ is called θ-*palindrome* if $w = \theta(w)$, see [7] and [9]. Now we say that a word $w \in \Sigma^+$ has a θ-*decomposition* if there exist a positive integer $k \geq 2$ and some words $t, w_1, \ldots, w_k \in \Sigma^+$ such that $w = w_1 \ldots w_k$ and $w_i \in \{t, \theta(t)\}$ for all $1 \leq i \leq k$. In this case, we say that w is θ-periodic, with θ-period $|t|$. We call a word $w \in \Sigma^+$ θ-primitive if it has no θ-decompositions, i.e., its least θ-period is $|w|$. We define the θ-primitive root of w, denoted by $\rho_\theta(w)$, as the shortest word t such that $w = w_1 \ldots w_k$ for some $k \geq 2$, $w_i \in \{t, \theta(t)\}$ for all $1 \leq i \leq k$, and $w_1 = t$. Note that if w is θ-primitive, then we can fix $\rho_\theta(w) = w$.

3   Properties of θ-Primitive Words

In this section, we consider $\theta : \Sigma^* \to \Sigma^*$ to be either a morphic or antimorphic involution, other than the identity function. We start by looking at some basic properties of θ-primitive words.

Proposition 3. *If a word $w \in \Sigma^+$ is θ-primitive, then it is also primitive. Moreover, the converse is not always true.*

Proof. Suppose that w is a θ-primitive word but not primitive. Then, there exists some $t \in \Sigma^+$ such that $w = t^n$ with $n \geq 2$. But then we can θ-decompose w as $w = w_1 \ldots w_n$, where $w_1 = \ldots = w_n = t$, which contradicts the θ-primitivity of w. For the converse, since θ is not the identity function, there exists a letter a such that $\theta(a) \neq a$. Then, if we take $w = a\theta(a)$, it is obvious that w is primitive, but not θ-primitive. □

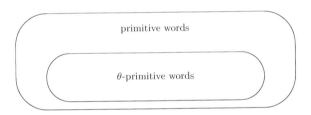

Fig. 1. The sets of primitive and θ-primitive words

Thus, the class of θ-primitive words is strictly included in the set of primitive ones, as illustrated in Fig. 1.

Proposition 4. *The θ-primitive root of a word is θ-primitive.*

Proof. Let $w \in \Sigma^+$ and $t = \rho_\theta(w)$ be its θ-primitive root. We can suppose, without loss of generality, that w is not θ-primitive; otherwise, $\rho_\theta(w) = w$ and thus the θ-primitive root is obviously θ-primitive. Then, we can write $w = w_1 \ldots w_n$, where $n \geq 2$ and $w_i \in \{t, \theta(t)\}$ for all $1 \leq i \leq n$. Suppose, now that t is not θ-primitive. Then, there exist a word $s \in \Sigma^+$ with $|s| < |t|$ and a positive integer $k \geq 2$, such that t has the θ-decomposition $t = t_1 \ldots t_k$, where for all $1 \leq i \leq k$, $t_i \in \{s, \theta(s)\}$. Thus, we obtain another θ-decomposition of w, i.e., $w = v_1 \ldots v_{kn}$, where all $v_i \in \{s, \theta(s)\}$ and $|s| < |t|$. But this contradicts the fact that t is the θ-primitive root of w. □

We also obtain the following result as an immediate consequence.

Corollary 1. *The θ-primitive root of a word is primitive.*

Contrary to the case of primitive words, a conjugate of a θ-primitive word need not be θ-primitive, as shown by the following two examples.

Example 1. Let $\theta : \{A, T, C, G\}^* \to \{A, T, C, G\}^*$ be the Watson-Crick antimorphic involution defined by $\theta(A) = T$, $\theta(T) = A$, $\theta(G) = C$, and $\theta(C) = G$. Then, the word $w = GCTA$ is θ-primitive, while its conjugate $w' = AGCT = AG\theta(AG)$ is not.

Example 2. Let $\theta : \{a, b, c, d\}^* \to \{a, b, c, d\}^*$ be a morphic involution defined by $\theta(a) = c$, $\theta(c) = a$, $\theta(b) = d$, and $\theta(d) = b$. Then, the word $w = abadcb$ is θ-primitive, while its conjugate $w' = babadc = (ba)^2\theta(ba)$ is not.

So, we can formulate the following result.

Proposition 5. *The class of θ-primitive words is not necessarily closed under circular permutations.*

Fine and Wilf's result on words, i.e., Theorem 4, constitutes one of the fundamental periodicity properties of words. Thus, a natural question is whether we can obtain an extension of this result when, instead of taking powers of two words u^n and v^m, we look at expressions over $\{u, \theta(u)\}$ and $\{v, \theta(v)\}$, respectively. In particular, since the mapping θ is an involution, we can suppose without loss of generality that the two expressions start with u and v, respectively. First, we analyze the case when θ is a morphic involution; it turns out that in this case we can obtain the same bound as in Theorem 4. However, since the proof of this result is analogous to the one for Theorem 4, see for instance [8], we will not include it here due to space limitations.

Theorem 5. *Let $\theta : \Sigma^* \to \Sigma^*$ be a morphic involution, $u, v \in \Sigma^+$ with $n = |u|$, $m = |v|$, and $d = gcd(n, m)$, $\alpha(u, \theta(u)) \in u\{u, \theta(u)\}^*$, and $\beta(v, \theta(v)) \in v\{v, \theta(v)\}^*$. If the two expressions $\alpha(u, \theta(u))$ and $\beta(v, \theta(v))$ have a common prefix of length at least $n + m - d$, then there exists a word $t \in \Sigma^+$ such that $u, v \in t\{t, \theta(t)\}^*$, i.e., $\rho_\theta(u) = \rho_\theta(v)$. Moreover, the bound $n + m - d$ is optimal.*

However, as illustrated by the following example, if the mapping θ is an antimorphic involution, then the bound given by Theorem 5 is not enough anymore.

Example 3. Let $\theta : \{a, b\}^* \to \{a, b\}^*$ be the mirror mapping defined as follows: $\theta(a) = a$, $\theta(b) = b$, and $\theta(w_1 \ldots w_n) = w_n \ldots w_1$, where $w_i \in \{a, b\}$ for all $1 \le i \le n$. Obviously, θ is an antimorphic involution on $\{a, b\}^*$. Let now $u = (ab)^k b$ and $v = ab$. Then, u^2 and $v^k\theta(v)^{k+1}$ have a common prefix of length $2|u| - 1 > |u| + |v| - gcd(|u|, |v|)$. However, $\rho_\theta(u) \ne \rho_\theta(v)$.

The next result gives a lower bound for the antimorphic case, for which we employ similar techniques as in [4], so we omit the proof here. As usual, $lcm(n, m)$ denotes the *least common multiple* of n and m.

Theorem 6. *Let* $\theta : \Sigma^* \to \Sigma^*$ *be an antimorphic involution,* $u, v \in \Sigma^+$, *and* $\alpha(u, \theta(u)) \in u\{u, \theta(u)\}^*$, $\beta(v, \theta(v)) \in v\{v, \theta(v)\}^*$ *be two expressions sharing a common prefix of length at least* $lcm(|u|, |v|)$. *Then, there exists a word* $t \in \Sigma^+$ *such that* $u, v \in t\{t, \theta(t)\}^*$, *i.e.,* $\rho_\theta(u) = \rho_\theta(v)$. *In particular, if* $\alpha(u, \theta(u)) = \beta(v, \theta(v))$, *then* $\rho_\theta(u) = \rho_\theta(v)$.

Note that, in many cases there is a big gap between the bounds given in Theorems 5 and 6. Moreover, Theorem 6 does not give the optimal bound for the general case when θ is an antimorphic involution. In Section 6, we show that this optimal bound for the general case is $2|u| + |v| - gcd(|u|, |v|)$, where $|u| > |v|$, while for some particular cases we obtain bounds as low as $|u| + |v| - gcd(|u|, |v|)$. As an immediate consequence of Theorems 5 and 6, we obtain the following result.

Corollary 2. *For any word* $w \in \Sigma^+$ *there exists a unique* θ-*primitive word* $t \in \Sigma^+$ *such that* $w \in t\{t, \theta(t)\}^*$, *i.e.,* $\rho_\theta(w) = t$.

Let us note now that, maybe even more importantly, just as in the case of primitive words, this result provides us with an alternative, equivalent way for defining the θ-primitive root of a word w, i.e., *the* θ-*primitive word* t *such that* $w \in t\{t, \theta(t)\}^*$. This proves to be a very useful tool in our future considerations.

Moreover, we also obtain the following two results as immediate consequences of Theorems 5 and 6.

Corollary 3. *Let* $u, v \in \Sigma^+$ *be two words such that* $\rho(u) = \rho(v) = t$. *Then,* $\rho_\theta(u) = \rho_\theta(v) = \rho_\theta(t)$.

Corollary 4. *If we have two words* $u, v \in \Sigma^+$ *such that* $u \in v\{v, \theta(v)\}^*$, *then* $\rho_\theta(u) = \rho_\theta(v)$.

4   Relations Imposing θ-Periodicity

It is well-known, due to Theorem 2, that any non-trivial equation over two distinct words forces them to be powers of a common word, i.e., to share a common period. Thus, a natural question is whether this would be the case also when we want two distinct words to have θ-decompositions depending on the same u and $\theta(u)$, i.e., to share a common θ-period. From [6], we already know that the equation $uv = \theta(v)u$ imposes $\rho_\theta(u) = \rho_\theta(v)$ only when θ is a morphic involution. In this section, we give several examples of equations over $\{u, \theta(u), v, \theta(v)\}$ forcing $\rho_\theta(u) = \rho_\theta(v)$ in the case when $\theta : \Sigma^* \to \Sigma^*$ is an antimorphic involution.

The first equation we look at is very similar to the commutation equation of two words, but it involves also the mapping θ.

Theorem 7. *Let* $\theta : \Sigma^* \to \Sigma^*$ *be an antimorphic involution over the alphabet* Σ *and* $u, v \in \Sigma^+$. *If* $uv\theta(v) = v\theta(v)u$, *then* $\rho_\theta(u) = \rho_\theta(v)$.

Proof. Since $uv\theta(v) = v\theta(v)u$, we already know, due to Theorem 2, that there exists a primitive word $t \in \Sigma^+$ such that $u = t^i$ and $v\theta(v) = t^j$, for some $i, j \geq 0$. If $j = 2k$ for some $k \geq 0$, then we obtain immediately that $v = \theta(v) = t^k$, i.e., $\rho(u) = \rho(v) = t$. Thus, $\rho_\theta(u) = \rho_\theta(t) = \rho_\theta(v)$. Otherwise, i.e., $j = 2k+1$, we can write $v = t^k t_1$ and $\theta(v) = t_2 t^k$, where $t = t_1 t_2$ and $|t_1| = |t_2| > 0$. Hence, $\theta(v) = \theta(t_1)\theta(t)^k = t_2 t^k$, which implies $t_2 = \theta(t_1)$. In conclusion, $u, v \in t_1\{t_1, \theta(t_1)\}^*$, for some word $t_1 \in \Sigma^+$, i.e., $\rho_\theta(u) = \rho_\theta(t_1) = \rho_\theta(v)$. $\square$

Next, we modify a little bit the previous equation, such that on one side, instead of $v\theta(v)$, we take its conjugate $\theta(v)v$.

Theorem 8. *Let $\theta : \Sigma^* \to \Sigma^*$ be an antimorphic involution over the alphabet Σ and $u, v \in \Sigma^+$. If $v\theta(v)u = u\theta(v)v$, then $\rho_\theta(u) = \rho_\theta(v)$.*

Proof. If we concatenate the word $\theta(v)$ to the right on both sides of the equation $v\theta(v)u = u\theta(v)v$, then we obtain $(v\theta(v))(u\theta(v)) = (u\theta(v))(v\theta(v))$. Due to Theorem 2, this means that there exists a primitive word $t \in \Sigma^+$ such that $v\theta(v) = t^i$ and $u\theta(v) = t^j$, for some $i, j \geq 0$, $j \geq \lceil i/2 \rceil$. If $i = 2k$ for some $k \geq 0$, then $\theta(v) = v = t^k$ and thus also $u = t^{j-k}$, i.e., $\rho(u) = \rho(v) = t$. Henceforth, $\rho_\theta(u) = \rho_\theta(t) = \rho_\theta(v)$. Otherwise, i.e., $j = 2k+1$, we can write $v = t^k t_1$ and $\theta(v) = t_2 t^k$, where $t = t_1 t_2$ and $|t_1| = |t_2| > 0$. Hence, we achieve again $t_2 = \theta(t_1)$, which implies that $v \in t_1\{t_1, \theta(t_1)\}^*$. Moreover, since $u\theta(v) = t^j$, we also obtain $u = t^{j-k-1}t_1 \in t_1\{t_1, \theta(t_1)\}^*$. Thus, $\rho_\theta(u) = \rho_\theta(t_1) = \rho_\theta(v)$. $\square$

Next, we look at the case when both uv and vu are θ-palindrome words, which also proves to be enough to impose that $u, v \in \{t, \theta(t)\}^*$ for some $t \in \Sigma^+$.

Theorem 9. *Let $u, v \in \Sigma^*$ be two words such that both uv and vu are θ-palindrome words and let $t = \rho(uv)$. Then, $t = \theta(t)$ and either $\rho(u) = \rho(v) = t$ or $u = (t_1\theta(t_1))^i t_1$ and $v = \theta(t_1)(t_1\theta(t_1))^j$, where $t = t_1\theta(t_1)$ and $i, j \geq 0$.*

Proof. The equality $uv = \theta(uv)$ immediately implies that $t = \theta(t)$. Moreover, if u and v commute, then $\rho(u) = \rho(v) = \rho(uv) = t$. Assume now that u and v do not commute. Since $\rho(u) \neq \rho(v)$ and $uv = t^n$ for some $n \geq 1$, we can write $u = t^i t_1$ and $v = t_2 t^{n-i-1}$ for some $i \geq 0$ and $t_1, t_2 \in \Sigma^+$ such that $t = t_1 t_2$. Thus, $vu = t_2 t^{n-1} t_1 = (t_2 t_1)^n$ and since $vu = \theta(vu)$ we obtain that also $t_2 t_1$ is θ-palindrome, i.e., $t_2 t_1 = \theta(t_2 t_1) = \theta(t_1)\theta(t_2)$. Now, if $|t_1| = |t_2|$, then $t_2 = \theta(t_1)$ and thus $t = t_1\theta(t_1)$, $u = t^i t_1$, and $v = \theta(t_1)t^{n-i-1}$. Otherwise, either $|t_1| > |t_2|$ or $|t_1| < |t_2|$. We consider next only the case $|t_1| > |t_2|$, the other one being similar. Since $t_2 t_1 = \theta(t_1)\theta(t_2)$, we can write $\theta(t_1) = t_2 x$ and $t_1 = x\theta(t_2)$ for some word $x \in \Sigma^+$ with $x = \theta(x)$. Then, since $t = \theta(t)$ we have that $t = t_1 t_2 = x\theta(t_2)t_2 = \theta(x\theta(t_2)t_2) = \theta(t_2)t_2 x$. Hence, x and $\theta(t_2)t_2$ commute, which contradicts the primitivity of t. $\square$

As an immediate consequence we obtain the following result.

Corollary 5. *For $u, v \in \Sigma^*$, if $uv = \theta(uv)$ and $vu = \theta(vu)$, then $\rho_\theta(u) = \rho_\theta(\theta(v))$. In particular, there exists some $t \in \Sigma^+$ such that $u, v \in \{t, \theta(t)\}^*$.*

5   On θ-Primitive and θ-Palindrome Words

In this section, we investigate two word equations under which a θ-primitive word must be θ-palindrome. Throughout this section we consider $\theta : \Sigma^* \to \Sigma^*$ to be an antimorphic involution over the alphabet Σ.

Theorem 10. *Let* $\theta : \Sigma^* \to \Sigma^*$ *be an antimorphic involution over the alphabet* Σ *and* $v \in \Sigma^+$ *be a* θ-*primitive word. If* $\theta(v)vx = yv\theta(v)$ *for some words* $x, y \in \Sigma^*$ *with* $|x|, |y| < |v|$, *then* v *is* θ-*palindrome and* $x = y = \epsilon$.

Proof. Assume there exist some words $x, y \in \Sigma^*$ with $|x|, |y| < |v|$, such that $\theta(v)vx = yv\theta(v)$, as illustrated in Fig. 2.

Then, we can write $v = v_1v_2 = v_2v_3$, with $v_1, v_2, v_3 \in \Sigma^*$, $y = \theta(v_2) = x$, $v_1 = \theta(v_1)$, $v_3 = \theta(v_3)$. Since $v_1v_2 = v_2v_3$, we can write $v_1 = pq$, $v_3 = qp$, $v_2 = (pq)^i p$, and $v = (pq)^{i+1}p$ for some words $p, q \in \Sigma^*$ and some $i \geq 0$. Thus, $pq = \theta(pq)$ and $qp = \theta(qp)$, which, due to Theorem 9, leads to one of the following two cases. First, if $p = t^k t_1$ and $q = \theta(t_1)t^j$, where $k, j \geq 0$ and $t = t_1\theta(t_1)$ is the primitive root of pq, then we obtain that $v = t^{(k+j+1)(i+1)+k}t_1$ with $(k+j+1)(i+1)+k \geq 1$, which contradicts the θ-primitivity of v. Second, if $\rho(p) = \rho(q) = t$, then also $v \in \{t\}^*$ where $t = \theta(t)$. Thus, $v = \theta(v)$, and

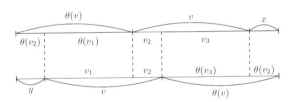

Fig. 2. The equation $\theta(v)vx = yv\theta(v)$

the initial identity becomes $v^2x = yv^2$. But, since v is θ-primitive and thus also primitive, we immediately obtain, due to Proposition 1, that $x = y = \epsilon$.   □

In other words, the previous result states that if v is a θ-primitive word, then $\theta(v)v$ cannot overlap with $v\theta(v)$ in a nontrivial way. However, the following example shows that this is not the case anymore if we look at the overlaps between $\theta(v)v$ and v^2, or between $v\theta(v)$ and v^2, respectively, even if we consider the larger class of primitive words.

Example 4. Let $\theta : \Sigma^* \to \Sigma^*$ be an antimorphic involution over the alphabet Σ, $p, q \in \Sigma^+$ such that $\rho(p) \neq \rho(q)$, $p = \theta(p)$, and $q = \theta(q)$, and let $v = p^2q^2p$ and $u = pq^2p^2$. It is easy to see that u and v are primitive words. In addition, if we take $\Sigma = \{a, b\}$, the mapping θ to be the mirror image, $p = a$, and $q = b$, then u and v are actually θ-primitive words. Since $\theta(v) = pq^2p^2$ and $\theta(u) = p^2q^2p$, we can write $xv^2 = v\theta(v)y$ and $y\theta(u)u = u^2z$ where $x = p^2q^2$, $y = pq^2p$, and $z = q^2p^2$. Thus, for primitive (resp. θ-primitive) words u and v, $v\theta(v)$ can overlap with v^2 and $\theta(u)u$ with u^2 in a nontrivial way.

Maybe even more surprisingly, the situation changes again if we try to fit v^2 inside $v\theta(v)v$, as shown by the following result.

Theorem 11. *Let $\theta : \Sigma^* \to \Sigma^*$ be an antimorphic involution over the alphabet Σ and $v \in \Sigma^+$ be a primitive word. If $v\theta(v)v = xv^2y$ for some words $x, y \in \Sigma^*$, then v is θ-palindrome and either $x = \epsilon$ and $y = v$ or $x = v$ and $y = \epsilon$.*

Proof. Suppose that $v\theta(v)v = xv^2y$ for some words $x, y \in \Sigma^*$, as illustrated in Fig. 3. If we look at this identity from left to right, then we can write $v = xv_1 = v_1v_2$, with $v_1, v_2 \in \Sigma^*$ such that $|x| = |v_2|$ and $\theta(v) = \theta(v_2)\theta(v_1)$. Now, if we look at the right sides of this identity, then we immediately obtain that $x = v_2$ and $v_1 = y$. Thus, $v = xy = yx$, implying that $x, y \in \{t\}^*$, for some primitive word t. However, since v is primitive, this means that either $x = \epsilon$ and $y = v$ or $x = v$ and $y = \epsilon$. Moreover, in both cases we also obtain $v = \theta(v)$.   □

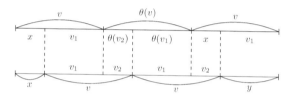

Fig. 3. The equation $v\theta(v)v = xv^2y$

6   An Optimal Bound for the Antimorphic Extension of the Fine and Wilf Theorem

Throughout this section we take $\theta : \Sigma^* \to \Sigma^*$ to be an antimorphic involution, $u, v \in \Sigma^+$ with $|u| > |v|$, $\alpha(u, \theta(u)) \in \{u, \theta(u)\}^+$, and $\beta(v, \theta(v)) \in \{v, \theta(v)\}^+$. Since θ is an involution, we can suppose, without loss of generality, that $\alpha(u, \theta(u))$ and $\beta(v, \theta(v))$ start with u and v, respectively. We start our analysis with the case when v is θ-palindrome.

Theorem 12. *Let u and v be two words with $|u| > |v|$ and $v = \theta(v)$. If there exist two expressions $\alpha(u, \theta(u)) \in u\{u, \theta(u)\}^*$ and $\beta(v, \theta(v)) \in v\{v, \theta(v)\}^*$ having a common prefix of length at least $|u| + |v| - gcd(|u|, |v|)$, then $\rho_\theta(u) = \rho_\theta(v)$.*

Proof. First, we can suppose, without loss of generality that $gcd(|u|, |v|) = 1$. Otherwise, i.e., $gcd(|u|, |v|) = d \geq 2$, we consider a new alphabet $\Sigma' = \Sigma^d$, where the new letters are words of length d in the original alphabet, and we look at the words u and v as elements of $(\Sigma')^+$. In the larger alphabet $gcd(|u|, |v|) = 1$, and if we can prove the theorem there it immediately gives the general proof.

Since $v = \theta(v)$, $\beta(v, \theta(v)) = v^n$ for some $n \geq 2$. Moreover, if $v \in \Sigma$, then trivially $u \in v\{v, \theta(v)\}^*$, i.e., $\rho_\theta(u) = \rho_\theta(v)$. So, suppose next that $|v| \geq 2$ and, since $gcd(|u|, |v|) = 1$, $u = v^iv_1$, where $i \geq 1$ and $v = v_1v_2$ with $v_1, v_2 \in \Sigma^+$.

If $\alpha(u, \theta(u)) = u^2\alpha'(u, \theta(u))$, then u^2 and v^n have a common prefix of length at least $|u| + |v| - gcd(|u|, |v|)$, which, due to Theorem 4, implies that $\rho(u) = \rho(v) = t$, for some primitive word $t \in \Sigma^+$, and thus $\rho_\theta(u) = \rho_\theta(t) = \rho_\theta(v)$.

Otherwise, $\alpha(u, \theta(u)) = u\theta(u)\alpha'(u, \theta(u))$ for some $\alpha'(u, \theta(u)) \in \{u, \theta(u)\}^*$. Now, we have two cases depending on $|v_1|$ and $|v_2|$. We present here only the case when $|v_1| \leq |v_2|$, see Fig. 4, the other one being symmetric. Now, since θ is an antimorphism, $\theta(suf_{|v|-1}(u)) = pref_{|v|-1}(\theta(u))$. So, we can write $v_2 = \theta(v_1)z$ for some $z \in \Sigma^*$, since $|v_1| \leq |v_2| \leq |v| - 1 = |v| - gcd(|u|, |v|)$. Now, to the

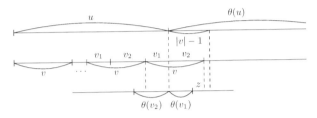

Fig. 4. The common prefix of $u\theta(u)$ and v^n of length $|u| + |v| - 1$

left of the border-crossing v there is at least one occurrence of another v, so we immediately obtain $z = \theta(z)$, as $v_2 = \theta(v_1)z$ and $\theta(v_2) = \theta(z)v_1$. Then, $v = v_1\theta(v_1)z = zv_1\theta(v_1) = \theta(v)$ implying, due to Theorem 7, $\rho_\theta(v_1) = \rho_\theta(z)$. So, since $v = v_1\theta(v_1)z$ and $u = v^iv_1 = (v_1\theta(v_1)z)^iv_1$, we obtain $\rho_\theta(u) = \rho_\theta(v)$.   □

Let us look next at the case when u is θ-palindrome.

Theorem 13. *Let u and v be two words with $|u| > |v|$ and $u = \theta(u)$. If there exist two expressions $\alpha(u, \theta(u)) \in u\{u, \theta(u)\}^*$ and $\beta(v, \theta(v)) \in v\{v, \theta(v)\}^*$ having a common prefix of length at least $|u| + |v| - gcd(|u|, |v|)$, then $\rho_\theta(u) = \rho_\theta(v)$.*

Proof. As before, we can suppose without loss of generality that $gcd(|u|, |v|) = 1$. Also, since $u = \theta(u)$, we actually have $\alpha(u, \theta(u)) = u^n$ for some $n \geq 2$. Moreover, since u starts with v and $u = \theta(u)$, we also know that u ends with $\theta(v)$. Now, if $v \in \Sigma$, then trivially $u \in v\{v, \theta(v)\}^*$, i.e., $\rho_\theta(u) = \rho_\theta(v)$. So, we can suppose next that $|v| \geq 2$ and thus, since $gcd(|u|, |v|) = 1$, we have $u = \beta'(v, \theta(v))v'$, where $\beta'(v, \theta(v))$ is a prefix of $\beta(v, \theta(v))$ and $v' \in \Sigma^+$, $v' \in Pref(v) \cup Pref(\theta(v))$.

Case 1: We begin our analysis with the case when the border between the first two u's falls inside a v, as illustrated in Fig. 5. Then, we can write $v = v_1v_2 = v_2v_3$ where $v_1, v_2, v_3 \in \Sigma^+$, implying that $v_1 = xy$, $v_3 = yx$, and $v_2 = (xy)^jx$ for some $j \geq 0$ and $x, y \in \Sigma^*$. Moreover, since u ends with $\theta(v)$, we also have $v_1 = \theta(v_1)$, i.e., $xy = \theta(y)\theta(x)$. If $x = \epsilon$, then $v_1, v_2, v_3, v \in \{y\}^*$, which implies that also $u \in y\{y, \theta(y)\}^*$, i.e., $\rho_\theta(u) = \rho_\theta(v) = \rho_\theta(y)$; moreover, since $gcd(|u|, |v|) = 1$ we actually must have $y \in \Sigma$. Similarly, we also obtain $\rho_\theta(u) = \rho_\theta(v)$ when $y = \epsilon$. So, from now on we can suppose that $x, y \in \Sigma^+$.

Let us consider next the case when, before the border-crossing v we have an occurrence of another v, as illustrated in Fig. 5. Then, we have that $v_2 = \theta(v_2)$,

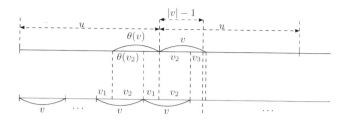

Fig. 5. The common prefix of u^2 and $\beta(v, \theta(v))$ of length $|u| + |v| - 1$

i.e., $(xy)^j x = (\theta(x)\theta(y))^j \theta(x)$. If $j \geq 1$, then this means that $x = \theta(x)$ and $y = \theta(y)$. Then, the equality $xy = \theta(y)\theta(x)$ becomes $xy = yx$. So, there exists a word $t \in \Sigma^+$ such that $x, y \in \{t\}^*$, and thus also $v \in \{t\}^+$ and $u \in t\{t, \theta(t)\}^*$, i.e., $\rho_\theta(u) = \rho_\theta(v)$. Otherwise, $j = 0$ and we have $x = \theta(x)$. But then, the equality $xy = \theta(y)\theta(x)$ becomes $xy = \theta(y)x$, implying that $x = p(qp)^n$ and $y = (qp)^m$ for some $m \geq 1$, $n \geq 0$, and some words p and q with $p = \theta(p)$ and $q = \theta(q)$, see [6]. Since u^2 and $\beta(v, \theta(v))$ share a common prefix of length at least $|u| + |v| - gcd(|u|, |v|) = |u| + |v| - 1$, v_3 and some $\beta'(v, \theta(v))$ share a prefix of length $|v_3| - 1$. Furthermore, as $v_3 = yx = (qp)^m p(qp)^n$, $v = v_1 v_2 = p(qp)^{m+n} p(qp)^n$, and $\theta(v) = (pq)^n p(pq)^{m+n} p$, this means that independently of what follows to the right the border-crossing v, either v or $\theta(v)$, we have two expressions over p and q sharing a common prefix of length at least $|p| + |q|$. Thus, from [2], we can conclude that $p, q \in \{t\}^*$ for some $t \in \Sigma^+$, which implies that also $x, y, v \in \{t\}^+$ and $u \in t\{t, \theta(t)\}^*$, i.e., $\rho_\theta(u) = \rho_\theta(v)$.

Now, suppose that before the border-crossing v we have an occurrence of $\theta(v)$. If $|u| < 2|v| + |v_1|$, then, since $\beta(v, \theta(v))$ starts with v, we must have $v = \theta(v)$, in which case we can use Theorem 12 to conclude that $\rho_\theta(u) = \rho_\theta(v)$. Otherwise, $|u| \geq 2|v| + |v_1|$ and since $u = \theta(u)$, u ends either with $v\theta(v)$ or with $\theta(v)\theta(v)$. In the first case, we obtain $v_3 = \theta(v_3)$, i.e., $yx = \theta(yx)$, which together with $xy = \theta(xy)$ imply, due to Corollary 5, that $x, y \in \{t, \theta(t)\}^*$, for some $t \in \Sigma^+$ and thus, $\rho_\theta(u) = \rho_\theta(v)$. In the second case, we obtain $v_1 = v_3$, i.e., $xy = yx$. So, $x, y \in \{t\}^*$, and thus also $v \in \{t\}^+$ and $u \in t\{t, \theta(t)\}^*$, i.e., $\rho_\theta(u) = \rho_\theta(v)$.
Case 2: The case when the border between the first two u's falls inside $\theta(v)$ is similar to the one above. So, due to page limitations, we omit it here. □

Although the previous two results give a very short bound, i.e., $|u| + |v| - gcd(|u|, |v|)$, this is not enough in the general case, as illustrated also in Example 3. However, we can prove that, independently of how the expression $\alpha(u, \theta(u))$ starts, $2|u| + |v| - gcd(|u|, |v|)$ is enough to impose θ-periodicity of u and v. The first case we consider is when $\alpha(u, \theta(u))$ starts with u^2. The proofs of Theorems 14 and 16 are rather complex and necessitate the analysis of many cases. Their inclusion would double the length of this paper and we therefore omit them here.

Theorem 14. *Given two distinct words $u, v \in \Sigma^+$ with $|u| > |v|$, if there exist two expressions $\alpha(u, \theta(u)) \in u\{u, \theta(u)\}^*$ and $\beta(v, \theta(v)) \in v\{v, \theta(v)\}^*$ hav-*

ing a common prefix of length at least $2|u| + |v| - gcd(|u|, |v|)$ *and, moreover,* $\alpha(u, \theta(u)) = u^2 \alpha'(u, \theta(u))$ *for some* $\alpha'(u, \theta(u)) \in \{u, \theta(u)\}^+$, *then* $\rho_\theta(u) = \rho_\theta(v)$.

The next result considers the case when $\alpha(u, \theta(u))$ starts with $u\theta(u)u$, which is an immediate consequence of Theorem 13.

Theorem 15. *Given two distinct words* $u, v \in \Sigma^+$ *with* $|u| > |v|$, *if there exist two expressions* $\alpha(u, \theta(u)) \in u\{u, \theta(u)\}^*$ *and* $\beta(v, \theta(v)) \in v\{v, \theta(v)\}^*$ *having a common prefix of length at least* $2|u| + |v| - gcd(|u|, |v|)$ *and, moreover,* $\alpha(u, \theta(u)) = u\theta(u)u\alpha'(u, \theta(u))$ *with* $\alpha'(u, \theta(u)) \in \{u, \theta(u)\}^*$, *then* $\rho_\theta(u) = \rho_\theta(v)$.

The only case left is when $\alpha(u, \theta(u))$ starts with $u\theta(u)\theta(u)$.

Theorem 16. *Let* $u, v \in \Sigma^+$ *be two words with* $|u| > |v|$. *If there exist two expressions* $\alpha(u, \theta(u)) \in u\{u, \theta(u)\}^*$ *and* $\beta(v, \theta(v)) \in v\{v, \theta(v)\}^*$ *having a common prefix of length at least* $2|u| + |v| - gcd(|u|, |v|)$, *and, moreover,* $\alpha(u, \theta(u)) = u\theta(u)\theta(u)\alpha'(u, \theta(u))$ *for some* $\alpha'(u, \theta(u)) \in \{u, \theta(u)\}^*$, *then* $\rho_\theta(u) = \rho_\theta(v)$.

To conclude, in this section we proved that if θ is an antimorphic involution, then we only need two expressions $\alpha(u, \theta(u))$ and $\beta(v, \theta(v))$ to share a common prefix of length $2|u|+|v|-gcd(|u|, |v|)$, where $|u| > |v|$, in order to impose $\rho_\theta(u) = \rho_\theta(v)$. Moreover, the following examples show that this bound is optimal.

Example 5. Let $\theta : \{a, b\}^* \to \{a, b\}^*$ be the mirror involution, $u_1 = a^2ba^3b$, $v_1 = a^2ba$, with $gcd(|u_1|, |v_1|) = 1$, and $u_2 = ba^2baba$, $v_2 = ba^2ba$, with $gcd(|u_2|, |v_2|) = 1$. Then, u_1^3 and $v_1^2\theta(v_1)^2v_1$ have a common prefix of length $2|u_1| + |v_1| - 2$, but $\rho_\theta(u_1) \neq \rho_\theta(v_1)$. Also, $u_2\theta(u_2)^2$ and v_2^4 have a common prefix of length $2|u_2| + |v_2| - 2$, but $\rho_\theta(u_2) \neq \rho_\theta(v_2)$.

Acknowledgments. This research was supported by Natural Sciences and Engineering Research Council of Canada Discovery Grant and Canada Research Chair Award of Lila Kari.

References

1. Berstel, J., Boasson, L.: Partial words and a theorem of Fine and Wilf. WORDS 1997 (Rouen), Theoret. Comput. Sci. 218(1), 135–141 (1999)
2. Choffrut, C., Karhumaki, J.: Combinatorics of words. In: Rozenberg, G., Salomaa, A. (eds.) Handbook of Formal Languages, vol. 1, pp. 329–438. Springer, Berlin (1997)
3. Constantinescu, S., Ilie, L.: Generalized Fine and Wilfs theorem for arbitrary number of periods. Theoret. Comput. Sci. 339(1), 49–60 (2005)
4. Constantinescu, S., Ilie, L.: Fine and Wilf's theorem for abelian periods. Bulletin of EATCS 89, 167–170 (2006)
5. Fine, N.J., Wilf, H.S.: Uniqueness theorem for periodic functions. Proc. Amer. Math. Soc. 16, 109–114 (1965)
6. Kari, L., Mahalingam, K.: Watson-Crick Conjugate and Commutative Words. In: Garzon, M., Yan, H. (eds.) Preproceedings of the International Conference on DNA 13, pp. 75–87 (2007)

7. Kari, L., Mahalingam, K.: Watson-Crick Palindromes (submitted)

8. Lothaire, M.: Combinatorics on words, Encyclopedia of Mathematics and its applications 17. Addison-Wesley Publishing Co, Reading (1983)

9. de Luca, A., De Luca, A.: Pseudopalindrome closure operators in free monoids. Theoret. Comput. Sci. 362, 282–300 (2006)

10. Mignosi, F., Restivo, A., Silva, P.S.: On Fine and Wilfs theorem for bidimensional words. Selected papers in honor of Jean Berstel, Theoret. Comput. Sci. 292(1), 245–262 (2003)

11. Shyr, H.J., Thierrin, G.: Disjunctive languages and codes. In: Proc. FCT77. LNCS, vol. 56, pp. 171–176. Springer, Heidelberg (1977)

12. Yu, S.S.: Languages and Codes, Lecture Notes, Department of Computer Science. National Chung-Hsing University, Taichung, Taiwan 402 (2005)

Complexity of Data Tree Patterns over XML Documents

Claire David

LIAFA, University Paris 7 and CNRS, France
cdavid@liafa.jussieu.fr

Abstract. We consider Boolean combinations of data tree patterns as a specification and query language for XML documents. Data tree patterns are tree patterns plus variable (in)equalities which express joins between attribute values. Data tree patterns are a simple and natural formalism for expressing properties of XML documents. We consider first the model checking problem (query evaluation), we show that it is DP-complete[1] in general and already NP-complete when we consider a single pattern. We then consider the satisfiability problem in the presence of a DTD. We show that it is in general undecidable and we identify several decidable fragments.

1 Introduction

The relational model and its popular query language SQL are widely used in database systems. However, it does not fit well in the ever changing Internet environment, since its structure is fixed by an initially specified schema which is difficult to modify. When exchanging and manipulating large amounts of data from different sources, a less structured and more flexible data model is preferable. This was the initial motivation for the Extensible Markup Language (XML) model which is now the standard for data exchange.

An XML document is structured as an unranked, labelled tree. The main difference with the relational model is that in XML, data is also extracted because of its position in the tree and not only because of its value. Consequently, all the tools manipulating XML data, like XML query languages and XML schema, combine navigational features with classical data extraction ones. XPath[2] is a typical example. It has a navigational core, known as Core-XPath and studied in [16], which is essentially a modal language that walks around in the tree. XPath also allows restricted tests on data attributes. It is the building block of most XML query languages (XQuery, XSLT...). Similarly, in order to specify integrity constraints in XML Schema, XML languages have navigational features for description of walks in the tree and selection of nodes. The nodes are for instance chosen according to a key or a foreign key [15].

---

[1] A problem DP is the intersection of a NP problem and a co-NP problem.
[2] In all the paper, XPath refers to XPath1.0.

E. Ochmański and J. Tyszkiewicz (Eds.): MFCS 2008, LNCS 5162, pp. 278–289, 2008.

In this paper, we study an alternative formalism as a building block for querying and specifying XML data. It is based on Boolean combinations of data tree patterns. A data tree pattern is essentially a tree with child or descendant edges, labelled nodes and (in)equality constraints on data values. Intuitively, a document satisfies a data tree pattern if there exists an injective mapping from the tree pattern into the tree that respects edges, node labels and data value constraints. Using patterns, one can express properties on trees in a natural, visual and intuitive way. These properties can express queries, as well as some integrity constraints.

At first glance, the injectivity requirement does not seem important; however, it has some consequences in terms of expressive power. As we do not consider horizontal order between siblings, without injectivity data tree patterns are invariant by bisimulation. Data tree patterns with injective semantics are strictly more expressive than with non-injective semantics. For example, it is not possible to express desirable properties such as a node has two a-labelled children without injectivity. Another consequence of injectivity appears when considering conjunctions of data tree patterns. With non-injective semantics, the conjunction of two patterns would be equivalent to a new pattern obtained by merging the two patterns at the root. With injectivity this no longer works and we have to consider conjunctions of tree patterns. This difference appears when we study the complexity of the satisfiability problem: for one pattern the problem is PTIME while it is untractable for a conjunction of patterns.

XPath and data tree patterns are incomparable in terms of expressiveness. Without data value, XPath queries are closed under bisimulation while data tree patterns are not. On the other hand, XPath allows negation of subformulas while we only allow negation of a full data tree pattern. For example XPath can check whether a node has a-labelled children but no b-labelled child. This is not possible with Boolean combinations of tree pattern. In terms of data comparison, Xpath allows very limited joins because XPath queries cannot compare more than two elements at a time, while a single pattern can compare simultaneously an arbitrary number of elements.

In this paper, to continue this comparison, we study the complexity of two questions related to data tree patterns: the model checking problem (query evaluation) and the satisfiability problem in the presence of schema.

The evaluation of XPath queries has been extensively studied (see [6] for a detailed survey). The evaluation problem is PTIME for general XPath queries. In our case, this problem is more difficult: the combined complexity of the model checking problem for Boolean combinations of data tree patterns is untractable. We prove that it is DP-complete in general and already NP-complete when considering only one tree pattern.

The satisfiability problem for XPath is undecidable in general [5]. However for many fragments the problem is decidable with a complexity ranging from NP to NEXPTIME. Similarly, for Boolean combinations of data tree patterns the satisfiability problem is undecidable in general. We identify several decidable fragments by restraining the expressivity of tree patterns or by bounding the depth of the documents. The corresponding complexities range from NP to 2EXPTIME.

Related Work: Tree patterns have already been investigated in a database context, often without data values [22,3,20]. The focus is usually optimisation techniques for efficient navigation [1,12,7]. In this work, we focus on the difficulty raised by data values and we are not interested in optimisation but in the worst case complexity for the model checking and satisfiability problems.

Several papers considered the non injective semantics of tree pattern with data constraints. First, [19] considered the satisfiability problem for one positive pattern while we consider Boolean combinations of tree pattern. Then, the authors of [2] consider the type checking problem which is more powerful that unsatisfiability but incomparable to the satisfiability problem.

Data tree patterns are used in [4] to specify data exchange settings. They study two problems: the first one is consistency of data exchange settings, the second one is query answering under data exchange settings. Given a conjunction of data tree patterns and a DTD, we can construct a data exchange setting such that the consistency of this setting is equivalent to the satisfiability of the conjunction of patterns in the presence of the DTD. However the data tree patterns they consider are less expressive than ours, in that they can not express inequality constraints on data values nor Boolean combinations of data tree patterns. The other problem considered in this paper is query answering. This problem seems related to our model checking problem. However it does not seem possible to use their result or their proof techniques.

Fragment of XPath: In [14], the authors consider an XPath fragment (simple XPath) allowing only vertical navigation but augmented with data comparisons. Negation is disallowed, both in the navigation part and in the comparison part. A simple XPath expression can be viewed as a pattern with non-injective semantics and only data equality. They study the inclusion problem of such expressions wrt special schemes (SXIC) containing integrity constraints like inclusion dependency. We cannot simulate inclusion dependency even with Boolean combinations of data tree patterns. Hence, their framework is incomparable to ours.

Conjunctive queries on trees: Conjunctive queries on trees can be expressed by tree patterns. They were considered in [17,8] without data values. Very recently [9], an extension by schema constraints is proposed and in very few cases they allow data comparison. Notice that, without sibling predicate, those conjunctive queries are strictly less expressive than our framework because they do not allow negation and do not have an injective semantic. It is shown that the query satisfiability problem is NP-complete, whereas the query validity problem is 2EXPTIME-complete. Moreover, the validity of a disjunction of conjunctive queries is shown to be undecidable. This last result corresponds to our undecidability result but the proof is different.

Logics over infinite alphabets: Another related approach is to consider logic for trees over an infinite alphabet. In [10,11], the authors study an extension of First Order Logic with two variables. In [13,18], the focus is on temporal logic and μ-calculus. These works are very elegant, but the corresponding complexities are non primitive recursive. Our work can be seen as a continuation of this work aiming for lower complexities.

Structure: Section 2 contains the necessary definitions. In Section 3, we consider the model-checking problem. In Section 4, we consider the satisfiability problem in general and the restricted cases. Section 5 contains a summary of our results and a discussion. Omitted proofs can be found in the appendix available at http://www.liafa.jussieu.fr/~cdavid/publi/mfcs08.pdf.

2   Preliminary

In this paper, we consider XML documents that are modeled as unordered, unranked *data trees*, as considered e.g. in [10].

Definition 1. A **data tree** over a finite alphabet Σ is an unranked, unordered, labelled tree with data values. Every node v has a *label* $v.l \in \Sigma$ and a *data value* $v.d \in \mathcal{D}$, where $\mathcal{D}$ is an infinite domain.

We only consider equality tests between data values. The data part of a tree can thus be seen as an equivalence relation $\sim$ on its nodes. In the following, we write $u \sim v$ for two nodes u, v, if $u.d = v.d$ and we use the term *class* without more precision to denote an equivalence class for the relation $\sim$.

The **data erasure** of a data tree t over Σ is the tree obtained from t by ignoring the data value $v.d$ of each node v of t.

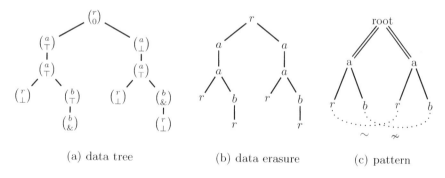

(a) data tree   (b) data erasure   (c) pattern

Fig. 1. Examples

Data tree patterns are a natural way to express properties of data trees, or to query such trees. They describe a set of nodes through their relative positions in the tree, and (in)equalities between their data values.

Definition 2. A **data tree pattern** $P = (p, C_\sim, C_\nsim)$ consists of:

- an unordered, unranked tree p, with nodes labelled either by Σ or by the wildcard symbol $*$, and edges labelled either by $|$ (child edges) or by $\|$ (descendant edge), and
- two binary relations $C_\sim$ and $C_\nsim$ on the set of nodes of p.

A data tree t satisfies a pattern $P = (p, C_\sim, C_\nsim)$, and we write $t \models P$, if there exists an *injective* mapping f from the nodes of p to the nodes of t that is consistent with the labelling, the relative positions of nodes, the branching structure and the data constraints. Formally, we require the following:

- for every node v from p with $v.l \in \Sigma$, we have $v.l = f(v).l$,
- for every pair of nodes (u,v) from p, if $(u, v) \in C_\sim$ (resp. $(u, v) \in C_\nsim$) then $f(u) \sim f(v)$ (resp. $f(u) \nsim f(v)$),
- for every pair of nodes (u,v) from p, if (u, v) is an edge of p labelled by $|$ (resp. by $\|$), then $f(v)$ is a child (resp. a descendant) of $f(u)$,
- for any nodes u,v,z from p, if (u, v) and (u, z) are both edges of p labelled by $\|$, then $f(v)$ and $f(z)$ are *not* related by the descendant relation in t.

A mapping f as above is called a **witness** of the pattern P in the data tree t.

Notice that the semantic does not preserve the least common ancestor and asks for an injective mapping between the nodes of a pattern and those of the tree. This enables patterns to express integrity constraints. We will discuss the impact of those choices in Section 5. Data tree patterns can describe properties that XPath cannot, see *e.g.* the pattern in Fig 1 (XPath cannot talk simultaneously about the two r-nodes and the two b-nodes).

We denote by $Ptn(\sim, |, \|)$ the set of data tree patterns and by $BC(\sim, |, \|)$ the set of Boolean combinations of data tree patterns. We will also consider restricted patterns, that do not use child relations or do not use descendant relations (denoted respectively by $Ptn(\sim, \|)$, $Ptn(\sim, |)$). From these, we derive the corresponding classes of Boolean combinations. Finally, BC^+ (resp. BC^-) denotes conjunctions of patterns (resp. negations of patterns).

In proofs, we consider the *parse tree* of a Boolean formula φ over patterns, denoted by $\mathcal{T}(\varphi)$. The leaves of this tree are labelled by (possibly negation of) patterns and inner nodes are labelled by conjunctions or disjunctions. Such trees are of linear size in the size of the formula and can be computed in PTIME.

Given a pattern formula from $BC(\sim, |, \|)$, the main problems we are interested in are the model-checking on a data tree (evaluation), the satisfiability problem, in the general case as well as for interesting fragments. Because the general structure of XML documents is usually constrained, we may consider DTDs as additional inputs. DTDs are essentially regular constraints on the finite structure of the tree. Since we work on unordered, unranked trees, we use as DTDs an unordered version of hedge automata. A DTD is a bottom-up automaton $\mathcal{A}$ where the transition to a state q' with label a is given by a Boolean combination of clauses of the form $\#q \leq k$ where q is a state and k a constant (unary encoded). A clause $\#q \leq k$ is satisfied if there are at most k children in state q. Adding a DTD constraint does not change the complexity results for the model-checking, since checking whether the data erasure of a tree satisfies a DTD is PTIME. Therefore, we do not mention DTDs in the model-checking part. We consider the following problems:

Problem 1. Given a data tree t and a pattern formula φ, the **model-checking** problem asks whether t satisfies φ.

Problem 2. Given a pattern formula φ and a DTD L, the **satisfiability** problem in the presence of a DTD asks whether φ is satisfied by some data tree whose data erasure belongs to L.

3 Model Checking

Patterns provide a formalism for expressing properties. In this section, we see how efficiently we can evaluate them. Our main result is the exact complexity of the model-checking problem for pattern formulas from $BC(\sim, |, \|)$.

Theorem 3. *The model-checking problem for $BC(\sim, |, \|)$ is DP-complete.*

The class of complexity DP is defined as the class of problems that are the conjunction of a NP problem and a co-NP problem [21]. In particular, DP includes both NP and co-NP. A typical DP-complete problem is SAT/UNSAT: given two propositional formulas φ_1, φ_2, it asks whether φ_1 is satisfiable, and φ_2 is unsatisfiable.

The key to the proof of Theorem 3 is the case where only one pattern is present. This problem is already NP-complete.

Proposition 4. The model-checking problem for a single pattern from $Ptn(\sim, |, \|)$ is NP-complete.

Proof. The upper bound is obtained by an algorithm guessing a witness for the pattern in the data tree and checking in PTIME whether the witness is correct. The lower bound is more difficult. It is obtained by a reduction of 3SAT.

Given a propositional formula φ in 3-CNF, we build a data tree t_φ and a pattern P_φ of polynomial size, such that $t_\varphi \vDash P_\varphi$ iff φ is satisfiable. Because we consider the model-checking problem, the data tree is fixed in the input. Thus, it must contain all possible valuations of the variables and at least all possible true valuations of each variable. Moreover, one positive data tree pattern should identify a true valuation of the formula and check its consistency. Hence, it does not seem possible to use previously published encodings of 3SAT into trees.

The pattern selects one valuation per variable and per clause. Its structure ensures that only one valuation per variable and per literal is selected. The constraints on data ensure the consistency of the selection. The data tree and the tree of the pattern depend only on the number of variables and clauses of the formula. Only the constraints on data of the pattern are specific to the formula. They encode the link between variables and clauses.

Let $\Sigma = \{r, \bar{r}, X, Y, Z, \#, \$\}$ be the finite alphabet. Assume that φ has k variables and n clauses. The data tree t_φ is composed of k copies of the tree t_v and n copies of the tree t_c as depicted in Figure 2. Even if we consider unordered trees, each copy of t_v corresponds to a variable of the formula and each copy of t_c to a clause. The tree t_φ involves exactly three classes, denoted as $0, \top, \bot$.

Each subtree t_v, see Figure 2(b), contains the two possible values for a variable. The left (right) branch of the tree represents true (resp. false).

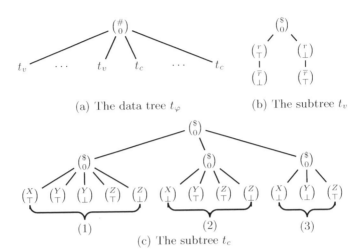

(a) The data tree t_φ (b) The subtree t_v

(1)   (2)   (3)

(c) The subtree t_c

Fig. 2. The data tree t_φ

A clause is viewed here as the disjunction of three literals, say X, Y, and Z. Each subtree t_c, see Figure 2(c), is formed by three subtrees. Each of them represents one of the three disjoint possibilities for a clause to be true: (1) X is true, or (2) X is false and Y is true, or (3) X and Y are false and Z is true.

We now turn to the definition of the tree pattern $P_\varphi = (tp_\varphi, C_\sim, C_\nsim)$, depicted in Figure 3. Similarly to t_φ, the tree tp_φ is formed by k copies of tp_v (each of them implicitly corresponding to a variable) and n copies of tp_c (each of them implicitly corresponding to a clause).

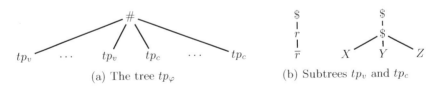

(a) The tree tp_φ (b) Subtrees tp_v and tp_c

Fig. 3. The tree tp_φ

The form of the data erasures of t_φ and tp_φ ensures that any witness of P_φ in t_φ selects exactly one value per variable and one (satisfying) valuation for each clause. Note that this is ensured by the definition of witness, since the witness mapping is injective.

It remains to define the data constraints $C_\sim$ and $C_\nsim$ in order to guarantee that each clause is satisfied. Assume that the first literal of clause c is a positive variable x (resp. the negation of x). Then we add in $C_\sim$ the r-position (resp. the $\bar{r}$-position) of the subtree tp_v corresponding to the variable x together with the X-position of the subtree tp_c corresponding to the clause c. The same can be done with the literals Y and Z. Figure 4 gives the example of the pattern for the formula φ with only the clause $a \vee \neg b \vee c$.

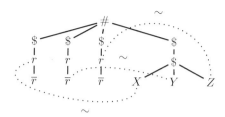

Fig. 4. Pattern for $\varphi = a \vee \neg b \vee c$

We now prove that $t_\varphi \vDash P_\varphi$ iff φ is satisfiable. Assume that the formula φ is satisfiable. From any satisfying assignment of φ we derive a mapping of P_φ into t_φ: the subtree p_v corresponding to the variable v is mapped on the left branch of the corresponding t_v if the value of v is true, and on the right branch otherwise. Since each clause is satisfied, one of the three cases represented by the subtree t_c happens, and we can map the tp_c corresponding to the clause on the branch of the corresponding t_c. The converse is similar.   □

In the proofs of Proposition 4 and Theorem 3, the patterns use only the child predicate. We can do the same with similar patterns using only the descendant predicate. As a consequence, we have:

Theorem 5. *The model-checking problem for both fragments* $BC(\sim, \|)$, $BC(\sim, |)$ *is DP-complete.*

Corollary 6. The model-checking problem for $BC^+(\sim, \|)$, $BC^+(\sim, |)$ and for $BC^+(\sim, |, \|)$ is NP-complete.

Similarly, we can see that the model-checking problem of a (conjunction of) negated pattern(s) is co-NP-complete. Notice that in the proof of Theorem 3, the pattern formula of the lower bound is a conjunction of one pattern and the negation of one pattern. Thus, the model-checking problem is already DP-complete for a conjunction of one pattern and one negated pattern.

The model checking problem for conjunctive queries is also exponential (NP) in relational databases. However, the algorithms work very well in practice, when models or queries are simple. In particular, when the query is acyclic, the problem becomes polynomial. The worst cases that lead to exponential behaviors do not appear often. It would be interesting to know how the algorithms following from our proofs behave on practical cases, and whether we can find some restriction on the patterns that would lead to efficient evaluation in practice.

4   Satisfiability

In this section, we study the satifiability problem in the presence of DTDs. Checking satisfiability of a query is useful for optimization of query evaluation or minimization techniques. In terms of schema design, satisfiability corresponds to checking the consistency of the specification.

We show that the satisfiability problem is undecidable in general. However the reduction needs the combination of negation, child and descendant operations. Indeed, removing any one of these features yields decidability, and we give the corresponding precise complexities.

4.1 Undecidability

Theorem 7. *The satisfiability problem for $BC(\sim, |, \|)$ in the presence of DTD is undecidable.*

Proof sketch. We prove the undecidability by a reduction from the acceptance problem of two-counter machines (or Minsky machines). Our reduction builds a DTD and a pattern formula of size polynomial in the size of the machine whose models are exactly the encodings of the accepting runs.

The encoding of a run can be split in three parts:

1. The general structure of the tree, which depends only on the data erasure, and is controlled by the DTD.
2. The internal consistency of a configuration.
3. The evolution of counter values between two successive configurations.

The global structure contains a branch that is labelled by the sequence of transitions. Ensuring that a tree is of this shape is done by the DTD. It recognizes the data erasure of sequences of configurations. In particular it checks that a counter is zero when this is required by the transition. It also ensures that the sequence of transitions respects the machine's rules (succession of control states, initial and final configurations).

The data values allow us to control the evolution of the counters between two consecutive configurations. In order to do so, we need to guarantee a certain degree of structure and continuity of the values through a run. The data structure and the evolution of counters are ensured by the pattern formula.   □

The proof uses only conjunctions of negated patterns. Thus, the satisfiability problem is already undecidable for the $BC^-(\sim, |, \|)$ fragment in the presence of a DTD. Alternatively, the DTD can be replaced by a pattern formula. To do so, we need a few positive patterns to constrain initial and final configurations in the coding. Thus, the satisfiability problem is undecidable for $BC(\sim, |, \|)$ without DTDs. It is interesting to notice that the satisfiability problem of $BC(\sim, |, \|)$ is undecidable on word models. We will discuss this in Section 5.

4.2 Decidable Restrictions

We can obtain decidability by restraining either the expressive power of pattern formulas or the data trees considered. For the first part, using only one kind of edge predicate ($|$ or $\|$) leads to decidability. For the second part, restricting the trees to bounded depth leads to decidability. We provide the exact complexities.

Restricted Fragments: The proof of undecidability uses both $\|$ and $|$ in the pattern to count unbounded values of the counters. If we restrict expressivity of patterns to use either $\|$ or $|$, we can't do this anymore and the problem becomes decidable. The key to both lower bounds is that patterns can still count up to a polynomial value and thus compare positions of a tree of polynomial depth. We use this idea to encode exponential size configurations of a Turing machine into the leaves of polynomial depth subtrees.

Theorem 8. *The satisfiability problem of $BC(\sim, |)$ in the presence of a DTD is* 2EXPTIME-*complete.*

Proof sketch. The upper bound is obtained by a small model property. We can prove that a pattern formula φ of $BC(\sim, |)$ is satisfiable in the presence of a given DTD iff it has a model with a number of classes that is doubly exponential in the size of the formula. We can recognize the data erasure of such small models with an automaton of size doubly exponential in the size of the formula. Because emptiness of such automata is PTIME, we have the 2EXPTIME upper bound.

The lower bound is obtained by a coding of accepting runs of AEXPSPACE Turing machines. We can build a DTD and a pattern formula from $BC(\sim, |)$ such that a data tree is a model on the pattern formula and respects the DTD iff it is the encoding of an accepting run of the machine.   □

Theorem 9. *The satisfiability problem of $BC(\sim, \|)$ in the presence of a DTD is* NEXPTIME-*complete.*

Bounded Depth restrictions: In the context of XML documents, looking at the satisfiability problem restricted to data trees of bounded depth is a crucial restriction. This restriction leads to decidability for $BC(\sim, <, +1)$.

Problem 3. Consider a pattern formula φ, an integer d and a DTD L. The problem of **bounded depth satisfiability in the presence of a DTD** asks whether φ is satisfiable by a data tree of depth smaller than d whose data erasure belongs to L.

Theorem 10. *If d is fixed, the bounded depth satisfiability problem in the presence of a DTD for $BC(\sim, \|, |)$ is Σ_2-complete.*

Theorem 11. *If d is part of the input, the bounded depth satisfiability problem in the presence of a DTD for $BC(\sim, \|, |)$ is* NEXPTIME-*complete.*

Other remarks: All the lower bound results of this section only use conjunctions of negated patterns. Thus these results hold for the BC^- fragments.

Proposition 12. *The satisfiability problem of a single pattern is* PTIME.

Proposition 13. *The satisfiability problem for $BC^+(\sim, |, \|)$ is NP-complete in the presence of DTD.*

5 Conclusion

The table below summarizes our results. bnd (resp. bnd_f) Sat stands for Bounded depth Satisfiability when the bound is part of the input (resp. fixed). The gray parts of the table gives complexity results for data words models. Data words are the linear model corresponding to data trees. This model is studied in the verification area [11,13]. Data patterns can also be considered for data words. The proofs are more complex and will be available in a longer version.

Fragments	Model-Checking	Satisfiability	bnd Sat	bnd_f Sat
$BC(\sim, \|, \mid)$	DP-complete	Undecidable	NExpTime-complete	Σ_2-complete
$BC(\sim, \mid)$	DP-complete	2ExpTime-complete	NExpTime-complete	Σ_2-complete
Data Word	PTime	PSpace-complete		
$BC(\sim, \|)$	DP-complete	NExpTime-complete	NExpTime-complete	Σ_2-complete
Data Word	DP-complete	Σ_2-complete		
$BC^-(\sim, \|, \mid)$	coNP-complete	Undecidable	NExpTime-complete	Σ_2-complete
Data Word	coNP-complete	undecidable		
$BC^+(\sim, \|, \mid)$	NP-complete	NP-complete	NP-complete	NP-complete
Data Word	NP-complete	NP-complete		

Discussion:

- In our framework we use the unordered version of trees. If we consider the next-sibling predicate, the situation is different. For the model checking problem all results hold with similar proofs. However, the complexity of the satisfiability problem can increase when negation is allowed. In particular the satisfiability problem for bounded depth tree becomes undecidable since we can encode data words.
- Recall that our pattern formalism does not preserve the least common ancestor. All results hold if we add the least common ancestor.
- An important issue of semi-structured databases is the containment problem. Given a DTD and two pattern formulas we want to know whether every tree satisfying the DTD and the first formula also satisfies the second one. When the set of formulas we consider is closed under negation, we can decide whether a formula φ_1 is more constraining than φ_2 by checking the satisfiability of $\varphi_2 \wedge \neg\varphi_1$. In Boolean combinations, we have closure under negation, hence the inclusion problem reduces to the satisfiability problem. For the positive fragment, the precise complexity seems harder to state and the question is left open.
- In terms of expressiveness, our pattern formalism is incomparable to XPath. In terms of tractability, evaluation of XPath queries is PTime whereas model-checking of one data tree pattern is already NP-hard. A question is to find good notions of constraints in order to isolate interesting fragments with lower complexity. Considering the complexity of the satisfiability problem, XPath and our pattern formalism behave similarly.
- In this paper, we only consider patterns as filters in order to define properties on the data trees. Defining a query language would be a natural extension of this work. To do this, some of the variables of the patterns can be chosen as output variables.

References

1. Al-Khalifa, S., Jagadish, H.V., Patel, J.M., Wu, Y., Koudas, N., Srivastava, D.: Structural Joins: A Primitive for Efficient XML Query Pattern Matching. In: ICDE, pp. 141–153. IEEE, Los Alamitos (2002)
2. Alon, N., Milo, T., Neven, F., Suciu, D., Vianu, V.: XML with data values: type-checking revisited. J. Comput. Syst. Sci. 66(4), 688–727 (2003)
3. Amer-Yahia, S., Cho, S., Lakshmanan, L.V.S., Srivastava, D.: Minimization of tree pattern queries. SIGMOD Rec. 30(2), 497–508 (2001)
4. Arenas, M., Libkin, L.: XML data exchange: consistency and query answering. In: PODS, pp. 13–24 (2005)
5. Benedikt, M., Fan, W., Geerts, F.: XPath satisfiability in the presence of DTDs. In: PODS, pp. 25–36 (2005)
6. Benedikt, M., Koch, C.: XPath Leashed. ACM Computing Surveys (to appear)
7. Benzaken, V., Castagna, G., Miachon, C.: CQL: a pattern-based query language for XML. In: BDA, pp. 469–490 (2004)
8. Björklund, H., Martens, W., Schwentick, T.: Conjunctive Query Containment over Trees. In: Arenas, M., Schwartzbach, M.I. (eds.) DBPL 2007. LNCS, vol. 4797, pp. 66–80. Springer, Heidelberg (2007)
9. Björklund, H., Martens, W., Schwentick, T.: Optimizing Conjunctive Queries over Trees using Schema Information. In: MFCS (to appear, 2008)
10. Bojanczyk, M., David, C., Muscholl, A., Schwentick, T., Segoufin, L.: Two-variable logic on data trees and XML reasoning. In: PODS, pp. 10–19 (2006)
11. Bojanczyk, M., Muscholl, A., Schwentick, T., Segoufin, L., David, C.: Two-Variable Logic on Words with Data. In: LICS, pp. 7–16. IEEE, Los Alamitos (2006)
12. Bruno, N., Koudas, N., Srivastava, D.: Holistic twig joins: optimal XML pattern matching. In: SIGMOD Conference, pp. 310–321. ACM, New York (2002)
13. Demri, S., Lazic, R.: LTL with the Freeze Quantifier and Register Automata. In: LICS, pp. 17–26. IEEE, Los Alamitos (2006)
14. Deutsch, A., Tannen, V.: Containment and integrity constraints for xpath. In: KRDB CEUR Workshop Proceedings. CEUR-WS.org, vol. 45 (2001)
15. Fan, W., Libkin, L.: On XML integrity constraints in the presence of DTDs. J. ACM 49(3), 368–406 (2002)
16. Gottlob, G., Koch, C., Pichler, R.: Efficient algorithms for processing XPath queries. ACM Trans. Database Syst. 30(2), 444–491 (2005)
17. Gottlob, G., Koch, C., Schulz, K.U.: Conjunctive queries over trees. J. ACM 53(2), 238–272 (2006)
18. Jurdzinski, M., Lazic, R.: Alternation-free modal mu-calculus for data trees. In: LICS, pp. 131–140. IEEE, Los Alamitos (2007)
19. Lakshmanan, L.V.S., Ramesh, G., Wang, H., Zhao, Z.J.: On Testing Satisfiability of Tree Pattern Queries. In: VLDB, pp. 120–131 (2004)
20. Neumann, A., Seidl, H.: Locating matches of tree patterns in forests. In: Arvind, V., Ramanujam, R. (eds.) FST TCS 1998. LNCS, vol. 1530, pp. 134–145. Springer, Heidelberg (1998)
21. Papadimitriou, C.H., Yannakakis, M.: The Complexity of Facets (and Some Facets of Complexity). J. Comput. Syst. Sci. 28(2), 244–259 (1984)
22. Wu, Y., Patel, J.M., Jagadish, H.V.: Structural Join Order Selection for XML Query Optimization. In: ICDE, pp. 443–454. IEEE, Los Alamitos (2003)

A PTAS for the Sparsest Spanners Problem on Apex-Minor-Free Graphs*

Feodor F. Dragan[1], Fedor V. Fomin[2], and Petr A. Golovach[2]

[1] Department of Computer Science, Kent State University, Kent, Ohio 44242, USA
dragan@cs.kent.edu
[2] Department of Informatics, University of Bergen, PB 7803, 5020 Bergen, Norway
{fedor.fomin,petr.golovach}@ii.uib.no

Abstract. A *t-spanner* of a graph G is a spanning subgraph S in which the distance between every pair of vertices is at most t times their distance in G. The SPARSEST t-SPANNER problem asks to find, for a given graph G and an integer t, a t-spanner of G with the minimum number of edges. On general n-vertex graphs, the problem is known to be NP-hard for all $t \geq 2$, and, even more, it is NP-hard to approximate it with ratio $O(\log n)$ for every $t \geq 2$. For $t \geq 5$, the problem remains NP-hard for planar graphs, and up to now the approximability status of the problem on planar graphs considered to be open. In this note, we resolve this open issue by showing that the SPARSEST t-SPANNER problem admits a *polynomial time approximation scheme (PTAS)* for every $t \geq 1$. Actually, our results hold for a much wider class of graphs, namely, on the class of *apex-minor-free graphs* which contains the classes of planar and bounded genus graphs.

1 Introduction

The concept of *sparse graph spanners* was introduced in [28] and [29] and has been studied since then in a number of papers, in the context of wired or wireless communication networks, distributed computing, robotics, computational geometry and biology [2,3,11,12,13,15,28,29]. A *t-spanner* of a graph G is a spanning subgraph S in which the distance between every pair of vertices is at most t times their distance in G. One is interested in finding a sparsest t-spanner for a graph G, i.e., a t-spanner with the minimum number of edges.

The original application of spanners was in the efficient simulation of synchronized protocols in unsynchronized networks [5,29]. Thereafter spanners were used in the design of low-stretch routing schemes using small routing tables (see [6,30] and the references therein), computing almost shortest paths in graphs [18], and in approximation algorithms for geometric spaces [27]. A recent application of spanners is in the design of approximate distance oracles and labeling schemes for arbitrary metrics; see [30,31] for further references. In all the applications cited above the quality of the solution is directly related to the quality of the

* Supported by Norwegian Research Council.

E. Ochmański and J. Tyszkiewicz (Eds.): MFCS 2008, LNCS 5162, pp. 290–298, 2008.
© Springer-Verlag Berlin Heidelberg 2008

underlying spanners. For example, in [29], close relationships were established between the quality of spanners (in terms of *stretch factor t* and the number of spanner edges), and the time and communication complexities of any synchronizer for the network based on this spanner.

Unfortunately, as it was shown in [28], the problem of determining, for a given graph G and integers t and m, whether G has a t-spanner with at most m edges is NP-complete. This indicates that it is unlikely to find in polynomial time an exact solution for the sparsest t-spanner problem in general graphs even for small values of t and m. Later, [24] showed that for every $t \geq 2$ there is a constant $c < 1$ such that it is NP-hard to approximate the sparsest t-spanner with the ratio $c \cdot \log n$, where n is the number of vertices in the graph. On the other hand, the problem admits a $O(\log n)$-ratio approximation for $t = 2$ [25,24] and a $O(n^{2/(t+1)})$-ratio approximation for $t > 2$ [21]. For some other inapproximability and approximability results for the sparsest t-spanner problem on general graphs we refer the reader to [19,20,21] and papers cited therein.

In this note, we consider the sparsest t-spanner problem on so-called *apex-minor-free graphs* which is a large class of graphs including all planar graphs and all graphs with bounded genus. Spanners for these graph classes were considered in [16]. Particularly, it was shown that for any fixed positive integer t and nonnegative integer r, it is possible to decide in a polynomial time whether a graph G has a t-spanner with at most $n - 1 + r$ edges. From another side, it is known that, on planar graphs, the problem of determining, for a given graph G and integers m and t, if G has a t-spanner with at most m edges is NP-complete for every fixed $t \geq 5$ (the case $2 \leq t \leq 4$ is open) [10]. This indicates that it is unlikely to find in polynomial time an exact solution for the sparsest t-spanner problem in planar graphs, too, and, consequently, a possible remaining course of action for investigating the problem is devising approximation algorithms for it.

Here, we show that the sparsest t-spanner problem admits a polynomial time approximation scheme (PTAS) on the class of apex-minor-free graphs for every $t \geq 1$ (and, hence, for the planar graphs and for the graphs with bounded genus). For NP-hard optimization problems, a PTAS is one of the best types of algorithm one can hope for. In proving our result, we employ the well known technique for solving NP-hard problems on planar graphs proposed by Baker [7] and generalized by Eppstein [22,23] (see also [14]) to graphs with bounded local treewidth (alias, apex-minor-free graphs). Previously, a PTAS was known only for the sparsest 2-spanner problem on 4-connected planar triangulations [17].

Our result also answers the following questions explicitly mentioned in in [10] and [17]:

– What is the approximability status of the sparsest t-spanner problem for planar graphs?
– Does a PTAS exist for the sparsest t-spanner problem for 4-connected planar triangulations and $t > 2$, or even for all planar graphs?

2   Preliminaries

In this section we present necessary definitions, notations and some auxiliary results.

Let $G = (V, E)$ be an undirected graph with the vertex set V and edge set E. We often will use notations $V(G) = V$ and $E(G) = E$. For $U \subseteq V$ by $G[U]$ is denoted the subgraph of G induced by U. The *distance* $\mathrm{dist}_G(u, v)$ between vertices u and v of a connected graph G is the length (the number of edges) of a shortest u, v-path in G.

Let t be a positive integer. A subgraph S of G, such that $V(S) = V(G)$, is called a *(multiplicative) t-spanner* of G, if $\mathrm{dist}_S(u, v) \leq t \cdot \mathrm{dist}_G(u, v)$ for every pair of vertices u and v. The parameter t is called the *stretch factor* of S. It is easy to see that the t-spanners can equivalently be defined as follows.

Proposition 1. *Let G be a connected graph, and t be a positive integer. A spanning subgraph S of G is a t-spanner of G if and only if for every edge (x, y) of G $\mathrm{dist}_S(x, y) \leq t$.*

Let $A \subseteq E(G)$. We call a subgraph S of G, such that for every edge $(x, y) \in A$ $\mathrm{dist}_S(x, y) \leq t$, a *partial t-spanner for A*. Clearly, if $A = E(G)$ then a partial t-spanner for this set is a t-spanner for G.

The SPARSEST t-SPANNER problem asks to find, for a given graph G and an integer t, a t-spanner of G with the minimum number of edges. Correspondingly, the SPARSEST PARTIAL t-SPANNER problem asks to find a partial t-spanner with the minimum number of edges for a given graph G, an integer t and a set $A \subset E(G)$.

A *tree decomposition* of a graph G is a pair (X, U) where U is a tree whose vertices we call *nodes* and $X = (\{X_i \mid i \in V(U)\})$ is a collection of subsets of $V(G)$ such that

1. $\bigcup_{i \in V(U)} X_i = V(G)$,
2. for each edge $(v, w) \in E(G)$, there is an $i \in V(U)$ such that $v, w \in X_i$, and
3. for each $v \in V(G)$ the set of nodes $\{i \mid v \in X_i\}$ forms a subtree of U.

The *width* of a tree decomposition $(\{X_i \mid i \in V(U)\}, U)$ equals $\max_{i \in V(U)} \{|X_i| - 1\}$. The *treewidth* of a graph G is the minimum width over all tree decompositions of G. We use notation $\mathbf{tw}(G)$ to denote the treewidth of a graph G.

It is said that a graph class $\mathcal{G}$ has bounded local treewidth if there is a function $f(r)$ (which depends only on r) such that for any graph G in $\mathcal{G}$, the treewidth of the subgraph of G induced by the set of vertices at distance at most r from any vertex is bounded above by $f(r)$. A graph class $\mathcal{G}$ has *linear local treewidth* if $f(r) = O(r)$. For example, it is known [9,1] that, for every planar graph G, $f(r) \leq 3r - 1$, and a corresponding tree decomposition of width at most $3r - 1$ of the subgraph induced by the set of vertices at distance at most r from any vertex can be found in time $O(rn)$.

Given an edge $e = (x, y)$ of a graph G, the graph G/e is obtained from G by contracting the edge e; that is, to get G/e we identify the vertices x and y

and remove all loops and replace all multiple edges by simple edges. A graph H obtained by a sequence of edge-contractions is said to be a *contraction* of G. H is a *minor* of G if H is a subgraph of a contraction of G. A graph class $\mathcal{G}$ is *minor-closed* if for every graph $G \in \mathcal{G}$ all minors of G are in $\mathcal{G}$, too.

We say that a graph G is H-*minor-free* when it does not contain H as a minor. We also say that a graph class $\mathcal{G}$ is H-*minor-free* (or, excludes H as a minor) when all its members are H-minor-free. Clearly, all minor-free graph classes are minor-closed.

An *apex graph* is a graph obtained from a planar graph G by adding a vertex and making it adjacent to some vertices of G. A graph class is *apex-minor-free* if it does not contain any graph with some fixed apex graph as a minor. For example, planar graphs (and bounded-genus graphs) are apex-minor-free graphs.

Eppstein [22,23] characterized all minor-closed graph classes that have bounded local treewidth. It was proved that they are exactly apex-minor-free graphs. These results were improved by Demaine and Hajiaghayi [14]. They proved that all apex-minor-free graphs have linear local treewidth.

3   Main Result

Many optimization problems can be solved efficiently for graphs of bounded treewidth by formulating the problem in a logical language, called *Monadic Second Order Logic* (abbr. MSOL). It is known that problems which can be expressed in this way can be solved in linear time for graphs with bounded treewidth [4]. We need such a result for the SPARSEST PARTIAL t-SPANNER problem.

Lemma 1. *Let k and t be positive integers. Let also G be a graph of treewidth at most k, and let $A \subseteq E(G)$. The* SPARSEST PARTIAL t-SPANNER *problem can be solved by a linear-time algorithm (the constant which is used in the bound of the running time depends only on k and t) if a corresponding tree decomposition of G is given.*

Proof. The SPARSEST PARTIAL t-SPANNER problem can formulated in MSOL as follows. We ask for a a subgraph S of G (i.e. a subset of edges) with the following property: for every edge $(x,y) \in A$ $\mathrm{dist}_S(x,y) \leq t$. This property is expressible in MSOL because $\mathrm{dist}_S(x,y) \leq t$ means that there are edges $(v_0,v_1),(v_1,v_2),\ldots,(v_{l-1},v_l) \in E(S)$ for some $l \leq t$ such that $x = v_0$ and $y = v_l$. Then the claim follows from the well known results of Arnborg et al.[4].   □

It should be noted also that the dynamic-programming algorithm for the case $A = E(G)$ was given by Makowsky and Rotics [26]. The algorithm of Makowsky and Rotics can be easily adapted to solve the problem for arbitrarily choice of A.

Let u be a vertex of a graph G. For $i \geq 0$ we denote by L_i the i-th level of breadth first search, i.e. the set of vertices at distance i from u. We call the partition of the vertex set $V(G)$ $\mathcal{L}(G,u) = \{L_0, L_1, \ldots, L_r\}$ *breadth first search (BFS) decomposition* of G. We assume for convenience that for BFS decomposition $\mathcal{L}(G,u)$ $L_i = \emptyset$ for $i < 0$ or $i > r$, and we use further negative indices and indices that are more than r. It can be easily seen that the BFS decomposition can be constructed by the breadth first search in a linear time.

Let G be a graph with BFS decomposition $\mathcal{L}(G, u) = (L_0, L_1, \ldots, L_r)$, and t be a positive integer. Suppose that $i \le j$ are integers. For $i \le j$ we define

$$G_{ij} = G[\bigcup_{k=i}^{j} L_k].$$

Graph G_{ij} is shown on Fig. 1.

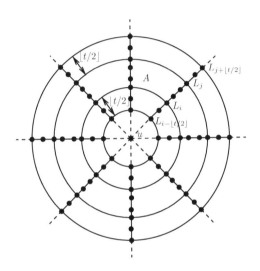

Fig. 1. Graphs G_{ij} and G'_{ij}

The following result is due to Demaine and Hajiaghayi [14] (see also the work of Eppstein [23]),

Lemma 2 ([14]). *Let G be an apex-minor-free graph. Then* $\mathbf{tw}(G_{ij}) = O(j-i)$.

Denote by $G'_{ij} = G_{i-\lfloor t/2 \rfloor, j+\lfloor t/2 \rfloor}$ (see Fig. 1), and let $A = E(G_{ij})$. Let S be a t-spanner of G and S' be the subgraph of S induced by $V(G'_{ij})$. We need the following claim.

Lemma 3. *S' is a partial t-spanner for A in G'_{ij}.*

Proof. Let $(x, y) \in A$. Note that $x, y \in V(G_{ij})$. Since S is a t-spanner for G, we have that there is a x, y-path P in S of length at most t. Suppose that some vertex v of this path does not belong to G'. Then $v \in L_l$ for some $l < i - \lfloor t/2 \rfloor$ or $l > j + \lfloor t/2 \rfloor$. By the definition of the BFS decomposition $\mathrm{dist}_G(x, v) > \lfloor t/2 \rfloor$ and $\mathrm{dist}_G(y, v) > \lfloor t/2 \rfloor$. But then P has length at least $\mathrm{dist}_G(x, v) + \mathrm{dist}_G(v, y) \ge 2\lfloor t/2 \rfloor + 2 > t$. So, all vertices of P are vertices of G'_{ij}, and this path is a path in S'.   □

Now we are ready to describe our algorithm. Let t, k be positive integers, $t < k$. For a given apex-minor-free graph G the BFS decomposition $\mathcal{L}(G, u) = (L_0, L_1, \ldots, L_r)$ is constructed for some vertex u.

If $r \leq k$ then a t-spanner S of G is constructed directly. We use the fact that $\mathbf{tw}(G) = O(k)$ and, for example, use Bodlaender's Algorithm [8] to construct in linear time a suitable tree decomposition of G. Then, by Lemma 1, a sparsest t-spanner of G can be found in linear time.

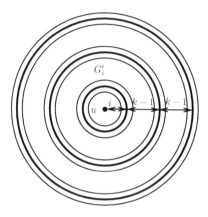

Fig. 2. Graphs G'_j

Suppose now that $r > k$. We consequently construct t-spanners S_i of G for $i = 1, 2, \ldots, k - 1$ as follows. Let

$$J_i = \{j \in \{2 - k, 3 - k, \ldots, r - 1\} : j \equiv i \ (\mod k - 1)\}.$$

For every $j \in J_i$ we consider graph $G'_j = G_{j - \lfloor t/2 \rfloor, j + k + \lfloor t/2 \rfloor - 1}$ and set of edges $A_j = E(G_{j, j+k-1})$. In other words, we "cover" graph G by graphs $G'_{i-(k-1)}, G'_i, G'_{i+(k-1)}, \ldots$, and two consecutive graphs "overlap" by $2\lfloor t/2 \rfloor + 1$ levels in the BFS decomposition (see Fig. 2). The union of all sets A_j is the set $E(G)$. By Lemma 2, $\mathbf{tw}(G'_j) = O(k + t)$. For every graph G'_j we construct a sparsest partial t-spanner S_{ij} for A_j in G'_j by making use of Lemma 1. We define

$$S_i = \bigcup_{j \in J_i} S_{ij}.$$

Finally, we choose among graphs $S_1, S_2, \ldots, S_{k-1}$ the graph with the minimum number of edges and denote it by S.

The following theorem describes properties of the graph S.

Theorem 1. *Let S be the subgraph of an apex-minor-free graph G, obtained by the algorithm described above. Then the following holds*

1. *S is a t-spanner of G.*
2. *For every t and $k > t$, S can be constructed by a linear-time algorithm.*
3. *S has at most $(1 + \frac{t+1}{k-1})\mathrm{OPT}(G)$ edges, where $\mathrm{OPT}(G)$ is the number of edges in the solution of the SPARSEST t-SPANNER problem on G.*

Proof. 1. Every S_i is a t-spanner of G. Indeed, for every $(x, y) \in E(G)$, there is $j \in J_i$ such that $(x, y) \in A_j$, and $\mathrm{dist}_{S_i}(x, y) \leq \mathrm{dist}_{S_{i_j}}(x, y) \leq t$.

2. The second claim yields by Lemmata 1 and 2.

3. If $k \geq r$ then the claim is obvious. Let $k < r$ and let T be a t-spanner of G with the minimum number of edges, $m = |E(T)| = \mathrm{OPT}(G)$. Assume that $i \in \{1, 2, \ldots, k-1\}$ and $j \in J_i$. Let $T_j = T[V(G'_j)]$. By Lemma 3, T_j is a partial t-spanner for the set A_j in T_j. Then

$$|E(T_j)| \geq |E(S_{ij})|,$$

and

$$|E(S_i)| \leq \sum_{j \in J_i} |E(T_j)|$$

$$= m + \sum_{j \in J_i} |E(T) \cap E(G_{j-\lfloor t/2 \rfloor, j+\lfloor t/2 \rfloor})|.$$

We have only to note that

$$|E(S)| = \min_{1 \leq i \leq k-1} |E(S_i)|$$

$$\leq m + \min_{1 \leq i \leq k-1} \sum_{j \in J_i} |E(T) \cap E(G_{j-\lfloor t/2 \rfloor, j+\lfloor t/2 \rfloor})|$$

$$\leq m + \min_{1 \leq i \leq k-1} \sum_{j \in J_i} |E(G_{j-\lfloor t/2 \rfloor, j+\lfloor t/2 \rfloor})|$$

$$\leq (1 + \frac{t+1}{k-1})m. \qquad \square$$

Finally, we have the following corollary.

Corollary 1. *For every $t \geq 1$, the SPARSEST t-SPANNER problem admits a PTAS with linear running time for the class of apex-minor-free graphs (and, hence, for the planar graphs and for the graphs with bounded genus).*

Note that, since the proof of Lemma 1 was not constructive, we can not claim that we have a very efficient algorithm. It would be interesting to find a more efficient solution to the problem at least for the class of planar graphs by utilizing the dynamic programming technique, the planarity of the graph and the specifics of the problem. For the planar graphs, the initial problem on G will be reduced to a subproblem of constructing a sparsest partial t-spanner for a subgraph of G with bounded outerplanarity.

References

1. Alber, J., Bodlaender, H.L., Fernau, H., Kloks, T., Niedermeier, R.: Fixed parameter algorithms for dominating set and related problems on planar graphs. Algorithmica 33, 461–493 (2002)
2. Althöfer, I., Das, G., Dobkin, D., Joseph, D., Soares, J.: On sparse spanners of weighted graphs. Discrete Comput. Geom. 9, 81–100 (1993)
3. Alzoubi, K., Li, X.-Y., Wang, Y., Wan, P.J., Frieder, O.: Geometric spanners for wireless ad hoc networks. IEEE Transactions on Parallel and Distributed Systems 14, 408–421 (2003)
4. Arnborg, S., Lagergren, J., Seese, D.: Easy problems for tree-decomposable graphs. J. Algorithms 12, 308–340 (1991)
5. Awerbuch, B.: Complexity of network synchronization. J. ACM 32, 804–823 (1985)
6. Awerbuch, B., Peleg, D.: Routing with polynomial communication-space trade-off. SIAM Journal on Discrete Mathematics 5, 151–162 (1992)
7. Baker, B.S.: Approximation algorithms for NP-complete problems on planar graphs. J. Assoc. Comput. Mach. 41, 153–180 (1994)
8. Bodlaender, H.L.: A linear-time algorithm for finding tree-decompositions of small treewidth. SIAM J. Comput. 25, 1305–1317 (1996)
9. Bodlaender, H.L.: A partial k-arboretum of graphs with bounded treewidth. Theoretical Computer Science 209, 1–45 (1998)
10. Brandes, U., Handke, D.: NP-completeness results for minimum planar spanners. Discrete Mathematics and Theoretical Computer Science 3, 1–10 (1998)
11. Cai, L.: Tree spanners: Spanning trees that approximate the distances, Ph.D. thesis, University of Toronto (1992)
12. Cai, L., Corneil, D.G.: Tree spanners. SIAM J. Disc. Math. 8, 359–387 (1995)
13. Chew, L.P.: There are planar graphs almost as good as the complete graph. J. of Computer and System Sciences 39, 205–219 (1989)
14. Demaine, E.D., Hajiaghayi, M.T.: Fast Algorithms for Hard Graph Problems: Bidimensionality, Minors, and Local Treewidth. In: Pach, J. (ed.) GD 2004. LNCS, vol. 3383, pp. 517–533. Springer, Heidelberg (2005)
15. Dobkin, D.P., Friedman, S.J., Supowit, K.J.: Delaunay graphs are almost as good as complete graphs. Discrete and Computational Geometry 5, 399–407 (1990)
16. Dragan, F.F., Fomin, F.V., Golovach, P.A.: Spanners in sparse graphs. In: Aceto, L., et al. (eds.) ICALP 2008, Part I. LNCS, vol. 5125, pp. 597–608. Springer, Heidelberg (2008)
17. Duckworth, W., Wormald, N.C., Zito, M.: A PTAS for the sparsest 2-spanner of 4-connected planar triangulations. J. Discrete Algorithms 1, 67–76 (2003)
18. Elkin, M.: Computing almost shortest paths. ACM Transactions on Algorithms 1, 283–323 (2005)
19. Elkin, M., Peleg, D.: Strong Inapproximability of the Basic t-Spanner Problem. In: Montanari, U., Rolim, J.D.P., Welzl, E. (eds.) ICALP 2000. LNCS, vol. 1853, pp. 636–647. Springer, Heidelberg (2000)
20. Elkin, M., Peleg, D.: The Hardness of Approximating Spanner Problems. In: Reichel, H., Tison, S. (eds.) STACS 2000. LNCS, vol. 1770, pp. 370–381. Springer, Heidelberg (2000)
21. Elkin, M., Peleg, D.: Approximating k-spanner problems for k > 2. Theoret. Comput. Sci. 337, 249–277 (2005)
22. Eppstein, D.: Subgraph isomorphism in planar graphs and related problems. In: SODA, pp. 632–640 (1995)

23. Eppstein, D.: Diameter and treewidth in minor-closed graph families. Algorithmica 27, 275–291 (2000)
24. Kortsarz, G.: On the Hardness of Approximating Spanners. Algorithmica 30, 432–450 (2001)
25. Kortsarz, G., Peleg, D.: Generating Sparse 2-Spanners. J. Algorithms 17, 222–236 (1994)
26. Makowsky, J.A., Rotics, U.: Optimal spanners in partial k-trees (manuscript)
27. Narasimhan, G., Smid, M.: Geometric Spanner Networks. Cambridge (2007)
28. Peleg, D., Schäffer, A.A.: Graph spanners. J. Graph Theory 13, 99–116 (1989)
29. Peleg, D., Ullman, J.D.: An optimal synchronizer for the hypercube. SIAM J. Comput. 18, 740–747 (1989)
30. Thorup, M., Zwick, U.: Compact routing schemes. In: Proceedings of the 13th Ann. ACM Symp. on Par. Alg. and Arch (SPAA 2001), pp. 1–10. ACM Press, New York (2001)
31. Thorup, M., Zwick, U.: Approximate distance oracles. J. ACM 52, 1–24 (2005)

Computational Complexity of Perfect-Phylogeny-Related Haplotyping Problems

Michael Elberfeld and Till Tantau

Institut für Theoretische Informatik
Universität zu Lübeck, 23538 Lübeck, Germany
{elberfeld,tantau}@tcs.uni-luebeck.de

Abstract. Haplotyping, also known as haplotype phase prediction, is the problem of predicting likely haplotypes based on genotype data. This problem, which has strong practical applications, can be approached using both statistical as well as combinatorial methods. While the most direct combinatorial approach, maximum parsimony, leads to NP-complete problems, the perfect phylogeny model proposed by Gusfield yields a problem, called PPH, that can be solved in polynomial (even linear) time. Even this may not be fast enough when the whole genome is studied, leading to the question of whether parallel algorithms can be used to solve the PPH problem. In the present paper we answer this question affirmatively, but we also give lower complexity bounds on its complexity. In detail, we show that the problem lies in Mod_2L, a subclass of the circuit complexity class NC^2, and is hard for logarithmic space and thus presumably not in NC^1. We also investigate variants of the PPH problem that have been studied in the literature, like the perfect path phylogeny haplotyping problem and the combined problem where a perfect phylogeny of maximal parsimony is sought, and show that some of these variants are TC^0-complete or lie in AC^0.

Keywords: bioinformatics, haplotyping, computational complexity, circuit classes, perfect phylogenies.

1 Introduction

We investigate the computational complexity of haplotype phase prediction problems. Haplotype phase prediction is an important preprocessing step in genomic disease and medical condition association studies. In these studies two groups of people are identified, where one group has a certain disease or medical condition while the other has not, and one tries to find correlations between group membership and the genomic data of the individuals in the groups. The genomic data typically consists of information about which bases are present in an individual's DNA at so-called SNP sites (single nucleotide polymorphism sites). While the DNA sequences of different individuals are mostly identical, at SNP sites there may be variations. Low-priced methods for large-scale inference

E. Ochmański and J. Tyszkiewicz (Eds.): MFCS 2008, LNCS 5162, pp. 299–310, 2008.

of genomic data can read out, separately for each SNP site, the bases present, of which there can be two since we inherit one chromosome from our father and one from our mother. However, since the bases at different sites are determined independently, we have no information on which chromosome a base belongs to. For *homozygous sites,* where the same base is present on both chromosomes, this is not a problem, but for *heterozygous sites* this information, called the *phase* of an SNP site, is needed for accurate correlations. The idea behind *haplotype phase prediction* or just *haplotyping* is to computationally predict likely phases based on the laboratory data (which misses this information). For an individual, the genomic input data without phase information is called the *genotype* while the two predicted chromosomes are called *haplotypes.*

There are both statistical [11,12,18] as well as combinatorial approaches to haplotyping. The present paper will treat only combinatorial approaches, of which there are two main ones: Given a set of observed genotypes, the maximum parsimony approach [5,6,13,16,19] tries to minimize the number of different haplotypes needed to explain the genotypes. The rationale behind this approach is that mutations producing new haplotypes are rare and, thus, genotypes can typically be explained by a small set of distinct haplotypes. Unfortunately, the computational problem resulting from the maximum parsimony approach, called MH for *maximum parsimony haplotyping,* is NP-complete [26]. This is remedied by Gusfield's perfect-phylogeny-based approach [17]. Here the rationale is that mutation events producing new haplotypes can typically be arranged in a perfect phylogeny (a sort of "optimal" evolutionary tree). The resulting problem, called PPH for *perfect phylogeny haplotyping,* can be solved in polynomial time as shown in Gusfield's seminal paper [17]. It is also possible to combine these two approaches, that is, to look for a minimal set of haplotypes whose mutation events form a perfect phylogeny, but the resulting problems are – not very surprisingly – NP-complete once more [1,27].

Due to the great practical importance of solving the PPH problem efficiently, a lot of research has been invested into finding quick algorithms for it and also for different variants. These efforts have culminated in the recent linear-time algorithms [8,23,24] for PPH. On the other hand, for a number of variants, in particular when missing data is involved, NP-completeness results can be obtained. (In the present paper we only consider the case where complete data is available.) These results have sparked our interest in, ideally, determining the exact computational complexity of these problems in complexity-theoretic terms. For instance, by Gusfield's result, PPH $\in$ P, but is it also hard for this class? Note that this question is closely linked to the question of whether we can find an efficient parallel algorithm.

Before we list the results obtained in the present paper, let us first describe the problems that we investigate (detailed definitions are given in the next section). The input is always a *genotype matrix,* whose rows represent individuals and whose columns represent SNP sites. An entry in this matrix encodes the measurement made for the given individual and the given SNP site. The question is always whether there exists a *haplotype matrix* with certain properties that *explains* the genotype matrix. A haplotype matrix explaining a genotype matrix has twice as many rows, namely two haplotypes for each individual, one from the father and one from the mother, and these two haplotypes taken together must explain exactly the observed genotype in the input matrix for this individual.

The *perfect phylogeny model* is an evolutionary model according to which mutations at a specific site can happen only once, in other words, there cannot be any "back-mutations." For haplotype matrices this means that there must exist an (evolutionary) tree whose nodes can be labeled with the haplotypes in the matrix in such a way that all haplotypes sharing a base at a given site form a connected subtree. In the perfect *path* phylogeny model [15] the phylogeny is restricted to be a very special kind of tree, namely a path. In the *directed* version of the perfect phylogeny model the tree is rooted and the root label is part of the input.

Problem	Question: Given a genotype matrix, an integer d (where applicable), and a root label (where applicable), does there exist an explaining haplotype matrix ...

MH   ...that has at most d different haplotypes?

PPH   ...that admits a perfect phylogeny?

DPPH   ...that admits a perfect phylogeny with the given root label?

MPPH   ...that admits a perfect phylogeny and has at most d different haplotypes?

PPPH   ...that admits a perfect path phylogeny?

DPPPH   ...that admits a perfect path phylogeny with the given root label?

MPPPH   ...that admits a perfect path phylogeny and has at most d different haplotypes?

Our Results. In this paper we show that PPH is hard for logarithmic space under first-order reductions. This is the first lower bound on the complexity of this problem. We also show PPH $\in$ Mod$_2$L, which is a close (but not matching) upper bound on the complexity of this central problem.

We show that the PPPH problem, where the tree topology of the perfect phylogeny is restricted to a path, is (provably) easier: This problem lies in FO, which is the same as the uniform circuit class AC^0 (constant depth, unbounded fan-in, polynomial-size circuits). This implies, in particular, that PPPH cannot be hard for logarithmic space – unlike PPH. To obtain this results we extend the partial order method introduced by Gramm et al. [15] for directed perfect path phylogenies to the undirected case. We also show that MPPPH is complete for the uniform threshold circuit complexity class TC^0, which is the same as FO(COUNT).

Restricting the tree topology to a path is one way of simplifying the PPH problem. Another approach that has also been studied in the literature is to restrict the number of heterozygous entries in the genotype matrices. We show that PPH is in FO when genotype matrices are restricted to contain at most two heterozygous entries per row or at most one heterozygous entry per column. In contrast, if we allow at most three heterozygous entries per row, PPH is still L-hard.

Related Work. Perfect phylogeny haplotyping was suggested by Gusfield [17]. The computational complexity of the PPH problem is quite intriguing since, at first sight, it is not even clear whether this problem is solvable in polynomial time. Gusfield showed that this is, indeed, the case and different authors soon proposed simplified algorithms with easier implementations [2,10]. A few years later three groups independently devised linear time algorithms for the problem [8,23,24]. All these papers are concerned with the time complexity of PPH and all these algorithms have at least linear space complexity.

Haplotyping with perfect path phylogenies was introduced by Gramm et al. [15] in an attempt to find faster algorithms for restricted versions of PPH. For instance, Gramm et al. present a simple and fast linear-time algorithm for DPPPH and they show that the version with incomplete data is fixed parameter tractable with respect to the number of missing entries per site. These results all suggested (but did not prove) that the PPPH problem is somehow "easier" than PPH. The results of the present paper, namely that PPPH ∈ FO while PPH is L-hard, settle this point.

Our result that MPPPH is TC^0-complete contrasts sharply with the NP- completeness of MPPH proved in [1,27]. One might try to explain these contrasting complexities by arguing that "considering perfect *path* phylogenies rather than perfect phylogenies makes the problems trivial anyway, so this is no surprise," but this is not the case: For instance, the incomplete perfect path phylogeny problem, IPPPH, is known to be NP-complete [15] just like IPPH.

Van Iersel et al. [27] have studied the complexity of MPPH for inputs with a bounded number of heterozygous entries and they show that for certain bounds the problem is still NP-complete while for other bounds it lies in P. These results are closely mirrored by the results of the present paper, only the classes we consider are much smaller: The basic PPH problem is L-hard and so are some restricted versions, while other restricted versions lie in FO. It turns out that, despite the different proof techniques, the restrictions that make MPPH lie in P generally also make PPH lie in FO, while restrictions that cause MPPH to be NP-hard also cause PPH to be L-hard.

Organization of This Paper. After the following preliminaries section, Section 3 presents our results on PPH, including the variants of PPH with restricted inputs matrices. In Section 4 we present the results on PPPH and the maximum parsimony variant MPPPH. Due to lack of space, most proofs are omitted; the missing proofs can be found in the technical report version [9].

2   Basic Notations and Definitions

2.1   Haplotypes, Genotypes, Perfect Phylogenies, Induced Sets

Conceptually, a haplotype describes genetic information from a single chromosome. When SNP sites are used for this purpose, a haplotype is a sequence of the bases A, C, G and T. In the human genome, at any given SNP site at most two different bases can be observed in almost all cases (namely an original base and a mutated version). Since these two bases are known and fixed for a given site, we can encode one base with 0 and the other with 1 (for this particular site). For example, the haplotypes AAGC and TATC might be encoded by 0110 and 1100. A genotype combines two haplotypes by joining their entries. For example, the haplotypes AAGC and TATC lead to the genotype $\{A, T\}\{A\}\{G, T\}\{C\}$ and so do the two haplotypes AATC and TAGC. Given a genotype, we call sites with only one observed base *homozygous* and sites with two bases *heterozygous*. It is customary to simplify the representation of genotypes by encoding a homozygous entry by 0 or 1 (depending on the single base present) and heterozygous entries with the number 2. Thus, we encode the above genotype by 2120.

Formally, a *haplotype* is a vector of 0's and 1's. A *genotype* is a vector of 0's, 1's and 2's. Given a vector c, let $c[i]$ denote the ith component. Two haplotypes $h_1, h_2 \in \{0, 1\}^n$ *explain* a genotype $g \in \{0, 1, 2\}^n$ if for each $i \in \{1, \ldots, n\}$ we have $g[i] = h_1[i] = h_2[i]$ whenever $g[i] \in \{0, 1\}$ and $h_1[i] \neq h_2[i]$ whenever $g[i] = 2$. To examine multiple genotypes or haplotypes, we arrange them in matrices. The rows of a *haplotype matrix* are haplotypes and the rows of a *genotype matrix* are genotypes. We call a column of a matrix *polymorphic* if it contains both 0-entries and 1-entries or a 2-entry. We say that a $2n \times m$ haplotype matrix B *explains* an $n \times m$ genotype matrix A if for each $i \in \{1, \ldots, n\}$ the rows $2i - 1$ and $2i$ of B explain the row i of A.

Perfect phylogenies for haplotype and genotype matrices are defined as follows:

Definition 1. *A haplotype matrix B admits a perfect phylogeny if there exists a rooted tree T_B, called a perfect phylogeny for B, such that:*

1. *Each row of B labels exactly one node of T_B.*
2. *Each column of B labels exactly one edge of T_B.*
3. *Each edge of T_B is labeled by at least one column of B.*
4. *For every two rows h_1 and h_2 of B and every column i, we have $h_1[i] \neq h_2[i]$ if, and only if, i lies on the path from h_1 to h_2 in T_B.*

A genotype matrix A admits a perfect phylogeny *if there exists a haplotype matrix B that explains A and admits a perfect phylogeny.*

We say that T_B is a *perfect path phylogeny* if the topology of T_B is a path, that is, T_B consists of at most two disjoint branches emanating from the root. If the root of T_B is labelled with a haplotype given beforehand, we call T_B *directed*. Since the roles of 0's and 1's can be exchanged individually for each column, we will always require the given haplotype to be the all-0-haplotype. Formally, we say that a haplotype matrix B *admits a directed perfect (path) phylogeny* if there exists a perfect (path) phylogeny as in Definition 1 with the root label 0^n.

The *four gamete property* is a well-known alternative characterization of perfect phylogenies: A haplotype matrix admits a perfect phylogeny if, and only if, no column pair contains the submatrix $\begin{bmatrix} 0 & 0 \\ 0 & 1 \\ 1 & 0 \\ 1 & 1 \end{bmatrix}$. By the four gamete property it is important to know for each pair of columns which combinations of 0's and 1's are (or must be) present in the pair. Following Eskin, Halperin, and Karp [10] we call these combinations the *induced set* of the columns. Formally, given a haplotype matrix B and a pair of columns (c, c') the *induced set* $\text{ind}_B(c, c') \subseteq \{00, 01, 10, 11\}$ contains all bitstrings $ab \in \{00, 01, 10, 11\}$ for which there is a row r in B such that the entry in column c is a and the entry in column c' is b. For a genotype matrix A and two columns, the *induced set* $\text{ind}_A(c, c') \subseteq \{00, 01, 10, 11\}$ is the intersection of all $\text{ind}_B(c, c')$ where B is any haplotype matrix that explains A. This means, for instance, that the induce of the two columns of the genotype matrix $\left[\begin{smallmatrix} 0 & 1 \\ 2 & 0 \end{smallmatrix}\right]$ is $\{01, 00, 10\}$, the induce of $\left[\begin{smallmatrix} 2 & 0 \\ 1 & 2 \end{smallmatrix}\right]$ is $\{00, 10, 11\}$, and the induce of $\left[\begin{smallmatrix} 0 & 0 \\ 2 & 2 \end{smallmatrix}\right]$ is $\{00\}$.

For a genotype with at least two heterozygous entries there exist multiple pairs of explaining haplotypes. If a genotype g contains $[2\ 2]$ in columns c and c', then two explaining haplotypes for g contain either $\left[\begin{smallmatrix} 0 & 0 \\ 1 & 1 \end{smallmatrix}\right]$ or $\left[\begin{smallmatrix} 0 & 1 \\ 1 & 0 \end{smallmatrix}\right]$ in c and c'. In the first case we say that g *is resolved equally in* (c, c') and in the second case we say that g *is resolved unequally in* (c, c'). By the four gamete property, when a genotype matrix admits a perfect phylogeny, for each column pair all genotypes are resolved either equally or unequally. The resolution of a column pair is often, but not always, determined by the induce: In order to obtain a perfect phylogeny, a column pair that induces 00 and 11 must be resolved equally and a column pair that induces 01 and 10 must be resolved unequally.

2.2 Complexity Classes, Circuit Classes, Descriptive Complexity Theory

The classes L, P, and NP denote logarithmic space, polynomial time, and nondeterministic polynomial time, respectively. The class Mod_2L contains all languages L for which there exists a nondeterministic logspace Turing machine such that $x \in L$ if, and only if, the number of accepting computation paths on input x is odd. The circuit classes AC^0, TC^0, NC^1, and NC^2, all of which are assumed to be uniform in the present paper, are defined as follows: AC^0 is the class of problems that can be decided by a logspace-uniform family of constant-depth and polynomial-size circuits over and-, or- and not-gates with an unbounded fan-in. For TC^0 we may additionally use threshold gates. The class NC^1 contains all languages that can be decided by a logspace-uniform family of polynomial-size, $O(\log n)$-depth circuits over and-, or- and not-gates with bounded fan-in. For the class NC^2 the depth only needs to be $O(\log^2 n)$.

We use several notions from descriptive complexity theory, which provides equivalent characterizations of the classes AC^0 and TC^0. In descriptive complexity theory inputs are encoded as logical structures. A genotype matrix is described by the logical structure $\mathcal{A} = (I^{\mathcal{A}}, A_0^{\mathcal{A}}, A_1^{\mathcal{A}}, A_2^{\mathcal{A}}, n_r^{\mathcal{A}}, n_c^{\mathcal{A}})$ as follows: $I^{\mathcal{A}}$ is a finite set of indices, which are used both for rows and columns, and $n_r^{\mathcal{A}}$ and $n_c^{\mathcal{A}}$ are elements from $I^{\mathcal{A}}$ that are the maximum row and column indices, respectively. The relation $A_0^{\mathcal{A}} \subseteq I^{\mathcal{A}} \times I^{\mathcal{A}}$ indicates 0-entries, that is, $(r, c) \in A_0^{\mathcal{A}}$ holds exactly if the row r has a 0-entry in column c. The relations $A_1^{\mathcal{A}}$ and $A_2^{\mathcal{A}}$ are defined similarly, only for 1- and 2-entries. We assume that the universe $I^{\mathcal{A}}$ is ordered (we always have free access to a predicate $<$), but will not need the bit-predicate (see [20] for a discussion of its importance).

Given a formula, the set of all finite logical structures satisfying the formula (that are models of the formula) can be regarded as a language. If the formula is a first-order formula, the described language is called *first-order definable*. The class of all such languages is denoted FO and equals AC^0. For example, the formula $(\exists r, c)[r \leq n_r \wedge c \leq n_c \wedge A_2(r, c)]$ is true for genotype matrices that contain a row with a heterozygous entry and, therefore, defines the set of genotype matrices with at least one heterozygous entry. The computational power of first-order logic can be increased by adding an additional number domain and counting quantifiers. This class, which is called FO(COUNT), equals TC^0.

To describe mappings between logical structures, one can use *first-order queries*, which are tuples of defining formulas for the relations of the image structure. Since L is closed under reductions that can be described by first-order queries, we use these *first-order reductions* to prove L-hardness.

The inclusion structure between the described classes is known to be the following:

$$FO = AC^0 \subsetneq FO(COUNT) = TC^0 \subseteq NC^1 \subseteq L \subseteq Mod_2L \subseteq NC^2 \subseteq P \subseteq NP.$$

3   Complexity of Haplotyping Via Perfect Phylogenies

In the present section we study the complexity of PPH as well as the variants where the number of heterozygous entries in the input is restricted. In Section 3.1 we show that PPH and its directed variant are L-hard and in Mod_2L. Thus, both problems are in NC^2 by the inclusion $Mod_2L \subseteq NC^2$ shown in [3], but not in NC^1, unless $L = NC^1$. In Section 3.2 we additionally take restrictions into account and show that the hardness result still holds when we restrict the input to genotype matrices with at most three heterozygous entries per row. In contrast, we show that PPH is first-order definable for genotype matrices with at most two heterozygous entries per row or at most one heterozygous entry per column.

We will focus on the directed version DPPH rather than PPH since Eskin, Halperin and Karp [10] have shown that PPH reduces to DPPH via an easy construction: In each column, search downward for the first homozygous entry and if it equals 1, exchange the roles of 0 and 1 in this column. Indeed, this construction is so simple that it can be implemented using a first-order query: for each homozygous entry we have to decide whether it is inverted and this depends on the value of the first homozygous entry in the same column, which in turn is easy to determine for an ordered universe (recall that we always have access to an ordering of the universe). Note that DPPH trivially reduces to PPH by adding a row with only 0-entries to the matrix. Interestingly, the path variants PPPH and DPPPH are also equivalent via first-order reductions, which is shown in Section 4, but this is harder to prove.

3.1   Complexity of the PPH Problem

In the present section we prove two theorems on the complexity of PPH. The first gives a lower bound, namely that PPH is hard for logarithmic space under first-order reductions. The second gives an upper bound, namely PPH ∈ Mod₂L. Since DPPH is first-order equivalent to PPH, the same results hold for the directed version, also. The bounds show that both problems are in NC^2, but not in NC^1 under common assumptions from complexity theory. We thank Arfst Nickelsen for hinting at the basic idea of the proof of Theorem 1 in a personal communication.

Theorem 1. PPH *is hard for* L *under first-order reductions.*

Proof (sketch). We show that there is a first-order reduction from the reachability problem for undirected graphs to the complement of PPH. Given a graph G and a start node s and a target node t we first modify the graph such that any path from s to t must have even length. Then we construct a genotype matrix such that the resolutions of certain column pairs are predetermined by induces. The matrix is setup in such a way that an edge joining two nodes enforces that two specific columns in the matrix are resolved unequally. This can be used to enforce that columns corresponding to s and to nodes at an even distance from s must be resolved equally. By adding further restrictions that enforce that the columns of s and t must be resolved unequally, we can ensure that (a) if s and t lie in different components there exists an explaining haplotype matrix that admits a perfect phylogeny and (b) otherwise every explaining haplotype matrix violates the four gamete property.   □

Theorem 2. PPH *is in* Mod₂L.

Proof (sketch). PPH can be logspace-many-one reduced to solving systems of linear equations over $\mathbb{Z}/2\mathbb{Z}$. This reduction is implicit in the construction of Theorem 1 of the paper by Eskin, Halperin, and Karp [10]. Solving systems of linear equations over $\mathbb{Z}/2\mathbb{Z}$ is in Mod₂L as shown in [3] and since Mod₂L is closed under logspace-many-one reductions, we get the claim.   □

3.2   Complexity of PPH for Restricted Instances

How do restrictions on the number of heterozygous sites influence the complexity of PPH? This question will be addressed in the present section. Following Sharan, Halldórsson, and Istrail [26] we say that a genotype matrix is (k, l)-*bounded* if each row contains at most k and each column at most l heterozygous entries. We use a star to indicate that a parameter is not bounded. We parametrize problems in the same way, so PPH$(3, *)$ denotes the set of all genotype matrices with at most three heterozygous entries per row that admit a perfect phylogeny.

In the literature (k, l)-bounded variants were first studied for the NP-complete problem MH. The hope was to find restrictions that hold in practice and that make the problem tractable. In different papers parameters k and l have been determined such that MH(k, l) is either efficiently solvable or NP-complete [4,21,22,26,27]. Bounded variants of MPPH have also been studied and the main results are the same as for the corresponding variants of MH.

We study the complexity of (k, l)-bounded variants of PPH. Our results, summarized in Theorem 3, show strong similarities to the complexity of bounded variants of MH and MPPH, but one complexity level further down. We show that PPH$(3, *)$ is L-hard; and it is known [27] that MH$(3, *)$ and MPPH$(3, *)$ are NP-complete. We show PPH$(2, *) \in$ FO and PPH$(*, 1) \in$ FO; and it is known [4,22,27] that MH$(2, *)$, MPPH$(2, *) \in$ P and MH$(*, 1)$, MPPH$(*, 1) \in$ P. We do not know the complexity of PPH$(*, 2)$; and the complexity MH$(*, 2)$ and MPPH$(*, 2)$ are also open.

Theorem 3

1. PPH$(3, *)$ *is* L-*hard.*
2. PPH$(2, *)$ *is first-order definable.*
3. PPH$(*, 1)$ *is first-order definable.*

4 Complexity of Haplotyping Via Perfect Path Phylogenies

In the present section we consider perfect *path* phylogenies, a problem variant first advocated in Gramm et al. [15] and later studied in [14]. In the first of these papers it is shown that PPPH is solvable in linear time. We show that it is first-order definable and, therefore, in AC^0. To obtain this result, we first prove DPPPH $\in$ FO and then reduce PPPH to DPPPH by a first-order reduction. While this reduction is similar to the reduction from PPH to DPPH in Section 3, the correctness proof for the path variant differs. Bafna et al. [1] have shown the NP-completeness of MPPH. For the path variant MPPPH we show that it can be described by a first-order formula with counting quantifiers and is, therefore, in TC^0. In addition, we prove that MPPPH is hard for TC^0.

4.1 Complexity of the Basic Decision Problem

In the present section we show that the set of all genotype matrices admitting a perfect path phylogeny can be described using a first-order formula; in other words, we show PPPH $\in$ FO, see Theorem 6. In order to prove this, we first show that the simpler problem DPPPH lies in FO and then show that PPPH can be first-order reduced to its directed version.

We start with some notations and a characterization for DPPPH from the literature. Then we present an alternative characterization that can be formalized with first-order logic and a first-order reduction from PPPH to DPPPH. While the construction of this reduction is easy, its correctness proof is not. Finally, we conclude that PPPH is first-order definable since the class of first-order definable problems is closed under first-order reduction.

We define a partial order on the columns of a genotype matrix as follows: Let $1 \succ 2 \succ 0$. For two columns c and c' we have $c \succeq c'$ if $c[i] \succeq c'[i]$ for each row i. We say that two columns c and c' are *comparable* if $c \succeq c'$ or $c' \succeq c$. If $c \succeq c'$ and $c \neq c'$, then we say that c *dominates* c'. In the following let C be a set of columns. A subset of C is called an *(anti)chain* if its elements are pairwise (in)comparable. An antichain $C' \subseteq C$ is *maximal* if it is not properly contained in any other antichain. A maximal antichain C' of size i is the *highest maximal antichain of size* i if there is no other antichain of size (exactly) i with an element that dominates an element from C'. For a set C of columns there exists at most one highest maximal antichain of size i, which we denote by $\text{hma}_i(C)$. Let $\text{hma}_i(C) = \emptyset$ if there is no such set.

We call two columns *separable* if each column has a 0-entry in the rows where the other has a 1-entry. Following [15] we say that a column set C *has the ppp-property* if there exist two (possibly empty) chains C_1 and C_2 that cover C, so that their maximal elements (if they exist) are separable. We call (C_1, C_2) a *ppp-cover* of C. The following fact characterizes DPPPH.

Fact 4 (Gramm et al. [15]). *A genotype matrix A admits a directed perfect path phylogeny if, and only if, the set of columns of A has the ppp-property.*

The above characterization does not readily yield a first-order description of DPPPH since we cannot quantify over chains (a second-order quantifier would be needed, lifting the complexity up to the polynomial hierarchy). What we need is a more "element-oriented" characterization such as the one given by the following lemma.

Lemma 1. *A column set C has the ppp-property if, and only if, the width of C is at most 2 and one of the following statements is true:*

1. $\text{hma}_1(C) = \{c^*\}$ *and* $\text{hma}_2(C) = \emptyset$.
2. $\text{hma}_1(C) = \emptyset$, $\text{hma}_2(C) = \{c_1, c_2\}$, *and* c_1 *and* c_2 *are separable.*
3. $\text{hma}_1(C) = \{c^*\}$, $\text{hma}_2(C) = \{c_1, c_2\}$, *and* c^* *and* c_1 *are separable or* c^* *and* c_2 *are separable.*

Theorem 5. DPPPH *is first-order definable.*

Theorem 6. PPPH *is first-order definable.*

To prove Theorem 5, it suffices to show that the characterization given in Lemma 1 can be tested using a first-order formula. Theorem 6 can be proved by a first-order reduction from PPPH to DPPPH.

4.2 Combining Perfect Path Phylogenies and Maximum Parsimony

In the present section we prove that MPPPH is TC^0-complete, in stark contrast to the fact that MPPH is NP-complete.

Theorem 7. MPPPH *is* TC^0-*complete under* AC^0-*reductions.*

Proof (sketch). First, we show that $(A, d) \in$ MPPPH if, and only if, $A \in$ PPPH and d is greater than the number of distinct polymorphic columns in A. Due to this characterization, MPPPH has nearly the same complexity as PPPH, we only need to add counting quantifiers, which are used to count the number of distinct polymorphic columns in an input matrix. This implies MPPPH $\in \text{TC}^0$. To prove the TC^0-hardness of MPPPH, we present an AC^0-reduction from MAJORITY, where a binary string is given and the question is whether at least half of the input bits are 1. We construct unique genotypes for bits that equal 0 and use an MPPPH oracle gate with an appropriate budget value to count them.   □

5 Conclusion

The three main results of the present paper are that (a) the complexity of PPH lies between L and Mod_2L, (b) while PPPH lies in AC^0 and MPPPH is TC^0-complete, and (c) restricted variants of PPH are either L-hard or they lie in AC^0. Concerning the latter results, the complexity of a few restricted variants is still open. In particular, what is the complexity of PPH(3, 2)?

A much broader, still largely open research field is the complexity of these problems when data may be missing. Typically, the resulting problems are NP-complete, so we need to look for approximation algorithms, fixed-parameter algorithms, or moderately exponential time algorithms. Specialized results are known in this context, but there are still only few precise complexity-theoretic results in this setting.

References

1. Bafna, V., Gusfield, D., Hannenhalli, S., Yooseph, S.: A note on efficient computation of haplotypes via perfect phylogeny. J. Comput. Biol. 11(5), 858–866 (2004)
2. Bafna, V., Gusfield, D., Lancia, G., Yooseph, S.: Haplotyping as perfect phylogeny: A direct approach. J. Comput. Biol. 10(3–4), 323–340 (2003)
3. Buntrock, G., Damm, C., Hertrampf, U., Meinel, C.: Structure and importance of logspace-MOD-classes. Math. Syst. Theor. 25(3), 223–237 (1992)
4. Tromp, J., Cilibrasi, R., van Iersel, L., Kelk, S.: On the Complexity of Several Haplotyping Problems. In: Casadio, R., Myers, G. (eds.) WABI 2005. LNCS (LNBI), vol. 3692, pp. 128–139. Springer, Heidelberg (2005)
5. Clark, A.G.: Inference of haplotypes from PCR-amplified samples of diploid populations. J. Mol. Biol. and Evol. 7(2), 111–122 (1990)
6. Daly, M., Rioux, J., Schaffner, S., Hudson, T., Ladner, E.: High-resolution haplotype structure in the human genome. Nat. Genet. 29, 229–232 (2001)
7. Dilworth, R.P.: A decomposition theorem for partially ordered sets. Ann. Math. 51(1), 161–166 (1950)
8. Ding, Z., Filkov, V., Gusfield, D.: A linear-time algorithm for the perfect phylogeny haplotyping (PPH)problem. J. Comput. Biol. 13(2), 522–553 (2006)
9. Elberfeld, M., Tantau, T.: Computational complexity of perfect-phylogeny-related haplotyping problems. Tech. Rep. SIIM-TR-A-08-02, Universitat zu Lubeck (2008)
10. Eskin, E., Halperin, E., Karp, R.M.: Efficient reconstruction of haplotype structure via perfect phylogeny. J. Bioinform. and Comput. Biol. 1(1), 1–20 (2003)
11. Excoffier, L., Slatkin, M.: Maximum-likelihood estimation of molecular haplotype frequencies in a diploid population. Mol. Biol. and Evol. 12(5), 921–927 (1995)
12. Fallin, D., Schork, N.: Accuracy of haplotype frequency estimation for biallelic loci via the expectation-maximization algorithm for unphased diploid genotype data. Am. J. Hum. Genet. 67, 947–959 (2000)
13. Friss, L., Hudson, R., Bartoszewicz, A., Wall, J., Donfalk, T., Di Rienzo, A.: Gene conversion and differential population histories may explain the contrast between polymorphism and linkage disequilibrium levels. Am. J. Hum. Genet. 69, 831–843 (2001)
14. Gramm, J., Hartman, T., Nierhoff, T., Sharan, R., Tantau, T.: On the complexity of SNP block partitioning under the perfect phylogeny model. Discrete Math (2008)

15. Gramm, J., Nierhoff, T., Sharan, R., Tantau, T.: Haplotyping with missing data via perfect path phylogenies. Discrete and Appl. Math. 155(6–7), 788–805 (2007)
16. Gusfield, D.: Inference of haplotypes from samples of diploid populations: complexity and algorithms. J. Comput. Biol. 8(3), 305–323 (2001)
17. Gusfield, D.: Haplotyping as perfect phylogeny: Conceptual framework and efficient solutions. In: Proc. RECOMB 2002, pp. 166–175. ACM Press, New York (2002)
18. Hawley, M., Kidd, K.: Haplo: A program using the EM algorithm to estimate the frequency of multi-site haplotypes. J. Hered. 86, 409–441 (1995)
19. Helmuth, L.: Map of the human genome 3.0. Science 293(5530), 582–585 (2001)
20. Immerman, N.: Descriptive Complexity. Springer, New York (1999)
21. Lancia, G., Pinotti, M.C., Rizzi, R.: Haplotyping populations by pure parsimony: Complexity of exact and approximation algorithms. INFORMS J. Comput. 16(4), 348–359 (2004)
22. Lancia, G., Rizzi, R.: A polynomial case of the parsimony haplotyping problem. Oper. Res. Lett. 34(3), 289–295 (2006)
23. Liu, Y., Zhang, C.-Q.: A linear solution for haplotype perfect phylogeny problem. In: Proc. Int. Conf. Adv. in Bioinfor. and Appl., pp. 173–184. World Scientific, Singapore (2005)
24. Vijaya Satya, R., Mukherjee, A.: An optimal algorithm for perfect phylogeny haplotyping. J. Comput. Biol. 13(4), 897–928 (2006)
25. Vijaya Satya, R., Mukherjee, A.: The undirected incomplete perfect phylogeny problem. IEEE/ACM T. Comput. Biol. and Bioinfor (to appear, 2008)
26. Sharan, R., Halldórsson, B.V., Istrail, S.: Islands of tractability for parsimony haplotyping. IEEE/ACM T. Comput. Biol. and Bioinfor. 3(3), 303–311 (2006)
27. van Iersel, L., Keijsper, J., Kelk, S., Stougie, L.: Shorelines of islands of tractability: Algorithms for parsimony and minimum perfect phylogeny haplotyping problems. IEEE/ACM T. Comput. Biol. and Bioinfor. 5(2), 301–312 (2008)

Sincere-Strategy Preference-Based Approval Voting Broadly Resists Control[*]

Gábor Erdélyi, Markus Nowak, and Jörg Rothe[**]

Institut für Informatik, Heinrich-Heine-Universität Düsseldorf, 40225 Düsseldorf, Germany

Abstract. We study sincere-strategy preference-based approval voting (SP-AV), a system proposed by Brams and Sanver [8], with respect to procedural control. In such control scenarios, an external agent seeks to change the outcome of an election via actions such as adding/deleting/partitioning either candidates or voters. SP-AV combines the voters' preference rankings with their approvals of candidates, and we adapt it here so as to keep its useful features with respect to approval strategies even in the presence of control actions. We prove that this system is computationally resistant (i.e., the corresponding control problems are NP-hard) to at least 16 out of 20 types of constructive and destructive control. Thus, for the 20 control types studied here, SP-AV has more resistances to control, by at least two, than is currently known for any other natural voting system with a polynomial-time winner problem.

Keywords: Complexity theory, artificial intelligence, approval voting, complexity of procedural control.

1 Introduction

Voting provides a particularly useful method for preference aggregation and collective decision-making. While voting systems were originally used in political science, economics, and operations research, they are now also of central importance in various areas of computer science, such as artificial intelligence (in particular, within multi-agent systems). In automated, large-scale computer settings, voting systems have been applied, e.g., for planning [11] and similarity search [14], and have also been used in the design of recommender systems [19] and ranking algorithms [10] (where they help to lessen the spam in meta-search web-page rankings). For such applications, it is crucial to explore the computational properties of voting systems and, in particular, to study the complexity of problems related to voting (see, e.g., the survey by Faliszewski et al. [15]).

The study of voting systems from a complexity-theoretic perspective was initiated by Bartholdi, Tovey, and Trick's series of seminal papers about the complexity of winner determination [2], manipulation [1] and procedural control [3] in elections. This paper contributes to the study of electoral control, where an external agent—traditionally

---

[*] URLs: ccc.cs.uni-duesseldorf.de/~{erdelyi, rothe} (G. Erdélyi and J. Rothe). Supported in part by DFG grant RO 1202/11-1, an ESF grant in the LogICCC program, and by the Humboldt Foundation's TransCoop program.
[**] Corresponding author.

E. Ochmański and J. Tyszkiewicz (Eds.): MFCS 2008, LNCS 5162, pp. 311–322, 2008.
© Springer-Verlag Berlin Heidelberg 2008

called *the chair*—seeks to influence the outcome of an election via procedural changes to the election's structure, namely via adding/deleting/partitioning either candidates or voters (see Section 2.2 and the full version [12] for the formal definitions of our control problems). We consider both *constructive* control (introduced by Bartholdi, Tovey, and Trick [3]), where the chair's goal is to make a given candidate the unique winner, and *destructive* control (introduced by Hemaspaandra, Hemaspaandra, and Rothe [22]), where the chair's goal is to prevent a given candidate from being a unique winner.

We investigate the same twenty types of constructive and destructive control that were studied for approval voting [22], and we do so for a voting system that was proposed by Brams and Sanver [8] as a combination of preference-based and approval voting. Approval voting was introduced by Brams and Fishburn ([4,5], see also [6]) as follows: Every voter either approves or disapproves of each candidate, and every candidate with the largest number of approvals is a winner. One of the simplest preference-based voting systems is plurality: All voters report their preference rankings of the candidates, and the winners are the candidates that are ranked first-place by the largest number of voters. The purpose of this paper is to show that Brams and Sanver's combined system (adapted here so as to keep its useful features even in the presence of control actions) combines the strengths, in terms of computational resistance to control, of plurality and approval voting.

Some voting systems are *immune* to certain types of control in the sense that it is never possible for the chair to reach his or her goal via the corresponding control action. Of course, immunity to any type of control is most desirable, as it unconditionally shields the voting system against this particular control type. Unfortunately, like most voting systems approval voting is *susceptible* (i.e., not immune) to many types of control, and plurality voting is susceptible to all types of control.[1] However, and this was Bartholdi, Tovey, and Trick's brilliant insight [3], even for systems susceptible to control, the chair's task of controlling a given election may be too hard computationally (namely, NP-hard) for him or her to succeed. The voting system is then said to be *resistant* to this control type. If a voting system is susceptible to some type of control, but the chair's task can be solved in polynomial time, the system is said to be *vulnerable* to this control type.

The quest for a natural voting system with an easy winner-determination procedure that is universally resistant to control lasts for more than 15 years now. Among the voting systems that have been studied with respect to control are plurality, Condorcet, approval, cumulative, Llull, and (variants of) Copeland voting [3,22,23,24,16,17]. Among these systems, plurality and Copeland voting (denoted Copeland$^{0.5}$ in [17]) display the broadest resistance to control, yet even they are not universally control-resistant. The only system currently known to be fully resistant—to the 20 types of constructive and destructive control studied in [22,23] and here—is a highly artificial system constructed via hybridization [23]. (We mention that this system was not designed for

---

[1] A related line of research has shown that, in principle, all natural voting systems can be manipulated by strategic voters. Most notable among such results is the classical work of Gibbard [20] and Satterthwaite [25]. The study of strategy-proofness is still an extremely active and interesting area in social choice theory (see, e.g., Duggan and Schwartz [9]) and in artificial intelligence (see, e.g., Everaere et al. [13]).

Table 1. Number of resistances, immunities, and vulnerabilities to our 20 control types

Number of	Condorcet	Approval	Llull	Plurality	Copeland	SP-AV
resistances	3	4	13	14	14	≥ 16
immunities	4	8	0	0	0	0
vulnerabilities	7	8	7	6	6	≤ 4

direct, real-world use as a "natural" system but rather was intended to rule out the existence of a certain impossibility theorem [23].)

While approval voting nicely distinguishes between each voter's acceptable and inacceptable candidates, it ignores the preference rankings the voters may have about their approved (or disapproved) candidates. This shortcoming motivated Brams and Sanver [8] to introduce a voting system that combines approval and preference-based voting, and they defined the related notions of sincere and admissible approval strategies, which are quite natural requirements. We adapt their sincere-strategy preference-based approval voting system in a natural way such that, for elections with at least two candidates, admissibility of approval strategies (see Definition 1) can be ensured even in the presence of control actions such as deleting candidates and partitioning candidates or voters. Note that in control by partition of voters (see Section 2.2) the run-off may have a reduced number of candidates.

The purpose of this paper is to study if, and to what extent, this hybrid system (where "hybrid" is not meant in the sense of [23]) inherits the control resistances of plurality (which is perhaps the simplest preference-based system) and approval voting. Denoting this system by SP-AV, we show that SP-AV does combine the resistances of plurality and approval voting (see also the full version [12] of this paper).

More specifically, we prove here that sincere-strategy preference-based approval voting is resistant to at least 16 and vulnerable to at most four of the 20 types of control considered here and in [22].[2] For comparison, Table 1 shows the number of resistances, immunities, and vulnerabilities to our 20 control types that are known for each of Condorcet,[3] approval, Llull, plurality, and Copeland voting (see [3,22,16,17]), and for SP-AV (see Theorem 1 and Table 2 in Section 3.1).

This paper is organized as follows. In Section 2, we define sincere-strategy preference-based approval voting, the types of control studied in this paper, and the notions of

---

[2] *Note added in final revision:* We state in this paper only those results that have been submitted to and peer-reviewed for MFCS-08. Note, however, that the full version [12] of this paper contains more results than are stated here. In particular, the two partition-of-voters cases left open in Table 2 have been resolved meanwhile: SP-AV is resistant to this control type in the constructive case and is vulnerable in the destructive case [12]. Thus, SP-AV is now known to have 17 resistances and three vulnerabilities to the 20 control types considered here, which gives a larger number of results for SP-AV than stated in Tables 1 and 2.

[3] As in [22], we consider two types of control by partition of candidates (namely, with and without run-off) and one type of control by partition of voters, and for each partition case we use the rules TE ("ties eliminate") and TP ("ties promote") for handling ties that may occur in the corresponding subelections (see the full version [12] of this paper). However, since Condorcet winners are always unique when they exist, the distinction between TE and TP is not made for Condorcet voting, which thus has only 14 instead of 20 types of control.

immunity, susceptibility, vulnerability, and resistance [3]. In Section 3, we provide our results on SP-AV and some proofs. Some other proofs are omitted due to space limitations, but can be found in the full and expanded version [12] of this paper. Finally, in Section 4 we give our conclusions.

2 Preliminaries

2.1 Preference-Based Approval Voting

An election $E = (C,V)$ is specified by a finite set C of candidates and a finite collection V of voters who express their preferences over the candidates in C, where distinct voters may of course have the same preferences. How the voter preferences are represented depends on the voting system used. In approval voting (AV, for short), every voter draws a line between his or her acceptable and inacceptable candidates (by specifying a 0-1 approval vector, where 0 represents disapproval and 1 represents approval), yet does not rank them. In contrast, many other important voting systems (e.g., Condorcet voting, Copeland voting, all scoring protocols including plurality, Borda count, veto, etc.) are based on voter preferences that are specified as tie-free linear orderings of the candidates.

Brams and Sanver [8] introduced a voting system that combines approval and preference-based voting. To distinguish this system from other systems that these authors introduced with the same purpose of combining approval and preference-based voting [7], we call the variant considered here (including the conventions and rules to be explained below) *sincere-strategy preference-based approval voting* (SP-AV, for short).

Definition 1 ([8]). *Let (C,V) be an election, where the voters both indicate approvals/ disapprovals of the candidates and provide a tie-free linear ordering of all candidates. For each voter $v \in V$, an AV strategy of v is a subset $S_v \subseteq C$ such that v approves of all candidates in S_v and disapproves of all candidates in $C - S_v$. The list of AV strategies for all voters in V is called an AV strategy profile for (C,V). (We sometimes also speak of V's AV strategy profile for C.) For each $c \in C$, let $score_{(C,V)}(c) = \|\{v \in V \mid c \in S_v\}\|$ denote the number of c's approvals. Every candidate c with the largest $score_{(C,V)}(c)$ is a winner of election (C,V).*

An AV strategy S_v of a voter $v \in V$ is said to be admissible *if S_v contains v's most preferred candidate and does not contain v's least preferred candidate. S_v is said to be* sincere *if for each $c \in C$, if v approves of c then v also approves of each candidate ranked higher than c (i.e., there are no gaps allowed in sincere approval strategies). An AV strategy profile for (C,V) is* admissible *(respectively,* sincere*) if the AV strategies of all voters in V are admissible (respectively, sincere).*

Admissibility and sincerity are quite natural requirements. In particular, requiring the voters to be sincere ensures that their preference rankings and their approvals/disapprovals are not contradictory. Note further that admissible AV strategies are not dominated in a game-theoretic sense [4], and that sincere strategies for at least two candidates are always admissible if voters are neither allowed to approve of everybody nor to disapprove of everybody (i.e., if we require voters v to have only AV strategies S_v with $\emptyset \neq S_v \neq C$), a convention adopted by Brams and Sanver [8] and also adopted here.[4] Henceforth, we will tacitly assume that only sincere AV strategy profiles are considered (which by the above convention, whenever there are at least two candidates,[5] necessarily are admissible), i.e., a vote with an insincere strategy will be considered void.

Preferences are represented by a left-to-right ranking (separated by a space) of the candidates (e.g., a b c), with the leftmost candidate being the most preferred one, and approval strategies are denoted by inserting a straight line into such a ranking, where all candidates left of this line are approved and all candidates right of this line are disapproved (e.g., "$a \mid b$ c" means that a is approved, while both b and c are disapproved). In our constructions, we sometimes also insert a subset $B \subseteq C$ into such approval rankings, where we assume some arbitrary, fixed order of the candidates in B (e.g., "$a \mid B$ c" means that a is approved, while all $b \in B$ and c are disapproved).

2.2 Control Problems for Preference-Based Approval Voting

The control problems considered here were introduced by Bartholdi et al. [3] for constructive control and by Hemaspaandra et al. [22] for destructive control. In constructive control scenarios the chair's goal is to make a favorite candidate win, and in destructive control scenarios the chair's goal is to ensure that a despised candidate does not win. As is common, the chair is assumed to have complete knowledge of the voters' preference rankings and approval strategies (see [22] for a detailed discussion of this assumption), and as in most papers on electoral control (exceptions are, e.g., [24,17]) we define the control problems in the unique-winner model.

To achieve his or her goal, the chair modifies the structure of a given election via adding/deleting/partitioning either candidates or voters. Such control actions— specifically, those with respect to control via deleting or partitioning candidates or via partitioning voters—may have an undesirable impact on the resulting election in that they might violate our conventions about admissible AV strategies. That is why we define the following rules that preserve (or re-enforce) our conventions under such control actions:

1. Whenever during or after a control action (such as deletion or partition of candidates or partition of voters) it happens that we obtain an election (C,V) with $\|C\| = 1$, then each voter $v \in V$ approves of the candidate in C (even if v originally did not approve of this candidate).
2. Whenever during or after a control action it happens that we obtain an election (C,V) with $\|C\| \geq 2$ and for some voter $v \in V$ we have $S_v = \emptyset$ or $S_v = C$, then each such voter's AV strategy is changed to approve of his or her top candidate and to disapprove of his or her bottom candidate. This rule re-enforces $\emptyset \neq S_v \neq C$ for each $v \in V$, as desired.

We now formally define our control problems, where each problem is defined by stating the problem instance together with two questions, one for the constructive and one for

---

[4] Brams and Sanver [8] actually preclude only the case $S_v = C$ for voters v. However, an AV strategy that disapproves of all candidates obviously is sincere, yet not admissible, which is why we also exclude the case of $S_v = \emptyset$.

[5] Note that an AV strategy is never admissible for less than two candidates. For elections with one candidate, we by convention require each voter to approve of this candidate.

the destructive case. These control problems are tailored to sincere-strategy preference-based approval voting by requiring every election occuring in these control problems (be it before, during, or after a control action—so, in particular, this also applies to the subelections in the partitioning cases) to have a sincere AV strategy profile and to satisfy the above conventions and rules. In particular, this means that when the number of candidates is reduced (due to deleting candidates or partitioning candidates or voters), approval lines may have to be moved in accordance with the above rules.

Due to space, we confine ourselves to defining, as an example, only control by deleting candidates. The remaining definitions of control problems can be found in, e.g., [3,16,22], and also in the full version [12] of this paper.

Control by Deleting Candidates: In this control scenario, the chair seeks to reach his or her goal by deleting (up to a given number of) candidates. Here it may happen that our conventions are violated by the control action, but will be re-enforced by the above rules (namely, by moving the line between some voter's acceptable and inacceptable candidates to behind the top candidate or to before the bottom candidate whenever necessary).

Name: Control by Deleting Candidates.
Instance: An election (C,V), a designated candidate $c \in C$, and a nonnegative integer ℓ.
Question (constructive): Is it possible to delete up to ℓ candidates from C such that c is the unique winner of the resulting election?
Question (destructive): Is it possible to delete up to ℓ candidates (other than c) from C such that c is not a unique winner of the resulting election?

2.3   Immunity, Susceptibility, Vulnerability, and Resistance

The following notions—which are due to Bartholdi, Tovey, and Trick [3]—will be central to our complexity analysis of the control problems for preference-based approval voting.

Definition 2 ([3]). *Let $\mathcal{E}$ be an election system and let Φ be some given type of control. $\mathcal{E}$ is said to be* immune *to Φ-control if (a) Φ is a constructive control type and it is never possible for the chair to turn a designated candidate from being not a unique winner into being the unique winner via exerting Φ-control, or (b) Φ is a destructive control type and it is never possible for the chair to turn a designated candidate from being the unique winner into being not a unique winner via exerting Φ-control. $\mathcal{E}$ is said to be* susceptible *to Φ-control if it is not immune to Φ-control. $\mathcal{E}$ is said to be* vulnerable *to Φ-control if $\mathcal{E}$ is susceptible to Φ-control and the control problem associated with Φ is solvable in polynomial time. $\mathcal{E}$ is said to be* resistant *to Φ-control if $\mathcal{E}$ is susceptible to Φ-control and the control problem associated with Φ is NP-hard.*

For example, approval voting is known to be immune to eight of the twelve types of candidate control considered here and in [22]. The proofs of these results crucially employ the links between immunity/susceptibility for various control types shown in [22] and the fact that approval voting satisfies the unique version of the Weak Axiom of Revealed Preference (denoted by Unique-WARP, see [22,3]): If a candidate c is the unique

Table 2. Overview of results. Key: I means immune, S means susceptible, R means resistant (i.e., R = S + NP-hard), V means vulnerable (i.e., V = S + P-membership), TE means ties-eliminate, and TP means ties-promote. Results for SP-AV are new. Results for AV, stated here to allow comparison, are due to Hemaspaandra, Hemaspaandra, and Rothe [22].

Control by	SP-AV		AV	
	Constr.	Destr.	Constr.	Destr.
Adding Candidates	R	R	I	V
Deleting Candidates	R	R	V	I
Partition	TE: R	TE: R	TE: V	TE: I
of Candidates	TP: R	TP: R	TP: I	TP: I
Run-off Partition	TE: R	TE: R	TE: V	TE: I
of Candidates	TP: R	TP: R	TP: I	TP: I
Adding Voters	R	V	R	V
Deleting Voters	R	V	R	V
Partition	TE: S	TE: S	TE: R	TE: V
of Voters	TP: R	TP: R	TP: R	TP: V

winner in a set C of candidates, then c is the unique winner in every subset of C that includes c. In contrast with approval voting, sincere-strategy preference-based approval voting does not satisfy Unique-WARP, and we will see later in Section 3.2 that it indeed is susceptible to each type of control considered here. See the full version [12] for the proof of Proposition 1.

Proposition 1. *Sincere-strategy preference-based approval voting does not satisfy Unique-WARP.*

3   Results for Sincere-Strategy Preference-Based Approval Voting

3.1   Overview

Theorem 1 below (see also Table 2) shows the complexity results regarding control of elections for SP-AV. As mentioned in the introduction, with at least 16 resistances and at most four vulnerabilities (note, however, Footnote 2), this system has more resistances and fewer vulnerabilities to control (for our 20 control types) than is currently known for any other natural voting system with a polynomial-time winner problem.

Theorem 1. *Sincere-strategy preference-based approval voting is susceptible, resistant, and vulnerable to the twenty types of control considered here as shown in Table 2.*

3.2   Susceptibility

By definition, all resistance and vulnerability results in particular require susceptibility. The following two lemmas (the proofs of which can be found in the full version [12] of this paper) show that SP-AV is susceptible to the twenty types of control considered here.

Lemma 1. *SP-AV is susceptible to constructive and destructive control by adding candidates, by deleting candidates, and by partition of candidates (with or without run-off and for each in both tie-handling models, TE and TP).*

Lemma 2. *SP-AV is susceptible to constructive and destructive control by adding voters, by deleting voters, and by partition of voters in both tie-handling models, TE and TP.*

3.3 Candidate Control

Theorems 2 and 3 below show that sincere-strategy preference-based approval voting is fully resistant to candidate control. This result should be contrasted with that of Hemaspaandra, Hemaspaandra, and Rothe [22], who proved immunity and vulnerability for all cases of candidate control within approval voting (see Table 2). In fact, SP-AV has the same resistances to candidate control as plurality, and we will show that the construction presented in [22] to prove plurality resistant also works for sincere-strategy preference-based approval voting in all cases of candidate control except one—namely, except for constructive control by deleting candidates. In the proof of Theorem 3, we will provide a novel construction that works for this one missing case.

All resistance results in this section follow via a reduction from the NP-complete problem Hitting Set (see, e.g., Garey and Johnson [18]): Given a set $B = \{b_1, b_2, \ldots, b_m\}$, a collection $\mathscr{S} = \{S_1, S_2, \ldots, S_n\}$ of subsets $S_i \subseteq B$, and a positive integer $k \leq m$, does $\mathscr{S}$ have a hitting set of size at most k, i.e., is there a set $B' \subseteq B$ with $\|B'\| \leq k$ such that for each i, $S_i \cap B' \neq \emptyset$?

Some of our proofs use constructions and arguments for SP-AV that are straightforward modifications of the constructions and arguments of the corresponding results for approval voting or plurality from [22], and we thus attribute them to Hemaspaandra et al. [22] (such as Theorem 2 below). Some other of our results require new insights to make the proof work for SP-AV (such as Theorem 3 below). Due to space limitations, we present here only the results on candidate control, and for voter control and other details omitted here we refer the interested reader to the full version [12] of this paper.

Theorem 2 ([22]). *SP-AV is resistant to all types of constructive and destructive candidate control considered here, except for constructive control by deleting candidates.*

The proof of Theorem 2 is similar to the corresponding proof for plurality in [22], except that only the arguments for *destructive* candidate control are given there (simply because plurality was shown resistant to all cases of constructive candidate control already by Bartholdi, Tovey, and Trick [3] via a different construction). We now provide a short proof sketch of Theorem 2 and the construction from [22] (slightly modified so as to be formally conform with the SP-AV voter representation) in order to (a) show that the same construction can be used to establish all but one resistances of SP-AV to *constructive* candidate control, and (b) explain why the one missing case (namely, *constructive control by deleting candidates*) does *not* follow from this construction.

Proof Sketch of Theorem 2. Susceptibility holds by Lemma 1 in each case. The resistance proofs are based on a reduction from Hitting Set and employ Construction 1 below, slightly modified so as to be formally conform with the SP-AV voter representation.

Construction 1 ([22]). *Let $(B, \mathscr{S}, k)$ be a given instance of Hitting Set, where $B = \{b_1, b_2, \ldots, b_m\}$ is a set, $\mathscr{S} = \{S_1, S_2, \ldots, S_n\}$ is a collection of subsets $S_i \subseteq B$, and $k \leq m$ is a positive integer. Define the election (C, V), where $C = B \cup \{c, w\}$ is the candidate set and where V consists of the following voters:*

1. *There are $2(m - k) + 2n(k + 1) + 4$ voters of the form: $c \mid w\ B$.*
2. *There are $2n(k + 1) + 5$ voters of the form: $w \mid c\ B$.*
3. *For each i, $1 \leq i \leq n$, there are $2(k + 1)$ voters of the form: $S_i \mid c\ w\ (B - S_i)$.*
4. *For each j, $1 \leq j \leq m$, there are two voters of the form: $b_j \mid w\ c\ (B - \{b_j\})$.*

Since $score_{(\{c,w\}, V)}(c) - score_{(\{c,w\}, V)}(w) = 2k(n - 1) + 2n - 1$ is positive (because of $n \geq 1$), c is the unique winner of election $(\{c, w\}, V)$. The key observation is the following proposition, which can be proven as in [22].

Proposition 2 ([22])

1. *If $\mathscr{S}$ has a hitting set B' of size k, then w is the unique SP-AV winner of election $(B' \cup \{c, w\}, V)$.*
2. *Let $D \subseteq B \cup \{w\}$. If c is not the unique SP-AV winner of election $(D \cup \{c\}, V)$, then there exists a set $B' \subseteq B$ such that*
   (a) $D = B' \cup \{w\}$,
   (b) w is the unique SP-AV winner of election $(B' \cup \{c, w\}, V)$, and
   (c) B' is a hitting set of $\mathscr{S}$ of size less than or equal to k.

As an example, the resistance of SP-AV to constructive and destructive control by adding candidates now follows immediately from Proposition 2, via mapping the Hitting Set instance $(B, \mathscr{S})$ to the set $\{c, w\}$ of qualified candidates and the set B of spoiler candidates, to the voter collection V, and by having c be the designated candidate in the destructive case and by having w be the designated candidate in the constructive case. The other cases of Theorem 2 can be proven similarly. □

Turning now to the one missing case mentioned above: Why does Construction 1 not work for constructive control by deleting candidates? Informally put, the reason is that c is the only serious rival of w in the election (C, V) of Construction 1, so by simply deleting c the chair could make w the unique SP-AV winner, regardless of whether $\mathscr{S}$ has a hitting set of size k. However, via a different construction, we can prove resistance also in this case.

Theorem 3. *SP-AV is resistant to constructive control by deleting candidates.*

Proof. Susceptibility holds by Lemma 1. To prove resistance, we provide a reduction from Hitting Set. Let $(B, \mathscr{S}, k)$ be a given instance of Hitting Set, where $B = \{b_1, b_2, \ldots, b_m\}$ is a set, $\mathscr{S} = \{S_1, S_2, \ldots, S_n\}$ is a collection of subsets $S_i \subseteq B$, and $k < m$ is a positive integer.[6]

Define the election (C, V), where $C = B \cup \{w\}$ is the candidate set and where V consists of the following $4n(k + 1) + 4m - 2k + 3$ voters:

---

[6] Note that if $k = m$ then B is always a hitting set of size at most k (provided that $\mathscr{S}$ contains only nonempty sets—a requirement that doesn't affect the NP-completeness of the problem), and we thus may require that $k < m$.

1. For each i, $1 \leq i \leq n$, there are $2(k+1)$ voters of the form: $S_i \mid (B - S_i) \; w$.
2. For each i, $1 \leq i \leq n$, there are $2(k+1)$ voters of the form: $(B - S_i) \; w \mid S_i$.
3. For each j, $1 \leq j \leq m$, there are two voters of the form: $b_j \mid w \; (B - \{b_j\})$.
4. There are $2(m-k)$ voters of the form: $B \mid w$.
5. There are three voters of the form: $w \mid B$.

Since for each $b_j \in B$, the difference

$$score_{(C,V)}(w) - score_{(C,V)}(b_j) = 2n(k+1) + 3 - (2n(k+1) + 2 + 2(m-k)) = 1 - 2(m-k)$$

is negative (due to $k < m$), w loses to each member of B and so does not win election (C,V).

We claim that $\mathscr{S}$ has a hitting set B' of size k if and only if w can be made the unique SP-AV winner by deleting at most $m - k$ candidates.

From left to right: Suppose $\mathscr{S}$ has a hitting set B' of size k. Then, for each $b_j \in B'$,

$$\begin{aligned} score_{(B' \cup \{w\}, V)}(w) &- score_{(B' \cup \{w\}, V)}(b_j) \\ &= 2n(k+1) + 2(m-k) + 3 - (2n(k+1) + 2 + 2(m-k)) \\ &= 1, \end{aligned}$$

since the approval line is moved for $2(m-k)$ voters of the third group, thus transferring their approvals from members of $B - B'$ to w. So w is the unique SP-AV winner of election $(B' \cup \{w\}, V)$. Since $B' \cup \{w\} = C - (B - B')$, it follows from $\|B\| = m$ and $\|B'\| = k$ that deleting $m - k$ candidates from C makes w the unique SP-AV winner.

From right to left: Let $D \subseteq B$ be any set such that $\|D\| \leq m - k$ and w is the unique SP-AV winner of election $(C - D, V)$. Let $B' = (C - D) - \{w\}$. Note that $B' \subseteq B$ and that we have the following scores in $(B' \cup \{w\}, V)$:

$$\begin{aligned} score_{(B' \cup \{w\}, V)}(w) &= 2(n - \ell)(k+1) + 2(m - \|B'\|) + 3, \\ score_{(B' \cup \{w\}, V)}(b_j) &\leq 2n(k+1) + 2(k+1)\ell + 2 + 2(m-k) \quad \text{for each } b_j \in B', \end{aligned}$$

where ℓ is the number of sets $S_i \in \mathscr{S}$ that are not hit by B', i.e., $B' \cap S_i = \emptyset$. Since w is the unique SP-AV winner of $(B' \cup \{w\}, V)$, w has more approvals than any candidate b_j in B':

$$\begin{aligned} score_{(B' \cup \{w\}, V)}(w) &- score_{(B' \cup \{w\}, V)}(b_j) \\ &\geq 2(n - \ell)(k+1) + 2(m - \|B'\|) + 3 - 2n(k+1) - 2\ell(k+1) - 2 - 2(m-k) \\ &= 1 + 2(k - \|B'\|) - 4\ell(k+1) > 0. \end{aligned}$$

Solving this inequality for ℓ, we obtain

$$0 \leq \ell < \frac{1 + 2(k - \|B'\|)}{4(k+1)} < \frac{4 + 4k}{4(k+1)} = 1.$$

Thus $\ell = 0$. It follows that $1 + 2(k - \|B'\|) > 0$, which implies $\|B'\| \leq k$. Thus, B' is a hitting set of size at most k.   ❑

4   Conclusions

We have shown that Brams and Sanver's sincere-strategy preference-based approval voting system [8] combines the resistances of approval and plurality voting to procedural control: SP-AV is here shown to be resistant to at least 16 of the 20 types of control studied here and in [22] (note, however, Footnote 2). Thus, for these 20 types of control, SP-AV has more resistances and fewer vulnerabilities to control than is currently known for any other natural voting system with a polynomial-time winner problem.

Acknowledgments. We thank the anonymous MFCS referees for their helpful comments on a preliminary version of this paper.

References

1. Bartholdi III, J., Tovey, C., Trick, M.: The computational difficulty of manipulating an election. Social Choice and Welfare 6(3), 227–241 (1989)
2. Bartholdi III, J., Tovey, C., Trick, M.: Voting schemes for which it can be difficult to tell who won the election. Social Choice and Welfare 6(2), 157–165 (1989)
3. Bartholdi III, J., Tovey, C., Trick, M.: How hard is it to control an election? Mathematical Comput. Modelling 16(8/9), 27–40 (1992)
4. Brams, S., Fishburn, P.: Approval voting. American Political Science Review 72(3), 831–847 (1978)
5. Brams, S., Fishburn, P.: Approval Voting. Birkhäuser, Boston (1983)
6. Brams, S., Fishburn, P.: Voting procedures. In: Arrow, K., Sen, A., Suzumura, K. (eds.) Handbook of Social Choice and Welfare, vol. 1, pp. 173–236. North-Holland, Amsterdam (2002)
7. Brams, S., Sanver, R.: Voting systems that combine approval and preference. In: Brams, S., Gehrlein, W., Roberts, F. (eds.) The Mathematics of Preference, Choice, and Order: Essays in Honor of Peter C. Fishburn. Springer, Heidelberg (to appear)
8. Brams, S., Sanver, R.: Critical strategies under approval voting: Who gets ruled in and ruled out. Electoral Studies 25(2), 287–305 (2006)
9. Duggan, J., Schwartz, T.: Strategic manipulability without resoluteness or shared beliefs: Gibbard–Satterthwaite generalized. Social Choice and Welfare 17(1), 85–93 (2000)
10. Dwork, C., Kumar, R., Naor, M., Sivakumar, D.: Rank aggregation methods for the web. In: Proc. WWW 2001, pp. 613–622. ACM Press, New York (2001)
11. Ephrati, E., Rosenschein, J.: Multi-agent planning as a dynamic search for social consensus. In: Proc. IJCAI 1993, pp. 423–429 (1993)
12. Erdélyi, G., Nowak, M., Rothe, J.: Sincere-strategy preference-based approval voting fully resists constructive control and broadly resists destructive control. Technical Report cs.GT/0806.0535, ACM Computing Research Repository (CoRR) (June 2008)
13. Everaere, P., Konieczny, S., Marquis, P.: The strategy-proofness landscape of merging. Journal of Artificial Intelligence Research 28, 49–105 (2007)
14. Fagin, R., Kumar, R., Sivakumar, D.: Efficient similarity search and classification via rank aggregation. In: Proc.ACM SIGMOD Intern.Conf.on Management of Data, pp. 301–312. ACM Press, New York (2003)
15. Faliszewski, P., Hemaspaandra, E., Hemaspaandra, L., Rothe, J.: A richer understanding of the complexity of election systems. In: Ravi, S., Shukla, S. (eds.) Fundamental Problems in Computing: Essays in Honor of Professor Daniel J. Rosenkrantz. Springer, Heidelberg (to appear, 2006); Available as Technical Report cs.GT/0609112, ACM Computing Research Repository (CoRR), September 2006

16. Faliszewski, P., Hemaspaandra, E., Hemaspaandra, L., Rothe, J.: Llull and Copeland voting broadly resist bribery and control. In: Proc.AAAI 2007, pp. 724–730. AAAI Press, Menlo Park (2007)

17. Faliszewski, P., Hemaspaandra, E., Hemaspaandra, L., Rothe, J.: Copeland voting fully resists constructive control. In: Proc. AAIM 2008, June 2008, pp. 165–176. Springer, Heidelberg (to appear, 2008)

18. Garey, M., Johnson, D.: Computers and Intractability: A Guide to the Theory of NP-Completeness. W.H. Freeman, New York (1979)

19. Ghosh, S., Mundhe, M., Hernandez, K., Sen, S.: Voting for movies: The anatomy of recommender systems. In: Proc.3rd Annual Conference on Autonomous Agents, pp. 434–435. ACM Press, New York (1999)

20. Gibbard, A.: Manipulation of voting schemes. Econometrica 41(4), 587–601 (1973)

21. Hemaspaandra, E., Hemaspaandra, L., Rothe, J.: Anyone but him: The complexity of precluding an alternative. In: Proc. AAAI 2005, pp. 95–101. AAAI Press, Menlo Park (2005)

22. Hemaspaandra, E., Hemaspaandra, L., Rothe, J.: Anyone but him: The complexity of precluding an alternative. Artificial Intelligence 171(5–6), 255–285 (2007)

23. Hemaspaandra, E., Hemaspaandra, L., Rothe, J.: Hybrid elections broaden complexity-theoretic resistance to control. In: Proc. IJCAI 2007, pp. 1308–1314. AAAI Press, Menlo Park (2007)

24. Procaccia, A., Rosenschein, J., Zohar, A.: Multi-winner elections: Complexity of manipulation, control, and winner-determination. In: Proc. IJCAI 2007, pp. 1476–1481. AAAI Press, Menlo Park (2007)

25. Satterthwaite, M.: Strategy-proofness and Arrow's conditions: Existence and correspondence theorems for voting procedures and social welfare functions. Journal of Economic Theory 10(2), 187–217 (1975)

Reversal-Bounded Counter Machines Revisited[*]

Alain Finkel[1] and Arnaud Sangnier[1,2]

[1] LSV, ENS Cachan, CNRS
[2] EDF R&D
61 av. du pdt Wilson 94230 Cachan, France
{finkel,sangnier}@lsv.ens-cachan.fr

Abstract. We extend the class of reversal-bounded counter machines by autho-rizing a finite number of alternations between increasing and decreasing mode over a given bound. We prove that extended reversal-bounded counter machines also have effective semi-linear reachability sets. We also prove that the property of being reversal-bounded is undecidable in general even when we fix the bound, whereas this problem becomes decidable when considering Vector Addition System with States.

1 Introduction

The *verification of infinite state systems* has shown in the last years to be an efficient technique to model and verify computer systems. Various models of infinite-state systems have also been proposed as for instance counter systems, lossy channel systems, pushdown automata, timed automata, etc, in order to obtain an automatic verification procedure. Among them, counter systems which consist in finite automata extended with operations on integer variables enjoy a central position for both theoretical results and maturity of tools like FAST [3], LASH [16] and TREX [1].

Reachability problem for counter systems. It has been proved in [20] that Minsky machines, which correspond to counter systems where each counter can be incremented, decremented or tested to zero, have an undecidable reachability problem, even when they manipulate only two counter variables. Because of that, different restrictions over counter systems have been proposed in order to obtain the decidability. For instance, Vector Addition Systems with States (or Petri nets) are a special class of counter systems, in which it is not possible to perform equality tests (equivalent to zero-tests), and for which the reachability problem is decidable [14,19].

Counter systems with semi-linear reachability sets. In many verification problems, it is convenient not only to have an algorithm for the reachability problem, but also to be able to compute effectively the reachability set. In the past, many classes of counter systems with a semi-linear reachability set have been found. Among the VASS (or Petri nets), we distinguish the BPP-nets [5], the cyclic Petri nets [2], the persistent Petri nets [15,18], the regular Petri nets [21], the 2-dimensional VASS [9]. In [10], the class of reversal-bounded counter machines is introduced as follows: each counter can only per-form a bounded number of alternations between increasing and decreasing mode. The author shows that reversal-bounded counter machines have a semi-linear reachability

---

[*] Partly supported by project AVERISS (ANR-06-SETIN-001).

E. Ochmański and J. Tyszkiewicz (Eds.): MFCS 2008, LNCS 5162, pp. 323–334, 2008.

set and these results have been extended in [11] authorizing more complex guards and restricting the way the alternations are counted. In [17], it has been shown that most of the counter systems with a semi-linear reachability set are in fact flattable, which means that their control graph can be replaced equivalently w.r.t. reachability, by another one with no nested loops. In fact, it has been proved in [6], that counter machines with no nested loops in their control structure have a semi-linear reachability set.

Our contribution. In this paper, we first propose an extension of the definition of reversal-bounded machines saying that a counter machine is k-reversal-b-bounded if each counter does at most k alternations between increasing and decreasing mode above a given bound b. We show that these new reversal-bounded counter machines do also have a semilinear reachability set, which can be effectively computed. We study the decidability of the reversal-boundedness of a given counter machine, proving that the only case, which is decidable, is the one when the two parameters b and k are provided. Finally, we study reversal-bounded VASS, showing that one can decide using the coverability graph whether a VASS is reversal-bounded or not. Doing so, we propose a new recursive class of VASS with semi-linear reachability sets which contains all the bounded VASS. Furthermore, to the best of our knowledge, it is not known whether one can or cannot decide if a VASS has a semi-linear reachability set or if it is flattable.

Due to lack of space, some details are omitted and can be found in [7].

2   Preliminaries

2.1   Useful Notions

Let $\mathbb{N}$ (resp. $\mathbb{Z}$) denotes the set of nonnegative integers (resp. integers). The usual total order over $\mathbb{Z}$ is written $\leq$. By $\mathbb{N}_\omega$, we denote the set $\mathbb{N} \cup \{\omega\}$ where ω is a new symbol such that $\omega \notin \mathbb{N}$ and for all $k \in \mathbb{N}_\omega$, $k \leq \omega$. We extend the binary operation $+$ and $-$ to $\mathbb{N}_\omega$ as follows: for all $k \in \mathbb{N}$, $k + \omega = \omega$ and $\omega - k = \omega$. For $k, l \in \mathbb{N}_\omega$ with $k \leq l$, we write $[k..l]$ for the interval of integers $\{i \in \mathbb{N} \mid k \leq i \leq l\}$.

Given a set X and $n \in \mathbb{N}$, X^n is the set of n-dim vectors with values in X. For any index $i \in [1..n]$, we denote by $\mathbf{v}(i)$ the ith component of a n-dim vector $\mathbf{v}$. We write $\mathbf{0}$ the vector such that $\mathbf{0}(i) = 0$ for all $i \in [1..n]$. The classical order on $\mathbb{Z}^n$ is also denoted $\leq$ and is defined by $\mathbf{v} \leq \mathbf{w}$ if and only if for all $i \in [1..n]$, we have $\mathbf{v}(i) \leq \mathbf{w}(i)$. We also define the operation $+$ over n-dim vectors of integers in the classical way (ie for $\mathbf{v}$, $\mathbf{v}' \in \mathbb{Z}^n$, $\mathbf{v} + \mathbf{v}'$ is defined by $(\mathbf{v} + \mathbf{v}')(i) = \mathbf{v}(i) + \mathbf{v}'(i)$ for all $i \in [1..n]$).

Let $n \in \mathbb{N}$. A subset $S \subseteq \mathbb{N}^n$ is *linear* if there exist $k + 1$ vectors $\mathbf{v}_0, \mathbf{v}_1, \ldots, \mathbf{v}_k$ in $\mathbb{N}^n$ such that $S = \{\mathbf{v} \mid \mathbf{v} = \mathbf{v}_0 + \lambda_1.\mathbf{v}_1 + \ldots + \lambda_k.\mathbf{v}_k$ with $\lambda_i \in \mathbb{N}$ for all $i \in [1..k]\}$. A *semi-linear set* is any finite union of linear sets. We extend the notion of semi-linearity to subsets of $Q \times \mathbb{N}^n$ where Q is a finite (non-empty) set.

For an alphabet Σ, we denote by Σ^* the set of finite words over Σ and ϵ represents the empty word.

2.2   Counter Machines

A *Minsky machine* is a finite control state automaton which manipulates integer variables, called counters. From each control state, the machine can do the following operations: 1) Increment a counter and go to another control state, 2) Test the value of a counter, if it is 0, it passes to a control state, and if not, it decrements the counter and goes to another control state. There is also a control state called the final state (or halting state) from which the machine cannot do anything. The Minsky machine is said to halt when it reaches this control state. We define here a slight extension of Minsky machines.

We call a n-*dim guarded translation* (shortly a translation) any function $t : \mathbb{N}^n \rightarrow \mathbb{N}^n$ such that there exist $\# \in \{=, \leq\}^n$, $\mu \in \mathbb{N}^n$ and $\delta \in \mathbb{Z}^n$ with $0 \leq \mu + \delta$ and $dom(t) = \{\mathbf{v} \in \mathbb{N}^n \mid \mu \# \mathbf{v}\}$ and for all $\mathbf{v} \in dom(t)$, $t(\mathbf{v}) = \mathbf{v} + \delta$. We will sometimes use the encoding $(\#, \mu, \delta)$ to represent a translation. In the following, T_n will denote the set of the n-dim guarded translations. Let $t = (\#, \mu, \delta)$ be a guarded translation in T_n. We define the vector $D_t \in \mathbb{Z}^n$ as follows, $\forall i \in [1..n]$, $D_t(i) = \delta(i)$. We extend this definition to words of guarded translations, recursively as follows, if $\sigma \in T_n^*$ and $t \in T_n$, we have $D_{t\sigma} = D_t + D_\sigma$ and by convention, $D_\epsilon = \mathbf{0}$.

Definition 1. *A n-dim counter machine (shortly counter machine) is a finite valuated graph $S = \langle Q, E \rangle$ where Q is a finite set of control states and E is a finite relation $E \subseteq Q \times T_n \times Q$.*

The semantics of a counter machine $S = \langle Q, E \rangle$ is given by its associated transition system $TS(S) = \langle Q \times \mathbb{N}^n, \rightarrow \rangle$ where $\rightarrow \subseteq Q \times \mathbb{N}^n \times T_n \times Q \times \mathbb{N}^n$ is a relation defined as follows:

$$(q, \mathbf{v}) \xrightarrow{t} (q', \mathbf{v}') \text{ iff } \exists (q, t, q') \in E \text{ such that } \mathbf{v} \in dom(t) \text{ and } \mathbf{v}' = t(\mathbf{v})$$

We write $(q, \mathbf{v}) \rightarrow (q', \mathbf{v}')$ if there exists $t \in T_n$ such that $(q, \mathbf{v}) \xrightarrow{t} (q', \mathbf{v}')$. The relation $\rightarrow^*$ represents the reflexive and transitive closure of $\rightarrow$. Given a configuration $(q, \mathbf{v})$ of $TS(S)$, $\text{Reach}(S, (q, \mathbf{v})) = \{(q', \mathbf{v}') \mid (q, \mathbf{v}) \rightarrow^* (q', \mathbf{v}')\}$. Furthermore, we extend the relation $\rightarrow$ to words in T_n^*. We have then $(q, \mathbf{v}) \xrightarrow{\epsilon} (q, \mathbf{v})$ and if $t \in T_n$ and $\sigma \in T_n^*$, $(q, \mathbf{v}) \xrightarrow{t\sigma} (q'', \mathbf{v}'')$ if $(q, \mathbf{v}) \xrightarrow{t} (q', \mathbf{v}') \xrightarrow{\sigma} (q'', \mathbf{v}'')$.

Given a counter machine $S = \langle Q, E \rangle$ and an initial configuration $c \in Q \times \mathbb{N}^n$, the pair (S, c) is an intialized counter machine. Since, the notations are explicit, in the following we shall write counter machine for both (S, c) and S.

It is true that any counter machine can be easily encoded into a Minsky machine. For instance to encode a test of the form $x_i = c$, the Minsky machine can decrement c times the counter, test to 0 and increment again c times the counter. Note that this encoding modifies the number of alternations between increasing and decreasing mode for the counters, which is the factor we are interested in when considering reversal-boundedness. That is the reason why we propose this extension of Minsky machine. We do not go further for instance extending the guards, because in [11], it is proved that the reachability problem for reversal-bounded counter machines with linear guards (of the form $x = y$ where x, y are two counters variables) is undecidable.

3   New Reversal-Bounded Counter Machines

3.1   Reversal-Bounded Counter Machines

We would like to extend the notion of reversal-bounded to capture and verify a larger class of counter machines. In fact, if we consider the counter machine represented by

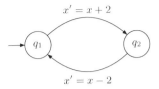

Fig. 1. A simple not reversal-bounded counter machine

the figure 1 with the initial configuration $(q_1, 0)$. Its reachability set is finite equal to $\{(q_1, 0), (q_2, 2)\}$ and consequently semi-linear but the counter machine is not reversal-bounded. We propose here an extension of the notion of reversal-bounded, which allows us to handle such cases and more generally every bounded counter machines.

Given an integer $b \in \mathbb{N}$, we now consider the number of alternations between increasing and decreasing mode when the value of a counter goes above the bound b. Let $S = \langle Q, E \rangle$ be a n-dim counter machine and $TS(S) = \langle Q \times \mathbb{N}^n, \rightarrow \rangle$. From it, we define another transition system $TS_b(S) = \langle Q \times \{\downarrow, \uparrow\}^n \times \mathbb{N}^n \times \mathbb{N}^n, \rightarrow_b \rangle$. Intuitively for a configuration $(q, \mathbf{m}, \mathbf{v}, \mathbf{r}) \in Q \times \{\downarrow, \uparrow\}^n \times \mathbb{N}^n \times \mathbb{N}^n$, the vector $\mathbf{m}$ is used to store the current mode of each counter -increasing ($\uparrow$) or decreasing ($\downarrow$)-, the vector $\mathbf{v}$ contains the values and the vector $\mathbf{r}$ the numbers of alternations performed over b. Formally, we have $(q, \mathbf{m}, \mathbf{v}, \mathbf{r}) \xrightarrow{t}_b (q', \mathbf{m}', \mathbf{v}', \mathbf{r}')$ if and only if the following conditions hold:

1. $(q, \mathbf{v}) \xrightarrow{t} (q', \mathbf{v}')$
2. for each $i \in [1..n]$, the relation expresses by the following array is verified:

$\mathbf{v}(i) - \mathbf{v}'(i)$	$\mathbf{m}(i)$	$\mathbf{m}'(i)$	$\mathbf{v}(i)$	$\mathbf{r}(i)$
> 0	$\downarrow$	$\downarrow$	$-$	$\mathbf{r}(i)$
> 0	$\uparrow$	$\downarrow$	$\leq b$	$\mathbf{r}(i)$
> 0	$\uparrow$	$\downarrow$	$> b$	$\mathbf{r}(i) + 1$
< 0	$\uparrow$	$\uparrow$	$-$	$\mathbf{r}(i)$
< 0	$\downarrow$	$\uparrow$	$\leq b$	$\mathbf{r}(i)$
< 0	$\downarrow$	$\uparrow$	$> b$	$\mathbf{r}(i) + 1$
$= 0$	$\downarrow$	$\downarrow$	$-$	$\mathbf{r}(i)$
$= 0$	$\uparrow$	$\uparrow$	$-$	$\mathbf{r}(i)$

We denote by $\rightarrow_b^*$ the reflexive and transitive closure of $\rightarrow_b$. Given a configuration $(q, \mathbf{m}, \mathbf{v}, \mathbf{r})$ of $TS_b(S)$, $\text{Reach}_b(S, (q, \mathbf{m}, \mathbf{v}, \mathbf{r})) = \{(q', \mathbf{m}', \mathbf{v}', \mathbf{r}') \mid (q, \mathbf{m}, \mathbf{v}, \mathbf{r}) \rightarrow_b^* (q', \mathbf{m}', \mathbf{v}', \mathbf{r}')\}$. We extend this last notation to the configurations of $TS(S)$, saying that if $(q, \mathbf{v}) \in Q \times \mathbb{N}^n$ is a configuration of $TS(S)$, then $\text{Reach}_b(S, (q, \mathbf{v}))$ is equal to the set $\text{Reach}_b(S, (q, \uparrow, \mathbf{v}, \mathbf{0}))$ where $\uparrow$ denotes here the vector with all components equal to $\uparrow$.

Definition 2. *Let $b, k \in \mathbb{N}$. A counter machine (S, c) is k-reversal-b-bounded if and only if for all $(q, \mathbf{m}, \mathbf{v}, \mathbf{r}) \in \text{Reach}_b(S, c)$ and for all $i \in [1..n]$, we have $\mathbf{r}(i) \leq k$.*

We then say that:

1. A counter machine is *reversal-bounded* if there exist $k, b \in \mathbb{N}$ such that it is k-reversal-b-bounded,

2. For a given $k \in \mathbb{N}$, a counter machine is *k-reversal-bounded*, if there exists $b \in \mathbb{N}$ such that it is k-reversal-b-bounded,
3. For a given $b \in \mathbb{N}$, a counter machine is *reversal-b-bounded*, if there exists $k \in \mathbb{N}$ such that it is k-reversal-b-bounded.

We remark that this definition includes the definition of reversal-bounded given in [10], which corresponds to reversal-0-bounded. In comparison to what is presented in [10], there is a slight difference because we do not have here accepting states and consequently we consider all the possible runs of the counter machine as accepted runs. We will see in section 4 that this difference can change some decidability results. Note that in later works [11], the counter machines are also defined without any accepting state.

3.2 Reachability Set

In [10], it has been proved that the reversal-0-bounded counter machines have an effectively computable semi-linear reachability set. We extend here this result to all the reversal-bounded counter machines. The idea consists in building from a k-reversal-b-

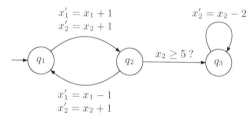

Fig. 2. A 1-reversal-1-bounded counter machine

bounded counter machine (S, c) a k-reversal-0-bounded counter machine (S', c') as it is done for the counter machine of the figure 2 (with the initial configuration $(q_1, (0, 0))$) from which we obtain the counter machine represented in the figure 3 (with the initial configuration $((q_1, 0, 0), (0, 0))$). We assume $S = \langle Q, E \rangle$ and $S' = \langle Q', E' \rangle$. First we introduce two symbols $\perp$ and ω_b which are not integers. ω_b represents a counter value strictly greater than b and $\perp$ a counter value for which it is not known whether it is greater or not than b. The location set Q' is then equal to $Q \times B^n$ where $B = \{0, \dots, b\} \cup \{\omega_b, \perp\}$. Intuitively, the counter machine S' encodes the run of S and when a counter value in S is under the bound b, its value is stored into the control state of S' and the corresponding value of the counter in S' is 0, but when the value goes above b in S then it is restored in the counter in S'. Furthermore (S', c') being k-reversal-0-bounded, we use the results of [10] to compute the reachability set $\text{Reach}(S', c')$ from which we deduce $\text{Reach}(S, c)$.

Theorem 3. *Given a reversal-bounded counter machine, its reachability set is an effectively computable semi-linear set.*

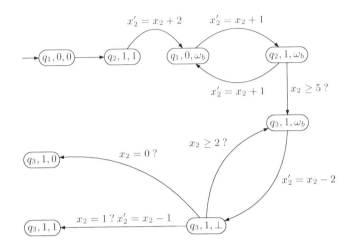

Fig. 3. A 1-reversal-0-bounded counter machine obtained from the counter machine of Fig. 2

4   Deciding Reversal-Boundedness

In this section, we will study the decidabilty of reversal-boundedness.

4.1   Undecidability

In [10], the author shows that it is not possible to decide whether a counter machine is reversal-0-bounded or not. We prove here that this theorem is still true when considering reversal-boundedness.

Theorem 4. *Verifying if a counter machine is reversal-bounded is undecidable.*

Proof. We reduce the halting problem for 2-counters deterministic Minsky Machines. We consider a deterministic Minsky Machine S with the initial configuration $(q_0, (0,0))$ working over two counter variables x_1 and x_2. "Deterministic" here means that there is a unique possible run starting on $(q_0, (0,0))$. From S, we build a counter machine S' working over three counter variables x_1, x_2 and x_3, such that for each $(q, t, q') \in E$, we add two control states q_1 and q_2 and the transitions (q, t_1, q_1), (q_1, t_2, q_2) and (q_2, t, q') where t_1 and t_2 only change the counter variable x_3 doing $x_3' = x_3 + 2$ for t_1 and $x_3' = x_3 - 1$ for t_2. Note that S' starting on $(q_0, (0,0,0))$ is also deterministic. Furthermore $(S', (q_0, (0,0,0)))$ is reversal-bounded if and only if its unique run is finite, which is equivalent to halting. Since S' starting with $(q_0, (0,0,0))$ halts if and only if S starting from $(q_0, (0,0))$ halts and since this last problem is undecidable, we conclude the theorem.   □

4.2   Fixing One Parameter

We will see here that fixing one of the parameters is not enough to obtain decidability for the reversal-boundedness.

Theorem 5. *Given* $b \in \mathbb{N}$, *verifying if a counter machine is reversal-b-bounded is un-decidable.*

Sketch of Proof. For each b in $\mathbb{N}$, we can reuse the same proof as for the theorem 4, we can show that the 3-counter machine $(S', (q_0, (0, 0, 0)))$ is reversal-b-bounded if and only if the deterministic Minsky machine (S, c) from which it is built halts.   □

Theorem 6. *Given* $k \in \mathbb{N}$, *verifying if a counter machine is k-reversal-bounded is undecidable.*

Sketch of Proof. To prove this result we again use the 3-counter machine S' with the initial configuration $(q_0, (0, 0, 0))$ that we complete so that each run can begin with doing at least k alternations between increasing and decreasing mode over any bound.   □

4.3   Fixing the Two Parameters

We will now prove that if the two parameters b and k are fixed, it is possible to decide if a counter machine is k-reversal-b-bounded. Let $b, k \in \mathbb{N}$ and (S, c) be a counter machine. The idea consists in building a counter machine (S', c') which will be $(k+1)$-reversal-b-bounded and which will reach a special control state q_{err} if and only if (S, c) is not k-reversal-b-bounded. Note that since (S', c') is reversal-bounded, it is possible to decide whether the control state q_{err} is reachable or not. In the control state of (S', c'), we store the mode -increasing ($\uparrow$) or decreasing ($\downarrow$)- for each counter and also the number of alternations already performed over b. We also add some control states to test at each step if each counter value is strictly greater (denoted by $b_>$) or smaller than b (denoted by $b_<$). The figure 4 gives an example of the counter machine we build to decide if the counter machine from figure 1 with the initial configuration $(q_1, 0)$ is 1-reversal-1-bounded.

Theorem 7. *Given* $b, k \in \mathbb{N}$, *verifying if a counter machine is k-reversal-b-bounded is decidable.*

This result contrasts with the one given in [10], which says that given $k \in \mathbb{N}$, verifying if a counter machine is k-reversal-0-bounded is undecidable. This is due to the fact that in [10], the considered counter machines have accepting control states, whereas our definition is equivalent to have all the control states as accepting. In fact, when we define the reversal-bounded counter machines, we consider all the possible runs and not only the one ending in an accepting state.

4.4   Computing the Parameters

When a counter machine is reversal-bounded, it could be useful to characterize the pairs (k, b) for which it is k-reversal-b-bounded, first because it gives us information on the behavior of the counter machine but also because these parameters are involved in the way the reachability set is built as one can see in the proof of theorem 3 and in [10].

Let (S, c) be a counter machine. We define the following set to talk about the parameters of reversal-bounded counter machines:

$$RB(S, c) = \{(k, b) \in \mathbb{N} \times \mathbb{N} \mid (S, c) \text{ is } k\text{-reversal-}b\text{-bounded}\}$$

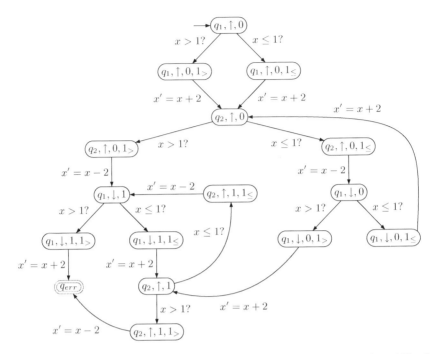

Fig. 4. A 2-reversal-1-bounded counter machine to decide if the counter machine of Fig. 2 is 1-reversal-1-bounded

Then $RB(S,c) = \emptyset$ if and only if (S,c) is not reversal-bounded, hence the non-emptiness problem for $RB(S,c)$ is in general not decidable, but this set is recursive (cf. theorem 7). Furthermore, if there exist (k,b) in $RB(S,c)$ and $(k',b') \in \mathbb{N} \times \mathbb{N}$ such that $(k,b) \leq (k',b')$ then we know, by definition of reversal-boundedness that (S,c) is also k'-reversal-b'-bounded, ie $(k',b') \in RB(S,c)$. Since the order relation $\leq$ on $\mathbb{N} \times \mathbb{N}$ is a well-ordering we can deduce:

Lemma 8. *Let (S,c) be a reversal-bounded counter machine. The set $RB(S,c)$ is upward-closed, it has a finite number of minimal elements, which can effectively be computed.*

Sketch of proof. The facts that $RB(S,c)$ is upward closed is a direct consequence of reversal-boundedness. And since $(\mathbb{N} \times \mathbb{N}, \leq)$ is a well-ordering, each of its upward-closed set has a finite number of minimal elements [8]. To compute the minimal elements, we add "reversal-bounded" counters either to count the number of alternations between increasing and decreasing mode over a bound b or to store the value of a counter each time it changes mode over a given b. □

5   Analysis of VASS

In this section, we recall the definition of Vector Addition System with States and show that the notion of reversal-boundedness we newly introduce is well-suited for the verification of these systems.

5.1   VASS and Their Coverability Graphs

Definition 9. *A n-dim counter machine $\langle Q, E \rangle$ is a* Vector Addition System with States *(shortly VASS) if and only if for all transitions $(q, t, q') \in E$, t is a guarded translation $(\#, \mu, \delta)$ such that $\# = (\leq, \ldots, \leq)$,*

Hence in VASS, it is not possible to test if a counter value is equal to a constant but only if it is greater than a constant.

In [13], the authors provide an algorithm to build from a VASS a labeled tree, the *Karp and Miller tree* (the algorithm is provided in [7]). The main idea of this construction is to cover in a finite way the reachable configurations using the symbol ω, when a counter is not bounded. They have shown that their algorithm always terminates and that it enjoys some good properties. In particular, this tree can be used to decide the boundedness of a VASS. In [21], the authors have proposed a further construction based on the Karp and Miller tree in order to test the regularity of the language of the unlabeled traces of a VASS. This last construction is known as the *coverability graph*. To obtain it, the nodes of the Karp and Miller tree with the same labels are grouped together. Formally if (S, c) is a n-dim initialized VASS, we denote by $CG(S, c)$ its coverability graph defined as follows, $CG(S, c) = \langle N, \Delta \rangle$ where:

- $N \subseteq Q \times \mathbb{N}_\omega^n$ is a finite set of nodes,
- $\Delta \subseteq N \times T_n \times N$ is a finite set of edges labeled with guarded transitions.

We call a *circuit* in the coverability graph a path ending in the starting node and a circuit will be said to be *elementary* if all nodes are different with the exception of the starting and ending nodes. For a vector $\mathbf{w} \in \mathbb{N}_\omega^n$, we denote by $\mathrm{Inf}(\mathbf{w})$ the set $\{i \in [1..n] \mid \mathbf{w}(i) = \omega\}$ and $\mathrm{Fin}(\mathbf{w}) = [1..n] \setminus \mathrm{Inf}(\mathbf{w})$. Using these notions, it has been proved that the coverability graph verifies the following properties.

Let (S, c) be a n-dim initialized VASS with $S = \langle Q, E \rangle$, $TS(S) = \langle Q \times \mathbb{N}^n, \to \rangle$ its associated transition system and $G = \langle N, \Delta \rangle$ its coverability graph.

Theorem 10. *[13,21]*

1. *If $(q, \mathbf{w})$ is a node in G, then for all $k \in \mathbb{N}$, there exists $(q, \mathbf{v}) \in \mathrm{Reach}(S, c)$ such that for all $i \in \mathrm{Inf}(\mathbf{w})$, $k \leq \mathbf{v}(i)$ and for all $i \in \mathrm{Fin}(\mathbf{w})$, $\mathbf{w}(i) = \mathbf{v}(i)$.*
2. *For $\sigma \in T_n^*$, if $c \xrightarrow{\sigma} (q, \mathbf{v})$ then there is a unique path in G labeled by σ and leading from c to a node $(q, \mathbf{w})$ and for all $i \in \mathrm{Fin}(\mathbf{w})$, $\mathbf{v}(i) = \mathbf{w}(i)$.*
3. *If $\sigma \in T_n^*$ is a word labeling a circuit in G and $(q, \mathbf{w})$ is the initial node of this circuit, then there exist $(q, \mathbf{v}) \in \mathrm{Reach}(S, c)$ and $(q', \mathbf{v}')$ such that $(q, \mathbf{v}) \xrightarrow{\sigma} (q, \mathbf{v}')$ and for all $i \in \mathrm{Fin}(\mathbf{w})$, $\mathbf{w}(i) = \mathbf{v}(i) = \mathbf{v}'(i)$.*

From this theorem, we deduce the following lemma, we will then use to decide the reversal-boundedness of a VASS:

Lemma 11. *If there exists an elementary circuit $((q_1, \mathbf{w}_1) \xrightarrow{t_1} (q_2, \mathbf{w}_2) \xrightarrow{t_2} \ldots \xrightarrow{t_f} (q_1, \mathbf{w}_1))$ in G, then for all $k, l \in \mathbb{N}$, there exist $\mathbf{v}_1, \ldots, \mathbf{v}_l \in \mathbb{N}^n$ such that:*

(i) *$c \to^* (q_1, \mathbf{v}_1) \xrightarrow{\sigma} (q_1, \mathbf{v}_2) \xrightarrow{\sigma} \ldots \xrightarrow{\sigma} (q_1, \mathbf{v}_l)$ in $TS(S)$ with $\sigma = t_1 \ldots t_f$, and,*
(ii) *for all $j \in [1..l]$, for all $i \in \mathrm{Inf}(\mathbf{w}_1)$, $k \leq \mathbf{v}_j(i)$ and for all $i \in \mathrm{Fin}(\mathbf{w}_1)$, $\mathbf{w}_1(i) = \mathbf{v}_j(i)$.*

5.2 Deciding If a VASS Is Reversal-b-Bounded

In this section, we show that its possible to decide if a VASS is reversal-b-bounded using a characterization over its coverability graph.

Let $S = \langle Q, E \rangle$ be a n-dim counter machine. We build a $2n$-dim counter machine $\widetilde{S} = \langle Q', E' \rangle$ adding for each counter another counter, whose role is to count the alternation of the first counter between increasing and decreasing mode. Formally, $Q' = Q \times \{\uparrow, \downarrow\}^n$ and T' is built as follows, for each $(q, (\#, \mu, \delta), q') \in E$ and $\mathbf{m}, \mathbf{m}' \in \{\uparrow, \downarrow\}^n$, we have $((q, \mathbf{m}), (\#', \mu', \delta'), (q', \mathbf{m}')) \in E'$ if and only if:

- for all $i \in [1..n]$, $\#'(i) = \#(i)$, $\mu'(i) = \mu_i$ and $\delta'(i) = \delta(i)$;
- for all $i \in [n+1..2n]$, $\#'(i) \in \{\leq\}$ and $\mu'(i) = 0$;
- δ, $\mathbf{m}$, $\mathbf{m}'$ and δ' satisfy for all $i \in [1..n]$ the conditions described in the following array:

$\delta(i)$	$\mathbf{m}(i)$	$\mathbf{m}'(i)$	$\delta'(n+i)$
$= 0$	$\uparrow$	$\uparrow$	0
$= 0$	$\downarrow$	$\downarrow$	0
> 0	$\uparrow$	$\uparrow$	0
> 0	$\downarrow$	$\uparrow$	1
< 0	$\downarrow$	$\downarrow$	0
< 0	$\uparrow$	$\downarrow$	1

By construction, we remark that if S is a VASS then $\widetilde{S}$ is a VASS too. We define then the relation $\sim \in (Q \times \{\uparrow, \downarrow\}^n \times \mathbb{N}^n \times \mathbb{N}^n) \times (Q \times \{\uparrow, \downarrow\}^n \times \mathbb{N}^{2n})$ between the configurations of $TS_0(S)$ and the ones of $TS(\widetilde{S})$ saying that $(q, \mathbf{m}, \mathbf{v}, \mathbf{r}) \sim (q', \mathbf{m}', \mathbf{v}')$ if and only if:

- $q = q'$,
- $\mathbf{m} = \mathbf{m}'$,
- for all $i \in [1..n]$, $\mathbf{v}(i) = \mathbf{v}'(i)$ and $\mathbf{r}(i) = \mathbf{v}'(n+i)$.

The relation $\sim$ is a bisimulation between $TS_0(S)$ and $TS(\widetilde{S})$. Given an initial configuration $c = (q, \mathbf{v})$, we have $(q, \uparrow, \mathbf{v}, \mathbf{0}) \sim (q, \uparrow, (\mathbf{v}, \mathbf{0}))$. Hence, if we denote by $\widetilde{c}$ the triple $(q, \uparrow, (\mathbf{v}, \mathbf{0}))$, we can deduce that the VASS (S, c) is reversal-0-bounded if and only if there exists $k \in \mathbb{N}$ such that for all $(q, \mathbf{m}, \mathbf{v}) \in \mathrm{Reach}(\widetilde{S}, \widetilde{c})$, for all $i \in [1..n]$, $\mathbf{v}(n+i) \leq k$. Using the coverability graph of $(\widetilde{S}, \widetilde{c})$, this last property is decidable for a VASS. Generalizing this method for any $b \in \mathbb{N}$, counting only the alternations that are done above b, we can deduce that:

Theorem 12. *Given $b \in \mathbb{N}$, verifying if a VASS is reversal-b-bounded is decidable.*

5.3 Deciding If a VASS Is Reversal-Bounded

We will now show that the analysis of the coverabilty graph of $(\widetilde{S}, \widetilde{c})$ allows us to decide if a VASS is reversal-bounded (without a fixed bound). Note that this is not a direct consequence of the previous theorem, because it is not possible to enumerate the different bounds b and test if the VASS is reversal-b-bounded, since this method never terminates when the VASS is not reversal-bounded.

Lemma 13. *A n-dim VASS (S, c) is reversal-b-bounded if and only if for all $i \in [1..n]$, all nodes $(q, \boldsymbol{w})$ belonging to an elementary circuit labeled by $\sigma \in T_n^*$ of $CG(\widetilde{S}, \widetilde{c})$ with $D_\sigma(n + i) > 0$ verify $\boldsymbol{w}(i) \leq b$.*

In other words, this last lemma states that (S, c) is reversal-b-bounded if and only if for all $i \in [1..n]$, there is no elementary circuit in the coverability graph $CG(\widetilde{S}, \widetilde{c})$ which strictly increases the $(n + i)$-th counter and which has a node, whose i-th component is strictly greater than b or equal to ω. In fact, applying the lemma 11, we deduce that if such an elementary circuit exists, we can build a run of the VASS (S, c) which does not respect the definition of reversal-b-boundedness. The details of the proof can be found in [7].

For a VASS (S, c), the lemma 13 gives us a necessary and sufficient condition over the coverability graph of $(\widetilde{S}, \widetilde{c})$, and this condition can effectively be tested. This allows us to deduce the following decidability result.

Theorem 14. *Verifying if a VASS is reversal-bounded is decidable.*

Unfortunately, the decision algorithm we propose here builds entirely the coverability graph of a VASS, and this building is known to be non-primitive-recursive in space (some details can be found in [12]).

6   Perspectives

In [4], the authors have proved that some liveness problems are decidable for reversal-0-bounded counter machines and others not. For instance, it is decidable to verify if a run of a reversal-bounded counter machine passes infinitely often through a semilinear set of possible configurations; but the same problem becomes undecidable when all the runs are considered. It seems that this result can easily be extended to the class of reversal-bounded counter machines, we have introduced. It would then pave the way to verify more complex properties than reachability over reversal-bounded counter machines. It could also be interesting to look at these liveness problems in the particular case of reversal-bounded VASS.

An other perspective for our work would be to use reversal-bounded counter machines to analyze counter machines which are not necessarily reversal-bounded. In fact, we have seen with the proof of theorem 7, that for any $k, b \in \mathbb{N}$ and from any counter machine, it is possible to build another counter machine, which is k-reversal-b-bounded and whose runs represent an under-approximation of the set of runs of the first one. We could consequently build a tool which given a counter machine would build successively, incrementing the parameters k and b, the corresponding k-reversal-b-bounded counter machines, and would test at each step if the reachability set of the initial counter machine has been built (this can be easily done, since this set is a fixpoint of the transition relation). This algorithm might never terminate, if the reachability set is not semilinear for instance, but it will refine at each step the under-approximation of the reachability set.

References

1. Annichini, A., Bouajjani, A., Sighireanu, M.: TReX: A tool for reachability analysis of complex systems. In: Berry, G., Comon, H., Finkel, A. (eds.) CAV 2001. LNCS, vol. 2102, pp. 368–372. Springer, Heidelberg (2001)
2. Araki, T., Kasami, T.: Decidable problems on the strong connectivity of Petri net reachability sets. Theor. Comput. Sci. 4(1), 99–119 (1977)
3. Bardin, S., Leroux, J., Point, G.: FAST extended release. In: Ball, T., Jones, R.B. (eds.) CAV 2006. LNCS, vol. 4144, pp. 63–66. Springer, Heidelberg (2006)
4. Ibarra, O.H., Dang, Z., San Pietro, P.: Liveness Verification of Reversal-Bounded Multicounter Machines with a Free Counter. In: Hariharan, R., Mukund, M., Vinay, V. (eds.) FSTTCS 2001. LNCS, vol. 2245, pp. 132–143. Springer, Heidelberg (2001)
5. Esparza, J.: Petri nets, commutative context-free grammars, and basic parallel processes. Fundam. Inform. 31(1), 13–25 (1997)
6. Finkel, A., Leroux, J.: How to compose Presburger-accelerations: Applications to broadcast protocols. In: Agrawal, M., Seth, A.K. (eds.) FSTTCS 2002. LNCS, vol. 2556, pp. 145–156. Springer, Heidelberg (2002)
7. Finkel, A., Sangnier, A.: Reversal-bounded counter machines revisited. Research report, Laboratoire Specification et Verification, ENS Cachan (2008)
8. Higman, G.: Ordering by divisibility in abstract algebras. Proc. London Math. Soc (3) 2, 326–336 (1952)
9. Hopcroft, J.E., Pansiot, J.-J.: On the reachability problem for 5-dimensional vector addition systems. Theor. Comput. Sci. 8, 135–159 (1979)
10. Ibarra, O.H.: Reversal-bounded multicounter machines and their decision problems. J. ACM 25(1), 116–133 (1978)
11. Ibarra, O.H., Su, J., Dang, Z., Bultan, T., Kemmerer, R.A.: Counter machines and verification problems. Theor. Comput. Sci. 289(1), 165–189 (2002)
12. Jantzen, M.: Complexity of place/transition nets. In: APN 1986. LNCS, vol. 254, pp. 413–434. Springer, Heidelberg (1987)
13. Karp, R.M., Miller, R.E.: Parallel program schemata: A mathematical model for parallel computation. In: FOCS 1967, pp. 55–61. IEEE Computer Society Press, Los Alamitos (1967)
14. Kosaraju, S.R.: Decidability of reachability in vector addition systems (preliminary version). In: STOC 1982, pp. 267–281. ACM Press, New York (1982)
15. Landweber, L.H., Robertson, E.L.: Properties of conflict-free and persistent Petri nets. J. ACM 25(3), 352–364 (1978)
16. Homepage of LASH,
   http://www.montefiore.ulg.ac.be/~boigelot/research/lash
17. Leroux, J., Sutre, G.: Flat counter almost everywhere! In: Peled, D.A., Tsay, Y.-K. (eds.) ATVA 2005. LNCS, vol. 3707, pp. 474–488. Springer, Heidelberg (2005)
18. Mayr, E.W.: Persistence of vector replacement systems is decidable. Acta Inf. 15, 309–318 (1981)
19. Mayr, E.W.: An algorithm for the general Petri net reachability problem. SIAM J. Comput. 13(3), 441–460 (1984)
20. Minsky, M.L.: Computation: finite and infinite machines. Prentice-Hall, Inc., Upper Saddle River, NJ, USA (1967)
21. Valk, R., Vidal-Naquet, G.: Petri nets and regular languages. J. Comput. Syst. Sci. 23(3), 299–325 (1981)

Iterative Compression and Exact Algorithms

Fedor V. Fomin[1,*], Serge Gaspers[1], Dieter Kratsch[2],
Mathieu Liedloff[2], and Saket Saurabh[1]

[1] Department of Informatics, University of Bergen,
N-5020 Bergen, Norway
{fomin,serge,saket}@ii.uib.no
[2] Laboratoire d'Informatique Théorique et Appliquée,
Université Paul Verlaine - Metz, 57045 Metz Cedex 01, France
{kratsch,liedloff}@univ-metz.fr

Abstract. Iterative Compression has recently led to a number of break-throughs in parameterized complexity. The main purpose of this paper is to show that iterative compression can also be used in the design of exact exponential time algorithms. We exemplify our findings with algorithms for the MAXIMUM INDEPENDENT SET problem, a counting version of k-HITTING SET and the MAXIMUM INDUCED CLUSTER SUBGRAPH problem.

1 Introduction

Iterative Compression is a tool that has recently been used successfully in solving a number of problems in the area of Parameterized Complexity. This technique was first introduced by Reed et al. to solve the ODD CYCLE TRANSVERSAL problem, where one is interested in finding a set of at most k vertices whose deletion makes the graph bipartite [20]. Iterative compression was used in obtaining faster FPT algorithms for FEEDBACK VERTEX SET, EDGE BIPARTIZATION and CLUSTER VERTEX DELETION on undirected graphs [6,12,14]. Recently this technique has led to an FPT algorithm for the DIRECTED FEEDBACK VERTEX SET problem [4], one of the longest open problems in the area of parameterized complexity.

The main idea behind iterative compression for parameterized algorithms is an algorithm which, given a solution of size $k+1$ for a problem, either compresses it to a solution of size k or proves that there is no solution of size k. This is known as the compression step of the algorithm. Based on this compression step, iterative (and incremental) algorithms for minimization problems are obtained. The most technical part of an FPT algorithm based on iterative compression is to show that the compression step can be carried out in time $f(k) \cdot n^{O(1)}$, where f is an arbitrary computable function, k is a parameter and n is the length of the input.

The presence of a solution of size $k + 1$ can provide important structural information about the problem. This is one of the reasons why the technique of iterative compression has become so powerful. Structures are useful in designing

---

* Partially supported by the Research Council of Norway.

E. Ochmański and J. Tyszkiewicz (Eds.): MFCS 2008, LNCS 5162, pp. 335–346, 2008.
© Springer-Verlag Berlin Heidelberg 2008

algorithms in most paradigms. By seeing so much success of iterative compression in designing fixed parameter tractable algorithms, it is natural and tempting to study its applicability in designing exact exponential time algorithms.

The goal of the design of moderately exponential time algorithms for NP-complete problems is to establish algorithms for which the worst-case running time is provably faster than the one of enumerating all prospective solutions, or loosely speaking, *algorithms better than brute-force enumeration*. For example, for NP-complete problems on graphs on n vertices and m edges whose solutions are either subsets of vertices or edges, the brute-force or trivial algorithms basically enumerate all subsets of vertices or edges. This mostly leads to algorithms of time complexity 2^n or 2^m, modulo some polynomial factors, based on whether we are enumerating vertices or edges. Almost all the iterative compression based FPT algorithms with parameter k have a factor of 2^{k+1} in the running time, as they all branch on all partitions (A, D) of a $k + 1$ sized solution S and look for a solution of size k with a restriction that it should contain all elements of A and none of D. This is why, at first thought, iterative compression is a quite useless technique for solving optimization problems because for $k = \Omega(n)$, we end up with an algorithm having a factor 2^n or 2^m in the worst-case running time, while a running time of 2^n or 2^m (up to a polynomial factor) often can be achieved by (trivial) brute force enumeration. Luckily, our intuition here appears to be wrong and with some additional arguments, iterative compression can become a useful tool in the design of moderately exponential time algorithms as well. We find it interesting because despite of several exceptions (like the works of Björklund et al. [1,2,16]), the area of exact algorithms is heavily dominated by branching algorithms, in particular, for subset problems. It is very often that an (incremental) improvement in the running time of branching algorithm requires an extensive case analysis, which becomes very technical and tedious. The analysis of such algorithms can also be very complicated and even computer based.

The main advantage of iterative compression is that it provides combinatorial algorithms based on problem structures. While the improvement in the running time compared to (complicated) branching algorithms is not so impressive, the simplicity and elegance of the arguments allow them to be used in a basic algorithm course.

To our knowledge, this paper is the first attempt to use iterative compression outside the domain of FPT algorithms. We exemplify this approach by the following results:

1. We show how to solve MAXIMUM INDEPENDENT SET for a graph on n vertices in time $\mathcal{O}(1.3196^n)$. While the running time of our iterative compression algorithm is slower than the running times of modern branching algorithms [10,21], this simple algorithm serves as an introductory example to more complicated applications of the method.

2. We obtain algorithms counting the number of minimum hitting sets of a family of sets of an n-element ground set in time $\mathcal{O}(1.7198^n)$, when the size of each set is at most 3 (#MINIMUM 3-HITTING SET). For #MINIMUM 4-HITTING SET we obtain an algorithm of running time $\mathcal{O}(1.8997^n)$. For

MINIMUM 4-HITTING SET similar ideas lead to an algorithm of running time $\mathcal{O}(1.8704^n)$. These algorithms are faster than the best algorithms known for these problems so far [9,19].

3. We provide an algorithm to solve the MAXIMUM INDUCED CLUSTER SUB-GRAPH problem in time $\mathcal{O}(1.6181^n)$. The only algorithm for this problem we were aware of before is the use of a very complicated branching algorithm of Wahlström [23] for solving 3-HITTING SET (let us note that MAXIMUM INDUCED CLUSTER SUBGRAPH is a special case of 3-HITTING SET, where every subset is a set of vertices inducing a path of length 3), which results in time $\mathcal{O}(1.6278^n)$.

2   Maximum Independent Set

MAXIMUM INDEPENDENT SET (MIS) is one of the well studied problems in the area of exact exponential time algorithms and many papers have been written on this problem [10,21,22]. It is customary that if we develop a new method then we first apply it to well known problems in the area. Here, as an introductory example, we consider the NP-complete problem MIS.

MAXIMUM INDEPENDENT SET (MIS): Given a graph $G = (V, E)$ on n vertices, find a maximum independent set of G. An *independent set* of G is a set of vertices $I \subseteq V$ such that no two vertices of I are adjacent in G. A *maximum independent set* is an independent set of maximum size.

It is well-known that I is an independent set of a graph G iff $V \setminus I$ is a vertex cover of G, i.e. every edge of G has at least one end point in $V \setminus I$. Therefore MINIMUM VERTEX COVER (MVC) is the complement of MIS in the sense that I is a maximum independent set of G iff $V \setminus I$ is a minimum vertex cover of G. This fact implies that when designing exponential time algorithms we may equivalently consider MVC. We proceed by defining a compression version of the MVC problem.

COMP-MVC: Given a graph $G = (V, E)$ with a vertex cover $S \subseteq V$, find a vertex cover of G of size at most $|S| - 1$ if one exists.

Note that if we can solve COMP-MVC efficiently then we can solve MVC efficiently by repeatedly applying an algorithm for COMP-MVC as follows. Given a graph $G = (V, E)$ on n vertices with $V = \{v_1, v_2, ..., v_n\}$, let $G_i = G[\{v_1, v_2, ..., v_i\}]$ and let C_i be a minimum vertex cover of G_i. By V_i we denote the set $\{v_1, v_2, ..., v_i\}$. We start with G_1 and put $C_1 = \emptyset$. Suppose that we already have computed C_i for the graph G_i for some $i \geq 1$. We form an instance of COMP-MVC with input graph G_{i+1} and $S = C_i \cup \{v_{i+1}\}$. In this stage we either *compress* the solution S which means that we find a vertex cover S' of G_{i+1} of size $|S| - 1$ and put $C_{i+1} = S'$, or (if there is no S') we put $C_{i+1} = S$.

Our algorithm is based on the following lemma.

Lemma 1. [⋆][1] *Let G_{i+1} and S be given as above. If there exists a vertex cover C_{i+1} of G_{i+1} of size $|S| - 1$, then it can be partitioned into two sets A and B such that*
(a) $A \subset S$, $|A| \leq |S| - 1$ and A is a minimal vertex cover of $G_{i+1}[S]$.
(b) $B \subseteq (V_{i+1} \setminus A)$ is a minimum vertex cover of the bipartite graph $G_{i+1}[V_{i+1} \setminus A]$.

Lemma 1 implies that the following algorithm solves COMP-MVC correctly.

Step 1: Enumerate all minimal vertex covers of size at most $|S| - 1$ of $G_{i+1}[S]$ as a possible candidate for A.
Step 2: For each minimal vertex cover A find a minimum vertex cover B of the bipartite graph $G_{i+1}[V_{i+1} \setminus A]$ (via the computation of a maximum matching in this bipartite graph [13]).
Step 3: If the algorithm finds a vertex cover $A \cup B$ of size $|S| - 1$ in this way, set $C_{i+1} = A \cup B$, else set $C_{i+1} = S$.

Steps 2 and **3** of the algorithm can be performed in polynomial time, and the running time of **Step 1**, which is exponential, dominates the running time of the algorithm. To enumerate all maximal independent sets or equivalently all minimal vertex covers of a graph in **Step 1**, one can use the polynomial-delay algorithm of Johnson et al. [15].

Proposition 1 ([15]). *All maximal independent sets of a graph can be enumerated with polynomial delay.*

For the running time analysis of the algorithm we need the following bounds on the number of maximal independent sets or minimal vertex covers due to Moon and Moser [17] and Byskov [3].

Proposition 2 ([17]). *A graph on n vertices has at most $3^{n/3}$ maximal independent sets.*

Proposition 3 ([3]). *The maximum number of maximal independent sets of size at most k in any graph on n vertices for $k \leq n/3$ is*

$$N[n, k] = \lfloor n/k \rfloor^{(\lfloor n/k \rfloor + 1)k - n} (\lfloor n/k \rfloor + 1)^{n - \lfloor n/k \rfloor k}.$$

Moreover, all such sets can be enumerated in time $\mathcal{O}^*(N[n, k])$.[2]

Since

$$\max \left\{ \max_{0 \leq \alpha \leq 3/4} (3^{\alpha n/3}), \max_{3/4 < \alpha \leq 1} (N[\alpha n, (1 - \alpha)n]) \right\} = O^*(2^{2n/5}),$$

---

[1] Proofs of results labeled with [⋆] will appear in the long version of the paper.
[2] Throughout this paper we use a modified big-Oh notation that suppresses all polynomially bounded factors. For functions f and g we write $f(n) = \mathcal{O}^*(g(n))$ if $f(n) = \mathcal{O}(g(n) poly(n))$, where $poly(n)$ is a polynomial. Furthermore, since $c^n \cdot poly(n) = \mathcal{O}((c + \epsilon)^n)$ for any $\epsilon > 0$, we omit polynomial factors in the big-Oh notation every time we round the base of the exponent.

we have that by Propositions 1, 2, and 3, all minimal vertex covers of $G_{i+1}[S]$ of size at most $|S| - 1$ can be listed in time $\mathcal{O}^*(2^{2n/5}) = \mathcal{O}(1.3196^n)$.

Thus, the overall running time of the algorithm solving COMP-MVC is $\mathcal{O}(1.3196^n)$. Since the rounding of the base of the exponent dominates the polynomial factor of the other steps of the iterative compression, we obtain the following theorem.

Theorem 1. MAXIMUM INDEPENDENT SET *and* MINIMUM VERTEX COVER *can be solved in time* $\mathcal{O}(1.3196^n)$ *on graphs of n vertices by a compression based algorithm.*

3   #k-Hitting Set

The HITTING SET problem is a generalization of VERTEX COVER. Here, given a family of sets over a ground set of n elements, the objective is to hit every set of the family with as few elements of the ground set as possible. We study a version of the hitting set problem where every set in the family has at most k elements.

> MINIMUM k-HITTING SET (MHS$_k$): Given a universe V of n elements and a collection $\mathcal{C}$ of subsets of V of size at most k, find a minimum hitting set of $\mathcal{C}$. A *hitting set* of $\mathcal{C}$ is a subset $V' \subseteq V$ such that every subset of $\mathcal{C}$ contains at least one element of V'.

A counting version of the problem is #MINIMUM k-HITTING SET (#MHS$_k$) that asks for the number of different minimum hitting sets. We denote an instance of #MHS$_k$ by $(V, \mathcal{C})$. Furthermore we assume that for every $v \in V$, there exists at least one set in $\mathcal{C}$ containing it.

We show how to obtain an algorithm to solve #MHS$_k$ using iterative compression which uses an algorithm for #MHS$_{k-1}$ as a subroutine. First we define the compression version of the #MHS$_k$ problem.

> COMP-#k-HITTING SET: Given a universe V of n elements, a collection $\mathcal{C}$ of subsets of V of size at most k, and a (not necessarily minimum) hitting set $H' \subseteq V$ of $\mathcal{C}$, find a minimum hitting set $\widehat{H}$ of $\mathcal{C}$ and compute the number of all minimum hitting sets of $\mathcal{C}$.

Lemma 2. *Let* $\mathcal{O}^*(a_{k-1}^n)$ *be the running time of an algorithm solving* #MHS$_{k-1}$, *where* $a_{k-1} > 1$ *is some constant. Then* COMP-#k-HITTING SET *can be solved in time*

$$\mathcal{O}^* \left(2^{|H'|} a_{k-1}^{|V| - |H'|} \right).$$

Moreover, if $|H'|$ *is greater than* $2|V|/3$ *and the minimum size of a hitting set in* $\mathcal{C}$ *is at least* $|H'| - 1$, *then* COMP-#k-HITTING SET *can be solved in time*

$$\mathcal{O}^* \left(\binom{|H'|}{2|H'| - |V|} a_{k-1}^{|V| - |H'|} \right).$$

Proof. To prove the lemma, we give an algorithm that, for each possible partition $(N, \bar{N})$ of H', computes a minimum hitting set H_N and the number h_N of minimum hitting sets subject to the constraint that these hitting sets contain all the elements of N and none of the elements of $\bar{N}$.

For every partition $(N, \bar{N})$ of H', we either reject it as invalid or we reduce the instance $(V, \mathcal{C})$ to an instance $(V', \mathcal{C}')$ by applying the following two rules in the given order.

(H) If there exists a set $C_i \in \mathcal{C}$ such that $C_i \subseteq \bar{N}$ then we refer to such a partition as *invalid* and reject it.

(R) For all sets C_i with $C_i \cap N \neq \emptyset$ put $\mathcal{C} = \mathcal{C} \setminus C_i$. In other words, all sets of $\mathcal{C}$, which are already hit by N, are removed.

If the partition $(N, \bar{N})$ of H' is not invalid based on rule **(R)** the instance $(V, \mathcal{C})$ can be reduced to the instance $I' = (V', \mathcal{C}')$, where $V' = V \setminus H'$ and $\mathcal{C}' = \{X \cap V' \mid X \in \mathcal{C} \text{ and } X \cap N = \emptyset\}$.

Summarizing, the instance I' is obtained by removing all the elements of V for which it has already been decided if they are part of H_N or not and all the sets that are hit by the elements in N. To complete H_N, it is sufficient to find a minimum hitting set of I' and to count the number of minimum hitting sets of I'. The crucial observation here is that I' is an instance of $\#\mathrm{MHS}_{k-1}$. Indeed, H' is a hitting set of $(V, \mathcal{C})$ and by removing it we decrease the size of every set at least by one. Therefore, we can use an algorithm for $\#\mathrm{MHS}_{k-1}$ to complete this step. When checking all partitions $(N, \bar{N})$ of H' it is straightforward to keep the accounting information necessary to compute a minimum hitting set $\widehat{H}$ and to count all minimum hitting sets.

Thus for every partition $(N, \bar{N})$ of H' the algorithm solving $\#\mathrm{MHS}_{k-1}$ is called for the instance I'. There are $2^{|H'|}$ partitions $(N, \bar{N})$ of the vertex set H'. For each such partition, the number of elements of the instance I' is $|V'| = |V \setminus H'| = |V| - |H'|$. Thus, the running time of the algorithm is $\mathcal{O}^*\left(2^{|H'|} a_{k-1}^{|V|-|H'|}\right)$.

If $|H'| > 2|V|/3$ and the minimum size of a hitting set in $\mathcal{C}$ is at least $|H'| - 1$, then it is not necessary to check all partitions $(N, \bar{N})$ of H' and in this case we can speed up the algorithm. Indeed, since

- $|H'| \geq |\widehat{H}| \geq |H'| - 1$, and
- $|\widehat{H} \cap (V \setminus H')| \leq |V| - |H'|$,

it is sufficient to consider only those partitions $(N, \bar{N})$ of H' such that

$$|N| \geq |H'| - 1 - (|V| - |H'|) = 2|H'| - |V| - 1.$$

In this case, the running time of the algorithm is $\mathcal{O}^*\left(\binom{|H'|}{2|H'|-|V|} a_{k-1}^{|V|-|H'|}\right)$. $\square$

Now we are ready to use iterative compression to prove the following theorem.

Theorem 2. *Suppose there exists an algorithm to solve $\#\mathrm{MHS}_{k-1}$ in time $\mathcal{O}^*(a_{k-1}^n)$, $1 < a_{k-1} \leq 2$. Then $\#\mathrm{MHS}_k$ can be solved in time*

$$\mathcal{O}^*\left(\max_{2n/3 \leq j \leq n} \left\{ \binom{j}{2j-n} a_{k-1}^{n-j} \right\}\right).$$

Proof. Let $(V, \mathcal{C})$ be an instance of $\#\mathrm{MHS}_k$, where $V = \{v_1, v_2, \cdots, v_n\}$. For $i = 1, 2, \ldots, n$, let $V_i = \{v_1, v_2, \cdots, v_i\}$ and $\mathcal{C}_i = \{X \in \mathcal{C} \mid X \subseteq V_i\}$. Then $I_i = (V_i, \mathcal{C}_i)$ constitutes an instance for the ith stage of the iteration. We denote by H_i and h_i, a minimum hitting set of an instance I_i and the number of different minimum hitting sets of I_i respectively.

If $\{v_1\} \in \mathcal{C}$, then $H_1 = \{v_1\}$ and $h_1 = 1$; otherwise $H_1 = \emptyset$ and $h_1 = 0$.

Consider the ith stage of the iteration. We have that $|H_{i-1}| \leq |H_i| \leq |H_{i-1}|+1$ because at least $|H_{i-1}|$ elements are needed to hit all the sets of I_i except those containing element v_i and $H_{i-1} \cup \{v_i\}$ is a hitting set of I_i. Now, use Lemma 2 with $H' = H_{i-1} \cup \{v_i\}$ to compute a minimum hitting set of I_i. If $|H'| \leq 2i/3$, its running time is $\mathcal{O}^* \left(\max_{0 \leq j \leq 2i/3} \left\{ 2^j a_{k-1}^{i-j} \right\} \right) = \mathcal{O}^* \left(2^{2i/3} a_{k-1}^{i/3} \right)$ (for $a_{k-1} \leq 2$). If $|H'| > 2i/3$, the running time is $\mathcal{O}^* \left(\max_{2i/3 < j \leq i} \left\{ \binom{j}{2j-i} a_{k-1}^{i-j} \right\} \right)$. Since for every fixed $j > 2i/3$, and $1 \leq i \leq n$,

$$\binom{j}{2j-i} a_{k-1}^{i-j} \leq \binom{j}{2j-n} a_{k-1}^{n-j},$$

the worst case running time of the algorithm is

$$\mathcal{O}^* \left(\max \left\{ \max_{1 \leq i \leq n} 2^{2i/3} a_{k-1}^{i/3}, \max_{2n/3 \leq j \leq n} \left\{ \binom{j}{2j-n} a_{k-1}^{n-j} \right\} \right\} \right).$$

Finally, $\binom{2n/3}{n/3} = 2^{2n/3}$ up to a polynomial factor, and thus the running time is $\mathcal{O}^* \left(\max_{2n/3 \leq j \leq n} \left\{ \binom{j}{2j-n} a_{k-1}^{n-j} \right\} \right)$. □

Based on the $\mathcal{O}(1.2377^n)$ algorithm for $\#\mathrm{MHS}_2$ [23], the worst-case running time of the algorithm of Theorem 2 is obtained for $0.7049n < j < 0.7050n$.

Corollary 1. $\#\mathrm{MHS}_3$ *can be solved in time* $\mathcal{O}(1.7198^n)$.

The same approach can be used design an algorithm for the optimization version MHS_k, assuming that an algorithm for MHS_{k-1} is available. Based on the $\mathcal{O}(1.6278^n)$ algorithm for MHS_3 [23] this leads to an $\mathcal{O}(1.8704^n)$ time algorithm for solving MHS_4 (in that case, the maximum is obtained for $0.6824n < j < 0.6825n$).

Corollary 2. MHS_4 *can be solved in time* $\mathcal{O}(1.8704^n)$.

In the following theorem we provide an alternative approach to solve $\#\mathrm{MHS}_k$. This is a combination of brute force enumeration (for sufficiently large hitting sets) with one application of the compression algorithm of Lemma 2. For large values of a_{k-1}, more precisely for $a_{k-1} \geq 1.6553$, this new approach gives faster algorithms than the one obtained by Theorem 2.

Theorem 3. *Suppose there exists an algorithm with running time* $\mathcal{O}^*(a_{k-1}^n)$, $1 < a_{k-1} \leq 2$, *solving* $\#\mathrm{MHS}_{k-1}$. *Then* $\#\mathrm{MHS}_k$ *can be solved in time*

$$\min_{0.5 \leq \alpha \leq 1} \max \left\{ \mathcal{O}^* \left(\binom{n}{\alpha n} \right), \mathcal{O}^* \left(2^{\alpha n} a_{k-1}^{n-\alpha n} \right) \right\}.$$

k	#MHS$_k$	MHS$_k$
2	$\mathcal{O}(1.2377^n)$ [23]	$\mathcal{O}(1.2108^n)$ [21]
3	$\mathcal{O}(1.7198^n)$	$\mathcal{O}(1.6278^n)$ [23]
4	$\mathcal{O}(1.8997^n)$	$\mathcal{O}(1.8704^n)$
5	$\mathcal{O}(1.9594^n)$	$\mathcal{O}(1.9489^n)$
6	$\mathcal{O}(1.9824^n)$	$\mathcal{O}(1.9781^n)$
7	$\mathcal{O}(1.9920^n)$	$\mathcal{O}(1.9902^n)$

Fig. 1. Running times of the algorithms for #MHS$_k$ and MHS$_k$

Proof. First the algorithm tries all subsets of V of size $\lfloor \alpha n \rfloor$ and identifies those that are a hitting set of I.

Now there are two cases. In the first case, there is no hitting set of this size. Then the algorithm verifies all sets of larger size whether they are hitting sets of I. It is straightforward to keep some accounting information to determine the number of hitting sets of the smallest size found during this enumeration phase. The running time of this phase is $\mathcal{O}^*\left(\sum_{i=\lfloor \alpha n \rfloor}^{n} \binom{n}{i}\right) = \mathcal{O}^*\left(\binom{n}{\alpha n}\right)$.

In the second case, there exists a hitting set of size $\lfloor \alpha n \rfloor$. Then count all minimum hitting sets using the compression algorithm of Lemma 2 with H' being a hitting set of size $\lfloor \alpha n \rfloor$ found by the enumeration phase. By Lemma 2, this phase of the algorithm has running time $\mathcal{O}^*\left(2^{\alpha n} a_{k-1}^{n-\alpha n}\right)$.   □

The best running times of algorithms solving #MHS$_k$ and MHS$_k$ are summarized in Figure 1. For #MHS$_{\geq 4}$ and MHS$_{\geq 5}$, we use the algorithm of Theorem 3. Note that the MHS$_2$ problem is equivalent to MVC and MIS.

4   Maximum Induced Cluster Subgraph

Clustering objects according to given similarity or distance values is an important problem in computational biology with diverse applications, e.g., in defining families of orthologous genes, or in the analysis of microarray experiments [5,8,11,14,18]. A graph theoretic formulation of the clustering problem is called CLUSTER EDITING. To define this problem we need to introduce the notion of a *cluster graph*. A graph is called a cluster graph if it is a disjoint union of cliques. In the most common parameterized version of CLUSTER EDITING, given an input graph $G = (V, E)$ and a positive integer k, the question is whether the input graph G can be transformed into a cluster graph by adding or deleting at most k edges in time $f(k) \cdot n^{O(1)}$, where f is an arbitrary computable function. This problem has been extensively studied in the realm of parameterized complexity [5,8,11,18]. In this section, we study a vertex version of CLUSTER EDITING. We study the following optimization version of the problem.

> MAXIMUM INDUCED CLUSTER SUBGRAPH (MICS): Given a graph $G = (V, E)$ on n vertices, find a maximum size subset $C \subseteq V$ such that $G[C]$, the subgraph of G induced by C, is a cluster graph.

Due to the following well-known observation, the MICS problem is also known as MAXIMUM INDUCED P_3-FREE SUBGRAPH.

Observation 1. *A graph is a disjoint union of cliques if and only if it contains no induced subgraph isomorphic to the graph P_3, the path on 3 vertices.*

Clearly, $C \subseteq V$ induces a cluster graph in $G = (V, E)$ (that is $G[C]$ is a disjoint union of cliques of G) iff $S = V \setminus C$ hits all induced paths on 3 vertices of G. Thus solving the MICS problem is equivalent to finding a minimum size set of vertices whose removal produces a maximum induced cluster subgraph of G. By Observation 1, this reduces to finding a minimum hitting set S of the collection of vertex sets of (induced) P_3's of G. Such a hitting set S is called a P_3-HS.

As customary when using iterative compression, we first define a compression version of the MICS problem.

COMP-MICS: Given a graph $G = (V, E)$ on n vertices and a P_3-HS $S \subseteq V$, find a P_3-HS of G of size at most $|S| - 1$ if one exists.

Lemma 3. COMP-MICS *can be solved in time* $\mathcal{O}(1.6181^n)$.

Proof. For the proof we distinguish two cases based on the size of S.

Case 1: If $|S| \leq 2n/3$ then the following algorithm which uses matching techniques is applied.

Step 1: Enumerate all partitions of $(N, \bar{N})$ of S.
Step 2: For each partition, compute a maximum set $C \subseteq V$ such that $G[C]$ is a cluster graph, subject to the constraints that $N \subseteq C$ and $\bar{N} \cap C = \emptyset$, if such a set C exists.

In **Step 2**, we reduce the problem of finding a maximum sized C to the problem of finding a maximum weight matching in an auxiliary bipartite graph. Independent of our work, Hüffner et al. [14] also use this natural idea of reduction to weighted bipartite matching to obtain FPT algorithm for the vertex weighted version of CLUSTER VERTEX DELETION using iterative compression. For completeness, we present the details of **Step 2**.

If $G[N]$ contains an induced P_3 then there is obviously no $C \subseteq V$ inducing a cluster graph that respects the partition $(N, \bar{N})$. We call such a partition *invalid*.

Otherwise, $G[N]$ is a cluster graph, and thus the goal is to find a maximum size subset C' of $\bar{S} = V \setminus S$ such that $G[C' \cup N]$ is a cluster graph. Fortunately, such a set C' can be computed in polynomial time by reducing the problem to finding a maximum weight matching in an auxiliary bipartite graph.

First we describe the construction of the bipartite graph. Consider the graph $G[N \cup \bar{S}]$ and note that $G[N]$ and $G[\bar{S}]$ are cluster graphs. Now the following reduction rule is applied to the graph $G[N \cup \bar{S}]$.

(R) Remove every vertex $b \in \bar{S}$ for which $G[N \cup \{b\}]$ contains an induced P_3.

Clearly all vertices removed by **(R)** cannot belong to any C' inducing a cluster subgraph of G. Let $\hat{S}$ be the subset of vertices of $\bar{S}$ which are not removed by

(R). Hence the current graph is $G[N \cup \hat{S}]$. Clearly $G[\hat{S}]$ is a cluster graph since $G[\overline{S}]$ is one. Further, note that no vertex of $\hat{S}$ has neighbors in two different maximal cliques of $G[N]$ and if a vertex of $\hat{S}$ has a neighbor in one maximal clique of $G[N]$ then it is adjacent to each vertex of this maximal clique. Thus, every vertex in $\hat{S}$ has either no neighbor in N or it is adjacent to all the vertices of exactly one maximal clique of $G[N]$.

Now we are ready to define the auxiliary bipartite graph $G' = (A, B, E')$. Let $\{\mathcal{C}_1, \mathcal{C}_2, \cdots, \mathcal{C}_r\}$ be the maximal cliques of the cluster graph $G[N]$. Let $\{\mathcal{C}'_1, \mathcal{C}'_2, \cdots, \mathcal{C}'_s\}$ be the maximal cliques of the cluster graph $G[\hat{S}]$. Let $A = \{a_1, a_2, \ldots, a_r, a'_1, a'_2, \ldots, a'_s\}$ and $B = \{b_1, b_2, \ldots, b_s\}$. Here, for all $i \in \{1, \ldots, r\}$, each maximal clique $\mathcal{C}_i$ of $G[N]$ is represented by $a_i \in A$; and for all $j \in \{1, 2, \ldots, s\}$, each maximal clique $\mathcal{C}'_j$ of $G[\hat{S}]$ is represented by $a'_j \in A$ and by $b_j \in B$.

Now there are two types of edges in G': $a_j b_k \in E'$ if there is a vertex $u \in \mathcal{C}'_k$ such that u has a neighbor in $\mathcal{C}_j$, and $a'_j b_j \in E'$ if there is a vertex $u \in \mathcal{C}'_j$ such that u has no neighbor in N. Finally we define the weights for both types of eges in the bipartite graph G'. For an edge $a_j b_k \in E'$, its weight $w(a_j b_k)$ is the number of vertices in $\mathcal{C}'_k$ being adjacent to all vertices of the maximal clique $\mathcal{C}_j$. For an edge $a'_j b_j$, its weight $w(a'_j b_j)$ is the number of vertices in $\mathcal{C}'_j$ without any neighbor in N.

This transformation is of interest due to the following claim that uses the above notation.

Claim. [⋆] The maximum size of a subset C' of $\hat{S}$ such that $G[N \cup C']$ is a cluster subgraph of the graph $G^* = G[N \cup \hat{S}]$ is equal to the maximum total weight of a matching in the bipartite graph $G' = (A, B, E')$.

Note that the construction of the bipartite graph G', including the application of **(R)** and the computation of a maximum weighted matching of G' can be performed in time $\mathcal{O}(n^3)$ [7]. Thus, the running time of the algorithm in Case 1 is the time needed to enumerate all subsets of S (whose size is bounded by $2n/3$) and this time is $\mathcal{O}^*(2^{2n/3}) = \mathcal{O}(1.5875^n)$.

Case 2: If $|S| > 2n/3$ then the algorithm needs to find a P_3-HS of G of size $|S| - 1$, or show that none exists.

The algorithm proceeds as in the first case. Note that at most $n - |S|$ vertices of $V \setminus S$ can be added to N. Therefore, the algorithm verifies only those partitions $(N, \bar{N})$ of S satisfying $|N| \geq |S| - 1 - (n - |S|) = 2|S| - n - 1$. In this second case, the worst-case running time is obtained for $0.7236 < \alpha < 0.7237$, and it is

$$\mathcal{O}^* \left(\max_{2/3 < \alpha \leq 1} \left\{ \binom{\alpha n}{(2\alpha - 1)n} \right\} \right) = \mathcal{O}(1.6181^n). \qquad \square$$

Now we are ready to prove the following theorem using iterative compression.

Theorem 4. MICS *can be solved in time* $\mathcal{O}(1.6181^n)$ *on a graph on n vertices.*

Proof. Given a graph $G = (V, E)$ with $V = \{v_1, \ldots, v_n\}$. Let $G_i = G[\{v_1, \ldots, v_i\}]$ and let C_i be a maximum induced cluster subgraph of G_i. Let $S_i = V_i \setminus C_i$.

The algorithm starts with G_1, $C_1 = \{v_1\}$ and $S_1 = \emptyset$. At the ith iteration of the algorithm, $1 \leq i \leq n$, we maintain the invariant that we have at our disposal C_{i-1} a maximum set inducing a cluster subgraph of G_{i-1} and S_{i-1} a minimum P_3-HS of G_{i-1}. Note that $S_{i-1} \cup \{v_i\}$ is a P_3-HS of G_i and that no P_3-HS of G_i has size smaller than $|S_{i-1}|$. Now use the algorithm of Lemma 3 to solve COMP-MICS on G_i with $S = S_{i-1} \cup \{v_i\}$. Then the worst-case running time is attained at the nth stage of the iteration and the run time is $\mathcal{O}(1.6181^n)$. $\square$

5   Conclusion

Iterative compression is a technique which is succesfully used in the design of FPT algorithms. In this paper we show that this technique can also be used to design exact exponential time algorithms. This suggests that it might be used in other areas of algorithms as well. For example, how useful can iterative compression be in the design of approximation algorithms?

Carrying over techniques from the design of FPT algorithms to the design of exact exponential time algorithms and vice-versa is a natural and tempting idea. A challenging question in this regard is whether Measure and Conquer, a method that has been succesfully used to improve the time analysis of simple exponential-time branching algorithms, can be adapted for the analysis of FPT branching algorithms.

References

1. Björklund, A., Husfeldt, T.: Inclusion–Exclusion Algorithms for Counting Set Partitions. In: Proceedings of FOCS 2006, pp. 575–582 (2006)
2. Björklund, A., Husfeldt, T., Kaski, P., Koivisto, M.: Fourier meets Mobius: Fast Subset Convolution. In: Proceedings of STOC 2007, pp. 67–74 (2007)
3. Byskov, J.M.: Enumerating maximal independent sets with applications to graph colouring. Oper. Res. Lett. 32(6), 547–556 (2004)
4. Chen, J., Liu, Y., Lu, S., Razgon, I., O'Sullivan, B.: A Fixed-Parameter Algorithm for the Directed Feedback Vertex Set Problem. In: Proceedings of STOC 2008, pp. 177–186 (2008)
5. Dehne, F., Langston, M.A., Luo, X., Pitre, S., Shaw, P., Zhang, Y.: The Cluster Editing Problem: Implementations and Experiments. In: Proceedings of IWPEC 2006, pp. 13–24 (2006)
6. Dehne, F., Fellows, M., Langston, M., Rosamond, F., Stevens, K.: An $O(2^{O(k)}n^3)$ FPT algorithm for the undirected feedback vertex set problem. In: Proceedings of COCOON 2005, pp. 859–869 (2005)
7. Edmonds, J., Karp, R.M.: Theoretical improvements in algorithmic efficiency for network flow problems. J. ACM 19(2), 248–264 (1972)
8. Fellows, M.R., Langston, M.A., Rosamond, F.A., Shaw, P.: Efficient Parameterized Preprocessing for Cluster Editing. In: Proceedings of FCT 2007, pp. 312–321 (2007)
9. Fernau, H.: Parameterized Algorithms for Hitting Set: The Weighted Case. In: Proceedings of CIAC 2006, pp. 332–343 (2006)
10. Fomin, F.V., Grandoni, F., Kratsch, D.: Measure and conquer: A simple $O(2^{0.288\,n})$ independent set algorithm. In: Proceedings of SODA 2006, pp. 18–25 (2006)

11. Guo, J.: A More Effective Linear Kernelization for Cluster Editing. In: Proceedings of ESCAPE 2007, pp. 36–47 (2007)

12. Guo, J., Gramm, J., Hüffner, F., Niedermeier, R., Wernicke, S.: Compression-based fixed-parameter algorithms for feedback vertex set and edge bipartization. J. Comput. Syst. Sci. 72(8), 1386–1396 (2006)

13. Hopcroft, J.E., Karp, R.M.: An $n^{5/2}$ algorithm for maximum matching in bipartite graphs. SIAM J. Computing 2(4), 225–231 (1973)

14. Hüffner, F., Komusiewicz, C., Moser, H., Niedermeier, R.: Fixed-parameter algorithms for cluster vertex deletion. In: Proceedings of LATIN 2008, pp. 711–722 (2008)

15. Johnson, D.S., Papadimitriou, C.H., Yannakakis, M.: On Generating All Maximal Independent Sets. Inf. Process. Lett. 27(3), 119–123 (1988)

16. Koivisto, M.: An $O(2^n)$ Algorithm for Graph Colouring and Other Partitioning Problems via Inclusion-Exclusion. In: Proceedings of FOCS 2006, pp. 583–590 (2006)

17. Moon, J.W., Mose, L.: On Cliques in Graphs. Israel J. Mathematics 3, 23–28 (1965)

18. Rahmann, S., Wittkop, T., Baumbach, J., Martin, M., Trub, A., Böcker, S.: Exact and Heuristic Algorithms for Weighted Cluster Editing. In: Proceedings of Comput. Syst. Bioinformatics Conference 2007, vol. 6(1), pp. 391–401 (2007)

19. Raman, V., Saurabh, S., Sikdar, S.: Efficient Exact Algorithms through Enumerating Maximal Independent Sets and Other Techniques. Theory Comput. Syst. 41(3), 1432–4350 (2007)

20. Reed, B.A., Smith, K., Vetta, A.: Finding odd cycle transversals. Oper. Res. Lett. 32(4), 299–301 (2004)

21. Robson, J.M.: Algorithms for maximum independent sets. J. Algorithms 7, 425–440 (1986)

22. Tarjan, R., Trojanowski, A.: Finding a maximum independent set. SIAM J. Computing 6(3), 537–546 (1977)

23. Wahlstrom, M.: Algorithms, measures and upper bounds for satisfiability and related problems, PhD thesis, Linkoping University, Sweden (2007)

Complexity and Limiting Ratio of Boolean Functions over Implication*

Hervé Fournier[1], Danièle Gardy[1], Antoine Genitrini[1],
and Bernhard Gittenberger[2]

[1] Laboratoire PRiSM
CNRS UMR 8144 and Université de Versailles St-Quentin en Yvelines
45 av. des États-Unis, 78035 Versailles, France
{herve.fournier,daniele.gardy,antoine.genitrini}@prism.uvsq.fr
[2] Technische Universität Wien
Wiedner Hauptstrasse 8-10/104, A-1040 Wien, Austria
gittenberger@dmg.tuwien.ac.at

Abstract. We consider the logical system of boolean expressions built on the single connector of implication and on positive literals. Assuming all expressions of a given size to be equally likely, we prove that we can define a probability distribution on the set of boolean functions expressible in this system. We then show how to approximate the probability of a function f when the number of variables grows to infinity, and that this asymptotic probability has a simple expression in terms of the complexity of f. We also prove that most expressions computing any given function in this system are "simple", in a sense that we make precise.

Keywords: Boolean functions; Implicational formulas; Complexity; Limiting ratio; Probability distribution; Analytic combinatorics.

1 Introduction

Write at random a boolean expression on given sets of boolean variables and of connectors: we obtain a boolean function. How random is this boolean function? E.g., what is the probability that we obtain a tautology? A literal? Any specified function? Is the probability of obtaining a given function related to the complexity of the function? Does the Shannon effect, i.e. the fact that "almost all" functions have maximal complexity, still hold for this probability distribution? These and some others are questions that we would like to investigate for general logic systems.

We present here a first step in this direction, with an in-depth study of the simple system obtained from the single connector of implication and positive

* This research was partially supported by the A.N.R. project *SADA* and by the P.H.C. Amadeus project *Probabilities and tree representations for boolean functions*. The last author's work has been supported by *ÖAD, project Amadée, grant 3/2006*, as well as by *FWF (Austrian Science Foundation), National Research Area S9600, grant S9604*.

E. Ochmański and J. Tyszkiewicz (Eds.): MFCS 2008, LNCS 5162, pp. 347–362, 2008.
© Springer-Verlag Berlin Heidelberg 2008

literals. Our interest for this system stems from its relation to intuitionistic logic [1,2] and from its relative simplicity, although not all boolean functions can be obtained in this system: The set of functions that can be obtained in this system is the Post class S_0, i.e. the set of functions that we can write as $x \vee g$ for suitable boolean variable x and function g.

Consider the ratio of the number of formulas of size n that compute a fixed boolean function f, among all formulas of size n, and let the size grow to infinity. It is possible to show that the limit of this ratio exists for a wide variety of logical systems [3], and that we can thus define a probability distribution on the set of boolean functions.

Some of us have shown in a former paper [4] that the tautologies in the implication system have the simple shape $(..., a, ...) \rightarrow a$ with high probability, and that, if the number k of boolean variables grows large enough, the probability of a tautology is asymptotically $1/k$. The next natural step is then to try and compute the probability that a random expression computes a literal, a function $x_i \vee x_j$, etc., and to check if the "average" expression computing, e.g., a literal, has a simple form. When studying the random expressions that compute a given boolean function f, one major parameter is the complexity $L(f)$, i.e. the size of the smallest expressions that represent f. We shall prove in the present paper that the probability of any given function f depends exponentially on its complexity; in passing we are also able to characterize the shape of a random expression computing f, and to show that these expressions are obtained quite simply from minimal trees.

The efforts to define non-uniform probability distributions, induced by random boolean expressions, or formulae, on the set of boolean functions, date back several years. The starting point is generally the description of formulae as *trees* of a suitable shape and suitably labelled. The first efforts in this direction were by Paris et al. [5] on And/Or trees (i.e. expressions built on the two connectors $\wedge$ and $\vee$); the underlying model was that of binary Catalan trees, suitably labelled. The study of these trees was further pursued by Lefman and Savický [6], who proved by a pruning argument the existence of a probability distribution induced by random expressions, and established important lower and upper bounds for the probability of any boolean function in terms of its complexity. At the same time, Woods [7] proved independently the existence of a limiting distribution for general formulae. Some of the authors of the present paper then gave an alternative construction of the probability distribution for And/Or trees, together with an improvement on the upper bound [8]. The survey paper [3] presents an overview of the probability distributions induced by random boolean expressions on boolean functions, and of the way we can obtain them using the tools of analytic combinatorics: enumeration of formulae/trees by generating functions, the Drmota-Lalley-Woods theorem for solving an algebraic system of equations and asymptotics.

We should also mention that several researchers have concentrated on the probability of tautologies, i.e. on the probability of the single constant function *True*. Let us mention the Polish school around Zaionc, who began a systematic investigation of the probability of a tautology in various logical frameworks

[9,10,11,12]; see also [13,14] for the expressions built on the single equivalence connector. For And/Or trees, we refer the reader to Woods's result that the tautologies have asymptotic probability $3/4k$, and that almost all of them have the simple form $l \vee \bar{l} \vee \ldots$ [15], and to Kozik [16] for a different, later proof.

Significant results have also been established for a different family of formulae/trees, namely balanced trees obtained by iteration of a single connector. The first result in this area is due to Valiant [17], whose aim was to compute a boolean expression for the function *Majority* with high enough probability. Then Boppana [18] and Gupta-Mahajan [19] improved Valiant's result for majority; Boppana went on to prove that iteration by a single, well-chosen connector gives a uniform distribution on the set of threshold functions. Savický [20] showed that iterating a non-linear balanced connector leads to the uniform distribution on the set of all boolean functions. Finally, Brodsky and Pippenger [21] present a systematic study of different classes of connectors and of the distributions induced on boolean functions; these distributions are either uniform on subsets of boolean functions, or concentrated on a single function.

The present paper is organized as follows. We show in Section 2 how all the trees computing a specific boolean function can be derived from a finite set of minimal trees by a few simple operations. Our main results are also given in this section, namely the asymptotic expression of the probability of the boolean function in terms of its complexity, and the (relatively) simple form of a random expression computing a boolean function. The rest of the paper is devoted to the proof of these results. We first recall in Section 3 basic facts and former results on tautologies, i.e. on the trees that compute the simplest boolean function in our system: the constant *True*. Next we give technical results on expansions and on the inverse operation of pruning in Section 4, before considering irreducible trees in Section 5. Finally we present possible extensions in Section 6.

2   Results: Limiting Ratio of Trees Computing a Given Function

We begin by a brief presentation of the formulas we consider, then give a couple of definitions in order to state the main result concerning the limiting ratio of trees computing a given function.

Trees over implication. We consider in this paper formulas built with the single connector of implication (denoted by $\rightarrow$) and k positive literals $\{x_1, \ldots, x_k\}$. These formulas can be represented by full binary trees whose internal nodes are all labelled by $\rightarrow$ and leaves by some literals. We denote by $\mathcal{F}_k$ this set of formulas. Each formula, or tree, is associated to a specific boolean function; we say that a tree is *computing* a specific boolean function. The boolean function computed by a tree A is denoted by $[A]$. Every tree A of $\mathcal{F}_k$ can be written in a unique way as

$$A = A_1 \rightarrow (A_2 \rightarrow (\ldots \rightarrow (A_p \rightarrow r(A)) \ldots))$$

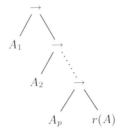

Fig. 1. The canonical decomposition of a tree

where $A_i \in \mathcal{F}_k$ and $r(A) \in \{x_1, \ldots, x_k\}$. We refer to this as the *canonical decomposition* of a tree A – see Figure 1. The subtrees $A_1, \ldots, A_p$ are called the *premises* of A, and the rightmost leaf $r(A)$ is called the *goal* of A. Analogously, premises and goal of any subtree of A is defined.

Limiting ratio. We define the size $|A|$ of a tree A as the number of its *leaves*. The *limiting ratio* of a subset $\mathcal{A}$ of trees is defined as

$$\mu_k(\mathcal{A}) = \lim_{n \to \infty} \frac{|\{A \in \mathcal{A}, |A| = n\}|}{|\{A \in \mathcal{F}_k, |A| = n\}|}$$

if this limit exists. We now define the *limiting ratio of a function* f as the limiting ratio of all trees computing f; that is, $\mu_k(f) = \mu_k(\{A \in \mathcal{F}_k \mid [A] = f\})$. Introducing the generating functions $\sum_n |\{A \in \mathcal{A}, |A| = n, [A] = f\}| z^n$, the results of Drmota [22], Lalley [23] and Woods [7] give us an easy way to prove that the limiting ratio of each boolean function is defined in the system $\mathcal{F}_k$ – i.e. for all boolean functions f, the limit defining $\mu_k(f)$ exists. These theorems are nicely described in Flajolet and Sedgewick [24,25].

Valid expansions of a tree. We now define three rules, called *expansion rules*, that allow, starting from a tree A, to obtain larger trees computing the same function as A. Let A be a tree and B one of its subtrees and let the root of B be denoted by ν.

The first expansion of A is called *valid expansion by a tautology*. We say that the tree A' obtained by replacing the subtree B with the subtree $C \to B$ in A, where C is a tautology, is a valid expansion of A by a tautology at node ν. Of course A' computes the same function as A since $[C \to B] = [B]$.

The second expansion of A is called *valid expansion by goal α*. If substituting B with $C \to B$ yields a tree A' computing the same function as A *for any tree C with goal α*, we say that any of these trees A' is obtained from A by a valid expansion of type "goal α" at node ν.

The third expansion of A is called *valid expansion by premise α*. If substituting B with $C \to B$ yields a tree A' computing the same function as A *for any tree C with a premise equal to α*, we say that any of these trees A' is obtained from A by a valid expansion of type "premise α" at node ν.

Figure 2 represents the shape of the tree obtained after a valid expansion at the root of B. Given a tree A, we define $E(A)$ to be the set of all trees obtained

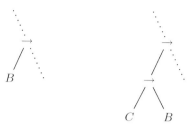

Fig. 2. Valid expansion with the subtree C in the root of B

from A by a single valid expansion of any of the three types defined above. Note that all trees in $E(A)$ compute the same function as A. We naturally extend E to any set of trees $\mathcal{A} \subseteq \mathcal{F}_k$ by letting $E(\mathcal{A}) = \bigcup_{A \in \mathcal{A}} E(A)$. In the same way we define $E^0(\mathcal{A}) = \mathcal{A}$, $E^i(\mathcal{A}) = E(E^{i-1}(\mathcal{A}))$ and $E^*(\mathcal{A}) = \bigcup_{i \in \mathbb{N}} E^i(\mathcal{A})$.

Given a tree A, we define $\lambda(A)$ as the number of types of valid expansions of A; more precisely, this is the number of pairs (ν, α), where ν is a node of A (either an internal node or a leaf) and $\alpha \in \{x_1, \ldots, x_k\}$, such that an expansion of type "goal α" is valid in the node ν, plus the number of couples (ν, α) such that an expansion of type "premise α" is valid in the node ν, plus $2|A| - 1$ (this is counting the tautology expansions in every of the $2|A| - 1$ nodes of A).

For a boolean function f depending on a finite number of variables of $\{x_i \mid i > 0\}$, we define its *complexity* $L(f)$ to be the size of the smallest trees (over implication) computing f. Trees of size $L(f)$ computing f are called *minimal trees* of f; their set is denoted by $\mathcal{M}(f)$. Given a boolean function f, we define $\lambda(f)$ as the sum of all $\lambda(M)$ when M runs over all minimal trees computing f. It will be proved that $\lambda(f)$ does not depend on the number k of ambient variables. We can now state the main result of this paper.

Theorem 1. *Let f be a boolean function different from* True. *Almost all trees computing f are obtained by a single expansion of a minimal tree of f:*

$$\mu_k(f) \sim \mu_k(E(\mathcal{M}(f))).$$

As a consequence, the limiting ratio of f is asymptotically (as $k \to \infty$) equal to:

$$\mu_k(f) = \frac{\lambda(f)}{4^{L(f)} \, k^{L(f)+1}} + O\left(\frac{1}{k^{L(f)+2}}\right).$$

A proof of this theorem is given at the end of Section 5, where bounds on $\lambda(f)$ are also provided – see Proposition 1.

3   Limiting Ratio and Structure of Tautologies

In this section, we recall some results from [4] on trees computing the constant function *True* (tautologies). Some of us proved there that the limiting ratio of all tautologies is equivalent to the limiting ratio of the family of *simple tautologies*:

formulas such that one premise is equal to the goal of the formula. The limiting ratio of the set G_k of simple tautologies is equal to:

$$\mu_k(G_k) = \frac{4k+1}{(2k+1)^2} = \frac{1}{k} - \frac{3}{4k^2} + O\left(\frac{1}{k^3}\right).$$

Moreover, the following bounds on the limiting ratio of all tautologies (denoted by Cl_k) were given for k tending to infinity:

$$\frac{1}{k} - \frac{47}{64k^2} + O\left(\frac{1}{k^3}\right) \leqslant \mu_k(Cl_k) \leqslant \frac{1}{k} + \frac{17}{4k^2} + O\left(\frac{1}{k^3}\right).$$

Finally we recall two facts on the structure of tautologies. For a node ν of a tree A, we define the *left depth* of the node ν as the number of left branches needed to reach ν from the root of A. We define in the same way the left depth of a subtree B of A as the left depth of its root. Let A be a tree and B one of its subtrees; B is called a *left subtree* of A if the root of B is the left son of its first ancestor. Let A be a tree which is a tautology: then the goal of A has a second occurrence at left depth 1 in A. Moreover, if A is a non-simple tautology, there exist in A either three occurrences of the same variable or two times two occurrences of two distinct variables among the leaves of left depth at most 3.

4   Expansion and Pruning

We now study some of the properties of the expansion rules defined in Section 2. Given a tree A and a left subtree B of A, we denote by $A \setminus B$ the tree obtained by removing B from A. More precisely, since B is a left subtree of A, it is the left son of a tree of the form $B \to C$ in A; the tree $A \setminus B$ is obtained by substituting the subtree $B \to C$ by C in A – see Figure 3. The following three lemmas give (necessary and) sufficient conditions for a tree to be a single expansion of a certain type of a smaller tree.

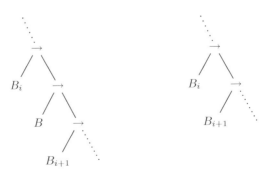

Fig. 3. Removing a left subtree B

Lemma 1. *Let A be a tree and B be a left subtree of A. If B is a tautology, then A is obtained by a single valid expansion of type "tautology" of $A \setminus B$.*

Proof. This is obvious from the definition of expansions by tautologies.   □

Lemma 2. *Let A be a tree and B be a left subtree of A. Let β be the goal of B. If substituting B by 1 or β in A yields a tree computing $[A]$ in both cases, then A is obtained by a single valid expansion of type "goal β" of $A \setminus B$.*

Proof. Let A_1 be the tree A where B is replaced with β, and A_2 be the tree A where B is replaced with 1. Let B' be any tree with goal β, and A' be the tree obtained from A by replacing B with B'. Of course $\beta \leqslant [B'] \leqslant 1$. Then by induction on the size of the formula, we obtain $[A] = [A_1] \leqslant [A'] \leqslant [A_2] = [A]$ or $[A] = [A_1] \geqslant [A'] \geqslant [A_2] = [A]$, depending whether the left depth of the root of B is even or odd. In any case, $[A'] = [A]$. Moreover, $[A \setminus B] = [A]$ since $[A \setminus B] = [A_2]$.   □

Lemma 3. *Let A be a tree and B be a left subtree of A. Suppose that B has a premise of size one β. If substituting B with 1 or $\bar{\beta}$ in A gives a tree computing $[A]$ in both cases, then A is obtained by a single valid expansion of type "premise β" of $A \setminus B$.*

Proof. The proof is similar to the previous one.   □

When going from A to $A \setminus B$ with one of the three lemmas above, we shall say $A \setminus B$ is obtained by *pruning* the left subtree B in A. Note the difference between pruning a subtree and cutting a subtree: if A is a tree and B one of its left subtrees, we will use the term "cutting the subtree B" when we remove B from A without any condition on B, and the term "pruning the subtree B" when we remove B from A while A is a valid expansion of $A \setminus B$. However, both final trees are denoted by $A \setminus B$.

A tree which cannot be pruned is called an *irreducible tree*. Of course all minimal trees computing a function f are irreducible. However, the converse is not true; indeed consider the function $f = x_1 \vee (\bar{x}_2 \wedge \bar{x}_3) \vee (\bar{x}_2 \wedge x_4)$. It can be checked that $(x_4 \rightarrow x_2) \rightarrow (((x_2 \rightarrow x_3) \rightarrow x_3) \rightarrow x_1)$ computes f, is irreducible, but not minimal since $((x_3 \rightarrow x_4) \rightarrow x_2) \rightarrow x_1$ is smaller and also computes f. We also remark that the system of pruning rules is not confluent.

We now define a new way of getting large trees from a smaller one. But this time, it does not preserve the function computed by the initial tree; its purpose is to establish some upper bounds on the limiting ratio of expansions. This new mapping X is called *extension* (it is different from expansion). The mapping X is defined recursively as follows: for a tree T consists of a single leaf α, $X(\alpha)$ is the set of all trees whose goal is labelled by α. If $T = L \rightarrow R$, we let

$$X(L \rightarrow R) = \left\{ A_1 \rightarrow \left(\ldots \rightarrow \left(A_p \rightarrow \left(\tilde{L} \rightarrow \tilde{R} \right) \right) \ldots \right) \mid A_1, \ldots, A_p \in \mathcal{F}_k, \ \tilde{L} \in X(L), \ \tilde{R} \in X(R) \right\}.$$

See Figure 4 for a graphical representation of the recursive definition of this mapping, and Figure 5 for the general shape of extensions of a given tree: the

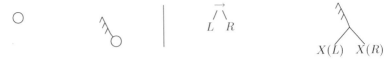

Fig. 4. The recursive definition of the *extension* mapping

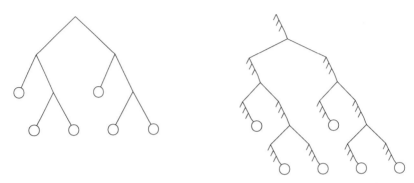

Fig. 5. A tree A on the left and the set $X(A)$ it defines, on the right

left sons of the combs are arbitrary trees (not represented not to overload the picture). We naturally extend X to a set of trees $\mathcal{A} \subseteq \mathcal{F}_k$ by letting $X(\mathcal{A}) = \bigcup_{A \in \mathcal{A}} X(A)$. Notice that $X(X(\mathcal{A})) = X(\mathcal{A})$ for any $\mathcal{A} \subseteq \mathcal{F}_k$. The relationship between extensions and expansions is given below.

Lemma 4. *Let A be a tree. Then $E^*(A) \subseteq X(A)$.*

Proof. Let A be a tree. Since $X(X(A)) = X(A)$, all is needed is to prove that $E(A) \subseteq X(A)$. Recall that any tree $A' \in E(A)$ is obtained by substituting a subtree B of A with a tree of the form $C \to B$. It is clear from the definition of extensions that $C \to B \in X(B)$, and it follows that $A' \in X(A)$. □

Let $\mathcal{V}$ be a fixed finite subset of the variables $\{x_i \mid i > 0\}$, independent of the number of variables k we consider. Let p and q be two integers. Let $\mathcal{B}_q^p(\mathcal{V})$ the set of trees $B \subseteq \mathcal{F}_k$ such that $p \leqslant |B| \leqslant pq + 1$ and which contain at least p leaves labelled in $\mathcal{V}$. Note that $\mathcal{B}_q^p(\mathcal{V})$ implicitly depends on the number k of variables considered in our system.

Lemma 5. *The limiting ratio of $X(\mathcal{B}_q^p(\mathcal{V}))$ satisfies the next equation:*

$$\mu_k \left(X \left(\mathcal{B}_q^p(\mathcal{V}) \right) \right) = O \left(\frac{1}{k^p} \right).$$

Proof of this lemma is omitted in this short abstract. Let $\mathcal{A}_q^p(\mathcal{V})$ be the set of trees of $\mathcal{F}_k$ which contain p leaves labelled in $\mathcal{V}$, all of them being of left depth at most q. Notice that $\mathcal{A}_q^p(\mathcal{V})$ is infinite – as opposed to $\mathcal{B}_q^p(\mathcal{V})$.

Lemma 6. *$\mathcal{A}_q^p(\mathcal{V}) \subseteq X(\mathcal{B}_q^p(\mathcal{V}))$ and consequently $E^*(\mathcal{A}_q^p(\mathcal{V})) \subseteq X(\mathcal{B}_q^p(\mathcal{V}))$.*

Proof. Let $A \in \mathcal{A}_q^p(\mathcal{V})$. Let $\nu_1, \ldots, \nu_p$ be p leaves of A labelled with variables from $\mathcal{V}$, all with left depth at most q. Let $C_1, \ldots, C_r$ be the set of maximal (w.r.t. inclusion) left subtrees of A not containing any of the nodes ν_i. Let B be the tree obtained from A by removing all C_i, i.e. $B = A \setminus \{C_1, \ldots, C_r\}$. Of course $A \in X(B)$, and it can be checked that $B \in \mathcal{B}_q^p(\mathcal{V})$: indeed the largest tree B that can be obtained is when all nodes ν_i have a left depth q and belong to distinct premises of A, and $|B| = pq + 1$ in this case. Thus $A \in X(\mathcal{B}_q^p(\mathcal{V}))$. The second part of the lemma follows from Lemma 4 and the fact that $X(X(\mathcal{B}_q^p(\mathcal{V}))) = X(\mathcal{B}_q^p(\mathcal{V}))$. □

Using Lemma 5 and 6 we obtain:

Corollary 1. *It holds that:*

$$\mu_k \left(E^* \left(\mathcal{A}_q^p(\mathcal{V}) \right) \right) = O \left(\frac{1}{k^p} \right).$$

5   Irreducible Trees and Their Expansions

Let f be a boolean function different from *True*. The variable x is called an *essential variable* of f if the two functions obtained by evaluating x to 0 and to 1 are two distinct functions. Otherwise, x is called an *inessential variable* of f.

Let A be a tree and ν one of its nodes of positive left depth (either an internal node or a leaf). We define $\Delta(\nu)$ to be the smallest left subtree of A containing ν. In the same way, for a node ν of left depth at least 2, we define $\Delta^2(\nu)$ to be the smallest left subtree strictly containing $\Delta(\nu)$ – see Figure 6. We shall also write $\Delta(B)$ for a subtree B as a shortcut for $\Delta(\nu)$, where ν is the root of B (and in the same way $\Delta^2(B)$ for $\Delta^2(\nu)$).

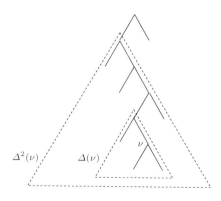

Fig. 6. The left subtrees $\Delta(\nu)$ and $\Delta^2(\nu)$ associated to a node ν of a tree

Lemma 7. *Any tree A computing a function $f \neq True$ contains at least $L(f)$ occurrences of essential variables of f.*

Proof. Let A be a tree computing $f \neq True$. First remark that the goal of A is an essential variable; otherwise A would compute the constant $True$ function. So $\Delta(\nu)$ is well defined for any leaf labelled with an inessential variable. Let $\{\Delta_1, \ldots, \Delta_p\}$ be the set of maximal $\Delta(\nu)$ (with respect to inclusion) when ν runs over all the leaves of A labelled with inessential variables. Of course all Δ_i are disjoint. If we assign the value 1 to all inessential variables, all Δ_i evaluate to 1, because their goals are inessential variables. Thus $A' = A \setminus \{\Delta_1, \ldots, \Delta_p\}$ computes f. Since A' only contains essential variables and $|A| \geqslant |A'| \geqslant L(f)$, it follows that there are at least $L(f)$ essential variables.   □

The size of a tree A computing $f \neq True$ can be written $|A| = L(f) + e + i$, where $L(f) + e$ is the number of leaves labelled with essential variables and i is the number of leaves labelled with inessential variables; notice that $e \geqslant 0$ by Lemma 7. Given a function f different from $True$, we decompose the set of irreducible trees computing f into the following disjoint sets:

- $\mathcal{M}(f)$ is the set of all minimal trees, i.e. trees of size $L(f)$ (case $e = i = 0$);
- $\mathcal{P}_1(f)$ is the set of irreducible trees of size greater than $L(f)$, with exactly $L(f)$ occurrences of essential variables and at least one occurrence of an inessential variable (case $e = 0$, $i > 0$);
- $\mathcal{P}_2(f)$ is the set of irreducible trees of size $L(f) + 1$, without any occurrence of inessential variables (case $e = 1$, $i = 0$);
- $\mathcal{P}_3(f)$ is the set of irreducible trees of size greater than $L(f) + 1$, with exactly $L(f) + 1$ occurrences of essential variables and $i > 0$ occurrences of *all distinct* inessential variables (case $e = 1$, $i > 0$, first part);
- $\mathcal{P}_4(f)$ is the set of irreducible trees of size greater than $L(f) + 2$, with exactly $L(f) + 1$ occurrences of essential variables and $i > 0$ occurrences of inessential variables such that at least one inessential variable is repeated (case $e = 1$, $i > 0$, second part);
- $\mathcal{P}_5(f)$ is the set of irreducible trees containing at least $L(f) + 2$ occurrences of essential variables (case $e \geqslant 2$, $i \geqslant 0$).

Of course any tree computing f falls in an iterated expansion of an irreducible tree computing f (obtained by repeated pruning). Theorem 1 relies on evaluating the limiting ratios of $E^*(\mathcal{C})$ for each of the classes $\mathcal{C}$ defined above. We first prove that the two sets $\mathcal{P}_1(f)$ and $\mathcal{P}_3(f)$ are empty.

Lemma 8. *For any boolean function f different from* True, *the set $\mathcal{P}_1(f)$ is empty.*

Proof. Suppose that $\mathcal{P}_1(f)$ is not empty, and let $A \in \mathcal{P}_1(f)$. The size A is $L(f)+i$, with exactly $L(f)$ occurrences of essential variables and $i > 0$ occurrences of inessential ones. Let $\{\Delta_1, \ldots, \Delta_p\}$ be the set of maximal $\Delta(\nu)$ (with respect to inclusion) when ν runs over all the leaves of A labelled by an inessential variable. If we assign the value 1 to all inessential variables, all Δ_i evaluate to 1, because their goals are inessential variables. Moreover, since they are left subtrees, the tree $A' := A \setminus \{\Delta_1, \ldots, \Delta_p\}$ computes f. Since A contains $L(f)$ occurrences of essential variables, no Δ_i contains an essential variable.

Suppose now that there does not exist any assignment of the inessential variables such that Δ_1 evaluates to 0: then Δ_1 is a tautology and A is reducible, this is absurd. Hence there exists an assignement a of all inessential variables such that Δ_1 evaluates to 0 under a. Notice that Δ_1 cannot be a premise of A because $f \neq True$, so $\Delta_1^2 := \Delta^2(\Delta_1)$ is well defined and evaluates to 1 under a (because its premise Δ_1 evaluates to 0). Let $S = \{\Delta_i \mid [\Delta_{i|a}] = 1\} \cup \{\Delta_i^2 \mid [\Delta_{i|a}] = 0\}$ – where $[C_{|a}]$ denotes the function computed by the subtree C under the assignement a. The set S is composed of left subtrees, all evaluating to 1 under a. Moreover S contains Δ_1^2 which contains at least one essential variable (its goal) – otherwise Δ_1 would not be maximal. Thus $A'' := A \setminus S$ is of size at most $L(f) - 1$ and computes f, this is absurd. □

Using similar ideas, we obtain the following:

Lemma 9. *For any boolean function f different from True, the set $\mathcal{P}_3(f)$ is empty.*

Proof is not given in this short abstract. It is easy to check that both $\mathcal{P}_4(f)$ and $\mathcal{P}_5(f)$ are non empty for any function $f \neq True$. On the other hand, $\mathcal{P}_2(f)$ may be empty or not, depending on f: $\mathcal{P}_2(f)$ is empty for $f = x_1$ while $\mathcal{P}_2(g)$ is not empty for $g = x_1 \vee (\bar{x}_2 \wedge x_3 \wedge x_4) \vee (\bar{x}_2 \wedge \bar{x}_5)$. Indeed, it can be checked that $L(g) = 6$ and $(x_3 \to (x_4 \to x_2)) \to (((x_5 \to x_2) \to x_2) \to x_1)$ belongs to $\mathcal{P}_2(g)$. Our next step is to prove that iterated expansions of $\mathcal{P}_2(f)$ yields a family with small limiting ratio. Thus we can assume that f is such that $\mathcal{P}_2(f) \neq \emptyset$ – otherwise the limiting ratio of $E^*(\mathcal{P}_2(f))$ is obviously 0. The following lemma establishes some restrictions on the possible types of expansions in $\mathcal{M}(f)$ and $\mathcal{P}_2(f)$.

Lemma 10. *Let f be a boolean function different from True, and $A \in \mathcal{M}(f) \cup \mathcal{P}_2(f)$. No valid expansion of type* goal *or* premise *with respect to an inessential variable is possible in A.*

Proof. We develop here a proof by contradiction. Let $A \in \mathcal{M}(f) \cup \mathcal{P}_2(f)$. Let ν be one of its nodes where we are able to perform a valid expansion of type *goal* or *premise* with respect to an inessential variable α. Notice that the left depth of ν is at least 1; otherwise f would be equal to *True*. Thus $\Delta(\nu)$ is well defined. We shall first prove that $|\Delta(\nu)| = 1$.

Suppose the valid expansion in ν is of type *goal* α. Let A' be the tree obtained by expanding A in ν with the left subtree reduced to α. Then we have $[A'_{|\alpha=0}] = [A \setminus \Delta(\nu)] = f$ and we conclude that $|\Delta(\nu)| = 1$ and $|A| = L(f) + 1$ (otherwise we would have a tree smaller than $L(f)$ computing f). Suppose now the valid expansion in ν is of type *premise* α. Let x be the goal of A, and let A' be the tree obtained by expanding A in ν with the left subtree $\alpha \to x$. Then we have $[A'_{|\alpha=1,\ x=0}] = [A \setminus \Delta(\nu)_{|x=0}] = f_{|x=0}$ and of course $[A \setminus \Delta(\nu)_{|x=1}] = 1 = f_{|x=1}$. Again we conclude that $[A \setminus \Delta(\nu)] = f$; it follows that $|\Delta(\nu)| = 1$ and $|A| = L(f) + 1$ in this case too.

So no matter the type of the valid expansion, we know that $\Delta(\nu)$ is a leaf. Let y be the label of $\Delta(\nu)$. In the tree A, the left subtree $\Delta(\nu)$ computes y. Moreover, for both types of expansion, we have shown that $A \setminus \Delta(\nu)$ still computes f; to put it otherwise, substituting $\Delta(\nu)$ with 1 or its goal y in A does not change the computed function. It follows by Lemma 2 that A is reducible, which is absurd.   □

Notice that Lemma 10 shows that $\lambda(f)$ defined in Section 2 does not depend on the number of variables k we consider. We are now ready to bound the limiting ratio of $E^*(\mathcal{P}_2(f))$.

Lemma 11. *For any boolean function f different from* True,

$$\mu_k(E^*(\mathcal{P}_2(f))) = O\left(\frac{1}{k^{L(f)+2}}\right).$$

Proof. Since $\mathcal{P}_2(f)$ is finite, $\mu_k(\mathcal{P}_2(f)) = 0$ and all we have to show is that $\mu_k(E^*(E(\mathcal{P}_2(f)))) = O(1/k^{L(f)+2})$. Let $A \in E(\mathcal{P}_2(f))$; we have $A \in E(I)$ for an irreducible tree $I \in \mathcal{P}_2(f)$. We will prove that A satisfies one of the following conditions:

- A contains at least $L(f)+2$ occurrences of essential variables with left depth at most $L(f)+2$;
- A contains $L(f)+1$ occurrences of essential variables and two occurrences of the same inessential variable, all with a left depth at most $L(f)+2$.

If A is obtained from I by an expansion of type *goal* or *premise*, then it must be with respect to an essential variable by Lemma 10. Now remark that I is of size $L(f)+1$, so all its node are of left depth at most $L(f)$ (there is at least a node per left depth). Since expansions preserve left depth of nodes present in the initial tree, we conclude that A satisfies the first condition above.

Suppose now that A is obtained from I by an expansion of type *tautology*. From the results recalled in Section 3, we know that this tautology contains two occurrences of some variable x in its nodes of left depth at most 1. If x is an essential variable of f, then A satisfies the first condition above. Otherwise, if x is an inessential variable of f, then A satisfies the second condition above.

Let us denote by Γ the set of essential variables of f. Let $\mathcal{N}_1 = \mathcal{A}_{L(f)+2}^{L(f)+2}(\Gamma)$ and

$$\mathcal{N}_2 = \bigcup_{\alpha \in \{x_1, \ldots, x_k\}} \mathcal{A}_{L(f)+2}^{L(f)+3}(\Gamma \cup \{\alpha\}).$$

We have just proved that $E(\mathcal{P}_2(f)) \subseteq \mathcal{N}_1 \cup \mathcal{N}_2$. It follows that $E^*(E(\mathcal{P}_2(f))) \subseteq E^*(\mathcal{N}_1) \cup E^*(\mathcal{N}_2)$. Corollary 1 yields

$$\mu_k(E^*(\mathcal{N}_1)) = O\left(\frac{1}{k^{L(f)+2}}\right).$$

Moreover, still with Corollary 1 we obtain that for any variable α:

$$\mu_k(E^*(\mathcal{A}_{L(f)+2}^{L(f)+3}(\Gamma \cup \{\alpha\}))) = O\left(\frac{1}{k^{L(f)+3}}\right).$$

It follows that

$$\mu_k(E^*(\mathcal{N}_2)) = O\left(\frac{1}{k^{L(f)+2}}\right).$$

Thus, we have proved that $\mu_k(E^*(\mathcal{P}_2(f))) = O(1/k^{L(f)+2})$.   □

We now turn our attention towards $\mathcal{P}_4(f)$ and $\mathcal{P}_5(f)$. In the same way as for the limiting ratio of iterated expansions of $\mathcal{P}_2(f)$, we can show the following:

Lemma 12. *For any boolean function f different from* True, *it holds that:*

$$\mu_k(E^*(\mathcal{P}_4(f) \cup \mathcal{P}_5(f))) = O\left(\frac{1}{k^{L(f)+2}}\right).$$

Proof is omitted in this short abstract. The last step towards the proof of Theorem 1 is to study limiting ratio of expansions of the minimal trees computing a given function. We show the following:

Lemma 13. *Let f be a boolean function different from* True. *Using a single expansion of the minimal trees, we get:*

$$\mu_k(E(\mathcal{M}(f))) = \frac{\lambda(f)}{4^{L(f)}k^{L(f)+1}} + O\left(\frac{1}{k^{L(f)+2}}\right).$$

Moreover,

$$\mu_k\left(E^*\left(\mathcal{M}(f)\right) \setminus E(\mathcal{M}(f))\right) = O\left(\frac{1}{k^{L(f)+2}}\right).$$

The proof is not given in this short abstract. The first part relies on simple calculations of limiting ratio using generating functions and the results on tautologies recalled in Section 3; the second part, in the spirit of Lemma 11, is more technical and relies on a case study with regard to the location of the second expansion with respect to the first one. Theorem 1 is now obtained easily.

Proof of Theorem 1. Of course each tree computing f falls in a set obtained by an arbitrary number of expansions of an irreducible tree computing f. That is, the set of trees computing f is exactly $\mathcal{A}(f) = E^*(\mathcal{M}(f) \cup \mathcal{P}_1(f) \cup \mathcal{P}_2(f) \cup \mathcal{P}_3(f) \cup \mathcal{P}_4(f) \cup \mathcal{P}_5(f))$. Now of course

$$E(\mathcal{M}(f)) \subseteq \mathcal{A}(f) \subseteq E(\mathcal{M}(f)) \cup (E^*(\mathcal{M}(f)) \setminus E(\mathcal{M}(f))) \cup \bigcup_{i \in \{1,\dots,5\}} E^*(\mathcal{P}_i(f)).$$

The result follows from Lemmas 8, 9, 11, 12 and 13.   □

Let us now provide some bounds on the integer $\lambda(f)$.

Proposition 1. *Let f be a boolean function different from* True, *and let n be its number of essential variables. It holds that*

$$2 \cdot (2L(f) - 1) \cdot |\mathcal{M}(f)| \leqslant \lambda(f) \leqslant (1 + 2n) \cdot (2L(f) - 1) \cdot |\mathcal{M}(f)|.$$

Proof. Let M be a minimal tree computing f and let ν be a node of M. Since M is of size $L(f)$, the tree M has $2L(f) - 1$ nodes in total. Thus all we have to show is that the number λ_ν of types of valid expansions in the node ν of M satisfies $2 \leqslant \lambda_\nu \leqslant 1 + 2n$.

The upper bound is simply obtained by counting all possible types of expansions: 1 for the tautology type, n of goal type, and n of premise type – recall from Lemma 10 that no expansion of type goal or premise with respect to an inessential variable is valid in M.

The lower bound is obtained by remarking that, besides the tautology type of expansion which is of course valid in ν, the expansion of type *premise* x is also valid in ν, where x is the goal of M. Indeed, let A be the tree obtained from M by replacing the subtree B rooted in ν by $C \to B$, where C is any tree with a premise equal to x. Of course $[A|_{x=0}] = [M|_{x=0}]$ because $[C|_{x=0}] = 1$. Moreover, $[A|_{x=1}] = [M|_{x=1}] = 1$ because x is the goal of M. Thus $[A] = [M]$ and we conclude that the expansion of type "goal x" is valid in ν.   □

When building the tree underlying the formula, we assumed that it is chosen uniformly among the trees of a given (large) size. Assume now that it is obtained by a critical branching process, so that its size itself is random (see [8] for the definition of this process in the case of And/Or trees). This gives a different probability distribution on the set of boolean functions; let us denote it by π. Then we can obtain a similar result for this new distribution.

Proposition 2. *Let f be a boolean function different from* True*; then*

$$\pi(f) = \frac{|\mathcal{M}(f)|}{2^{2L(f)-1} \, k^{L(f)}} + O\left(\frac{1}{k^{L(f)+1}}\right).$$

6   Conclusion

When considering the limiting ratio of a boolean function, e.g., in the system of implication, it may not be enough to know that the limiting ratio exists, and one may naturally wish for some numerical information. For a fixed, (very) small number of boolean variables, explicit computation of the limiting ratios is feasible by writing, then solving, an algebraic system; see [3] for an overview of the mathematical technology involved and [8] for the application to And/Or trees. However, the fact that size of the system grows exponentially in k severely restricts hand-made evaluation. For a moderate number of variables, very recent results on explicit solving of algebraic systems [26] give us hope to extend the numerical computations a little bit farther. But exact computation will eventually fail, even for a "reasonable" number of boolean variables. Then we turn to asymptotic analysis; this is where our result comes in. Although it is likely that no general, easy-to-use expression of the constant factor $\lambda(f)$ holds for all boolean functions, we can still hope to obtain results for well-defined classes. Consider for example *read-once* functions, i.e., functions with $L(f)$ essential variables. An alternative definition is as functions whose minimal trees contain no repetition of variables. We can prove that the average number of expansions of read-once functions of complexity c is

$$\bar{\lambda}_{r.o.}(c) \sim \frac{\sqrt{\pi}}{2\sqrt{2}} \, c^{3/2} \left(\frac{4}{e}\right)^c .$$

We should also mention that Theorem 1 requires us to specify the boolean function, and does not hold uniformly over *all* boolean functions; hence we are still unable to compute the average complexity of a boolean function chosen according to this probability distribution. Further work is required before we can either verify or invalidate the Shannon effect for this non-uniform probability distribution.

Acknowledgements. We are grateful to Jakub Kozik for fruitful discussions about this problem.

References

1. Zaionc, M.: Statistics of implicational logic. Electronic Notes in Theoretical Computer Science 84 (2003)
2. Kostrzycka, Z., Zaionc, M.: Statistics of intuitionnistic versus classical logic. Studia Logica 76(3), 307–328 (2004)
3. Gardy, D.: Random boolean expressions. In: Colloquium on Computational Logic and Applications (2006)
4. Fournier, H., Gardy, D., Genitrini, A., Zaionc, M.: Classical and intuitionistic logic are asymptotically identical. In: CSL, pp. 177–193 (2007)
5. Paris, J.B., Vencovská, A., Wilmers, G.M.: A natural prior probability distribution derived from the propositional calculus. Annals of Pure and Applied Logic 70, 243–285 (1994)
6. Lefmann, H., Savický, P.: Some typical properties of large And/Or Boolean formulas. Random Structures and Algorithms 10, 337–351 (1997)
7. Woods, A.R.: Coloring rules for finite trees, and probabilities of monadic second order sentences. Random Struct. Algorithms 10(4), 453–485 (1997)
8. Chauvin, B., Flajolet, P., Gardy, D., Gittenberger, B.: And/Or trees revisited. Combinatorics, Probability and Computing 13(4-5), 475–497 (2004)
9. Moczurad, M., Tyszkiewicz, J., Zaionc, M.: Statistical properties of simple types. Mathematical Structures in Computer Science 10(5), 575–594 (2000)
10. Kostrzycka, Z.: On the density of truth of implicational parts of intuitionistic and classical logics. J. of Applied Non-Classical Logics 13(2) (2003)
11. Zaionc, M.: On the asymptotic density of tautologies in logic of implication and negation. Reports on Mathematical Logic 39, 67–87 (2005)
12. Kostrzycka, Z.: On the density of truth in modal logics. In: Mathematics and Computer Science, Nancy (France); Proceedings in DMTCS (September 2006) (to appear)
13. Matecki, G.: Asymptotic density for equivalence. Electronic Notes in Theoretical Computer Science 140, 81–91 (2005)
14. Kostrzycka, Z.: On asymptotic divergency in equivalential logics. Mathematical Structures in Computer Science 18, 311–324 (2008)
15. Woods, A.: On the probability of absolute truth for and/or formulas. Bulletin of Symbolic Logic 12(3) (2005)

16. Kozic, J.: Subcritical pattern languages for and/or trees. Technical report (2008)
17. Valiant, L.: Short monotone formulae for the majority function. Journal of Algorithms 5, 363–366 (1984)
18. Boppana, R.B.: Amplification of probabilistic boolean formulas. In: Proceedings of the 26th Annual IEEE Symposium on Foundations of Computer Science, pp. 20–29 (1985)
19. Gupta, A., Mahajan, S.: Using amplification to compute majority with small majority gates. Comput. Complex. 6(1), 46–63 (1997)
20. Savicky, P.: Random Boolean formulas representing any Boolean function with asymptotically equal probability. Discrete Mathematics 83, 95–103 (1990)
21. Brodsky, A., Pippenger, N.: The boolean functions computed by random boolean formulas or how to grow the right function. Random Structures and Algorithms 27, 490–519 (2005)
22. Drmota, M.: Systems of functional equations. Random Structures and Algorithms 10(1-2), 103–124 (1997)
23. Lalley, S.P.: Finite range random walk on free groups and homogeneous trees. The Annals of Probability 21 (1993)
24. Flajolet, P., Sedgewick, R.: Analytic combinatorics: Functional equations, rational and algebraic functions. Technical Report 4103, INRIA (January 2001)
25. Flajolet, P., Sedgewick, R.: Analytic Combinatorics. Cambridge University Press, Cambridge (to appear, 2008) Available from the authors' web page
26. Pivoteau, C., Salvy, B., Soria, M.: Combinatorial Newton iteration to compute Boltzmann oracle. Technical report, Journées ALÉA (March 2008), http://www-calfor.lip6.fr/pivoteau

Succinctness of Regular Expressions with Interleaving, Intersection and Counting

Wouter Gelade[*]

Hasselt University and Transnational University of Limburg
School for Information Technology
wouter.gelade@uhasselt.be

Abstract. Studying the impact of operations, such as intersection and interleaving, on the succinctness of regular expressions has recently received renewed attention [12,13,14]. In this paper, we study the succinctness of regular expressions (REs) extended with interleaving, intersection and counting operators. We show that in a translation from REs with interleaving to standard regular expressions a double exponential size increase can not be avoided. We also consider the complexity of translations to finite automata. We give a tight exponential lower bound on the translation of REs with intersection to NFAs, and, for each of the three classes of REs, we show that in a translation to a DFA a double exponential size increase can not be avoided. Together with known results, this gives a complete picture of the complexity of translating REs extended with interleaving, intersection or counting into (standard) regular expressions, NFAs, and DFAs.

1 Introduction

Regular expressions are used in many applications such as text processors, programming languages [30], and XML schema languages [5,28]. These applications, however, usually do not restrict themselves to the standard regular expression using disjunction ($+$), concatenation ($\cdot$) and star (*), but also allow the use of additional operators. Although these operators mostly do not increase the expressive power of the regular expressions, they can have a drastic impact on succinctness, thus making them harder to handle. For instance, it is well known that expressions extended with the complement operator can describe certain languages non-elementary more succinct than standard regular expressions or finite automata [29].

In this paper, we study the succinctness of regular expressions extended with counting (RE(#)), intersection (RE($\cap$)), and interleaving (RE(&)) operators. The counting operator allows for expressions such as $a^{[2,5]}$, specifying that there must occur at least two and at most 5 a's. These RE(#)s are used in egrep [16] and Perl [30] patterns and in the XML schema language XML Schema [28]. The class RE($\cap$) is a well studied extension of the regular expressions, and is often

---

[*] Research Assistant of the Fund for Scientific Research - Flanders (Belgium).

E. Ochmański and J. Tyszkiewicz (Eds.): MFCS 2008, LNCS 5162, pp. 363–374, 2008.
© Springer-Verlag Berlin Heidelberg 2008

referred to as the semi-extended regular expressions. The interleaving operator allows for expressions such as $a \& b \& c$, specifying that a, b, and c may occur in any order, and is used, for instance, in the XML schema language Relax NG [5].

A problem we consider, is the translation of extended regular expressions into (standard) regular expressions. For RE(#) and RE(∩) the complexity of this translation has already been settled and is exponential [18] and double exponential [12], respectively. We show that also in constructing an expression for the interleaving of a set of expressions (an hence also for an RE(&)) a double exponential size increase can not be avoided. This is the main technical result of the paper. Apart from a pure mathematical interest, the latter result has two important consequences. First, it prohibits an efficient translation from Relax NG (which allows interleaving) to XML Schema Definitions (which does not). However, as XML Schema is the widespread W3C standard, and Relax NG is a more flexible alternative, such a translation would be more than desirable. A second consequence concerns the automatic discovery of regular expression describing a set of given strings. The latter problem occurs in the learning of XML schema languages [1,2,3]. At present these algorithms do not take into account the interleaving operator, but for Relax NG this would be wise as this would allow to learn significantly smaller expressions.

We recently learned that Gruber and Holzer independently obtained a similar result [Personal communication]. They show that any regular expression defining the language $(a_1 b_1)^* \& \cdots \& (a_n b_n)^*$ must be of size at least double exponential in n. Compared to the result in this paper, this gives a tighter bound ($2^{2^{\Omega(n)}}$ instead of $2^{2^{\Omega(\sqrt{n})}}$), and shows that the double exponential size increase already occurs for very simple expressions. On the other, the alphabet of the counterexamples grows linear with n, whereas the alphabet size is constant for the languages in this paper. Hence, the two results nicely complement each other.

We also consider the translation of extended regular expressions to NFAs. For the standard regular expressions, it is well known that such a translation can be done efficiently [4]. Therefore, when considering problems such as membership, equivalence, and inclusion testing for regular expressions the first step is almost invariantly a translation to a finite automaton. For extended regular expressions, such an approach is less fruitful. We show that an RE(&, ∩, #) can be translated in exponential time into an NFA. However, it has already been shown by Kilpelainen and Tuhkanen [18] and Mayer and Stockmeyer [20] that such an exponential size increase can not be avoided for RE(#) and RE(&), respectively. For the translation from RE(∩) to NFAs, a $2^{\Omega(\sqrt{n})}$ lower bound is reported in [25], which we here improve to $2^{\Omega(n)}$.

As the translation of extended regular expressions to NFAs already involves an exponential size increase, it is natural to ask what the size increase for DFAs is. Of course, we can translate any NFA into a DFA in exponential time, thus giving a double exponential translation, but can we do better? For instance, from the results in [12] we can conclude that given a set of regular expressions, constructing an NFA for their intersection can not avoid an exponential size increase. However, it is not too hard to see that also a DFA of exponential size accepting their intersection can be constructed. In the present paper, we show that this is not possible for the classes $\mathrm{RE}(\#)$, $\mathrm{RE}(\cap)$, and $\mathrm{RE}(\&)$. For each class we show that in a translation to a DFA, a double exponential size increase can not be avoided. An overview of all results is given in Figure 1(a).

	NFA	DFA	RE
$\mathrm{RE}(\#)$	$2^{\Omega(n)}$ [18]	$2^{2^{\Omega(n)}}$ (Th. 3)	$2^{\theta(n)}$ [18]
$\mathrm{RE}(\cap)$	$2^{\Omega(n)}$ (Pr. 2)	$2^{2^{\Omega(n)}}$ (Th. 4)	$2^{2^{\Omega(\sqrt{n})}}$ [12]
$\mathrm{RE}(\&)$	$2^{\Omega(n)}$ [20]	$2^{2^{\Omega(\sqrt{n})}}$ (Th. 5)	$2^{2^{\Omega(\sqrt{n})}}$ (Th. 6)
$\mathrm{RE}(\&, \cap, \#)$	$2^{\mathcal{O}(n)}$ (Pr. 1)	$2^{2^{\mathcal{O}(n)}}$ (Pr. 3)	$2^{2^{\mathcal{O}(n)}}$ (Pr. 4)

	RE
$\mathrm{RE} \cap \mathrm{RE}$	$2^{\Omega(n)}$ [13]
$\bigcap \mathrm{RE}$	$2^{2^{\Omega(\sqrt{n})}}$ [12]
$\mathrm{RE} \,\&\, \mathrm{RE}$	$2^{\Omega(n)}$ [13]

(a)   (b)

Fig. 1. Table (a) gives the complexity of translating extended regular expressions into NFAs, DFAs, and regular expressions. Proposition and theorem numbers are given in brackets. Table (b) lists some related results obtained in [12] and [13].

Related work. The different classes of regular expressions considered here have been well studied. In particular, the $\mathrm{RE}(\cap)$ and its membership [17,19,25] and equivalence and emptiness [10,24,26] problems, but also the classes $\mathrm{RE}(\#)$ [18,23] and $\mathrm{RE}(\&)$ [11,20] have received interest. Succinctness of regular expressions has been studied by Ehrenfeucht and Zeiger [8] and, more recently, by Ellul et. al [9], Gelade and Neven [12], Gruber and Holzer [13,14], and Gruber and Johannsen [15]. Some relevant results of these papers are listed in Figure 1(b). Schott and Spehner give lower bounds for the translation of the interleaving of words to DFAs [27]. Also related, but different in nature, are the results on state complexity [32], in which the impact of the application of different operations on finite automata is studied.

Outline. In Section 2 we give the necessary definitions and present some basic results. In Sections 3, 4, and 5 we study the translation of extended regular expressions to NFAs, DFAs, and regular expressions, respectively. A version of this paper containing all proofs is available from the webpage of the author.

2   Definitions and Basic Results

2.1   Regular Expressions

By $\mathbb{N}$ we denote the natural numbers without zero. For the rest of the paper, Σ always denotes a finite alphabet. A Σ-*string* (or simply string) is a finite sequence $w = a_1 \cdots a_n$ of Σ-symbols. We define the length of w, denoted by $|w|$, to be n. We denote the empty string by ε. The set of *positions of* w is $\{1, \ldots, n\}$ and the *symbol of* w *at position* i is a_i. By $w_1 \cdot w_2$ we denote the *concatenation* of two strings w_1 and w_2. As usual, for readability, we denote the concatenation of w_1 and w_2 by $w_1 w_2$. The set of all strings is denoted by Σ^*. A *string language* is a subset of Σ^*. For two string languages $L, L' \subseteq \Sigma^*$, we define their concatenation

$L \cdot L'$ to be the set $\{ww' \mid w \in L, w' \in L'\}$. We abbreviate $L \cdot L \cdots L$ (i times) by L^i. By $w_1 \,\&\, w_2$ we denote the set of strings that is obtained by *interleaving* w_1 and w_2 in every possible way. That is, for $w \in \Sigma^*$, $w \,\&\, \varepsilon = \varepsilon \,\&\, w = \{w\}$, and $aw_1 \,\&\, bw_2 = (\{a\}(w_1 \,\&\, bw_2)) \cup (\{b\}(aw_1 \,\&\, w_2))$. The operator $\&$ is then extended to languages in the canonical way.

The set of *regular expressions* over Σ, denoted by RE, is defined in the usual way: $\emptyset$, ε, and every Σ-symbol is a regular expression; and when r_1 and r_2 are regular expressions, then $r_1 \cdot r_2$, $r_1 + r_2$, and r_1^* are also regular expressions. By RE($\&, \cap, \#$) we denote the class of *extended regular expressions*, that is, REs extended with interleaving, intersection and counting operators. So, when r_1 and r_2 are RE($\&, \cap, \#$)-expressions then so are $r_1 \,\&\, r_2$, $r_1 \cap r_2$, and $r_1^{[k,\ell]}$ for $k, \ell \in \mathbb{N}$ with $k \leq \ell$. By RE($\&$), RE($\cap$), and RE($\#$), we denote RE extended solely with the interleaving, intersection and counting operator, respectively.

The language defined by an extended regular expression r, denoted by $L(r)$, is inductively defined as follows: $L(\emptyset) = \emptyset$; $L(\varepsilon) = \{\varepsilon\}$; $L(a) = \{a\}$; $L(r_1 r_2) = L(r_1) \cdot L(r_2)$; $L(r_1 + r_2) = L(r_1) \cup L(r_2)$; $L(r^*) = \{\varepsilon\} \cup \bigcup_{i=1}^{\infty} L(r)^i$; $L(r_1 \,\&\, r_2) = L(r_1) \,\&\, L(r_2)$; $L(r_1 \cap r_2) = L(r_1) \cap L(r_2)$; and $L(r^{[k,\ell]}) = \bigcup_{i=k}^{\ell} L(r)^i$.

By r^+, $\bigcup_{i=1}^{k} r_i$, and r^k, with $k \in \mathbb{N}$, we abbreviate the expression rr^*, $r_1 + \cdots + r_k$, and $rr \cdots r$ (k-times), respectively. For a set $S = \{a_1, \ldots, a_n\} \subseteq \Sigma$, we abbreviate by S the regular expression $a_1 + \cdots + a_n$. When $r^{[k,l]}$ is used in a standard regular expression, this is an abbreviation for $r^k (r + \varepsilon)^{l-k}$.

We define the *size* of an extended regular expression r over Σ, denoted by $|r|$, as the number of Σ-symbols and operators occurring in r plus the sizes of the binary representations of the integers. Formally, $|\emptyset| = |\varepsilon| = |a| = 1$, for $a \in \Sigma$, $|r_1 r_2| = |r_1 \cap r_2| = |r_1 + r_2| = |r_1 \,\&\, r_2| = |r_1| + |r_2| + 1$, $|r^*| = |r| + 1$, and $|r^{[k,\ell]}| = |r| + \lceil \log k \rceil + \lceil \log \ell \rceil$.

Intuitively, the *star height* of a regular expression r, denoted by $\mathrm{sh}(r)$ equals the number of nested stars in r. Formally, $\mathrm{sh}(\emptyset) = \mathrm{sh}(\varepsilon) = \mathrm{sh}(a) = 0$, for $a \in \Sigma$, $\mathrm{sh}(r_1 r_2) = \mathrm{sh}(r_1 + r_2) = \max\{\mathrm{sh}(r_1), \mathrm{sh}(r_2)\}$, and $\mathrm{sh}(r^*) = \mathrm{sh}(r) + 1$. The star height of a regular language L, denoted by $\mathrm{sh}(L)$, is the minimal star height among all regular expressions defining L.

The latter two concepts are related through the following theorem due to Gruber and Holzer [13], which will allow us to reduce our questions about the size of regular expressions to questions about the star height of regular languages.

Theorem 1 ([13]). *Let L be a regular language. Then any regular expression defining L is of size at least $2^{\frac{1}{3}(\mathrm{sh}(L)-1)} - 1$.*

2.2 Finite Automata and Graphs

A non-deterministic finite automaton (NFA) A is a 4-tuple (Q, q_0, δ, F) where Q is the set of states, q_0 is the initial state, F is the set of final states and $\delta \subseteq Q \times \Sigma \times Q$ is the transition relation. As usual, we denote by $\delta^* \subseteq Q \times \Sigma^* \times Q$ the reflexive-transitive closure of δ. Then, w is accepted by A if $(q_0, w, q_f) \in \delta^*$ for some $q_f \in F$. The set of strings accepted by A is denoted by $L(A)$. The size of an NFA is $|Q|+|\delta|$. An NFA is *deterministic* (or a DFA) if for all $a \in \Sigma, q \in Q$, $|\{(q, a, q') \in \delta \mid q' \in Q\}| \leq 1$.

A state $q \in Q$ is *useful* if there exist strings $w, w' \in \Sigma^*$ such that $(q_0, w, q) \in \delta^*$, and $(q, w', q_f) \in \delta^*$, for some $q_f \in F$. An NFA is *trim* if it only contains useful states. For $q \in Q$, let symbols$(q) = \{a \mid \exists p \in Q, (p, a, q) \in \delta\}$. Then, A is *state-labeled* if for any $q \in Q$, $|\text{symbols}(q)| \leq 1$, i.e., all transitions to a single state are labeled with the same symbol. In this case, we also denote this symbol by symbol(q). Further, A is *non-returning* if symbols$(q_0) = \emptyset$, i.e., q_0 has no incoming transitions. A language L is *bideterministic* if there exists a DFA A, accepting L, such that the inverse of A is again deterministic. That is, A may have at most one final state and the automaton obtained by inverting every transition in A, and exchanging the initial and final state, is again deterministic.

A (directed) *graph* G is a tuple (V, E), where V is the set of *vertices* and $E \subseteq V \times V$ is the set of *edges*. A graph (U, F) is a *subgraph* of G if $U \subseteq V$ and $F \subseteq E$. For a set of vertices $U \subseteq V$, the *subgraph of G induced by U*, denoted by $G[U]$, is the graph (U, F), where $F = \{(u, v) \mid u, v \in U \wedge (u, v) \in E\}$.

A graph $G = (V, E)$ is *strongly connected* if for every pair of vertices $u, v \in V$, both u is reachable from v, and v is reachable from u. A set of edges $V' \subseteq V$ is a *strongly connected component (SCC)* of G if $G[V']$ is strongly connected and for every set V'', with $V' \subsetneq V''$, $G[V'']$ is not strongly connected.

We now introduce the *cycle rank* of a graph $G = (V, E)$, denoted by cr(G), which is a measure for the structural complexity of G. It is inductively defined as follows: (1) if G is acyclic or empty, then cr$(G) = 0$, otherwise (2) if G is strongly connected, then cr$(G) = \min_{v \in V} \text{cr}(G[V \setminus \{v\}]) + 1$, and otherwise (3) cr$(G) = \max_{V' \text{ SCC of } G} \text{cr}(G[V'])$.

Let $A = (Q, q_0, \delta, F)$ be an NFA. The *underlying graph* G of A is the graph obtained by removing the labels from the transition edges of A, or more formally $G = (Q, E)$, with $E = \{(q, q') \mid \exists a \in \Sigma, (q, a, q') \in \delta\}$. In the following, we often abuse notation and for instance say the cycle rank of A, referring to the cycle rank of its underlying graph.

There is a strong connection between the star height of a regular language, and the cycle rank of the NFAs accepting it, as witnessed by the following theorem. Theorem 2(1) is known as Eggan's Theorem [7] and proved in its present form by Cohen [6]. Theorem 2(3) is due to McNaughton [21].

Theorem 2. *For any regular language L,*

1. *$sh(L) = \min\{cr(A) \mid A$ is an NFA accepting $L\}$. [7,6].*
2. *$sh(L) \cdot |\Sigma| \geq \min\{cr(A) \mid A$ is a non-returning state-labeled NFA accepting $L\}$.*
3. *if L is bideterministic, then $sh(L) = cr(A)$, where A is the minimal trim DFA accepting L. [21]*

3   Succinctness w.r.t. NFAs

In this section, we study the complexity of translating extended regular expressions into NFAs. We show that such a translation can be done in exponential time, by constructing the NFA by induction on the structure of the expression.

Proposition 1. *Let r be a RE(&,$\cap$,#). An NFA A with at most $2^{|r|}$ states, such that $L(r) = L(A)$, can be constructed in time $2^{\mathcal{O}(|r|)}$.*

This exponential size increase can not be avoided for any of the classes. For RE(#) this is witnessed by the expression $a^{[2^n,2^n]}$ and for RE(&) by the expression $a_1 \& \cdots \& a_n$, as already observed by Kilpelainen and Tuhkanen [18] and Mayer and Stockmeyer [20], respectively. For RE($\cap$), a $2^{\Omega(\sqrt{n})}$ lower bound has already been reported in [25]. The present tighter statement, however, will follow from Theorem 4 and the fact that any NFA with n states can be translated into a DFA with 2^n states [31].

Proposition 2. *For any $n \in \mathbb{N}$, there exist an RE(#) $r^{\#}$, an RE($\cap$) $r^{\cap}$, and an RE(&) $r^{\&}$, each of size $\mathcal{O}(n)$, such that any NFA accepting $r^{\#}$, $r^{\cap}$, or $r^{\&}$ contains at least 2^n states.*

4   Succinctness w.r.t. DFAs

In this section, we study the complexity of translating extended regular expressions into DFAs. First, from Proposition 1 and the fact that any NFA with n states can be translated into a DFA with 2^n states in exponential time [31], we can immediately conclude the following.

Proposition 3. *Let r be a RE(&,$\cap$,#). A DFA A with at most $2^{2^{|r|}}$ states, such that $L(r) = L(A)$, can be constructed in time $2^{2^{\mathcal{O}(|r|)}}$.*

We show that, for each of the classes RE(#), RE($\cap$), or RE(&), this double exponential size increase can not be avoided. For RE(#), this is witnessed by the expression $(a + b)^* a(a + b)^{[2^n,2^n]}$ which is of size $\mathcal{O}(n)$, but for which any DFA accepting it must contain at least 2^{2^n} states.

Theorem 3. *For any $n \in \mathbb{N}$ there exists an RE(#) r_n of size $\mathcal{O}(n)$ such that any DFA accepting $L(r_n)$ contains at least 2^{2^n} states.*

We now move to regular expressions extended with the intersection operator. The succinctness of RE($\cap$) with respect to DFAs can be obtained along the same lines as the simulation of exponential space turing machines by RE($\cap$) in [10].

Theorem 4. *For any $n \in \mathbb{N}$ there exists an RE($\cap$) $r_n^{\cap}$ of size $\mathcal{O}(n)$ such that any DFA accepting $L(r_n^{\cap})$ contains at least 2^{2^n} states.*

Proof. Let $n \in \mathbb{N}$. We start by describing the language $\mathcal{G}_n$ which will be used to establish the lower bound. This will be a variation of the following language over the alphabet $\{a,b\}$: $\{ww \mid |w| = 2^n\}$. It is well known that this language is hard to describe by a DFA. However, to define it very succinct by an RE($\cap$), we need to add some additional information to it.

Thereto, we first define a *marked number* as a string over the alphabet $\{0, 1, \bar{0}, \bar{1}\}$ defined by the regular expression $(0 + 1)^* \bar{1} 0^* + \bar{0}^*$, i.e., a binary number in which the rightmost 1 and all following 0's are marked. Then, for any $i \in [0, 2^n - 1]$ let $\text{enc}(i)$ denote the n-bit marked number encoding i. These marked numbers were introduced by Fürer in [10], where the following is observed: if $i, j \in [0, 2^n - 1]$ are such that $j = i + 1 (\text{mod } 2^n)$, then the bits of i and j which are different are exactly the marked bits of j. For instance, for $n = 2$, $\text{enc}(1) = 0\bar{1}$ and $\text{enc}(2) = \bar{1}\bar{0}$ and they differ in both bits as both bits of $\text{enc}(2)$ are marked. Further, let $\text{enc}^R(i)$ denote the reversal of $\text{enc}(i)$.

Now, for a string $w = a_0 a_1 \ldots a_{2^n - 1}$ define

$$\text{enc}(w) = \text{enc}^R(0) a_0 \text{enc}(0)\$\text{enc}^R(1) a_1 \text{enc}(1)\$ \cdots \text{enc}^R(2^n - 1) a_{2^n - 1} \text{enc}(2^n - 1)$$

and, finally, define

$$\mathcal{G}_n = \{\#\text{enc}(w)\#\text{enc}(w) \mid w \in L((a + b)^*) \wedge |w| = 2^n\}$$

For instance, for $n = 2$, and $w = abba$, $\text{enc}(w) = \bar{0}\bar{0}a\bar{0}\bar{0}\$\bar{1}\bar{0}b0\bar{1}\$0\bar{1}b\bar{1}\bar{0}\$\bar{1}\bar{1}a\bar{1}\bar{1}$ and hence $\#\bar{0}\bar{0}a\bar{0}\bar{0}\$\bar{1}\bar{0}b0\bar{1}\$0\bar{1}b\bar{1}\bar{0}\$\bar{1}\bar{1}a\bar{1}\bar{1}\#\bar{0}\bar{0}a\bar{0}\bar{0}\$\bar{1}\bar{0}b0\bar{1}\$0\bar{1}b\bar{1}\bar{0}\$\bar{1}\bar{1}a\bar{1}\bar{1} \in \mathcal{G}_2$.

It can be shown that any DFA accepting $\overline{\mathcal{G}_n}$, the complement of $\mathcal{G}_n$, must contain at least 2^{2^n} states. Furthermore, we can construct an expression $r_n^\cap$ of size $\mathcal{O}(n)$ defining $\overline{\mathcal{G}_n}$. Here, $r_n^\cap$ is the disjunction of many expressions, each describing some mistake a string can make in order not to be in $\mathcal{G}_n$. □

We can now extend the results for RE($\cap$) to RE($\&$). We do this by using a technique of Mayer and Stockmeyer [20] which allows, in some sense, to simulate an RE($\cap$) by an RE($\&$). To formally define this, we need some notation. Let $w = a_0 \cdots a_n$ be a string over an alphabet Σ, and let c be a symbol not in Σ. Then, for any $i \in \mathbb{N}$, define $\text{pump}_i(w) = a_0^i c^i a_1^i c^i \cdots a_k^i c^i$. Now, they proved the following:

Lemma 1 ([20]). *Let r be an RE($\cap$) containing k $\cap$-operators. Then, there exists an RE($\&$) s of size at most $|r|^2$ such that for any $w \in \Sigma^*$, $w \in L(r)$ iff $\text{pump}_k(w) \in L(s)$.*

That is, the expression s constructed in this lemma may define additional strings, but the set of valid pumped string it defines, corresponds exactly to $L(r)$. Using this lemma, we can now prove the following theorem.

Theorem 5. *For any $n \in \mathbb{N}$ there exists an RE($\&$) $r_n^\&$ of size $\mathcal{O}(n^2)$ such that any DFA accepting $L(r_n^\&)$ contains at least 2^{2^n} states.*

5   Succinctness w.r.t. Regular Expressions

In this section, we study the translation of extended regular expressions to (standard) regular expressions. First, for the class RE($\#$) it has already been shown by Kilpelainen and Tuhkanen [18] that this translation can be done in exponential time, and that an exponential size increase can not be avoided. Furthermore, from Proposition 1 and the fact that any NFA with n states can be translated into a regular expression in time $2^{\mathcal{O}(n)}$ [9] it immediately follows that:

Proposition 4. *Let r be a $RE(\&, \cap, \#)$. A regular expression s equivalent to r can be constructed in time $2^{2^{O(|r|)}}$.*

Furthermore, from the results in [12] (see also Figure 1(b)) it follows that in a translation from $RE(\cap)$ to standard regular expressions, a double exponential size increase can not be avoided.

Hence, it only remains to show a double exponential lower bound on the translation from $RE(\&)$ to standard regular expressions, which is exactly what we will do in the rest of this section. Thereto, we proceed in several steps and define several families of languages. First, we introduce the family of languages $(\mathcal{K}_n)_{n \in \mathbb{N}}$, on which all following languages will be based, and establish its star height. The star height of languages will be our tool for proving lower bounds on the size of regular expressions defining these languages. Then, we define the family $(\mathcal{L}_n)_{n \in \mathbb{N}}$ which is a binary encoding of $(\mathcal{K}_n)_{n \in \mathbb{N}}$ and show that these languages can be defined as the intersection of small regular expressions.

Finally, we define the family $(\mathcal{M}_n)_{n \in \mathbb{N}}$ which is obtained by simulating the intersection of the previously obtained regular expressions by the interleaving of related expressions, similar to the simulation of $RE(\cap)$ by $RE(\&)$ in Section 4. Bringing everything together, this then leads to the desired result: a double exponential lower bound on the translation of $RE(\&)$ to RE.

As an intermediate corollary of this proof, we also obtain a double exponential lower bound on the translation of $RE(\cap)$ to RE, similar to a result in [12]. We note, however, that the succinctness results for $RE(\&)$ can not be obtained by using the results in [12], and that, hence, the different lemmas which prove the succinctness of $RE(\cap)$ are necessary to obtain the subsequent results on $RE(\&)$.

5.1 $\mathcal{K}_n$: The Basic Language

We first introduce the family $(\mathcal{K}_n)_{n \in \mathbb{N}}$ defined by Ehrenfeucht and Zeiger over an alphabet whose size grows quadratically with the parameter n [8]:

Definition 1. *Let $n \in \mathbb{N}$ and $\Sigma_n = \{a_{i,j} \mid 0 \le i, j \le n-1\}$. Then, $\mathcal{K}_n$ contains exactly all strings of the form $a_{0,i_1} a_{i_1,i_2} \cdots a_{i_k, n-1}$ where $k \in \mathbb{N} \cup \{0\}$.*

An alternative definition of $\mathcal{K}_n$ is through the minimal DFA accepting it. Thereto, let $A_n^{\mathcal{K}} = (Q, q_0, \delta, F)$ be defined as $Q = \{q_0, \dots, q_{n-1}\}$, $F = \{q_{n-1}\}$, and for all $i, j \in [0, n-1]$, $(q_i, a_{i,j}, q_j) \in \delta$. That is, $A_n^{\mathcal{K}}$ is the complete DFA on n states where the transition from state i to j is labeled by $a_{i,j}$.

We now determine the star height of $\mathcal{K}_n$. This is done by observing that $\mathcal{K}_n$ is bideterministic, such that, by Theorem 2(3), $sh(\mathcal{K}_n) = cr(A_n^{\mathcal{K}})$, and subsequently showing that $cr(A_n^{\mathcal{K}}) = n$.

Lemma 2. *For any $n \in \mathbb{N}$, $sh(\mathcal{K}_n) = n$.*

5.2 $\mathcal{L}_n$: Succinctness of $RE(\cap)$

In this section we want to construct a set of small regular expressions such that any expression defining their intersection must be large (that is, of double exponential size). Ideally, we would like to use the family of languages $(\mathcal{K}_n)_{n \in \mathbb{N}}$ for this as we have shown in the previous section that they have a large star height, and thus by Theorem 1 can not be defined by small expressions. Unfortunately, this is not possible as the alphabet of $(\mathcal{K}_n)_{n \in \mathbb{N}}$ grows quadratically with n.

Therefore, we will introduce in this section the family of languages $(\mathcal{L}_n)_{n \in \mathbb{N}}$ which is a binary encoding of $(\mathcal{K}_n)_{n \in \mathbb{N}}$ over a fixed alphabet. Thereto, let $n \in \mathbb{N}$ and recall that $\mathcal{K}_n$ is defined over the alphabet $\Sigma_n = \{a_{i,j} \mid i, j \in [0, n-1]\}$. Now, for $a_{i,j} \in \Sigma_n$, define the function ρ_n as

$$\rho_n(a_{i,j}) = \#\mathrm{enc}(j)\$\mathrm{enc}(i)\triangle\mathrm{enc}(i+1)\triangle \cdots \triangle\mathrm{enc}(n-1)\triangle,$$

where $\mathrm{enc}(k)$, for $k \in \mathbb{N}$, denotes the $\lceil \log(n) \rceil$-bit marked number encoding k as defined in the proof of Theorem 4. So, the encoding starts by the encoding of the second index, followed by an ascending sequence of encodings of all numbers from the first index to $n-1$. We extend the definition of ρ_n to strings in the usual way: $\rho_n(a_{0,i_1} \cdots a_{i_{k-1},n-1}) = \rho_n(a_{0,i_1}) \cdots \rho_n(a_{i_k,n-1})$.

We are now ready to define $\mathcal{L}_n$.

Definition 2. Let $\Sigma = \{0, 1, \bar{0}, \bar{1}, \$, \#, \triangle\}$. For $n \in \mathbb{N}$, $\mathcal{L}_n = \{\rho_n(w) \mid w \in \mathcal{K}_n\}$.

For instance, for $n = 3$, $a_{0,1}a_{1,2} \in \mathcal{K}_3$ and hence $\rho_3(a_{0,1}a_{1,2}) = \#0\bar{1}\$\bar{0}0\triangle0\bar{1}\triangle\bar{1}0\triangle$ $\#\bar{1}0\$0\bar{1}\triangle\bar{1}0\triangle \in \mathcal{L}_3$. We now show that this encoding does not affect the star height. This is done by observing that, due to the specific encoding of $\mathcal{K}_n$, $\mathcal{L}_n$ is still bideterministic. Then, we obtain the star height of $\mathcal{L}_n$ by determining the cycle rank of the minimal DFA accepting it.

Lemma 3. For any $n \in \mathbb{N}$, $sh(\mathcal{L}_n) = n$.

Further, it can be shown that $\mathcal{L}_n$ can be described as the intersection of a set of small regular expressions.

Lemma 4. For every $n \in \mathbb{N}$, there are regular expressions $r_1, \ldots, r_m$, with $m = 4n + 3$, each of size $\mathcal{O}(n)$, such that $\bigcap_{i \leq m} L(r_i) = \mathcal{L}_{2^n}$.

Although it is not our main interest, we can now obtain the following by combining Theorem 1, and Lemmas 3 and 4.

Corollary 1. For any $n \in \mathbb{N}$, there exists an $RE(\cap)$ r of size $\mathcal{O}(n^2)$ such that any (standard) regular expression defining $L(r)$ is of size at least $2^{\frac{1}{3}(2^n - 1)} - 1$.

5.3 $\mathcal{M}_n$: Succinctness of RE(&)

In this section we will finally show that RE(&) are double exponentially more succinct than standard regular expressions. We do this by simulating the intersection of the regular expressions obtained in the previous section, by the interleaving of related expressions, similar to the simulation of RE(∩) by RE(&) in Section 4. This approach will partly yield the following family of languages. For any $n \in \mathbb{N}$, define

$$\mathcal{M}_n = \{\mathrm{pump}_{4\lceil \log n \rceil + 3}(w) \mid w \in \mathcal{L}_n\}$$

As $\mathcal{M}_n$ is very similar to $\mathcal{L}_n$, we can easily extend the result on the star height of $\mathcal{L}_n$ (Lemma 3) to $\mathcal{M}_n$:

Lemma 5. *For any $n \in \mathbb{N}$, $sh(\mathcal{M}_n) = n$.*

However, the language we will eventually define will not be exactly $\mathcal{M}_n$. Therefore, we need an additional lemma, for which we first introduce some notation. For $k \in \mathbb{N}$, and an alphabet Σ, we define $\Sigma^{(k)}$ to be the language defined by the expression $(\bigcup_{\sigma \in \Sigma} \sigma^k)^*$, i.e., all strings which consist of a sequence of blocks of identical symbols of length k. Further, for a language L, define index$(L) = \max \{i \mid i \in \mathbb{N} \wedge \exists w, w' \in \Sigma^*, a \in \Sigma \text{ such that } wa^i w' \in L\}$. Notice that index$(L)$ can be infinite. However, we will only be interested in languages for which it is finite, as in the following lemma.

Lemma 6. *Let L be a regular language, and $k \in \mathbb{N}$, such that index$(L) \leq k$. Then, $sh(L) \cdot |\Sigma| \geq sh(L \cap \Sigma^{(k)})$.*

This lemma is proved by combining Theorem 2(2) with an algorithm that transforms any non-returning state-labeled NFA A, with index$(L(A)) \leq k$, into an NFA accepting $L(A) \cap \Sigma^{(k)}$, without increasing its cycle rank.

Now, we are finally ready to prove the desired theorem:

Theorem 6. *For every $n \in \mathbb{N}$, there are regular expressions $s_1, \ldots, s_m$, with $m = 4n + 3$, each of size $\mathcal{O}(n)$, such that any regular expression defining $L(s_1) \& L(s_2) \& \cdots \& L(s_m)$ is of size at least $2^{\frac{1}{24}(2^n - 8)} - 1$.*

Proof. Let $n \in \mathbb{N}$, and let $r_1, \ldots, r_m$, with $m = 4n + 3$, be the regular expressions obtained in Lemma 4 such that $\bigcap_{i \leq m} L(r_i) = \mathcal{L}_{2^n}$.

Now, it is shown in [20], that given $r_1, \ldots, r_m$, it is possible to construct regular expressions $s_1, \ldots, s_m$ such that (1) for all $i \in [1, m]$, $|s_i| \leq 2|r_i|$, and if we define $\mathcal{N}_{2^n} = L(s_1) \& \cdots \& L(s_m)$, then (2) index$(\mathcal{N}_{2^n}) \leq m$, and (3) for every $w \in \Sigma^*$, $w \in \bigcap_{i \leq m} L(r_i)$ iff pump$_m(w) \in \mathcal{N}_{2^n}$. Furthermore, it follows immediately from the construction in [20] that any string in $\mathcal{N}_{2^n} \cap \Sigma^{(m)}$ is of the form $a_1^m c^m a_2^m c^m \cdots a_l^m c^m$, i.e., pump$_m(w)$ for some $w \in \Sigma^*$.

Since $\bigcap_{i \leq m} L(r_i) = \mathcal{L}_{2^n}$, and $\mathcal{M}_{2^n} = \{$pump$_m(w) \mid w \in \mathcal{L}_{2^n}\}$, it hence follows that $\mathcal{M}_{2^n} = \mathcal{N}_{2^n} \cap \Sigma^{(m)}$. As furthermore, by Lemma 5, $sh(\mathcal{M}_{2^n}) = 2^n$ and index$(\mathcal{N}_{2^n}) \leq m$, it follows from Lemma 6 that $sh(\mathcal{N}_{2^n}) \geq \frac{sh(\mathcal{M}_{2^n})}{|\Sigma|} = \frac{2^n}{8}$. So, $\mathcal{N}_{2^n}$ can be described by the interleaving of the expressions s_1 to s_m, each of size $\mathcal{O}(n)$, but any regular expression defining $\mathcal{N}_{2^n}$ must, by Theorem 1, be of size at least $2^{\frac{1}{24}(2^n - 8)} - 1$. This completes the proof. $\square$

Corollary 2. *For any $n \in \mathbb{N}$, there exists an RE(&) r_n of size $\mathcal{O}(n^2)$ such that any regular expression defining $L(r_n)$ must be of size at least $2^{\frac{1}{24}(2^n - 8)} - 1$.*

This completes our paper. As a final remark, we note that all lower bounds in this paper make use of a constant size alphabet and can furthermore easily be extended to a 2-letter alphabet. For any language over an alphabet $\Sigma = \{a_1, \ldots, a_k\}$, we obtain a new language by replacing, for any $i \in [1, k]$, every symbol a_i by $b^i c^{k-i+1}$. Obviously, the size of a regular expression for this new language is at most $k + 1$ times the size of the original expression, and the lower bounds on the number of states of DFAs trivially carry over. Furthermore, it is shown in [22] that this transformation does not affect the star height, and hence the lower bounds on the sizes of the regular expression also carry over.

Acknowledgement. I thank Frank Neven and the anonymous referees for helpful suggestions, and Hermann Gruber for informing me about their results on the succinctness of the interleaving operator.

References

1. Bex, G., Gelade, W., Neven, F., Vansummeren, S.: Learning deterministic regular expressions for the inference of schemas from XML data. In: WWW, pp. 825–834 (2008)
2. Bex, G., Neven, F., Schwentick, T., Tuyls, K.: Inference of concise DTDs from XML data. In: VLDB, pp. 115–126 (2006)
3. Bex, G., Neven, F., Vansummeren, S.: Inferring XML Schema Definitions from XML data. In: VLDB, pp. 998–1009 (2007)
4. Bruggemann-Klein, A.: Regular expressions into finite automata. Theoretical Computer Science 120(2), 197–213 (1993)
5. Clark, J., Murata, M.: RELAX NG Specification. OASIS (December 2001)
6. Cohen, R.S.: Rank-non-increasing transformations on transition graphs. Information and Control 20(2), 93–113 (1972)
7. Eggan, L.C.: Transition graphs and the star height of regular events. Michigan Mathematical Journal 10, 385–397 (1963)
8. Ehrenfeucht, A., Zeiger, H.: Complexity measures for regular expressions. Journal of Computer and System Sciences 12(2), 134–146 (1976)
9. Ellul, K., Krawetz, B., Shallit, J., Wang, M.: Regular expressions: New results and open problems. Journal of Automata, Languages and Combinatorics 10(4), 407–437 (2005)
10. Fürer, M.: The complexity of the inequivalence problem for regular expressions with intersection. In: ICALP, pp. 234–245 (1980)
11. Gelade, W., Martens, W., Neven, F.: Optimizing schema languages for XML: Numerical constraints and interleaving. In: Schwentick, T., Suciu, D. (eds.) ICDT 2007. LNCS, vol. 4353, pp. 269–283. Springer, Heidelberg (2006)
12. Gelade, W., Neven, F.: Succinctness of the complement and intersection of regular expressions. In: STACS, pp. 325–336 (2008)
13. Gruber, H., Holzer, M.: Finite automata, digraph connectivity, and regular expression size. In: ICALP (to appear, 2008)
14. Gruber, H., Holzer, M.: Language operations with regular expressions of polynomial size. In: DCFS (to appear, 2008)
15. Gruber, H., Johannsen, J.: Optimal lower bounds on regular expression size using communication complexity. In: Amadio, R.M. (ed.) FOSSACS 2008. LNCS, vol. 4962, pp. 273–286. Springer, Heidelberg (2008)
16. Hume, A.: A tale of two greps. Software, Practice and Experience 18(11), 1063–1072 (1988)
17. Jiang, T., Ravikumar, B.: A note on the space complexity of some decision problems for finite automata. Information Processing Letters 40(1), 25–31 (1991)
18. Kilpelainen, P., Tuhkanen, R.: Regular expressions with numerical occurrence indicators — preliminary results. In: SPLST 2003, pp. 163–173 (2003)

19. Kupferman, O., Zuhovitzky, S.: An improved algorithm for the membership problem for extended regular expressions. In: Diks, K., Rytter, W. (eds.) MFCS 2002. LNCS, vol. 2420, pp. 446–458. Springer, Heidelberg (2002)
20. Mayer, A.J., Stockmeyer, L.J.: Word problems-this time with interleaving. Information and Computation 115(2), 293–311 (1994)
21. McNaughton, R.: The loop complexity of pure-group events. Information and Control 11(1/2), 167–176 (1967)
22. McNaughton, R.: The loop complexity of regular events. Information Sciences 1(3), 305–328 (1969)
23. Meyer, A.R., Stockmeyer, L.J.: The equivalence problem for regular expressions with squaring requires exponential space. In: FOCS, pp. 125–129 (1972)
24. Petersen, H.: Decision problems for generalized regular expressions. In: DCAGRS, pp. 22–29 (2000)
25. Petersen, H.: The membership problem for regular expressions with intersection is complete in LOGCFL. In: Alt, H., Ferreira, A. (eds.) STACS 2002. LNCS, vol. 2285, pp. 513–522. Springer, Heidelberg (2002)
26. Robson, J.M.: The emptiness of complement problem for semi extended regular expressions requires c^n space. Information Processing Letters 9(5), 220–222 (1979)
27. Schott, R., Spehner, J.C.: Shuffle of words and araucaria trees. Fundamenta Informatica 74(4), 579–601 (2006)
28. Sperberg-McQueen, C.M., Thompson, H.: XML Schema (2005), http://www.w3.org/XML/Schema
29. Stockmeyer, L.J., Meyer, A.R.: Word problems requiring exponential time: Preliminary report. In: STOC, pp. 1–9 (1973)
30. Wall, L., Christiansen, T., Orwant, J.: Programming Perl, 3rd edn. OReilly, Sebastopol (2000)
31. Yu, S.: Regular languages. In: Rozenberg, G., Salomaa, A. (eds.) Handbook of formal languages, ch. 2, vol. 1, pp. 41–110. Springer, Heidelberg (1997)
32. Yu, S.: State complexity of regular languages. Journal of Automata, Languages and Combinatorics 6(2), 221–234 (2001)

Nilpotency and Limit Sets of Cellular Automata[*]

Pierre Guillon[1] and Gaétan Richard[2]

[1] Université Paris-Est
Laboratoire d'Informatique Gaspard Monge, UMR CNRS 8049
5 bd Descartes, 77454 Marne la Vallée Cedex 2, France
pierre.guillon@univ-mlv.fr
[2] Laboratoire d'Informatique Fondamentale de Marseille
Aix-Marseille Université, CNRS
39 rue Joliot-Curie, 13013 Marseille, France
gaetan.richard@lif.univ-mrs.fr

Abstract. A one-dimensional cellular automaton is a dynamical system which consisting in a juxtaposition of cells whose state changes over discrete time according to that of their neighbors. One of its simplest behaviors is nilpotency: all configurations of cells are mapped after a finite time into a given "null" configuration. Our main result is that nilpotency is equivalent to the condition that all configurations converge towards the null configuration for the Cantor topology, or, equivalently, that all cells of all configurations asymptotically reach a given state.

Keywords: cellular automata, nilpotency, dynamical systems, attractors.

1 Introduction

Discrete dynamical systems aim at representing evolutions of objects in astronomy, chemistry, cellular biology, zoology, computing networks... Evolutions of these objects can often be described by iterations of a continuous function. The sequence of values obtained is called orbit.

A long-standing issue in dynamical systems is the distinction between limit behavior and finitely-reached behavior: in which case does convergence of orbits imply reaching the limit in finite time? Of course, there are obvious examples in which all orbits converge towards the same limit without ever reaching it, such as the division by 2 on segment $[0, 1]$.

Here, we limit our study to some particular systems: *cellular automata* (CA). A CA consists in an infinite number of identical *cells* arranged on a regular lattice. All cells evolve synchronously according to their own state and those of their neighbors. It is thus a dynamical system on the set of *configurations* (which map each cell of the lattice to some state).

Endowing the set of configurations with the product topology allows the following restatement of the above-mentioned issue: in which case can local behavior (evolution of a particular cell) be uniformed into global behavior (evolution

---

[*] This work has been supported by the ANR Blanc "Projet Sycomore".

of the whole configuration)? This question can also be seen as a comparison between the limit set (configurations that can be reached arbitrarily late) and the ultimate set (adjacent values of orbits).

In this article, we study in detail the nilpotency of CA, in which all configurations eventually reach a given uniform configuration. Even though it represents the simplest behavior a dynamical system can have, its relevance grew in the CA field when Jarkko Kari made it the prototype of the undecidable problem on dynamics of CA [1], which has been widely used in reductions since then. This paper is dedicated to the equivalence between nilpotency and convergence toward a given configuration, or nilpotency of the trace subshift, which is a dynamical system linked to the CA, studied in [2,3], and which represents the observation of the evolution of a cell by a fixed observer. In other words, in that case, the local behavior can be uniformed.

The restriction to the case of CA with a spreading state is simpler and was already useful in some reductions (see for instance [3]). Here, we extend the result to all one-dimensional CA.

Section 2 is devoted to definitions and preliminary results. In Sect. 3, we recall some results about the nilpotency, and prove that uniform configurations are isolated in the limit set only if the CA is nilpotent. Sect. 4 is devoted to proving our new characterization of nilpotency and Sect. 5 gives some new tracks of generalization.

2   Preliminaries

Let $\mathbb{N}^* = \mathbb{N} \setminus \{0\}$. For $i, j \in \mathbb{N}$ with $i \leq j$, $[i, j]$ (resp. $]i, j[$) denotes the set of integers between i and j inclusive (resp. exclusive). For any function F from $A^{\mathbb{Z}}$ into itself, F^n denotes the n-fold composition of F with itself.

Words. Let A be a finite *alphabet* with at least two *letters*. A *word* is a finite sequence of letters $w = w_0 \ldots w_{|w|-1} \in A^*$, where $|w|$ is the *length* of w. A *factor* of a word $w = w_0 \ldots w_{|w|-1} \in A^*$ is a word $w_{[i,j]} = w_i \ldots w_j$, for $0 \leq i \leq j < |w|$.

2.1   Dynamical Systems

A (discrete) *dynamical system* (DS for short) is a couple (X, F), where X is a compact metric space, and F is a continuous function. When no confusion is possible (especially when $X = A^{\mathbb{Z}}$), X will be omitted. The *orbit* of initial point $x \in X$ is the sequence of points $F^j(x)$ when *generation* $j \in \mathbb{N}$ grows (where F^j denotes the j-fold composition of F with itself). We note $\mathcal{O}_F(x) = \left\{ F^j(x) \mid j \in \mathbb{N} \right\}$. We say that $Y \subset X$ is F-*stable* and that (Y, F) is a *subsystem* of (X, F) if $F(Y) \subset Y$.

Asymptotic behavior of orbits is represented by two particular sets:

- the (Ω-) *limit* set of a DS (X, F) is the set $\Omega_F = \bigcap_{j \in \mathbb{N}} F^j(X)$ of all configurations that can appear arbitrarily late in orbits.

– The *ultimate set* is the set $\omega_F = \bigcup_{x \in X} \bigcap_{J \in \mathbb{N}} \overline{\mathcal{O}_F(F^J(x))}$ of all adhering values of orbits.

One can notice that both are F-stable sets, and that, by definition, $\omega_F \subset \Omega_F$, whereas the converse is generally false. Moreover, compactness implies the non-emptiness of those two sets, and that the limit and ultimate sets of any DS F are equal to those of any of its iterated F^k for $k \in \mathbb{N}^*$.

2.2   Space $A^{\mathbb{Z}}$

Let A be a finite alphabet. The set $A^{\mathbb{Z}}$ (resp. $A^{\mathbb{N}}$) is the set of bi-infinite (resp. infinite) sequences over alphabet A. An element $x \in A^{\mathbb{Z}}$ is called *configuration* (see Fig. 1a). A *factor* of x is a word $x_{[i,j]} = x_i \ldots x_j$, for $i \leq j$. This notion of factor can be extended to infinite intervals (such as $] - \infty, j]$ or $[i, +\infty[$). The notation $x_{<i>}$ stands for the central factor $x_{[-i,i]}$.

Topology. $A^{\mathbb{Z}}$ is endowed with the *product* (or *Cantor*) topology, corresponding to the distance

$$d : \begin{array}{l} A^{\mathbb{Z}} \times A^{\mathbb{Z}} \to \mathbb{R}_+ \\ (x, y) \mapsto 2^{-\min_{x_i \neq y_i} |i|} \end{array} ,$$

which makes it compact, perfect, totally disconnected. A similar distance can be defined on $A^{\mathbb{N}}$. An example of two configuration at distance 2^{-2} is depicted in Fig. 1b.

(a) Configuration $c \in A^{\mathbb{Z}}$ and $c_{[-2,5]}$ (b) Two configurations at distance 2^{-i}

Fig. 1. Elements of $A^{\mathbb{Z}}$

Cylinders. For $j, k \in \mathbb{N}$ and a finite set W of words of length j, we note $[W]_k$ the set $\{x \in A^{\mathbb{Z}} | x_{[k,k+j[} \in W\}$. Such a set is called a *cylinder*. Cylinders form a countable base of clopen sets.

We note $[W]_k^C$ the complement of the cylinder $[W]_k$, $[W]$ the center cylinder $[W]_{\lfloor \frac{j}{2} \rfloor}$, if $K \subset A^{\mathbb{N}}$, $[W]_k K = \{x \in A^{\mathbb{Z}} | x_{[k,k+j[} \in W$ and $x_{[k+j,\infty[} \in K\}$, and, if $K \subset A^{-\mathbb{N}}$, $K[W]_k = \{x \in A^{\mathbb{Z}} | x_{[k,k+j[} \in W$ and $x_{]-\infty,k[} \in K\}$. Similarly to what is done for languages, we will assimilate a singleton with its unique element; for instance, $z[u]_i z'$ will denote the configuration $x \in A^{\mathbb{Z}}$ such that $x_{]-\infty,i[} = z$, $x_{[i,i+|u|[} = u$ and $x_{[i+|u|,\infty[} = z'$. Finally, note that notations such as $z[]z'$ will be used for the configuration $x \in A^{\mathbb{Z}}$ such that $x_{]-\infty,0[} = z$ and $x_{[0,\infty[} = z'$.

Finite configurations. If $q \in A$, then q^{ω} is the infinite word of $A^{\mathbb{N}}$ consisting in periodic repetitions of q, ${}^{\omega}q^{\omega}$ is the (spatially) *periodic* configuration of $A^{\mathbb{Z}}$ consisting in repetitions of q, ${}^{\omega}q^{\omega}$ is the *q-uniform* configuration, and any configuration ${}^{\omega}q[u]q^{\omega}$, where $u \in A^*$, is a *q-finite* configuration. A *q-semi-finite* configuration is a configuration $x = {}^{\omega}qz$ for some $z \in A^{\mathbb{N}} \backslash \{q^{\omega}\}$. Note that the set of q-finite configurations is dense, since any finite word u can be extended as ${}^{\omega}quq^{\omega} \in [u]$.

2.3 Cellular Automata

A (one-dimensional two-sided) *cellular automaton* (CA for short) is a discrete dynamical system consisting in cells distributed over the regular lattice $\mathbb{Z}$. Each *cell* $i \in \mathbb{Z}$ of the configuration $x \in A^{\mathbb{Z}}$ has a state x_i in the finite alphabet A. That state evolves according to the state of their neighbors: $F(x)_i = f(x_{[i-r,i+r[})$, where $f : A^{2r+1} \to A$ is the *local rule* and $r \in \mathbb{N}$ the *radius* of the CA. By abuse of notation, we assimilate the CA to its *global function* $F : A^{\mathbb{Z}} \to A^{\mathbb{Z}}$, which is a DS on the configurations space. Usually, an orbit is graphically represented by a two-dimensional *space-time diagram*, such as in Figure 2.

Example 1. A simple example of CA is the min CA, defined on alphabet $A = \{0, 1\}$ by radius 1 and local rule

$$f : \begin{array}{c} A^3 \to A \\ (x_{-1}, x_0, x_1) \mapsto \end{array} \left| \begin{array}{l} 1 \text{ if } x_{-1} = x_0 = x_1 = 1 \\ 0 \text{ otherwise} \end{array} \right. .$$

The typical evolution makes all finite bunches of 1s disappear progressively; only the infinite configuration ${}^{\omega}1^{\omega}$ will not lose any 1 and nearly all configurations tend towards ${}^{\omega}0^{\omega}$. The ultimate set is $\omega_F = \{{}^{\omega}0^{\omega}, {}^{\omega}1^{\omega}\}$ and the limit set Ω_F is the set of configurations that do not contain patterns of the form 10^k1 for any $k \in \mathbb{N}^*$.

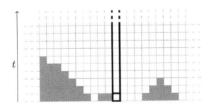

Fig. 2. A space-time diagram of the min CA with highlighted r-blocking word

Quiescent states. A state $0 \in A$ is said to be *quiescent* for the CA F with local rule f if $f(0, \ldots, 0) = 0$. Equivalently, $F({}^{\omega}0^{\omega}) = {}^{\omega}0^{\omega}$, which is called a *quiescent uniform configuration*. For instance, both 0 and 1 are quiescent states for the min CA.

We can see that the set of uniform configurations form a finite, hence ultimately periodic, subsystem of the cellular automaton; hence $\forall a \in A, \exists j < |A|, F^{|A|}({}^{\omega}a^{\omega}) = F^{|A|-j-1}({}^{\omega}a^{\omega})$, which gives the following remark:

Remark 1. For any CA F, F^j admits a quiescent state for some $j \in [1, |A|]$.

It can also easily be seen that the ultimate set of any CA contains all quiescent configurations. Since $\omega_{F^j} = \omega_F$, the previous remark gives, in particular, that for any CA F, there is at list one uniform configuration in ω_F (and therefore Ω_F).

Blocking words. A word $w \in A^*$ is (i,k)-*blocking* (or simply k-*blocking*) for CA $F : A^{\mathbb{Z}} \to A^{\mathbb{Z}}$ if $\forall x, y \in [w]_{-i}, \forall j \in \mathbb{N}, F^j(x)_{[0,k[} = F^j(y)_{[0,k[}$. If r is the radius of F, note that if a word w is (i,r)-blocking, and $x \in [w]_{-i}$, then for all configurations $y \in A^{\mathbb{Z}}$ such that $y_{[-i,\infty[} = x_{[-i,\infty[}$ (resp. $y_{]-\infty,|w|-i[} = x_{]-\infty,|w|-i[}$), and for all generation $j \in \mathbb{N}$, we have $F^j(y)_{[0,\infty[} = F^j(x)_{[0,\infty[}$ (resp. $F^j(y)_{]-\infty,r[} = F^j(x)_{]-\infty,r[}$), *i.e.* information cannot pass through an r-blocking word. For instance, in the min CA, the word 0 is $(0,1)$-blocking, since $\forall x \in [0], \forall j \in \mathbb{N}, F^j(x)_0 = 0$ (see Fig.2), hence space-time diagrams containing one 0 can be separated into two independent evolutions.

2.4   Symbolic Dynamics

The *shift map* $\sigma : A^{\mathbb{Z}} \to A^{\mathbb{Z}}$ is a particular CA defined by $\sigma(x)_i = x_{i+1}$ for every $x \in A^{\mathbb{Z}}$ and $i \in \mathbb{Z}$ which shifts configurations to the left. Hedlund [4] has shown that cellular automata are exactly the DS on $A^{\mathbb{Z}}$ which commute with the shift.

This characterization leads us to study shift-invariant subsets of $A^{\mathbb{Z}}$, among which, for instance, the image or limit sets of any CA. With the help of topology, we shall give some useful properties on those sets.

Subshifts. The *one-sided shift* (or simply shift when no confusion is possible), also noted σ by abuse of notation, is the self-map of $A^{\mathbb{N}}$ such that $\sigma(z)_i = z_{i+1}$, for every $z \in A^{\mathbb{N}}$ and $i \in \mathbb{N}$. A *one-sided subshift* $\Sigma \subset A^{\mathbb{N}}$ is a σ-stable closed set of infinite words. The following statements concern a two-sided version of subshifts.

Proposition 1. *The only closed shift-stable subset of $A^{\mathbb{Z}}$ of nonempty interior is $A^{\mathbb{Z}}$.*

Proof. Let Σ be such a set. Having nonempty interior, it must contain some cylinder $[u]_i$ with $u \in A^*$ and $i \in \mathbb{Z}$. Being closed and shift-stable, it must also contain $\overline{\bigcup_{j \leq i} \sigma^j([u]_i)} = A^{\mathbb{Z}}$.   □

Corollary 1. *If $A^{\mathbb{Z}} = \bigcup_{j \in \mathbb{N}} \Sigma_j$ where the Σ_j are closed and shift-stable, then $A^{\mathbb{Z}}$ is equal to some Σ_{j_0}.*

Proof. $A^{\mathbb{Z}}$ being complete, and of nonempty interior, the Baire Theorem states for some $j_0 \in \mathbb{N}$, Σ_{j_0} has nonempty interior too. By Proposition 1, $\Sigma_{j_0} = A^{\mathbb{Z}}$.   □

2.5   Trace

Rather than observing the whole configuration, trace consists in observing a specific portion of configuration during evolution of a cellular automaton. This notion establishes a link between the theory of cellular automata and symbolic dynamics.

Definition 1 (Trace). *Given a CA F, the* trace *of F with initial condition $x \in A^{\mathbb{Z}}$ is $T_F(x) = (F^j(x)_0)_{j \in \mathbb{N}}$. In other words, it is the central column of the space-time diagram of initial configuration x (see Figure 3). The* trace subshift *of F is the image set $\tau_F = T_F(A^{\mathbb{Z}})$. Similarly, we define $T_F^k(x) = (F^j(x)_{[0,k[})_{j \in \mathbb{N}}$, the sequence of the words at positions in $[0, k[$ in the space-time diagram of initial configuration x, and $\tau_F^k = T_F^k(A^{\mathbb{Z}})$.*

For instance, if $A = \{0,1\}$, then the trace subshift is $(0+1)^\omega$ for the shift, $0^\omega + 1^\omega$ for the identity and $(0 + 1)0^\omega$ for the CA that maps every cell to 0 (see [2] for more examples).

Note that $T_F F = \sigma T_F$ and, as T_F is continuous, we say it is a *factorization* between the CA and the subshift (τ_F, σ) which means that their two dynamics are very closely related. For more about factorizations, see [5]. In the following, τ_F and τ_F^k may stand for the dynamical systems (τ_F, σ) and (τ_F^k, σ).

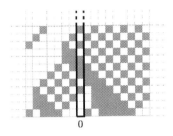

0

Fig. 3. Trace seen on the space-time diagram

In this paper, we try to obtain links between the set of possible traces and the global behavior of the cellular automaton. That is try to "globalize" a local observation of the behavior.

3   Nilpotent Cellular Automata

In this section, we recall the definition and characterizations of cellular automata nilpotency, and its known link with the limit set. We present a short proof emphasizing on the fact that a CA is nilpotent as soon as some uniform configuration is isolated in its limit set.

Definition 2. *A DS (X, F) is* weakly nilpotent *if there is a particular "null" configuration $z \in X$ such that for every point $x \in X$, there is a generation $J \in \mathbb{N}$ such that $\forall j > J, F^j(x) = z$. The system is said* nilpotent *if the generation J does not depend on the point, i.e. $\exists J \in \mathbb{N}, \forall x \in X, \forall j > J, F^j(x) = z$.*

Let us now consider the specific case of cellular automata. If a CA F is weakly nilpotent, then, from the definition, the null configuration must be F-invariant. Moreover, for any $z \in A^{\mathbb{Z}}$, there is some j for which, on the one hand $F^j(\sigma(z)) = z$, and on the other hand, $F^j(\sigma(z)) = \sigma(F^j(z)) = \sigma(z)$. Hence $z = \sigma(z)$ is a *uniform quiescent configuration*, i.e. $z = {}^\omega 0^\omega$ for some quiescent state 0. One can also make the following easy remarks:

Remark 2

- Suppose F is a CA and $j \in \mathbb{N}$ a generation such that for all configurations $x \in A^{\mathbb{Z}}$, $F^j(x)_0 = 0$. Then shift-invariance of $A^{\mathbb{Z}}$ gives $A^{\mathbb{Z}} = \bigcap_{k \in \mathbb{Z}} \sigma^k F^{-j}([0])$, and by commuting F and the shift, we obtain $A^{\mathbb{Z}} = F^{-j}({}^\omega 0^\omega)$. Hence F is nilpotent.
- On the other hand, if F is not nilpotent, then $F^{-j}([0^C])$ is a nonempty open set; in particular, it contains some q-finite configurations, for any state $q \in A$.
- Note that if F is a non-nilpotent CA, from Remark 1, there is some $J \in \mathbb{N}$ such that some $0 \in A$ is quiescent for F^J. We can see that the set of 0-finite configurations is F^J-stable. Hence, from the previous point, for all $j \in \mathbb{N}$, $F^j(A^{\mathbb{Z}})$ contains some non-uniform 0-finite configuration $z = {}^\omega 0[u]0^\omega$ with $u \in A^*$ and $z_0 \neq 0$.

Here is a precise characterization of the limit set of nilpotent CA.

Theorem 1 (Čulik, Pachl, Yu [6]). *A CA whose limit set is a singleton is nilpotent.*

Proof. If F is a CA such that $\Omega_F \cap [0^C] = \emptyset$, the intersection $\bigcap_{j \in \mathbb{N}}(F^j(A^{\mathbb{Z}}) \cap [0^C])$ of closed sets is empty. By compactness, $F^j(A^{\mathbb{Z}}) \cap [0^C]$ is empty for some $j \in \mathbb{N}$. $F^j(A^{\mathbb{Z}})$ being a subshift, we conclude that $F^j(A^{\mathbb{Z}}) \cap (\bigcup_{i \in \mathbb{Z}}[0^C]_i) = F^j(A^{\mathbb{Z}}) \setminus {}^\omega 0^\omega$ is empty. □

With this characterization and the fact that $\Omega_F = \Omega_{F^j}$, we have the following corollary.

Corollary 2. *A CA F is nilpotent if and only if so is some (so are all) of its iterated CA F^n, for $n \in \mathbb{N}$.* □

The next allows to prove the well-known equivalence between nilpotency and weak nilpotency.

Proposition 2. *If for every configuration $x \in A^{\mathbb{Z}}$, there is a generation $j \in \mathbb{N}$ and a cell $i \in \mathbb{Z}$ such that $F^j(x)_{[i,\infty[} = 0^\omega$ (resp. $F^j(x)_{]-\infty,i]} = {}^\omega 0$), then F is nilpotent.*

Proof. Assume $A^{\mathbb{Z}} = \bigcup_{j \in \mathbb{N}, i \in \mathbb{Z}} \sigma^i(F^{-j}({}^\omega A[0^\omega]))$ (the other case is symmetric). From Corollary 1, $A^{\mathbb{Z}} = \sigma^i(F^{-j}({}^\omega A[0^\omega]))$ for some generation $j \in \mathbb{N}$ and some $i \in \mathbb{Z}$. In particular, $\sigma^i(A^{\mathbb{Z}}) = F^{-j}([0])$, and Remark 2 allows to conclude the claim. □

Corollary 3. *A weakly nilpotent CA is nilpotent.* □

There is indeed some "gap" between the possible evolutions of nilpotent and non-nilpotent CA. This is also stressed by the following result on limit set: it is sufficient that the it admits an isolated uniform configuration for the CA to be nilpotent.

Proposition 3. *If F is a non-nilpotent CA, then Ω_F contains some semi-finite configuration $^\omega 0[]z$, for some $0 \in A$ and some $z \in A^{\mathbb{N}} \backslash \{0^\omega\}$.*

Proof. Let F be a non-nilpotent CA. Should we iterate it, we can suppose it has a quiescent state $0 \in A$, thanks to Remark 1 and Corollary 2. By Remark 2, for every generation $j \in \mathbb{N}$, $F^j(A^{\mathbb{Z}}) \cap {}^\omega 0[0^C] \neq \emptyset$, and compactness gives that $\Omega_F \cap {}^\omega 0[0^C] \neq \emptyset$.   □

Theorem 2. *The limit set of any non-nilpotent CA does not have any isolated uniform configurations.*

Proof. Let F a non-nilpotent CA, $0 \in A$, $k \in \mathbb{N}$ and $x \in {}^\omega 0[] \cap \Omega_F \backslash \{{}^\omega 0^\omega\}$ as in Proposition 3. Then, for any $k \in \mathbb{N}$, $\sigma^{-k}(x)$ is a configuration of Ω_F distinct from $^\omega 0^\omega$, and which is in the cylinder $[0^{2k+1}]$ around $^\omega 0^\omega$.   □

Remark that, on the contrary, ω_F can have isolated uniform configurations even in the non-nilpotent case. For instance, the ultimate set of the min CA is $\{{}^\omega 0^\omega, {}^\omega 1^\omega\}$.

The previous result allows state that limit sets are either a singleton or infinite.

Corollary 4 (Čulik, Pachl, Yu [6]). *The limit set of any non-nilpotent CA is infinite.*   □

Note that the limit set can be numerable, as for the min CA, or innumerable, as for the shift CA.

To sum up, the limit set of a CA has a very precise structure, and, as will be seen more deeply in the following section, constraining a little can make it collapse to a singleton, *i.e.* make the CA nilpotent.

4   Nilpotent Traces

We have seen in the previous section that there is a "gap" between nilpotent and non-nilpotent CA in terms of the limit set. The aim of this section is to prove that it is also the case for the ultimate set, or, equivalently, for the trace subshift: a constraint such as the ultimate set being a singleton is restrictive enough to imply nilpotency. We first give a directly proved characterization of that notion for trace subshifts.

Remark 3. Similarly to the case of CA, we can notice that a subshift $\Sigma \in A^{\mathbb{N}}$ is weakly nilpotent if and only if there is some state $0 \in A$ such that for every infinite word $w \in \Sigma$, there is some $j \in \mathbb{N}$ such that $w_{[j,\infty[} = 0^\omega$.

The previous remark will justify the concept of 0-*nilpotency*, where 0 is a quiescent state. We also say that some configuration x is 0-*mortal* for the CA F if there exists $j \in \mathbb{N}$ such that $F^j(x) = 0^\omega$.

Proposition 4. *Let F a CA. The following statements are equivalent:*

1. τ_F *is weakly 0-nilpotent.*
2. τ_F^k *is weakly 0^k-nilpotent, for any integer $k \in \mathbb{N}^*$.*
3. *For every configuration $x \in A^{\mathbb{Z}}$, $F^n(x)$ tends, as $n \to \infty$, to the same "null" configuration $^\omega 0^\omega$.*
4. $\omega_F = \{^\omega 0^\omega\}$.   □

Let us prove now that, in the same way as weakly nilpotent CA are nilpotent, weakly nilpotent traces are nilpotent, *i.e.* the corresponding CA is nilpotent (at least in dimension 1).

The proof is organized the following way: we first prove that a CA with weakly nilpotent trace has an r-blocking word, and then that there exists a bound on the converging time for mortal finite configurations. At last, we exhibit a contradiction with a non-nilpotency hypothesis.

Lemma 1. *Let F a CA whose trace is weakly 0-nilpotent; then for every $k \in \mathbb{N}$, there is a generation $J \in \mathbb{N}$ such that for every configuration $x \in A^{\mathbb{Z}}$, $\exists j \le J, F^j(x) \in [0^k]$.*

Proof. Let F a CA such that τ_F is weakly nilpotent and $k \in \mathbb{N}$. By Point 2 of Proposition 4 and Remark 3, we have $A^{\mathbb{Z}} = \bigcup_{j \in \mathbb{N}} F^{-j}([0^k])$. By compactness, we can extract a finite covering $A^{\mathbb{Z}} = \bigcup_{j \le J} F^{-j}([0^k])$ for some $J \in \mathbb{N}$.   □

Lemma 2. *If F is a CA with a nilpotent trace, then it admits an r-blocking word.*

Proof. Let F such a CA. By Point 2 of Proposition 4, $A^{\mathbb{Z}} = \bigcup_{J \in \mathbb{N}} \bigcap_{j>J} F^{-j}([0^r])$ has nonempty interior, for some $0 \in A$; hence the Baire Theorem states that there is some $J \in \mathbb{N}$ for which $\bigcap_{j>J} F^{-j}([0^r])$ has nonempty interior; it contains a cylinder $[u]_{-i}$ for some $u \in A^*$ and $i \in \mathbb{N}$. We can assume without loss of generality that $i \ge Jr$ and $|u| \ge (J+1)r + i$, which gives $\forall x, y \in [u]_{-i}, \forall j \le J, F^j(x)_{[0,r[} = F^j(y)_{[0,r[}$, and we already have by construction that $\forall x, y \in [u]_{-i}, \forall j > J, F^j(x)_{[0,r[} = 0^r = F^j(y)_{[0,r[}$. Hence, u is an (i,r)-blocking word.   □

Lemma 3. *Let F a CA whose trace is weakly 0-nilpotent; then there is a generation j for which any 0-finite 0-mortal configuration x satisfies $F^j(x) = {}^\omega 0^\omega$.*

Proof. Consider F a CA such that τ_F is weakly nilpotent, and for all $j \in \mathbb{N}$, there is a finite mortal configuration y such that $F^j(y) \ne {}^\omega 0^\omega$. Should it be shifted, we can suppose $F^j(y)_0 \ne 0$.

Claim. For every $k \in \mathbb{N}$, there is a 0-finite 0-mortal configuration $y \in [0^{2k+1}]$ and a generation $j \ge k$ such that $F^j(y)_0 \ne 0$.

Proof. Let $k \in \mathbb{N}$. From Lemma 1, $\exists J \in \mathbb{N}, \forall y \in A^{\mathbb{Z}}, \exists j \leq J, F^j(y) \in [0^{2k+1}]$. By hypothesis, there exists a finite mortal configuration x for which $F^J(x)_0 \neq 0$. Hence, $\exists j \leq J, F^j(x) \in [0^{2k+1}]$. Just take the configuration $y = F^j(x)$; we have $F^{J-j}(y)_0 = F^J(x)_0 \neq 0$. Of course, $\forall j < k, F^j(y)_0 = 0$.   □

Claim. Let x a 0-finite 0-mortal configuration and $k \in \mathbb{N}$. Then, there is a 0-finite 0-mortal configuration $y \in [x_{<rk>}]$ and

$$\left\{ j \in \mathbb{N} \middle| F^j(y)_0 \neq 0 \right\} \supsetneq \left\{ j \in \mathbb{N} \middle| F^j(x)_0 \neq 0 \right\} .$$

Proof. Let x a finite configuration and $n \in \mathbb{N}$ such that $F^n(x) = {}^\omega 0^\omega$. Should we take a larger k, we can assume that $x \in {}^\omega 0[A^{2r(k-2n)}]0^\omega$. In particular, note that $k \geq 2n$. By the previous claim, there exists a finite mortal configuration $x' \in [0^{2k+1}]$ and a generation $j \geq k$ such that $F^j(x)_0 \neq 0$. Let $y = x'_{]-\infty,-rk[}[x_{<rk>}]x'_{]rk,\infty[}$. We can easily see by induction on $j \leq n$ that $F^j(y) \in F^j(x')_{]-\infty,-r(k-2n+j)[}[A^{2rj}F^j(x)_{<r(k-j)>}A^{2rj}]F^j(x')_{]r(k-2n+j),\infty[}$. In particular, on the one hand, $\left\{ j \in \mathbb{N} \middle| F^j(y)_0 \neq 0 \right\} \cap [0,n] = \left\{ j \in \mathbb{N} \middle| F^j(x)_0 \neq 0 \right\} \cap [0,n] = \left\{ j \in \mathbb{N} \middle| F^j(x)_0 \neq 0 \right\}$. On the other hand, since $F^n(x)_{<r(k-n)>} = 0^{2r(k-n)} = F^n(x')_{<r(k-n)>}$, we get $F^n(y) = F^n(x')$. By construction, there is a generation $j \geq k > n$ such that $F^j(y)_0 = F^j(x')_0 \neq 0$. To sum up, $\left\{ j \in \mathbb{N} \middle| F^j(y)_0 \neq 0 \right\} \supsetneq \left\{ j \in \mathbb{N} \middle| F^j(x)_0 \neq 0 \right\}$.   □

We can now build by induction a sequence $(y^k)_{k \in \mathbb{N}}$ of finite mortal configurations, with $x^0 = {}^\omega 0^\omega$, and such that for $k \in \mathbb{N}$, $x^{k+1} \in [x^k_{<r(k+1)>}]$ and $\left\{ j \in \mathbb{N} \middle| F^j(x^{k+1})_0 \neq 0 \right\} \supsetneq \left\{ j \in \mathbb{N} \middle| F^j(x^k)_0 \neq 0 \right\}$. That Cauchy-Bolzano sequence of finite mortal configurations tends to some configuration $x \in A^{\mathbb{Z}}$ such that $\left\{ j \in \mathbb{N} \middle| F^j(x)_0 \neq 0 \right\}$ is infinite (by continuity of the trace), *i.e.* τ_F is not nilpotent.   □

Lemma 4. *Let F a one-dimensional non-nilpotent CA admitting a weakly 0-nilpotent trace and an r-blocking word. Then for every generation $j \in \mathbb{N}$, there exists some 0-finite 0-mortal configuration y such that $F^j(y) \neq {}^\omega 0^\omega$.*

Proof. Let F such a CA, u a (i,r)-blocking word and $j \in \mathbb{N}$. By blockingness and Point 2 of Proposition 4, there is a generation $k \in \mathbb{N}$ such that $\forall n \geq k, \forall z \in [u]_{-i}, F^n(z) \in [0^r]_0$. Should we take a strict superword for u, we can assume $|u| = 2i + 1$. Remark 2 states that there is some finite configuration x such that $F^{j+k}(x)_0 \neq 0$. Consider the configuration $x' = {}^\omega 0[ux_{<r(j+k)>}u]0^\omega$, and the configuration $y = {}^\omega 0[F^k(x')_{<r(j+k)+i>}]0^\omega$. By construction, $F^j(y)_0 = F^{j+k}(x)_0 \neq 0$. By induction on the generation $n \in \mathbb{N}$, $F^{n+k}(x') \in [0^r A^{2r(j+k)+2i+1}0^r]$, and thus $F^n(y) \in {}^\omega 0[A^{2r(j+k)+2i+1}]0^\omega$. By hypothesis and Point 2 of Proposition 4, there is some generation $l \in \mathbb{N}$ such that $F^l(y) \in [0^{2r(j+k)+2i+1}]$, which gives $F^l(y) = {}^\omega 0^\omega$.   □

Theorem 3. *Any one-dimensional CA whose trace is weakly nilpotent is nilpotent.*

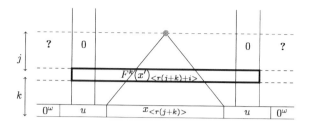

Fig. 4. Construction of mortal finite configurations with arbitrary slow convergence

Proof. Suppose there is a non-nilpotent CA F such that τ_F is nilpotent. Then, by Lemma 2, we know that there exists an r-blocking word. Lemmas 4 and 3 give a contradiction. □

Corollary 5. *The subshift $(\varepsilon + 0^*1)0^\omega$ is not the trace of any one-dimensional CA.* □

5 Conclusion

To sum up, for a one-dimensional CA F, the following statements are equivalent:

F is nilpotent $\Leftrightarrow F$ is weakly nilpotent
$\Leftrightarrow \tau_F^k$ is nilpotent $\Leftrightarrow \tau_F^k$ is weakly nilpotent
$\Leftrightarrow \Omega_F$ is finite $\Leftrightarrow \Omega_F$ has some isolated uniform configuration
$\Leftrightarrow \Omega_F$ is a singleton $\Leftrightarrow \omega_F$ is a singleton

We can remark that the nilpotency of a CA associated to a local rule does not depend on the anchor. Hence, all characterizations work up to a shift of the rule.

The main open question is of course the generalization of these result to CA in upper dimensions (configurations in $A^{\mathbb{Z}^d}$, $d > 1$). Juxtaposing blocking words (Lemma 4) is the crucial part which is difficult to transpose in dimension 2.

Another one is whether it is sufficient to take as a hypothesis that all orbits have $^\omega 0^\omega$ as an adhering value, instead of a limit - *i.e.* all space-time diagrams walk through cylinders $[0^k]$ for any $k \in \mathbb{N}$ - to conclude the CA is nilpotent.

A natural question is whether the characterization of nilpotency in terms of trace can be generalized, for instance to preperiodicity (*i.e.* ultimate periodicity), in the same way as the finite-time characterization of nilpotency (Corollary 3; see for instance [5]). This is untrue; some CA F may have all the infinite words of its trace preperiodic (*i.e.* $\forall z \in \tau_F, \exists p, q \in \mathbb{N}, \sigma^{p+q}(z) = \sigma^q(z)$), but with an unbounded preperiod, such as the min CA, whose trace subshift $1^\omega + 1^*0^\omega$ has only preperiodic words, but with an arbitrarily high preperiod.

Note that our result can be seen as a constraint on the structure of limit sets of non-nilpotent CA. In that idea, Proposition 3 allows to state that a CA is nilpotent as soon as its limit set contains only periodic configurations. Note that this is untrue for ω_F: for instance, the min CA has only spatially periodic configurations in its ultimate set. We can also mention another misleadingly simple issue, which concerns the presence of infinite configurations.

Conjecture 1. If F is a CA and $0 \in A$ such that Ω_F contains only 0-finite configurations, then F is 0-nilpotent.

Acknowledgement

We gratefully thank Guillaume Theyssier, François Blanchard, Jarkko Kari for our very profitable discussions on that matter. Thanks also to our referees, who criticized constructively a rather unrefined version.

References

1. Kari, J.: The nilpotency problem of one-dimensional cellular automata. SIAM Journal on Computing 21, 571–586 (1992)
2. Cervelle, J., Formenti, E., Guillon, P.: Sofic trace of a cellular automaton. In: Cooper, S.B., Löwe, B., Sorbi, A. (eds.) CiE 2007. LNCS, vol. 4497, pp. 152–161. Springer, Heidelberg (2007)
3. Cervelle, J., Guillon, P.: Towards a Rice theorem on traces of cellular automata. In: Kučera, L., Kučera, A. (eds.) MFCS 2007. LNCS, vol. 4708, pp. 310–319. Springer, Heidelberg (2007)
4. Hedlund, G.A.: Endomorphisms and automorphisms of the shift dynamical system. Mathematical Systems Theory 3, 320–375 (1969)
5. Kůrka, P.: Topological and symbolic dynamics. Société Mathématique de France (2003)
6. Čulik II, K., Pachl, J., Yu, S.: On the limit sets of cellular automata. SIAM Journal on Computing 18, 831–842 (1989)

A Note on k-Colorability of P_5-Free Graphs

Chính T. Hoàng[1,*], Marcin Kamiński[2,**], Vadim Lozin[3,***],
Joe Sawada[4,†], and Xiao Shu[4]

[1] Physics and Computer Science, Wilfrid Laurier University, Canada
[2] Department of Computer Science, Université Libre de Bruxelles, Belgium
Marcin.Kaminski@ulb.ac.be
[3] DIMAP and Mathematics Institute, University of Warwick, United Kingdom
[4] Computing and Information Science, University of Guelph, Canada

Abstract. We present a polynomial-time algorithm determining whether or not, for a fixed k, a P_5-free graph can be k-colored. If such a coloring exists, the algorithm will produce one.

Keywords: P_5-free graphs, graph coloring, dominating clique.

1 Introduction

Graph coloring is among the most important and applicable graph problems. The *k-colorability problem* is the question of whether or not the vertices of a graph can be colored with one of k colors so that no two adjacent vertices are assigned the same color. In general, the k-colorability problem is NP-complete [10]. Even for planar graphs with no vertex degree exceeding 4, the problem is NP-complete [5]. However, for other classes of graphs, like perfect graphs [8], the problem is polynomial-time solvable. For the following special class of perfect graphs, there are efficient polynomial time algorithms for finding optimal colorings: chordal graphs [6], weakly chordal graphs [9], and comparability graphs [4]. For more information on perfect graphs, see [1], [3], and [7].

Another interesting class of graphs are those that are P_t-free, that is, graphs with no chordless paths $v_1, v_2, \ldots, v_t$ of length $t-1$ as induced subgraph for some fixed t. If $t = 3$ or $t = 4$, then there exists efficient algorithms to answer the k-colorability question (see [3]). However, it is known that CHROMATIC NUMBER for P_5-free graphs is NP-complete [11]. Thus, it is of some interest to consider the problem of k-coloring a P_t-free graph for some fixed $k \geq 3$ and $t \geq 5$. Taking this parameterization into account, a snapshot of the known complexities for the k-colorability problem of P_t-free graphs is given in Table 1. From this chart we can see that there is a polynomial algorithm for the 3-colorability of P_6-free graphs [13].

In this paper we focus on P_5-free graphs. Notice that when $k = 3$, the colorability question for P_5-free graphs can be answered in polynomial time (see [14]). The authors

* Research supported by NSERC.
** Corresponding author.
*** Supported by DIMAP – Centre for Discrete Mathematics and its Applications at the University of Warwick.
† Research supported by NSERC.

E. Ochmański and J. Tyszkiewicz (Eds.): MFCS 2008, LNCS 5162, pp. 387–394, 2008.
© Springer-Verlag Berlin Heidelberg 2008

Table 1. Known complexities for k-colorability of P_t-free graphs

$k\backslash t$	3	4	5	6	7	8	$\ldots$	12	$\ldots$
3	$O(m)$	$O(m)$	$O(n^\alpha)$	$O(mn^\alpha)$	?	?	?	?	$\ldots$
4	$O(m)$	$O(m)$	??		?	?	?	NP_c	$\ldots$
5	$O(m)$	$O(m)$	??	?	?	NP_c	NP_c	NP_c	$\ldots$
6	$O(m)$	$O(m)$	??	?	?	NP_c	NP_c	NP_c	$\ldots$
7	$O(m)$	$O(m)$	??	?	?	NP_c	NP_c	NP_c	$\ldots$
$\ldots$	$\ldots$	$\ldots$	$\ldots$	$\ldots$	$\ldots$	$\ldots$	$\ldots$	$\ldots$	$\ldots$

of [12] proved that the 4-colorability of (P_5, C_5)-free graphs can be decided in polynomial time. We obtain a theorem (Theorem 2) on the structure of P_5-free graphs and use it to design a polynomial-time algorithm that determines whether a P_5-free graph can be k-colored. If such a coloring exists, then the algorithm will yield a valid k-coloring.

The remainder of the paper is presented as follows. In Section 2 we present relevant definitions, concepts, and notations. Then in Section 3, we present our recursive polynomial-time algorithm that answers the k-colorability question for P_5-free graphs.

2   Background and Definitions

In this section we provide the necessary background and definitions used in the rest of the paper. For starters, we assume that $G = (V, E)$ is a simple undirected graph where $|V| = n$ and $|E| = m$. If A is a subset of V, then we let $G(A)$ denote the subgraph of G induced by A.

Definition 1. *A set of vertices A is said to* dominate *another set B, if every vertex in B is adjacent to at least one vertex in A.*

The following structural result about P_5-free graphs is from Bacsó and Tuza [2]:

Theorem 1. *Every connected P_5-free graph has either a dominating clique or a dominating P_3.*

Definition 2. *Given a graph G, an integer k and for each vertex v, a list $l(v)$ of k colors, the k-list coloring problem asks whether or not there is a coloring of the vertices of G such that each vertex receives a color from its list.*

Definition 3. *The* restricted k-list coloring problem *is the k-list coloring problem in which the lists $l(v)$ of colors are subsets of $\{1, 2, \ldots, k\}$.*

Our general approach is to take an instance of a specific coloring problem Φ for a given graph and replace it with a polynomial number of instances $\phi_1, \phi_2, \phi_3, \ldots$ such that the answer to Φ is "yes" if and only if there is some instance ϕ_k that also answers "yes".

For example, consider a graph with a dominating vertex u where each vertex has color list $\{1, 2, 3, 4, 5\}$. This listing corresponds to our initial instance Φ. Now, by considering different ways to color u, the following set of four instances will be equivalent to Φ:

ϕ_1: $l(u) = \{1\}$ and the remaining vertices have color lists $\{2,3,4,5\}$,
ϕ_2: $l(u) = \{2\}$ and the remaining vertices have color lists $\{1,3,4,5\}$,
ϕ_3: $l(u) = \{3\}$ and the remaining vertices have color lists $\{1,2,4,5\}$,
ϕ_4: $l(u) = \{4,5\}$ and the remaining vertices have color lists $\{1,2,3,4,5\}$.

In general, if we recursively apply such an approach we would end up with an exponential number of equivalent coloring instances to Φ.

3 The Algorithm

Let G be a connected P_5-free graph. This section describes a polynomial time algorithm that decides whether or not G is k-colorable. The algorithm is outlined in 3 steps. Step 2 requires some extra structural analysis and is presented in more detail in the following subsection.

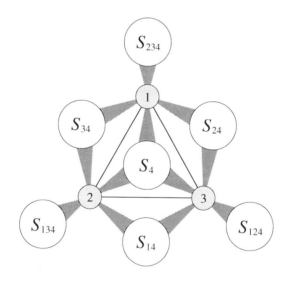

Fig. 1. The fixed sets in a P_5-free graph with a dominating K_3 where $k = 4$

1. Identify and color a maximal dominating clique or a P_3 if no such clique exists (Theorem 1). This partitions the vertices into **fixed sets** indexed by available colors. For example, if a P_5-free graph has a dominating K_3 (and no dominating K_4) colored with $\{1,2,3\}$ and $k = 4$, then the fixed sets would be given by: $S_{124}, S_{134}, S_{234}, S_{14}, S_{24}, S_{34}$. For an illustration, see Figure 1. Note that all the vertices in S_{124} are adjacent to the vertex colored 3 and thus have color lists $\{1,2,4\}$. This gives rise to our original restricted list-coloring instance Φ. Although the illustration in Figure 1 does not show it, it is possible for there to be edges between any two fixed sets.

2. Two vertices are *dependent* if there is an edge between them and the intersection of their color lists is non-empty. In this step, we remove all dependencies between each pair of fixed sets. This process, detailed in the following subsection, will create a polynomial number of coloring instances $\{\phi_1, \phi_2, \phi_3, \ldots\}$ equivalent to Φ.

3. For each instance ϕ_i from Step 2 the dependencies between each pair of fixed sets has been removed which means that the vertices within each fixed set can be colored independently. Thus, for each instance ϕ_i we recursively see if each fixed set can be colored with the corresponding restricted color lists (the base case is when the color lists are a single color). If *one* such instance provides a valid k-coloring then return the coloring. Otherwise, the graph is not k-colorable.

As mentioned, the difficult part is reducing the dependencies between each pair of fixed sets (Step 2).

3.1 Removing the Dependencies between Two Fixed Sets

Let S_{list} denote a fixed set of vertices with color list given by $list$. We partition each such fixed set into **dynamic sets** that each represent a unique subset of the colors in $list$. For example: $S_{123} = P_{123} \cup P_{12} \cup P_{13} \cup P_{23} \cup P_1 \cup P_2 \cup P_3$. Initially, $S_{123} = P_{123}$ and the remaining sets in the partition are empty. However, as we start removing dependencies, these sets will dynamically change. For example, if a vertex u is initially in P_{123} and one of its neighbors gets colored 2, then u will be removed from P_{123} and added to P_{13}.

Recall that our goal is to remove the dependencies between two fixed sets S_p and S_q. To do this, we remove the dependencies between each pair (P, Q) where P is a dynamic subset of S_p and Q is a dynamic subset of S_q. Let $col(P)$ and $col(Q)$ denote the color lists for the vertices in P and Q respectively. By visiting these pairs in order from largest to smallest with respect to $|col(P)|$ and then $|col(Q)|$, we ensure that we only need to consider each pair once. Applying this approach, the crux of the reduction process is to remove the dependencies between a pair (P, Q) by creating at most a polynomial number of equivalent colorings.

Now, observe that there exists a vertex v from the dominating set found in Step 1 of the algorithm that dominates every vertex in one set, but is not adjacent to any vertex in the other. This is because P and Q are subsets of different fixed sets. WLOG assume that v dominates Q.

Theorem 2. *Let H be a P_5-free graph partitioned into three sets P, Q and $\{v\}$ where v is adjacent to every vertex in Q but not adjacent to any vertex in P. If we let Q' denote all components of $H(Q)$ that are adjacent to some vertex in P then one of the following must hold.*

1. *There exists exactly one special component C in $G(P)$ that contains two vertices a and b such that a is adjacent to some component $Y_1 \in G(Q)$ but not adjacent to another component $Y_2 \in G(Q)$ while b is adjacent to Y_2 but not Y_1.*
2. *There is a vertex $x \in P$ that dominates every component in Q', except at most one (call it T).*

PROOF: Suppose that there are two unique components $X_1, X_2 \in G(P)$ with $a, b \in X_1$ and $c, d \in X_2$ and components $Y_1 \neq Y_2$ and $Y_3 \neq Y_4$ from $G(Q)$ such that:

- a is adjacent to Y_1 but not adjacent to Y_2,
- b is adjacent to Y_2 but not adjacent to Y_1,
- c is adjacent to Y_3 but not adjacent to Y_4,
- d is adjacent to Y_4 but not adjacent to Y_3.

Let y_1 (respectively, y_2, y_3, y_4) be a vertex in Y_1 (respectively, Y_2, Y_3, Y_4) that is adjacent to a (respectively, b, c, d) and not b (respectively, a, d, c). Since H is P_5-free, there must be edges (a, b) and (c, d), otherwise a, y_1, v, y_2, b or c, y_3, v, y_4, d would be P_5s. An illustration of these vertices and components is given in Figure 2.

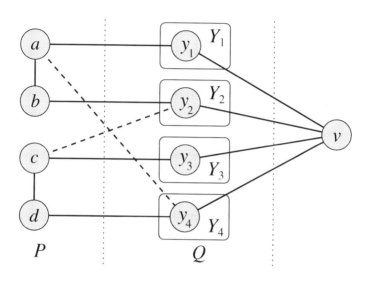

Fig. 2. Illustration for proof of Theorem 2

Suppose $Y_2 = Y_3$. Then b is not adjacent to y_3, for otherwise there exists a P_5 induced by vertices a, b, y_3, c, d. Now, there exists a P_5 y_1, a, b, y_2, y_3 (if y_2 is adjacent to y_3) or a P_5 a, b, y_2, v, y_3 (if y_2 is not adjacent to y_3). Thus, Y_2 and Y_3 must be unique components. Similarly, we have $Y_2 \neq Y_4$. Now since b, y_2, v, y_3, c cannot be a P_5, either b is adjacent to y_3 or c must be adjacent to y_2. WLOG, suppose the latter. Now a, b, y_2, v, y_4 implies that either a or b is adjacent to y_4. If y_4 is adjacent to b but not a, then a, b, y_4, d, c would be a P_5 which implies that a must be adjacent to y_4 anyway. Thus, we end up with a P_5 a, y_4, v, y_2, c which is a contradiction to the graph being P_5-free. Thus there must be at most one special component C.

Now suppose that there is no special component C. Let Q' denote all components in Q that are adjacent to some vertex in P. Let x be a vertex in P that is adjacent to the largest number of components in Q'. Suppose that x is not adjacent to a component T of Q'. Thus there is some other vertex $x' \in P$ adjacent to T. The maximality of x implies there is a component S of Q such that x is adjacent to S but x' is not. If x is not adjacent to x', then there is a P_5 with x, s, v, r, x' with some vertices $s \in S, r \in T$.

Thus x and x' belongs to a special component C of P - a contradiction. Thus, x must be adjacent to all components of Q'.

If there are two components A, B of Q' that are not dominated by x, then there are adjacent vertices $a, b \in A$, adjacent vertices $c, d \in B$ such that x is adjacent to a, c but not to b, d; but now the five vertices b, a, x, c, d form a P_5. □

Given a list-coloring instance ϕ of a P_5-free graph, we will at some points need to reduce the color lists for a given connected component C. This can be done by considering all possible ways to color C's dominating clique or P_3 (Theorem 1). Since there are a constant number of vertices in such a dominating set, we obtain a constant number of new instances that together are equivalent to ϕ. For future reference, we call this function that returns this set of equivalent instances ReduceComponent(C, ϕ). If C is empty, the function simply returns ϕ.

Procedure RemoveDependencies(P', Q, φ)

    if no dependencies between P' and Q
        then output φ
        else find x, T from Theorem 2
            for each $c \in col(P) \cap col(Q)$ **do**
                output ReduceComponent(T, φ with x colored c)
            RemoveDependencies($P'-\{x\}, Q, \varphi$ with $l(x) = col(P)-col(Q)$)

Fig. 3. Algorithm to remove dependencies between two dynamic sets P' and Q (with no special component C) by creating an equivalent set of coloring instances with the dependences removed

Using this procedure along with Theorem 2, we can remove the dependencies between two dynamic sets P and Q for a given list-coloring instance ϕ. First, we find the special component C if it exists, and set $C = \emptyset$ otherwise. Then we call ReduceComponent(C, ϕ) which will effectively remove all vertices in C from P as their color lists change. Then, for each resulting coloring instance φ we remove the remaining dependencies between $P' = P - C$ and Q by applying procedure RemoveDependencies(P', Q, φ) shown in Figure 3. In this procedure we find a vertex x and component T from Theorem 2, since we know that the special component C has already been handled. If T does not exist, then we set $T = \emptyset$. Now by considering each color in $col(P') \cap col(Q)$ along with the list $col(P') - col(Q)$ we can create a set of equivalent instances to φ (as described in Section 2). If we modify φ by assigning x a color from $col(P') \cap col(Q)$, then all vertices in Q adjacent to x will have their color list reduced by the color of x. Thus, only the vertices in T may still have dependencies with the original set P - but these dependencies can be removed by a single call to Reduce-Component. In the single remaining instance where we modify φ by assigning x the color list $col(P) - col(Q)$, we simply repeat this process (at most $|P|$ times) by setting $P' = P' - \{x\}$ until there are no remaining dependencies between P and Q. Thus, each iteration of RemoveDependencies produces a constant number of instances with no dependencies between P and Q and one instance in which the size of P' is reduced by at least one.

The output of this step is $O(n)$ list-coloring instances (that are obtained in polynomial time), with no dependencies between P and Q, that together are equivalent to the original instance ϕ. Since there are a constant number of pairs of dynamic sets for each pair of fixed sets, and since there are a constant number of pairs of fixed sets, this proves the following theorem:

Theorem 3. *Determining whether or not a P_5-free graph can be colored with k-colors can be decided in polynomial time.*

4 Summary

In this paper, we obtain a theorem (Theorem 2) on the structure of P_5-free graphs and use it to design a polynomial-time algorithm that determines whether a P_5-free graph can be k-colored. The algorithm recursively uses list coloring techniques and thus its complexity is high even though it is polynomial, as is the case with all list coloring algorithms. In a related paper (in preparation), we will give a slightly faster algorithm also based on list coloring techniques, however this algorithm provides less insight into the structure of P_5-free graphs. It would be of interest to find a polynomial-time algorithm to k-color a P_5-free graph without using list coloring techniques.

Continuing with this vein of research, the following open problems are perhaps the next interesting avenues for future research:

- Does there exist a polynomial time algorithm to determine whether or not a P_7-free graph can be 3-colored?
- Does there exist a polynomial time algorithm to determine whether or not a P_6-free graph can be 4-colored?
- Is the problem of k-coloring a P_7-free graph NP-complete for any $k \geq 3$?

Two other related open problems are to determine the complexities of the MAXIMUM INDEPENDENT SET and MINIMUM INDEPENDENT DOMINATING SET problems on P_5-free graphs.

References

1. Ramirez Alfonsin, J.L., Reed, B.A.: Perfect Graphs. John Wiley & Sons, LTD, Chichester (2001)
2. Bacsó, G., Tuza, Z.: Dominating cliques in P_5-free graphs. Period. Math. Hungar. 21(4), 303–308 (1990)
3. Berge, C., Chvátal, V. (eds.): Topics on perfect graphs. North-Holland, Amsterdam (1984)
4. Even, S., Pnueli, A., Lempel, A.: Permutation graphs and transitive graphs. J. Assoc. Comput. Mach. 19, 400–410 (1972)
5. Garey, M.R., Johnson, D.S., Stockmeyer, L.: Some simplified NP-complete problems. Theoretical Computer Science 1, 237–267 (1976)
6. Gavril, F.: Algorithms for minimum coloring, maximum clique, minimum coloring by cliques, and maximum independent set of a chordal graph. SIAM J. Comput. 1, 180–187 (1972)
7. Golumbic, M.C.: Algorithmic graph theory and perfect graphs. Academic Press, New York (1980)
8. Grötschel, M., Lovász, L., Schrjver, A.: The ellipsoid method and its consequences in combinatorial optimization. Combinatorica 1, 169–197 (1981)

9. Hayward, R., Hoàng, C.T., Maffray, F.: Optimizing weakly triangulated graphs. Graphs and Combinatorics 5, 339–349 (1989)
10. Karp, R.M.: Reducibility among combinatorial problems. In: Miller, R.E., Thatcher, J.W. (eds.) Complexity of Computer Computations, pp. 85–103. Plenum Press, New York (1972)
11. Kral, D., Kratochvil, J., Tuza, Z., Woeginger, G.J.: Complexity of Coloring Graphs without Forbidden Induced Subgraphs. In: Brandstädt, A., Van Le, B. (eds.) WG 2001. LNCS, vol. 2204, pp. 254–262. Springer, Heidelberg (2001)
12. Le, V.B., Randerath, B., Schiermeyer, I.: On the complexity of 4-coloring graphs without long induced paths. Theoretical Computer Science 389(1-2), 330–335 (2007)
13. Randerath, B., Schiermeyer, I.: 3-Colorability $\in P$ for P_6-free graphs. Discrete Applied Mathematics 136(2-3), 299–313 (2004)
14. Sgall, J., Woeginger, G.J.: The complexity of coloring graphs without long induced paths. Acta Cybernet 15(1), 107–117 (2001)

Combinatorial Bounds and Algorithmic Aspects of Image Matching under Projective Transformations*

Christian Hundt[1],** and Maciej Liśkiewicz[2],***

[1] Institut für Informatik, Universität Rostock, Germany
Christian.Hundt@uni-rostock.de
[2] Institut für Theoretische Informatik, Universität zu Lübeck, Germany
liskiewi@tcs.uni-luebeck.de

Abstract. Image matching is an important problem in image processing and arises in such diverse fields as video compression, optical character recognition, medical imaging, watermarking etc. Given two images, image matching determines a transformation that changes the first image such that it most closely resembles the second. Common approaches require either exponential time, or find only an approximate solution, even when only rotations and scalings are allowed. This paper provides the first general polynomial time algorithm to find the exact solution to the image matching problem under projective, affine or linear transformations. Subsequently, nontrivial lower bounds on the number of different transformed images are given which roughly induce the complexity of image matching under the three classes of transformations.

1 Introduction

Image matching research is strongly motivated by various practical applications. In computer vision, e.g., image matching searches digital camera images for distorted versions of objects with known shape, like latin letters [15] or human silhouettes [21]. In mpeg video compression image matching can be used to reduce redundancy by computing the similarity between successive video frames (see e.g. [22] and the references therein). One challenge in medical imaging is to match images of one object taken in different times, from different perspectives or using different medical image devices (see e.g. [5,19]). Finally, image matching can be applied to digital watermarking robust against some geometrical distortions, like e.g. the print-and-scan-process [6,23].

Informally, matching two given digital images A and B is the optimization problem of computing an admissible distortion f which changes A closest to B.

---

* Supported by DFG research grant RE 672/5-1.
** The work on this paper was done during the stay of the first author at the University of Lübeck.
*** On leave from Instytut Informatyki, Uniwersytet Wrocławski, Poland.

E. Ochmański and J. Tyszkiewicz (Eds.): MFCS 2008, LNCS 5162, pp. 395–406, 2008.

For admissibility we require f to be contained in a given set $\mathcal{F}$ of injective functions $f : \mathbb{R}^2 \to \mathbb{R}^2$ that we call transformations. The IMAGE MATCHING PROBLEM (IMP, for short) is intractable for elastic transformations [16] and remains hard for other, more complex transformation classes. On the other hand research in combinatorial pattern matching has succeeded in showing the tractability of the problem under very small subsets of elastic transformations like rotation and scaling (see e.g. [3,4]). However, there has been very little progress in understanding the discrete algorithmic aspects of image matching for the numerous interesting classes of transformations between the both extreme cases.

Let, as usually, $f(A)$ denote the image which results from transforming an image A according to f. We denote by $\mathcal{D}(A, \mathcal{F})$ the set of images $f(A)$ for all f in $\mathcal{F}$. In this paper we present a new discretization technique to reduce certain uncountable sets of transformations $\mathcal{F}$ to finite subsets $\{f_1, f_2, \ldots, f_t\}$ such that the union $\cup_{i=1}^t \{f_i(A)\}$ coincides with $\mathcal{D}(A, \mathcal{F})$. Thus, we provide a finite (and comparatively small) subset of $\mathcal{F}$ sufficient to generate all images that can be obtained by transforming A according to $\mathcal{F}$. Consequently, the transformation of A which is closest to B can be found in $f_1(A)$ to $f_t(A)$. We apply our discretization approach to projective transformations $\mathcal{F}_p$, affine transformations $\mathcal{F}_a$, and linear transformations $\mathcal{F}_1$. Through the rest of the paper we will use the placeholder $\star$ to indicate that any symbol p, a or 1 is allowed instead of repeatedly enumerating the three cases.

Using the discretization approach we show first non-trivial lower and upper polynomial bounds on the cardinalities of transformations $\{f_1, f_2, \ldots, f_t\}$ needed and sufficient to generate the complete sets $\mathcal{D}(A, \mathcal{F}_\star)$. Moreover, we provide a general polynomial time image matching algorithm for the classes $\mathcal{F}_p$, $\mathcal{F}_a$ and $\mathcal{F}_1$ and thus, narrow the gap between known tractable and intractable cases.

Through the rest of this section we give a short overview on previous work and an informal discussion of our results and techniques.

Previous Work. Image matching has been studied both experimentally and theoretically by using different approaches ranging from techniques based on continuous analysis to discrete methods. For an overview we refer to [6,5,15,19,18,2] and the references therein.

The most common approach in continuous analysis is to look at the images A and B in an analogue way and interpret them as mappings over the real plane $\mathbb{R}^2$. Then the function $\Delta(f, A, B) = \int_{x,y \in \mathbb{R}} |A(f(x,y)) - B(x,y)| \, dx \, dy$ gives for any transformation f the difference between the f-transformed version of A and image B. The image match is found by minimizing Δ over the (typically uncountable) set of transformations. Such approaches are mostly considered in medical imaging (see e.g. [20]). The disadvantage is the high complexity to find the global optimum due to the continuous nature of the considered objects.

The feature based matching approach is another technique to solve IMP (see e.g. [1,17]). In this setting, one extracts salient features like points, lines, regions, etc. from images A and B and subsequently, one tries to transform the geometrical objects from A closest to the ones from B. This approach relies heavily on the quality of feature extraction and feature matching, two highly non-trivial tasks even for simple points (for this well studied case, called *optimized point matching* or *geometric point set matching*, see [14] for a survey). Moreover, due to the underlying continuous problem setting, known algorithms using this approach give again only approximate solutions. Remarkably, to reduce the trouble with continuous object spaces Hagedoorn [11] introduced a similar discretization method, like proposed in this paper for IMP. Still, his approach does not provide efficient algorithms even for very small subclasses of affine transformations.

Using purely discrete approaches the matchings between images are mostly required to be exact and globally optimal. In fact, the few known exact polynomial time matching algorithms were developed in combinatorial pattern matching (CPM, for short) [18,9,10,2,4,3]. The problem statement of CPM and IMP are such similar that most techniques can be easily transferred in both directions. Thus, CPM research provides efficient algorithms for image matching under rotations $\mathcal{F}_r$ or scalings $\mathcal{F}_s$. After a series of improving results, the best known algorithm for image matching on rotations is derived from a result of Amir et al. [4] and can solve IMP in time $O(n^4)$, for images of size $n \times n$. The best algorithm for image matching with scaling can immediately be derived from a result of Amir and Chencinski [3] running in time $O(n^3)$. However, the combination of both algorithms does not solve the image matching on $\mathcal{F}_{sr}$, the set of transformations combining rotation and scaling. In this setting the discrete nature of digital images makes their transformations neither commutative nor transitive. Recently, we have presented the first efficient CPM algorithm for $\mathcal{F}_{sr}$ which allows to solve the image matching under $\mathcal{F}_{sr}$ in $O(n^6)$ time [13]. On the other hand no efficient image matching algorithms are known for more general classes of transformations. Their complexity limitations are well understood for elastic transformations, where Keysers and Unger [16] have shown NP-hardness results. A systematic study of the inherent complexity of IMP for functions inside elastic transformations was launched in [12].

Our Contributions. In this paper we present a new technique to select finite subsets $\{f_1, f_2, \ldots, f_t\}$ of $\mathcal{F}_\star$ which completely render the sets $\mathcal{D}(A, \mathcal{F}_\star)$ of all images gained by transforming A with transformations in $\mathcal{F}_\star$. Due to our discretization method only a polynomial sized subset of uncountable $\mathcal{F}_\star$ is sufficient to compute all images in $\mathcal{D}(A, \mathcal{F}_\star)$. Enumerating them solves the IMP under $\mathcal{F}_\star$ in polynomial time with respect to the image size for any $\star \in \{p, a, 1\}$.

In our setting, each transformation f in $\mathcal{F}_\star$ can be represented by an appropriate point p in a parameter space $\mathbb{R}^d$. Thus, considering an image A one can imagine a natural correspondence between the point p and the image $f(A)$ obtained from transforming A according to f. The proposed discretization essentially arises from an appropriate partition of the parameter space into a finite number of convex subspaces $\varphi_1, \varphi_2 \ldots, \varphi_t$. The crucial property of the partition is that for any subspace φ_i all points of φ_i correspond to the same transformed image. Hence, if two transformations f and f' are represented by points p and p' that belong to the same subspace φ_i then the transformations are not distinguished by the images $f(A)$ and $f'(A)$ (meaning $f(A) = f'(A)$).

The partition of the parameter space is determined by a hyperplane arrangement of a specific set $\mathcal{H}$ of hyperplanes. Thus, we get that the corresponding subspaces $\varphi_1, \varphi_2 \ldots, \varphi_t$ are exactly the faces of a hyperplane arrangement. To enumerate the set of transformations $\{f_1, f_2, \ldots, f_t\}$ we traverse the faces φ_i choosing for each one a point $p_i \in \varphi_i$ that determines a transformation f_i. Due to the equivalence between all points in φ_i we get that, whatever selection strategy is used, f_i is a correct representative for all images $f(A)$ with f described by a point $p \in \varphi_i$. However, in our setting transformations are injective. To choose a point p_i inside the face φ_i which really represents an (injective) transformation in $\mathcal{F}_\star$ emerges to a technical difficulty. Additionally, the point should be encoded by rational numbers of length $O(\log n)$.

To argue the polynomial running time of our proposed matching method we need to estimate for any $\mathcal{F}_\star$ the number of transformations in $\{f_1, f_2, \ldots, f_t\}$. The geometrical properties of hyperplane arrangements allows us to obtain upper and lower bounds on the cardinalities of these sets which in turn indicate upper and lower worst case bounds on $|\mathcal{D}(A, \mathcal{F}_\star)|$. In fact, we get that $|\mathcal{D}(A, \mathcal{F}_p)|$ is bounded from above by $O(n^{24})$ and from below by $\Omega(n^{12})$. Analogously, we can show bounds $O(n^{18})$ and $\Omega(n^{12})$ for $|\mathcal{D}(A, \mathcal{F}_a)|$ and $O(n^{12})$ and $\Omega(n^{10})$ for $|\mathcal{D}(A, \mathcal{F}_1)|$. Our results incorporate into a general polynomial time algorithm for image matching under $\mathcal{F}_\star$ which enumerates the set $\{f_1, f_2, \ldots, f_t\}$ by traversing the faces of the hyperplane arrangement. Computing the similarities between B and succeeding $f_i(A)$ we finally test all images in $\mathcal{D}(A, \mathcal{F}_\star)$ against B. To improve efficiency, we traverse the faces in an order implied by their geometrical incidence. This effects in minimal updates between successively tested images. In turn the worst-case running time of the presented algorithm for the IMAGE MATCHING PROBLEM under $\mathcal{F}_\star$ meets the announced upper bounds on the cardinalities of $\mathcal{D}(A, \mathcal{F}_\star)$. Moreover the algorithm needs only integer arithmetic which avoids technical problems due to the use of floating point arithmetic.

2   Preliminaries

In this paper a digital image A is a two dimensional array of pixels, i.e, of unit squares covering a certain area of the real plane $\mathbb{R}^2$. The pixels are indexed over the set $\mathcal{N} = \{(i,j) \mid i,j \in \mathbb{Z} \text{ with } -n \leq i,j \leq n\}$ and we assume that the pixel with index $(i,j) \in \mathcal{N}$ has its geometric center point at coordinates (i,j). We call n the size and $\mathcal{N}$ the support of the image A. Each pixel (i,j) has a color $A\langle i,j\rangle$ from a finite set $\Sigma = \{0, 1, \ldots, \sigma\}$ of colors. For the sake of simplicity we let $A\langle i,j\rangle = 0$ if $(i,j) \notin \mathcal{N}$. For two images A and B of the same support the distortion $\Delta(A, B)$ between A and B is measured by $\sum \delta(A\langle i,j\rangle, B\langle i,j\rangle)$ where $\delta(a,b)$ is a function charging mismatches, for example, $\delta(a,b) = |a - b|$.

Throughout this paper transformations are injective functions $f : \mathbb{R}^2 \to \mathbb{R}^2$. Applying a transformation $f : \mathbb{R}^2 \to \mathbb{R}^2$ to an image A we get the image $f(A)$ which has the same support as A. Define for $g = f^{-1}$ the mapping $\gamma_g : \mathcal{N} \to \mathbb{Z}^2$ which determines for any pixel (i,j) in $f(A)$ the corresponding pixel (i', j') in A. We let $\gamma_g(i,j) = [g(i,j)]$, where $[(x,y)] := ([x], [y])$ denotes rounding all

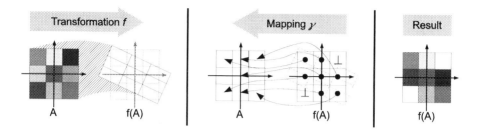

Fig. 1. The image A is distorted by a continuous transformation $f : \mathbb{R}^2 \to \mathbb{R}^2$. However image $f(A)$ can represent transformed color values only at integer coordinates. Thus, the mapping γ determines by nearest neighbor interpolation the discrete version of f.

components of a vector $(x, y) \in \mathbb{R}^2$. Then the color value of pixel (i, j) in $f(A)$ is defined as the color value of the pixel $(i', j') = \gamma_g(i, j)$ in A. Hence, we choose the pixel which geometrically contains the point $f^{-1}(i, j)$ in its square area. For an example see Fig. 1.

For given image A and a prescribed $\mathcal{F}$ we define the set of transformed images $\mathcal{D}(A, \mathcal{F}) = \{f(A) \mid f \in \mathcal{F}\}$. Then we call the following optimization problem the IMAGE MATCHING PROBLEM for $\mathcal{F}$:

Problem: For a given reference image A and a distorted image B, both of the same size n, find in the search space $\mathcal{D}(A, \mathcal{F})$ an image A' minimizing the distortion $\Delta(A', B)$ and return $f \in \mathcal{F}$ with $f(A) = A'$.

For the analysis of complexity aspects we will apply the unit cost model for arithmetic operations. Therefore, we assume that mathematical basic operations can be done in constant time disregarding the length of input numbers.

In this paper we are interested in $\mathcal{F}_p$ the set of projective transformations. Any transformation f in $\mathcal{F}_p$ can be uniquely described by $f(x, y) = (\frac{a}{c}, \frac{b}{c})^T$ where

$$\begin{pmatrix} a \\ b \\ c \end{pmatrix} = \begin{pmatrix} p_1 & p_2 & p_3 \\ p_4 & p_5 & p_6 \\ p_7 & p_8 & 1 \end{pmatrix} \cdot \begin{pmatrix} x \\ y \\ 1 \end{pmatrix} \tag{1}$$

for some $p_1, \ldots, p_8 \in \mathbb{R}$. For the set $\mathcal{F}_p$ all transformations can be characterized by the parameters p_1 to p_8. Hence, each such transformation f can be characterized by a vector $(p_1, \ldots, p_8)^T$ in $\mathbb{R}^8$. However, there are some points in $\mathbb{R}^8$ which belong to non-injective transformations.

In addition to $\mathcal{F}_p$ we will also regard its subsets of affine transformations $\mathcal{F}_a$ and linear transformations $\mathcal{F}_l$. The transformations in $\mathcal{F}_a$ are exactly the projective transformations with $p_7 = p_8 = 0$ and in turn $\mathcal{F}_l$ is the subset of $\mathcal{F}_a$ containing transformations with $p_3 = p_6 = 0$. Due to this restrictions every affine transformation can be characterized by a point in $\mathbb{R}^6$ and every linear transformation by a point in $\mathbb{R}^4$. According to our definition all considered classes of transformations are closed under inversion, i.e., for all $\star \in \{p, a, l\}$ it is true, if $f \in \mathcal{F}_\star$ then $f^{-1} \in \mathcal{F}_\star$.

For our approach being connected to combinatorial geometry we need some further definitions: We denote by H a set of linear equations h of the form

$h : c_1p_1 + \ldots + c_8p_8 = c_9$. Let $p = (p_1, \ldots, p_8) \in \mathbb{R}^8$. Then we define for each $h \in H$ the value $h(p) = c_1p_1 + \ldots + c_8p_8 - c_9$. Each equation h describes a hyperplane $\ell = \{p \mid h(p) = 0\}$ in $\mathbb{R}^8$. Notice the difference: h is a mathematical expression whereas ℓ is a subspace of $\mathbb{R}^8$. Denote by $\mathcal{H}$ the set of all hyperplanes defined by the equations in H. Now define for all $h \in H$ the following additional subspaces of $\mathbb{R}^8$:

$$\ell^+ = \{p \mid h(p) > 0\}, \quad \text{and} \quad \ell^- = \{p \mid h(p) < 0\}.$$

For a finite set of equations $H = \{h_1, \ldots, h_t\}$ consider the following partition of $\mathbb{R}^8$ into subspaces:

$$\mathcal{A}(H) = \left\{\varphi \subseteq \mathbb{R}^8 \mid \varphi = \bigcap_{w=1}^{t} \ell_w^{s_w} \text{ for some } s_1, \ldots, s_t \in \{+, -, 0\}\right\},$$

where ℓ_w is the line corresponding to h_w and ℓ_w^0 denotes just ℓ_w. In literature the set $\mathcal{A}(H)$ is called the hyperplane arrangement given by the hyperplanes $\mathcal{H}$. See [7] for detailed information on hyperplane arrangements.

We call the elements of $\mathcal{A}(H)$ faces. We say that a face φ is a d-face if its dimension is d for $d \in \{0, \ldots, 8\}$. In particular, a 0-face is a point, a 1-face is a line, half-line or line segment, a 2-face is a convex region on a plane, etc. A face φ' is a subface of another face φ if the dimension of φ' is one less than of φ and φ' is contained in the boundary of φ. We also say that φ and φ' are incident and that φ is a superface of φ'.

The incidence graph $\mathcal{I}(H)$ of $\mathcal{A}(H)$ contains a node $v(\varphi)$ for each face φ and $v(\varphi)$ and $v(\varphi')$ are connected by an edge if the faces φ and φ' are incident. The incidence graph is described in detail in [8] (see also [7]).

3   Structural Properties of the Search Spaces

In this section we first present the structure of the sets $\mathcal{D}(A, \mathcal{F}_p)$, $\mathcal{D}(A, \mathcal{F}_a)$ and $\mathcal{D}(A, \mathcal{F}_1)$. Afterwards we will show how to estimate the worst case number of images in the sets. For the three cases of $\star \in \{1, a, p\}$ we define the following sets $H_{\star,n}$ of equations

$\forall (i, j) \in \mathcal{N}$ and $\forall k \in \{-n, \ldots, n + 1\}$:

$$X_{ijk} : ip_1 + jp_2 + p_3 + (0.5i - ik)p_7 + (0.5j - jk)p_8 + (0.5 - k) = 0 \in H_{p,n}$$
$$Y_{ijk} : ip_4 + jp_5 + p_6 + (0.5i - ik)p_7 + (0.5j - jk)p_8 + (0.5 - k) = 0 \in H_{p,n}$$

$$X_{ijk} : ip_1 + jp_2 + p_3 + 0.5 - k = 0 \in H_{a,n}$$
$$Y_{ijk} : ip_4 + jp_5 + p_6 + 0.5 - k = 0 \in H_{a,n}$$

$$X_{ijk} : ip_1 + jp_2 + 0.5 - k = 0 \in H_{1,n} \text{ if } (i,j) \neq (0,0)$$
$$Y_{ijk} : ip_4 + jp_5 + 0.5 - k = 0 \in H_{1,n} \text{ if } (i,j) \neq (0,0)$$

Each of these equations $h \in H_{\star,n}$ describes a hyperplane ℓ which partitions the parameter space into three parts ℓ^+, ℓ and ℓ^-. Then any point p representing a transformation g is in exactly one of the three subspaces for all equations in $H_{\star,n}$.

The specific location of p within the subspaces determines how A is transformed to $f(A)$ by $f = g^{-1}$. Particularly, for $\mathcal{F}_1$ and $\mathcal{F}_a$ the following holds:

Lemma 1. *Let* $\star \in \{1, a\}$. *Assume* $f \in \mathcal{F}_\star$, $g = f^{-1}$, *and* p *is the point representing* g. *Then for all* $(i, j) \in \mathcal{N}$ *and* $k \in \{-n, \ldots, n+1\}$, *for any hyperplane* ℓ *described by* $X_{ijk} \in H_{\star, n}$, *and for the coordinates* (i', j') *in* A *which correspond to* (i, j) *in* $f(A)$, *i.e.* $(i', j') = \gamma_g(i, j)$, *it is true that* $i' < k$ *if* $p \in \ell^-$ *and* $i' \geq k$ *if* $p \in \ell^+ \cup \ell$. *Analogously, for any hyperplane* ℓ *described by* $Y_{ijk} \in H_{\star, n}$, *it holds that* $j' < k$ *if* $p \in \ell^-$ *and* $j' \geq k$ *if* $p \in \ell^+ \cup \ell$.

In case of projective transformations a more involved relationship is true:

Lemma 2. *Assume* $g = f^{-1}$ *for* $f \in \mathcal{F}_p$ *and let* p *be the point representing* g. *Furthermore let* $(i, j) \in \mathcal{N}$ *and* $k_1, k_2 \in \{-n, \ldots, n+1\}$ *such that* $k_1 < k_2$ *and denote by* ℓ_1 *and* ℓ_2 *the hyperplanes described by* $H_{p,n}$-*equations* X_{ijk_1} *and* X_{ijk_2}, *respectively. Consider* $\gamma_g(i, j) = (i', j')$. *Then* $k_1 \leq i' < k_2$ *iff* $p \in (\ell_1 \cup \ell_1^+) \cap \ell_2^-$ *or* $p \in (\ell_1^- \cup \ell_1) \cap \ell_2^+$.

Analogously, if we consider Y_{ijk_1} *and* Y_{ijk_2} *in* $H_{p,n}$ *to describe the hyperplanes* ℓ_1 *and* ℓ_2, *then* $k_1 \leq j' < k_2$ *iff* $p \in ((\ell_1^+ \cup \ell_1) \cap \ell_2^-) \cup ((\ell_1^- \cup \ell_1) \cap \ell_2^+)$.

With the help of the equations $H_{\star, n}$ and by the use of the above lemmas we are now ready to provide for all $\star \in \{1, a, p\}$ the relation between the set $\mathcal{D}(A, \mathcal{F}_\star)$ and the set $\mathcal{A}(H_{\star, n})$ of faces.

Theorem 1. *For all* $\star \in \{1, a, p\}$ *and all digital images* A *of size* n *there exist surjective mappings* $\Gamma_{\star, n} : \mathcal{A}(H_{\star, n}) \to \mathcal{D}(A, \mathcal{F}_\star)$.

By the theorem it suffices to estimate the number of faces in $\mathcal{A}(H_{\star, n})$ to get a bound on the cardinality of $\mathcal{D}(A, \mathcal{F}_\star)$. Due to the surjective mapping $\Gamma_{\star, n}$ we know that $\mathcal{D}(A, \mathcal{F}_\star)$ cannot contain a larger number of images than the number of faces in $\mathcal{A}(H_{\star, n})$. Furthermore, the surjective mappings $\Gamma_{\star, n}$ enable a simple method to enumerate the images in $\mathcal{D}(A, \mathcal{F}_\star)$. One simply has to construct $\mathcal{A}(H_{\star, n})$, traverse its elements φ in an appropriate way and each time compute $\Gamma_{\star, n}(\varphi)$ to obtain another transformed image of $\mathcal{D}(A, \mathcal{F}_\star)$. Since $\Gamma_{\star, n}$ is not bijective it may happen for certain images A that $\mathcal{D}(A, \mathcal{F}_\star)$ is smaller than $\mathcal{A}(H_{\star, n})$. However, it is hard to identify such images and thus, our algorithm will search the whole set $\mathcal{A}(H_{\star, n})$. An important task is to determine tight lower and upper bounds on the cardinalities $|\mathcal{A}(H_{\star, n})|$ that can provide good estimation of the running time of our algorithm.

Theorem 2. 1. $|\mathcal{A}(H_{p,n})| \in \Omega(n^{12}) \cap O(n^{24})$,   2. $|\mathcal{A}(H_{a,n})| \in \Omega(n^{12}) \cap O(n^{18})$,   3. $|\mathcal{A}(H_{1,n})| \in \Omega(n^{10}) \cap O(n^{12})$.

4   Computing Representatives

Theorem 1 of the previous section presents the main structural property used in our image matching algorithm. However, the possibility of enumerating the images of $\mathcal{D}(A, \mathcal{F}_\star)$ by traversing the faces of $\mathcal{A}(H_{\star, n})$ does not solve entirely IMP as defined in this paper. For the solution the algorithm has to return not only the image $A' \in \mathcal{D}(A, \mathcal{F}_\star)$ minimizing the distance to B but also a transformation $f \in \mathcal{F}_\star$ such that $A' = f(A)$. Therefore, we compute for all faces φ a point $p(\varphi)$ contained in φ. To this aim we proceed as follows: If φ is a 0-face, then $p(\varphi)$ is the only contained point. Otherwise, if $\varphi_1, \ldots, \varphi_t$ are the subfaces of φ then $p(\varphi) := \frac{1}{t} \sum_{w=1}^{t} p(\varphi_w)$. Accordingly, we get the transformation $g(\varphi)$ which is associated to $p(\varphi)$. Thus, for a face φ_{opt}, associated to the optimum image A', we are able to get the solution $f = g(\varphi_{\text{opt}})^{-1}$. The issue with this approach is the existence of transformations g represented by points of the parameter space which are not invertible and hence not included in $\mathcal{F}_\star$. In fact, certain faces do not represent invertible transformations at all.

In this section we show for $\star \in \{\mathsf{p}, \mathsf{a}, \mathsf{l}\}$ how to detect faces φ which either miss invertible transformations or which are associated to the same image as some other face. These faces will be marked as "irrelevant" in a preprocessing phase, before the main searching algorithm starts. Moreover, we introduce how to correct $g(\varphi)$ to an invertible transformation for the remaining relevant faces.

Generally, projective transformations are invertible iff the matrix defined by the parameters p_1 to p_8 is regular. The points of parameter space which violate regularity can be described by the following non-linear surface:

$$Z_\mathsf{p} : p_1 p_5 + p_2 p_6 p_7 + p_3 p_4 p_8 - p_1 p_6 p_8 - p_2 p_4 - p_3 p_5 p_7 = 0,$$

that is, by those parameters p_1 to p_8 which represent a matrix with determinant zero. A structural rewarding property of Z_p is its limited degree of freedom. In particular, to describe any point p in Z_p, it is sufficient to fix any combination of only seven parameters from p_1 to p_8. However, in every 8-face of $\mathcal{A}(H_{\mathsf{p},n})$ there are infinitely many points which are equal on parameters p_1 to p_7 and distinguish only in p_8. At most one of them can be part of Z_p. It follows that every 8-face in $\mathcal{A}(H_{\mathsf{p},n})$ has at least one point p which represents an invertible transformation and thus, can represent the face. For affine and linear transformations the surface of non-invertible transformations becomes less complex:

$$Z_\mathsf{a}, Z_\mathsf{l} : p_1 p_5 - p_2 p_4 = 0.$$

Still, any combination of five or respectively three parameters suffices to determine a point in Z_a or Z_l. Hence, every 6-face in $\mathcal{A}(H_{\mathsf{a},n})$ and every 4-face in $\mathcal{A}(H_{\mathsf{l},n})$ has at least one point p to represent the face.

The following lemma provides properties of the faces in $\mathcal{A}(H_{\star,n})$ which enable the computation of representative points by the use of the above characterization:

Lemma 3

1. *For every q-face φ' in $\mathcal{A}(H_{\mathsf{p},n})$ with $q < 8$ there is an 8-face φ in $\mathcal{A}(H_{\mathsf{p},n})$ such that the images $\Gamma_\mathsf{p}(\varphi')$ and $\Gamma_\mathsf{p}(\varphi)$ are equal.*
2. *For every q-face φ' in $\mathcal{A}(H_{\mathsf{a},n})$ with $q < 6$ there is a 6-face φ in $\mathcal{A}(H_{\mathsf{a},n})$ such that the images $\Gamma_\mathsf{a}(\varphi')$ and $\Gamma_\mathsf{a}(\varphi)$ are equal.*
3. *Every q-face φ in $\mathcal{A}(H_{\mathsf{l},n})$ with $0 < q \leq 4$ contains at least one point which is not in Z_l.*

By the lemma all q-faces with $q < 8$ may be marked "irrelevant" for projective transformations. This has the convenient side effect that faces are marked "irrelevant", if they are contained in one of the subspaces described by the equations:

$$Z_{ij} : ip_7 + jp_8 + 1 = 0 \quad \text{for all } (i,j) \in \mathcal{N}, \text{ with } (i,j) \neq (0,0),$$

that characterize transformations for which equation (1) results with $c = 0$. Such faces do not represent invertible transformations because neither of them maps the pixel (i,j). Again, since these subspaces are seven-dimensional only q-faces with $q < 8$ can be properly contained in them.

For affine transformations we have to mark exactly all q-faces with $q < 6$ "irrelevant". However, for linear transformations it is not that easy to exclude faces from the search. Nevertheless, Lemma 3 states at least that all but 0-faces contain points associated to invertible transformations. Hence, in this case a face φ should be marked "irrelevant" if (1) it is a 0-face and (2) the only contained point $p(\varphi)$ is in Z_1, which is easy to detect.

It remains to check, and if necessary to correct, for all relevant faces whether the point $p(\varphi)$ represents an invertible transformation $g(\varphi)$. If predefined $p(\varphi)$ is not contained in the surface $Z_\star$ then we are done. Otherwise, let $dZ_\star(p(\varphi))$ be the gradient of $Z_\star$ at $p(\varphi)$. Then we add to $p(\varphi)$ the vector $p(\varphi) + \epsilon \cdot dZ_\star(p(\varphi))$, where $\epsilon > 0$ is a constant chosen such that $p(\varphi)$ (1) remains in φ and (2) can be stored by $O(\log n)$ bits. Therewith we move $p(\varphi)$ out of Z_p and obtain an invertible transformation $g(\varphi)$. The existence of ϵ is straightforward for the projective and affine case. For the linear one it may however happen that the construction leads to $p(\varphi)$ outside φ, if φ is not a 4-face. Still, in this case φ is a proper subspace of some hyperplanes ℓ_1 to ℓ_t. To ensure that $p(\varphi)$ remains in φ we project $dZ_1(p(\varphi))$ onto the intersection of ℓ_1 to ℓ_t. Then the approach works again.

5   The Algorithm

To solve IMP under $\mathcal{F}_\star$ we use the mapping $\Gamma_{\star,n}$ introduced in Theorem 1 to enumerate $\mathcal{D}(A, \mathcal{F}_\star)$. For the implementation of the mapping we search $\mathcal{A}(H_{\star,n})$ and choose for each encountered face φ the representative transformation $g(\varphi)$. Then, the image A' associated to φ can be computed using $\gamma_{g(\varphi)}$. Theorem 1 guarantees that in this way all images in $\mathcal{D}(A, \mathcal{F}_\star)$ will be tested.

This straightforward approach of visiting all faces φ, computing A' by $g(\varphi)$ and estimating its distortion against B, has at least a time complexity of $|\mathcal{A}(H_{\star,n})|$ times $O(n^2)$, where the last term describes the cost of distortion estimation. Using our approach we can improve this complexity by computing A' incrementally. It turns out that incident faces correspond to very similar images. In fact, Lemmas 1 and 2 and Theorem 1 imply the following:

Corollary 1. *Let $\varphi, \varphi' \in \mathcal{A}(H_{\star,n})$ be two incident faces and $(i,j) \in \mathcal{N}$ a pixel. Then the two images $\Gamma_{\star,n}(\varphi)$ and $\Gamma_{\star,n}(\varphi')$ can differ at the pixel (i,j), only if there exists k in $\{-n, \ldots, n+1\}$ such that the hyperplane ℓ described by X_{ijk} or Y_{ijk} contains as a subspace the face φ or the face φ'.*

Thus, it is profitable to enumerate faces in the order implied by geometrical incidence to guarantee minimal changes in A' when going from one face to the next one. Our algorithm achieves linear running time with respect to $|\mathcal{A}(H_{\star,n})|$

Algorithm. ImageMatching   /* Image Matching for $\mathcal{F}_\star$ with $\star \in \{1, \mathtt{a}, \mathtt{p}\}$ */

Input: Images A and B of size n; $\star \in \{1, \mathtt{a}, \mathtt{p}\}$.

Output: $f = \arg\min_{f' \in \mathcal{F}_\star} \{\Delta(f'(A), B)\}$.

---

1. **Procedure** SEARCH($v(\varphi)$);   /* Depth first searching */
2. **begin**
3.   mark node $v(\varphi)$ as *seen*;
4.   **for all** *unseen* neighbors $v(\varphi')$ of $v(\varphi)$ **do begin**
5.   **for all** (i, j) in $Update(\varphi) \cup Update(\varphi')$ **do begin**
6. $\Delta = \Delta - \delta(A'\langle i, j\rangle, B\langle i, j\rangle)$; $A'\langle i, j\rangle = A\langle \gamma_{g(\varphi')}(i, j)\rangle$; $\Delta = \Delta + \delta(A'\langle i, j\rangle, B\langle i, j\rangle)$;
7.   **end**;
8.   **if** (φ' is not irrelevant) and ($\Delta < \Delta_{\mathrm{opt}}$) **then**
9.   **begin** $\Delta_{\mathrm{opt}} = \Delta$; $\varphi_{\mathrm{opt}} = \varphi'$; **end**;
10.   call SEARCH(φ');
11.   **for all** (i, j) in $Update(\varphi) \cup Update(\varphi')$ **do begin**
12. $\Delta = \Delta - \delta(A'\langle i, j\rangle, B\langle i, j\rangle)$; $A'\langle i, j\rangle = A\langle \gamma_{g(\varphi)}(i, j)\rangle$; $\Delta = \Delta + \delta(A'\langle i, j\rangle, B\langle i, j\rangle)$;
13.   **end**;
14.   **end**;
15. **end**;

16. **begin**   /* ImageMatching */
17.   construct the incidence graph $\mathcal{I}(H_{\star,n})$;
18.   PREPROCESS($\mathcal{I}(H_{\star,n})$);
19.   let φ_{id} be the face of $\mathcal{A}(H_{\star,n})$ which corresponds to the identity mapping;
20. $\varphi_{\mathrm{opt}} = \varphi_{id}$; $\Delta_{\mathrm{opt}} = \Delta = \Delta(A, B)$; $A' = A$;
21.   call SEARCH($v(\varphi_{id})$);   /* find φ_{opt} */
22.   return $f = g^{-1}(\varphi_{\mathrm{opt}})$;
23. **end**.

Fig. 2. The image matching algorithm. The main procedure prepares the DFS-search of $\mathcal{I}(H_{\star,n})$. The search itself is realized recursively by the SEARCH procedure. With each call one face φ becomes *seen*. Then the neighborhood of φ is processed by updating the pixels which have possibly changed and estimating the new distortion.

because it enumerates $\mathcal{A}(H_{\star,n})$ according to the geometrical incidence by DFS-traversing the incidence graph $\mathcal{I}(H_{\star,n})$.

In the following we introduce our image matching algorithm, which works for all three classes $\mathcal{F}_\mathtt{p}$, $\mathcal{F}_\mathtt{a}$ and $\mathcal{F}_1$. Furthermore, we present its running times.

The algorithm starts with the construction of $\mathcal{I}(H_{\star,n})$. Then $\mathcal{I}(H_{\star,n})$ is passed to PREPROCESS which (1) marks all nodes $v(\varphi)$ "irrelevant", if φ should not become the solution φ_{opt} and (2) labels all nodes $v(\varphi)$ with auxiliary information like (a) a set $Update(\varphi)$ of pixel coordinates, (b) the point $p(\varphi)$, and (c) a representative transformation $g(\varphi)$. The node labels $p(\varphi)$ and $g(\varphi)$ as well as the attribute "irrelevant" are computed according to the discussion of Section 4. Notice that in case of an "irrelevant" node $v(\varphi)$ the transformation $g(\varphi)$ may not be invertible. However, by $g(\varphi)$ we can still compute $\gamma_{g(\varphi)}$ and in turn the image A', which is associated to all points of the face φ. Because $v(\varphi)$ is "irrelevant" it cannot become the solution φ_{opt} and we will never attempt to invert $g(\varphi)$. Subsequently, PREPROCESS computes for all nodes $Update(\varphi)$, sets used to keep the number of incremental pixels updated as small as possible during the DFS-traversal of $\mathcal{I}(H_{\star,n})$. For all d-faces φ we let $Update(\varphi) := \emptyset$. If φ is a $(d-1)$-face, i.e., a subspace of one hyperplane $\ell \in \mathcal{H}_{\star,n}$, then

$$Update(\varphi) := \{(i,j) \mid \exists k \in \{-n, \ldots, n+1\} : \ X_{ijk} \text{ or } Y_{ijk} \text{ describes } \ell\}.$$

Finally, if $\varphi_1, \ldots, \varphi_t$ are the superfaces of a q-face φ with $q < d-1$, we let

$$Update(\varphi) := \bigcup_{w=1}^{t} Update(\varphi_w).$$

The definition of d depends on $\star$. For $\star = \mathsf{p}$ we have $d = 8$, for $\star = \mathsf{a}$ it is $d = 6$ and we set $d = 4$ for $\star = \mathsf{l}$.

After PREPROCESS our algorithm performs DFS on the graph $\mathcal{I}(H_{\star,n})$. Visiting a node $v(\varphi)$ the algorithm stores the current distortion between A', the image determined by $\gamma_{g(\varphi)}$, and B. Next, when traversing from φ to an incident (sub or super)face φ' the algorithm has to compute from A' the image A'', determined by $\gamma_{g(\varphi')}$. According to Corollary 1 it suffices to update only the pixel values, coordinates of which are elements in $Update(\varphi)$ or $Update(\varphi')$. For input images A and B we use the algorithm ImageMatching listed in Figure 2.

As a conclusion of this section we give a bound on the running time of the image matching algorithm listed in Figure 2.

Theorem 3. *For given A and B of size n, the* IMAGE MATCHING PROBLEM *can be solved by the algorithm listed in Figure 2 in time*
  1. $O(n^{12})$ for $\mathcal{F}_{\mathsf{l}}$,
  2. $O(n^{18})$ for $\mathcal{F}_{\mathsf{a}}$, and
  3. $O(n^{24})$ for $\mathcal{F}_{\mathsf{p}}$.

6   Conclusions and Future Work

In this work we analyzed the Image Matching Problem with respect to projective, affine and linear transformations. We introduced a general polynomial time searching strategy which takes advantage of the search space structure common among the covered classes of transformations.

To provide precise bounds for the running time of the searching algorithm we examined the complexity of the search space structure for each class of transformations. We showed narrow bounds for linear transformations $\Omega(n^{10}) \cap O(n^{12})$ and gave non trivial bounds for affine transformations $\Omega(n^{12}) \cap O(n^{18})$ and projective transformations $\Omega(n^{12}) \cap O(n^{24})$. A challenging task is to close the gaps between the corresponding lower and upper bounds.

References

1. Bovik, A. (ed.): Handbook of Image and Video Processing. Academic Press, San Diego, California (2000)
2. Amir, A., Butman, A., Crochemore, M., Landau, G., Schaps, M.: Two dimensional pattern matching with rotations. Theor. Comput. Sci. 314(1-2), 173–187 (2004)

3. Amir, A., Chencinski, E.: Faster 2-Dimensional Scaled Matching. In: Lewenstein, M., Valiente, G. (eds.) CPM 2006. LNCS, vol. 4009, pp. 200–210. Springer, Heidelberg (2006)

4. Amir, A., Tsur, D., Kapah, O.: Faster Two Dimensional Pattern Matching with Rotations. In: Sahinalp, S.C., Muthukrishnan, S.M., Dogrusoz, U. (eds.) CPM 2004. LNCS, vol. 3109, pp. 409–419. Springer, Heidelberg (2004)

5. Brown, L.G.: A survey of image registration techniques. ACM Computing Surveys 24(4), 325–376 (1992)

6. Cox, I.J., Bloom, J.A., Miller, M.L.: Digital Watermarking, Principles and Practice. Morgan Kaufmann, San Francisco, California (2001)

7. Edelsbrunner, H.: Algorithms in Combinatorial Geometry. Springer, Berlin (1987)

8. Edelsbrunner, H., O'Rourke, J., Seidel, R.: Constructing arrangements of lines and hyperplanes with applications. SIAM J. Comput. 15, 341–363 (1986)

9. Fredriksson, K., Ukkonen, E.: A rotation invariant filter for two dimensional string matching. In: Proc. CPM. LNCS, vol. 1448, pp. 118–125 (1998)

10. Ukkonen, E., Navarro, G., Fredriksson, K.: Optimal Exact and Fast Approximate Two Dimensional Pattern Matching Allowing Rotations. In: Apostolico, A., Takeda, M. (eds.) CPM 2002. LNCS, vol. 2373. Springer, Heidelberg (2002)

11. Hagedoorn, M.: Pattern matching using similarity measures, PhD Thesis, Univ. Utrecht (2000)

12. Hundt, C., Liśkiewicz, M.: On the Complexity of Affine Image Matching. In: Thomas, W., Weil, P. (eds.) STACS 2007. LNCS, vol. 4393, pp. 284–295. Springer, Heidelberg (2007)

13. Hundt, C., Liśkiewicz, M.: Two-dimensional Pattern Matching with Combined Scaling and Rotation. In: Proc. CPM 2008. LNCS, vol. 5029 (2008)

14. Indyk, P.: Algorithmic aspects of geometric embeddings. In: Proc. FOCS, pp. 10–33. IEEE Computer Society Press, Los Alamitos (2001)

15. Kasturi, R., Jain, R.C.: Computer Vision: Principles. IEEE Computer Society Press, Los Alamitos (1991)

16. Keysers, D., Unger, W.: Elastic image matching is NP-complete. Pattern Recogn. Letters 24, 445–453 (2003)

17. Kropatsch, W.G., Bischof, H. (eds.): Digital Image Analysis - Selected Techniques and Applications. Springer, Berlin (2001)

18. Landau, G.M., Vishkin, U.: Pattern matching in a digitized image. Algorithmica 12(3/4), 375–408 (1994)

19. Maintz, J.B.A., Viergever, M.A.: A survey of medical image registration. Medical Image Analysis 2, 1–36 (1998)

20. Modersitzki, J.: Numerical Methods for Image Registration. Oxford University Press, Oxford (2004)

21. Moeslund, T.B., Hilton, A., Krüger, V.: A survey of advances in vision-based human motion capture and analysis. Computer Vision and Image Understanding 104(2-3), 90–126 (2006)

22. Shi, Y.Q., Sun, H.: Image and video Compression for multimedia engineering. CRC Press, Boca Raton (2000)

23. Solanki, K., Madhow, U., Manjunath, B.S., Chandrasekaran, S., El-Khalil, I.: 'Print and Scan' Resilient Data Hiding in Images. IEEE Transactions on Information Forensics and Security 1(4), 464–478 (2006)

Lower Bounds for Syntactically Multilinear Algebraic Branching Programs

Maurice J. Jansen

Centre for Theory in Natural Science, University of Aarhus
Department of Computer Science, University of Aarhus, IT-Parken, Aabogade 34,
DK-8200 Aarhus N, Denmark
Phone: +45 8942 5600
mjjansen@daimi.au.dk

Abstract. It is shown that any weakly-skew circuit can be converted into a skew circuit with constant factor overhead, while preserving either syntactic or semantic multilinearity. This leads to considering syntactically multilinear algebraic branching programs (ABPs), which are defined by a natural read-once property. A $2^{n/4}$ size lower bound is proven for ordered syntactically multilinear ABPs computing an explicitly constructed multilinear polynomial in $2n$ variables. Without the ordering restriction a lower bound of level $\Omega(n^{3/2}/\log n)$ is observed, by considering a generalization of a hypercube covering problem by Galvin [1].

Keywords. Computational complexity, arithmetical circuits, lower bounds, multilinear polynomials, algebraic branching programs.

1 Introduction

It is not known whether polynomial size arithmetical circuits (VP) are computationally more powerful than polynomial size arithmetical formulas (VP_e). For the former, we have a surprising construction by Valiant, Skyum, Berkowitz and Rackoff, which shows that $\mathrm{VP} = \mathrm{VNC}_2$ [2]. For the latter, we know by a result of Brent that $\mathrm{VP}_e = \mathrm{VNC}_1$ [3]. Recently, Raz made a breakthrough by showing that polynomial size *multilinear* circuits are strictly more powerful than polynomial size *multilinear* formulas. Raz proved that

Theorem 1 ([4]). s-mlin-$\mathrm{VNC}_1 \neq$ s-mlin-VNC_2.

Here "s-mlin-" denotes *syntactic* multilinearity. Technically, multilinearity comes in two flavors: *syntactic* and *semantic* (See Section 2). For formulas these two notions are equivalent, but this is not known to be true for circuits.

Intermediate between circuits and formulas we have so-called *weakly-skew circuits* (See [5]). Letting VP_s and VP_{ws} stand for the classes of p-families of polynomials that have skew circuits or weakly-skew circuits of polynomial size, respectively, Malod and Portier prove that $\mathrm{VP}_s = \mathrm{VP}_{ws}$. The situation can be summarized as follows:

Theorem 2 ([5,2]). $\mathrm{VNC}_1 \subseteq \mathrm{VP}_s = \mathrm{VP}_{ws} \subseteq \mathrm{VNC}_2 = \mathrm{VP}$.

E. Ochmański and J. Tyszkiewicz (Eds.): MFCS 2008, LNCS 5162, pp. 407–418, 2008.

A priori it is not clear whether the equality $VP_s = VP_{ws}$ holds up when passing to the multilinear variants of these classes, as the proofs in [5] appeal to the completeness of the determinant or trace of iterated matrix multiplication for the class VP_{ws}. For the determinant, currently no polynomial size multilinear circuits are known. Furthermore, multilinearity is not necessarily preserved under Valiant projections.

1.1   Results

It will be demonstrated one can convert any weakly-skew circuit into a skew-circuit with constant factor overhead, while maintaining either syntactic or semantic multilinearity. Further, it will be observed that the conversion without multilinearity restrictions can be done with constant factor overhead as well[1], which improves [5]. There an appeal is made to either polynomial size skew circuits for the determinant [6], or for the trace of iterated matrix multiplication. Both families are shown to be complete for VP_{ws}. One obtains that

Theorem 3

$$\text{s-mlin-VNC}_1 \subseteq \text{s-mlin-VP}_s = \text{s-mlin-VP}_{ws} \subseteq \text{s-mlin-VNC}_2 = \text{s-mlin-VP}.$$

In the above, the rightmost equality was proven in [7]. Looking at Theorem 3, the question is raised whether perhaps the techniques used to prove Theorem 1 can be strengthened to show that s-mlin-$VP_s \neq$ s-mlin-VNC_2, or that we can prove s-mlin-$VNC_1 \neq$ s-mlin-VP_s, by showing, in the terminology of [7,8], some *full rank polynomial* has polynomial size skew circuits.

Without multilinearity, $VP_s \neq VNC_2$ iff the determinant is not complete for VP under poly-size Valiant projections, due to [5]. Also, $VNC_1 \neq VP_s$ iff the determinant does not have poly-size formulas. Both statements are major outstanding problems in algebraic complexity theory.

A skew circuit can be transformed into an *algebraic branching program* (ABP, See [9]) with relatively little overhead. If the initial skew circuit is syntactically multilinear, this results in an ABP B which is syntactically multilinear in the following natural sense: on any directed path in B, any variable can appear at most once. This can be thought of as the algebraic analog of a Boolean read-once branching program. In the latter model we know of tight exponential lower bounds [10]. Also exponential lower bounds are known for ABPs in the non-commutative case [9]. Bryant introduced so-called *ordered binary decision diagrams* (OBDDs), for which he proved exponential lower bounds [11]. These are read-once Boolean branching programs in which variables are restricted to appear in the same order on all directed paths. This restriction can naturally also be considered for ABPs, leading to the following result:

Theorem 4. *Let X be a set of $2n$ variables. For any field F, there exists an extension field G of F and explicitly constructed multilinear polynomial $f \in G[X]$,*

---

[1] This observation has simultaneously been made by Kaltofen and Koiran in a so far unpublished paper, which was unknown to the author at the time of this research.

*such that any ordered algebraic branching program over X and G computing f
has at least $2^{n/4}$ nodes.*

Finally, the problem of proving lower bounds for unrestricted ABPs and (un-ordered) multilinear ABPs is explored. For any fixed constant $0 < \alpha < 1$, it will be shown that any unrestricted ABP requires size $\Omega(n^{3/2}\alpha\sqrt{1-\alpha})$ to compute the elementary symmetric polynomial of degree $\lfloor \alpha n \rfloor$ in n variables. Next a relation between proving lower bounds for multilinear ABPs and the generalization of a hypercube covering problem by Galvin will be established [1]. By straightforward counting this yields a lower bound for multilinear ABPs of level $\Omega(\frac{n^{3/2}}{\log n})$, for computing any full rank polynomial. Potentially however, this technique yields up to quadratic lower bounds, provided linear lower bounds can be proven for certain generalizations of Galvin's Problem.

2   Preliminaries

For non-negative integer n, $[n]$ denotes the set $\{1, 2, \ldots, n\}$. Let F be a field and $X = \{x_1, x_2, \ldots, x_n\}$ be a set of variables. An arithmetical circuit Φ over F and X is a directed acyclic graph with nodes of in-degree zero or two. Nodes with in-degree zero are called *inputs* and are labeled by variables or field elements. Nodes with in-degree two are called *gates* and have labels $\in \{\times, +\}$. For each gate g in Φ, one has associated the *polynomial computed by g*, denoted by Φ_g, which is defined in the obvious manner. Also let Φ_g stand for the subcircuit rooted at gate g. It will be made clear from the context which meaning is intended. Denote by X_g the set of variables used by the subcircuit Φ_g. The size of Φ, denoted by $|\Phi|$, is taken to be the number gates. If the underlying graph of Φ is a tree, Φ is called a *formula*. For a polynomial f, $C(f)$ and $L(f)$ denote the smallest size of a circuit or formula, respectively, computing f.

An arithmetical circuit Φ is called *weakly-skew* if at every multiplication gate g with inputs g_1 and g_2, one of Φ_{g_1} and Φ_{g_2} is disjoint from the rest of Φ. Φ is called *skew* if for each multiplication gate at least one g_1 and g_2 is an input gate (See [5]). For a polynomial f, $C_{ws}(f)$ and $C_s(f)$ denote the smallest size of a weakly-skew or skew circuit computing f, respectively.

A polynomial f is called *multilinear* if for any monomial of f, every variable has degree at most one. A circuit Φ is *semantically multilinear* if every polynomial computed at any gate of Φ is multilinear. Φ is called *syntactically multilinear* if for each multiplication gate g with inputs g_1 and g_2, X_{g_1} and X_{g_2} are disjoint. For a polynomial f, $C_{syn}(f)$ and $C_{sem}(f)$ denote syntactic and semantic multilinear circuit size, respectively. Similarly, $L_{syn}(f)$ and $L_{sem}(f)$ denote syntactic and semantic multilinear formula size. These definitions will be combined in the obvious manner. For example, $C_{syn,ws}(f)$ denotes the smallest size of a syntactically multilinear weakly-skew circuit computing f. Standard notation for arithmetical circuit classes will be followed (See e.g. [12]). For any class $\mathcal{C} \in \{VNC_1, VNC_2, VP_s, VP_{ws}, VP\}$, let syn-mlin-$\mathcal{C}$ and mlin-$\mathcal{C}$ stand for the class of p-families of polynomials that have $\mathcal{C}$-circuits which additionally are required to be syntactically or semantically multilinear, respectively.

Definition 1 (See [9]). *An algebraic branching program (ABP) over a field F and a set of variables X is a 4-tuple $B = (G, w, s, t)$, where $G = (V, E)$ is a directed acyclic graph for which V can be partitioned into levels $L_0, L_1, \ldots, L_d$, where $L_0 = \{s\}$ and $L_d = \{t\}$. Vertices s and t are called the source and sink of B, respectively. Edges may only go between consecutive levels L_i and L_{i+1}. The weight function $w : E \to F[X]$ assigns homogeneous linear forms to the edges of G. For a path p in G, we extend the weight function by $w(p) = \prod_{e \in p} w(e)$. For $i, j \in V$, let $P_{i,j}$ be the collection of all path in G from i to j. The program B computes the polynomial $\sum_{p \in P_{s,t}} w(p)$. The size of B is taken to be $|V|$.*

For a linear form $L(x) = \sum_{i=1}^{n} c_i x_i$, define $\operatorname{coef}[L, x_i] = c_i$. An ABP B is called *syntactically multilinear*, if for any directed path p in B, for all i, there is at most one edge e on p with $\operatorname{coef}[w(e), x_i] \neq 0$. $B(f)$ denotes the smallest size of an ABP computing f, and $B_{syn}(f)$ denotes such size with the additional restriction of syntactic multilinearity.

3   Circuit Transformations

It is convenient to work with the following data structure: a *skew schedule* is a directed acyclic graph G with weights on the edges $\in F \cup X$, where the out-degree of a vertex is either zero, one or two, and where for any vertex v with distinct edges $e_1 = (v, w)$ and $e_2 = (v, u)$, the labels of e_1 and e_2 equal to 1. For a directed acyclic graph G with node $s \in V[G]$, a path p in G is called a *maximal path with starting point s*, if the first vertex of p is s and the last vertex of p has no outgoing edges.

Lemma 1. *For any polynomial f, $C_{syn,s}(f) \leq 5C_{syn,ws}(f)$.*

Proof. First process Φ so that any addition gate has its input coming in from different gates by inserting dummy addition gates. This at most doubles the size. Let e' be the new size. Let $g_1, g_2, \ldots, g_{e'}$ be a topological sort of the gates of Φ, where wlog. we assume $\Phi_{g_{e'}} = f$, and that $g_{e'}$ is the only gate with out-degree zero. Let $g_{-m+1}, g_{-m+2}, \ldots, g_0$ be the set of inputs of Φ. Sequentially for stages $k = 1, 2, \ldots, e'$, we construct a skew schedule G_k from G_{k-1}. To initialize, let G_0 consists of m distinct directed edges. For each input g of Φ, we select a unique edge among $E[G_0]$ and put the label of g on it. Let B be the set of vertices in G_0 with out-degree zero. The set B will remain as a subset of vertices in each G_k. We will never change the in-degree of vertices in B. At the beginning of stage k, the skew schedule G_{k-1} will satisfy:

1. Each node $g = g_{k'}$ with $k' < k$ will correspond one-to-one with a vertex $v_g \in V[G_{k-1}] \backslash B$.
2. Let $\mathcal{G}_{k-1}$ be the set of nodes $g_{k'}$ with $k' < k$, that are used by gates $g_{k''}$ for some $k'' \geq k$. For any $g \in \mathcal{G}_{k-1}$, $\sum_{p \in \mathcal{P}} \prod_{e \in p} w(e) = \Phi_g$, where $\mathcal{P}$ is the collection of all maximal paths with starting point v_g in G_{k-1},
3. On any directed path in G_{k-1} no variable appears more than once,

4. For any node $g = g_{k'}$ with $k' < k$, the set of nodes not in B reachable in G_{k-1} from v_g, is precisely $\{v_{g'} : g' \in \Phi_g\}$.

At the beginning of stage $k = 1$, we have that $\mathcal{G}_0$ is the set of all input gates. For each input gate g, v_g is defined to be the starting vertex of the unique edge we have selected for it. Properties (1)-(4) can now be verified to hold. At stage k we do the following:

Case I: $g_k = +(g_i, g_j)$. We have that $g_i, g_j \in \mathcal{G}_{k-1}$. We construct G_k from G_{k-1} by adding one new vertex w with edges of weight 1 from w to v_{g_i} and v_{g_j}. No parallel edges are created since $i \neq j$. Let us verify the needed properties. Property (3) is clear. It is also clear that if we let $\mathcal{P}$ be the collection of all maximal paths starting in w that $\sum_{p \in \mathcal{P}} \prod_{e \in p} w(e) = \Phi_{g_k}$. If we are at the last iteration, i.e. $k = e'$, then this is all we are required to verify. Otherwise, g_k will be used later on, i.e. $g_k \in \mathcal{G}_k$. Observe that $\mathcal{G}_k = \mathcal{G}_{k-1} \cup \{g_k\} - S$, where $S \subseteq \{g_i, g_j\}$. We define $v_{g_k} = w$. By our above observation for the vertex w, and the fact that we do not modify connectivity for the other vertices, Property (2) holds. Property (4) is clear.

Case II: $g_k = \times(g_i, g_j)$. Wlog. assume Φ_{g_j} is disjoint from the rest of Φ. We have that $g_i, g_j \in \mathcal{G}_{k-1}$. For $s \in \{i, j\}$, let W_s be the set of vertices in G_{k-1} reachable from v_{g_s}. W_i and W_j are disjoint. Namely, suppose $w \in W_i \cap W_j$. If $w \notin B$ then Property (4) implies there exists a shared node in Φ_{g_i} and Φ_{g_j}, which is a contradiction. In case $w \in B$, then since we do not add edges into w, we have a vertex $w' = v_{g'}$ for some input gate g' with $w' \in W_i \cap W_j$. Hence we again have a contradiction.

Let $E \subseteq W_j$ be the set of vertices reachable from v_{g_j} with out-degree zero. We define G_k to be the graph G_{k-1} modified by adding an edge (v, v_{g_i}) of weight 1 for each $v \in E$. We add[2] a new vertex w and edge (w, v_{g_j}) with weight 1, and let $v_{g_k} = w$. Since $W_i \cap W_j = \emptyset$, no vertex from E is reachable from v_{g_i}. Hence G_k is an acyclic graph. Observe G_k is a skew schedule. We will now verify Properties (1)-(4).

Let $\mathcal{P}$ be the set of maximal paths starting in v_{g_k} in G_k. Let $\mathcal{P}_s$ be the set of maximal paths in G_{k-1} starting in v_{g_s}, for $s \in \{i, j\}$. For $p \in \mathcal{P}_i$ and $q \in \mathcal{P}_j$, let $q \# p$ denote the path in G_k that is (v_{g_k}, v_{g_j}), followed by q, followed by the edge with weight 1 into v_{g_i}, followed by p. We have that $\mathcal{P} = \{q \# p : q \in \mathcal{P}_j, p \in \mathcal{P}_i\}$. This means $\sum_{r \in \mathcal{P}} \prod_{e \in r} w(e) = \sum_{p \in \mathcal{P}_i} \prod_{e \in p} w(e) \cdot \sum_{q \in \mathcal{P}_j} \prod_{e \in q} w(e) = \Phi_{g_i} \cdot \Phi_{g_j} = \Phi_{g_k}$. In case this was the last iteration, i.e. $k = e'$, this is all we need together with Property (3) to be verified below. Otherwise, since g_k will be used later again, $g_k \in \mathcal{G}_k$. Observe that $\mathcal{G}_k = \mathcal{G}_{k-1} - S \cup \{g_k\}$, where S is the set of nodes in Φ_{g_j}.

By what we observed above, Property (2) holds for g_k. For $g \neq g_k$ in $\mathcal{G}_k$, the only way Property (2) can be disturbed is if some vertex $w \in E$ is reachable from v_g in G_{k-1}. This means some $w' \notin B$ is reachable from both v_g and v_{g_j} in

---

[2] This is not strictly necessary, but we do so to simplify the proof.

G_{k-1}, but then Φ_g and Φ_{g_j} share a vertex by 'previous' Property (4). Since Φ is weakly-skew, g must be a node in Φ_{g_j}, but that is a contradiction since $g \in \mathcal{G}_k$.

To check Property (3), we note that the only edges with variables are of the form (v, b) with $b \in B$ and $v = v_g$, for some input $g \in \Phi$. Property (3) can be violated only, if in G_{k-1} we have for such (v, b) that some vertex in E can be reached from b, and that in G_{k-1} there exists a path starting in v_{g_i} going over an edge (v', b') with $b' \in B$ and $v' = v_{g'}$, for some input gate g', that has the same variable label as (v, b). This means that $g' \in \Phi_{g_i}$. Similar as above, by Property (4), it must be that $g \in \Phi_{v_{g_j}}$. By syntactic multilinearity, we conclude the labels of (v, b) and (v', b') must be different. Property (4) clearly holds.

This completes the description the graphs $G_1, G_2, \ldots, G_{e'}$. $G_{e'}$ is a skew schedule of size at most $2m + e' \leq 5e'$. We can evaluate it node by node in a bottom-up fashion. This yields a syntactically multilinear skew circuit computing f of size at most $5e' \leq 10e$ gates. To optimize the constant to be 5 instead of 10, we observe adding dummy addition gates is not required.   □

Lemma 2. *For any polynomial f, $C_{sem,s}(f) \leq 5C_{sem,ws}(f)$.*

Proof. Modify the multiplication case in the proof of Lemma 1 as follows. For each variable x_i appearing in the *polynomial Φ_{g_i}*, replace any edge weight x_i in the induced subgraph $G_{k-1}[W_j]$ by zero. This does not alter the polynomial represented at vertex v_{g_i}, since it cannot contain variable x_i. Polynomials represented at other vertices in $G_k[W_j]$ can have changed, but cannot be used at later stages. The substitution has forced all these polynomials to be multilinear.   □

Removing Property (3) in the proof of Lemma 1 immediately yields a proof that for any polynomial f, $C_s(f) \leq 5C_{ws}(f)$. This reproves $\text{VP}_{ws} = \text{VP}_s$, but without an underlying cubic blow-up in size as in [5]. Let us put the basic facts together about the measures that are considered. The proof is left to the reader.

Lemma 3. *For a homogeneous polynomial of degree d,*

1. $C(f) \leq C_{ws}(f) \leq L(f)$ and $C_{syn}(f) \leq C_{syn,ws}(f) \leq L_{syn}(f)$.
2. $C_{ws}(f) \leq C_s(f) \leq 5C_{ws}(f)$ and $C_{syn,ws}(f) \leq C_{syn,s}(f) \leq 5C_{syn,ws}(f)$.
3. $B(f) \leq d \cdot (4C_s(f) + 2)$ and $B_{syn}(f) \leq d \cdot (4C_{syn,s}(f) + 2)$.
4. $\alpha\sqrt{C_s(f)/n} \leq B(f)$ and $\alpha\sqrt{C_{syn,s}(f)/n} \leq B_{syn}(f)$, for some constant $\alpha > 0$.

Note that the above lemma implies Item 1 of Lemma 1 in [9], modulo constants. Also note that proving super-polynomial lower bounds on $B_{syn}(f)$ for some homogeneous $f \in$ s-mlin-VP is equivalent to showing s-mlin-$\text{VP}_{ws} \neq$ s-mlin-VP.

4   Ordered ABPs

Definition 2. *Let $B = (G, w, s, t)$ be an ABP over a field F and set of variables $X = \{x_1, x_2, \ldots, x_n\}$. Say a directed path p from s to t respects a permutation $\pi : [n] \to [n]$, if whenever an edge e_1 appears before an edge e_2 on p and $coef[w(e_1), x_{\pi(i)}] \neq 0$ and $coef[w(e_2), x_{\pi(j)}] \neq 0$, one has that $i < j$. B is called ordered, if there exists a permutation π that is respected by all directed s, t-paths.*

For a polynomial f, we denote ordered ABP size by $B_{ord}(f)$. Note an ordered ABP is syntactically multilinear. We observe that lower bounds for non-commutative algebraic branching programs of Nisan [9] can be transferred to the ordered model.

Theorem 5. *Any ordered algebraic branching program $B = (G, w, s, t)$ computing the permanent or determinant of an $n \times n$ matrix of variables has size at least 2^n.*

Proof. (sketch) First suppose that the branching program B respects a row-by-row ordering of the variables, i.e. if x_{ij} comes before x_{kl} on a directed s, t-path in B, then $i < k$. Interpreting B as defining an ABP over non-commuting variables, we have that B computes the 'ordered' permanent or determinant as defined in Definition 3 in [9], which is shown to require 2^n nodes to be computed. In case B respects some arbitrary permutation π, it can be observed that evaluating B non-commutatively yields a polynomial, which, in Nisan's terms, is *weakly equivalent* to the permanent or determinant, and thus also requires size 2^n. ☐

The above bound for the permanent and determinant is of level $2^{\Omega(\sqrt{N})}$, where N is the number of variables. In [9] a bound of $2^{\Omega(N)}$ is proven for the non-commutative model, but this is for a polynomial that is not multilinear. In order to obtain a bound of level $2^{\Omega(N)}$, we now turn to the aforementioned full rank polynomials [4,7,8].

4.1 Full Rank Polynomials

Let $X = \{x_1, x_2, \ldots, x_{2n}\}$, $Y = \{y_1, y_2, \ldots, y_n\}$ and $Z = \{z_1, z_2, \ldots, z_n\}$ be sets of indeterminates and let F be a field. Following [4,7,8], for a multilinear polynomial $g \in F[Y, Z]$, we define the $2^n \times 2^n$ *partial derivatives matrix* M_g with entries from F by taking $M_g(m_1, m_2) =$ coefficient of monomial $m_1 m_2$ in g, where m_1 and m_2 range over all multilinear monic monomials in Y and Z variables, respectively. A *partition* of X into Y and Z is any bijection $A : X \to Y \cup Z$. For a partition A and a polynomial $f \in F[X]$ denote by f^A the polynomial obtained from f by substitution of x_i by $A(x_i)$, for all $i \in [2n]$. The polynomial f is said to be of *full rank* if for every partition A, rank $M_{f^A} = 2^n$. For a multilinear polynomial $g \in F[Y, Z]$, let Y_g and Z_g be the sets of Y and Z variables appearing in g with exponent one. The rank of the partial derivatives matrix enjoys the following elementary properties:

Proposition 1 ([8]). *Let $g, g_1, g_2 \in F[Y, Z]$ be multilinear polynomials. Then*

1. *rank $M_g \leq 2^{\min(|Y_g|, |Z_g|)}$,*
2. *rank $M_{g_1+g_2} \leq$ rank $M_{g_1} +$ rank M_{g_2}, and*
3. *rank $M_{g_1 \cdot g_2} =$ rank $M_{g_1} \cdot$ rank M_{g_2}, provided $Y_{g_1} \cap Y_{g_2} = \emptyset$ and $Z_{g_1} \cap Z_{g_2} = \emptyset$.*

4.2 Lower Bound

Theorem 6. *Let X be a set of $2n$ variables, and let F be a field. For any full rank homogeneous polynomial f of degree n over X and F, $B_{ord}(f) \geq 2^{n/4}$.*

Proof. Let $B = (G, w, s, t)$ be an ordered ABP computing f. Let $L_0, L_1, \ldots, L_n$ be the levels of B. For $v, w \in V[G]$ such that v is on a lower level than w, let $f_{v,w}$ denote the polynomial computed by the subprogram of B with source v and sink w. Let $X_{v,w}$ denote the set of all variables appearing on directed paths from v to w.

Suppose that B respects the permutation $\pi : [2n] \to [2n]$. Consider any partition A that assign all n y-variables to $\{x_{\pi(1)}, x_{\pi(2)}, \ldots, x_{\pi(n)}\}$ and all n z-variables to $\{x_{\pi(n+1)}, x_{\pi(n+2)}, \ldots, x_{\pi(2n)}\}$. Let $0 < i < n$, then we can write

$$f = f_{s,t} = \sum_{v \in L_i} f_{s,v} f_{v,t}. \tag{1}$$

Consider a node $v \in L_i$.

Case I: $X_{s,v}$ contains a variable $x_{\pi(k)}$ with $k > n$. In this case, since B respects π, each variable in $X_{v,t}$ is assigned a z-variable by A. Hence *rank* $M_{f_{v,t}^A} \leq 1$. Paths from v to t are of length $n - i$, so $|X_{v,t}| \geq n - i$. None of the variables in $X_{v,t}$ can appear on paths from s to v, so $|X_{s,v}| \leq 2n - |X_{v,t}| \leq n + i$. By Item 1 of Proposition 1, we get

$$rank \ M_{f_{s,t}^A} \leq 2^{|X_{s,v}|/2} \leq 2^{(n+i)/2}.$$

Using multiplicativity (Proposition 1, Item 3), we conclude that *rank* $M_{f_{s,v}^A f_{v,t}^A}$ is at most $2^{(n+i)/2}$.

Case II: $X_{s,v}$ does not contain a variable $x_{\pi(k)}$ with $k > n$. In this case all of $X_{s,v}$ is assigned y-variables. Note $|X_{s,v}| \geq i$. Hence $|X_{v,t}| \leq 2n - i$. Arguing as above, we obtain that *rank* $M_{f_{s,v}^A f_{v,t}^A} \leq 2^{n-i/2}$.

Combining both cases, we conclude that for any $v \in L_i$, for $i \leq n/2$, *rank* $M_{f_{s,v}^A f_{v,t}^A} \leq 2^{\max(n-i/2,(n+i)/2)} \leq 2^{n-i/2}$. Using subadditivity (Proposition 1, Item 2) and Equation (1) we get that that for $i \leq n/2$, *rank* $M_{f^A} \leq |L_i| 2^{n-i/2}$. Since f is of full rank, *rank* $M_{f^A} = 2^n$. Hence, for even n, $|L_{n/2}| \geq 2^{n/4}$. For odd n, one can observe similarly that $|L_{(n-1)/2}| + |L_{(n+1)/2}| \geq 2^{n/4}$. $\qquad\square$

If the construction of a full rank polynomial from [7] can be modified to yield a *homogeneous* full rank polynomial, we can apply Theorem 6 to obtain lower bounds. We next verify this can be done, provided we work over a suitable extension field of the underlying field F.

4.3   Constructing a Homogeneous Full Rank Polynomial

Let $\mathcal{W} = \{\omega_{i,l,j}\}_{i,l,j \in [2n]}$ be sets of variables. For each interval $[i,j] \subseteq [2n]$ of even length, we define a polynomial $f_{i,j} \in F[X, \mathcal{W}]$ inductively as follows: if $|[i,j]| = 0$, then define $f_{i,j} = 1$. If $|[i,j]| > 0$, define $f_{i,j} = (x_i + x_j) f_{i+1,j-1} + \sum_l \omega_{i,l,j} f_{i,l} f_{l+1,j}$, where we sum over all l such that $|[i,l]|$ is even. Finally, f is defined to be $f_{1,2n}$. It can be verified inductively that $f_{i,j}$ is homogeneous of degree $|[i,j]|/2$ in the X-variables. Our definition differs from [7] in that we have the term $(x_i + x_j) f_{i+1,j-1}$ instead of $(1 + x_i x_j) f_{i+1,j-1}$. So we also immediately know each $f_{i,j}$ is multilinear. Call a partition A *balanced* on an interval $[i,j]$, if $|\{A(x_k) : k \in [i,j]\} \cap Y| = |\{A(x_k) : k \in [i,j]\} \cap Z|$. We have the following adaption of Lemma 4.3 in [7]:

Lemma 4. *Let $A : X \to Y \cup Z$ be a partition. Let $G = F(\mathcal{W})$ be the field of rational functions over the field F and set of variables $\mathcal{W}$. Then for any interval $[i,j]$ of length $2m$ that is balanced on A, we have that rank $M_{f^A_{i,j}} = 2^m$, where the rank is measured over the field G.*

Proof. The proof of Lemma 4.3 of [7] goes through for the modified polynomial $f_{i,j}$ after noting that for the case $\mathcal{D}_{i,i} = 1$ and $\mathcal{D}_{j,j} = -1$ of their proof, one still has rank $M_{A(x_i)+A(x_j)} = 2$. Namely, in this case $A(x_i) \in Y$ and $A(x_j) \in Z$, and rank $M_{y_s+z_t} = 2$, for any s and t. $\qquad\square$

Since the interval $[1, 2n]$ is balanced on any partition A, one concludes Theorem 4 follows from Lemma 4 and Theorem 6.

5   Unrestricted and Multilinear ABPs

Consider the following observation by Kristoffer Arnsfelt Hansen: if for a homogeneous algebraic branching program with linear forms on the edges, the number of edges between any two consecutive levels L_d and L_{d+1} is at most $K < n$, then the polynomial $f(x_1, x_2, \ldots, x_n)$ computed by the branching program vanishes on a linear space of dimension at least $n - K$. Provided this is a contradiction for f, one concludes $\max(|L_d|, |L_{d+1}|) \geq \sqrt{K}$.

For example, working this out for the elementary symmetric polynomial of degree d in n variables defined by $\sum_{S \subset [n], |S|=d} \prod_{i \in S} x_i$, it is known that if S_n^d vanishes on an affine linear space A, then $\dim(A) < (n + d)/2$ (See Theorem 1.1 in [13]). Applying above reasoning yields the following lower bound:

Theorem 7. *Let α be a constant with $0 < \alpha < 1$ and assume αn is integer. Over fields of characteristic zero, for the elementary symmetric polynomial $S_n^{\alpha n}$ of degree αn in n variables, it holds that $B(S_n^{\alpha n}) = \Omega(n^{3/2} \alpha \sqrt{1 - \alpha})$.*

For above argument to work, K must be smaller than n. Thus lower bounds obtained this way will be no stronger than $\Omega(n^{3/2})$, for polynomials of degree $\Theta(n)$ in n variables. In order to overcome this limitation, we turn to a hypercube covering problem.

Consider the $2n$-dimensional hypercubes $H_{2n} = \{-1, 1\}^{2n}$ and $B_{2n} = \{0, 1\}^{2n}$ over the real numbers. Let $x \cdot y$ denote the standard inner product defined by $x \cdot y = \sum_{i=1}^{2n} x_i y_i$. Let $\mathbf{1}$ denote the vector 1^{2n}. Let $H_{2n}^e = \{x \in H_{2n} : x \cdot \mathbf{1} = 0\}$. Of interest are minimal size coverings of H_{2n}^e by hyperplanes, where coefficients for the defining equations are taken from particular subsets of B_{2n}. More precisely, for $W \subset [2n]$, define $B_{2n}^W = \{b \in B_{2n} : \mathrm{wt}(b) \in W\}$, where $\mathrm{wt}(b) = b \cdot \mathbf{1}$. Define

$$Q(n, W, d) = \min\{|E| : E \subseteq B_{2n}^W, (\forall x \in H_{2n}^e), (\exists e \in E), |x \cdot e| \leq d\}.$$

Finding the value of $Q(n, W, d)$ becomes interesting only for certain weight sets W. For example, if $2n \in W$, then $E = \{1\}$ trivially covers all of H_{2n}^e. The *discrepancy* parameter d should also be small w.r.t. $\min(W)$, e.g. trivially $Q(n, W, \min(W)) = 1$. Also, in case W does not contain an even number and $d = 0$, we have that $Q(n, W, d)$ is ill-defined. In all other cases $Q(n, W, d)$ is well-defined. Namely, say $2\ell \in W$. By taking E to be all $2n$ cyclic shifts of $1^{2\ell}0^{2n-2\ell}$, we have that for each $x \in H_{2n}^e$, there must exist some $e \in E$, $x \cdot e = 0$. Similarly, this set E works, in case W contains no even numbers, but $d \geq 1$. The crucial question is whether for cases that avoid trivialities, one can do significantly better than $|E|$ being linear in n.

The special case of finding $m(k) := Q(2k, \{2k\}, 0)$ is a problem posed originally by Galvin (See [14]). For an upper bound, note one requires only half of all $4k$ cyclic shifts, i.e. $m(k) \leq 2k$. For odd k, Frankl and Rödl proved the linear lower bound $m(k) > \epsilon k$, for fixed $\epsilon > 0$ [1]. The proof relies on a strong result in extremal set theory they proved, which resolved a \$250 problem of Erdős. Later the bound was improved to $m(k) \geq 2k$, for odd k [15].

Consider $Q(n, [\epsilon_0 n, (1 + \epsilon_1)n], 2 \log n)$, for fixed $0 < \epsilon_0 < \epsilon_1 < 1$. From Theorem 8 below it will follow, that proving an $\widetilde{\Omega}(n)$ lower bounds on this quantity would yield an $\widetilde{\Omega}(n^2)$ multilinear ABP lower bound. In light of the result by Frankl and Rödl such a linear lower bound appears plausible. Also note the linear lower bound by Alon et al. for covering the entire hypercube, in case the defining equations have coefficients in $\{-1, 1\}$ instead of in $\{0, 1\}$ [14]. They define for $n \equiv d \pmod 2$,

$$K(n, d) = \min\{|E| : E \subseteq H_n, (\forall x \in H_n), (\exists e \in E), |x \cdot e| \leq d\},$$

and prove $K(n, d) = \lceil n/(d+1) \rceil$. The relation between $Q(n, W, d)$ and multilinear ABPs is expressed in the following theorem:

Theorem 8. *Let X be a set of $2n$ variables, and let F be a field. For any full rank homogeneous polynomial f of degree n over X and F,*

$$B_{syn}(f) = \Omega\left(\sum_{r=1}^{n-1} \min(n, Q(n, [r, n + r], 2 \log n))\right).$$

Proof. Let $B = (G, w, s, t)$ be a multilinear ABP computing f. Let $L_0, L_1, \ldots, L_n$ be the levels of B. For $v, w \in V[G]$, let $f_{v,w}$ denote the polynomial computed by the subprogram of B with source v and sink w. Let $X_{v,w}$ denote the set of all variables appearing on directed paths from v to w. Consider r such that $0 < r < n$. Write $f = f_{s,t} = \sum_{v \in L_r} f_{s,v} f_{v,t}$.

By syntactic multilinearity, we have that $|X_{s,v}| \geq r$ and $|X_{v,t}| \geq n - r$. The latter implies $|X_{s,v}| \leq n + r$, again by syntactic multilinearity. Let $\chi(X_{s,v}) \in B_{2n}^{[r,n+r]}$ denote the characteristic vector of $X_{s,v}$.

Suppose that $|L_r| < Q(n, [r, n + r], 2 \log n)$. Then there exists $\gamma \in H_{2n}^e$ such that for every $b \in \{\chi(X_{s,v}) : v \in L_r\}$, $|\gamma \cdot b| > 2 \log n$. Let $A : X \to Y \cup Z$ be any partition that assigns a Y variable to x_i, if $\gamma_i = 1$, and a Z variable otherwise, for all $i \in [2n]$.

Let B' be the branching program obtained from B by substituting according to A. For nodes v and w, we let $Y_{v,w}$ and $Z_{v,w}$ denote the sets of y and z variables, respectively, appearing on paths from v to w in B'. Then for any $v \in L_r$, $\min(Y_{s,v}, Z_{s,v}) \leq (Y_{s,v} + Z_{s,v})/2 - \log n \leq |X_{s,v}|/2 - \log n$. Hence, by Item 1 of Proposition 1, we have that $rank\ M_{f_{s,v}^A} \leq 2^{|X_{s,v}|/2 - \log n}$. By syntactic multilinearity, none of the variables appearing on paths from s to v can appear on paths from v to t. So $|X_{v,t}| \leq 2n - |X_{s,v}|$. By Item 1 of Proposition 1 we get $rank\ M_{f_{v,t}^A} \leq 2^{n - |X_{s,v}|/2}$. Using multiplicativity (Proposition 1, Item 3), we conclude $rank\ M_{f_{s,v}^A f_{v,t}^A} \leq 2^{n - \log n}$, and thus using subadditivity (Proposition 1, Item 2) and since $f = \sum_{v \in L_r} f_{s,v} f_{v,t}$ that $rank\ M_{f^A} \leq |L_r| 2^{n - \log n}$. Since f is of full rank, $rank\ M_{f^A} = 2^n$. We conclude that $|L_r| \geq n$. $\qquad\square$

One derives a lower bound as follows: assuming for simplicity ℓ is even, for an individual vector $e \in B_{2n}$ of weight ℓ, with $r \leq \ell \leq n + r$, the number of vectors $x \in H_{2n}^e$ with $|x \cdot e| \leq d$, is given by

$$\sum_{\substack{i=-d \\ i \text{ even}}}^{d} \binom{\ell}{\ell/2 - i/2}\binom{2n - \ell}{n - \ell/2 + i/2}. \tag{2}$$

One can bound (2) by $O(\frac{(d+1)2^{2n}}{\sqrt{\ell(2n-\ell)}}) = O(\frac{(d+1)2^{2n}}{\sqrt{r(n-r)}})$. This is an $O((d+1)\sqrt{\frac{2n}{r(n-r)}})$ fraction of H_{2n}^e. Hence $Q(n, [r, n+r], d) = \Omega((d+1)^{-1}\sqrt{\frac{r(n-r)}{2n}})$. Applying Theorem 8 and Lemma 4 and summing for r in the range $[\epsilon_0 n, \epsilon_1 n]$, for fixed $0 < \epsilon_0 < \epsilon_1 < 1$ yields the following theorem:

Theorem 9. *Let X be a set of $2n$ variables. For any field F, there exists an extension field G of F and explicitly constructed multilinear polynomial $f \in G[X]$, such that any multilinear algebraic branching program over X and G computing f has $\Omega(\frac{n\sqrt{n}}{\log n})$ nodes.*

6   Conclusions

The author believes that full rank polynomials cannot be computed by polynomial size multilinear ABPs. If true, this unfortunately would rule out separating s-mlin-VNC$_1$ and s-mlin-VP$_s$ by means of "merely" supplying an algorithm. Hence the attention has been on using full rank polynomials to prove lower bounds for multilinear ABPs. As the constructions in this paper show, proving super-polynomial lower bounds on the size of a syntactically multilinear ABP computing a polynomial $f \in$ s-mlin-VNC$_2$ is equivalent to separating s-mlin-VP$_{ws}$ and s-mlin-VP.

It should be noted the ABP-model can be quite powerful. It is possible for certain polynomials to have branching program size sublinear in the number of input variables. The prime example being that, given two $n \times n$ matrices X and Y of variables, one can compute $f = \sum_{i,j \in [n]}(XY)_{ij}$ with a syntactically multilinear ABP with $O(\sqrt{N})$ many nodes, where $N = 2n^2$ is the number of input variables. This is an example of a polynomial, for which the multilinear ABP-model is at least "quadratically more efficient" than the syntactic multilinear circuit model. In the latter model, the currently best know lower bound for an explicit function is of level $\Omega(\frac{n^{4/3}}{\log^2 n})$ [8]. Theorem 8 supplies a lower bound strategy, which yielded an $\Omega(\frac{n^{3/2}}{\log n})$ lower bound for multilinear ABPs. By resolving a certain generalization of Galvin's Problem this method can yield an $\tilde{\Omega}(n^2)$ lower bound for syntactically multilinear ABPs.

Acknowledgments. I thank Peter Bro Miltersen, Kristoffer Arnsfelt Hansen and Oded Lachish for helpful discussions.

References

1. Frankl, P., Rödl, V.: Forbidden intersections. Trans. Amer. Math. Soc. 300(1), 259–286 (1987)
2. Valiant, L., Skyum, S., Berkowitz, S., Rackoff, C.: Fast parallel computation of polynomials using few processors. SIAM J. Comput. 12, 641–644 (1983)
3. Brent, R.: The parallel evaluation of general arithmetic expressions. J. Assn. Comp. Mach. 21, 201–206 (1974)
4. Raz, R.: Separation of multilinear circuit and formula size. Theory of Computing 2(6), 121–135 (2006)
5. Malod, G., Portier, N.: Characterizing Valiant's algebraic complexity classes. In: MFCS, pp. 704–716 (2006)
6. Berkowitz, S.: On computing the determinant in small parallel time using a small number of processors. Inf. Proc. Lett. 18, 147–150 (1984)
7. Raz, R., Yehudayoff, A.: Balancing syntactically multilinear arithmetical circuits. Journal of Computational Complexity (to appear)
8. Raz, R., Shpilka, A., Yehudayoff, A.: A lower bound for the size of syntactically multilinear arithmetic circuits. In: Proc. 48th.Annual IEEE Symposium on Foundations of Computer Science, pp. 438–448 (2007)
9. Nisan, N.: Lower bounds for non-commutative computation: extended abstract. In: Proc. 23rd Annual ACM Symposium on the Theory of Computing, pp. 410–418 (1991)
10. Andreev, A., Baskov, J., Clementi, E., Rolim, R.: Small Pseudo-Random Sets Yield Hard Functions: New Tight Explicit Lower Bounds for Branching Programs. In: Wiedermann, J., Van Emde Boas, P., Nielsen, M. (eds.) ICALP 1999. LNCS, vol. 1644. Springer, Heidelberg (1999)
11. Bryant, R.E.: On the complexity of vlsi implementations and graph representations of boolean functions with application to integer multiplication. IEEE Trans. Computers 40(2), 205–213 (1991)
12. Bürgisser, P., Claussen, M., Shokrollahi, M.A.: Algebraic Complexity Theory. Springer, Heidelberg (1997)
13. Shpilka, A.: Affine projections of symmetric polynomials. In: Proc. 16th Annual IEEE Conference on Computational Complexity, pp. 160–171 (2001)
14. Alon, N., Bergmann, E., Coppersmith, D., Odlyzko, A.: Balancing sets of vectors (1988)
15. Enomoto, H., Frankl, P., Ito, N., Nomura, K.: Codes with given distances. Graphs and Combinatorics 3, 25–38 (1987)

Periodicity and Immortality
in Reversible Computing[*]

Jarkko Kari[1] and Nicolas Ollinger[2]

[1] Department of Mathematics, FIN-20014 University of Turku, Finland
jkari@utu.fi
[2] Laboratoire d'informatique fondamentale de Marseille (LIF),
Aix-Marseille Université, CNRS,
39 rue Joliot-Curie, 13 013 Marseille, France
Nicolas.Ollinger@lif.univ-mrs.fr

Abstract. We investigate the decidability of the periodicity and the immortality problems in three models of reversible computation: reversible counter machines, reversible Turing machines and reversible one-dimensional cellular automata. Immortality and periodicity are properties that describe the behavior of the model starting from arbitrary initial configurations: immortality is the property of having at least one non-halting orbit, while periodicity is the property of always eventually returning back to the starting configuration. It turns out that periodicity and immortality problems are both undecidable in all three models. We also show that it is undecidable whether a (not-necessarily reversible) Turing machine with moving tape has a periodic orbit.

Introduction

Reversible computing is the classical counterpart of quantum computing. Reversibility refers to the fact that there is an inverse process to retrace the computation back in time, i.e., the system is time invertible and no information is ever lost. Much of the research on reversible computation is motivated by the Landauer's principle which states a strict lower bound on the amount of energy dissipation which must take place for each bit of information that is erased [1]. Reversible computation can, in principle, avoid this generation of heat.

Reversible Turing machine (RTM) was the earliest proposed reversible computation model [2,3]. Since then, reversibility has been investigated within other common computation models such as Minsky's counter machines [4,5] and cellular automata [6]. In particular, reversible cellular automata (RCA) have been extensively studied due to the other physics-like attributes of cellular automata such as locality, parallelism and uniformity in space and time of the update rule.

All three reversible computation models are Turing complete: they admit simulations of universal Turing machines, which naturally leads to various undecidability results for reachability problems. In this work we view the systems,

---

[*] Work supported by grants of the French ANR and Academy of Finland # 211967.

E. Ochmański and J. Tyszkiewicz (Eds.): MFCS 2008, LNCS 5162, pp. 419–430, 2008.

however, rather differently by investigating their behavior from arbitrary starting configurations. This is more a dynamical systems approach. Each device is understood as a transformation $F : X \longrightarrow X$ acting on its configuration space X. In all cases studied here (counter machines, two Turing machine models – with moving head and with moving tape – and cellular automata) space X is endowed a topology under which F is continuous. In the cases of Turing machines with moving tape and cellular automata, it is the compact and metrizable topology obtained as the enumerable infinite product of the discrete topology on each finite component of a configuration. The action F may be partial, so that it is undefined for some elements of X. Configurations on which F is undefined are called *halting*. We call F *immortal* if there exists a configuration $x \in X$ that never evolves into a halting configuration, that is, $F^n(x)$ is defined for all positive integers n. In contrast, a *mortal* system eventually halts, regardless of the starting configuration. We call F *uniformly mortal* if a uniform time bound n exists such that $F^n(x)$ is not defined for any $x \in X$. If F is continuous, X compact, and the set of halting configurations open then mortality and uniform mortality are equivalent concepts. This means that mortal Turing machines and cellular automata are automatically uniformly mortal. In contrast, a counter machine may be mortal without being uniformly mortal. (A simple example is a one-counter machine where the counter value is repeatedly decremented until it becomes zero and the machine halts.)

Periodicity, on the other hand, is defined for *complete* systems: systems without halting configurations. We call total $F : X \longrightarrow X$ *uniformly periodic* if there is a positive integer n such that F^n is the identity map. *Periodicity* refers to the property that every configuration is periodic, that is, for every $x \in X$ there exists time n such that $F^n(x) = x$. Periodicity and uniform periodicity are equivalent concepts in the cases of cellular automata (Section 3.3) and Turing machines under both modes (Section 2.1), while a counter machine can be periodic without being uniformly periodic (Example 1 in Section 1.1).

In this work we are mainly concerned with decidability of these concepts. Immortality of unrestricted (that is, not necessarily reversible) Turing machines was proved undecidable already in 1966 by Hooper [7]. Our main result (Theorem 7) is a reversible variant of Hooper's approach where infinite searches during counter machine simulations by a Turing machine are replaced by recursive calls to the counter machine simulation itself with empty initial counters. Using reversible counter machines, the recursive calls can be unwound once the search is complete. In a sense this leads to a simpler construction than in Hooper's original article.

Our result also answers an open problem of control theory from [8]. That paper pointed out that if the immortality problem for reversible Turing machines is undecidable, then so is observability for continuous rational piecewise-affine planar homeomorphisms.

As another corollary we obtain the undecidability of the periodicity of Turing machines (Theorem 8). The related problem of determining if a given Turing machine has at least one periodic orbit (under the moving tape mode) is proved undecidable for reversible, non-complete Turing machines, and for non-reversible, complete Turing machines. The problem remains open under reversible and complete machines. The existence of periodic orbits in Turing machines and counter machines have been investigated before in [9,10]. Article [9] formulated a conjecture that every complete Turing machine (under the moving tape mode) has at least one periodic orbit, while [10] refuted the conjecture by providing an explicit counter example. The counter example followed the general idea of [7] in that recursive calls were used to prevent unbounded searches. In [10] is was also shown that it is undecidable if a given complete counter machine has a periodic orbit. We show that this is the case even under the additional constraint of reversibility (Theorem 6).

In Theorem 12 we reduce the periodicity problem of reversible Turing machine into the periodicity problem of one-dimensional cellular automata. The immortality problem of reversible cellular automata has been proved undecidable in [11]. Our proofs for the undecidability of immortality (Theorem 1) and periodicity (Theorem 3) among reversible counter machines follow the techniques of [5]. Interestingly, the uniform variants of both immortality and periodicity problems are decidable for counter machines (Theorems 2 and 4).

The paper is organized into three parts dealing with RCM (section 1), with RTM (section 2) and with RCA (section 3). Each part consists of four subsections on (1) definitions, (2) the immortality problem, (3) the periodicity problem, and (4) the existence of periodic orbits. Due to page constraints most proofs are short sketches of the main idea.

1   Reversible Counter Machines

1.1   Definitions

Following [5], we define special counter machine instructions for a simpler syntactic characterization of local reversibility and forget about initial and accepting states as we are only interested in dynamical properties.

Let $\Upsilon = \{0, +\}$ be the set of test values and $\Phi = \{-, 0, +\}$ be the set of counter operations whose reverse are defined by $-^{-1} = +$, $0^{-1} = 0$ and $+^{-1} = -$. For all $j \in \mathbb{Z}_k$ and $\phi \in \Phi$, testing τ and modifying $\theta_{j,\phi}$ actions are defined for all $k \in \mathbb{Z}$, $i \in \mathbb{Z}_k$ and $v \in \mathbb{N}^k$ as:

$$\tau(k) = \begin{cases} 0 \text{ if } k = 0 \\ + \text{ if } k > 0 \end{cases} \qquad \theta_{j,\phi}(v)(i) = \begin{cases} v(i) - 1 \text{ if } v(i) > 0, \ i = j \text{ and } \phi = - \\ v(i) \qquad \text{if } i \neq j \text{ or } \phi = 0 \\ v(i) + 1 \text{ if } i = j \text{ and } \phi = + \end{cases}$$

A k-counter machine M is a triple (S, k, T) where S is a finite set of states, $k \in \mathbb{N}$ is the number of counters, and $T \subseteq S \times \Upsilon^k \times \mathbb{Z}_k \times \Phi \times S$ is the transition table of the machine. Instruction $(s, u, i, -, t)$ is not allowed in T if $u(i) = 0$. A configuration $\mathfrak{c}$ of the machine is a pair (s, v) where $s \in S$ is a state and $v \in \mathbb{N}^k$ is the value of the counters. The machine can transform a configuration $\mathfrak{c}$ in a configuration $\mathfrak{c}'$ in one step, noted as $\mathfrak{c} \vdash \mathfrak{c}'$, by applying an instruction $\iota \in T$.

An instruction $(s, u, i, \phi, t) \in T$ can be applied to any configuration (s, v) where $\tau(v) = u$ leading to the configuration $(t, \theta_{i,\phi}(v))$. The transitive closure of $\vdash$ is noted as $\vdash^*$.

A counter machine (S, k, T) is a *deterministic k-counter machine (k-DCM)* if at most one instruction can be applied from any configuration. Formally, the transition table must satisfy the following condition:

$$(s, u, i, \phi, t) \in T \wedge (s, u, i', \phi', t') \in T \Rightarrow (i, \phi, t) = (i', \phi', t').$$

The *transition function* of a deterministic counter machine is the function $G : S \times \mathbb{N}^k \to S \times \mathbb{N}^k$ which maps a configuration to the unique transformed configuration, that is for all $(s, v) \in S \times \mathbb{Z}^k$,

$$G(s, v) = \begin{cases} (t, \theta_{i,\phi}(v)) \text{ if } (s, u, i, \phi, t) \in T \text{ and } \tau(v) = u \\ \bot \qquad\qquad \text{otherwise} \end{cases}$$

The set of reverse instructions of an instruction is defined as follows:

$(s, u, i, 0, t)^{-1} = \{(t, u, i, 0, s)\},$
$(s, u, i, +, t)^{-1} = \{(t, u', i, -, s)\},$ where $u'(i) = +, u'(j) = u(j)$ for $j \neq i$,
$(s, u, i, -, t)^{-1} = \{(t, u, i, +, s), (t, u', i, +, s)\},$ where $u'(i) = 0, u'(j) = u(j)$ for $j \neq i$.

The reverse T^{-1} of a transition table T is defined as $T^{-1} = \bigcup_{\iota \in T} \iota^{-1}$. The *reverse* of counter machine $M = (S, T)$ is the machine $M^{-1} = (S, T^{-1})$. A *reversible k-counter machine (k-RCM)* is a deterministic k-counter machine whose reverse is deterministic.

Example 1. The complete DCM $(\{l, l', r, r'\}, 2, T)$ with the following T is periodic but not uniformly periodic (*: any value): $\{ (l, (0, *), 0, 0, r), (r, (*, 0), 1, 0, l), (l, (+, *), 0, -, l'), (r, (*, +), 1, -, r'), (l', (*, *), 1, +, l), (r', (*, *), 0, +, r) \}$. In l, l' tokens are moved from the first counter to the second, and in states r, r' back to the first counter. Its reverse is obtained by swapping $l \leftrightarrow r$ and $l' \leftrightarrow r'$. $\qquad\square$

1.2   Undecidability of the Immortality Problem

Theorem 1. *It is undecidable whether a given 2-RCM is immortal.*

Proof sketch. By [7] the immortality problem is undecidable among 2-CM, while [5] provides an effective immortality/mortality preserving conversion of an arbitrary k-CM into a 2-RCM. $\qquad\blacksquare$

Remark. The 2-RCM constructed in the proof through Morita's construction [5] can be forced to have mortal reverse. This is obtained by adding in the original CM an extra counter that is being continuously incremented.

Theorem 2. *It is decidable whether a given k-CM is uniformly mortal.*

Proof sketch. Induction on k: The claim is trivial for $k = 0$. For the inductive step, let M be a k-CM, $k \geq 1$. For $i = 1, 2, \ldots, k$ set counter i to be always positive and test whether the so obtained $(k-1)$-CM M_i is uniformly mortal. If all k recursive calls return a positive answer, set n to be a common uniform mortality time bound for all k machines M_i. Since counters can be decremented by one at most, we know that configurations of M with some counter value $\geq n$ are mortal. Immortality hence occurs only if there is a period within the finite number of configurations with all counters $< n$.   ∎

1.3   Undecidability of the Periodicity Problem

Theorem 3. *It is undecidable whether a given 2-RCM is periodic.*

Proof sketch. Let $M = (S, 2, T)$ be a given 2-RCM whose reverse is mortal. In particular, there are no periodic configurations in M. According to the remark after Theorem 1 it is enough to effectively construct a complete 2-RCM M' that is periodic if and only if M is mortal. Machine M' has state set $S \times \{+, -\}$ where states $(s, +)$ and $(s, -)$ represent M in state s running forwards or backwards in time, respectively. In a halting configuration the direction is switched.   ∎

Analogously to Theorem 2 one can prove the following result.

Theorem 4. *It is decidable whether a given k-CM is uniformly periodic.*

1.4   Periodic Orbits

Theorem 5 ([10]). *It is undecidable whether a given complete 2-DCM admits a periodic configuration.*

Theorem 6. *It is undecidable whether a given complete 3-RCM admits a periodic configuration, and it is undecidable whether a given (not necessarily complete) 2-RCM admits a periodic configuration.*

Proof sketch. We first prove the result for complete 3-RCM. The construction in [5] shows that it is undecidable for a given 2-RCM $M = (S, 2, T)$ without periodic configurations and two given states s_1 and s_2 whether there are counter values n_1, n_2, m_1 and m_2 such that $(s_1, n_1, m_1) \vdash^* (s_2, n_2, m_2)$. By removing all transitions from state s_2 and all transitions into state s_1 we can assume without loss of generality that all configurations (s_1, n_1, m_1) and (s_2, n_2, m_2) are halting in M^{-1} and M, respectively. Using a similar idea as in the proof of Theorem 3 we effectively construct a 3-RCM $M' = (S \times \{+, -\}, 3, T')$ that simulates M forwards and backwards in time using states $(s, +)$ and $(s, -)$, respectively, and counters 1 and 2. The direction is switched at halting configurations. In addition, counter 3 is incremented at halting configurations, except when the state is s_1 or s_2.

Machine M' is clearly reversible and complete. Moreover, since M has no periodic configurations, the only periodic configurations of M' are those where M is simulated back and forth between states s_1 and s_2. This completes the proof for 3-RCM.

Using the construction of [5] a three counter RCM can be converted into a 2-RCM and that conversion preserves periodic orbits.   ∎

The 2-RCM provided by the construction in [5] is not complete. It seems likely that it can be modified to give a complete 2-RCM, but details remain to be worked out:

Conjecture 1. It is undecidable whether a given complete 2-RCM admits a periodic configuration.

2 Reversible Turing Machines

2.1 Definitions

The classical model of Turing machines consider machines with a moving head (a configuration is a triple $(s, z, c) \in S \times \mathbb{Z} \times \Sigma^{\mathbb{Z}}$). Following Kůrka [9], we consider machines with a moving tape as our base model to endow the space of configurations with a compact topology. Following [5], we define two kinds of instructions for a simpler syntactic characterization of local reversibility.

Let $\Delta = \{\leftarrow, \rightarrow\}$ be the set of directions with inverses $(\leftarrow)^{-1} = \rightarrow$ and $(\rightarrow)^{-1} = \leftarrow$. For all $\delta \in \Delta$ and $a \in \Sigma$, moving σ_δ and writing μ_a actions are defined for all $c \in \Sigma^{\mathbb{Z}}$ and $z \in \mathbb{Z}$ as:

$$\sigma_\delta(c)(z) = \begin{cases} c(z+1) & \text{if } \delta = \rightarrow \\ c(z-1) & \text{if } \delta = \leftarrow \end{cases} \qquad \mu_a(c)(z) = \begin{cases} a & \text{if } z = 0 \\ c(z) & \text{if } z \neq 0 \end{cases}$$

A *Turing machine* M is a triple (S, Σ, T) where S is a finite set of states, Σ is a finite set of symbols, and $T \subseteq (S \times \Delta \times S) \cup (S \times \Sigma \times S \times \Sigma)$ is the transition table of the machine. A configuration $\mathfrak{c}$ of the machine is a pair (s, c) where $s \in S$ is a state and $c \in \Sigma^{\mathbb{Z}}$ is the content of the tape. The machine can transform a configuration $\mathfrak{c}$ in a configuration $\mathfrak{c}'$ in one step, noted as $\mathfrak{c} \vdash \mathfrak{c}'$, by applying an instruction $\iota \in T$. An instruction $(s, \delta, t) \in T \cap (S \times \Delta \times S)$ is a *move instruction* of the machine, it can be applied to any configuration (s, c), leading to the configuration $(t, \sigma_\delta(c))$. An instruction $(s, a, t, b) \in T \cap (S \times \Sigma \times S \times \Sigma)$ is a *matching instruction* of the machine, it can be applied to any configuration (s, c) where $c(0) = a$, leading to the configuration $(t, \mu_b(c))$.

A Turing machine (S, Σ, T) is a *deterministic Turing machine (DTM)* if at most one instruction can be applied from any configuration. Formally, the transition table must satisfy the following conditions:

$$(s, \delta, t) \in T \wedge (s', a', t', b') \in T \Rightarrow s \neq s'$$
$$(s, \delta, t) \in T \wedge (s, \delta', t') \in T \Rightarrow \delta = \delta' \wedge t = t'$$
$$(s, a, t, b) \in T \wedge (s, a, t', b') \in T \Rightarrow t = t' \wedge b = b'$$

The *local transition function* of a DTM is the function $f : S \times \Sigma \to S \times \Delta \cup S \times \Sigma \cup \{\bot\}$ defined for all $(s, a) \in S \times \Sigma$ as follows. The associated partial *global transition function* $G : S \times \Sigma^{\mathbb{Z}} \to S \times \Sigma^{\mathbb{Z}}$ maps a configuration to the unique transformed configuration, that is for all $(s, c) \in S \times \Sigma^{\mathbb{Z}}$,

$$f(s, a) = \begin{cases} (t, \delta) & \text{if } (s, \delta, t) \in T \\ (t, b) & \text{if } (s, a, t, b) \in T \\ \bot & \text{otherwise} \end{cases} \qquad G(s, c) = \begin{cases} (t, \sigma_\delta(c)) & \text{if } f(s, c(0)) = (t, \delta) \\ (t, \mu_b(c)) & \text{if } f(s, c(0)) = (t, b) \end{cases}$$

Lemma 1. *If all configurations of a DTM are periodic or mortal then there is a uniform bound n such that for all configurations (s, c) either $G^n(s, c)$ is undefined or $G^t(s, c) = (s, c)$ for some $0 < t < n$. In particular, a periodic DTM is uniformly periodic and a mortal DTM is uniformly mortal.*

Proof. For every $n > 0$ let $U_n = \{(s, c) \mid G^n(s, c) = (s, c) \text{ or } G^n(s, c) \text{ undef}\}$ be the set of configurations that are mortal or periodic at time n. Sets U_n are open so $U_1, U_2, \ldots$ is an open cover of the compact set of all configurations. It has a finite subcover. ∎

One might think that periodicity characterizes a different set of machines if one considers Turing machines with a moving head instead of a moving tape but it is not the case. The global transition function with moving head $H : S \times \mathbb{Z} \times \Sigma^{\mathbb{Z}} \to S \times \mathbb{Z} \times \Sigma^{\mathbb{Z}}$ is defined so that for each $(s, z, c) \in S \times \mathbb{Z} \times \Sigma^{\mathbb{Z}}$, $H(s, z, c) = (s', z', c')$ where $G(s, \sigma_\to^z(c)) = (s', \sigma_\to^{z'}(c'))$. A DTM is periodic with moving head if for each configuration $\mathfrak{c}$, there exists $t \in \mathbb{N}$ such that $H^t(\mathfrak{c}) = \mathfrak{c}$ or equivalently if there exists some $t \in \mathbb{N}$ such that $H^t = \mathrm{Id}$.

Lemma 2. *A DTM is periodic if and only if it is periodic with moving head.*

Proof. Assume that Σ has at least two elements. For each $t \in \mathbb{N}$ and $(s, z, c) \in S \times \mathbb{Z} \times \Sigma^{\mathbb{Z}}$, $H^t(s, z, c) = (s', z', c')$ where $G^t(s, \sigma_\to^z(c)) = (s', \sigma_\to^{z'}(c'))$. Thus, if $H^t = \mathrm{Id}$ then $G^t = \mathrm{Id}$. Conversely, let $G^t = \mathrm{Id}$. By definition, $H^t(s, z, c) = (s, z', c')$ for some z' such that $\sigma_\to^z(c) = \sigma_\to^{z'}(c')$. Moreover, as the machine acts locally, for all d and k such that $c_{|[z-t, z+t]} = d_{|[k-t, k+t]}$, $H^t(s, k, d) = (s, k + z' - z, d')$ where $d' = \sigma_\to^{z'-z}(d')$. If $z' - z \neq 0$, one might choose d such that $d(k + t(z' - z)) \neq d(k + (t + 1)(z' - z))$, contradicting the hypothesis. Thus, $H^t = \mathrm{Id}$. ∎

The reverse of an instruction is defined as follows: $(s, \delta, t)^{-1} = (t, \delta^{-1}, s)$ and $(s, a, t, b)^{-1} = (t, b, s, a)$. The reverse T^{-1} of a transition table T is defined as $T^{-1} = \{\iota^{-1} \mid \iota \in T\}$. The *reverse* of Turing machine $M = (S, \Sigma, T)$ is the machine $M^{-1} = (S, \Sigma, T^{-1})$. A *reversible Turing machine (RTM)* is a deterministic Turing machine whose reverse is deterministic.

Lemma 3. *It is decidable whether a given Turing machine is reversible.*

Proof. It is sufficient to syntactically check the transition table. ∎

Lemma 4. *The reverse of a mortal RTM is mortal.*

Proof. The uniform bound is valid for both the mortal RTM and its reverse. ∎

Lemma 5. *The reverse of a complete RTM is a complete RTM. In particular, a complete RTM is surjective.*

Proof. A DTM is complete if and only if $n|\Sigma| + m = |S||\Sigma|$ where n and m are the numbers of move and matching instructions, respectively. The claim follows from the fact that M and M^{-1} always have the same numbers of move and matching instructions. ∎

2.2   Undecidability of the Immortality Problem

Theorem 7. *It is undecidable whether a given RTM is immortal.*

Proof sketch. For a given 2-RCM without periodic configurations, and given initial state s_0, we effectively construct a reversible Turing machine that is mortal if and only if the 2-RCM halts from the initial configuration $(s_0, 0, 0)$. The Theorem then follows from [5], where it was shown that the halting problem is undecidable for 2-RCM. Note that our additional constraint that the 2-RCM has no periodic configurations can be easily established by having an extra counter that is incremented on each step of the counter machine. This counter can then be incorporated in the existing two counters with the methods of [5].

As a first step we do a fairly standard simulation of a 2-CM by a TM. Configuration (s, a, b) where s is a state and $a, b \in \mathbb{N}$ is represented as a block "@1^{a}x2^{b}y" of length $a + b + 3$, and the Turing machine is positioned on the symbol "@" in state s. A simulation of one move of the CM consists of (1) finding delimiters "x" and "y" on the right to check if either of the two counters is zero, and (2) incrementing or decrementing the counters as determined by the CM. The TM is then returned to the beginning of the block in the new state of the CM. If the CM halts then also the TM halts. All this can be done reversibly if the simulated CM is reversible.

The TM constructed as outline above has the problem that it has immortal configurations even if the CM halts. These are due to the unbounded searches for delimiter symbols "@", "x" or "y". Searches are needed when testing whether the second counter is zero, as well as whenever either counter is incremented or decremented.

Unbounded searches lead to infinite searches if the symbol is not present in the configuration. (For example, searching to the right for symbol "x" when the tape contains "@111...".) To prevent such infinite searches we follow the idea of [7], also employed in [10]. Instead of a straightforward search using a loop, the search is done by performing a recursive call to the counter machine from its initial configuration $(s_0, 0, 0)$. More precisely, we first make a bounded search of length three to see if the delimiter is found within next three symbols. If the delimiter is not found, we start a recursive simulation of the CM by writing "@xy" over the next three symbols, step on the new delimiter symbol "@", and enter the initial state s_0. This begins a nested simulation of the CM.

In order to be able to continue the higher level execution after returning from the recursive search, the present state of the TM needs to be written on the tape when starting the recursive call. For this purpose we increase the tape alphabet by introducing several variants "@$_\alpha$" of the start delimiter "@". Here α is the Turing machine state at the time the search was begun. When returning from a successful recursive search, the higher level computation can pick up from where it left off by reading the state α from the delimiter "@$_\alpha$".

If the recursive search procedure finds the delimiter this is signalled by reversing the search. Once returned to the beginning, the three symbol initial segment "@xy" is moved three positions to the right and the process is repeated. The repeated applications of recursive searches, always starting the next search three positions further right, will eventually bring the machine on the delimiter it was looking for, and the search is completed.

On the other hand, if the CM halts during a recursive search then the TM halts. This always happens when a sufficiently long search is performed using a CM that halts from its initial configuration.

With some additional tricks one can make the TM outlined above reversible, provided the CM is reversible. Now we reason as follows: If the initial configuration $(s_0, 0, 0)$ is immortal in the CM then the TM has a non-halting simulation of the CM. So the TM is not mortal. Conversely, suppose that the CM halts in k steps but the TM has an immortal configuration. The only way for the TM not to halt is to properly simulate the CM from some configuration (s, a, b), where the possibilities $a = \infty$ and $b = \infty$ have to be taken into account. Since the CM has no periodic configurations, one of the two counters necessarily obtains arbitrarily large values during the computation. But this leads to arbitrarily long recursive searches, which is not possible since each such search halts within k steps. ∎

Remarks. (1) The RTM constructed in the proof has no periodic configurations. So the undecidability of the immortality problem holds among RTM without any periodic configurations. (2) Add to the 2-RCM a new looping state s_1 in which the first counter is incremented indefinitely. We can also assume without loss of generality that the 2-RCM halts only in state s_2. Then the RTM constructed in the proof has computation $(s_1, c_1) \vdash^* (s_2, c_2)$ for some $c_1, c_2 \in \Sigma^{\mathbb{Z}}$ if and only if the 2-RCM halts from the initial configuration $(s_0, 0, 0)$.

These detailed observations about the proof will be used later in the proofs of Theorems 8 and 9.

2.3   Undecidability of the Periodicity Problem

Theorem 8. *It is undecidable whether a given complete RTM is periodic.*

Proof sketch. For a given RTM $A = (S, \Sigma, T)$ without periodic configurations we effectively construct a complete RTM $A' = (S \times \{+, -\}, \Sigma, T')$ that is periodic if and only if every configuration of A is mortal. States $(s, +)$ and $(s, -)$ of A' are used to represent A in state s running forwards or backwards in time, respectively. In a halting configuration the direction is switched. The result now follows from Theorem 7 and the first remark after its proof. ∎

2.4   Periodic Orbits

Theorem 9. *It is undecidable whether a given (non-complete) RTM admits a periodic configuration.*

Proof. Remark (2) after the proof of Theorem 7 pointed out that it is undecidable for a given RTM $A = (S, \Sigma, T)$ without periodic configurations, and two given states $s_1, s_2 \in S$ whether there are configurations (s_1, c_1) and (s_2, c_2) such that $(s_1, c_1) \vdash^* (s_2, c_2)$. By removing all transitions from state s_2 and all transitions into state s_1 we can assume without loss of generality that all configurations (s_1, c_1) and (s_2, c_2) are halting in A^{-1} and A, respectively. Using a similar idea as in the proof of Theorem 8 we effectively construct an RTM $A' = (S \times \{+, -\}, \Sigma, T')$ in which A is simulated forwards and backwards in time using states $(s, +)$ and $(s, -)$, respectively. But now the direction is swapped from "-" to "+" only in state s_1, and from "+" to "-" in state s_2. In other halting situations of A, also A' halts. Clearly $((s_1, +), c_1)$ is periodic in A' if and only if $(s_1, c_1) \vdash^* (s_2, c_2)$ for some $c_2 \in \Sigma^{\mathbb{Z}}$. No other periodic orbits exist in A'. ∎

Theorem 10. *It is undecidable whether a given complete DTM admits a periodic configuration.*

Proof. In [10] a complete DTM over the binary tape alphabet was provided that does not have any periodic configurations. This easily gives an analogous DTM for any bigger tape alphabet. For a given RTM $A = (S, \Sigma, T)$ we effectively construct a complete DTM that has a periodic configuration if and only if A has a periodic configuration. The result then follows from Theorem 9. Let $B = (S', \Sigma, T')$ be the fixed complete DTM without periodic configurations from [10], $S \cap S' = \emptyset$. The complete DTM we construct has state set $S \cup S'$ and its transitions includes $T \cup T'$, and in addition a transition into a state $s' \in S'$ whenever A halts. It is clear that the only periodic configurations are those that are periodic already in A. ∎

Conjecture 2. A complete RTM without a periodic point exists. Moreover, it is undecidable whether a given complete RTM admits a periodic configuration.

3   Reversible Cellular Automata

3.1   Definitions

A *one-dimensional cellular automaton* A is a triple (S, r, f) where S is a finite state set, $r \in \mathbb{N}$ is the *neighborhood radius* and $f : S^{2r+1} \longrightarrow S$ is the *local update rule* of A. Elements of $\mathbb{Z}$ are called cells, and a configuration of A is an element of $S^{\mathbb{Z}}$ that assigns a state to each cell. Configuration c is turned into configuration c' in one time step by a simultaneous application of the local update rule f in the radius r neighborhood of each cell:

$$c'(i) = f(c(i - r), c(i - r + 1), \ldots, c(i + r - 1), c(i + r)) \text{ for all } i \in \mathbb{Z}.$$

Transformation $G : c \mapsto c'$ is the *global transition function* of A. The Curtis-Hedlund-Lyndom -theorem states that a function $S^{\mathbb{Z}} \longrightarrow S^{\mathbb{Z}}$ is a global transition function of some CA if and only if it is continuous and commutes with the shift σ, defined by $\sigma(c)_i = c_{i+1}$ for all $c \in S^{\mathbb{Z}}$ and $i \in \mathbb{Z}$.

Cellular automaton A is called *reversible* if the global function G is bijective and its inverse G^{-1} is a CA function. We call A *injective, surjective* and *bijective* if G is injective, surjective and bijective, respectively. Injectivity implies surjectivity, and bijectivity implies reversibility. See [6] for more details on these classical results.

3.2 Undecidability of the Immortality Problem

Let some states of a CA be identified as halting. Let us call a configuration c *halting* if $c(i)$ is a halting state for some i. We call c *locally halting* if $c(0)$ is a halting state. These two definitions reflect two different ways that one may use to define an accepting computation in CA: either acceptance happens when a halting state appears somewhere, in an unspecified cell, or one waits until a halting state shows up in a fixed, predetermined cell. A configuration c is immortal (locally immortal) for G if $G^n(c)$ is not halting (locally halting, respectively) for any $n \geq 0$. CA function G is immortal (locally immortal) if there exists an immortal (locally immortal) configuration.

Theorem 11 ([11]). *It is undecidable whether a given reversible one-dimensional CA is immortal (locally immortal).*

3.3 Undecidability of the Periodicity Problem

In cellular automata periodicity and uniform periodicity are equivalent. Indeed, suppose that a period n that is common to all configurations does not exist. Then for every $n \geq 1$ there is $c_n \in S^{\mathbb{Z}}$ such that $G^n(c_n) \neq c_n$. Each c_n has a finite segment p_n of length $2rn + 1$ that is mapped in n steps into a state that is different from the state in the center of p_n. Configuration c that contains a copy of p_n for all n, satisfies $G^n(c) \neq c$ for all n, and hence such c is not periodic.

Theorem 12. *It is undecidable whether a given one-dimensional CA is periodic.*

Proof sketch. For a given complete reversible Turing machine $M = (S, \Sigma, T)$ we effectively construct a one-dimensional reversible CA $A = (Q, 2, f)$ that is periodic if and only if M is periodic. The result then follows from Theorem 8. The state set

$$Q = \Sigma \times ((S \times \{+, -\}) \cup \{\leftarrow, \rightarrow\})$$

consists of two tracks: The first track stores elements of the tape alphabet Σ and it is used to simulate the content of the tape of the Turing machine, while the second track stores the current state of the simulated machine at its present location, and arrows $\leftarrow$ and $\rightarrow$ in other positions pointing towards the position of the Turing machine on the tape. The arrows are needed to prevent several Turing machine heads accessing the same tape location and interfering with each other's computation. The state is associated a symbol '+' or '-' indicating whether the reversible Turing machine is being simulated forwards or backwards in time. The direction is switched if the Turing machine sees a local error, i.e., an arrow pointing away from the machine.

It follows from the reversibility of M that A is a reversible CA. If M has a non-periodic configuration c then A has a non-periodic configuration which simulates the computation from c. Conversely, if M is periodic it is uniformly periodic under the moving head mode. It easily follows that all configurations of A are periodic. ∎

A one-dimensional RCA is equicontinuous if and only if it is periodic, so we have

Corollary 1. *It is undecidable whether a given one-dimensional reversible CA is equicontinuous.*

3.4   Periodic Orbits

Every cellular automaton has periodic orbits so the existence of periodic orbits is trivial among cellular automata.

References

1. Landauer, R.: Irreversibility and heat generation in the computing process. IBM Journal of Research and Development 5, 183–191 (1961)
2. Lecerf, Y.: Machines de turing réversibles. C. R. Acad. Sci. Paris 257, 2597–2600 (1963)
3. Bennett, C.B.: Logical reversibility of computation. IBM Journal of Research and Development 17(6), 525–532 (1973)
4. Minsky, M.: Computation: Finite and Infinite Machines. Prentice Hall, Englewoods Cliffs (1967)
5. Morita, K.: Universality of a reversible two-counter machine. Theor. Comput. Sci. 168(2), 303–320 (1996)
6. Kari, J.: Theory of cellular automata: a survey. Theor. Comput. Sci. 334, 3–33 (2005)
7. Hooper, P.K.: The undecidability of the turing machine immortality problem. J. Symb. Log. 31(2), 219–234 (1966)
8. Collins, P., van Schuppen, J.H.: Observability of hybrid systems and turing machines. In: 43rd IEEE Conference on Decision and Control, vol. 1, pp. 7–12. IEEE Press, Los Alamitos (2004)
9. Kůrka, P.: On topological dynamics of turing machines. Theor. Comput. Sci. 174(1-2), 203–216 (1997)
10. Blondel, V.D., Cassaigne, J., Nichitiu, C.M.: On the presence of periodic configurations in turing machines and in counter machines. Theor. Comput. Sci. 289(1), 573–590 (2002)
11. Kari, J., Lukkarila, V.: Some undecidable dynamical properties for one-dimensional reversible cellular automata. In: Condon, A., Harel, D., Kok, J.N., Salomaa, A., Winfree, E. (eds.) Algorithmic Bioprocesses. Springer, Heidelberg (to appear, 2008)

Step-Out Ring Signatures*

Marek Klonowski, Łukasz Krzywiecki, Mirosław Kutyłowski, and Anna Lauks

Institute of Mathematics and Computer Science, Wrocław University of Technology

Abstract. We propose a version of ring signatures for which the set of potential signers may be reduced: the real signer can prove that he or she has created the signature, while every other member of the ring can prove not to be the signer. Possibility to run these protocols is triggered by publishing certain secret information. The proposed scheme is an intermediate solution between the classical ring and group signatures, and can be used for instance for e-auction schemes.

1 Introduction

Recent development of ring and group signatures provide tools for authenticating digital data so that the signer remains anonymous within a set of users - members of a group or a ring.

A ring signature scheme [1] enables a signer to sign a message and remain hidden within an arbitrary group of people, called a *ring*. The real signer uses his secret key and the public keys of other ring members. So once the public keys are published, one can be included in a ring even against her or his will. Moreover, one is unable to prove that he was not the actual signer of a particular message. Group signatures [2] allow any group member to sign anonymously on behalf of the group: a verifier can check that a group member has signed the message, but he cannot indicate who. However, identity of the signer of a particular message can be revealed. Depending on a particular scheme, it can be done only by a group manager (e.g. [2]) or by group members (e.g. [3]).

The primary goal of ring signatures is to prove that a message comes from a member of a certain group of people without revealing from whom exactly. However, this is not the only functionality of ring signatures. Ring signatures has been expanded to deniable ring authentication [4,5]. Linkable ring signature schemes allow to link signatures, if they were signed by the same person. Short versions of linkable ring signatures were proposed in [6,7]. There are identity based ring signature schemes, which allow ring construction across different identity-based master domains [8,9,10,11,12,13]. For a confessible threshold ring signature [14] the actual signer can prove that he has created the signature.

In this paper we introduce *step-out ring signature scheme* that serves as an intermediate solution between ring and group signatures. We make it possible to change the anonymity status of the creator of a signature in the following way:

---

* Partially supported by Polish Ministry of Science and Higher Education, grant N N206 2701 33.

E. Ochmański and J. Tyszkiewicz (Eds.): MFCS 2008, LNCS 5162, pp. 431–442, 2008.

Confession procedure: the signer can reveal himself and prove that he has created the signature.

Step-out procedure: a member of a ring that has not created the signature can prove that he <u>is not</u> the signer. Depending on a version of the scheme, it can be possible only if some requirements are fulfilled.

So the anonymity set of a signer can be reduced to just one element by the signer himself -as for confessible ring signatures. However, the potential signers can leave the ring one by one. Additionally, step-out procedure cannot be started, unless some triggering event has occurred (revealing certain secret material).

Let us mention that similar functionality of signatures have been concerned in [15,16]. However, these papers focus on possibility of confirming to be the real signer.

Applications. We mention just two examples. The first one is an electronic auction scheme where the ID's of the participants are known – this kind of auction is used by authorities in some countries for selling. Every auction participant is obliged to pay a deposit, which is returned except for the winner. The deposit paid by the winner is assumed to be a part of the price paid. If the winner does not sign the contract then the deposit is retained by the seller as a compensation.

In the scenario described step-out ring signatures can be used to sign the bids. The bid with the highest value is kept by the organizer of the auction. If somebody wishes to leave the auction, then he or she executes the step-out procedure and the deposit can be returned immediately. When the auction is closed, the winner can reveal its identity to the seller. When the winner fails to do it, the other participants prove that they are not the winners for returning their deposits, and the effect is finally the same.

The second case is a joint bank account run for two joint owners, say Alice and Bob. Each of their orders can be signed by a step-out ring signature with ring of Alice and Bob. As long as everything is fine, neither bank nor anybody else should know details about who of the joint owners is communicating with the bank. However, in some situations revealing the signer becomes necessary - it might be a divorce case or dissolving a common enterprise. Last not least, due to inheritance procedures a court has to determine if the decendent conducted certain operations.

2   Scheme Description

Preliminaries. Let p, q be prime, $q|p - 1$, and $G = \langle g \rangle$ be a cyclic subgroup of $\mathbb{Z}_p^*$. For the sake of simplicity we shall skip "mod p" if it follows from the context. We shall consider rings with n participants, which will be denoted by $\mathcal{U}_1, \ldots, \mathcal{U}_n$. We assume that user $\mathcal{U}_i$ holds a private key x_i; the corresponding public key is $y_i = g^{x_i}$. The key y_i is known to all other participants.

We assume that the following assumptions are fulfilled in G:

Definition 1 (Decisional Diffie-Hellman Assumption). *Let G be a cyclic group generated by g of order q. Let $\mathcal{A}^{DDH}$ be an algorithm that has to distinguish $c_0 = (g, g^a, g^b, g^{ab})$ from $c_1 = (g, g^a, g^b, g^c)$ for randomly chosen $a, b, c \in \mathbb{Z}_q$. Let*

$\mathbf{Adv}_{\mathcal{A}}^{ddh} = \Pr[\mathcal{A}(c_d) = d]$ be called the advantage of $\mathcal{A}$ in breaking the DDH problem.

The DDH assumption holds for G, if advantage $\mathbf{Adv}_{\mathcal{A}}^{ddh}$ is negligible for each probabilistic polynomial-time algorithm $\mathcal{A}$, i.e. $\mathbf{Adv}_{\mathcal{A}}^{ddh} < \epsilon_{ddh}$ where ϵ_{ddh} is negligible.

Definition 2 (Discrete Logarithm (DL) Assumption). *Let G be a cyclic group generated by g of order q. Let $\mathcal{A}$ be an algorithm such that on input g^a, where $a \in \mathbb{Z}_q$, $\mathcal{A}$ should output a. Let $\mathbf{Adv}_{\mathcal{A}}^{dl} = \Pr[\mathcal{A}(g^a) = a]$ be the advantage of $\mathcal{A}$ in breaking the DL problem.*

The DL assumption holds in G, if for each probabilistic polynomial-time algorithm $\mathcal{A}$ advantage $\mathbf{Adv}_{\mathcal{A}}^{dl}$ is negligible, i.e. $\mathbf{Adv}_{\mathcal{A}}^{dl} < \epsilon_{dl}$ where ϵ_{dl} is negligible.

Our scheme is based upon a scheme SEQDL $\begin{bmatrix} n \\ 1 \end{bmatrix}$ $(\hat{g}, g, \hat{y}, y_1, \ldots, y_n, m)$ briefly described in [17]. It is a signature of knowledge and equality of discrete logarithms $\log_{\hat{g}} \hat{y}$ and $\log_g y_i$, where y_i is some element from the list $\{y_1, \ldots, y_n\}$ (with i unrevealed to the prover), for a message m. We use here a slightly modified version of this scheme:

Definition 3. *Signature of knowledge* SEQDL $\begin{bmatrix} n \\ 1 \end{bmatrix}$ $(\hat{g}, g, x_j, r_j, \hat{y}_w, y_1, \ldots, y_n, w_1, \ldots, w_n, m)$ *(with public parameters $\hat{g}, g, \hat{y}_w, y_1, \ldots, y_n, w_1, \ldots, w_n, m$) is a tuple* $(c_1, \ldots, c_n, s_1, \ldots, s_n)$ *such that*

$$\sum_{i=1}^n c_i = \mathcal{H}(\hat{g}||g||\hat{y}_w||y_1|| \ldots \tag{1}$$
$$\ldots ||y_n||w_1|| \ldots ||w_n||\hat{g}^{s_1}\hat{y}_w^{c_1}||g^{s_1}(y_1 w_1)^{c_1}|| \ldots ||\hat{g}^{s_n}\hat{y}_w^{c_n}||g^{s_n}(y_n w_n)^{c_n}||m) \ ,$$

where $\mathcal{H}$ denotes a secure hash function. It is a signature of knowledge and equality of discrete logarithms of group element $\hat{y}_w$ with respect to $\hat{g}$ and discrete logarithm of one element out of the list $\{y_1 w_1, \ldots, y_n w_n\}$ with respect to g, for message m.

The main difference with the original scheme is that we use factors $w_1, \ldots, w_n$ that play an important blinding role in our ring signature scheme.

SEQDL Proof of Knowledge. Now we present a construction of a signature of knowledge and equality of discrete logarithms:

SEQDL $\begin{bmatrix} n \\ 1 \end{bmatrix}$ $(\hat{g}, g, x_j, r_j, \hat{y}_w, y_1, \ldots, y_n, w_1, \ldots, w_n, m) = (c_1, \ldots, c_n, s_1, \ldots, s_n)$

where $y_j = g^{x_j}$, $w_j = g^{r_j}$, and $\hat{g}^{x_j + r_j} = \hat{y}_w$, that is, signature SEQDL $\begin{bmatrix} n \\ 1 \end{bmatrix}$ proves that $\log_{\hat{g}} \hat{y}_w = \log_g (y_i w_i)$. The signature is created as follows:

1. User $\mathcal{U}_j$ generates at random the elements $r \in \mathbb{Z}_q$ and $c_i, s_i \in \mathbb{Z}_q$ for $i \in \{1, \ldots, n\} \setminus \{j\}$.
2. For all $i \in \{1, \ldots, n\} \setminus \{j\}$ user $\mathcal{U}_j$ computes:

$$t_i \leftarrow \hat{g}^{s_i}\hat{y}_w^{c_i} \ , \quad u_i \leftarrow g^{s_i}(y_i w_i)^{c_i} \ , t_j \leftarrow \hat{g}^r \ , \quad u_j \leftarrow g^r \ . \tag{2}$$

3. User $\mathcal{U}_j$ computes:

$$c_j \leftarrow \mathcal{H}(\hat{g}\|g\|\hat{y}_w\|y_1\|\cdots \tag{3}$$
$$\cdots \|y_n\|w_1\|\cdots\|w_n\|t_1\|u_1\|\cdots\|t_n\|u_n\|m) - \sum_{i<n,i\neq j} c_i \ ,$$
$$s_j \leftarrow r - (x_j + r_j)c_j \bmod q \ . \tag{4}$$

Such a signature is verified positively, since

$$g^{s_j}(w_j y_j)^{c_j} = g^{r-(x_j+r_j)c_j}(g^{r_j}g^{x_j})^{c_j} = g^r = u_j \ ,$$
$$\hat{g}^{s_j}(\hat{y}_w)^{c_j} = \hat{g}^{r-(x_j+r_j)c_j}(\hat{g}^{r_j}\hat{g}^{x_j})^{c_j} = \hat{g}^r = t_j$$

and so the condition (1) and equality of the discrete log are fulfilled.

Note that knowledge of the private key x_j as well as discrete logarithm r_j of w_j are necessary to split the exponent r from chosen values $t_j = \hat{g}^r$ and $u_j = g^r$, so that $t_j = \hat{g}^{s_j}\hat{y}_w^{c_j}$, and $u_j = g^{s_j}(y_j w_j)^{c_j}$. Since c_j is fixed by expression (3), s_j is determined uniquely, and its appropriate value is given by (4).

2.1 Scheme Description

Outline. Let us assume that $\mathcal{U}_j$ is the real signer and $\mathcal{U}_1,\ldots,\mathcal{U}_n$ are all ring members. Let the private and public key of user $\mathcal{U}_i$ be, respectively, x_i and $y_i = g^{x_i}$. For Step-out Ring Signatures (SRS) we have the following procedures:

Signing procedure. $\mathcal{S}_{\mathrm{SRS}}$ is a randomized algorithm that takes generator g and a random element $\hat{g} \in \langle g \rangle$, $\hat{g} \neq 1$, the secret key x_j, the set of public keys $\{y_1,\ldots,y_n\} \subset \langle g \rangle$ and a message m. It returns a signature σ.

Verification procedure. $\mathcal{V}_{\mathrm{SRS}}$ is a deterministic algorithm that takes a message m, and a signature σ for m. It returns a bit: 1 or 0 to indicate whether σ is valid, i.e. someone having a public key in a set Y indicated by σ has signed m.

Confession procedure. $\mathcal{C}_{\mathrm{SRS}}$ is a deterministic algorithm executed by signer $\mathcal{U}_j$ of a signature σ for m: it yields σ' - another SRS signature for the message m. It returns a bit 1 or 0 to indicate, if $\mathcal{U}_j$ is really the signer.

Step-out procedure. $\mathcal{D}_{\mathrm{SRS}}$ is a deterministic algorithm that takes from a user $\mathcal{U}_i$ SRS signatures σ'' and σ''' for a message $\tilde{m} =$"I have not signed m" and a signature σ for m. It returns a bit 1 or 0 to confirm that $\mathcal{U}_i$ has not created σ.

Details. In order to sign, first an element $\hat{g} \neq 1$ is chosen at random from G. Since the order q of G is prime, $\hat{g}$ is a generator of G as well. The signature has a form of a non–interactive zero knowledge proof that the exponent hidden in $\hat{y}\hat{w}$ equals one of the exponents hidden in $y_1 w_1, \ldots, y_n w_n$.

```
Algorithm S_SRS(g, ĝ, x_j, y_1, ..., y_n, m)
    repeat
        r_1, ..., r_n ←_R Z*_p
        w_i ← g^{r_i}, for each i = 1, ..., n
    until (y_i w_i ≠ y_j w_j for each i ≠ j)
```

$\hat{w} \leftarrow \hat{g}^{r_j}, \ \hat{y} \leftarrow \hat{g}^{x_j}, \ \hat{y}_w \leftarrow \hat{y}\hat{w}$

$(c_1, \ldots c_n, s_1, \ldots, s_n) \leftarrow \text{SEQDL} \begin{bmatrix} n \\ 1 \end{bmatrix} (\hat{g}, g, x_j, r_j, \hat{y}_w, y_1, \ \ldots$
$\ldots, y_n, w_1, \ \ldots, \ w_n, m)$

$Y \leftarrow \{y_1, \ldots, y_n\}$

$W \leftarrow \{w_1, \ldots, w_n\}$

$\sigma \leftarrow (g, \hat{g}, \hat{y}, \hat{w}, Y, W, c_1, \ldots, c_n, s_1, \ldots, s_n)$

return (m, σ)

In order to verify a signature (m, σ) a verifier computes $\hat{y}_w = \hat{y}\hat{w}$ and simply checks validity of $\text{SEQDL} \begin{bmatrix} n \\ 1 \end{bmatrix} (\hat{g}, g, x_j, r_j, \hat{y}_w, y_1, \ldots, y_n, w_1, \ldots, w_n, m)$.

Algorithm $\mathcal{V}_{\text{SRS}}(\sigma, m)$

    $\hat{y}_w \leftarrow \hat{y}\hat{w}$

    $d \leftarrow \mathcal{V}_{\text{SEQDL}[\begin{smallmatrix} n \\ 1 \end{smallmatrix}]}(\hat{g}, g, \hat{y}_w, y_1, \ldots, y_n, w_1, \ldots, w_n, c_1, \ldots, c_n, s_1, \ldots, s_n, m)$

    if $d = 1$ then return 1 else return 0

Note that, if $\mathcal{U}_j$ is the real signer of σ and still holds r_j, he can create $\sigma' = (g, \hat{g}, \hat{y}, \hat{w}, Y', W, \text{SEQDL} \begin{bmatrix} n \\ 1 \end{bmatrix} (\hat{g}, g, x_j, r_j, \hat{y} \cdot \hat{w}, y'_1, \ldots, y'_n, w_1, \ldots, w_n, m))$, a new signature for the same m, with the same parameters $g, \hat{g}$, the same set W and some new set of potential signers Y' such that $Y \cap Y' = \{y_j\}$. Moreover, y_j stands on the same position in both sequences: $y_j = y'_j$. Additionally, the signer chooses new ring members so that for $i_1 \neq i_2$, we have $y_{i_1} w_{i_1} \neq y'_{i_2} w_{i_2}$.

Algorithm $\mathcal{C}_{\text{SRS}}(\sigma, \sigma', y_j, m)$

    if(the same $g, \hat{g}, \hat{y}, \hat{w}, W$ were used in σ and σ') then

        $d_1 \leftarrow \mathcal{V}_{\text{SRS}}(\sigma, m), \ d_2 \leftarrow \mathcal{V}_{\text{SRS}}(\sigma', m)$

        if $(d_1 = d_2 = 1$ and $\{y_j\} = Y \cap Y'$

                   and y_j stands on position j in Y') then

      return 1 else return 0

    else return 0

Using m, σ and σ', a verifier can check, if $\mathcal{U}_j$ has really created σ: the verifier checks whether appropriate parameters used in σ and σ' are the same, whether verification for SRS signatures σ and σ' gives a positive answer and $\{y_j\} = Y \cap Y'$, i.e. $\mathcal{U}_j$ was the only potential signer of both σ and σ'. In Sect. 3 (Proposition 1) we shall see that nobody but the real signer can present such a proof.

When the secret r_i is given to the user $\mathcal{U}_i$, he can prove that he has not created signature σ. Namely, in such a situation he can create two control signatures:

1. $\sigma'' = (g, \hat{g}, \hat{y}'', \hat{w}'', Y'', W, \text{SEQDL} \begin{bmatrix} n \\ 1 \end{bmatrix} (\hat{g}, g, x_i, r_i, \hat{y}'' \cdot \hat{w}'', y''_1, \ldots, y''_n, w_1, \ldots, w_n, \tilde{m}))$ - a SRS signature with the same parameters $g, \hat{g}, W$ as in σ and $\hat{y}'' = \hat{g}^{x_i}, \ \hat{w}'' = \hat{g}^{r_i}$, some new set of potential signers Y'', for the control message $\tilde{m} = $ "I have not signed m".

2. $\sigma''' = (g, \hat{g}, \hat{y}'', \hat{w}'', Y''', W, \text{SEQDL} \begin{bmatrix} n \\ 1 \end{bmatrix} (\hat{g}, g, x_i, r_i, \hat{y}'' \cdot \hat{w}'', y'''_1, \ldots, y'''_n, w_1, \ldots, w_n, \tilde{m}))$ - a SRS signature for the same the control message $\tilde{m}$ with the same $g, \hat{g}, \hat{y}''\hat{w}'', W$ and Y''' such that $Y'' \cap Y''' = \{y_i\}$ and y_i stands on the same position in Y'' and Y'''. Moreover, $y''_{i_1} w_{i_1} \neq y'''_{i_2} w_{i_2}$ for $i_1 \neq i_2$.

To verify that $\mathcal{U}_i$ was not a real signer, the verifier checks a few conditions:

```
Algorithm D_SRS(σ, m, σ'', σ''', y_i, m̃)
    if(the same g, ĝ, W were used in σ, σ'', σ'''
                and the same ŷ'', ŵ'' were used in σ'', σ''') then
        d_1 ← V_SRS(σ, m),  d_2 ← V_SRS(σ'', m̃),  d_3 ← V_SRS(σ''', m̃)
        if (d_1 = d_2 = d_3 = 1 and {y_i} = Y'' ∩ Y''',
            and y_i stands at the same position in Y'' and Y''',
            and ŷŵ ≠ ŷ''ŵ'') then
                return 1 else return 0
    else return 0
```

We shall see in Section 3 that the real signer cannot create signatures σ' and σ'' so that $\hat{y}\hat{w} \neq \hat{y}''\hat{w}''$.

Parameters r_i. During construction of SEQDL signature of knowledge (as a part of signing, confession and step-out procedures) the signer has to know r_i corresponding to the factor w_i used. Depending on the application aimed, we can apply different scenarios:

- The numbers r_i are created by the signer at random. They are kept secret unless the signer enables a member of a ring to step out.
- The numbers r_i are given together with the signature. In this case the ring participants can immediately step out.
- $\mathcal{U}_i$ generates r_i herself and publishes w_i. Moreover, each w_i can be a kind of time stamp - a signature generated with w_i has to be created no earlier than at the time of creating w_i.

In the first case the signer would have to remember all numbers r_i, which might be inconvenient. A simple solution to this problem is to set r_i as an output from a pseudorandom function of other elements of the signature. For instance, we can put $r_i = H'(w_{i-1}^{z_j})$, where z_j is a secret exponent used for this purpose only and H' is a hash function with the range $[0, q-1]$.

3   Algorithm Analysis

Throughout this section we shall use the following notation and parameters:

1. g and $\hat{g}$ are generators used for constructing SRS signatures;
2. $x_1, \ldots, x_n, x'_1, \ldots, x'_n, x''_1, \ldots, x''_n, x'''_1, \ldots, x'''_n$ denote private keys; the corresponding public keys are $y_1 = g^{x_1}, \ldots, y_n = g^{x_n}, y'_1 = g^{x'_1}, \ldots, y'_n = g^{x'_n}, y''_1 = g^{x''_1}, \ldots, y''_n = g^{x''_n}, y'''_1 = g^{x'''_1}, \ldots, y'''_n = g^{x'''_n}$; we shall consider the sets of keys such that $\{y_1, \ldots, y_n\} \cap \{y'_1, \ldots, y'_n\} = \{y_j\} = \{y'_j\}$ and $\{y''_1, \ldots, y''_n\} \cap \{y'''_1, \ldots, y'''_n\} = \{y_i\}$, where $y_i = y''_i = y'''_i$,
3. $m \in \{0,1\}^*$ is a message, $\tilde{m}$ is a message "I have not signed m",
4. $\sigma = (g, \hat{g}, \hat{y}, \hat{w}, y_1, \ldots, y_n, w_1, \ldots, w_n, c_1, \ldots, c_n, s_1, \ldots, s_n)$ where $\sigma = \mathcal{S}_{SRS}(g, \hat{g}, x_j, y_1, \ldots, y_n, m)$,

5. $\sigma' = (g, \hat{g}, \hat{y}, \hat{w}, y_1', \ldots, y_n', w_1, \ldots, w_n, \text{SEQDL} \begin{bmatrix} n \\ 1 \end{bmatrix} (\hat{g}, g, x_j, r_j, \hat{y}\hat{w}, y_1', \ldots, y_n', w_1, \ldots, w_n, m))$,

6. for $x_i \neq x_j$, $\hat{y}'' \neq \hat{y}$, $\hat{w}'' \neq \hat{w}$:
$\sigma'' = (g, \hat{g}, \hat{y}'', \hat{w}'', y_1'', \ldots, y_n'', w_1, \ldots, w_n, \text{SEQDL} \begin{bmatrix} n \\ 1 \end{bmatrix} (\hat{g}, g, x_i, r_i, \hat{y}''\hat{w}'', y_1'', \ldots, y_n'', w_1, \ldots, w_n, \tilde{m}))$,
$\sigma''' = (g, \hat{g}, \hat{y}'', \hat{w}'', y_1''', \ldots, y_n''', w_1, \ldots, w_n, \text{SEQDL} \begin{bmatrix} n \\ 1 \end{bmatrix} (\hat{g}, g, x_i, r_i, \hat{y}''\hat{w}'', y_1''', \ldots, y_n''', w_1, \ldots, w_n, \tilde{m}))$.

Correctness. It follows directly from the construction that the signatures created according to the described procedures are verified positively.

Proposition 1. *If the confession procedure is executed by $\mathcal{U}_j$ according to the protocol, then the outcome of the procedure is positive and this is an evidence that $\mathcal{U}_j$ is the signer.*

Proof. Since $\mathcal{V}_{\text{SEQDL}\begin{bmatrix} n \\ 1 \end{bmatrix}}(\hat{g}, g, \hat{y}\hat{w}, y_1, \ldots, y_n, w_1, \ldots, w_n, c_1, \ldots c_n, s_1, \ldots s_n, m) \to 1$, there exist α such that $g^\alpha \in \{y_1 w_1, \ldots, y_n w_n\}$ and $\hat{g}^\alpha = \hat{y}\hat{w}$. Moreover, if σ' was created like in point 5 and $\mathcal{V}_{\text{SRS}}(\sigma', m) \to 1$, then $g^\alpha \in \{y_1' w_1, \ldots, y_n' w_n\}$ as well. So $g^\alpha \in \{y_1 w_1, \ldots, y_n w_n\} \cap \{y_1' w_1, \ldots, y_n' w_n\}$. Since $\{y_1, \ldots, y_n\} \cap \{y_1', \ldots, y_n'\} = \{y_j\}$, and $y_{i_1} w_{i_1} \neq y_{i_2}' w_{i_2}$ for $i_1 \neq i_2$, we know that $g^\alpha = y_j w_j$, so in this case user $\mathcal{U}_j$ was a creator of σ and $\mathcal{C}_{\text{SRS}}(\sigma, \sigma', y_j, m) \to 1$. $\square$

Proposition 2. *If the step-out procedure is executed by a non-signer $\mathcal{U}_i$ according to the protocol, then the outcome of the step-out procedure is positive.*

Proof. Let us assume that $\mathcal{D}_{\text{SRS}}(\sigma, m, \sigma'', \sigma''', y_i, \tilde{m}) \to 0$. It happens if $\hat{y}\hat{w} = \hat{y}''\hat{w}''$. As in the proof of Proposition 1, we can see that the signatures σ'' and σ''' guarantee that there exists α' such that $g^{\alpha'} = y_i w_i$ and $\hat{g}^{\alpha'} = \hat{y}''\hat{w}''$. So $\alpha' = \log_g (y_i w_i) = \log_{\hat{g}} (\hat{y}''\hat{w}'') = \log_{\hat{g}} (\hat{y}\hat{w}) = \log_g (y_j w_j)$, where $\mathcal{U}_j$ is the signer of σ. We have got that $y_i w_i = y_j w_j$, but this contradicts the assumption about generating secrets r_i and computing w_i during the signing procedure $\mathcal{S}_{\text{SRS}}$. $\square$

Proposition 3. *If a signer $\mathcal{U}_j$ of σ executes the step-out procedure, then the outcome is negative.*

Proof. When performing the step-out procedure and generating signatures σ', σ'', the user $\mathcal{U}_j$ has to generate $y'' \cdot w'' = \hat{g}^{x_j + r_j}$. However, this product is the same as in σ, so this would lead to a failure of the test of the steps-out procedure. $\square$

Unforgeability. It is based on Discrete Logarithm Assumption and the Forking Lemma (see [18]). Recall that the Forking Lemma can be applied to a signature that takes the form of a tuple (σ_1, h, σ_2), where σ_1 depends only on values chosen at random, h is a hash value that depends on the message m to be signed and σ_1, and σ_2 depends only on σ_1, m, and h.

We consider a *chosen-message* scenario for a forger $\mathcal{F}_{\text{SRS}}$ that is given only public parameters of the scheme. The forger can query, up to q_{max} times, some real signers for valid signatures of messages of his choice. Messages can be asked for being signed more than once in an adaptive manner, i.e. $\mathcal{F}_{\text{SRS}}$ can adapt his queries according to previous message-signature pairs. Forger's goal is to produce a signature σ over a document m - not previously signed in the query stage - such that $\mathcal{V}_{\text{SRS}}(m, \sigma) = 1$, the set of potential signers in σ equals $\{y_1, \ldots, y_n\}$, but m did not come from any of users $\mathcal{U}_i$ having public key y_i. We say that $\mathcal{F}_{\text{SRS}}$ succeeds, if $\mathcal{F}_{\text{SRS}}$ can forge a message in this way with a non-negligible probability.

In order to take advantage from the Forking Lemma for attacks in adaptive *chosen-message* scenario we show that the proposed SRS signature is of the form (σ_1, h, σ_2) and can be simulated without the knowledge of the corresponding secret signing key and with indistinguishable distribution probability.

Lemma 1. *SRS signatures can be simulated in the random oracle model and under DDH assumption without knowing the corresponding secret signing key and with distribution probability indistinguishable from SRS signatures produced by a legitimate signer.*

Proof. First, let us consider the simulation algorithm S_0 that requires knowledge of the secret signing key x_j:

Algorithm $S_0(g, \hat{g}, m, x_j, y_1, \ldots, y_n)$
    $r_1, \ldots, r_n \leftarrow_R \mathbb{Z}_p^*$
    $w_i = g^{r_i}$ for all $i \in \{1, \ldots, n\}$
    $\hat{y} \leftarrow \hat{g}^{x_j}$, $\hat{w} \leftarrow \hat{g}^{r_j}$
    $c_i \leftarrow_R \mathbb{Z}_p^*$, for all $i \in \{1, \ldots, n\}$
    $s_i \leftarrow_R \mathbb{Z}_p^*$, for all $i \in \{1, \ldots, n\}$
    $\mathcal{H}(\hat{g}||g||\hat{y}\hat{w}||y_1||\ldots||y_n||w_1||\ldots||w_n||\hat{g}^{s_1}(\hat{y}\hat{w})^{c_1}||g^{s_1}(y_1 w_1)^{c_1}||\ldots$
        $\ldots||\hat{g}^{s_n}(\hat{y}\hat{w})^{c_n}||g^{s_n}(y_n w_n)^{c_n}||m) \leftarrow \sum_{i=1}^{n} c_i$
    return $\sigma_{S_0} = (m, g, \hat{g}, \hat{y}, \hat{w}, y_1, \ldots, y_n, w_1, \ldots, w_n, c_1, \ldots, c_n, s_1, \ldots, s_n)$

Note that the simulator S_0 can take the value of $\mathcal{H}$ to be equal to $\sum_{i=1}^{n} c_i$ in the random oracle model. Apart from that, S_0 mimics the signature creation procedure. Therefore, the returned tuple σ_{s_0} is indeed a valid signature of the message m signed by the user $\mathcal{U}_j$ in the random oracle model.

Now, let us consider a slightly modified simulation algorithm S_1 creating signatures with no knowledge on the corresponding secret signing key:

Algorithm $S_1(g, \hat{g}, m, y_1, \ldots, y_n)$
    $\alpha, \beta, r_1, \ldots, r_n \leftarrow_R \mathbb{Z}_p^*$
    $w_i = g^{r_i}$ for all $i \in \{1, \ldots, n\}$
    $\hat{y} \leftarrow \hat{g}^{\alpha}$, $\hat{w} \leftarrow \hat{g}^{\beta}$
    $c_i \leftarrow_R \mathbb{Z}_p^*$, for all $i \in \{1, \ldots, n\}$
    $s_i \leftarrow_R \mathbb{Z}_p^*$, for all $i \in \{1, \ldots, n\}$
    $\mathcal{H}(\hat{g}||g||\hat{y}\hat{w}||y_1||\ldots||y_n||w_1||\ldots||w_n||\hat{g}^{s_1}(\hat{y}\hat{w})^{c_1}||g^{s_1}(y_1 w_1)^{c_1}||\ldots$
        $\ldots||\hat{g}^{s_n}(\hat{y}\hat{w})^{c_n}||g^{s_n}(y_n w_n)^{c_n}||m) \leftarrow \sum_{i=1}^{n} c_i$
    return $\sigma_{S_1} = (m, g, \hat{g}, \hat{y}, \hat{w}, y_1, \ldots, y_n, w_1, \ldots, w_n, c_1, \ldots, c_n, s_1, \ldots, s_n)$

We claim that the outputs of S_1 are indistinguishable from the outputs of S_0. Conversely, assume that there exists an algorithm $Dist^S$ that for a simulated signature σ_{S_b} returns a bit b that indicates (with a fair probability) the type of the input (S_0 or S_1). Then we use $Dist^S(S_b)$ to construct another algorithm $Dist^{DDH}(t)$ that solves an instance of the DDH problem $t = (g, g^a, g^b, g^c)$. The idea is that the only difference between the simulations is that in the second case the parameter $\hat{y}\hat{w}$ is set at random, while in the first case it has the value $\hat{g}^{x_1+r_1}$ related to $g^{x_1+r_1}$, which is another parameter of the simulation. The algorithm solving DDH first creates a simulated signature σ_{S_b} such that: $\hat{g} \leftarrow g^a$, $\hat{y}\hat{w} \leftarrow g^c$, $y_1 w_1 \leftarrow g^b$. Note here that if $c = ab$, then $\hat{y}\hat{w} = \hat{g}^{x_1+r_1}$ and σ_{S_b} is indeed a simulation of type S_0 for the user y_1:

```
Algorithm S_b(g, ĝ = g^a, m, y_1, ..., y_n, g, g^b, g^c)
    r_1, r_2, ..., r_n ←_R Z*_p
    w_i = g^{r_i} for all i ∈ {2, ..., n}
    w_1 ← g^b / y_1
    ŵ ← ĝ^{r_1} ,  ŷ ← g^c / ŵ
    c_i ←_R Z*_p, for all i ∈ {1, ..., n}
    s_i ←_R Z*_p, for all i ∈ {1, ..., n}
    H(ĝ||g||ŷŵ||y_1|| ... ||y_n||w_1|| ... ||w_n||ĝ^{s_1}(ŷŵ)^{c_1}||g^{s_1}(y_1 w_1)^{c_1}|| ...
        ... ||ĝ^{s_n}(ŷŵ)^{c_n}||g^{s_n}(y_n w_n)^{c_n}||m) ← ∑_{i=1}^{n} c_i
    return σ_{S_b} = (m, g, ĝ, ŷ, ŵ, y_1, ..., y_n, w_1, ..., w_n, c_1, ..., c_n, s_1, ..., s_n)
```

Then it inputs σ_{S_b} to $Dist^S(\sigma_{S_b})$:

```
Algorithm Dist^{DDH}((g, g^a, g^b, g^c))
    σ_{S_b} ← S_b(g, ĝ = g^a, m, y_1, ..., y_n, g, g^b, g^c)
    if Dist^S(σ_{S_b}) = 0 then c = ab
    if Dist^S(σ_{S_b}) = 1 then c ≠ ab
```

Hence σ_{s_2} is indistinguishable from σ_{s_1} and σ_{s_1} is indistinguishable from a regular signature σ, thus σ_{s_2} is also is indistinguishable from a regular signature σ.

Now we formalize the attacks of a forger $\mathcal{F}_{\text{SRS}}$ in the *chosen-message* scenario.

Definition 4. *We consider the following experiment of running a forger* $\mathcal{F}_{\text{SRS}}$:

```
Experiment Exp_{F_SRS}
    for k = 1 to q_{max}
        query for (m_k, σ_k), such that V_SRS(σ_k, m_k) = 1
    let (m, σ) ← F_SRS(g, ĝ, y_1, ..., y_n, m, (m_1, σ_1), ..., (m_k, σ_k))
    if V_SRS(σ, m) = 1 return 1
    else return 0
```

Then we define the advantage $\mathbf{Adv}_{\mathcal{F}_{\text{SRS}}}$ *of the forger* $\mathcal{F}_{\text{SRS}}$ *as the probability* $\Pr[\mathbf{Exp}_{\mathcal{F}_{\text{SRS}}} = 1]$.

Theorem 1. *Step-out Ring Signatures* (SRS) *are secure against forgery, i.e.* $\mathbf{Adv}_{\mathcal{F}_{\text{SRS}}}$ *is negligibly small.*

Proof. In order to prove the theorem we show that an adversary algorithm $\mathcal{A}$ that runs a forger $\mathcal{F}_{\text{SRS}}$ as a subroutine can be used to break DL assumption.

To utilize the Forking Lemma we abbreviate a signature σ of a message m from the signing algorithm as $\sigma = (\sigma_1, h, \sigma_2)$ with the following parameters:

- $\sigma_1 = (\hat{g}, \hat{y}, \hat{w}, w_1, \ldots, w_n, u_1, \ldots, u_n, t_1, \ldots, t_n)$, where u_i, t_i are constructed like in (2),
- $h = \mathcal{H}(\hat{g}||g||\hat{y}\hat{w}||y_1||\ldots||y_n||w_1||\ldots||w_n||u_1||t_1||\ldots||u_n||t_n||m)$,
- $\sigma_2 = (c_1, \ldots, c_n, s_1, \ldots, s_n)$.

In order to break DL assumption $\mathcal{A}$ runs twice the forger algorithm, yielding two signatures of a message m. For both cases the "random" parts are the same - the random oracle inserts the same value for $\hat{y}\hat{w}$, which is the element for which we seek its discrete logarithm. As seen in Lemma 1, for running the forger $\mathcal{A}$ does not really need to involve the real signers, the first signatures can be simulated as well. After acquiring two valid signatures (σ_1, h, σ_2) and $(\sigma_1, h', \sigma_2')$, such that $h \neq h'$ and $\sigma_2 \neq \sigma_2'$, adversary $\mathcal{A}$ can compute the secret $\alpha = \log_{\hat{g}} \hat{y}\hat{w}$. Namely $\mathcal{A}$ computes $\alpha_j = (s_j' - s_j)/(c_j - c_j')$ for $j = 1, \ldots, n$. For one of these j the adversary gets $\hat{g}^{\alpha_j} = \hat{y}\hat{w}$. Since we assume that the DL assumption holds, the above algorithm must have negligible probability of success, therefore $\mathbf{Adv}_{\mathcal{F}_{\mathrm{SRS}}}$ must be negligible, too. □

Anonymity. Another fundamental issue regarding security of our scheme is anonymity of the real signer. We need to ensure that a ring signature leaks no information about who has signed the message. More formally, we need to show that the signatures for a message m created by the owner of a key x_0 are indistinguishable from the signatures for m created by using x_1 provided that both corresponding public keys y_0 and y_1 belong to the ring of the signatures concerned. That is, we need to prove that the signature scheme meets the conditions from the following definition:

Definition 5. *Let $\mathcal{A}^{DIST}$ be a probabilistic polynomial time algorithm that can distinguish between $\sigma_0 = \mathcal{S}_{\mathrm{SRS}}(g, \hat{g}, x_0, y_0, y_1, m)$ and $\sigma_1 = \mathcal{S}_{\mathrm{SRS}}(g, \hat{g}, x_1, y_0, y_1, m)$ for an arbitrary message m.*

Let advantage of $\mathcal{A}^{DIST}$ be defined as $\mathbf{Adv}_{\mathcal{A}^{DIST}} = \Pr[\mathcal{A}(\sigma_b) = b]$. We say that the scheme provides anonymity, if for any efficient algorithm $\mathcal{A}^{DIST}$ the value of $\mathbf{Adv}_{\mathcal{A}^{DIST}}$ is at most negligibly greater than $\frac{1}{2}$.

We show that breaking anonymity of our scheme is not easier than breaking DDH problem. Namely, we construct an algorithm $\mathcal{A}^{ddh}$ for breaking instances of DDH problem built on the top of any $\mathcal{A}^{DIST}$. Moreover, we show that $\mathcal{A}^{ddh}$ calls $\mathcal{A}^{DIST}$ as a subprocedure a limited number T of times and $\mathbf{Adv}_{\mathcal{A}^{ddh}} = 1 - 1/e > \frac{1}{2}$.

Let us describe the algorithm $\mathcal{A}^{ddh}$. For an instance of DDH problem (g, g^a, g^b, g^c) we will construct simulated signatures σ_i, by the means of simulator S_b described above, and treat this simulations as inputs for algorithm $\mathcal{A}^{DIST}$. As assumed this algorithm gives 1 for the input of the form (g, g^a, g^b, g^c) for $c = ab$ with probability $\mathbf{Adv}_{\mathcal{A}^{DIST}} = \epsilon > \frac{1}{2}$. Moreover, we assume that it gives 0 for input of the form (g, g^a, g^b, g^c) for $c \neq ab$ with probability $\frac{1}{2}$.

```
Algorithm A^ddh(g, g^a, g^b, g^c)
    p_1, ..., p_T ←_r Z*_p
```

```
d_1, ..., d_T ←_r {0, 1}
m, y_0, y_1 ←_r Z_p*
For i = 1 to T {
    if (d_i = 1) σ_i ← S_b(m, y_0, y_1, g, g^{ap_i}, g^{bp_i}, g^{cp_i})
    else σ_i ← S_b(m, y_0, y_1, g, g^{bp_i}, g^{ap_i}, g^{cp_i})
    d'_i ← A^{DIST}(σ_i)
}
X = {i ≤ T | d_i = d'_i}
if (X < (ε/2 + 1/4) T) return 0
else return 1
```

We can consider the above algorithm as T independent Bernoulli trials, where $T = \frac{2}{(\epsilon - 1/2)^2}$, and with probability of success in each trial equal to $\frac{1}{2}$ if $c \neq ab$, or equal to $\epsilon > \frac{1}{2}$ if $c = ab$. Therefore $X \sim B(T, \epsilon)$ or $X \sim B(T, \frac{1}{2})$.

Let us recall the following well-known variant of Chernoff inequality:

Lemma 2. Let $X \sim B(T, p)$ be a random variable with the binomial distribution. Then $\Pr(X > EX + t) \leq \exp\left(-\frac{2t^2}{T}\right)$ and $\Pr(X \leq EX - t) \leq \exp\left(-\frac{2t^2}{T}\right)$ for any $t > 0$.

Let us analyze probability that the algorithm gives correct answer, if $c \neq ab$. In this case we can treat X as a random variable with binomial distribution $B(T, \frac{1}{2})$. That probability of failure equals

$$\Pr\left(X > \left(\tfrac{\epsilon}{2} + \tfrac{1}{4}\right) T\right) = \Pr\left(X > \tfrac{1}{2}T + \tfrac{\epsilon - 1/2}{2}T\right) .$$

Since $EX = \frac{1}{2}T$ and $T = \frac{2}{(\epsilon - 1/2)^2}$, using Lemma 2 this probability equals $\Pr\left(X > EX + \frac{\epsilon - 1/2}{2}T\right) \leq \frac{1}{e}$. Using exactly the same reasoning we can show that if $c = ab$, then $\Pr\left(X \leq \left(\frac{\epsilon}{2} + \frac{1}{4}\right) T\right) \leq \frac{1}{e}$. Indeed, it is enough to remember that $EX = \epsilon T$ and apply the second inequality in Lemma 2.

We can see that for each input A^{ddh} gives the correct answer with probability exceeding $1 - 1/e > \frac{1}{2}$. The runtime of A^{ddh} is polynomial provided that $\epsilon - 1/2 = \frac{1}{\text{poly}(n)}$. We have constructed an appropriate algorithm for a basic case of a distinguishing two signers, however such an approach can be easily generalized to the case of several potential users. Let us also note that in this analysis only some parameters of the signature were taken into account. One can easily see that rest of them cannot be used for recognizing the real signer since their distribution does not depend on the signer identity and therefore cannot reveal any information on the signer.

References

1. Rivest, R.L., Shamir, A., Tauman, Y.: How to Leak a Secret. In: Boyd, C. (ed.) ASIACRYPT 2001. LNCS, vol. 2248. pp. 552–556. Springer, Heidelberg (2001)
2. Chaum, D., van Heyst, E.: Group Signatures. In: Davies, D.W. (ed.) EUROCRYPT 1991. LNCS, vol. 547, pp. 257–265. Springer, Heidelberg (1991)

3. Manulis, M.: Democratic Group Signatures: on an Example of Joint Ventures. In: Proceedings of the 2006 ACM Symposium on Information, Computer and Communications Security, ASIACCS 2006, p. 365. ACM Press, New York (2006)

4. Naor, M.: Deniable Ring Authentication. In: Yung, M. (ed.) CRYPTO 2002. LNCS, vol. 2442, pp. 481–498. Springer, Heidelberg (2002)

5. Susilo, W., Mu, Y.: Deniable Ring Authentication Revisited. In: Jakobsson, M., Yung, M., Zhou, J. (eds.) ACNS 2004. LNCS, vol. 3089, pp. 149–163. Springer, Heidelberg (2004)

6. Wei, V.K., Tsang, P.P.: Short Linkable Ring Signatures for E-Voting, E-Cash and Attestation. In: Deng, R.H., Bao, F., Pang, H., Zhou, J. (eds.) ISPEC 2005. LNCS, vol. 3439, pp. 48–60. Springer, Heidelberg (2005)

7. Au, M.H., Chow, S.S.M., Susilo, W., Tsang, P.P.: Short Linkable Ring Signatures Revisited. In: Atzeni, A.S., Lioy, A. (eds.) EuroPKI 2006. LNCS, vol. 4043, pp. 101–115. Springer, Heidelberg (2006)

8. Zhang, F., Kim, K.: ID-Based Blind Signature and Ring Signature from Pairings. In: Zheng, Y. (ed.) ASIACRYPT 2002. LNCS, vol. 2501, pp. 533–547. Springer, Heidelberg (2002)

9. Herranz, J., Sáez, G.: A Provably Secure Id-based Ring Signature Scheme. Cryptology ePrint Archive, Report 2003/261 (2003), http://eprint.iacr.org/

10. Lin, C.Y., Wu, T.C.: An Identity-based Ring Signature Scheme from Bilinear Pairings. Cryptology ePrint Archive, Report 2003/117 (2003), http://eprint.iacr.org/

11. Chow, S.S.M., Yiu, S.M., Hui, L.C.K.: Efficient Identity Based Ring Signature. In: Ioannidis, J., Keromytis, A., Yung, M. (eds.) ACNS 2005. LNCS, vol. 3531, pp. 499–512. Springer, Heidelberg (2005)

12. Awasthi, A.K., Lal, S.: ID-based Ring Signature and Proxy Ring Signature Schemes from Bilinear Pairings. ArXiv Computer Science e-prints (2005), http://arxiv.org/abs/cs/0504097

13. Au, M.H., Liu, J.K., Yuen, T.H., Wong, D.S.: Id-based Ring Signature Scheme Secure in the Standard Model. In: Yoshiura, H., Sakurai, K., Rannenberg, K., Murayama, Y., Kawamura, S.-i. (eds.) IWSEC 2006. LNCS, vol. 4266, pp. 1–16. Springer, Heidelberg (2006)

14. Chen, Y.S., Lei, C.L., Chiu, Y.P., Huang, C.Y.: Confessible Threshold Ring Signatures. In: International Conference on Systems and Networks Communications (ICNSC). IEEE Computer Society, Los Alamitos (2006), http://doi.ieeecomputersociety.org/10.1109/ICSNC.2006.29

15. Komano, Y., Ohta, K., Shimbo, A., Ichi Kawamura, S.: Toward the Fair Anonymous Signatures: Deniable Ring Signatures. In: Pointcheval, D. (ed.) CT-RSA 2006. LNCS, vol. 3860, pp. 174–191. Springer, Heidelberg (2006)

16. Mu, Y., Susilo, W., Zhang, F., Wu, Q.: Ad Hoc Group Signatures. In: Yoshiura, H., Sakurai, K., Rannenberg, K., Murayama, Y., Kawamura, S.-i. (eds.) IWSEC 2006. LNCS, vol. 4266, pp. 120–135. Springer, Heidelberg (2006)

17. Camenisch, J.L.: Efficient and Generalized Group Signatures. In: Fumy, W. (ed.) EUROCRYPT 1997. LNCS, vol. 1233, pp. 465–479. Springer, Heidelberg (1997)

18. Pointcheval, D., Stern, J.: Security Arguments for Digital Signatures and Blind Signatures. J. Cryptology 13(3), 361–396 (2000)

The Height of Factorization Forests

Manfred Kufleitner

Institut für Formale Methoden der Informatik,
Universität Stuttgart, Germany
manfred.kufleitner@fmi.uni-stuttgart.de

Abstract. We show that for every homomorphism from A^+ to a finite semigroup S there exists a factorization forest of height at most $3\,|S| - 1$. Furthermore, we show that for every non-trivial group, this bound is tight. For aperiodic semigroups, we give an improved upper bound of $2\,|S|$ and we show that for every $n \geq 2$ there exists an aperiodic semigroup S with n elements which reaches this bound.

1 Introduction

Factorization forests where introduced by Simon [8,10]. An important property of finite semigroups is that they admit factorization forests of finite height. This fact is called the Factorization Forest Theorem. It can be considered as a Ramsey-type property of finite semigroups. There exist different proofs of this fact of different difficulty and with different bounds on the height. The first proof of the Factorization Forest Theorem is due to Simon [10]. He showed that for every finite semigroup S there exists a factorization forest of height $\leq 9\,|S|$. The proof relies on several different techniques. It uses graph colorings, Green's relations, and a decomposition technique inspired by the Rees-Suschkewitsch Theorem on completely 0-simple semigroups. In [11] Simon gave a simplified proof relying on the Krohn-Rhodes decomposition. The bound shown is $2^{|S|+1} - 2$. A concise proof has been given by Chalopin and Leung [2]. The proof relies on Green's relations and yields the bound $7\,|S|$ on the height. Independently of this work, Colcombet has also shown a bound of $3\,|S|$ for the height of factorization forests [4]. He uses a generalization of the Factorization Forest Theorem in terms of *Ramseyan splits*. The proof also relies on Green's relations. A variant of our proof for the special case of aperiodic monoids has been shown in [5] with a bound of $3\,|S|$. The benefit of that proof on the one hand is its little machinery, on the other hand it is strong enough for application in first-order logic. The proof in this paper can be seen as a refinement of that proof. Again, the main tool are Green's relations. We only require basic results from the theory of finite semigroups which can be found in standard textbooks such as [6].

A lower bound of $|S|$ was shown for rectangular bands in [9] and also in [2]. The same bound has been shown for every finite group [2]. In the same paper, a lower bound of $|S| + 1$ has been shown for an infinite class of semilattices S.

There exist many different applications of the Factorization Forest Theorem, see e.g. [1,7,12]. In addition, Colcombet presented extensions of the Factorization Forest Theorem to trees [3] and to infinite words [4].

E. Ochmański and J. Tyszkiewicz (Eds.): MFCS 2008, LNCS 5162, pp. 443–454, 2008.

The paper is structured as follows. In Section 2, we introduce some notation as well as some basic properties. In particular, a generalization of factorization forests involving Green's relations is presented. Section 3 describes algorithms for constructing factorization forests for Green's relations. In Section 4 we show that every finite semigroup S admits factorization forests of height at most $3|S| - 1$. Furthermore, we show that this bound is tight for every non-trivial group. Section 5 contains an upper bound of $2|S|$ for aperiodic semigroups. Furthermore, for every $n \geq 2$ we give a semilattice S with n elements which reaches this bound. Therefore, the bound $2|S|$ is tight for aperiodic semigroups.

2   Preliminaries

A *word* over some finite alphabet A is a finite sequence $w = a_1 \cdots a_n$ of letters $a_i \in A$. The set of all words over the alphabet A is denoted by A^*. It is the free monoid over the set A. The empty word is its neutral element and it is denoted by ε. The set of non-empty words $A^+ = A^* \setminus \{\varepsilon\}$ is the free semigroup over A. A word u is a *factor* of a word w if there exist $x, y \in A^*$ such that $xuy = w$.

For a semigroup S we let $S^1 = S \cup \{1\}$ where $S^1 = S$ if S has a neutral element 1. The multiplication of S^1 extends the multiplication of S by defining $x \cdot 1 = 1 \cdot x = x$ for all $x \in S^1$. With this multiplication, S^1 is a monoid with 1 as the neutral element. Let $u, v \in S$. Those of Green's relations which play a role here are defined by:

$$u \mathcal{J} v \Leftrightarrow S^1 u S^1 = S^1 v S^1, \qquad u \leq_{\mathcal{J}} v \Leftrightarrow S^1 u S^1 \subseteq S^1 v S^1,$$

$$u \mathcal{R} v \Leftrightarrow u S^1 = v S^1, \qquad u \leq_{\mathcal{R}} v \Leftrightarrow u S^1 \subseteq v S^1,$$

$$u \mathcal{L} v \Leftrightarrow S^1 u = S^1 v, \qquad u \leq_{\mathcal{L}} v \Leftrightarrow S^1 u \subseteq S^1 v,$$

$$u \mathcal{H} v \Leftrightarrow u \mathcal{R} v \text{ and } u \mathcal{L} v.$$

Let $\mathcal{G} \in \{\mathcal{J}, \mathcal{R}, \mathcal{L}\}$ be one of Green's relations. We write $u <_{\mathcal{G}} v$ if $u \leq_{\mathcal{G}} v$ but not $u \mathcal{G} v$. We use the notation $\mathcal{G}_u = \{v \in S \mid v \mathcal{G} u\}$ for the $\mathcal{G}$-class of u. A semigroup S is *aperiodic* if for all $u \in S$ there exists some $n \in \mathbb{N}$ such that $u^n = u^{n+1}$. An element $u \in S$ is *idempotent* if $u^2 = u$. A *semilattice* S is a commutative semigroup such that all elements in S are idempotent.

Lemma 1. *Let S be a finite semigroup. The $\mathcal{R}$-class of $u \in S$ is uniquely determined by some arbitrary prefix x of u such that $u \mathcal{J} x$, i.e., $u = xy$ and $u \mathcal{J} x$ implies $u \mathcal{R} x$. A left-right symmetric statement (involving $\mathcal{L}$-classes and suffixes) also holds.*

Proof. By definition, we have $u \leq_{\mathcal{R}} x$. Now, $u \mathcal{J} x$ implies $u \mathcal{R} x$, see e.g. [5, Lemma 7].   □

Let S be a finite semigroup. A *factorization forest* of a homomorphism $\varphi : A^+ \to S$ is a function d which maps every word $w \in A^+$ with length $|w| \geq 2$ to a factorization $d(w) = (w_1, \ldots, w_n)$ of $w = w_1 \cdots w_n$ with $n \geq 2$ and $|w_i| < |w|$

for all $1 \leq i \leq n$. Moreover, $n \geq 3$ implies $\varphi(w_1) = \cdots = \varphi(w_n)$ is idempotent in S. The *height* h_d of a word w is defined as

$$h_d(w) = \begin{cases} 0 & \text{if } |w| = 1, \\ 1 + \max\{h_d(w_1), \ldots, h_d(w_n)\} & \text{if } d(w) = (w_1, \ldots, w_n). \end{cases}$$

The height of a factorization forest is the supremum over the heights of all words. We call the tree defined by the "branching" d for the word w the *factorization tree* of w. The height $h_d(w)$ is the height of this tree.

Let $\mathcal{G} \in \{\mathcal{J}, \mathcal{R}, \mathcal{L}, \mathcal{H}, =\}$ be one of Green's relations or equality and let $w \in A^+$. A *$\mathcal{G}$-factorization tree* of a factorization $w = w_1 \ldots w_n$ is an ordered tree satisfying the following conditions:

(a) The root of the tree is w.
(b) Only factors of w of the form $w_i \cdots w_k$ with $i \leq k$ occur as vertices.
(c) The children $u_1, \ldots, u_n$ of every inner node u satisfy $u = u_1 \cdots u_n$.
(d) If a node u has more than two children $u_1, \ldots, u_n$ (i.e., u has degree more than two), then $\varphi(u) \; \mathcal{G} \; \varphi(u_1) \; \mathcal{G} \cdots \mathcal{G} \; \varphi(u_n)$ in S.
(e) Only words w_i occur as leafs.

An =-factorization tree is simply called a *factorization tree*. If we omit the factorization of w, then we usually use the trivial factorization $w = a_1 \cdots a_n$ into letters $a_i \in A$. Suppose for every word $w \in A^+$ there exists a factorization tree, then we can construct a factorization forest d by choosing an optimal tree for every word w (by minimizing over all occurrences of w as the root or as an inner node of some tree). For every word w we define $h_\varphi(w)$ as the minimal height of some tree with the above properties. Now, instead of showing that there exists a factorization forest d such that for every word w we have $h_d(w) \leq n$ for some bound $n \in \mathbb{N}$, it suffices to show $h_\varphi(w) \leq n$, i.e., for every word w there exists a factorization tree of height at most n. Replacing the quantification "there exists d such that for every word w" by "for every word w there exists a tree" simplifies the proofs. Note that these two points of view are equivalent since for every factorization forest d, we have $h_\varphi(w) \leq h_d(w)$. On the other hand, for every homomorphism φ there exists a factorization forest d such that $h_d(w) = h_\varphi(w)$.

We will need the following graph-theoretic terminology. We say that a node x is the *left neighbor* of the node y in a tree, if the parent z of y has two children such that its left child is x and its right child is y. A *branch* is a simple path from the root to some leaf.

3   Factorization Trees and Green's Relations

Let $\varphi : A^+ \to S$ be a homomorphism to a finite semigroup S.

Lemma 2. *Let $w \in A^+$ and let j be the number of $\mathcal{J}$-classes in S. Consider an arbitrary factorization $w = w_1 \cdots w_n$. There exists a $\mathcal{J}$-factorization tree for $w = w_1 \cdots w_n$ of height at most $2j$. Moreover, every $\mathcal{J}$-class of S occurs at most once per branch at some node of degree more than two.*

Proof. Let $j(w)$ be the number of $\mathcal{J}$-classes which are $\leq_{\mathcal{J}}$-below $\varphi(w)$, i.e., $j(w) = |\{\mathcal{J}_u \mid u \leq_{\mathcal{J}} \varphi(w)\}|$. By induction on $j(w)$, we show that there exists a $\mathcal{J}$-factorization tree for $w = w_1 \cdots w_n$ of height at most $2 \cdot j(w) - 1$ with at most $j(w)$ nodes per branch of degree more than two. During the induction, we preserve the following invariant: The left neighbor of every vertex to which we are applying the induction hypothesis is a leaf. In order to establish the invariant, the very first factorization of height 1 is

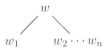

From now on, we can assume that the invariant holds and we continue with $w_2 \cdots w_n$ for which we reuse the name w. The $\mathcal{J}$-*factorization* of $w = w_1 \cdots w_n$ is $w = v_0 u_1 \cdots u_m$ where $u_i = w_{j_i} v_i$ and $v_i = w_{j_i+1} \cdots w_{j_{i+1}-1}$ (with $j_0 = 0$) such that

- $v_i \in \{\varepsilon\} \cup \{v \in A^+ \mid \varphi(w) <_{\mathcal{J}} \varphi(v)\}$ for all $0 \leq i \leq m$, and
- $\varphi(w) \, \mathcal{J} \, \varphi(u_i)$ for all $1 \leq i \leq m$.

The $\mathcal{J}$-factorization of $w = w_1 \cdots w_n$ is unique. It can be obtained by reading the word w from right to left, and as soon as the word one has read is $\mathcal{J}$-equivalent to w (under the homomorphism φ), then this word is the next u_i. Therefore, the words v_i are $<_{\mathcal{J}}$-above w (under φ). By definition of $\leq_{\mathcal{J}}$, we have $\varphi(w) \leq_{\mathcal{J}} \varphi(u)$ for every factor u of w. Note that $m \geq 1$.

First, suppose that some of the v_i are nonempty. W.l.o.g. we can assume that all v_i are nonempty, since $v_i = \varepsilon$ implies that u_i is a leaf. The construction of a factorization tree for w consists of three phases. In the first phase, we use the invariant for 'rotating' v_0 to the left (if $v_0 = \varepsilon$ then this phase can be omitted). The second phase is the factorization into the u_i which might yield a degree more than two. This is possible since $\varphi(w) \, \mathcal{J} \, \varphi(u_1 \cdots u_m) \, \mathcal{J} \, \varphi(u_i)$. Finally, the third phase is the factorization of the u_i into the leaf w_{j_i} and the word v_i. To summarize, we replace the tree

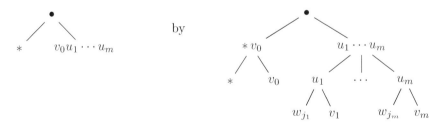

In the above pictures, $\bullet$ is the parent of w and $*$ is a leaf. In the resulting tree, the left neighbor of every v_i is a leaf. By induction hypothesis, every v_i admits a $\mathcal{J}$-factorization tree of height at most $2j(v_i) - 1$ with at most $j(v_i)$ nodes per branch of degree more than two. Only phase two and phase three each add $+1$

to the height of the trees for v_i and since only phase two might yield a branching of degree more than two, the resulting tree for $u_1 \cdots u_m$ is of height at most $\max \{2j(v_i) - 1 \mid 0 \leq i \leq m\} + 2 \leq 2j(w) - 1$. Moreover, vertex $u_1 \cdots u_m$ is the only vertex of degree more than two which is $\mathcal{J}$-equivalent to $\varphi(w)$. The first phase does not cost any additional resources. Hence, the resulting tree for w has the desired properties.

Let now all v_i be empty (which includes the case where $j(w) = 1$). This is how the induction stops. We have $u_i = w_i$ for all $1 \leq i \leq m = n$ and we can use the following tree of height 1:

This tree trivially satisfies the desired constraints on the height and the number of $\mathcal{J}$-classes at branchings of degree more than two.

Together with the (additional) very first factorization in which we established the invariant, the above induction proves the lemma.   □

Lemma 3. *Consider a factorization of a word* $w = w_1 \cdots w_n$ *such that* $\varphi(w)$ *and all* $\varphi(w_i)$ *belong to the same* $\mathcal{J}$-*class* J. *Let* ℓ *be the number of* $\mathcal{L}$-*classes in* J. *There exists an* $\mathcal{L}$-*factorization tree for* $w = w_1 \cdots w_n$ *of height at most* $2\ell - 1$ *such that there are at most* ℓ *nodes per branch of degree more than two.*

Proof. Let $\ell(w)$ be the number of $\mathcal{L}$-classes of prefixes of w under φ, i.e., $\ell(w) = |\{\mathcal{L}_{w_i} \mid 1 \leq i \leq n\}|$. By Lemma 1, we have $\varphi(w_1 \cdots w_i) \mathcal{L} \varphi(w_i) \mathcal{L} \varphi(w_j \cdots w_i)$ for every $j \leq i$. Using induction on $\ell(w)$, we show that there exists an $\mathcal{L}$-factorization tree for $w = w_1 \cdots w_n$ of height at most $2 \cdot \ell(w) - 1$ with at most $\ell(w)$ nodes per branch of degree more than two. The lemma then follows from this induction.

If $\ell(w) = 1$, then we can use the following tree of height 1:

This tree trivially satisfies the desired constraints on the height and the number of branchings of degree more than two.

Let now $\ell(w) > 1$ and let $j_1 < \cdots < j_m$ be the sequence of all indices such that $\varphi(w_{j_i}) \mathcal{L} \varphi(w_n)$. In particular, $j_m = n$. We can apply the following tree of height 2:

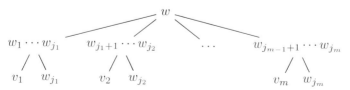

in which we used the abbreviation $v_i = w_{j_{i-1}+1} \cdots w_{j_i - 1}$ for $1 \le i \le m$ (where $j_0 = 0$). First, we factorized after every w_{j_i} and second, we chopped of the w_{j_i}. In the remaining words v_i, the $\mathcal{L}$-class of $\varphi(w)$ has been eliminated and hence, $\ell(v_i) < \ell(w)$ for all $1 \le i \le m$. Note that $\varphi(w) \; \mathcal{L} \; \varphi(w_{j_{i-1}} \cdots w_{j_i}) \; \mathcal{L} \; \varphi(w_n)$. By induction, there exists an $\mathcal{L}$-factorization tree for every v_i of height at most $2\ell(v_i) - 1$ with at most $\ell(v_i)$ nodes per branch of degree more than two. Therefore, the resulting tree for w is of height $\max\{2\ell(v_i) - 1 \mid 1 \le i \le m\} + 2 \le 2\ell(w) - 1$ and on every branch there are at most $\max\{\ell(v_i) \mid 1 \le i \le m\} + 1 \le \ell(w)$ nodes of degree more than two. $\qquad\square$

Lemma 4. *Consider a factorization of a word $w = w_1 \cdots w_n$ such that $\varphi(w)$ and all $\varphi(w_i)$ belong to the same $\mathcal{J}$-class J. Let r be the number of $\mathcal{R}$-classes in J. There exists an $\mathcal{R}$-factorization tree for $w = w_1 \cdots w_n$ of height at most $2r - 1$ such that there are at most r nodes per branch of degree more than two.*

Proof. The lemma and its proof are left-right symmetric to Lemma 3. $\qquad\square$

Lemma 5. *Consider a factorization of a word $w = w_1 \cdots w_n$ such that $\varphi(w)$ and all $\varphi(w_i)$ belong to the same $\mathcal{J}$-class J. Let h be the number of $\mathcal{H}$-classes in J. There exists an $\mathcal{H}$-factorization tree for $w = w_1 \cdots w_n$ of height at most $2h - 1$ such that there are at most h nodes per branch of degree more than two.*

Proof. We now combine Lemma 3 and Lemma 4 in order to construct $\mathcal{H}$-factorization trees. First, by Lemma 3 there exists an $\mathcal{L}$-factorization tree for $w = w_1 \cdots w_n$. In this tree we can substitute to any of its nodes v with more than two children $v_1, \ldots, v_m$ the $\mathcal{R}$-factorization tree given by Lemma 4 for $v = v_1 \cdots v_m$. The resulting tree is an $\mathcal{H}$-factorization tree for $w = w_1 \cdots w_n$.

Let us compute its height. Let ℓ be the number of $\mathcal{L}$-classes and let r be the number of $\mathcal{R}$-classes in J. We have $\ell r = h$: see e.g. [5, Lemma 6] for a proof of the fact that the intersection of an $\mathcal{R}$-class and an $\mathcal{L}$-class within the same $\mathcal{J}$-class is nonempty; a simple reflection shows that an $\mathcal{H}$-class is uniquely determined by its $\mathcal{L}$-class and its $\mathcal{R}$-class. Consider a branch in the constructed tree. Its length is at most $(2\ell - 1) - \ell + \ell(2r - 1) = 2\ell r - 1 = 2h - 1$. An upper bound for the number of nodes of degree more than two per branch is $\ell r = h$. $\qquad\square$

Lemma 6. *Consider a factorization of a word $w = w_1 \cdots w_n$ such that $\varphi(w)$ and all $\varphi(w_i)$ belong to the same $\mathcal{H}$-class H. There exists a factorization tree for $w = w_1 \cdots w_n$ of height at most $3|H| - 1$ such that there are at most $|H|$ nodes per branch of degree more than two.*

Proof. By Lemma 1, we see that $\varphi(w_1)\varphi(w_2) \; \mathcal{H} \; \varphi(w)$. From $\varphi(w) \; \mathcal{H} \; \varphi(w_1) \; \mathcal{H} \; \varphi(w_2)$ it follows that H is a group, see e.g. [6, Corollary 1.7, p.49]. Let $P(w) = \{\varphi(w_1 \cdots w_i) \mid 1 \le i \le m\}$ be the set of prefixes of w under φ. By induction on $|P(w)|$, we show that there is a factorization tree for w of height $3|P(w)| - 1$ with at most $|P(w)|$ nodes per branch of degree more than two. Since $P(w) \subseteq H$, the lemma then follows.

If $|P(w)| = 1$, then we have that $\varphi(w_i \cdots w_j) = \varphi(w_2 \cdots w_n) = \varphi(w_1)^{-1}\varphi(w)$, $2 \le i \le j \le n$, is the neutral element of the group H. Hence, we can use the following tree of height 2:

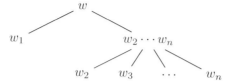

Let now $|P(w)| > 1$ and let $j_1 < \cdots < j_k$ be the sequence of all occurrences of the prefix $\varphi(w) \in P(w)$, i.e., all indices j_i such that $\varphi(w) = \varphi(w_1 \cdots w_{j_i})$. In particular, $j_k = n$. As before, we see that $\varphi(w_{j_s+1} \cdots w_{j_t})$, $1 \leq s \leq t \leq k$, is the neutral element of H. This gives the following tree of height 3:

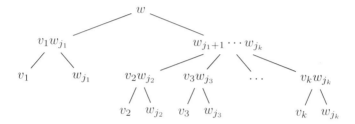

where $v_i = w_{j_{i-1}+1} \cdots w_{j_i-1}$ with $j_0 = 0$. We have $P(v_1) \subseteq P(w) \setminus \{\varphi(w)\}$. For $i \geq 2$ we have $\varphi(w_1 \cdots w_{j_{i-1}}) \cdot P(v_i) \subseteq P(w) \setminus \{\varphi(w)\}$. Therefore, for all $1 \leq i \leq k$ we have $|P(v_i)| < |P(w)|$. By induction, there exists a factorization tree for $v_i = w_{j_{i-1}+1} \cdots w_{j_i-1}$ of height at most $3|P(v_i)| - 1$ such that there are at most $|P(v_i)|$ nodes per branch of degree more than two. We compose those trees with the above tree for w. The height of the resulting tree is $3 \cdot \max\{|P(v_i)| \mid 1 \leq i \leq k\} - 1 + 3 \leq 3|P(w)| - 1$. It has at most $\max\{|P(v_i)| \mid 1 \leq i \leq k\} + 1 \leq |P(w)|$ nodes per branch of degree more than two.   □

Lemma 7. *Consider a factorization of a word $w = w_1 \cdots w_n$ such that $\varphi(w)$ and all $\varphi(w_i)$ belong to the same $\mathcal{J}$-class J. There exists a factorization tree for $w = w_1 \cdots w_n$ of height at most $3|J| - 1$ such that there are at most $|J|$ nodes per branch of degree more than two.*

Proof. The proof uses the same substitution principle as the proof of Lemma 5. By Lemma 5 there exists an $\mathcal{H}$-factorization tree for $w = w_1 \cdots w_n$. If we consider a node v of this tree with more than two children $v_1, \ldots, v_m$, then $\varphi(v)$ and all $\varphi(v_i)$ are $\mathcal{H}$-equivalent. Hence, we can apply Lemma 6 to the factorization $v = v_1 \cdots v_n$. We substitute every branching with more than two children by the respective tree given by Lemma 6. The resulting tree is a factorization tree for $w = w_1 \cdots w_n$.

We compute its height. Let h be the number of $\mathcal{H}$-classes in J. All $\mathcal{H}$-classes H within J have the same cardinality $|H|$, see e.g. [6, Green's Lemma]. Hence, $h \cdot |H| = |J|$. Consider a branch in the constructed tree. Its length is at most $(2h-1) - h + h(3|H| - 1) = 3h \cdot |H| - 1 = 3|J| - 1$. An upper bound for the number of nodes of degree more than two per branch is $h \cdot |H| = |J|$.   □

4   Arbitrary Semigroups

In this section, we show that, in general, $3|S| - 1$ is the optimal height of a factorization forest. First, we combine Lemma 2 and Lemma 7 in order to show that $3|S| - 1$ is an upper bound and second, we show that every factorization forest for every non-trivial group S reaches this bound. This closes the gap between the upper and the lower bound.

Theorem 1. *Let S be a finite semigroup. For every homomorphism $\varphi : A^+ \to S$ we have $h_\varphi(w) \leq 3|S| - 1$ for all $w \in A^+$.*

Proof. For simplicity, we only show the slightly weaker bound $h_\varphi(w) \leq 3|S|$. The improved bound $3|S| - 1$ can be obtained from the current proof by some additional rotation technique as in the proof of Lemma 2. We postpone this proof to a forthcoming journal version of this article.

By Lemma 2 there exists a $\mathcal{J}$-factorization tree for w. If we consider a node v of this tree with more than two children $v_1, \ldots, v_m$, then $\varphi(v)$ and all $\varphi(v_i)$ are $\mathcal{J}$-equivalent. Hence, we can apply Lemma 7 to the factorization $v = v_1 \cdots v_n$. We substitute every branching with more than two children by the respective tree given by Lemma 7. The resulting tree is a factorization tree for w.

We compute its height. Let $J_1, \ldots, J_j$ be the (j-many) $\mathcal{J}$-classes of S. Since every element of S is contained in exactly one $\mathcal{J}$-class, we have $|J_1| + \cdots + |J_j| = |S|$. The length of a branch in the constructed tree is at most $2j - j + ((3|J_1| - 1) + \cdots + (3|J_j| - 1)) = 3(|J_1| + \cdots + |J_j|) = 3|S|$.   □

Theorem 2. *For every non-trivial finite group G there exists a word $w \in G^+$ such that $h_\varphi(w) \geq 3|G| - 1$ where $\varphi : G^+ \to G$ is the evaluation homomorphism.*

Proof. By 1 we denote the neutral element of G. For a word $w = g_1 \cdots g_n \in G^n$ of length n we let

$$\omega = g_1\, \varphi(g_1 g_2) \cdots \varphi(g_1 \cdots g_n) \in G^n$$

be the sequence of all prefixes. Given a prefix sequence $\omega = h_1 \cdots h_n \in G^n$ one can reconstruct w by $w = h_1\, \varphi(h_1^{-1} h_2) \cdots \varphi(h_{n-1}^{-1} h_n) \in G^n$. Two prefix sequences $g_1 \cdots g_n, h_1 \cdots h_n \in G^n$ are *similar* if for all $1 \leq i \leq j \leq n$ we have $g_i = g_j$ if and only if $h_i = h_j$. Two words $w, v \in G^n$ are *similar* if their respective prefix sequences $\omega, \nu \in G^n$ are similar. Similarity forms an equivalence relation on the set of words in G^n. The concept of similarity has been introduced by [2, Section 6.1]. Its purpose is to ensure large factors in factorizations of degree more than two. We will construct words $w_1, \ldots, w_{|G|} \in G^+$ such that for all words $x, y, v_i \in G^*$ where v_i is similar to w_i we have

$$h_\varphi(x v_i y) \geq 3i - 2.$$

Let $G = \{a_1, \ldots, a_{|G|}\}$. The prefix sequence ω_i of the word w_i will be a word over the alphabet $\{a_1, \ldots, a_i\}$. By $\omega_{i-1}^{[j]}$ we denote the prefix sequence obtained from ω_{i-1} by replacing all occurrence of the letter a_j by the new letter a_i. Hence, $\omega_{i-1}^{[i]} = \omega_{i-1}$. The words w_i are given by the prefix sequences

$$\omega_1 = a_1^{12},$$

$$\omega_i = \left(\left(\omega_{i-1}^{[1]}\right)^{12} \cdots \left(\omega_{i-1}^{[i-1]}\right)^{12} \left(\omega_{i-1}^{[i]}\right)^{12} \right)^{12} \quad \text{for } i > 1.$$

The factor $\left(\omega_{i-1}^{[j]}\right)^{12}$, $1 \le j \le i$, does not contain the letter a_j. Hence, every prefix sequence ν_i which is similar to ω_i can be obtained by the very same construction using some other enumeration of the group elements. Therefore, it is sufficient to show $h_\varphi(x w_i y) \ge 3i - 2$. For $i = 1$ this is trivially true. Note that this lower bound is also the upper bound if and only if $a_1 = 1$. Let now $i > 1$. For technical reasons we need additional words w_i' and w_i'' defined by prefix sequences

$$\omega_i' = \left(\left(\omega_{i-1}^{[1]}\right)^{12} \cdots \left(\omega_{i-1}^{[i-1]}\right)^{12} \left(\omega_{i-1}^{[i]}\right)^{12} \right)^{6},$$

$$\omega_i'' = \left(\left(\omega_{i-1}^{[1]}\right)^{12} \cdots \left(\omega_{i-1}^{[i-1]}\right)^{12} \left(\omega_{i-1}^{[i]}\right)^{12} \right)^{3}.$$

We will show $h_\varphi(x v_i' y) \ge 3i - 3$ and $h_\varphi(x v_i'' y) \ge 3i - 4$ for all $x, y, v_i', v_i'' \in G^*$ such that v_i' is similar to w_i' and v_i'' is similar to w_i''. Again, it is sufficient to show $h_\varphi(x w_i' y) \ge 3i - 3$ and $h_\varphi(x w_i'' y) \ge 3i - 4$.

We start with w_i''. If $d(x w_i'' y) = (u_1, u_2)$ then u_1 or u_2 contains a factor v_{i-1} which is similar to w_{i-1}. W.l.o.g. let v_{i-1} be a factor of u_1. By induction, $h_\varphi(u_1) \ge 3(i-1) - 2 = 3i - 5$, and hence, the height of this factorization tree for w_i'' is at least $3i - 4$. If $d(x w_i'' y) = (u_1, \ldots, u_m)$ with $\varphi(u_i) = 1$ in G, then w.l.o.g. we can assume $|u_1| > |x|$. Let $g = \varphi(x)^{-1} \varphi(u_1) = \varphi(x)^{-1}$. If $g \notin \{a_1, \ldots, a_i\}$ then, by construction of w_i'', we see that w_i'' is a factor of u_1. Since some word v_{i-1} which is similar to w_{i-1} is a factor of w_i'', we conclude that the height of this factorization tree for $x w_i'' y$ is at least $3i - 4$. If $g = a_j$ for some $j \in \{1, \ldots, i\}$, then the factor of w_i'' which corresponds to $\omega_{i-1}^{[j]}$ is a factor of some u_ℓ, and therefore, u_ℓ contains a factor v_{i-1} which is similar to w_{i-1}. By induction, $h_\varphi(u_\ell) \ge 3i - 5$. Again, we conclude that the height of this factorization tree for $x w_i'' y$ is at least $3i - 4$. This shows $h_\varphi(x w_i'' y) \ge 3i - 4$.

Next, we consider w_i'. If $d(x w_i' y) = (u_1, u_2)$ then u_1 or u_2 contains a factor v_i'' which is similar to w_i''. Hence, the height of this tree for $x w_i' y$ is at least $3i - 3$. If $d(x w_i' y) = (u_1, \ldots, u_m)$ with $\varphi(u_j) = 1$ in G, then w.l.o.g. we can assume $|u_1| > |x|$. Let $g = \varphi(x)^{-1}$. If $g \notin \{a_1, \ldots, a_i\}$ then, by construction of w_i', we see that w_i' – and hence w_i'' – is a factor of u_1. We conclude that the height of this factorization tree for $x w_i' y$ is at least $3i - 3$. If $g = a_j$ for some $j \in \{1, \ldots, i\}$, then the factor of w_i' which corresponds to $\left(\omega_{i-1}^{[j]}\right)^{12}$ is a factor of some u_ℓ. If $d(u_\ell) = (u_1', u_2')$ then u_1' or u_2' contains a factor which is similar to w_{i-1}. Therefore, the height of this factorization tree for $x w_i' y$ is at least $3i - 3$. If $d(u_\ell) = (u_1', \ldots, u_k')$ with $\varphi(u_j') = 1$, then we could replace the factorization of $u w_i' x$ by $d(u w_i' x) = (u_1, \ldots, u_{\ell-1}, u_1', \ldots, u_k', u_{\ell+1}, \ldots, u_m)$. Therefore, this last case is not possible in an optimal factorization tree for $x w_i' y$. This shows $h_\varphi(x w_i' y) \ge 3i - 3$.

Finally, we consider w_i. If $d(xw_iy) = (u_1, u_2)$ then u_1 or u_2 contains a factor v_i' which is similar to w_i'. Hence, the height of this tree for xw_iy is at least $3i - 2$. If $d(xw_iy) = (u_1, \ldots, u_m)$ with $\varphi(u_j) = 1$ in G, then w.l.o.g. we can assume $|u_1| > |x|$. Let $g = \varphi(x)^{-1}$. If $g \notin \{a_1, \ldots, a_i\}$ then, by construction of ω_i, we see that w_i – and hence w_i' – is a factor of u_1. We conclude that the height of this factorization tree for xw_iy is at least $3i - 2$. If $g = a_j$ for some $j \in \{1, \ldots, i\}$, then the factor of w_i which corresponds to $\left(\omega_{i-1}^{[j]}\right)^{12}$ is a factor of some u_ℓ. As before, we do not need to consider the case $d(u_\ell) = (u_1', \ldots, u_k')$ with $\varphi(u_j') = 1$, since it is not optimal. Let $d(u_\ell) = (u_1', u_2')$. The factorization $d(u_1') = (u_1'', \ldots, u_k'')$ with $\varphi(u_j'') = 1$ is also not optimal, since this implies $\varphi(u_1') = 1$ and $\varphi(u_2') = 1$. Similarly, $d(u_2') = (u_1'', \ldots, u_j'')$ with $\varphi(u_j'') = 1$ does not yield an optimal tree. Hence, we can assume that $d(u_1') = (u_1'', u_2'')$ and $d(u_2') = (u_3'', u_4'')$. It follows that some u_j'' contains a factor which is similar to w_{i-1}. Therefore, the height of this factorization tree for xw_iy is at least $3i - 2$.

Let $w = w_{|G|}w_{|G|}a \in G^+$ where $a \in G$ is chosen such that $\varphi(w) \neq 1$. This is possible since G is non-trivial. Every factorization tree for w starts with a factorization $d(w) = (u_1, u_2)$. Now, u_1 or u_2 contains a factor $w_{|G|}$. It follows that $h_\varphi(w) \geq 3|G| - 1$. □

5   Aperiodic Semigroups

In this section we show that the height $2|S|$ is optimal for factorization forests over aperiodic semigroups S. In Theorem 3, we are using Lemma 2 and Lemma 5 in order to show that $2|S|$ is an upper bound and in Theorem 4 we show that there exists an infinite family of semilattices S for which $2|S|$ is also a lower bound. The factor 2 is surprising, since it is natural to try to find factorizations $w = xu_1 \cdots u_n y$ such that $u_1, \ldots, u_n$ are mapped to the same idempotent element of S. This yields the following tree of height 3:

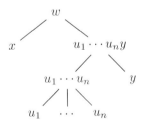

Theorem 3. *For every homomorphism* $\varphi : A^+ \to S$ *where* S *is a finite aperiodic semigroup we have* $h_\varphi(w) \leq 2|S|$ *for all* $w \in A^+$.

Proof. By Lemma 2 there exists a $\mathcal{J}$-factorization tree for w. If we consider a node v of this tree with more than two children $v_1, \ldots, v_m$, then $\varphi(v)$ and all $\varphi(v_i)$ are $\mathcal{J}$-equivalent. Hence, we can apply Lemma 5 to the factorization $v = v_1 \cdots v_n$. We substitute every branching with more than two children by the respective tree given by Lemma 5. The resulting tree is an $\mathcal{H}$-factorization tree for w. Since S is aperiodic, it is $\mathcal{H}$-trivial (see e.g. [5, Lemma 1]), and hence every $\mathcal{H}$-factorization tree is a factorization tree.

We compute the height of the resulting tree for w. Let $J_1, \ldots, J_j$ be the (j-many) $\mathcal{J}$-classes of S. Since every element of S is contained in exactly one $\mathcal{J}$-class, we have $|J_1| + \cdots + |J_j| = |S|$. The length of a branch in the constructed tree is at most $2j - j + \big((2\,|J_1| - 1) + \cdots + (2\,|J_j| - 1)\big) = 2(|J_1| + \cdots + |J_j|) = 2\,|S|$.   □

Theorem 4. *For every $n \geq 2$ there exists a semilattice M with n elements and a word $w \in M^+$ such that $h_\varphi(w) \geq 2\,|M|$ where $\varphi : M^+ \to M$ is the evaluation homomorphism.*

Proof. Let $M = \{a_1, \ldots, a_n\}$ equipped with the composition $a_i a_j = a_{\max(i,j)}$ for $a_i, a_j \in M$. These monoids have also been used in [2] in order to show a lower bound of $|M| + 1$. Note that a_1 is the neutral element of M. The monoid M has the following property. If a_i occurs in the alphabet of $w \in M^+$ and if $w = u_1 \cdots u_n$ such that $\varphi(u_1) = \cdots = \varphi(u_n) = a_j$, then $j \geq i$. We construct words $w_i \in \{a_1, \ldots, a_i\}^+$ such that $h_\varphi(x w_i y) \geq 2i - 1$ for all $x, y \in M^*$. Let

$$
\begin{aligned}
w_1 &= a_1^{12}, \\
w_i &= (w_{i-1}^{12} a_i)^{12} \quad \text{for } i > 1.
\end{aligned}
$$

The intuition is that if one wants to eliminate the letter a_{i+1} from w_{i+1}, one first has to use an idempotent factorization and then a simple factorization in order to cut off the letter a_{i+1}. For $i = 1$, we trivially have $h_\varphi(x w_1 y) \geq 2 - 1 = 1$. Let now $i > 1$. We have to distinguish several cases.

First, suppose $d(x w_i y) = (u_1, u_2)$. W.l.o.g. we can assume that u_1 contains a factor $w_{i-1}^{12} a_i$. If $d(u_1) = (u'_1, \ldots, u'_m)$ then some u'_ℓ contains a factor w_{i-1}: if $m > 2$, then $\varphi(u'_\ell) = a_j$ with $j \geq i$ for all $1 \leq \ell \leq m$. By induction, we see that the height of this factorization tree is at least $2(i - 1) - 1 + 2 = 2i - 1$.

Now, suppose $d(x w_i y) = (u_1, \ldots, u_m)$ with $\varphi(u_1) = \cdots = \varphi(u_m) = a_j$. We have $j \geq i$ since $\varphi(x w_i y) = a_j$ with $j \geq i$. Therefore, some u_ℓ contains a factor $w_{i-1}^6 a_i$ or a factor $a_i w_{i-1}^6$. In any case, if $d(u_\ell) = (u'_1, \ldots, u'_k)$, then some u'_t contains a factor w_{i-1} (in both cases $k = 2$ and $k > 2$, since $k > 2$ implies $\varphi(u'_1) = \cdots = \varphi(u'_k) = a_{j'}$ where $j' \geq i$). By induction, the height of this factorization tree is at least $2i - 1$. This shows that $h_\varphi(x w_i y) \geq 2i - 1$.

Let $w = w_{n-1}^{12}\big(a_n w_{n-1}^{12}\big)^{24}$. If $d(w) = (u_1, u_2)$, then u_1 or u_2 contains w_n as a factor and hence, the height of this factorization tree for w is at least $2n$. If $d(w) = (u_1, \ldots, u_m)$ with $\varphi(u_1) = \cdots = \varphi(u_m) = a_n$ then some u_j contains a factor $w_{n-1}^6 a_n w_{n-1}^6$. If $\varphi(u_j) = (u'_1, \ldots, u'_k)$ then some u'_ℓ contains a factor $w_{n-1}^3 a_n$ or a factor $a_n w_{n-1}^3$. In any case, if $d(u'_\ell) = (u''_1, \ldots, u''_t)$, then some u''_s contains a factor w_{i-1} and hence, the height of this factorization tree for w is at least $2n$.   □

Acknowledgments. The author would like to thank the anonymous referees for their comments and suggestions. They helped to improve the presentation of the results – in particular of those in Section 3.

References

1. Bojańczyk, M., Colcombet, T.: Bounds in ω-regularity. In: 21th IEEE Symposium on Logic in Computer Science (LICS 2006), 12-15 August 2006, pp. 285–296. IEEE Computer Society, Los Alamitos (2006)
2. Chalopin, J., Leung, H.: On factorization forests of finite height. Theoretical Computer Science 310(1-3), 489–499 (2004)
3. Colcombet, T.: A combinatorial theorem for trees. In: ICALP 2007. LNCS, pp. 901–912. Springer, Heidelberg (2007)
4. Colcombet, T.: Factorisation forests for infinite words. In: Csuhaj-Varjú, E., Ésik, Z. (eds.) FCT 2007. LNCS, vol. 4639, pp. 226–237. Springer, Heidelberg (2007)
5. Diekert, V., Gastin, P., Kufleitner, M.: A survey on small fragments of first-order logic over finite words. International Journal of Foundations of Computer Science 19(3), 513–548 (2008)
6. Pin, J.-É.: Varieties of Formal Languages. North Oxford Academic, London (1986)
7. Pin, J.-É., Weil, P.: Polynomial closure and unambiguous product. Theory of Computing Systems 30(4), 383–422 (1997)
8. Simon, I.: Factorization forests of finite height. Technical Report 87-73, Laboratoire d'Informatique Théorique et Programmation, Paris (1987)
9. Simon, I.: Properties of factorization forests. In: Pin, J.-É. (ed.) Formal Properties of Finite Automata and Applications: LITP Spring School on Theoretical Computer Science. LNCS, vol. 386, pp. 65–72. Springer, Heidelberg (1988)
10. Simon, I.: Factorization forests of finite height. Theoretical Computer Science 72(1), 65–94 (1990)
11. Simon, I.: A short proof of the factorization forest theorem. In: Nivat, M., Podelski, A. (eds.) Tree Automata and Languages, pp. 433–438. Elsevier, Amsterdam (1992)
12. Simon, I.: On semigroups of matrices over the tropical semiring. RAIRO – Informatique Theorique et Applications 28(3-4), 277–294 (1994)

Arithmetic Circuits, Syntactic Multilinearity, and the Limitations of Skew Formulae

Meena Mahajan and B.V. Raghavendra Rao

The Institute of Mathematical Sciences, Chennai 600 113, India
{meena,bvrr}@imsc.res.in

Abstract. Functions in arithmetic NC^1 are known to have equivalent constant width polynomial degree circuits, but the converse containment is unknown. In a partial answer to this question, we show that syntactic multilinear circuits of constant width and polynomial degree can be depth-reduced, though the resulting circuits need not be syntactic multilinear. We then focus specifically on polynomial-size syntactic multilinear circuits, and study relationships between classes of functions obtained by imposing various resource (width, depth, degree) restrictions on these circuits. Along the way, we obtain a characterisation of NC^1 (and its arithmetic counterparts) in terms of log width restricted planar branching programs. We also study the power of skew formulae, and show that even exponential sums of these are unlikely to suffice to express the determinant function.

1 Introduction

Among the parallel complexity classes, the class NC^1 of boolean functions computed by logarithmic depth polynomial size circuits has several equivalent characterisations, in the form of bounded width branching programs, polynomial size formulae and bounded width circuits of polynomial size. Its subclass AC^0, consisting of polynomial size constant depth unbounded fan-in circuits, has also been characterised via restricted branching programs.

However, when we consider the counting and arithmetic versions of those classes which are equivalent to NC^1, they seem to represent different classes of functions. In [10], it was shown that if inputs take values from $\{0,1\}$, and only the constants $-1,0,1$ are allowed, then counting the total weights of paths in a bounded width branching program is equivalent to the functions computable by log depth polynomial size arithmetic circuits, *i.e.* $GapBWBP = GapNC^1$. In [12], this study was extended to bounded width circuits of polynomial degree and size, sSC^0, showing that $GapNC^1 \subseteq GapsSC^0$, but it left open the question of equality of these classes.

The question of whether $GapsSC^0$ is in $GapNC^1$ can be seen as a depth reduction problem for bounded width circuits. We do not have an answer for this general question. So it is natural to ask if there are any restrictions on the circuit so that depth reduction is possible.

E. Ochmański and J. Tyszkiewicz (Eds.): MFCS 2008, LNCS 5162, pp. 455–466, 2008.
© Springer-Verlag Berlin Heidelberg 2008

Syntactic multilinearity is a restriction which has been studied in the literature. Syntactic multilinear circuits are those in which every multiplication gate operates on disjoint set of variables. The syntactic multilinear restriction is very fruitful in the sense that there are known unconditional separations and lower bounds for these classes (see [13,14,15]).

We show that depth reduction for small width circuits is possible if the circuit is syntactic multilinear; however, the depth-reduced circuit may not be syntactic multilinear or even multilinear. The setting we consider is more general than that of [10] and [12]; here the input variables are allowed to take arbitrary values from the underlying ring $\mathbb{K}$. The main result (Theorem 1) is that polynomial size, constant width syntactic multilinear circuits can be simulated (non-uniformly) by log depth bounded fan-in circuits of polynomial size, but this construction need not preserve the syntactic multilinearity property.

Once we take up the restriction of syntactic multilinearity for these arithmetic circuits, it is worthwhile to explore the relationships among the syntactic multilinear arithmetic circuit classes close to arithmetic NC^1.

In the model of branching programs, syntactic multilinearity is a well-studied notion, referred to as read-once branching programs (see *e.g.* [6]). There are several known lower bounds for syntactic multilinear branching programs.

For formulae, syntactic multilinearity is defined exactly as for circuits. A careful observation of the depth reduction for poly size arithmetic formula as given in [7] shows that it preserves syntactic multilinearity. Also some of the constructions in [10,11,12], relating branching programs and formulae, can be shown to preserve syntactic multilinearity.

In [3], the class of bounded depth arithmetic circuits is characterised in terms of a restricted version of grid programs, rGP, of bounded width BWrGP. We observe that this construction can be extended to show a new (non-uniform) characterisation of (arithmetic) NC^1 in terms of restricted planar branching programs of log width LWrGP. In addition, this can be shown to preserve syntactic multilinearity, for arithmetic NC^1 as well as arithmetic AC^0.

We also study the class of polynomial size skew formulas, denoted SkewF. The motivation for this study arises from Valiant's characterisations of the classes VP and VNP (see [18]; also, for more exposure on algebraic complexity theory, the reader is referred to [8,9]). Valiant proved that every polynomial $p(X) \in \mathsf{VNP}_{\mathbb{K}}$ (where $\mathbb{K}$ is an arbitrary ring), and in particular every polynomial in $\mathsf{VP}_{\mathbb{K}}$, can be written as $p(X) = \sum_{e \in \{0,1\}^m} \phi(X, e)$, where the polynomial ϕ has an arithmetic formula of polynomial size. So we ask if we can prove a similar equivalence in the case of skew circuits. That is, can we write polynomials computed by skew circuits as an exponential sum of polynomials computed by skew formulae? We show that this is highly unlikely, by showing that any polynomial which is expressible as an exponential sum of skew formulae belongs to the class VNC^1.

The existing and new relationships amongst the arithmetic classes (prefix a-) can be seen in Figure 1; Figure 2 shows the corresponding picture for the syntactic multilinear classes (prefix sma-). Our main depth-reduction result straddles the two figures, and along with [12] gives sma-$\mathsf{sSC}^0 \subseteq$ a-$\mathsf{NC}^1 \subseteq$ a-sSC^0.

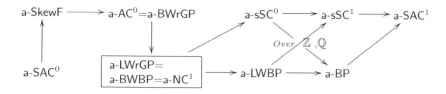

Fig. 1. Arithmetic classes around NC^1

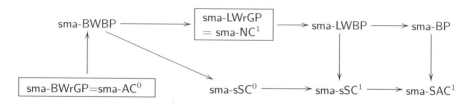

Fig. 2. Relationship among syntactic multilinear classes

The rest of the paper is organised as follows. Section 2 introduces basic definitions. In Section 3 we prove that small-width syntactic multilinear circuits can be depth-reduced. In Section 4, we establish the containments among the syntactic multilinear classes and obtain a new characterisation for NC^1 in terms of a restricted class of grid branching programs. In Section 5 we describe our results concerning skew formulae.

2   Preliminaries

We use standard notation for Boolean circuits and their size, width, depth and degree; see *e.g.* [12],[21]. Unless otherwise stated, fan-in is assumed to be bounded. NC^1 denotes the class of boolean functions which can be computed by boolean circuits of depth $O(\log n)$ and size $\mathsf{poly}(n)$. SC^i denotes the class of boolean functions computed by $\mathsf{poly}(n)$ size circuits of width $O(\log^i n)$. sSC^i is the class of boolean functions computed by $\mathsf{poly}(n)$ degree, $\mathsf{poly}(n)$ size circuits of width $O(\log^i n)$. SAC^i denotes the class of boolean functions computed by polynomial circuits of size $\mathsf{poly}(n)$ and depth $O(\log^i n)$, where $\vee$ gates can have unbounded fan-in. AC^0 denotes the class of boolean functions which can be computed by unbounded fan-in constant depth boolean circuits of size $\mathsf{poly}(n)$.

A *formula* is a circuit where every non-input gate has fan-out bounded by one. F and LWF denote the set of boolean functions which can be computed by polynomial size formulae of unbounded and log width respectively. Without loss of generality, NC^1, AC^0 and SAC^0 circuits can be assumed to be formulae.

A branching program (BP) is a directed acyclic layered graph with edges labelled from $\{x_1, \ldots, x_n, \neg x_1, \ldots, \neg x_n, 0, 1\}$, and with two designated nodes s and t. A BP is said to accept its input if and only if there exists an s-t path, in which every edge label evaluates to 1. A BP can also be viewed as a skew-circuit, *i.e.* a circuit where every $\wedge$ gate has at most one non-circuit input.

Let BWBP and LWBP denote the functions computed by constant width and log width branching programs of polynomial size respectively.

G-graphs are the graphs that have planar embeddings where vertices are embedded on a rectangular grid, and all edges are between adjacent columns from left to right. Let BWGP denote the class of boolean functions accepted by constant width polynomial size branching programs which are G-graphs. In these graphs, the node s is fixed as the leftmost bottom node and t is the rightmost top node. In [3], a restriction of G-graphs is considered where the width of the grid is a constant, and only certain kinds of connections are allowed between any two layers. Namely, for width $2k + 2$, the connecting pattern at any layer is one of the graphs $G_{k,i}$ shown alongside for $0 \le i \le 2k + 2$. Let BWrGP denote the class of boolean functions accepted by constant width polynomial size branch-

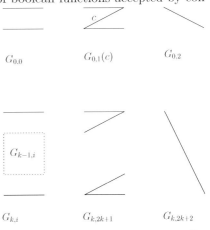

ing programs that are restricted G-graphs, and LWrGP the class corresponding to log width polynomial size programs that are restricted G-graphs. (see [3]).

The following proposition summaries the known relationships among the boolean complexity classes defined above; see for instance [21].

Proposition 1 ([20,4,16,11,3]). SAC^1 = Circuit Size, Deg(poly(n), poly(n));

$$NC^1 = BWBP = SC^0 = sSC^0 = F = LWF; \qquad AC^0 = BWrGP$$

An arithmetic circuit over a ring $\langle \mathbb{K}, +, -, \times, 0, 1 \rangle$ is a circuit with internal nodes labelled from $\{\times, +\}$, and leaves labelled by input variables $x_1, \ldots, x_n$ that take values in $\mathbb{K}$ or by one of the constants from $\{-1, 0, 1\}$.

The arithmetic circuit classes corresponding to the above defined boolean classes consist of functions $f : \mathbb{K}^* \to \mathbb{K}$, and are defined as follows.

BWBP$[\mathbb{K}] = \{f : \mathbb{K}^* \to \mathbb{K} \mid f = \text{sum of weights of all } s \rightsquigarrow t \text{ paths in a BWBP}\}$

Here the weight of a path is the product of the labels of edges along the path.

$$NC^1[\mathbb{K}] = \{f \mid f \text{ has a poly size, } O(\log n) \text{ depth, fan-in 2 circuit.}\}$$

sSC$^i[\mathbb{K}] = \{f \mid f \text{ has a poly size, } O(\log^i n) \text{ width, poly}(n) \text{ degree circuit.}\}$
For notational convenience we drop the $\mathbb{K}$ where understood from context to be any (or a specific) ring. To distinguish from the boolean case, we write $\mathcal{C}[\mathbb{K}]$ as a-$\mathcal{C}$. The following proposition summarises the known relationships among the arithmetic classes.

Proposition 2 ([3,10,12]). a-BWrGP = a-AC$^0 \subseteq$ a-BWBP = a-NC$^1 \subseteq$ a-sSC^0

Multilinear and syntactic multilinear circuits are as defined in [14]. Let C be an arithmetic circuit over the ring $\mathbb{K}$. For a gate g in C, let $P_g \in \mathbb{K}[X]$ be the polynomial represented at g. Let $X_g \subseteq X$ denote the set of variables that occur in the sub-circuit rooted at g. We call C *multilinear* if for every gate $g \in C$, P_g is a multilinear polynomial, and *syntactic multilinear* if for every multiplication gate $g = h \times f$ in C, $X_h \cap X_f = \emptyset$.

In the case of formulae, the notion of multilinearity and syntactic multilinearity are (non-uniformly) equivalent. Viewing branching programs as skew-circuits, a multilinear branching program P is one which computes multilinear polynomials at every node, and P is syntactic multilinear if in every path of the program (not just s-to-t paths), no variable appears more than once; *i.e.* the branching program is syntactic read-once.

For any arithmetic complexity class a-$\mathcal{C}$, we denote by ma-$\mathcal{C}$ and sma-$\mathcal{C}$ respectively the functions computed by multilinear and syntactic multilinear versions of the corresponding circuits.

In [15] it is shown that the depth reduction of [19] preserves syntactic multilinearity; thus

Proposition 3 ([15]). *Any function computed by a syntactic multilinear polynomial size polynomial degree arithmetic circuit is in* sma-SAC^1.

3   Depth Reduction in Small Width Sm-Circuits

This entire section is devoted to a proof of Theorem 1 below, which says that a circuit width bound can be translated to a circuit depth bound, provided the given small-width circuit is syntactic multilinear.

Theorem 1. *Let C be a* syntactic multilinear *circuit of length l and width w and circuit degree d, with $X = \{x_1, \ldots, x_n\}$ as the input variables, and constants $\{-1, 0, 1\}$ from the ring $\mathbb{K}$. Then there is an equivalent circuit E of depth $O(w^2 \log l + \log d)$ and size $O(2^{w^2 + 3w} l^{25w} + 4lwd)$.*

Corollary 1. sma-sSC$^0 \subseteq$ a-NC^1.

Corollary 2. sma-Size, Width, Deg($2^{\text{poly(log)}}$, poly(log), $2^{\text{poly(log)}}$)
$\subseteq$ a-Size, Depth($2^{\text{poly(log)}}$, poly(log))

We first give a brief outline of the technique used. The main idea is to first cut the circuit C at length $\lceil \frac{l}{2} \rceil$, to obtain circuits A (the upper part) and B (the lower part). Let $M = \{h_1, \ldots, h_w\}$ be the output gates of C at level $\lceil \frac{l}{2} \rceil$. (Note that output gates are at the topmost layer of the circuit.) A is obtained from C by replacing the gates in M by a set $Z = \{z_1, \ldots, z_w\}$ of new variables. Each gate g of A (or B) represents a polynomial $p_g \in \mathbb{K}[X, Z]$, and can also be viewed as a polynomial in $K[Z]$, where $K = \mathbb{K}[X]$. Since A and B are circuits of length bounded by $\lceil \frac{l}{2} \rceil$, if we can prove inductively that the coefficients of the polynomials at the output gates of A and B can be computed by small depth circuits (say $O(w \log(l/2))$, then, since p_g has at most 2^w monomials in variables from Z, we can substitute for the z_i's by the value at the output gate g_i of B (*i.e.* polynomials in $\mathbb{K}[X]$). This requires an additional depth of $O(w)$.

The first difficulty in the above argument can be seen even when $w = O(1)$. Though C is syntactic multilinear, the circuit A need not be multilinear in the new dummy variables from Z. This is because there can be gates which compute large constants from $\mathbb{K}$ (*i.e.* without involving any of the variables), and hence have large degree (bounded by the degree of the circuit). This means that the polynomials in the new variables Z at the output gates of A can have non-constant degree, and the number of monomials can be large. Thus the additional depth needed to compute the monomials will be non-constant; hence the argument fails.

To overcome this difficulty, we first transform the circuit C into a new circuit C', where no gates compute "large" constants in $\mathbb{K}$. Assume without loss of generality that every gate in C has a maximum fan-out of 2. Let $G = \{g \in C \mid \mathsf{leaf}(g) \cap X = \emptyset\}$, where for a gate $g \in C$, we define

$$\mathsf{leaf}(g) = \{h \in C \mid h \text{ is a leaf node in } C, \text{ and } g \text{ is reachable from } h \text{ in } C\}$$

Thus G is exactly the nodes that syntactically compute constants. Now define C' as a new circuit which is the same as C except that for all $g \in G$, we replace the ith ($i = 1, 2$) outgoing wire of g by a new variable y_{g_i}. Note that the number of such new variables introduced is at most $4lw$. Let $Y = \{y_{g_i} \mid g \in G, 1 \leq i \leq 2\}$. We show that C' is syntactic multilinear in the variables $X \cup Y$.

Lemma 1. *The circuit C' constructed above is syntactic multilinear in the variables $X \cup Y$. Further, C' does not have any constants.*

Next we show, in Lemma 2, how to achieve depth reduction for syntactic multilinear bounded width circuits which have no constants. Then we complete the proof of Theorem 1 by explicitly computing the constants (*i.e.* the actual values represented by variables in Y) computed by the circuit C.

Lemma 2. *Let C' be a width w, length l syntactic multilinear arithmetic circuit with leaves labelled from $X \cup Y$ (no constants). Then there is an equivalent arithmetic circuit C'' of size $O(2^{w^2+3w}l^{25w})$ and depth $O(w^2 \log l)$ which computes the same function as C'.*

To establish lemma 2, we use the intuitive idea sketched in the beginning of the section; namely, slice the circuit horizontally, introduce dummy variables along the slice, and proceed inductively on each part.

Now the top part has three types of variables: circuit inputs X, variables representing constants Y as introduced in Lemma 1, and variables along the slice Z. The variables Z appear only at the lowest level of this circuit. Note that this circuit for the top part is syntactic multilinear in Z as well (because there are no constants at the leaves).

To complete an inductive proof for Lemma 2, we need to show depth reduction for such circuits. We use Lemma 3 below, which tells us that viewing each gate as computing a polynomial in Z, with coefficients from $K = \mathbb{K}[X, Y]$, there are small-depth circuits representing each of the coefficients. We then combine these circuits to evaluate the original circuit.

Formally, let D be a width w, length l, syntactic multilinear circuit, with leaves labelled from $X \cup Y \cup Z$ (no constants), where variables from $Z = \{z_1, \ldots z_w\}$ appear only at the lowest level of the circuit. Let $h_1, \ldots, h_w$ be the set of output gates of D. Let $p_{h_i} \in \mathbb{K}[X, Y, Z]$ denote the multilinear polynomial computed at h_i. Note that p_{h_i} can also be viewed as a polynomial in $K[Z]$, i.e. a multilinear polynomial with variables from Z and polynomials from $\mathbb{K}[X, Y]$ as its coefficients; we use this viewpoint below. For $T \subseteq \{1, \ldots, w\}$, let $[p_{h_i}, T] \in \mathbb{K}[X, Y]$ denote the coefficient of the monomial $m_T = \prod_{j \in T} z_j$ in p_{h_i}. The following lemma tells us how to evaluate these coefficients $[p_{h_i}, T]$.

Lemma 3. *With circuit D as above, $\forall h \in \{h_1, \ldots, h_w\}$ and $T \subseteq \{1, \ldots, w\}$, there is a bounded fan-in arithmetic circuit $D^{h,T}$ of size bounded by $2^{w^2+2w}l25w$ and depth $O(w^2 \log l)$, with leaves labelled from $X \cup Y \cup \{0, 1\}$, such that the value computed at its output gate is exactly the value of $[p_h, T]$ evaluated at the input setting to $X \cup Y$.*

Proof Sketch. We proceed by induction on the length l of the circuit D. The base case, when $l = 1$, can be handled appropriately. Assume that the lemma holds for all circuits D' of length $l' < l$ and width w.

Now let D be the given circuit of length l, syntactic multilinear in $X \cup Y \cup Z$, where variables from Z appear only at the lowest level. Let $\{h_1, \ldots, h_w\}$ be the output gates of D. Let $\{g_1, \ldots, g_w\}$ be the gates of D at level $l' = \lceil \frac{l}{2} \rceil$. Denote by A the circuit resulting from replacing gates g_i with new variables z'_i for $1 \leq i \leq w$, and removing all the gates below level l', and denote by B the circuit with $\{g_1, \ldots, g_w\}$ as output gates, i.e. gates above the g_i's are removed. We rename the output gates of A as $\{f_1, \ldots, f_w\}$. Both A and B are syntactic multilinear circuits of length bounded by l' and width w, and of a form where the inductive hypothesis is applicable. For $i \in \{1, \ldots, w\}$, p_{f_i} is a polynomial in $K[Z']$ and p_{g_i} is a polynomial in $K[Z]$, where $K = \mathbb{K}[X, Y]$.

Applying induction on A and B, for all $S, Q \subseteq \{1, \ldots, w\}$, $[p_{f_i}, S]$ and $[p_{g_i}, Q]$ have circuits $A^{f_i, S}$ and $B^{g_i, Q}$. Note that $p_{h_i}(Z) = p_{f_i}(p_{g_1}(Z), \ldots, p_{g_w}(Z))$. But due to multilinearity,

$$p_{f_i}(Z') = \sum_{S \subseteq [w]} \left([p_{f_i}, S] \prod_{j \in S} z'_j \right) \qquad p_{g_j}(Z) = \sum_{Q \subseteq [w]} \left([p_{g_j}, Q] \prod_{s \in Q} z_s \right)$$

Using this expression for p_{f_i} in the formulation for p_{h_i}, we have

$$p_{h_i}(Z) = \sum_{S \subseteq [w]} \left([p_{f_i}, S] \prod_{j \in S} p_{g_j}(Z) \right)$$

Hence, we can extract coefficients of p_{h_i} as follows. The coefficient of the monomial m_T, for any $T \subseteq [w]$ in p_{h_i} is given by

$$[p_{h_i}, T] = \sum_{S \subseteq [w]} [p_{f_i}, S] \left(\text{coefficient of } m_T \text{ in } \prod_{j \in S} p_{g_j}(Z) \right)$$

If S has t elements, then the monomial m_T is built up in t disjoint parts (not necessarily non-empty), where the kth part is contributed by the kth polynomial of the form p_g in the above expression. So the coefficient of m_T is the product of the corresponding coefficients. Hence

$$[p_{h_i}, T] = \sum_{S = \{j_i, \dots, j_t\} \subseteq [w]} \left([p_{f_i}, S] \sum_{\substack{Q_1, \dots, Q_t : \\ \text{partition of } T}} \prod_{k=1}^{t} [p_{g_{j_k}}, Q_k] \right)$$

We use this expression to compute $[p_{h_i}, T]$. We first compute $[p_{f_i}, S]$ and $[p_{g_j}, Q]$ for all $i, j \in [w]$ and all $S, Q \subseteq [w]$ using the inductively constructed sub-circuits. Then a circuit on top of these does the required combination. Since the number of partitions of T is bounded by w^w, while the number of sets S is 2^w, this additional circuitry has size at most 2^{w^2} (for $w \geq 2$) and depth $O(w^2)$.

We can show that this construction satisfies the required bounds. □

Using Lemma 3, we can establish Lemma 2 and hence Theorem 1.

4 Relationships among Syntactic Multilinear Classes

This section explores the relationships among the syntactic multilinear versions of the arithmetic classes which are related to NC^1.

A classical result from [7] shows that for every arithmetic formula F of size s, there is an equivalent arithmetic formula F' which has depth $O(\log s)$ and size $\mathsf{poly}(s)$. A careful observation of this proof shows that if we start with a syntactic multilinear formula F, then the depth-reduced formula F' is also syntactic multilinear.

Theorem 2. *Every syntactic multilinear formula with n leaves has an equivalent syntactic multilinear circuit of depth $O(\log n)$ and size $O(n)$.*

In particular, $\mathsf{sma}\text{-}\mathsf{F} \subseteq \mathsf{sma}\text{-}\mathsf{NC}^1$.

It is easy to see that the path-preserving simulation of a constant width branching program by a log depth circuit preserves syntactic multilinearity:

Lemma 4. *For any syntactic multilinear branching program P of width w and size s over ring $\mathbb{K}$, there is an equivalent syntactic multilinear circuit C of depth $O(\log s)$ and size $O(s)$ with fan-in of $+$ gate bounded by w (or alternatively, depth $O(\log w \log s)$ and bounded fan-in).*

In particular, $\mathsf{sma}\text{-}\mathsf{BWBP} \subseteq \mathsf{sma}\text{-}\mathsf{NC}^1$ and $\mathsf{sma}\text{-}\mathsf{BP} \subseteq \mathsf{sma}\text{-}\mathsf{SAC}^1$.

It is also easy to see that the construction of [11], staggering a small-depth formula into a small-width one, preserves syntactic multilinearity. Thus

Lemma 5. *Let Φ be any sm-formula with depth d and size s. Then there is an equivalent syntactic multilinear formula Φ' of length $2s$ and width d. In particular, sma-NC^1 $\subseteq$ sma-LWF.*

From Lemma 5 and Theorem 2, we have the following equivalence.

Corollary 3. *Over any ring $\mathbb{K}$,*
sma-F= sma-LWF= sma-NC^1= sma-Formula-Depth,Size(log, poly).

A straightforward inductive construction of a branching program from a log depth formula results in a log width BP and preserves syntactic multilinearity. But the reverse containment may not hold. However, by restricting the branching program as in [3], we can obtain a characterisation for a-NC^1 which also preserves syntactic multilinearity. In [3] a characterisation for bounded depth arithmetic circuits in terms of counting number of paths in a restricted version of bounded width grid graphs is presented. We note that the characterisation given in [3] works for bounded depth arithmetic circuits over arbitrary rings, showing that a-BWrGP = a-AC^0. By closely examining the parameters in [3], we obtain a characterisation for a-NC^1 in terms of the restricted version of log width grid branching programs. We also note that these constructions preserve syntactic multilinearity. In the statements and proofs below, we use the notion of alternation-depth: a circuit C has alternation depth a if on every root-to-leaf path, the number of maximal segments of gates of the same type is at most a. Also, for an rGP (and in fact any branching program) P, we denote by $\mathsf{Var}(P)$ the set of variables that appear on some s-to-t path in P. For a formula F, $\mathsf{Var}(F)$ denotes the variables appearing anywhere in the formula F; if h is the root of F, then without loss of generality $\mathsf{Var}(F) = X_h$.

Lemma 6. *Let Φ be an arithmetic formula of size s (i.e. number of wires) and alternation-depth $2d$ over $\mathbb{K}$ and with input variables $X \in \mathbb{K}^n$. Then there is a restricted grid program P of length $s^2 + 2s$ (i.e. the number of edge layers) and width $\max\{2, 2d\}$, where the edges are labelled from $\mathsf{Var}(\Phi) \cup \mathbb{K}$, such that the weighted sum of s-to-t paths in P is equal to the function computed by Φ. Further, if Φ is syntactic multilinear, then so is P.*

Lemma 7. *Let P be an arithmetic rGP of length l (number of edge layers) and of width $2w + 2$ with variables from $X \in \mathbb{K}$. Then there exists an equivalent arithmetic formula Φ over $\mathbb{K}$, with alternation depth at most $2w+2$, size (number of wires) at most $2l$, and $\mathsf{Var}(\Phi) = \mathsf{Var}(P)$. Further, if P is syntactic multilinear, then so is Φ.*

Corollary 4. sma-AC^0 = sma-BWrGP.
     sma-NC^1 = sma-LWrGP;   a-NC^1 = a-LWrGP.

The above construction also holds in the case of boolean circuits, giving

Corollary 5. NC^1 = LWrGP.

Thus we get a characterisation for NC^1 and a-NC^1 in terms of a restricted class of log width polynomial size planar branching programs.

In [5] it is shown that any $O(\log n)$ depth polynomial size formula has an equivalent 3-register straight line program. This proves that a-NC^1 ⊆ a-BWBP. Can the same be stated for sma-NC^1 and sma-BWBP? It turns out that applying the construction of [5] does not preserve syntactic multilinearity; in fact the resulting program need not even be multilinear.

5   Skew Formulae

In this section, we consider a question motivated by the setting of Valiant's algebraic complexity classes defined in [18]. VP is the class of polynomials of polynomial degree, computable by polynomial-sized circuits. Similarly one can define VF, VNC^1, and so on. VNP is the class of polynomials expressible as $p(x_1, \ldots, x_n) = \sum_{e \in \{0,1\}^m} g(X, e)$ where $m \in O(\mathsf{poly}(n))$ and the polynomial g is in VP. Thus, loosely speaking, VNP equals $\sum \cdot$VP. See [8,9] for more details.

It is well known that the complexity class NP is equivalent to $\exists \cdot$P and in fact even to $\exists \cdot$F. A similar result holds in the case of Valiant's algebraic complexity classes too. Valiant has shown that VNP $= \sum \cdot$VF, and thus the polynomial g in the expression above can be assumed to be computable by a formula of polynomial size and polynomial degree.

Noting that VNP is the class of polynomials which are projection equivalent to the "permanent" polynomial, a natural question arises about the polynomials which are equivalent to determinant. Since the determinant exactly characterises the class of polynomials which are computable by skew arithmetic circuits ([17]), the question one could ask is: can the determinant be written as an exponential sum of partial instantiations of a polynomial that can be computed by *skew formula* of poly size, SkewF? Recall that a circuit is said to be skew if every $\times$ (or $\wedge$) gate has at most one child that is not a circuit input. Skew circuits are essentially equivalent to branching programs. Thus one could ask the related question: since VP ⊆ $\sum \cdot$VP $= \sum \cdot$VF, can we show that VSkew ⊆ $\sum \cdot$VSkewF?

We show that this is highly unlikely. We first give a characterisation of polynomials computed by skew formulae (Lemma 8) in terms of their degree and number of monomials. (As a corollary, this places a-SkewF inside a-AC^0.) We then use this to show that $\sum \cdot$VSkewF is in fact contained in VNC^1 (Theorem 3). Thus placing VSkew in $\sum \cdot$VSkewF is analogous to the statement that GapL equals GapNC^1, which we believe is quite unlikely.[1]

Lemma 8. *Let $f \in \mathbb{Z}[X]$ have degree d, where m monomials have non-zero coefficients. Then f can be computed by a skew formula Φ of size $O(md)$. Further, if all coefficients in f are bounded by c, then f can be computed by a skew formula Φ' that uses as constants only $-1, 0, 1$ and has size $O(md + mc)$. Conversely, let $f \in \mathbb{Z}[X]$ be computed by a skew formula Φ of size s. Then the degree and number of monomials in f are bounded by s. Further, if $-1, 0, 1$ are the only constants in Φ, then the absolute values of coefficients in f are bounded by s.*

Corollary 6. a-SAC^0 ⊂ a-SkewF ⊂ a-AC^0.

---

[1] For $\mathcal{C} \in \{\mathsf{SkewF}, \mathsf{NC}^1, \mathsf{BP}\}$, a-$\mathcal{C}$ is essentially the same as V$\mathcal{C}$ except that V$\mathcal{C}$ allows arbitrary constants from $\mathbb{K}$.

Theorem 3. *Let $f \in \mathbb{Z}[X]$ be expressible as $f(X) = \sum_{e \in \{0,1\}^m} \phi(X, e)$, where ϕ has a poly size skew formula. Then $f \in \mathsf{VNC}^1$. (i.e. , $\sum \cdot \mathsf{VSkewF} \subseteq \mathsf{VNC}^1$.)*

Proof Sketch. Since $\phi(X, Y)$ has a poly size skew formula, by Lemma 8 we know that the number of non-zero monomials in ϕ is bounded by some polynomial $q(n, m)$. Hence the number of non-zero monomials in $\phi(X, Y)|_X$, is also bounded by $q(n, m)$.

For any $\alpha \in \mathbb{N}^n$, consider the monomial $X^\alpha = \prod_{\alpha_i} X_i^{\alpha_i}$. Define the set S_α as $S_\alpha = \{\beta \in \{0, 1\}^m \mid X^\alpha Y^\beta \text{ has a non-zero coefficient } a_{\alpha,\beta} \text{ in } \phi\}$. Clearly, for each α, $|S_\alpha| \le q(n, m)$. Since $\phi(X, Y)$ is evaluated only at Boolean settings of Y, we can assume, w.l.o.g., that it is multilinear in Y. Hence

$$\phi(X, Y) = \sum_{\alpha \in \mathbb{N}^n} \sum_{\beta \in \{0,1\}^m} a_{\alpha,\beta} X^\alpha Y^\beta$$

Hence we can show that

$$f(X) = \sum_{\alpha \in \mathbb{N}^n} \left(X^\alpha \sum_{\beta \in S_\alpha} a_{\alpha,\beta} 2^{m-l_\beta} \right)$$

where $l_\beta = $ number of 1's in the bit vector $\beta \in \{0, 1\}^m$.

Now it is easy to see that the above expression can be computed in VNC^1.   □

Thus, if the Determinant polynomial is expressible as $\sum \cdot \mathsf{VSkewF}$ then it belongs to VNC^1.

We briefly consider (syntactic) multilinear versions of these classes. From Lemma 8, we know that a-SkewF is characterised by polynomials with polynomially many coefficients. The construction yields, for any multilinear polynomial computed by a skew formula, an equivalent skew formula which is syntactic multilinear. Hence the notion of multilinearity and syntactic multilinearity are the same for skew formulae. Since any multilinear polynomial that can be computed by an a-SAC^0 circuit has a small number of monomials, the containments of corollary 6 hold in the syntactic multilinear case too. Also, note that the polynomial $\prod_i (x_i + y_i)$ is multilinear, and can be computed by a sma-AC^0 circuit.

Corollary 7. sma-SAC$^0 \subset$ sma-SkewF $=$ ma-SkewF $\subset$ sma-AC^0.

6   Conclusion

This work came out of an attempt to close the gap in a-NC$^1 \subseteq$ a-sSC^0. We have not been able to do this; we can only show that sma-sSC$^0 \subseteq$ a-NC^1. Can the depth-reduction be pushed to all of a-sSC^0? At least ma-sSC^0? Alternatively, can the depth-reduced circuit be made multilinear?

Another unsettled question is to better understand the Boolean containments NC$^1 =$ LWrGP $\subseteq$ LWGP $\subseteq$ LWBP $\subseteq$ sSC$^1 \subseteq$ SC$^1 =$ L. Where exactly does the power of the classes actually jump from NC^1 to L?

Making the constructions described here uniform would also be of interest.

Acknowledgements. The referees' comments are gratefully acknowledged.

References

1. Agrawal, M., Allender, E., Datta, S.: On TC^0, AC^0, and arithmetic circuits. Journal of Computer and System Sciences 60(2), 395–421 (2000)
2. Allender, E.: Arithmetic circuits and counting complexity classes. In: Krajicek, J. (ed.) Complexity of Computations and Proofs, Quaderni di Matematica, vol. 13, pp. 33–72. Seconda Universita di Napoli (2004)
3. Allender, E., Ambainis, A., Barrington, D.A., Datta, S., LêThanh, H.: Bounded depth arithmetic circuits: Counting and closure. In: International Colloquium on Automata, Languages, and Programming ICALP, ICALP 1999, pp. 149–158 (1999)
4. Barrington, D.: Bounded-width polynomial-size branching programs recognize exactly those languages in NC^1. Journal of Computer and System Sciences 38(1), 150–164 (1989)
5. Ben-Or, M., Cleve, R.: Computing algebraic formulas using a constant number of registers. SIAM J. Comput. 21(1), 54–58 (1992)
6. Borodin, A., Razborov, A.A., Smolensky, R.: On lower bounds for read-k-times branching programs. Computational Complexity 3, 1–18 (1993)
7. Brent, R.P.: The parallel evaluation of arithmetic expressions in logarithmic time. In: Complexity of sequential and parallel numerical algorithms (Proc. Sympos. Carnegie-Mellon Univ., Pittsburgh, Pa., 1973), pp. 83–102. Academic Press, New York (1973)
8. Bürgisser, P.: Completeness and Reduction in Algebraic Complexity Theory. In: Algorithms and Computation in Mathematics. Springer, Heidelberg (2000)
9. Bürgisser, P., Clausen, M., Shokrollahi, M.: Algebraic Complexity Theory. Springer, Heidelberg (1997)
10. Caussinus, H., McKenzie, P., Thérien, D., Vollmer, H.: Nondeterministic NC^1 computation. Journal of Computer and System Sciences 57, 200–212 (1998)
11. Istrail, S., Zivkovic, D.: Bounded width polynomial size Boolean formulas compute exactly those functions in AC^0. Information Processing Letters 50, 211–216 (1994)
12. Limaye, N., Mahajan, M., Rao, B.V.R.: Arithmetizing classes around NC^1 and L. ECCC TR07-087 (2007). Preliminary version appeared in STACS 2007
13. Raz, R.: Multi-linear formulas for permanent and determinant are of superpolynomial size. In: STOC, pp. 633–641 (2004)
14. Raz, R.: Multilinear-NC^1 $\neq$ multilinear-NC^2. In: FOCS, pp. 344–351 (2004)
15. Raz, R., Yehudayoff, A.: Balancing syntactically multilinear arithmetic circuits. In: Computational Complexity (to appear)
16. Spira, P.M.: On time hardware complexity tradeoffs for boolean functions. In: Fourth Hawaii International Symposium on System Sciences, pp. 525–527 (1971)
17. Toda, S.: Counting problems computationally equivalent to the determinant. Technical Report CSIM 91-07, Dept. Comp. Sci. and Inf. Math. Univ. of Electro-Communications, Tokyo (1991)
18. Valiant, L.G.: Completeness classes in algebra. In: Symposium on Theory of Computing STOC, pp. 249–261 (1979)
19. Valiant, L.G., Skyum, S., Berkowitz, S., Rackoff, C.: Fast parallel computation of polynomials using few processors. SIAM J. Comput. 12(4), 641–644 (1983)
20. Venkateswaran, H.: Circuit definitions of nondeterministic complexity classes. SIAM Journal on Computing 21, 655–670 (1992)
21. Vollmer, H.: Introduction to Circuit Complexity: A Uniform Approach. Springer, New York (1999)

Smoothed Analysis of Binary Search Trees and Quicksort under Additive Noise

Bodo Manthey[1,*] and Till Tantau[2]

[1] Saarland University, Computer Science
Postfach 151150, 66041 Saarbrücken, Germany
manthey@cs.uni-sb.de
[2] Universität zu Lübeck, Institut für Theoretische Informatik
Ratzeburger Allee 160, 23538 Lübeck, Germany
tantau@tcs.uni-luebeck.de

Abstract. Binary search trees are a fundamental data structure and their height plays a key role in the analysis of divide-and-conquer algorithms like quicksort. We analyze their smoothed height under additive uniform noise: An adversary chooses a sequence of n real numbers in the range $[0, 1]$, each number is individually perturbed by adding a value drawn uniformly at random from an interval of size d, and the resulting numbers are inserted into a search tree. An analysis of the smoothed tree height subject to n and d lies at the heart of our paper: We prove that the smoothed height of binary search trees is $\Theta(\sqrt{n/d} + \log n)$, where $d \geq 1/n$ may depend on n. Our analysis starts with the simpler problem of determining the smoothed number of left-to-right maxima in a sequence. We establish matching bounds, namely once more $\Theta(\sqrt{n/d} + \log n)$. We also apply our findings to the performance of the quicksort algorithm and prove that the smoothed number of comparisons made by quicksort is $\Theta(\frac{n}{d+1}\sqrt{n/d} + n \log n)$.

1 Introduction

To explain the discrepancy between average-case and worst-case behavior of the simplex algorithm, Spielman and Teng introduced the notion of *smoothed analysis* [14]. Smoothed analysis interpolates between average-case and worst-case analysis: Instead of taking a worst-case instance or, as in average-case analysis, choosing an instance completely at random, we analyze the complexity of (possibly worst-case) objects subject to slight random perturbations. On the one hand, perturbations model that nature is not (or not always) adversarial. On the other hand, perturbations reflect the fact that data is often subject to measurement or rounding errors; even if the instance at hand was initially a worst-case instance, due to such errors we would probably get a less difficult instance in practice. Spielman and Teng [15] give a comprehensive survey on results and open problems in smoothed analysis.

* Work done at the Department of Computer Science at Yale University, supported by the Postdoc-Program of the German Academic Exchange Service (DAAD).

E. Ochmański and J. Tyszkiewicz (Eds.): MFCS 2008, LNCS 5162, pp. 467–478, 2008.

Binary search trees are one of the most fundamental data structures in computer science and they are the building blocks for a large variety of data structures. One of the most important parameter of binary search trees is their *height*. The worst-case height of a binary tree for n numbers is n. The average-case behavior has been the subject of a considerable amount of research, culminating in the result that the average-case height is $\alpha \ln n - \beta \ln \ln n + O(1)$, where $\alpha \approx 4.311$ is the larger root of $\alpha \ln(2e/\alpha) = 1$ and $\beta = 3/(2 \ln(\alpha/2)) \approx 1.953$ [12]. Furthermore, the variance of the height is bounded by a constant, as was proved independently by Drmota [6] and Reed [12], and also all higher moments are bounded by constants [6]. Drmota [7] gives a recent survey.

Beyond being an important data structure, binary search trees play a central role in the analysis of divide-and-conquer algorithms like quicksort [9, Section 5.2.2]. While quicksort needs $\Theta(n^2)$ comparisons in the worst case, the average number of comparisons is $2n \log n - \Theta(n)$ with a variance of $(7 - \frac{2}{3}\pi^2) \cdot n^2 - 2n \log n + O(n)$ as mentioned by Fill and Janson [8]. Quicksort and binary search trees are closely related: The height of the tree $T(\sigma)$ obtained from a sequence σ is equal to the number of levels of recursion required by quicksort to sort σ. The number of comparisons, which corresponds to the total path length of $T(\sigma)$, is at most n times the height of $T(\sigma)$.

Binary search trees are also related to the number of left-to-right maxima of a sequence, which is the number of new maxima seen while scanning a sequence from left to right. The number of left-to-right maxima of σ is equal to the length of the rightmost path of the tree $T(\sigma)$, which means that left-to-right maxima provide an easy-to-analyze lower bound for the height of binary search trees. In the worst-case, the number of left-to-right maxima is n, while it is $\sum_{i=1}^{n} 1/i \in \Theta(\log n)$ on average. The study of left-to-right maxima is also of independent interest. For instance, the number of times a data structure for keeping track of a bounding box of moving object needs to be updated is closely related to the number of left-to-right maxima (and minima) of the coordinate components of the objects. (See Basch et al. [2] for an introduction to data structures for mobile data.) Left-to-right maxima also play a role in the analysis of quicksort [13].

Given the discrepancies between average-case and worst-case behavior of binary search trees, quicksort, and the number of left-to-right maxima, the question arises of what happens in between when the randomness is limited.

Our results. We continue the smoothed analysis of binary search trees and quicksort begun by Banderier et al. [1] and Manthey and Reischuk [10]. However, we return to the original idea of smoothed analysis that input numbers are perturbed by adding random numbers. The perturbation model introduced by Spielman and Teng for the smoothed analysis of continuous problems like linear programming is appropriate for algorithms that process real numbers. In their model, each of the real numbers in the adversarial input is perturbed by adding a small Gaussian noise. This model of perturbation favors instances in the neighborhood of the adversarial input for a fairly natural and realistic notion of "neighborhood."

In our model the adversarial input sequence consists of n real numbers in the interval $[0, 1]$. Then, each of the real numbers is individually perturbed by adding a random number drawn uniformly from an interval of size d, where $d = d(n)$ may depend on n. If $d < 1/n$, then the sorted sequence $(1/n, 2/n, 3/n, \ldots, n/n)$ stays a sorted sequence and the smoothed height of binary search trees (as well as the performance of quicksort and the number of left-to-right maxima) is the same as in the worst-case. We always assume $d \geq 1/n$ in the following.

We study the smoothed height of binary search trees, the smoothed number of comparisons made by quicksort, and the smoothed number of left-to-right maxima under additive noise. In each case we prove tight upper and lower bounds:

1. The smoothed number of left-to-right maxima is $\Theta(\sqrt{n/d} + \log n)$ as shown in Section 3. This result will be exploited in the subsequent sections.
2. The smoothed height of binary search trees is $\Theta(\sqrt{n/d} + \log n)$ as shown in Section 4.
3. The smoothed number of comparisons made by quicksort is $\Theta(\frac{n}{d+1}\sqrt{n/d} + n \log n)$ as shown in Section 5. Thus, the perturbation effect of $d \in \omega(1)$ is stronger than for $d \in o(1)$.

Already for $d \in \omega(1/n)$, we obtain bounds that are asymptotically better than the worst-case bounds. For constant values of d, which correspond to a perturbation by a constant percentage like 1%, the height of binary search trees drops from the worst-case height of n to $O(\sqrt{n})$, and quicksort needs only $O(n^{3/2})$ comparisons.

It is tempting to assume that results such as the above will hold in the same way for other distributions, such as the Gaussian distribution, with d replaced by the standard deviation. We contribute a surprising result in Section 6: We present a well-behaved probability distribution (symmetric, monotone on the positive reals, smooth) for which sorting sequences can *decrease* the expected number of left-to-right maxima. This effect is quite counter-intuitive and the literature contains the claim that one can restrict attention to sorted sequences since they are the worst-case sequences also in the smoothed setting [4,3]. Our distribution refutes this claim.

Related work. The first smoothed analysis of quicksort, due to Banderier, Beier, and Mehlhorn [1], uses a perturbation model different from the one used in the present paper, namely a *discrete perturbation model*. Such models take discrete objects like permutations as input and again yield discrete objects like another permutation. Banderier et al. used *p-partial permutations*, which work as follows: An adversary chooses a permutation of the numbers $\{1, \ldots, n\}$ as sequence, every element of the sequence is marked independently with a probability of p, and then the marked elements are randomly permuted. Banderier et al. showed that the number of comparisons subject to p-partial permutations is $O(\frac{n}{p} \cdot \log n)$. Furthermore, they proved bounds on the smoothed number of left-to-right maxima subject to this model.

Manthey and Reischuk [10] analyzed the height of binary search trees under p-partial permutations. They proved a lower bound of $0.8 \cdot (1 - p) \cdot \sqrt{n/p}$ and an asymptotically matching upper bound of $6.7 \cdot (1 - p) \cdot \sqrt{n/p}$ for the smoothed tree height. For the number of left-to-right maxima, they showed a lower bound of $0.6 \cdot (1 - p) \cdot \sqrt{n/p}$ and an upper bound of $3.6 \cdot (1 - p) \cdot \sqrt{n/p}$.

Special care must be taken when defining perturbation models for discrete inputs: The perturbation should favor instances in the neighborhood of the adversarial instance, which requires a suitable definition of neighborhood in the first place, and the perturbation should preserve the global structure of the adversarial instance. Partial permutations have the first feature [10, Lemma 3.2], but destroy much of the global order of the adversarial sequence.

The smoothed number of left-to-right maxima for the additive noise model of the present paper was already considered by Damerow et al. [4,5,3]. They considered so-called kinetic data structures, which keep track of properties of a set of moving points like a bounding box for them or a convex hull, and they introduced the notion of *smoothed motion complexity*. They also considered left-to-right maxima since left-to-right maxima provide upper and lower bounds on the number of times a bounding box needs to be updated for moving points. For left-to-right maxima, Damerow et al. show an upper bound of $O(\sqrt{n \log n/d} + \log n)$ and a lower bound of $\Omega(\sqrt{n/d})$. In the present paper, we show that the exact bound is $\Theta(\sqrt{n/d} + \log n)$.

2   Preliminaries

Intervals are denoted by $[a, b] = \{x \in \mathbb{R} \mid a \le x \le b\}$. To denote an interval that does not include an endpoint, we replace the square bracket next to the endpoint by a parenthesis. We denote *sequences* of real numbers by $\sigma = (\sigma_1, \ldots, \sigma_n)$, where $\sigma_i \in \mathbb{R}$. For $U = \{i_1, \ldots, i_\ell\} \subseteq \{1, \ldots, n\}$ with $i_1 < i_2 < \cdots < i_\ell$ let $\sigma_U = (\sigma_{i_1}, \sigma_{i_2}, \ldots, \sigma_{i_\ell})$ denote the *subsequence* of σ of the elements at positions in U. We denote probabilities by $\mathbb{P}$ and expected values by $\mathbb{E}$.

Throughout the paper, we will assume for the sake of clarity that numbers like $\sqrt{d}$ are integers and we do not write down the tedious floor and ceiling functions that are actually necessary. Since we are interested in asymptotic bounds, this does not affect the validity of the proofs.

Due to lack of space, some proofs are omitted. For complete proofs, we refer to the full version of this paper [11].

2.1   Binary Search Trees, Left-to-Right Maxima, and Quicksort

Let σ be a sequence of length n consisting of pairwise distinct elements. For the following definitions, let $G = \{i \in \{1, \ldots, n\} \mid \sigma_i > \sigma_1\}$ be the set of positions of elements greater than σ_1, and let $S = \{i \in \{1, \ldots, n\} \mid \sigma_i < \sigma_1\}$ be the set of positions of elements smaller than σ_1.

From σ, we obtain a *binary search tree* $T(\sigma)$ by iteratively inserting the elements $\sigma_1, \ldots, \sigma_n$ into the initially empty tree as follows: The root of $T(\sigma)$ is σ_1. The left subtree of the root σ_1 is $T(\sigma_S)$, and the right subtree of σ_1 is $T(\sigma_G)$. The *height of* $T(\sigma)$ is the maximum number of nodes on any root-to-leaf path of

$T(\sigma)$: Let $\mathrm{height}(\sigma) = 1 + \max\{\mathrm{height}(\sigma_S), \mathrm{height}(\sigma_G)\}$, and let $\mathrm{height}(\sigma) = 0$ when σ is the empty sequence.

The number of *left-to-right maxima* of σ is the number of maxima seen when scanning σ from left to right: let $\mathrm{ltrm}(\sigma) = 1 + \mathrm{ltrm}(\sigma_G)$, and let $\mathrm{ltrm}(\sigma) = 0$ when σ is the empty sequence. The number of left-to-right maxima of σ is equal to the length of the rightmost path of $T(\sigma)$, so $\mathrm{ltrm}(\sigma) \leq \mathrm{height}(\sigma)$.

Quicksort is the following sorting algorithm: Given σ, we construct σ_S and σ_G. To do this, all elements of $(\sigma_2, \ldots, \sigma_n)$ have to be compared to σ_1, which is called the *pivot element*. Then we sort σ_S and σ_G recursively to obtain τ_S and τ_G, respectively. Finally, we output $\tau = (\tau_S, \sigma_1, \tau_G)$. The number $\mathrm{qs}(\sigma)$ of comparisons needed to sort σ is thus $\mathrm{qs}(\sigma) = (n-1) + \mathrm{qs}(\sigma_S) + \mathrm{qs}(\sigma_G)$ if σ has a length of $n \geq 1$, and $\mathrm{qs}(\sigma) = 0$ when σ is the empty sequence.

2.2   Perturbation Model

The perturbation model of *additive noise* is defined as follows: Let $d = d(n) \geq 0$ be the perturbation parameter (d may depend on n). Given a sequence σ of n numbers chosen by an adversary from the interval $[0, 1]$, we draw a *noise* ν_i for each $i \in \{1, \ldots, n\}$ uniformly and independently from each other at random from the interval $[0, d]$. Then we obtain the perturbed sequence $\overline{\sigma} = (\overline{\sigma}_1, \ldots, \overline{\sigma}_n)$ by adding ν_i to σ_i, that is, $\overline{\sigma}_i = \sigma_i + \nu_i$. Note that $\overline{\sigma}_i$ need no longer be an element of $[0, 1]$, but $\overline{\sigma}_i \in [0, d+1]$. For $d > 0$ all elements of $\overline{\sigma}$ are distinct with a probability of 1.

For this model, we define the random variables $\mathrm{height}_d(\sigma)$, $\mathrm{qs}_d(\sigma)$, as well as $\mathrm{ltrm}_d(\sigma)$, which denote the smoothed search tree height, smoothed number of quicksort comparisons, and smoothed number of left-to-right maxima, respectively, when the sequence σ is perturbed by d-noise. Since the adversary chooses σ, our goal are bounds for $\max_{\sigma \in [0,1]^n} \mathbb{E}(\mathrm{height}_d(\sigma))$, $\max_{\sigma \in [0,1]^n} \mathbb{E}(\mathrm{qs}_d(\sigma))$, and $\max_{\sigma \in [0,1]^n} \mathbb{E}(\mathrm{ltrm}_d(\sigma))$. In the following, we will sometimes write $\mathrm{height}(\overline{\sigma})$ instead of $\mathrm{height}_d(\sigma)$ if d is clear from the context. Since $\overline{\sigma}$ is random, $\mathrm{height}(\overline{\sigma})$ is also a random variable. Similarly, we will use $\mathrm{ltrm}(\overline{\sigma})$ and $\mathrm{qs}(\overline{\sigma})$.

The choice of the interval sizes is arbitrary since the model is invariant under scaling if we scale the perturbation parameter accordingly. This is summarized in the following lemma, which we will exploit a couple of times in the following.

Lemma 1. *Let $b > a$ and $d > 0$ be arbitrary real numbers, and let $d' = d/(b-a)$. Then $\max_{\sigma \in [a,b]^n} \mathbb{E}(\mathrm{height}_d(\sigma)) = \max_{\sigma \in [0,1]^n} \mathbb{E}(\mathrm{height}_{d'}(\sigma))$. For quicksort and the number of left-to-right maxima, we have analogous equalities.*

As argued earlier, if $d < 1/n$, the adversary can specify $\sigma = (1/n, \ldots, n/n)$ and adding the noise terms does not affect the order of the elements. This means that we get the worst-case height, number of comparisons, and number of left-to-right maxima. Because of this observation we will restrict our attention to $d \geq 1/n$.

If d is large, the noise will swamp out the original instance, and the order of the elements of $\overline{\sigma}$ will depend only on the noise rather than the original instance. For intermediate d, additive noise interpolates between average and worst case.

3   Smoothed Number of Left-to-Right Maxima

We start our analyses with the smoothed number of left-to-right maxima, which provides us with a lower bound on the height of binary search trees as well. Our aim for the present section is to prove the following theorem.

Theorem 1. *For $d \geq 1/n$, we have*

$$\max_{\sigma \in [0,1]^n} \mathbb{E}\big(\text{ltrm}_d(\sigma)\big) \in \Theta\big(\sqrt{n/d} + \log n\big).$$

The lower bound of $\Omega\big(\sqrt{n/d} + \log n\big)$ is already stated without proof in [4] and a proof can be found in [3], so we prove only the upper bound. The following notations will be helpful: For $j \leq 0$, let $\sigma_j = \nu_j = 0$. This allows us to define $\delta_i = \sigma_i - \sigma_{i-\sqrt{nd}}$ for all $i \in \{1, \ldots, n\}$. We define $I_i = \{j \in \{1, \ldots, n\} \mid i - \sqrt{nd} \leq j < i\}$ to be the set of the $|I_i| = \min\{i-1, \sqrt{nd}\}$ positions that precede i.

To prove the upper bound for the smoothed number of left-to-right maxima, we proceed in two steps: First, a "bubble-sorting argument" is used to show that the adversary should choose a sorted sequence. Note that this is not as obvious as it may seem since in Section 6 we show that this bubble-sorting argument does not apply to all distributions. Second, we prove that the expected number of left-to-right maxima of sorted sequences is $O(\sqrt{n/d} + \log n)$, which improves the bound of $O(\sqrt{n \log n/d} + \log n)$ [3,4].

Lemma 2. *For every σ and its sorted version τ, $\mathbb{E}\big(\text{ltrm}_d(\sigma)\big) \leq \mathbb{E}\big(\text{ltrm}_d(\tau)\big)$.*

Lemma 3. *For all σ of length n and all $d \geq 1/n$, we have $\mathbb{E}\big(\text{ltrm}_d(\sigma)\big) \in O\big(\sqrt{n/d} + \log n\big)$.*

Proof. By Lemma 2 we can restrict ourselves to proving the lemma for sorted sequences σ. We estimate the probability that a given $\overline{\sigma}_i$ for $i \in \{1, \ldots, n\}$ is a left-to-right maximum. Then the bound follows by the linearity of expectation. To bound the probability that $\overline{\sigma}_i$ is a left-to-right maximum (ltrm), consider the following computation:

$$\mathbb{P}\big(\overline{\sigma}_i \text{ is an ltrm}\big) \leq \mathbb{P}\big(\forall j \in I_i : \nu_j < \overline{\sigma}_i - \sigma_{i-\sqrt{nd}}\big) \tag{1}$$

$$\leq \mathbb{P}\big(d < \overline{\sigma}_i - \sigma_{i-\sqrt{nd}}\big) + \int_0^{d-\delta_i} \mathbb{P}\big(\forall j \in I_i : \nu_j < \sigma_i + x - \sigma_{i-\sqrt{nd}}\big) \cdot \tfrac{1}{d} \, dx \tag{2}$$

$$\leq \tfrac{\delta_i}{d} + \int_0^d \mathbb{P}\big(\forall j \in I_i : \nu_j < x\big) \cdot \tfrac{1}{d} \, dx \tag{3}$$

$$\leq \tfrac{\delta_i}{d} + \mathbb{P}\big(\forall j \in I_i : \nu_j < \nu_i\big) \;=\; \tfrac{\delta_i}{d} + \tfrac{1}{|I_i|+1}. \tag{4}$$

To see that (1) holds, assume that $\overline{\sigma}_i$ is a left-to-right maximum. Then $\overline{\sigma}_i - \sigma_{i-\sqrt{nd}}$ must be larger than the noises of all the elements in the index range I_i, for if the noise ν_j of some element σ_j were larger than $\overline{\sigma}_i - \sigma_{i-\sqrt{nd}}$, then $\overline{\sigma}_j = \sigma_j + \nu_j$ would be larger than $\sigma_j + \overline{\sigma}_i - \sigma_{i-\sqrt{nd}}$. Since the sequence is sorted, we would get $\sigma_j + \overline{\sigma}_i - \sigma_{i-\sqrt{nd}} \geq \overline{\sigma}_i$, and $\overline{\sigma}_i$ would not be a left-to-right maximum.

For (2), first observe that $\nu_j < \overline{\sigma}_i - \sigma_{i-\sqrt{nd}}$ is surely the case for all $j \in I_i$ if $d < \overline{\sigma}_i - \sigma_{i-\sqrt{nd}}$. So, consider the case $d \geq \overline{\sigma}_i - \sigma_{i-\sqrt{nd}} = \delta_i + \nu_i$. Then $\nu_i \in [0, d-\delta_i]$ and we can rewrite $\mathbb{P}(\forall j \in I_i : \nu_j < \delta_i + \nu_i)$ as $\int_0^{d-\delta_i} \mathbb{P}(\forall j \in I_i : \nu_j < \delta_i + x) \cdot \frac{1}{d} \, dx$, where $1/d$ is the density of ν_i. For (3) observe that $d < \overline{\sigma}_i - \sigma_{i-\sqrt{nd}}$ is equivalent to $d - \delta_i < \nu_i$ and the probability of this is δ_i/d. Furthermore, we performed an index shift in the integral. In (4), we replaced the integral by a probability once more and get the final result.

We have $\sum_{i=1}^n \delta_i = \sum_{i=1}^n (\sigma_i - \sigma_{i-\sqrt{nd}}) = \sum_{i=n-\sqrt{nd}+1}^n \sigma_i \leq \sqrt{nd}$. The second equality holds since most σ_i cancel themselves out and $\sigma_i = 0$ for $i \leq 0$. The inequality holds since there are $\sqrt{nd}$ summands. We bound $1/(|I_i| + 1) = 1/\min\{i, \sqrt{nd}+1\}$ by $1/i + 1/\sqrt{nd}$ and sum over all i: $\mathbb{E}(\mathrm{ltrm}_d(\sigma)) \leq \sum_{i=1}^n \left(\frac{\delta_i}{d} + \frac{1}{|I_i|+1}\right) \leq \frac{\sqrt{nd}}{d} + \sum_{i=1}^n \frac{1}{i} + \frac{n}{\sqrt{nd}} \in O(\sqrt{n/d} + \log n)$.   □

4   Smoothed Height of Binary Search Trees

In this section we prove our first main result, an exact bound on the smoothed height of binary search trees under additive noise. The bound is the same as for left-to-right maxima, as stated in the following theorem.

Theorem 2. For $d \geq 1/n$, we have

$$\max_{\sigma \in [0,1]^n} \mathbb{E}(\mathrm{height}_d(\sigma)) \in \Theta(\sqrt{n/d} + \log n).$$

In the rest of this section, we prove this theorem. We have to prove an upper and a lower bound, but the lower bound follows directly from the lower bound of $\Omega(\sqrt{n/d} + \log n)$ for the smoothed number of left-to-right maxima (the number of left-to-right maxima in a sequence is the length of the rightmost path of the sequence's search tree). Thus, we only need to focus on the upper bound. To prove the upper bound of $O(\sqrt{n/d} + \log n)$ on the smoothed height of binary search trees, we need some preparations. In the next subsection we introduce the concept of *increasing and decreasing runs* and show how they are related to binary search tree height. As we will see, bounding the length of these runs implicitly bounds the height of binary search trees. This allows us to prove the upper bound on the smoothed height of binary search trees in the main part of this section.

4.1   Increasing and Decreasing Runs

In order to analyze the smoothed height of binary search trees, we introduce a related measure for which an upper bound is easier to obtain. Given a sequence σ, consider a root-to-leaf path of the tree $T(\sigma)$. We extract two subsequences $\alpha = (\alpha_1, \ldots, \alpha_k)$ and $\beta = (\beta_1, \ldots, \beta_\ell)$ from this path according to the following algorithm: We start at the root. When we are at an element σ_i of the path, we look at the direction in which the path continues from σ_i. If it continues with the right child of σ_i, we append σ_i to α; if it continues with the left child, we append σ_i to β; and if σ_i is a leaf (has no children), then we append σ_i to both α

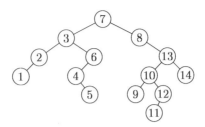

Fig. 1. The tree $T(\sigma)$ obtained from the sequence $\sigma = (7, 8, 13, 3, 2, 10, 9, 6, 4, 12, 14, 1, 5, 11)$. We have $\text{height}(\sigma) = 6$. The root-to-leaf path ending at 11 yields the increasing run $\alpha = (7, 8, 10, 11)$ and the decreasing run $\beta = (13, 12, 11)$.

and β. This construction ensures $\alpha_1 < \cdots < \alpha_k = \beta_\ell < \cdots < \beta_1$ and the length of σ is $k + \ell - 1$. Figure 1 shows an example of how α and β are constructed.

A crucial property of the sequence α is the following: Let $\alpha_i = \sigma_{j_i}$ for all $i \in \{1, \ldots, k\}$ with $j_1 < j_2 < \cdots < j_k$. Then none of $\sigma_1, \ldots, \sigma_{j_i - 1}$ lies in the interval (α_i, α_{i+1}), for otherwise α_i and α_{i+1} cannot be on the same root-to-leaf path. A similar property holds for the sequence β: No element of σ prior to β_i lies in the interval (β_{i+1}, β_i). We introduce a special name for sequences with this property.

Definition 1. *An* increasing run *of a sequence* σ *is a subsequence* $(\sigma_{i_1}, \ldots, \sigma_{i_k})$ *with the following property: For every* $j \in \{1, \ldots, k-1\}$, *no element of* σ *prior to* σ_{i_j} *lies in the interval* $(\sigma_{i_j}, \sigma_{i_{j+1}})$. *Analogously, a* decreasing run *of* σ *is a subsequence* $(\sigma_{i_1}, \ldots, \sigma_{i_\ell})$ *with* $\sigma_{i_1} > \cdots > \sigma_{i_\ell}$ *such no element prior to* σ_{i_j} *lies in the interval* $(\sigma_{i_{j+1}}, \sigma_{i_j})$.

Let $\text{inc}(\sigma)$ and $\text{dec}(\sigma)$ denote the length of the longest increasing and decreasing run of σ, respectively. Furthermore, let $\text{dec}_d(\sigma)$ and $\text{inc}_d(\sigma)$ denote the length of the longest runs under d-noise. In Figure 1, we have $\text{inc}(\sigma) = 4$ because of $(7, 8, 10, 12)$ or $(7, 8, 13, 14)$ and $\text{dec}(\sigma) = 4$ because of $(7, 3, 2, 1)$.

Since every root-to-leaf path can be divided into an increasing and a decreasing run, we immediately obtain the following lemma.

Lemma 4. *For every sequence* σ *and all* d *we have*

$$\text{height}(\sigma) \leq \text{dec}(\sigma) + \text{inc}(\sigma),$$
$$\mathbb{E}\big(\text{height}_d(\sigma)\big) \leq \mathbb{E}\big(\text{dec}_d(\sigma) + \text{inc}_d(\sigma)\big).$$

In terms of upper bounds, $\text{dec}(\sigma)$ and $\text{inc}(\sigma)$ as well as $\text{dec}_d(\sigma)$ and $\text{inc}_d(\sigma)$ behave equally. The reason is that given a sequence σ, the sequence τ with $\tau_i = 1 - \sigma_i$ has the properties $\text{dec}(\sigma) = \text{inc}(\tau)$ and $\mathbb{E}\big(\text{dec}_d(\sigma)\big) = \mathbb{E}\big(\text{inc}_d(\tau)\big)$. This observation together with Lemma 4 proves the next lemma.

Lemma 5. *For all* d, *we have*

$$\max_{\sigma \in [0,1]^n} \mathbb{E}\big(\text{height}_d(\sigma)\big) \leq 2 \cdot \max_{\sigma \in [0,1]^n} \mathbb{E}\big(\text{inc}_d(\sigma)\big).$$

The lemma states that in order to bound the smoothed height of search trees from above we can instead bound the smoothed length of increasing or decreasing runs. To simplify the analysis even further, we show that we can once more restrict our attention to sorted sequences.

Lemma 6. *For every σ and its sorted version τ, $\mathbb{E}\big(\mathrm{inc}_d(\sigma)\big) \leq \mathbb{E}\big(\mathrm{inc}_d(\tau)\big)$.*

4.2   Upper Bound on the Smoothed Height of Binary Search Trees

We are now ready to prove the upper bound for binary search trees by proving an upper bound on the smoothed length of increasing runs of sorted sequences. For this, we prove four lemmas, the last of which claims exactly the desired upper bound. Lemma 7 deals with $d = 1$ and states that $\mathbb{E}\big(\mathrm{height}_1(\sigma)\big) \in O(\sqrt{n})$ for every sequence σ. Lemma 8 states that in order to bound tree heights, we can divide sequences into (possibly overlapping) parts and consider the height of the trees induced by the subsequences individually. A less general form of the lemma has already been shown by Manthey and Reischuk [10, Lemma 4.1]. Lemma 9 establishes that if $d = n/\log^2 n$, a perturbed sequence behaves the same way as a completely random sequence with respect to the smoothed length of its longest increasing run. The core idea is to partition the sequence into a set of "completely random" elements, which behave as expected, and two sets of more bothersome elements lying in a small range. As we will see, the number of bothersome elements is roughly $\log^2 n$ and since the range of values of these elements is small, we can use the result $\mathbb{E}\big(\mathrm{height}_1(\sigma)\big) \in O(\sqrt{n})$ to show that their contribution to the length on an increasing run is just $O(\log n)$. Finally, in Lemma 10 we allow general $d \geq 1/n$. This case turns out to be reducible to the case $d = n/\log^2 n$ by scaling the numbers according to Lemma 1.

For the proofs of the lemmas, two technical terms will be helpful: For a given real interval $I = [a, b]$, we say that a position i of σ is *eligible* for I if $\overline{\sigma}_i$ can assume any value in I. In other words, i is eligible for $[a, b]$ if $\sigma_i \leq a$ and $\sigma_i + d \geq b$. Furthermore, we say that i is *regular* if $\overline{\sigma}_i$ actually lies inside I.

Lemma 7. *For all σ, we have $\mathbb{E}\big(\mathrm{inc}_1(\sigma)\big) \in O(\sqrt{n})$.*

Lemma 8. *For every sequence σ, all d, and every covering $U_1, U_2, \ldots, U_k$ of $\{1, \ldots, n\}$ (which means $\bigcup_{i=1}^{k} U_i = \{1, \ldots, n\}$), we have*

$$\mathrm{height}(\sigma) \leq \sum_{i=1}^{k} \mathrm{height}(\sigma_{U_i}),$$
$$\mathbb{E}\big(\mathrm{height}_d(\sigma)\big) \leq \sum_{i=1}^{k} \mathbb{E}\big(\mathrm{height}_d(\sigma_{U_i})\big).$$

Lemma 9. *For all sequences σ, and $d = n/\log^2 n$, we have $\mathbb{E}\big(\mathrm{height}_d(\sigma)\big) \in O(\log n)$.*

Lemma 10. *For every sequence σ and all $d \geq 1/n$ we have $\mathbb{E}\big(\mathrm{height}_d(\sigma)\big) \in O\big(\sqrt{n/d} + \log n\big)$.*

Proof. If $d \in \Omega\big(n/\log^2 n\big)$, then $\mathbb{E}\big(\mathrm{height}_d(\sigma)\big) \in O(\log n)$ by Lemma 9.

To prove the theorem for smaller values of d, we divide the sequence into subsequences. Let N be the largest real root of the equation $N^2/\log^2 N = nd$. Then $\log N \in \Theta(\log(nd))$, and thus $N = c \cdot \sqrt{nd} \cdot \log(nd)$ for some $c \in \Theta(1)$. Let n_j be the number of elements of σ with $\sigma_i \in [(j-1) \cdot N/n, j \cdot N/n]$. Choose $k_j \in \mathbb{N}$ such that $(k_j - 1) \cdot N < n_j \le k_j N$. We divide the n_j elements of the interval $[(j-1) \cdot N/n, j \cdot N/n]$ into k_j subsequences $\sigma^{j,1}, \ldots, \sigma^{j,k_j}$ such that no subsequence contains more than N elements. Since

$$\sum_{j=1}^{n/N} k_j \le \sum_{j=1}^{n/N} \frac{n_j + N}{N} \le 2n/N,$$

we obtain at most $2n/N$ such subsequences. Each subsequence spans at most an interval of length N/n and contains at most N elements. Thus, by Lemma 9, we have $\mathbb{E}\big(\mathrm{height}_d(\sigma^{j,\ell})\big) \in O(\log(N))$. Finally, Lemma 8 yields

$$\mathbb{E}\big(\mathrm{height}_d(\sigma)\big) \le \sum_{j=1}^{n/N} \sum_{\ell=1}^{k_j} \mathbb{E}\big(\mathrm{height}_d(\sigma^{j,\ell})\big) \in O\left(\frac{n \log N}{N}\right) = O\left(\sqrt{n/d}\right). \quad \square$$

5   Smoothed Number of Quicksort Comparisons

In this section, we apply our results on binary search trees and left-to-right maxima to the performance of the quicksort algorithm. The following theorem summarizes the findings.

Theorem 3. *For $d \ge 1/n$ we have*

$$\max_{\sigma \in [0,1]^n} \mathbb{E}\big(\mathrm{qs}_d(\sigma)\big) \in \Theta\big(\tfrac{n}{d+1} \sqrt{n/d} + n \log n\big).$$

In other words, for $d \in O(1)$, we have at most $O(n\sqrt{n/d})$ comparisons, while for $d \in \Omega(1)$, we have at most $O(\frac{n}{d}\sqrt{n/d})$ comparisons. This means that d has a stronger influence for $d \in \Omega(1)$.

To prove the upper bound, we first need a lemma similar to Lemma 8 that allows us to estimate the number of comparisons of subsequences.

Lemma 11. *For every sequence σ, all d, and every covering $U_1, U_2, \ldots, U_k$ of $\{1, \ldots, n\}$, we have*

$$\mathrm{qs}(\sigma) \le \sum_{i=1}^k \mathrm{qs}(\sigma_{U_i}) + Q,$$

$$\mathbb{E}\big(\mathrm{qs}_d(\sigma)\big) = \mathbb{E}\big(\mathrm{qs}(\overline{\sigma})\big) \le \sum_{i=1}^k \mathbb{E}\big(\mathrm{qs}(\overline{\sigma}_{U_i})\big) + \mathbb{E}(\overline{Q}),$$

where Q is the number of comparisons of elements of σ_{U_i} with elements of $\sigma_{\{1,\ldots,n\} \setminus U_i}$ for any i and the random variable $\overline{Q}$ is defined analogously for $\overline{\sigma}$.

Lemma 12. *For every sequence σ and all $d \ge 1/n$, we have $\mathbb{E}\big(\mathrm{qs}_d(\sigma)\big) \in O\big(\tfrac{n}{d+1} \sqrt{n/d} + n \log n\big)$.*

Our upper bound is tight. The standard sorted sequence provides a worst case, but we use a sequence that is slightly easier to handle technically.

Lemma 13. *For $\sigma = (1/n, 2/n, \ldots, \tfrac{n}{2}/n, 1, 1, \ldots, 1)$ and all $d \ge 1/n$, we have $\mathbb{E}\big(\mathrm{qs}_d(\sigma)\big) \in \Omega\big(\tfrac{n}{d+1} \sqrt{n/d} + n \log n\big)$.*

6   Sorting Decreases the Number of Left-to-Right Maxima

Lemma 2 states that sorting a sequence can never decrease the expected number of left-to-right maxima – at least when the noise is drawn uniformly from a single interval. Intuitively, this should not only hold for this kind of noise, but for *any* kind of noise – at least if the noise distribution is reasonably well-behaved. We demonstrate that this intuition is wrong and there exist a simple distribution and a sequence for which the sorted version has a lower expected number of left-to-right maxima than the original sequence.

Theorem 4. *The exist a sequence σ and a symmetric probability distribution $f \colon \mathbb{R} \to \mathbb{R}$ that is monotonically decreasing on $\mathbb{R}_+$ such that sorting σ to obtain τ decreases the expected number of left-to-right maxima after perturbation.*

To prove the theorem, we use the sequence $\sigma = (0, \ldots, 0, 1 + \frac{1}{\varepsilon}, \frac{1}{\varepsilon})$. The probability distribution has most of its mass in the interval $[-1, 1]$ with tails of length $1/\varepsilon$ to either side: $f(x) = 1 - 2\varepsilon$ for $x \in [-1, 1]$ and $f(x) = \varepsilon^2$ for $|x| \in [1, 1+1/\varepsilon]$. This distribution can easily be made smooth without changing the claim of the theorem.

7   Conclusion

The smoothed height of binary search trees and also the smoothed number of left-to-right maxima are $\Theta(\sqrt{n/d} + \log n)$; the smoothed number of quicksort comparisons is $\Theta(\frac{n}{d+1}\sqrt{n/d} + n\log n)$. While we obtain the average-case height of $\Theta(\log n)$ for binary search trees only for $d \in \Omega(n/\log^2 n)$ – which is large compared to the interval size $[0, 1]$ from which the numbers are drawn –, for the quicksort algorithm $d \in \Omega(\sqrt[3]{n/\log^2 n})$ suffices so that the expected number of comparisons equals the average-case number of $\Theta(n\log n)$. On the other hand, the recursion depth of quicksort, which is equal to the tree height, can be as large as $\Omega(\sqrt{n/d})$. Thus, although the average number of comparisons is already reached at $d \in \Omega(\sqrt[3]{n/\log^2 n})$, the recursion depth remains asymptotically larger than its average value for $d \in o(n/(\log n)^2)$.

A natural question arising from our results is, what happens when the noise is drawn according to distributions other than the uniform distribution? As we have demonstrated, we cannot expect all distributions to behave in the same way as the uniform distribution. A natural candidate for closer examination is the normal distribution, for which first results on left-to-right maxima have already been obtained [4]. We conjecture that if $\max_{x \in \mathbb{R}} f(x) = \phi$, where f is the noise distribution, then the expected tree height and the expected number of left-to-right maxima are $O(\sqrt{n\phi} + \log n)$ while the expected number of quicksort comparisons is $O(\frac{\phi n}{\phi+1}\sqrt{n\phi} + n\log n)$. These bounds would be in compliance with our bounds for uniformly distributed noise, where $\phi = 1/d$.

In our study of the quicksort algorithm we used the first element as the pivot element. This choice simplifies the analysis but one would often use the median

of the first, middle, and last element. Nevertheless, we conjecture that the same
bounds as for the simple pivot strategy also hold for this pivot strategy.

References

1. Mehlhorn, K., Banderier, C., Beier, R.: Smoothed Analysis of Three Combinatorial Problems. In: Rovan, B., Vojtáš, P. (eds.) MFCS 2003. LNCS, vol. 2747, pp. 198–207. Springer, Heidelberg (2003)
2. Basch, J., Guibas, L.J., Hershberger, J.: Data structures for mobile data. J. Algorithms 31(1), 1–28 (1999)
3. Damerow, V.: Average and Smoothed Complexity of Geometric Structures. PhD thesis, Universität Paderborn (2006)
4. Damerow, V., auf der Heide, F.M., Räcke, H., Scheideler, C., Sohler, C.: Smoothed motion complexity. In: Di Battista, G., Zwick, U. (eds.) ESA 2003. LNCS, vol. 2832, pp. 161–171. Springer, Heidelberg (2003)
5. Damerow, V., Sohler, C.: Extreme Points Under Random Noise. In: Albers, S., Radzik, T. (eds.) ESA 2004. LNCS, vol. 3221, pp. 264–274. Springer, Heidelberg (2004)
6. Drmota, M.: An analytic approach to the height of binary search trees II. J.ACM 50(3), 333–374 (2003)
7. Drmota, M.: Profile and height of random binary search trees. J. Iranian Statistical Society 3(2), 117–138 (2004)
8. Fill, J.A., Janson, S.: Quicksort asymptotics. J. Algorithms 44(1), 4–28 (2002)
9. Knuth, D.E.: Sorting and Searching, 2nd edn. The Art of Computer Programming, vol. 3. Addison-Wesley, Reading (1998)
10. Manthey, B., Reischuk, R.: Smoothed analysis of binary search trees. Theoret. Comput. Sci. 378(3), 292–315 (2007)
11. Manthey, B., Tantau, T.: Smoothed analysis of binary search trees and quicksort under additive noise. Report 07-039, Electronic Colloquium on Computational Complexity (ECCC) (2007)
12. Reed, B.: The height of a random binary search tree. J.ACM 50(3), 306–332 (2003)
13. Sedgewick, R.: The analysis of quicksort programs. Acta Inform. 7(4), 327–355 (1977)
14. Spielman, D.A., Teng, S.-H.: Smoothed analysis of algorithms: Why the simplex algorithm usually takes polynomial time. J.ACM 51(3), 385–463 (2004)
15. Spielman, D.A., Teng, S.-H.: Smoothed analysis of algorithms and heuristics: Progress and open questions. In: Foundations of Computational Mathematics, Santander 2005, pp. 274–342. Cambridge University Press, Cambridge (2006)

From λ-Calculus to Universal Algebra and Back

Giulio Manzonetto[1] and Antonino Salibra[1,2]

[1] Laboratoire PPS, Université Paris 7
[2] Dip. Informatica, Università Ca'Foscari di Venezia

Abstract. We generalize to universal algebra concepts originating from λ-calculus and programming to prove a new result on the lattice λT of λ-theories, and a general theorem of pure universal algebra which can be seen as a meta version of the Stone Representation Theorem. In this paper we introduce the class of *Church algebras* (which includes all Boolean algebras, combinatory algebras, rings with unit and the term algebras of all λ-theories) to model the "if-then-else" instruction of programming. The interest of Church algebras is that each has a Boolean algebra of central elements, which play the role of the idempotent elements in rings. Central elements are the key tool to represent any Church algebra as a weak Boolean product of indecomposable Church algebras and to prove the meta representation theorem mentioned above. We generalize the notion of easy λ-term of lambda calculus to prove that any Church algebra with an "easy set" of cardinality n admits (at the top) a lattice interval of congruences isomorphic to the free Boolean algebra with n generators. This theorem has the following consequence for the lattice of λ-theories: for every recursively enumerable λ-theory ϕ and each n, there is a λ-theory $\phi_n \supseteq \phi$ such that $\{\psi : \psi \supseteq \phi_n\}$ "is" the Boolean lattice with 2^n elements.

Keywords: Lambda calculus, Universal Algebra, Church Algebras, Stone Representation Theorem, Lambda Theories.

1 Introduction

Lambda theories are equational extensions of the untyped λ-calculus closed under derivation. They arise by syntactical or semantic considerations. Indeed, a λ-theory may correspond to some operational semantics of λ-calculus, as well as it may be induced by a λ-model, which is a particular *combinatory algebra* (CA, for short) [1, Sec. 5.2]. The set of λ-theories is naturally equipped with a structure of complete lattice (see [1, Ch. 4]), whose bottom element is the least λ-theory $\lambda\beta$, and whose top element is the inconsistent λ-theory. The lattice λT of λ-theories is a very rich and complex structure of cardinality $2^{\aleph_0}$ [1,10,11,2].

The interest of a systematic study of the lattice λT of λ-theories grows out of several open problems on λ-calculus. For example, Selinger's order-incompleteness problem asks for the existence of a λ-theory not arising as the equational theory of a non-trivially partially ordered model of λ-calculus. This problem can be proved equivalent to that of the existence of a recursively enumerable (r.e., for short) λ-theory ϕ whose term algebra generates an n-permutable variety of algebras for

E. Ochmański and J. Tyszkiewicz (Eds.): MFCS 2008, LNCS 5162, pp. 479–490, 2008.
© Springer-Verlag Berlin Heidelberg 2008

some $n \geq 2$ (see [15] and the remark after [16, Thm. 3.4]). Lipparini [9] has found out interesting non-trivial lattice identities that hold in the congruence lattices of all algebras living in an n-permutable variety. The failure of some Lipparini's lattice identities in λT would imply that Selinger's problem has a negative answer.

Techniques of universal algebra were applied in [14,10,3] to study the structure of λT. In this paper we validate the inverse slogan: λ-calculus can be fruitfully applied to universal algebra. By generalizing to universal algebra concepts originating from λ-calculus and programming, we create a zigzag path from λ-calculus to universal algebra and back. All the algebraic properties we have shown in [11] for CAs, hold for a wider class of algebras, that we call *Church algebras*. Church algebras include, beside CAs, all BAs (Boolean algebras) and all rings with unit, and model the "if-then-else" instruction by two constants $0, 1$ and a ternary term $q(x, y, z)$ satisfying the following identities:

$$q(1, x, y) = x; \quad q(0, x, y) = y.$$

The interest of Church algebras is that each has a BA of central elements, which can be used to represent the Church algebra as a weak Boolean product of directly indecomposable algebras (i.e., algebras which cannot be decomposed as the Cartesian product of two other non-trivial algebras).

We generalize the notion of easy λ-term from λ-calculus and use central elements to prove that: (i) any Church algebra with an "easy set" of cardinality κ admits a congruence ϕ such that (the lattice reduct of) the free BA with κ generators embeds into the lattice interval $[\phi)$ of all congruences greater than ϕ; (ii) If κ is a finite cardinal, this embedding is an isomorphism. This theorem applies directly to all BAs and rings with units. For λT it has the following consequence: for every r.e. λ-theory ϕ and each natural number n, there is a λ-theory $\phi_n \supseteq \phi$ such that the lattice interval $[\phi_n)$ is the finite Boolean lattice with 2^n elements. It is the first time that it is found an interval of λT whose cardinality is not $1, 2$ or $2^{\aleph_0}$.

Our contribution to general Universal Algebra is the following: using Church algebras we prove a meta version of the Stone Representation Theorem that applies to all varieties of algebras and not only to the classic ones. Indeed, we show that any variety of algebras can be decomposed as a weak Boolean product of directly indecomposable subvarieties. This means that, given a variety $\mathcal{V}$, there exists a family of "indecomposable" subvarieties $\mathcal{V}_i$ $(i \in I)$ of $\mathcal{V}$ for which every algebra of $\mathcal{V}$ is isomorphic to a weak Boolean product of algebras of $\mathcal{V}_i$ $(i \in I)$.

2 Preliminaries

We will use the notation of Barendregt's classic work [1] for λ-calculus and combinatory logic, and the notation of McKenzie et al. [12] for universal algebra.

A lattice L is *bounded* if it has a top element 1 and a bottom element 0. An element $x \in L$ is an *atom* (*coatom*) if it is a minimal element in $L - \{0\}$ (maximal element in $L - \{1\}$). For $x \in L$, we set $L_x = \{y \in L - \{0\} : x \wedge y = 0\}$. L is called: *lower semicomplemented* if $L_x \neq \emptyset$ for all $x \neq 1$; *pseudocomplemented* if each $L_x \cup \{0\}$ has a greatest element (called the *pseudocomplement* of x); *complemented* if, for every $x \in L$, there exists y such that $x \wedge y = 0$ and $x \vee y = 1$; *atomic* if, for every $x \in L$, there exists an atom $y \leq x$; *coatomic* if, for every $x \in L$, there exists a coatom $y \geq x$; *Boolean* if it is distributive and complemented.

An element x of a complete lattice L is *completely join-prime* if, for every $X \subseteq L$, $x \leq \bigvee X$ implies $x \leq y$ for some $y \in X$.

We write $[x)$ for $\{y : x \leq y \leq 1\}$.

An *algebraic similarity type* Σ is constituted by a non-empty set of operator symbols together with a function assigning to each operator $f \in \Sigma$ a finite *arity*.

A *Σ-algebra* $\mathbf{A}$ is determined by a non-empty set A together with an operation $f^{\mathbf{A}} : A^n \to A$ for every $f \in \Sigma$ of arity n. $\mathbf{A}$ is *trivial* if $|A| = 1$, where $|A|$ denotes the cardinality of A.

A compatible equivalence relation ϕ on a Σ-algebra $\mathbf{A}$ is called a *congruence*. We often write $x\phi y$ for $(x, y) \in \phi$. The set $\{y : x\phi y\}$ is denoted by $[x]_\phi$.

If $\phi \subseteq \psi$ are congruences on $\mathbf{A}$, then $\psi/\phi = \{([x]_\phi, [y]_\phi) : x\psi y\}$ is a congruence on the quotient $\mathbf{A}/\phi$.

If $X \subseteq A \times A$, then we write $\theta(X)$ for the least congruence including X. We write $\theta(x, y)$ for $\theta(\{(x, y)\})$. If $x \in A$ and $Y \subseteq A$, then we write $\theta(x, Y)$ for $\theta(\{(x, y) : y \in Y\})$.

We denote by $\mathrm{Con}(\mathbf{A})$ the algebraic complete lattice of all congruences of $\mathbf{A}$, and by $\nabla = \{(x, y) : x, y \in A\}$ and $\Delta = \{(x, x) : x \in A\}$ the top and the bottom element of $\mathrm{Con}(\mathbf{A})$. We recall that the join of two congruences ϕ and ψ is the least equivalence relation containing the union $\phi \cup \psi$. A congruence ϕ on $\mathbf{A}$ is called: *trivial* if it is equal to ∇ or Δ; *consistent* if $\phi \neq \nabla$; *compact* if $\phi = \theta(X)$ for some finite set $X \subseteq A \times A$. Two congruences ϕ and ψ are *compatible* if $\phi \vee \psi \neq \nabla$; otherwise, they are *incompatible*.

An algebra $\mathbf{A}$ is *directly decomposable* if there exist two non-trivial algebras $\mathbf{B}, \mathbf{C}$ such that $\mathbf{A} \cong \mathbf{B} \times \mathbf{C}$, otherwise it is called *directly indecomposable*.

An algebra $\mathbf{A}$ is a *subdirect product* of the algebras $(\mathbf{B}_i)_{i \in I}$, written $\mathbf{A} \leq \Pi_{i \in I} \mathbf{B}_i$, if there exists an embedding f of $\mathbf{A}$ into the direct product $\Pi_{i \in I} \mathbf{B}_i$ such that the projection $\pi_i \circ f : \mathbf{A} \to \mathbf{B}_i$ is onto for every $i \in I$.

A non-empty class $\mathcal{V}$ of algebras is a *variety* if it is closed under subalgebras, homomorphic images and direct products or, equivalently, if it is axiomatizable by a set of equations. A variety $\mathcal{V}'$ is a *subvariety* of the variety $\mathcal{V}$ if $\mathcal{V}' \subseteq \mathcal{V}$. We will denote by $\mathcal{V}(\mathbf{A})$ the variety generated by an algebra $\mathbf{A}$, i.e., $\mathbf{B} \in \mathcal{V}(\mathbf{A})$ if every equation satisfied by $\mathbf{A}$ is also satisfied by $\mathbf{B}$.

Let $\mathcal{V}$ be a variety. We say that $\mathbf{A}$ is *the free $\mathcal{V}$-algebra over the set X of generators* iff $\mathbf{A} \in \mathcal{V}$, $\mathbf{A}$ is generated by X and for every $g : X \to \mathbf{B} \in \mathcal{V}$, there is a unique homomorphism $f : \mathbf{A} \to \mathbf{B}$ that extends g (i.e., $f(x) = g(x)$ for every $x \in X$). A free algebra in the class of all Σ-algebras is called *absolutely free*.

Given two congruences σ and τ on $\mathbf{A}$, we can form their relative product: $\tau \circ \sigma = \{(x, z) : \exists y \in A \; x\sigma y\tau z\}$. We have $\tau \cup \sigma \subseteq \tau \circ \sigma \subseteq \tau \vee \sigma$; in general $\tau \circ \sigma$ is not a congruence.

Definition 1. *A congruence ϕ on an algebra* $\mathbf{A}$ *is a* factor congruence *if there exists another congruence ψ such that $\phi \cap \psi = \Delta$ and $\phi \circ \psi = \nabla$. In this case we call (ϕ, ψ) a* pair of complementary factor congruences *or* cfc-pair, *for short.*

If (ϕ, ψ) is a cfc-pair, then, for all x and y, there is just one element u satisfying $x \phi u \psi y$.

Under the hypotheses of the above definition the homomorphism $f : \mathbf{A} \to \mathbf{A}/\phi \times \mathbf{A}/\psi$ defined by $f(x) = ([x]_\phi, [x]_\psi)$ is an isomorphism. So, the existence of factor congruences is just another way of saying "this algebra is a direct product of simpler algebras".

The set of factor congruences of $\mathbf{A}$ is not, in general, a sublattice of $\mathrm{Con}(\mathbf{A})$. Δ and ∇ are the *trivial* factor congruences, corresponding to $\mathbf{A} \cong \mathbf{A} \times \mathbf{B}$, where $\mathbf{B}$ is a trivial algebra. An algebra $\mathbf{A}$ is directly indecomposable if, and only if, $\mathbf{A}$ has no non-trivial factor congruences.

It is possible to characterize cfc-pairs in terms of certain algebra homomorphisms called *decomposition operators* (see [12, Def. 4.32] for more details).

Definition 2. *A decomposition operator for an algebra* $\mathbf{A}$ *is an algebra homomorphism $f : \mathbf{A} \times \mathbf{A} \to \mathbf{A}$ such that $f(x, x) = x$ and $f(f(x, y), z) = f(x, z) = f(x, f(y, z))$.*

There exists a bijection between cfc-pairs and decomposition operators, and thus, between decomposition operators and factorizations like $\mathbf{A} \cong \mathbf{B} \times \mathbf{C}$.

Proposition 1. *[12, Thm. 4.33] Given a decomposition operator f, the relations ϕ, ψ defined by $x \phi y$ iff $f(x, y) = y$; and $x \psi y$ iff $f(x, y) = x$, form a cfc-pair. Conversely, given a cfc-pair (ϕ, ψ), the map f defined by $f(x, y) = u$ iff $x \phi u \psi y$, is a decomposition operator.*

The Boolean product construction allows us to transfer numerous fascinating properties of BAs into other varieties of algebras (see [5, Ch. IV]). We recall that a Boolean space is a compact, Hausdorff and totally disconnected topological space, and that *clopen* means "open and closed".

Definition 3. *A weak Boolean product of a family $(\mathbf{A}_i)_{i \in I}$ of algebras is a subdirect product $\mathbf{A} \leq \Pi_{i \in I} \mathbf{A}_i$, where I can be endowed with a Boolean space topology such that: (i) the set $\{i \in I : x_i = y_i\}$ is open for all $x, y \in A$, and (ii) if $x, y \in A$ and N is a clopen subset of I, then the element z, defined by $z_i = x_i$ for every $i \in N$ and $z_i = y_i$ for every $i \in I - N$, belongs to A. A Boolean product is a weak Boolean product such that the set $\{i \in I : x_i = y_i\}$ is clopen for all $x, y \in A$.*

A λ-*theory* is any congruence (w.r.t. the binary operator of application and the lambda abstractions) on the set of λ-terms including (α)- and (β)-conversion (see [1, Ch. 2]). We use for λ-theories the same notational convention as for congruences. The set of all λ-theories, ordered by inclusion, is naturally equipped with a structure of complete lattice, denoted by λT, with intersection as meet. The least element of λT is denoted by $\lambda\beta$, while the top element of λT is the inconsistent λ-theory ∇. The *term algebra of a λ-theory ϕ*, denoted by Λ_ϕ, has the equivalence classes of λ-terms modulo ϕ as elements, and is equipped with the natural operations induced by application and λ-abstraction. The lattice λT is isomorphic to the congruence lattice $\mathrm{Con}(\Lambda_{\lambda\beta})$, while its interval sublattice $[\phi)$ is isomorphic to $\mathrm{Con}(\Lambda_\phi)$.

As a matter of notation, Ω denotes the λ-term $(\lambda x.xx)(\lambda x.xx)$.

The variety CA of *combinatory algebras* [1, Sec. 5.1] consists of algebras $\mathbf{C} = (C, \cdot, \mathbf{k}, \mathbf{s})$, where $\cdot$ is a binary operation and $\mathbf{k}, \mathbf{s}$ are constants, satisfying $\mathbf{k}xy = x$ and $\mathbf{s}xyz = xz(yz)$ (as usual, the symbol "$\cdot$" is omitted and association is made on the left).

3   Church Algebras

Our key observation is that many algebraic structures, such as CAs, BAs etc., have in common the fact that all are *Church algebras*. In this section we study the algebraic properties of this class of algebras. Applications are given in Section 5 and in Section 6.

Definition 4. *An algebra $\mathbf{A}$ is called* a Church algebra *if there are two constants $0, 1 \in A$ and a ternary term $q(e, x, y)$ such that $q(1, x, y) = x$ and $q(0, x, y) = y$. A variety $\mathcal{V}$ is called a* Church variety *if every algebra in $\mathcal{V}$ is a Church algebra with respect to the same term $q(e, x, y)$ and constants $0, 1$.*

Note that the top element ∇ of the congruence lattice $\mathrm{Con}(\mathbf{A})$ of a Church algebra $\mathbf{A}$ is a compact element because obviously $\nabla = \theta(0, 1)$.

Example 1. The following are easily checked to be Church algebras:
1. Combinatory algebras: $q(e, x, y) \equiv (e \cdot x) \cdot y$; $1 \equiv \mathbf{k}$; $0 \equiv \mathbf{sk}$
2. Boolean algebras: $q(e, x, y) \equiv (e \vee y) \wedge (e^- \vee x)$
3. Heyting algebras: $q(e, x, y) \equiv (e \vee y) \wedge ((e \to 0) \vee x)$
4. Rings with unit: $q(e, x, y) \equiv (y + e - ey)(1 - e + ex)$

Let $\mathbf{A} = (A, +, \cdot, 0, 1)$ be a commutative ring with unit. Every idempotent element $e \in A$ (i.e., satisfying $e \cdot e = e$) induces a cfc-pair $(\theta(1, e), \theta(e, 0))$. In other words, the ring $\mathbf{A}$ can be decomposed as $\mathbf{A} \cong \mathbf{A}/\theta(1, e) \times \mathbf{A}/\theta(e, 0)$. $\mathbf{A}$ is directly indecomposable iff 0 and 1 are the unique idempotent elements. Vaggione [18] generalized the notion of idempotent to any universal algebra whose top congruence ∇ is compact, and called them *central elements*. Central elements were used to investigate the closure of varieties of algebras under Boolean products. Here we give a new characterization based on decomposition operators (see Def.2). Hereafter, we set $\theta_e \equiv \theta(1, e)$ and $\bar\theta_e \equiv \theta(e, 0)$.

Definition 5. *We say that an element e of a Church algebra $\mathbf{A}$ is* central, *and we write $e \in \mathrm{Ce}(\mathbf{A})$, if $(\theta_e, \bar\theta_e)$ is a cfc-pair. A central element e is called* nontrivial *if $e \neq 0, 1$.*

We now show that, in a Church algebra, factor congruences can be internally represented by central elements. The following lemma is easy to check.

Lemma 1. *Let* $\mathbf{A}$ *be a Church algebra and* $e \in A$. *Then we have, for all* $x, y \in A$:

(a) $x \; \theta_e \; q(e, x, y) \; \bar{\theta}_e \; y$.

(b) $x\theta_e y$ *iff* $q(e, x, y) \; (\theta_e \cap \bar{\theta}_e) \; y$.

(c) $x\bar{\theta}_e y$ *iff* $q(e, x, y) \; (\theta_e \cap \bar{\theta}_e) \; x$.

(d) $\theta_e \circ \bar{\theta}_e = \bar{\theta}_e \circ \theta_e = \nabla$.

Proposition 2. *Let* $\mathbf{A}$ *be a Church* Σ-*algebra and* $e \in A$. *Then the following conditions are equivalent:*

(i) e *is central;*

(ii) $\theta_e \cap \bar{\theta}_e = \Delta$;

(iii) *For all* x *and* y, $q(e, x, y)$ *is the unique element such that* $x\theta_e \; q(e, x, y) \; \bar{\theta}_e y$;

(iv) e *satisfies the following identities:*

    1. $q(e, x, x) = x$.

    2. $q(e, q(e, x, y), z) = q(e, x, z) = q(e, x, q(e, y, z))$.

    3. $q(e, f(\bar{x}), f(\bar{y})) = f(q(e, x_1, y_1), \ldots, q(e, x_n, y_n))$, *for every* $f \in \Sigma$.

    4. $e = q(e, 1, 0)$.

(v) *The function* f_e *defined by* $f_e(x, y) = q(e, x, y)$ *is a decomposition operator such that* $f_e(1, 0) = e$.

Thus a Church algebra $\mathbf{A}$ is directly indecomposable iff $\mathrm{Ce}(\mathbf{A}) = \{0, 1\}$ iff $\theta_e \cap \bar{\theta}_e \neq \Delta$ for all $e \neq 0, 1$.

Example 2. (i) All elements of a BA are central by Prop. 2(iv) and Example 1.

(ii) An element is central in a commutative ring with unit iff it is idempotent. This characterization does not hold for non-commutative rings with unit.

(iii) Let $\Omega \equiv (\lambda x.xx)(\lambda x.xx)$ be the usual looping term of λ-calculus. It is well-known that the λ-theories $\theta_\Omega = \theta(\Omega, \lambda xy.x)$ and $\bar{\theta}_\Omega = \theta(\Omega, \lambda xy.y)$ are consistent (see [1, Prop. 15.3.9]). Then by Lemma 5 below the term Ω is a non-trivial central element in the term algebra of $\theta_\Omega \cap \bar{\theta}_\Omega$.

We now show that the partial ordering on $\mathrm{Ce}(\mathbf{A})$, defined by:

$$e \leq d \text{ if, and only if, } \bar{\theta}_e \subseteq \bar{\theta}_d$$

is a Boolean ordering and that the meet, join and complementation operations are internally representable. 0 and 1 are respectively the bottom and top element of this ordering.

Theorem 1. *Let* $\mathbf{A}$ *be a Church algebra. The algebra* $(\mathrm{Ce}(\mathbf{A}), \wedge, \vee, ^-, 0, 1)$, *with operations defined by* $e \wedge d = q(e, d, 0)$, $e \vee d = q(e, 1, d)$, $e^- = q(e, 0, 1)$, *is a* BA, *which is isomorphic to the* BA *of factor congruences of* $\mathbf{A}$.

Next we turn to the Stone representation theorem for Church algebras. It is a corollary of Thm. 1 and of theorems by Comer [6] and by Vaggione [18].

Let $\mathbf{A}$ be a Church algebra. If I is an ideal of the Boolean algebra $\mathrm{Ce}(\mathbf{A})$, then $\phi_I = \cup_{e \in I} \bar{\theta}_e$ is a congruence. In the next theorem $\mathcal{S}$ is the Boolean space of maximal ideals of $\mathrm{Ce}(\mathbf{A})$.

Theorem 2. (The Stone Representation Theorem) *Let* **A** *be a Church algebra. Then, for all $I \in \mathcal{S}$ the quotient algebra $\mathbf{A}/\phi_I$ is directly indecomposable and the map $f : A \to \Pi_{I \in \mathcal{S}}(A/\phi_I)$, defined by $f(x) = ([x]_{\phi_I} : I \in \mathcal{S})$, gives a weak Boolean product representation of* **A**.

Note that, in general, Thm. 2 does not give a (non-weak) Boolean product representation. This was shown in [11] for combinatory algebras.

4   The Main Theorem

In λ-calculus there are *easy λ-terms*, i.e., terms, like Ω, that can be consistently equated with any other closed λ-term. In this section we generalize the notion of easiness to Church algebras and show that any Church algebra with an easy set of cardinality n admits a congruence ϕ such that the lattice interval of all congruences greater than ϕ is isomorphic to the free BA with n generators.

Let **A** be a Church algebra and $a \subseteq A$. As a matter of notation, for every $b \subseteq a$, we define

$$\delta_b \equiv \theta(1, b) \vee \theta(0, a - b). \tag{1}$$

By definition $\theta(1, \emptyset) = \theta(0, \emptyset) = \Delta$.

Definition 6. *Let* **A** *be a Church algebra. We say that $a \subseteq A$ is an* easy set *if $\delta_b \neq \nabla$ for every $b \subseteq a$.*

Note that, if a is an easy set, then the set of all δ_b ($b \subseteq a$) consists of $2^{|a|}$ pairwise incompatible congruences.

We say that an element x is *easy* if $\{x\}$ is an easy set. Thus, x is easy iff the congruences θ_x and $\overline{\theta}_x$ are both different from ∇.

Example 3. A finite subset a of a BA is an easy set if it holds: (i) $\bigvee a \neq 1$; (ii) $\bigwedge a \neq 0$; (iii) for all $b \subset a$, $\bigvee b \not\geq \bigwedge(a - b)$. Thus, for example, $\{\{1, 2\}, \{2, 3\}\}$ is an easy set in the powerset of $\{1, 2, 3, 4\}$.

The following lemmas are used in the proof of the main theorem.

Lemma 2. *The congruences of a Church algebra permute with its factor congruences, i.e., $\phi \circ \psi = \psi \circ \phi$ for every congruence ϕ and factor congruence ψ.*

Proof. Let $\psi = \theta_e$ for a central e and let $a \; \phi \; b \; \theta_e \; c$ for some b. We get the conclusion if $a \; \theta_e \; q(e, a, c) \; \phi \; c$. First $a \; \theta_e \; q(e, a, c)$ by Lemma 1(a). From $a \; \phi \; b$ we have $q(e, a, c) \; \phi \; q(e, b, c)$. Finally, $q(e, b, c) = c$ by $b \; \theta_e \; c$ and by Prop. 2(iii). $\quad\blacksquare$

Definition 7. *A bounded lattice L with top 1 and bottom 0 satisfies the* Zipper *condition if, for every set I and for every $x_i, y, z \in L$ $(i \in I)$, we have:*

$$\bigvee_{i \in I} x_i = 1, \; x_i \wedge y = z \; (i \in I) \implies y = z.$$

Lemma 3. *If* **A** *is a Church algebra, then* $\mathrm{Con}(\mathbf{A})$ *satisfies the Zipper condition.*

Proof. By [8] the congruence lattice of every algebra having a binary term with a right unit and a right zero satisfies the Zipper condition.

Lemma 4. *Let* $\mathbf{A}$ *be a Church algebra and* $\phi \in \mathrm{Con}(\mathbf{A})$. *Then,* $\mathbf{A}/\phi$ *is also a Church algebra and the map* $c_\phi : \mathrm{Ce}(\mathbf{A}) \to \mathrm{Ce}(\mathbf{A}/\phi)$, *defined by* $c_\phi(x) = [x]_\phi$ *is a homomorphism of* BAs.

Lemma 5. *Let* $\mathbf{A}$ *be a Church algebra,* $e \in A$ *and* $\phi \in \mathrm{Con}(\mathbf{A})$. *Then,*
(i) $[e]_\phi$ *is central in* $\mathbf{A}/\phi$ *iff* $\phi \supseteq \theta_e \cap \bar{\theta}_e$.
(ii) $[e]_{\theta_e \cap \bar{\theta}_e}$ *is a non-trivial central element in* $\mathbf{A}/\theta_e \cap \bar{\theta}_e$ *iff* $\theta_e \neq \nabla, \Delta$.

Theorem 3. *Let* $\mathbf{A}$ *be a Church algebra,* $a \subseteq A$ *be an easy set and* $\mathbf{B}(a)$ *be the free Boolean algebra over the set* a *of generators. Then there exists a congruence* ϕ_a *satisfying the following conditions:*

1. *The lattice reduct of* $\mathbf{B}(a)$ *can be embedded into the interval sublattice* $[\phi_a)$ *of* $\mathrm{Con}(\mathbf{A})$;
2. *If* a *has finite cardinality* n, *then the above embedding is an isomorphism and, hence,* $[\phi_a)$ *has cardinality* 2^{2^n}.

Proof. Let $\eta = \cap_{b \subseteq a} \delta_b$ (see (1) above for the definition of δ_b). We define ϕ_a as any maximal element of the set of all congruences ϕ which contain η and are compatible with each δ_b (i.e., $\phi \vee \delta_b \neq \nabla$). Note that ϕ_a exists by Zorn Lemma.

Claim 1. $[x]_{\phi_a}$ *is central in* $\mathbf{A}/\phi_a$ *for every* $x \in a$.

Proof. If we prove that $[x]_\eta$ is central in $\mathbf{A}/\eta$, then by $\eta \subseteq \phi_a$ and by Lemma 4 we get the conclusion of the claim. Since $x \in a$ is equivalent either to 1 or to 0 in each congruence δ_b, then $[x]_{\delta_b}$ is a trivial central element in $\mathbf{A}/\delta_b$, so that $[x]_\eta$ is central in $\mathbf{A}/\eta$ by Lemma 5(i) and by $\eta = \cap_{b \subseteq a} \delta_b$.

Let now $f_a : \mathbf{B}(a) \to \mathrm{Ce}(\mathbf{A}/\phi_a)$ be the unique Boolean homomorphism satisfying $f_a(x) = [x]_{\phi_a}$ $(x \in a)$.

Claim 2. f_a *is an embedding.*

Proof. Let $b \subseteq a$. Recall that $\phi_a \vee \delta_b \neq \nabla$. By Lemma 4 there exists a Boolean homomorphism (denoted by h_b in this proof) from $\mathrm{Ce}(\mathbf{A}/\phi_a)$ into $\mathrm{Ce}(\mathbf{A}/\phi_a \vee \delta_b)$. Since $(x, 1) \in \phi_a \vee \delta_b$ for every $x \in b$, and $(y, 0) \in \phi_a \vee \delta_b$ for every $y \in a - b$, then the kernel of $h_b \circ f_a$ is an ultrafilter of $\mathbf{B}(a)$. By the arbitrariness of $b \subseteq a$, every ultrafilter of $\mathbf{B}(a)$ can be the kernel of a suitable $h_b \circ f_a$. This is possible only if f_a is an embedding.

This concludes the proof of (1) of Thm. 3.

Hereafter, we assume that a *is finite and we let* $n = |a|$. Then $\mathbf{B}(a)$ is finite, atomic, has n generators, 2^n atoms, 2^n coatoms, and $|B(a)| = 2^{2^n}$. Recall that $\mathrm{Con}(\mathbf{A}/\phi_a)$ is isomorphic to $[\phi_a)$.

Let At_a be the set of atoms of $\mathrm{Con}(\mathbf{A}/\phi_a)$.

Claim 3. $\bigvee\{\beta \in At_a : \beta \text{ is a factor congruence}\} = \nabla$.

Proof. Let $v \in Ce(\mathbf{A}/\phi_a)$ such that $v = f_a(u)$ for some atom $u \in \mathbf{B}(a)$. Consider $\tau \in Con(\mathbf{A})$ such that $\tau/\phi_a = \theta(v, 0) \in Con(\mathbf{A}/\phi_a)$. We claim that $\tau/\phi_a \in At_a$. By the way of contradiction, let $\sigma \in Con(\mathbf{A})$ such that $\phi_a \subset \sigma \subset \tau$. By Lemma 4 we have a sequence of Boolean homomorphisms:

$$\mathbf{B}(a) \xrightarrow{f_a} Ce(\mathbf{A}/\phi_a) \xrightarrow{c_\sigma} Ce(\mathbf{A}/\sigma) \xrightarrow{c_{\tau/\sigma}} Ce(\mathbf{A}/\tau)$$

and a Boolean homomorphism $c_\tau : Ce(\mathbf{A}/\phi_a) \xrightarrow{c_\tau} Ce(\mathbf{A}/\tau)$ such that $c_\tau = c_{\tau/\sigma} \circ c_\sigma$. Since u is an atom of $\mathbf{B}(a)$, then the set $\{0, u\}$ is the Boolean ideal associated with the kernel of $c_\tau \circ f_a$. If $c_\sigma(v) = 0$, then σ/ϕ_a contains the pair $(v, 0)$, i.e., $\sigma = \tau$. It follows that $c_\sigma(v) \neq 0$ and the map $c_\sigma \circ f_a : \mathbf{B}(a) \to Ce(\mathbf{A}/\sigma)$ is an embedding. Since $\mathbf{B}(a)$ is free over a, for every $b \subseteq a$ there exists an atom $w \in \mathbf{B}(a)$ such that $w = (\bigwedge b) \wedge (\bigwedge \{x^- : x \in a - b\})$. Let $w' = c_\sigma(f_a(w)) \in \mathbf{A}/\sigma$ the corresponding nontrivial central element. By definition of w, the non-triviality of the factor congruence $\theta(w', 1) \in Con(\mathbf{A}/\sigma)$ is equivalent to $\sigma \vee \delta_b \neq \nabla$. The arbitrariness of b and the strict inclusion $\phi_a \subset \sigma$ contradict the maximality of ϕ_a. In conclusion $\tau/\phi_a \in At_a$. Finally, the claim follows because the join of all atoms of $\mathbf{B}(a)$ is the top element.

Let Co_a be the set of coatoms of $Con(\mathbf{A}/\phi_a)$. We say that the coatoms form a finite irredundant decomposition of Δ if Co_a is finite, $\cap Co_a = \Delta$, and $\cap(Co_a - \{\psi\}) \neq \Delta$ for every $\psi \in Co_a$.

Claim 4. $Con(\mathbf{A}/\phi_a)$ *is pseudocomplemented, complemented, atomic, and the coatoms form a finite irredundant decomposition of* Δ.

Proof. $Con(\mathbf{A}/\phi_a)$ satisfies the Zipper condition (by Lemma 3) and $\bigvee At_a = \nabla$ (by Claim 3). Then by [7, Prop. 2] $Con(\mathbf{A}/\phi_a)$ is complemented, atomic and every coatom has a complement which is an atom. It is also pseudocomplemented by [7, Prop. 1]. Since the top element ∇ is compact, by [7, Prop. 3] we get that the coatoms form a finite irredundant decomposition of Δ.

Claim 5. Let $\xi \in Con(\mathbf{A}/\phi_a)$ be a non-trivial congruence and $\gamma = \bigvee\{\delta \in At_a : \delta \subseteq \xi\}$. If $\beta \in At_a$ is a factor congruence and $\beta \nsubseteq \xi$, then $\xi \cap (\beta \vee \gamma) = \gamma$.

Proof. We always have $\gamma \subseteq \xi \cap (\beta \vee \gamma)$. We show the opposite direction. Let $(x, y) \in \xi \cap (\beta \vee \gamma)$, i.e., $x \, \xi \, y$ and $x(\beta \vee \gamma)y$. We have to show that $x \, \gamma \, y$. Since β is a factor congruence, by Lemma 2 we have $\beta \vee \gamma = \beta \circ \gamma$. Then $x \, \beta \, z \, \gamma \, y$ for some z. Since $\gamma \subseteq \xi$ and $z \, \gamma \, y$ then $z \, \xi \, y$, that together with $x \, \xi \, y$ implies $x \, \xi \, z$. Then $x(\xi \cap \beta)z$. Since β is an atom and $\beta \nsubseteq \xi$, then $\xi \cap \beta = \Delta$, so that $x = z$. This last equality and $z \, \gamma \, y$ imply $x \, \gamma \, y$. In other words, $\xi \cap (\beta \vee \gamma) = \gamma$.

Claim 6. Every $\xi \in Con(\mathbf{A}/\phi_a)$ is a join of atoms.

Proof. Let At_ξ be the set of atoms included in ξ. We will show that $\xi = \bigvee At_\xi$ by applying the Zipper condition of Def. 7. Let $\gamma = \bigvee At_\xi$. By Claim 5 we have: $\bigvee\{\nu : \xi \cap \nu = \gamma\} \supseteq \bigvee\{\beta \vee \gamma : \beta \in At_a, \beta \nsubseteq \xi, \beta$ is a factor congruence$\} \supseteq \bigvee\{\beta : \beta \in At_a$ is a factor congruence$\}$. By Claim 3 this last element is equal to ∇, so that $\bigvee\{\nu : \xi \cap \nu = \gamma\} = \nabla$. By the Zipper condition this entails $\xi = \gamma$.

Claim 7. $\mathrm{Con}(\mathbf{A}/\phi_a)$, *and hence* $[\phi_a)$, *is isomorphic to the power set of* At_a.

Proof. $\mathrm{Con}(\mathbf{A}/\phi_a)$ is atomic and pseudocomplemented (by Claim 4), so that each atom is completely join-prime. By this and by Claim 6 every element is univocally represented as a join of atoms. The conclusion follows because every join of atoms exists by completeness.

Claim 8. $\mathrm{Con}(\mathbf{A}/\phi_a)$, *and hence* $[\phi_a)$, *has* 2^n *coatoms and* 2^n *atoms.*

Proof. Since $\phi_a \vee \delta_b \neq \nabla$ for every $b \subseteq a$, $[\phi_a)$ has at least 2^n coatoms. For every $b \subseteq a$, let c_b be a coatom including $\phi_a \vee \delta_b$. Assume now that there is a coatom ξ distinct from each c_b for every $b \subseteq a$. Consider the intersection $\cap(Co_a - \{\xi\})$, where Co_a denotes the set of coatoms of $[\phi_a)$. By Claim 4 we have that $\cap(Co_a - \{\xi\}) \neq \phi_a$. This contradicts the maximality of ϕ_a among the congruences which contains $\cap_{b \subseteq a} \delta_b$ and are compatible with δ_b. In conclusion, we have 2^n coatoms, and hence 2^n atoms.

This concludes the proof of the main theorem.

The next proposition, which follows from [7, Prop. 4], says that the main theorem cannot be improved.

Proposition 3. *Let* $\mathbf{A}$ *be a Church algebra. Then there exists no congruence* ϕ *such that the interval sublattice* $[\phi)$ *is isomorphic to an infinite Boolean lattice.*

5   The Lattice of λ-Theories

The term algebra of a λ-theory ϕ is a Church algebra. This easy remark has the interesting consequence that the lattice λT of all λ-theories admits (at the top) Boolean lattice intervals of cardinality 2^k for every k.

Lemma 6. *For every r.e.* λ-*theory* ϕ, *the term algebra of* ϕ *admits an infinite easy set.*

Proof. The set consisting of all λ-terms $\Omega\hat{n}$, where $\hat{n}$ is the n-th Church numeral, is an easy set in the term algebra of $\lambda\beta$. This follows from the easiness of Ω and a compactness argument, and appears as [1, Ex. 15.4.3]. More generally, the term algebra of each r.e. λ-theory has an easy element [1, Prop. 17.1.9], and hence it has an infinite easy set, by the same compactness argument.

Theorem 4. *For every r.e.* λ-*theory* ϕ *and each natural number* k, *there is a* λ-*theory* $\phi_k \supseteq \phi$ *such that the lattice interval* $[\phi_k)$ *is isomorphic to the finite Boolean lattice with* 2^k *elements.*

Proof. By Lemma 6 and by Thm. 3 there exists a congruence ψ_k such that $\psi_k \supseteq \phi$ and $[\psi_k)$ is isomorphic to the free Boolean algebra with 2^{2^k} elements. The congruence ϕ_k of the theorem can be defined by using ψ_k and the following facts: (a) Every filter of a finite Boolean algebra is a Boolean lattice; (b) The free Boolean algebra with 2^{2^k} elements has filters of arbitrary cardinality 2^i ($i \leq 2^k$).

Note that the λ-theory ϕ_k of Thm. 4 is not r.e. because otherwise the lattice interval $[\phi_k)$ would have a continuum of elements by [1, Cor. 17.1.11].

6   Lattices of Equational Theories

We say that L is a *lattice of equational theories* iff L is isomorphic to the lattice $L(T)$ of all equational theories containing some equational theory T (or dually, the lattices of all subvarieties of some variety of algebras). Such lattice is algebraic and coatomic, possessing a compact top element; but no stronger property was known before Lampe's discovery [8] that any lattice of equational theories obeys the Zipper condition (see Def. 7).

In this section we show the existence of Boolean lattice intervals in the lattices of equational theories as well as a meta version of the Stone representation theorem that holds for all varieties of algebras.

It is well known that any lattice of equational theories is isomorphic to a congruence lattice (see [5,12]). Indeed, the lattice $L(T)$ of all equational theories containing T is isomorphic to the congruence lattice of the algebra $(\mathbf{F}_T, f)_{f \in \mathrm{End}(\mathbf{F}_T)}$, where $\mathbf{F}_T$ is the free algebra over a countable set of generators in the variety axiomatized by T, and $\mathrm{End}(\mathbf{F}_T)$ is the set of its endomorphisms.

We expand the algebra $(\mathbf{F}_T, f)_{f \in \mathrm{End}(\mathbf{F}_T)}$ (without changing the congruence lattice) by the operation q defined as follows (x_1, x_0 are two fixed variables) $q(t, s_1, s_0) = t[s_1/x_1, s_0/x_0]$, where $t[s_1/x_1, s_0/x_0]$ is the term obtained by substituting the term s_i for the variable x_i ($i = 0, 1$) within t. The algebra $(\mathbf{F}_T, f, q)_{f \in \mathrm{End}(\mathbf{F}_T)}$ was defined by Lampe in the proof of McKenzie Lemma in [8].

If we define $1 \equiv x_1$ and $0 \equiv x_0$, from the identities $q(x_1, s_1, s_0) = s_1$ and $q(x_0, s_1, s_0) = s_0$ we get that $(\mathbf{F}_T, f, q)_{f \in \mathrm{End}(\mathbf{F}_T)}$ is a Church algebra. It will be denoted by $\mathbf{C}_T$ and called hereafter the *Church algebra of T*.

In the following lemma we characterize the central elements of $\mathbf{C}_T$.

Lemma 7. *Let T be an equational theory and $\mathcal{V}$ be the variety of Σ-algebras axiomatized by T. Then the following conditions are equivalent, for every $e \in \mathbf{C}_T$ and term $t(x_1, x_0) \in e$:*

- (i) *e is a central element.*
- (ii) *T contains the identities $t(x, x) = x$; $t(x, t(y, z)) = t(x, z) = t(t(x, y), z)$ and $t(f(\overline{x}), f(\overline{y})) = f(t(x_1, y_1), \ldots, t(x_n, y_n))$, for $f \in \Sigma$.*
- (iii) *For every $\mathbf{A} \in \mathcal{V}$, the function $t^{\mathbf{A}} : A \times A \to A$ is a decomposition operator.*
- (iv) *$T = T_1 \cap T_0$, where T_i is the theory axiomatized (over T) by $t(x_1, x_0) = x_i$ ($i = 0, 1$).*

If e and t satisfy the above conditions and e is nontrivial as central element, then by Lemma 7(iii)-(iv) every algebra $\mathbf{A} \in \mathcal{V}$ can be decomposed as $\mathbf{A} \cong \mathbf{A}/\phi \times \mathbf{A}/\overline{\phi}$, where $(\phi, \overline{\phi})$ is the cfc-pair associated with the decomposition operator $t^{\mathbf{A}}$; moreover, the algebras $\mathbf{A}/\phi$ and $\mathbf{A}/\overline{\phi}$ satisfy respectively the equational theories T_1 and T_0. In this case, we say that $\mathcal{V}$ is *decomposable* as a product of the two subvarieties axiomatized respectively by T_1 and T_0 (see [17]); otherwise, we say that $\mathcal{V}$ is *indecomposable*.

Proposition 4. *Let T be an equational theory. Assume there exist n binary terms $t_0, \ldots, t_{n-1}$ such that, for every function $k : n \to \{0, 1\}$, the theory axiomatized (over T) by $t_i(x_1, x_0) = x_{k(i)}$ ($i = 0, \ldots, n-1$) is consistent. Then* *there exists a theory $T' \supseteq T$ such that $L(T')$ is isomorphic to the free Boolean lattice with 2^{2^n} elements.*

The set of all factor congruences of an algebra **A** does not constitute in general a sublattice of the congruence lattice of **A**. We now show that in every algebra there is a subset of factor congruences which always constitutes a Boolean sublattice of the congruence lattice.

We say that a variety $\mathcal{V}$ is *decomposable as a weak Boolean product of directly indecomposable subvarieties* if there exists a family $\langle \mathcal{V}_i : i \in X \rangle$ of indecomposable subvarieties $\mathcal{V}_i$ of $\mathcal{V}$ such that every algebra $\mathbf{A} \in \mathcal{V}$ is isomorphic to a weak Boolean product $\Pi_{i \in X} \mathbf{B}_i$ of algebras $\mathbf{B}_i \in \mathcal{V}_i$.

Theorem 5. (Meta-Representation Theorem) *Every variety $\mathcal{V}$ of algebras is decomposable as a weak Boolean product of directly indecomposable subvarieties.*

References

1. Barendregt, H.P.: The λ-calculus: Its syntax and semantics. North-Holland Publishing Co., Amsterdam (1984)
2. Berline, C.: Graph models of λ-calculus at work, and variations. Math. Struct. in Comp. Science 16, 185–221 (2006)
3. Berline, C., Salibra, A.: Easiness in graph models. Theo. Comp. Sci. 354, 4–23 (2006)
4. Bigelow, D., Burris, S.: Boolean algebras of factor congruences. Acta Sci. Math. 54, 11–20 (1990)
5. Burris, S., Sankappanavar, H.P.: A course in universal algebra. Springer, Berlin (1981)
6. Comer, S.: Representations by algebras of sections over boolean spaces. Pacific J. Math. 38, 29–38 (1971)
7. Diercks, V., Ernè, M., Reinhold, J.: Complements in lattices of varieties and equational theories. Algebra Universalis 31(4), 506–515 (1994)
8. Lampe, W.A.: A property of the lattice of equational theories. Algebra Universalis 23, 61–69 (1986)
9. Lipparini, P.: n-permutable varieties satisfy nontrivial congruence identities. Algebra Universalis 33(2), 159–168 (1995)
10. Lusin, S., Salibra, A.: The lattice of lambda theories. Journal of Logic and Computation 14, 373–394 (2004)
11. Manzonetto, G., Salibra, A.: Boolean algebras for λ-calculus. In: 21th Annual IEEE Symposium on Logic in Computer Science (LICS 2006) (2006)
12. McKenzie, R.N., McNulty, G.F., Taylor, W.F.: Algebras, Lattices, Varieties, vol. I. Wadsworth Brooks (1987)
13. Pierce, R.S.: Modules over commutative regular rings. Memoirs Amer. Math. Soc. (1967)
14. Salibra, A.: On the algebraic models of λ-calculus. Theo. Comp. Sci. 249, 197–240 (2000)
15. Salibra, A.: Topological incompleteness and order incompleteness of the λ-calculus. ACM Trans. on Computational Logic 4, 379–401 (2003) (LICS 2001 Special Issue)
16. Selinger, P.: Order-incompleteness and finite lambda reduction models. Theo. Comp. Sci. 309, 43–63 (2003)
17. Taylor, W.: The fine spectrum of a variety. Algebra Universalis 5, 263–303 (1975)
18. Vaggione, D.: Varieties in which the Pierce stalks are directly indecomposable. Journal of Algebra 184, 424–434 (1996)

A Complete Axiomatic System for a Process-Based Spatial Logic

Radu Mardare[1] and Alberto Policriti[2,3]

[1] Microsoft Research-University of Trento, CoSBi Centre, Trento, Italy
[2] Department of Mathematics and Computer Science, University of Udine, Italy
[3] Applied Genomics Institute, Udine, Italy

Abstract. The process-based Spatial Logics are multi-modal logics developed for semantics on Process Algebras and designed to specify concurrent properties of dynamic systems. On the syntactic level, they combine modal operators similar to operators of Hennessy-Milner logic, dynamic logic, arrow logic, relevant logic, or linear logic. This combination generates expressive logics, sometimes undecidable, for which a wide range of applications have been proposed.

In the literature, there exist some sound proof systems for spatial logics, but the problem of completeness against process-algebraic semantics is still open. The main goal of this paper is to identify a sound-complete axiomatization for such a logic. We focus on a particular spatial logic that combines the basic spatial operators with dynamic and classical operators. The semantics is based on a fragment of CCS calculus that embodies the core features of concurrent behaviors. We prove the logic decidable both for satisfiability/validity and mode-checking, and we propose a sound-complete Hilbert-style axiomatic system for it.

1   Introduction

Process algebras [2] are calculi designed for modelling complex systems of *processes*[1] organised in a modular way, which run in a decentralised manner and are able to interact, collaborate and communicate. Starting with Robin Milner's classical work on a *Calculus of Communicating Systems* [17], a plethora of process calculi have been developed and successfully applied to a multitude of issues in concurrent computing, e.g. modelling computer networks, cellular/molecular/chemical networks, and a wide class of problems related to them. This success raises the necessity to define query languages able to express complex properties of systems and, eventually, to develop model-verification techniques. The dual nature of these calculi - algebraical/equational syntax versus coalgebraical operational semantics, makes them appropriate for a modal logic-based approach.

---

[1] In this paradigm, the processes are understood as spatially localised and independently observable units of behaviour and computation (e.g. programs or processors running in parallel).

E. Ochmański and J. Tyszkiewicz (Eds.): MFCS 2008, LNCS 5162, pp. 491–502, 2008.

In this context were proposed the process semantics for modal logics, that can be considered as a special case of Kripke semantics: it involves structuring a class of processes as a Kripke model, by endowing it with accessibility relations and then using the standard clauses of Kripke semantics. The most obvious accessibility relations on processes are the ones induced by action transitions $\alpha.P \xrightarrow{\alpha} P$, and thus the corresponding (Hennessy-Milner) logic [13] was the first process-based modal logic to be developed. Later, temporal [21], mobile or concurrent features were added [10,18]. A relatively new type of process logics are *spatial logics* [8,3], which are particularly tailored for capturing spatial and concurrent properties of processes. Among the various spatial operators we mention: the *parallel operator*[2] $\phi|\psi$ and its adjoint - the *guarantee operator* $\phi \triangleright \psi$; the *location operators* characterize ambient logic[3] [8]; for semantics based on calculi with name passing and name restrictions other specific operators have been proposed, e.g. *placement*, *revelation* and *hiding* operators etc [3]. In addition, most of these logics include transition-based modalities and quantifiers.

The modal operators of spatial logics are similar to modal operators studied in other contexts. The parallel operator, for instance, is just a modal operator of arity 3 that satisfies the axioms of associativity, commutativity and modal distribution, as will be proved latter. Operators such as this have been studied, e.g., in the context of *Arrow Logic* [1] where it entails undecidability for Kripke semantics, as proved in [11]. The parallel operator and the guarantee operator of spatial logics are similar to two operators used in *Relevant* and *Substructural Logics* [22] - the *intentional conjunction* and *relevant implication* respectively. But, as in the case of Arrow Logic, Relevant Logic has a semantics in terms of Kripke structures. Consequently, not many known results can be projected over the process semantics. Some spatial logics are using dynamic operators [12] for expressing the transitions. There are also other relations between spatial logics and well studied modal logics[4].

On the other hand, there are many peculiarities of spatial logics that make them interesting from a modal perspective. For example, the spatial logic we study in this paper allows us to define characteristic formulas for processes. Such a formula identifies a process up to structural congruence, i.e. we have formulas f_P that names a particular state P of the system, thus giving to the logic the expressivity of Hybrid Logics [19]. Another peculiarity is that we can define a universal modality $\circ\phi$ and thus, we can express syntactically meta properties such as validity and satisfiability of a formula. The guarantee operator can be used to translate any satisfiability/validity problem of spatial logic into a model checking problem for the null process, as $\models \phi$ can be proved equivalent with $0 \models \top \triangleright \phi$, [9]. In this way, decidability of satisfiability and validity is directly related with the decidability of model checking. All these peculiarities of spatial

---

[2] A process P has the property $\phi|\psi$, if it can be split into two disjoint parts $P \equiv Q|R$ s.t. Q satisfies ϕ and R satisfies ψ.

[3] Ambient logic is a spatial logic defined over ambient calculus.

[4] See e.g. [8] for a detailed description of the connection between Ambient logic and Linear Logic.

logics emerge mainly from the structure of their models, which are not just labelled graphs, but processes with a structure bound by the rigid rules of the operational semantics of process calculi.

The challenge we take in this paper is to find a sound and complete Hilbert-style axiomatic system for spatial logic that will reveal the nature of the spatial operators, as well as the interrelation between them and the dynamic or classical operators. The axioms we propose are sometimes similar with the axioms of the related modal logics and these similarities are useful in placing the spatial logics in the general context of modal logics. To the best of our knowledge, the problem of completeness for this class of logics has not been approached in the literature, even if the problem of defining sound sequence calculi for them has been considered [6,8,4]. Related to static ambient logic, for instance, there exists a sound-complete sequent calculus [6], but its syntax differs from the syntax of ambient logics. It is done for atomic construction of type $P : \phi$ for a process P and a logic formula ϕ, that encodes the satisfiability relation $P \models \phi$ of ambient logic; the sequent rules just rewrite the semantics of ambient logic. In this context, the soundness and completeness are proved as $P \models \phi$ iff $\vdash P : \phi$, result that does not clarify the axiomatics of spatial logics, the syntactic behavior of the spatial operators, or the relation with other logics. Our previous work [14,15] present some completeness results from a modal perspective, but for only for epistemic versions of spatial logics without the guarantee operator.

A second achievement of the paper is a decidability result that is essential in the completeness proof. The particular spatial logic studied in this paper (that extends the Hennessy-Milner logic with the parallel and guarantee operators) is proved decidable for both satisfiability/validity and model checking against a fragment of CCS calculus that embodies the core features of finite concurrent behaviors. The decidability proof goes on the lines of decidability proofs in [7,6] and consist in proving the bound model property for the logic. As for the semantics, the same fragment of CCS yields undecidability for other spatial logics, e.g. with a modality encoding communication-based transitions [5].

2   Preliminaries on Process Algebra

In this section we recall a number of basic notions of process algebra, mainly to establish some basic terminology and notations for this paper. We introduce a fragment of CCS calculus that will be latter used as semantics for the logic. The novelty of the section is the *structural bisimulation*, a special relation on processes that will be latter used for proving the bounded model property for the spatial logic.

Definition 1 (CCS processes). *Let Σ be a denumerable set of elements called actions and $0 \notin \Sigma$ a special object called the* null *process. The* class of CCS processes *is introduced inductively, for arbitrary $\alpha \in \Sigma$, as follows.*
    $P := 0 \mid \alpha.P \mid P|P$

We denote by $\mathbb{P}$ the class of CCS processes.

Definition 2 (Structural congruence). *The structural congruence is the smallest congruence relation* $\equiv \subseteq \mathbb{P} \times \mathbb{P}$ *such that* $(\mathbb{P}, |, 0)$ *is an abelian monoid with respect to* $\equiv$*, i.e.*

1. $(P|Q)|R \equiv P|(Q|R)$ 2. $P|0 \equiv 0|P \equiv P$ 3. $P|Q \equiv Q|P$

Definition 3 (Operational semantics). *Let* $\tau \notin \Sigma \cup \mathbb{P}$ *and consider a function on* Σ *that associates to each* $\alpha \in \Sigma$ *its* complementary action $\overline{\alpha}$*, such that* $\overline{\overline{\alpha}} = \alpha$*. The operational semantics on* $\mathbb{P}$ *defines a labeled transition system* $\mathbb{T} : \mathbb{P} \to (\Sigma \cup \{\tau\}) \times \mathbb{P}$ *by means of the rules in Table 1, where* $\mathbb{T}(P) = (\alpha, Q)$ *is denoted by* $P \xrightarrow{\alpha} Q$ *for any* $\alpha \in \Sigma$*,* $\mathbb{T}(P) = (\tau, Q)$ *is denoted by* $P \xrightarrow{\tau} Q$*, and* μ *is used to denote arbitrary elements in* $\Sigma \cup \{\tau\}$*.*

Table 1. The transition system

$$\frac{}{\alpha.P \xrightarrow{\alpha} P}, \alpha \in \Sigma \qquad\qquad \frac{}{\alpha.P|\overline{\alpha}.Q \xrightarrow{\tau} P|Q}, \alpha \in \Sigma$$

$$\frac{P \equiv Q \quad P \xrightarrow{\mu} P'}{Q \xrightarrow{\mu} P'}, \mu \in \Sigma \cup \{\tau\} \qquad\qquad \frac{P \xrightarrow{\mu} P'}{P|Q \xrightarrow{\mu} P'|Q}, \mu \in \Sigma \cup \{\tau\}$$

Hereafter, we call a process P *guarded* if $P \equiv \alpha.Q$ for some $\alpha \in \Sigma$ and we use the notation $P^k \stackrel{def}{=} \underbrace{P|...|P}_{k}$ for $k \leq 1$.

Definition 4. *The set of actions* $Act(P) \subset \Sigma$ *of an arbitrary process* $P \in \mathbb{P}$ *is defined, inductively, as follows.*

1. $Act(0) \stackrel{def}{=} \emptyset$ 2. $Act(\alpha.P) \stackrel{def}{=} \{\alpha\} \cup Act(P)$ 3. $Act(P|Q) \stackrel{def}{=} Act(P) \cup Act(Q)$.

For a set $\Omega \subseteq \Sigma$ and a pair h, w of nonnegative integers we define the class $\mathbb{P}^{\Omega}_{(h,w)}$ of processes having the actions from Ω and the syntactic trees bound by two dimensions - the *depth* h of the tree and the width w that represents the maximum number of congruent processes that can be found in a node of the tree. $\mathbb{P}^{\Omega}_{(h,w)}$ is introduced inductively on h.

$\mathbb{P}^{\Omega}_{(0,w)} = \{0\}$;

$\mathbb{P}^{\Omega}_{(h+1,w)} = \{(\alpha_1.P_1)^{k_1}|...|(\alpha_i.P_i)^{k_i}, \text{ for } k_j \leq w, \alpha_j \in \Omega, P_j \in \mathbb{P}^{\Omega}_{(h,w)}, \forall j = 1..i\}$.

Lemma 1. *If* $\Omega \subseteq \Sigma$ *is a finite set, then* $\mathbb{P}^{\Omega}_{(h,w)}$ *is a finite set of processes.*

2.1 Structural Bisimulations

In this subsection we introduce the *structural bisimulation*, a relation on processes indexed by a subclass $\Omega \subseteq \Sigma$ of actions and by two nonnegative integers h, w. This relation is similar to the pruning relation proposed for trees (static ambients) in [6]. Intuitively, two processes are Ω-structural bisimilar on size (h, w) if they look indistinguishable for an external observer that sees only the actions in Ω, does not following a process for more than h transition steps and cannot distinguish more than w cloned subprocesses of a process.

Definition 5 (Ω-Structural Bisimulation). *Let $\Omega \subseteq \Sigma$ and h, w two non-negative integers. The Ω-structural bisimulation on $\mathbb{P}$ is denoted by $\approx^{\Omega}_{(h,w)}$ and is defined inductively as follows.*
If $P \equiv Q \equiv 0$, then $P \approx^{\Omega}_{(h,w)} Q$;
If $P \not\equiv 0$ and $Q \not\equiv 0$, then
  $P \approx^{\Omega}_{(0,w)} Q$ always.
  $P \approx^{\Omega}_{(h+1,w)} Q$ iff for any $i \in 1..w$ and any $\alpha \in \Omega$:

  $-$ $P \equiv \alpha.P_1|...|\alpha.P_i|P'$ implies $Q \equiv \alpha.Q_1|...|\alpha.Q_i|Q'$, $P_j \approx^{\Omega}_{(h,w)} Q_j$, $j = 1..i$;
  $-$ $Q \equiv \alpha.Q_1|...|\alpha.Q_i|Q'$ implies $P \equiv \alpha.P_1|...|\alpha.P_i|P'$, $Q_j \approx^{\Omega}_{(h,w)} P_j$, $j = 1..i$.

Hereafter we present some results about Ω-structural bisimulation.

Lemma 2 (Equivalence). *For a set $\Omega \subseteq \Sigma$ and nonnegative integers h, w, $\approx^{\Omega}_{(h,w)}$ is an equivalence relations on $\mathbb{P}$.*

Lemma 3 (Congruence). *Let $\Omega \subseteq \Sigma$ be a set of actions.*
1. If $P \approx^{\Omega}_{(h,w)} Q$, then $\alpha.P \approx^{\Omega}_{(h+1,w)} \alpha.Q$.
2. If $P \approx^{\Omega}_{(h,w)} P'$ and $Q \approx^{\Omega}_{(h,w)} Q'$, then $P|Q \approx^{\Omega}_{(h,w)} P'|Q'$.

For nonnegative integers h, h', w, w' we convey to write $(h', w') \leq (h, w)$ iff $h' \leq h$ and $w' \leq w$.

Lemma 4. *Let $\Omega' \subseteq \Omega \subseteq \Sigma$ and $(h', w') \leq (h, w)$. If $P \approx^{\Omega}_{(h,w)} Q$, then $P \approx^{\Omega'}_{(h',w')} Q$.*

Lemma 5 (Split). *If $P'|P'' \approx^{\Omega}_{(h,w_1+w_2)} Q$ for some $\Omega \subseteq \Sigma$, then there exists $Q, Q' \in \mathbb{P}$ such that $Q \equiv Q'|Q''$ and $P' \approx^{\Omega}_{(h,w_1)} Q'$, $P'' \approx^{\Omega}_{(h,w_2)} Q''$.*

Lemma 6 (Step-wise propagation). *If $P \approx^{\Omega}_{(h,w)} Q$ and $P \xrightarrow{\alpha} P'$ for some $\alpha \in \Omega \subseteq \Sigma$, then there exists a transition $Q \xrightarrow{\alpha} Q'$ such that $P' \approx^{\Omega}_{(h-1,w-1)} Q'$.*

As Σ is a denumerable set, assume a lexicographic order $\ll \subseteq \Sigma \times \Sigma$ on it. Then, any element $\alpha \in \Sigma$ has a successor denoted by $succ(\alpha)$ and any finite subset $\Omega \subset \Sigma$ has a maximum element denoted by $sup(\Omega)$. We define $\Omega^+ = \Omega \cup \{succ(sup(\Omega))\}$.

All the previous results can be used to prove the next theorem. It states that for any finite set Ω of actions and any nonnegative integers h, w, the equivalence relation $\approx^{\Omega}_{(h,w)}$ divides $\mathbb{P}$ in equivalence classes such that each equivalence class has a representative in the set $\mathbb{P}^{\Omega^+}_{(h,w)}$. This set, by Lemma1, is finite. This observation will be the key for proving, latter, the bounded model property.

Lemma 7 (Pruning Theorem). *For any finite set $\Omega \subseteq \Sigma$, any nonnegative integers h, w and any process $P \in \mathbb{P}$, there exists a process $Q \in \mathbb{P}^{\Omega^+}_{(h,w)}$ such that $P \approx^{\Omega}_{(h,w)} Q$.*

3 Spatial Logic

In this section we introduce the spatial logic SL that contains only one atomic proposition[5] 0, a class of dynamic operators $\langle \alpha \rangle$ indexed by a denumerable set $\Sigma \ni \alpha$, the parallel operator and its adjoint together with the Boolean operators.

Definition 6 (Syntax of Spatial Logics). *Let Σ be a denumerable alphabet. The class $\mathcal{L}$ of well formed formulas of SL is introduced inductively as follows.*

$$\phi := 0 \mid \neg\phi \mid \phi \wedge \phi \mid \langle \alpha \rangle \phi \mid \phi|\phi \mid \phi \triangleright \phi.$$

Definition 7 (Semantics of SL). *The semantics of SL is given by the satisfiability operator, $P \models \phi$ that relates a process $P \in \mathbb{P}$ with the formula $\phi \in \mathcal{L}$, inductively by.*

$P \models 0$ *iff* $P \equiv 0$.
$P \models \neg\phi$ *iff* $P \not\models \phi$.
$P \models \phi \wedge \psi$ *iff* $P \models \phi$ *and* $P \models \psi$.
$P \models \langle \alpha \rangle \phi$ *iff there exists a transition* $P \xrightarrow{\alpha} P'$ *and* $P' \models \phi$.
$P \models \phi|\psi$ *iff* $P \equiv Q|R$, $Q \models \phi$ *and* $R \models \psi$.
$P \models \phi \triangleright \psi$ *iff for any* Q, $Q \models \phi$ *implies* $P|Q \models \psi$.

For arbitrary $\phi, \psi \in \mathcal{L}$ and $\alpha \in \Sigma$ we introduce some derived operators[6].

$$\top \stackrel{def}{=} 0 \vee \neg 0 \qquad\qquad \bot \stackrel{def}{=} \neg\top \qquad\qquad \phi \parallel \psi \stackrel{def}{=} \neg(\neg\phi|\neg\psi)$$

$$\circ\phi \stackrel{def}{=} (\neg\phi) \triangleright \bot \qquad\qquad 1 \stackrel{def}{=} \neg 0 \wedge (0 \parallel 0) \qquad\qquad \alpha.\phi \stackrel{def}{=} 1 \wedge \langle \alpha \rangle \phi$$

$$\bullet\phi \stackrel{def}{=} \neg(\circ\neg\phi)$$

The derived operators can be characterized semantically by:

$P \models \top$ *always*.
$P \models \bot$ *never*.
$P \models \phi \parallel \psi$ *iff* $P \equiv P_1|P_2$, *then either* $P_i, v \models \phi$ *or* $P_j, v \models \psi$, $\{i,j\} = \{1,2\}$.
$P \models \circ\phi$ *iff for any process* Q, $Q \models \phi$.
$P \models \bullet\phi$ *iff there exists a process* Q, $Q \models \phi$.
$P \models 1$ *iff there exists* $\alpha \in \Sigma$ *and* $P \equiv \alpha.Q$.
$P \models \alpha.\phi$ *iff there exists* $\alpha \in \Sigma$ *s.t.* $P \equiv \alpha.P'$ *and* $P' \models \phi$.

Notice, from the semantics, that $\circ$ is a universal modality as the satisfiability of $\circ\phi$ is equivalent with the validity of ϕ, while $\bullet$ is its dual.

Definition 8. *A formula $\phi \in \mathcal{L}$ is satisfiable if there exists a process $P \in \mathbb{P}$ such that $P \models \phi$. A formula $\phi \in \mathcal{L}$ is valid (a validity), denoted by $\models \phi$, if for any process $P \in \mathbb{P}$, $P \models \phi$.*

---

[5] In spatial logics the symbol 0 it is used both in syntax for representing the atomic proposition and in semantics to represent the null process in CCS.

[6] We also assume all the boolean operators.

4   Decidability of *SL*

In what follows we show that satisfiability, validity and model checking are decidable for *SL* against process semantics. The proof is based on the bounded model property technique which consists in showing that, given a formula $\phi \in \mathcal{L}$, we can identify a finite class of processes bound by the dimension of the formula, $\mathbb{P}_\phi$ such that if ϕ has a model in $\mathbb{P}$, then it has a model in $\mathbb{P}_\phi$. Thus, the satisfiability problem in $\mathbb{P}$ is equivalent with the satisfiability in $\mathbb{P}_\phi$. This result can be further used to prove the decidability of satisfiability. Indeed, as $\mathbb{P}_\phi$ is finite, checking the satisfiability of a formula can be done by investigating, one by one, all the processes in $\mathbb{P}_\phi$.

Definition 9 (Size of a formula). *The sizes of a formula of $\mathcal{L}$, denoted by $(\!|\phi|\!) = (h, w)$, is defined inductively on the structure of a formula. In what follows, suppose that $(\!|\phi|\!) = (h, w)$ and $(\!|\psi|\!) = (h', w')$.*

1. $(\!|0|\!) \overset{def}{=} (1, 1)$.

2. $(\!|\neg\phi|\!) \overset{def}{=} (\!|\phi|\!)$.

3. $(\!|\phi \wedge \psi|\!) \overset{def}{=} (max(h, h'), max(w, w'))$.

4. $(\!|\langle\alpha\rangle\phi|\!) \overset{def}{=} (h + 1, w + 1)$.

5. $(\!|\phi \triangleright \psi|\!) \overset{def}{=} (max(h, h'), w + w')$.

6. $(\!|\phi|\psi|\!) \overset{def}{=} (max(h, h'), w + w')$.

Definition 10. *The set of actions of a formula ϕ, $act(\phi) \subseteq \Sigma$ is given by:*

1. $act(0) \overset{def}{=} \emptyset$

2. $act(\neg\phi) = act(\phi)$

3. $act(\phi \wedge \psi) \overset{def}{=} act(\phi) \cup act(\psi)$

4. $act(\langle\alpha\rangle\phi) \overset{def}{=} \{\alpha\} \cup act(\phi)$

5. $act(\phi \triangleright \psi) \overset{def}{=} act(\phi) \cup act(\psi)$

6. $act(\phi|\psi) \overset{def}{=} act(\phi) \cup act(\psi)$

The next Lemma states that a formula $\phi \in \mathcal{L}$ expresses a property of a process P up to $\approx_{(\!|\phi|\!)}^{act(\phi)}$. This means that ϕ expresses a property that involves only its actions and is bounded by its size.

Lemma 8. *If $P \approx_{(\!|\phi|\!)}^{act(\phi)} Q$, then $P \models \phi$ iff $Q \models \phi$.*

This result guarantees the bounded model property.

Theorem 1 (Bound model property). *If $P \models \phi$, then there exists $Q \in \mathbb{P}_{(\!|\phi|\!)}^{act(\phi)^+}$ such that $Q \models \phi$.*

Proof. The result is a direct consequence of Lemma 7 and Lemma 8.

Theorem 2 (Decidability). *For SL validity, satisfiability and model checking are decidable against process semantics.*

Proof. The decidability of satisfiability derives from the bounded model property. Indeed, if ϕ has a model, by Lemma1, it has a model in $\mathbb{P}_{(\!|\phi|\!)}^{act(\phi)^+}$. As $act(\phi)$ is finite, by Lemma 1, $\mathbb{P}_{(\!|\phi|\!)}^{act(\phi)^+}$ is finite, hence checking for membership is decidable.

The decidability of validity derives from the fact that ϕ is valid iff $\neg\phi$ is not satisfiable.

5   Characteristic Formulas

In this section we use the peculiarities of $\mathcal{L}$ to define characteristic formulas for processes. Consider the subclass $\overline{\mathcal{F}} \subseteq \mathcal{L}$ of well formed formulas of SL given, for arbitrary $\alpha \in \Sigma$ by $f := 0 \mid \alpha.f \mid f|f$. Let $* : \overline{\mathcal{F}} \to \overline{\mathcal{F}}$ be the function defined by:
$0^* = 0$; $(\alpha.f)^* = \alpha.f^*$; $(f|0)^* = f^*$; $(f_1|f_2)^* = f_1^*|f_2^*$, for $f_1 \neq 0 \neq f_2$.
Denote by $\mathcal{F} \subseteq \overline{\mathcal{F}}$ the set of fixed points of function $*$ called *proper formulas*, i.e., the set of formulas $f \in \overline{\mathcal{F}}$ s.t. $f^* = f$. For arbitrary positive integers h, w and arbitrary $S \subseteq \Sigma$, let

$$\mathcal{F}_{(h,w)}^S = \{f \in \mathcal{F} \mid (\!(f)\!) \leq (h, w), act(f) \subseteq S\}.$$

Observe that $\mathcal{F} \subseteq \mathcal{L}$ and for a finite set $S \subseteq \Sigma$, $\mathcal{F}_{(h,w)}^S$ is finite. In what follows, we use Greek letters (sometime with indexes) ϕ, ψ, ϕ_1, etc. to denote arbitrary formulas of $\mathcal{L}$ and f, f', f'', f_1, f_2, etc. to denote arbitrary proper formulas of $\mathcal{F}$.

The next Lemma proves that the $\equiv$-equivalence classes of $\mathbb{P}$ can be characterized by formulas of $\mathcal{F}$. For this reason, in what follows, we will use sometime the notation f_P to denote a proper formula $f \in \mathcal{F}$ that characterizes the $\equiv$-equivalence class of $P \in \mathbb{P}$.

Lemma 9. *1. Let $f \in \mathcal{F}$, $P, Q \in \mathbb{P}$. Then $P \models f$ and $Q \models f$, iff $P \equiv Q$.*
2. For any $P \in \mathbb{P}$ there exists $f \in \mathcal{F}$ such that $P \models f$.
3. For any $f \in \mathcal{F}$ there exists $P \in \mathbb{P}$ such that $P \models f$.

Proof. The function $[\,] : \mathcal{F} \to \mathbb{P}$ given by the next rules defines the relation between the formulas in $\mathcal{F}$ and the $\equiv$-equivalence classes in $\mathbb{P}$.
   $[0] = 0$; $[\alpha.f] = \alpha.[f]$; $[f_1|f_2] = [f_1]|[f_2]$.

6   A Hilbert-Style Axiomatic System of SL

In table 2 is proposed a Hilbert-style axiomatic system for SL. We assume the axioms and the rules of propositional logic. In addition we have axioms and rules that characterize the spatial and dynamic operators and their interrelations. Recall that we use Greek letters to specify arbitrary formulas of $\mathcal{L}$ and f, f_1, f_2 to specify arbitrary proper formulas (of $\mathcal{F}$).

Due to the way the proper formulas are defined, the axioms $(S1) - (S4)$ guarantees that for any formula $f \in \mathcal{F}$ the set $\{(f', f'') \in \mathcal{F} \times \mathcal{F} \mid \vdash f \leftrightarrow f'|f''\}$ is finite. This proves that the disjunction in axiom $(S6)$ is finitary.

Observe that the rules $(GR1)$ and $(GR2)$ depicts the adjunction between the two spatial operators $|$ and $\rhd$.

The condition $\alpha.f, f|f' \in \mathcal{F}_{(\!(\phi)\!)}^{act(\phi)^+}$ reflects the finite model property and guarantees that (Ind) can be based on a finite number of premises.

Definition 11. *A formula $\phi \in \mathcal{L}$ is provable in SL, denoted by $\vdash \phi$ if ϕ is an axiom or it can be derived, as a theorem, from the axioms of SL using the rules of SL. A formula $\phi \in \mathcal{L}$ is consistent in SL if $\neg\phi$ is not provable in SL.*

Table 2. The axiomatic system of SL

Spatial axioms
(S1): $\vdash (\phi|\psi)|\rho \rightarrow \phi|(\psi|\rho)$
(S2): $\vdash \phi|0 \leftrightarrow \phi$
(S3): $\vdash \phi|\psi \rightarrow \psi|\phi$
(S4): $\vdash \top|\bot \rightarrow \bot$
(S5): $\vdash \phi|(\psi \vee \rho) \rightarrow (\phi|\psi) \vee (\phi|\rho)$
(S6): $\vdash (f \wedge \phi|\psi) \rightarrow \bigvee_{f \leftrightarrow f'|f''}(f' \wedge \phi)|(f'' \wedge \psi)$

Spatial rules

(SR1): If $\vdash \phi \rightarrow \psi$ then $\vdash \phi|\rho \rightarrow \psi|\rho$

Dynamic axioms
(D1): $\vdash \langle \alpha \rangle \phi|\psi \rightarrow \langle \alpha \rangle (\phi|\psi)$
(D2): $\vdash [\alpha](\phi \rightarrow \psi) \rightarrow ([\alpha]\phi \rightarrow [\alpha]\psi)$
(D3): $\vdash 0 \vee \alpha.\top \rightarrow [\beta]\bot$, for $\alpha \neq \beta$
(D4): $\vdash \alpha.\phi \rightarrow [\alpha]\phi$

Dynamic rules

(DR1): If $\vdash \phi$ then $\vdash [\alpha]\phi$
(DR2): If $\vdash \phi_1 \rightarrow [\alpha]\phi_1'$ and $\vdash \phi_2 \rightarrow [\alpha]\phi_2'$
  $then \vdash \phi_1|\phi_2 \rightarrow [\alpha](\phi_1'|\phi_2 \vee \phi_1|\phi_2')$

Guarantee axiom

(G1): $\vdash \circ(f \rightarrow \phi) \rightarrow \bullet\phi$

Guarantee rules

(GR1): $\vdash \phi_1 \rightarrow (\phi_2 \triangleright \psi)$ iff $\vdash \phi_1|\phi_2 \rightarrow \psi$
(GR2): $\vdash \phi_1 \rightarrow \neg(\phi_2 \triangleright \psi)$ iff $\vdash \bullet(\phi_1|\phi_2 \wedge \neg\psi)$

Induction rule
(Ind): If for any $\alpha.f, f|f' \in \mathcal{F}_{(\!(\phi)\!)}^{act(\phi)^+}$
  $\vdash 0 \rightarrow \phi$
  $\vdash \circ(f \rightarrow \phi) \rightarrow \circ(\alpha.f \rightarrow \phi)$
  $\vdash (\circ(f \rightarrow \phi) \wedge \circ(f' \rightarrow \phi)) \rightarrow \circ(f|f' \rightarrow \phi)$
  then $\vdash \phi$

All the axioms and the rules of our axiomatic system depict true facts about processes. This is proved by the next soundness theorem.

Theorem 3 (Soundness). *The axiomatic system of SL is sound with respect to the process semantics, i.e. if $\vdash \phi$ then $\models \phi$.*

Before continuing with the completeness proof, we list some theorems of SL that will be useful further. Recall that, in what follows, we denote by $f_P \in \mathcal{F}$ any proper formula that characterizes the process P.

Lemma 10 (Spatial corollaries). *The next assertions are theorems of SL.*
1. $\vdash \phi|(\psi \wedge \rho) \rightarrow (\phi|\psi) \wedge (\phi|\rho)$
2. *If $\vdash \phi \rightarrow \rho$ and $\vdash \psi \rightarrow \theta$, then $\vdash \phi|\psi \rightarrow \rho|\theta$.*
3. *If $P \not\equiv Q$, then $\vdash f_P \rightarrow \neg f_Q$.*
4. *If for any Q, R s.t. $P \equiv Q|R$, $\vdash f_Q \rightarrow \neg\phi$ or $\vdash f_R \rightarrow \neg\psi$, then $\vdash f_P \rightarrow \neg(\phi|\psi)$.*

Lemma 11 (Dynamic corollaries). *The next assertions are theorems of SL.*
1. *If $\vdash \phi \rightarrow \psi$, then $\vdash \langle \alpha \rangle \phi \rightarrow \langle \alpha \rangle \psi$.*
2. *If $\vdash \phi \rightarrow \psi$, then $\vdash [\alpha]\neg\psi \rightarrow [\alpha]\neg\phi$.*
3. $\vdash f_P \rightarrow [\alpha]\bigvee\{f_Q \mid P \xrightarrow{\alpha} Q\}$.
4. *If $\vdash \bigvee\{f_Q \mid P \xrightarrow{\alpha} Q\} \rightarrow \phi$, then $\vdash f_P \rightarrow [\alpha]\phi$.*

Lemma 12 (Guarantee corollary). *The next assertions are SL-theorems.*

1. *If* $\vdash \bigvee_{f \in \mathcal{F}^{act(\phi)+}_{(\![\phi]\!)}} f \to \phi$, *then* $\vdash \phi$.

2. *If* $\vdash \phi$, *then* $\vdash \circ\phi$.

Now we approach the completeness problem. We begin with the next lemma stating that a process P satisfies a property ϕ iff its characteristic formula f_P implies the property ϕ and this implication is a theorem in SL system.

Lemma 13. *If $P \in \mathbb{P}$ and $f_P \in \mathcal{F}$ characterizes P, then $P \models \phi$ iff $\vdash f_P \to \phi$.*

Proof. ($\Longrightarrow$:) If $P \models \phi$, then $\vdash f_P \to \phi$. We prove it by induction on the syntactical structure of ϕ. We show here only the cases that require a more complex analysis.

The case $\phi = \phi_1|\phi_2$: $P \models \phi$ iff $P \equiv Q|R$, $Q \models \phi_1$ and $R \models \phi_2$. Using the inductive hypothesis, $\vdash f_Q \to \phi_1$ and $\vdash f_R \to \phi_2$. The case 2 of Lemma 10 implies further $\vdash f_Q|f_R \to \phi_1|\phi_2$, i.e. $\vdash f_P \to \phi$.

The case $\phi = \psi \triangleright \rho$: $P \models \psi \triangleright \rho$ iff for any process Q, $Q \models \psi$ implies $P|Q \models \rho$. The inductive hypothesis gives that for any Q, $\vdash f_Q \to \psi$ implies $\vdash f_P|f_Q \to \rho$. But Rule $(GR1)$ gives the equivalence of $\vdash f_P|f_Q \to \rho$ and $\vdash f_Q \to (f_P \triangleright \rho)$. Hence, for any Q, $\vdash f_Q \to (\phi \to f_P \triangleright \rho)$. Then, for any Q with $f_Q \in \mathcal{F}^{act(\phi \to f_P \triangleright \rho)+}_{(\![\phi \to f_P \triangleright \rho]\!)}$, $\vdash f_Q \to (\phi \to f_P \triangleright \rho)$. Hence, $\vdash \bigvee_{f \in \mathcal{F}^{act(\phi \to f_P \triangleright \rho)+}_{(\![\phi \to f_P \triangleright \rho]\!)}} f \to (\phi \to f_P \triangleright \rho)$ where from, using Lemma 12, $\vdash \phi \to f_P \triangleright \rho$ that is equivalent with $\vdash f_P \to \phi \triangleright \rho$.

The case $\phi = \neg(\psi_1|\psi_2)$: $P \models \neg(\psi_1|\psi_2)$ means that for any parallel decomposition of $P \equiv Q|R$, $Q \models \neg\psi_1$ or $R \models \neg\psi_2$, i.e., $\vdash f_Q \to \neg\psi_1$ or $\vdash f_R \to \neg\psi_2$. Then, the case 4 of Lemma 10 gives $\vdash f_P \to \neg\psi$.

The case $\psi = \neg(\phi_1 \triangleright \phi_2)$: $P \models \neg(\phi_1 \triangleright \phi_2)$ is equivalent with $P \not\models \phi_1 \triangleright \phi_2$. Hence, there exists $Q \models \phi_1$ such that $P|Q \models \neg\phi_2$, i.e., $\vdash f_Q \to \phi_1$ and $\vdash f_P|f_Q \to \neg\phi_2$. Hence, $\vdash f_P|f_Q \to (f_P|\phi_1 \wedge \neg\phi_2)$. Further, Lemma 12 implies $\vdash \circ(f_P|f_Q \to (f_P|\phi_1 \wedge \neg\phi_2))$, Axiom $(G1)$, $\vdash \bullet(f_P|\phi_1 \wedge \neg\phi_2)$ and Rule $(GR2)$, $\vdash f_P \to \neg(\phi_1 \triangleright \phi_2)$.

($\Longleftarrow$) Let $\vdash f_P \to \phi$. Suppose that $P \not\models \phi$. Then, $P \models \neg\phi$. Using the reversed implication we obtain $\vdash f_P \to \neg\phi$, thus, $\vdash f_P \to \bot$. But $P \models f_P$ which, using the soundness, gives $P \models \bot$ impossible! Hence, $P \models \phi$.

Using the result of the previous lemma we can prove that consistency implies satisfiability, as stated in the next lemma.

Lemma 14. *If ϕ is SL-consistent then there exists a process $P \in \mathbb{P}$ such that $P \models \phi$.*

Proof. Suppose that for any process P we do not have $P \models \phi$, i.e., $P \models \neg\phi$. Using Lemma 13, we obtain $\vdash f_P \to \neg\phi$, i.e. $\vdash \circ(f_P \to \neg\phi)$. as this is happening for all processes, implies that for any $f \in \mathcal{F}$ we have $\vdash f \to \neg\phi$, i.e. $\vdash f \to \neg\phi$. But then $\vdash 0 \to \neg\phi$, $\vdash \circ(f \to \neg\phi) \to \circ(\alpha.f \to \neg\phi)$ and $\vdash (\circ(f \to \neg\phi) \wedge \circ(f' \to \neg\phi)) \to \circ(f|f' \to \neg\phi)$. Further, the rule (Ind) gives $\vdash \neg\phi$ wich contradicts the consistency of ϕ.

At this point we have all the results needed to prove the completeness of our axiomatic system.

Theorem 4 (Completeness). *The axiomatic system of SL is complete with respect to process semantics, i.e. if $\models \phi$ then $\vdash \phi$.*

Proof. Suppose that ϕ is a valid formula with respect to our semantics, but ϕ is not provable from our the axiomatic system. Then neither is $\neg\neg\phi$, so, by definition, $\neg\phi$ is SL-consistent. It follows, from Lemma 14, that $\neg\phi$ is satisfiable with respect to process semantics, contradicting the validity of ϕ.

Consequently, the axiomatic system of SL proposed in Table 2 is sound and complete with respect to process semantics. This means that any fact about CCS processes that can be expressed in $\mathcal{L}$ has the properties:

- if it is true, then either it is stated in the axioms or it can be proved from the axioms;
- if it is stated in the axioms or if it can be proved from the axioms, then it true about processes.

These two characteristics of the axiomatic system, the soundness and completeness, present SL as a powerful tool for expressing and analysing properties of CCS processes.

7   Conclusion and Future Works

The achievements of this paper can be summarized as follows. We identified an interesting multi-modal logic, SL, with semantics on CCS calculus able to express dynamic and concurrent properties of distributed systems. The language of SL is expressive enough to characterize the CCS processes up to structural congruence, quality that reveal for SL an expressivity comparable with the expressivity of hybrid logics. In SL we can also define universal modalities that allow us to express meta properties such as validity and satisfiability. In spite of this level of expressivity, we proved the bounded model property for SL against a fragment of CCS for which other spatial logics are undecidable. The bounded model property entails decidability for satisfiability, validity, and model checking.

The main result of the paper is the sound-complete axiomatic system that we propose for SL. Some of the axioms and rules are similar with axioms and rules known from other modal logics, and this peculiarity can help in better understanding the modal face of the concurrency and in placing spatial logics in the general context of modal logics.

References

1. van Benthem, J.: Language in action. Categories, Lambdas and Dynamic Logic. Elsevier Science Publisher, Amsterdam (1991)
2. Bergstra, J.A., Ponse, A., Smolka, S.A. (eds.): Handbook of Process Algebra. North Holland, Elsevier (2001)

3. Caires, L., Cardelli, L.: A Spatial Logic for Concurrency (Part I), Information and Computation, vol. 186(2) (2003)

4. Caires, L., Cardelli, L.: A Spatial Logic for Concurrency (Part II). In: Brim, L., Jančar, P., Křetínský, M., Kucera, A. (eds.) CONCUR 2002. LNCS, vol. 2421. Springer, Heidelberg (2002)

5. Caires, L., Lozes, E.: Elimination of Quantifiers and Decidability in Spatial Logics for Concurrency. In: Gardner, P., Yoshida, N. (eds.) CONCUR 2004. LNCS, vol. 3170. Springer, Heidelberg (2004)

6. Calcagno, C., Cardelli, L., Gordon, A.D.: Deciding validity in a spatial logic for trees. Journal of Functional Programming 15 (2005)

7. Calcagno, C., et al.: Computability and complexity results for a spatial assertion language for data structures. In: Hariharan, R., Mukund, M., Vinay, V. (eds.) FSTTCS 2001. LNCS, vol. 2245. Springer, Heidelberg (2001)

8. Cardelli, L., Gordon, A.D.: Anytime, Anywhere: Modal Logics for Mobile Ambients. In: Proc. 27th ACM Symposium on Principles of Programming Languages (2000)

9. Charatonik, W., Talbot, J.M.: The decidability of model checking mobile ambients. In: Fribourg, L. (ed.) CSL 2001 and EACSL 2001. LNCS, vol. 2142. Springer, Heidelberg (2001)

10. Dam, M.: Model checking mobile processes. Information and Computation 129(1) (1996)

11. Gyuris, V.: Associativity does not imply undecidability without the axiom of Modal Distribution. In: Marx, M., et al. (eds.) Arrow Logic and Multi-Modal Logic, CSLI and FOLLI (1996)

12. Harel, D., et al.: Dynamic Logic. MIT Press, Cambridge (2000)

13. Hennessy, M., Milner, R.: Algebraic laws for Nondeterminism and Concurrency. Journal of J. ACM 32(1) (1985)

14. Mardare, R.: Observing distributed computation. In: Mossakowski, T., Montanari, U., Haveraaen, M. (eds.) CALCO 2007. LNCS, vol. 4624. Springer, Heidelberg (2007)

15. Mardare, R., Priami, C.: Decidable extensions of Hennessy-Milner Logic. In: Najm, E., Pradat-Peyre, J.-F., Donzeau-Gouge, V.V. (eds.) FORTE 2006. LNCS, vol. 4229. Springer, Heidelberg (2006)

16. Mardare, R., Polocriti, A.: Towards a complete axiomatization for Spatial Logics, TechRep. CoSBi, TR-03-2008, www.cosbi.eu

17. Milner, R.: A Calculus of Communicating Systems. Springer, New York (1982)

18. Milner, R., Parrow, J., Walker, D.: Modal logics for mobile processes. TCS 114 (1993)

19. Prior, A.: Past, Present and Future. Clarendon Press, Oxford (1967)

20. Sangiorgi, D.: Extensionality and Intensionality of the Ambient Logics. In: Proc. of the 28th ACM Annual Symposium on Principles of Programming Languages (2001)

21. Stirling, C.: Modal and temporal properties of processes. Springer, New York (2001)

22. Urquhart, A.: Semantics for Relevant Logics. Journal of Symbolic Logic 37(1) (1972)

Voronoi Games on Cycle Graphs[*]

Marios Mavronicolas[1], Burkhard Monien[2], Vicky G. Papadopoulou[1],
and Florian Schoppmann[2,3]

[1] Department of Computer Science, University of Cyprus, Nicosia CY-1678, Cyprus
{mavronic,viki}@ucy.ac.cy
[2] Faculty of Computer Science, Electrical Engineering and Mathematics,
University of Paderborn, Fürstenallee 11, 33102 Paderborn, Germany
{bm,fschopp}@uni-paderborn.de
[3] International Graduate School of Dynamic Intelligent Systems

Abstract. In a Voronoi game, each of a finite number of players chooses
a point in some metric space. The utility of a player is the total measure
of all points that are closer to him than to any other player, where
points equidistant to several players are split up evenly among the closest
players. In a recent paper, Dürr and Thang (2007) considered discrete
Voronoi games on graphs, with a particular focus on pure Nash equilibria.
They also looked at *Voronoi games on cycle graphs* with n nodes and
k players. In this paper, we prove a new characterization of all Nash
equilibria for these games. We then use this result to establish that Nash
equilibria exist if and only if $k \leq \frac{2n}{3}$ or $k \geq n$. Finally, we give exact
bounds of $\frac{9}{4}$ and 1 for the prices of anarchy and stability, respectively.

1 Introduction

1.1 Motivation and Framework

In a *Voronoi game*, there is a finite number of players and an associated metric
measurable space. Each player has to choose a point in the space, and all choices
are made simultaneously. The utility of a player is the measure of all points
that are closer to him than to any other player, plus an even share of the points
that are equidistant (and closest) to him and others. Voronoi games belong to
the huge class of *competitive location* games, which provide models of rivaling
sellers seeking to maximize their market share by strategic positioning in the
market.

The foundations of competitive location were laid by a seminal paper of
Hotelling [6]; he studied two competing merchants in a linear market with con-
sumers spread evenly along the line (also known as the ice-cream vendor problem). Since the unique *Nash equilibrium* of this duopoly is reached when both
merchants are located at the center, Hotelling's results were later described as
the "principle of minimum differentiation" [2]. Recall here that Nash equilibria

---

[*] This work was partially supported by the IST Program of the European Union under
contract number IST-15964 (AEOLUS).

E. Ochmański and J. Tyszkiewicz (Eds.): MFCS 2008, LNCS 5162, pp. 503–514, 2008.
© Springer-Verlag Berlin Heidelberg 2008

are the stable states of the game in which no player can improve his utility by unilaterally switching to a different strategy.

In subsequent works, Hotelling's model was shown to be very sensitive to his original assumptions; in fact, the principle of minimum differentiation cannot even be maintained if just a third player enters the market (see, e.g., [4]). Particularly since the 1970's, a myriad of different competitive location models have been studied. An extensive taxonomy with over 100 bibliographic references can be found in [5]; the authors classify the various models according to (i) the underlying metric measurable space, (ii) the number of players, (iii) the pricing policy (if any), (iv) the equilibrium concept, and (v) customers' behavior. They point out that competitive location "has become one of the truly interdisciplinary fields of study" with interest stemming from economists and geographers, as well as "operations researchers, political scientists and many others" [5].

1.2   Related Work

Eaton and Lipsey [4] studied Voronoi games on a (continuous) circle and observed that its Nash equilibria allow for a very easy characterization ("no firm's whole market is smaller than any other firm's half market"). They defined social cost as the total *transport cost*, i.e., the average distance, over all points of the circle, to the nearest player (i.e., firm). Using this measure, they pointed out that there is always an equilibrium configuration with optimal social cost, whereas the cost-maximizing equilibrium (all n firms are paired, n is even, all pairs are equidistantly located) incurs twice the optimum social cost.

Extending Hotelling's model to graphs has been suggested by Wendell and McKelvey [8]. Yet, to the best of our knowledge, Dürr and Thang [3] were the first to study Nash equilibria of Voronoi games on (undirected) graphs with more than just two players. They established several fundamental results: There is a relatively simple graph that does not allow for a Nash equilibrium even if there are only two players. Even more, deciding the existence of a Nash equilibrium for general graphs and arbitrary many players is NP-hard. Dürr and Thang [3] also introduced the *social cost discrepancy* as the maximum ratio between the social costs of any two Nash equilibria. For connected graphs, they showed an upper bound on the social cost discrepancy of $O(\sqrt{kn})$, and gave a construction scheme for graphs with social cost discrepancy of at least $\Omega(\sqrt{n/k})$. Finally, they considered *Voronoi games on cycle graphs* and gave a characterization of all Nash equilibria. However, it turns out that their characterization is not correct and requires some non-trivial modifications.

1.3   Contribution and Significance

The contribution of this paper and its structure are as follows:

- In Section 2, we prove for Voronoi games on cycle graphs that a strategy profile is a Nash equilibrium if and only if no more than two players have the same strategy, the distance between two strategies is at most twice the minimum utility of any player, and three other technical conditions hold.

We remark that an algebraic characterization of Nash equilibria on cycle graphs was already given in [3, Lemma 2]. However, their result contains mistakes. Fixing these is non-trivial and leads to a different set of conditions.

- In Section 3, we show that a Voronoi game on a cycle graph with n nodes and $k \leq n$ players has a Nash equilibrium if and only if $k \leq \frac{2n}{3}$ or $k = n$. If that condition is fulfilled, then any strategy profile that locates all players equidistantly on the cycle (up to rounding) is a Nash equilibrium.
- In Section 4, we prove that profiles with (almost) equidistantly located players have optimal social cost. Furthermore, no Nash equilibrium has social cost greater than $\frac{9}{4}$ times the optimal cost. If $\frac{1}{2} \cdot \lfloor \frac{2n}{k} \rfloor$ is not an odd integer, then the upper bound improves to 2. To obtain these results, we devise and employ carefully constructed optimization problems so that best and worst Nash equilibria coincide with global minima or maxima, respectively. We give families of Voronoi games on cycle graphs where the aforementioned ratios are attained exactly. Hence, these factors are also exact bounds on the price of anarchy. Clearly, the price of stability is 1.

We believe that our combinatorial constructions and techniques will spawn further interest; we hope they will be applicable to other Voronoi games on graphs. Note that, due to lack of space, we had to omit some of the smaller proofs.

1.4   The Model

Notation. For $n \in \mathbb{N}_0$, let $[n] := \{1, \ldots, n\}$ and $[n]_0 := [n] \cup \{0\}$. Given a vector $\boldsymbol{v}$, we denote its components by $\boldsymbol{v} = (v_1, v_2, \ldots)$. We write $(\boldsymbol{v}_{-i}, v_i')$ to denote the vector equal to $\boldsymbol{v}$ but with the i-th component replaced by v_i'.

Definition 1. *A Voronoi game on a connected undirected graph is specified by a graph $G = (V, E)$ and the number of players $k \in \mathbb{N}$. The strategic game is completed as follows:*

- *The strategy set of each player is V, the set of strategy profiles is $\mathscr{S} := V^k$.*
- *The utility, $u_i : \mathscr{S} \to \mathbb{R}$, of a player $i \in [k]$ is defined as follows: Let the distance, $\mathrm{dist} : V \times V \to \mathbb{N}_0$, be defined such that $\mathrm{dist}(v, w)$ is the length of a shortest path connecting v, w in G. Moreover, for any node $v \in V$, the function $F_v : \mathscr{S} \to 2^{[k]}$, $F_v(\boldsymbol{s}) := \arg\min_{i \in [k]} \{\mathrm{dist}(v, s_i)\}$, maps a strategy profile to the set of players closest to v. Then, $u_i(\boldsymbol{s}) := \sum_{v \in V : i \in F_v(\boldsymbol{s})} \frac{1}{|F_v(\boldsymbol{s})|}$.*

The "quality" of a strategy profile $\boldsymbol{s} \in \mathscr{S}$ is measured by the *social cost*, $\mathrm{SC}(\boldsymbol{s}) := \sum_{v \in V} \min_{i \in [k]} \{\mathrm{dist}(v, s_i)\}$. The *optimum social cost* (or just the *optimum*) associated to a game is $\mathrm{OPT} := \inf_{\boldsymbol{s} \in \mathscr{S}} \{\mathrm{SC}(\boldsymbol{s})\}$.

We are interested in profiles called *Nash equilibria*, where no player has an incentive to unilaterally deviate. That is, $\boldsymbol{s} \in \mathscr{S}$ is a Nash equilibrium if and only if for all $i \in [k]$ and all $s_i' \in V$ it holds that $u_i(\boldsymbol{s}_{-i}, s_i') \leq u_i(\boldsymbol{s})$. If such a profile exists in a game, to what degree can social cost deteriorate due to player's selfish behavior? Several metrics have been proposed to capture this question: The *price of anarchy* [7] is the worst-case ratio between a Nash equilibrium and the optimum, i.e., PoA $= \sup_{\boldsymbol{s} \text{ is NE}} \frac{\text{SC}(\boldsymbol{s})}{\text{OPT}}$. The *price of stability* [1] is the best-case ratio between a Nash equilibrium and the optimum, i.e., PoS $= \inf_{\boldsymbol{s} \text{ is NE}} \frac{\text{SC}(\boldsymbol{s})}{\text{OPT}}$. Finally, the *social cost discrepancy* [3] measures the maximum ratio between worst and best Nash equilibria, i.e., SCD $= \sup_{\boldsymbol{s},\boldsymbol{s}' \text{ are NE}} \frac{\text{SC}(\boldsymbol{s})}{\text{SC}(\boldsymbol{s}')}$. For these ratios, $\frac{0}{0}$ is defined as 1 and, for any $x > 0$, $\frac{x}{0}$ is defined as ∞.

In this paper, we consider *Voronoi games on cycle graphs*. A cycle graph is a graph $G = (V, E)$ where $V = \mathbb{Z}_n$ is the set of congruence classes modulo n, for some $n \in \mathbb{N}$, and $E := \{(x, x + 1) : x \in \mathbb{Z}_n\}$. Clearly, a Voronoi game on a cycle graph is thus fully specified by the number of nodes n and the number of players k. As an abbreviation we use $\mathcal{C}(n, k)$. We will assume $k \leq n$ throughout the rest of this paper as otherwise the games have a trivial structure. (In particular, whenever all nodes are used and the difference in the number of players on any two nodes is at most 1, this profile is a Nash equilibrium with zero social cost.)

We use a representation of strategy profiles that is convenient in the context of cycle graphs and which was also used in [3]. Define the *support* of a strategy profile $\boldsymbol{s} \in \mathscr{S}$ as the set of all chosen strategies, i.e., supp $: \mathscr{S} \to 2^V$, supp$(\boldsymbol{s}) :=$ $\{s_1, \ldots, s_k\}$. Now fix a profile $\boldsymbol{s}$. Then, define $\ell := |\text{supp}(\boldsymbol{s})|$ and $\theta_0 < \cdots < \theta_{\ell-1}$ such that $\{\theta_i\}_{i \in \mathbb{Z}_\ell} = \text{supp}(\boldsymbol{s})$. Now, for $i \in \mathbb{Z}_\ell$:

- Let $d_i := (n + \theta_{i+1} - \theta_i) \bmod n$; so, d_i is the distance from θ_i to θ_{i+1}.
- Let c_i be the number of players on node θ_i.
- Denote by v_i the utility of each player with strategy θ_i.
- Similar to [3], we define $a_i \in \mathbb{N}_0$, $b_i \in \{0, 1\}$ by $d_i - 1 = 2 \cdot a_i + b_i$.

Up to rotation and renumbering of the players, $\boldsymbol{s}$ is uniquely determined by ℓ, $\boldsymbol{d} = (d_i)_{i \in \mathbb{Z}_\ell}$, and $\boldsymbol{c} = (c_i)_{i \in \mathbb{Z}_\ell}$. With the above definitions, the utility of a player with strategy θ_i is obviously

$$v_i = \frac{b_{i-1}}{c_{i-1} + c_i} + \frac{a_{i-1} + 1 + a_i}{c_i} + \frac{b_i}{c_i + c_{i+1}}.$$

Throughout, we use $\mathbb{Z}_\ell$ for indexing; i.e., $c_i = c_{i+\ell}$ and $d_i = d_{i+\ell}$ for all $i \in \mathbb{Z}$. For better readability, we do not reflect the dependency between $\boldsymbol{s}$ and $\ell, \boldsymbol{d}, \boldsymbol{c}, \boldsymbol{a}, \boldsymbol{b}$ in our notation. This should be always clear from the context.

Example 1. Figure 1 shows a flattened segment of a cycle graph to illustrate our notation: On node θ_1, there are $c_1 = 2$ players, the distance from node θ_0 to θ_1 is $d_0 = 6$. Hence, $a_0 = 2$ and $b_0 = 1$, i.e., there is a middle node between θ_0 and θ_1. The players on node θ_1 share the shaded Voronoi area, i.e., $v_1 = \frac{1}{2} \cdot 4\frac{2}{3} = \frac{7}{3}$.

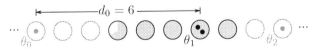

Fig. 1. Illustration of our notation

2   Characterization of Nash Equilibria

In this section, we prove an exact characterization of all Nash equilibria for Voronoi games on a cycle graph with $n \in \mathbb{N}$ nodes and $k \in [n]$ players.

Theorem 1 (Strong characterization). *Consider* $\mathcal{C}(n, k)$ *where* $n \in \mathbb{N}$, $k \in [n]$. *A strategy profile* $s \in \mathscr{S}$ *with minimum utility* $\gamma := \min_{i \in \mathbb{Z}_\ell} \{v_i\}$ *is a Nash equilibrium if and only if the following holds for all* $i \in \mathbb{Z}_\ell$:

S1. $c_i \leq 2$
S2. $d_i \leq 2\gamma$
S3. $c_i \neq c_{i+1} \Longrightarrow \lfloor 2\gamma \rfloor$ *odd*
S4. $c_i = 1, d_{i-1} = d_i = 2\gamma \Longrightarrow 2\gamma$ *odd*
S5. $c_i = c_{i+1} = 1, d_{i-1} + d_i = d_{i+1} = 2\gamma \Longrightarrow 2\gamma$ *odd*
$\quad\;\; c_i = c_{i-1} = 1, d_{i-1} = d_i + d_{i+1} = 2\gamma \Longrightarrow 2\gamma$ *odd*

Condition (S1) requires that no node may be shared by more than two players. Otherwise, by definition of γ, the neighboring strategies would be "far away". However, by condition (S2), the distance between two used nodes must not be too large because any player moving in between them could then achieve a utility larger than γ. Conditions (S3)–(S5) deal with strategies whose neighboring strategies are played by a different number of players. The necessity of these conditions is illustrated in Figure 2 for the case when $\gamma = 1$. Here, a double outline indicates the new (shared) Voronoi area after the respective player moves.
We need the following lemma, the proof of which is given in the full version:

Lemma 1. *If property (S2) of Theorem 1 is fulfilled then* $\forall i \in \mathbb{Z}_\ell, c_i = 2 : d_{i-1} = d_i = \lfloor 2\gamma \rfloor$. *If additionally (S1) and (S3) are fulfilled then also* $2\gamma = \lfloor 2\gamma \rfloor$ *and* $\forall i \in \mathbb{Z}_\ell, c_i = 2 : v_i = \gamma$.

Proof (of Theorem 1). We start with a weak characterization that essentially states the definition of a Nash equilibrium in the context of Voronoi games on cycle graphs. In order to deal with parity issues, we find it convenient to mix in Boolean arithmetic and identify $1 \equiv true$ and $0 \equiv false$. For instance, if $b, b' \in \{0, 1\}$, then $b \leftrightarrow b' = 1$ if $b = b'$, and 0 otherwise. Similarly, $b \vee b' = 1$ if $b = 1$ or $b' = 1$, and 0 otherwise.

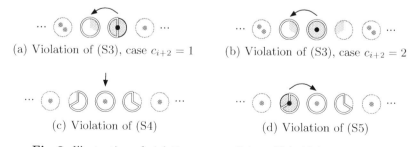

(a) Violation of (S3), case $c_{i+2} = 1$ (b) Violation of (S3), case $c_{i+2} = 2$

(c) Violation of (S4)   (d) Violation of (S5)

Fig. 2. Illustration of violations to conditions (S3)–(S5) when $\gamma = 1$

Claim 1 (Weak characterization). The strategy profile s is a Nash equilibrium if and only if the following holds:

W1. No player alone on a node can improve by moving to a neighboring node not in the support (which would swap parities of the distances to neighboring strategies), i.e., $\forall i \in \mathbb{Z}_\ell, c_i = 1 : (b_{i-1} = b_i = 1 \implies c_{i-1} = c_{i+1} = 1) \wedge (b_{i-1} = 1, b_i = 0 \implies c_{i-1} \le c_{i+1}) \wedge (b_{i-1} = 0, b_i = 1 \implies c_{i-1} \ge c_{i+1})$.

W2. No player can improve by moving to a node that is not in the support (for the cases not covered by (W1)), i.e., $\forall i, j \in \mathbb{Z}_\ell : v_i \ge a_j + \frac{\neg b_j}{\min\{c_{j-1}, c_j\}+1} + b_j$.

W3. No player can improve to an arbitrary non-neighboring strategy, i.e., $\forall i, j \in \mathbb{Z}_\ell, j \notin \{i-1, i+1\} : v_i \ge \frac{b_{j-1}}{c_{j-1}+c_j+1} + \frac{a_{j-1}+1+a_j}{c_j+1} + \frac{b_j}{c_j+c_{j+1}+1}$.

W4. No player sharing a node can improve to a neighboring strategy, i.e., $\forall i \in \mathbb{Z}_\ell, c_i \ge 2 : v_i \ge \frac{b_i}{c_i+c_{i+1}} + \frac{a_i+1+a_{i+1}}{c_{i+1}+1} + \frac{b_{i+1}}{c_{i+1}+c_{i+2}+1}$, with a corresponding inequality for moving to θ_{i-1} instead of θ_{i+1}.

W5. No player alone on a node can improve to a neighboring strategy, i.e., $\forall i \in \mathbb{Z}_\ell, c_i = 1 : v_i \ge \frac{b_{i-1} \leftrightarrow b_i}{c_{i-1}+1+c_{i+1}} + \frac{a_{i-1}+a_i+a_{i+1}+b_{i-1} \vee b_i+1}{c_{i+1}+1} + \frac{b_{i+1}}{c_{i+1}+c_{i+2}+1}$, with a corresponding inequality for moving to θ_{i-1} instead of θ_{i+1}.

Proof (of claim). Conditions (W1)–(W5) are exhaustive. ∎

We now continue by proving necessity ("$\implies$"). Note that (S1) and (S2) have also been stated in [3, Lemma 2 (i), (ii)]. For completeness and since their proof contained mistakes, we reestablish the claims here.

(S1) Assume by way of contradiction that there is some $i \in \mathbb{Z}_\ell$ with $c_i \ge 3$. W.l.o.g., assume $d_i \ge d_{i-1}$, i.e., also $a_i \ge a_{i-1}$ and $(b_{i-1} > b_i \implies a_{i-1} < a_i)$. Since $v_i \ge 1$, it must hold that $a_i \ge 1$. Consider now the move by some player with strategy θ_i to node $\theta_i + 1$.
  Since $\frac{b_{i-1}}{c_{i-1}+c_i} + \frac{a_{i-1}}{c_i} \le \frac{b_i}{c_{i-1}+c_i} + \frac{a_i}{c_i}$ and $2a_i + 1 \le c_i a_i$, his old utility is at most $v_i = \frac{b_{i-1}}{c_{i-1}+c_i} + \frac{a_{i-1}+1+a_i}{c_i} + \frac{b_i}{c_i+c_{i+1}} \le a_i + \frac{b_i}{2}$, whereas his new utility is $v' = a_i + b_i + \frac{\neg b_i}{1+c_{i+1}} > v_i$. This is a contradiction to the profile being a Nash equilibrium.

(S2) We first show that $d_i \le \lfloor 2\gamma \rfloor + 1$: Otherwise, if $d_i \ge \lfloor 2\gamma \rfloor + 2$, then a player with utility γ could move to node $\theta_i + 1$ and thus improve his utility to at least $\lfloor \frac{d_i}{2} \rfloor \ge \lfloor \frac{\lfloor 2\gamma \rfloor}{2} \rfloor + 1 = \lfloor \gamma \rfloor + 1 > \gamma$.
  Now assume $d_i = \lfloor 2\gamma \rfloor + 1$. Then, $c_i = 2$ because otherwise, if $c_i = 1$, a player with utility γ could change his strategy to $\theta_{i+1} - 1$ and thus achieve a new utility of $\frac{d_i}{2} = \frac{\lfloor 2\gamma \rfloor}{2} + \frac{1}{2} > \gamma$. The argument can be repeated correspondingly to obtain $c_{i+1} = 2$. Now note that $v_{i+1} \ge \frac{d_i}{2}$ because this is what a player with strategy θ_{i+1} could otherwise improve to, when moving to $\theta_i + 1$. It follows that $d_{i+1} = \lfloor 2\gamma \rfloor + 1 = d_i$. Inductively, we get for all $j \in \mathbb{Z}_\ell$ that $d_j = d_i$ and $c_j = c_i$. Then, n has to be a multiple of d_i, and for all $j \in \mathbb{Z}_\ell$ it holds that $v_j = \frac{d_j}{2} > \gamma$. Clearly, a contradiction.

(S3) W.l.o.g., assume that $c_i = 2$, $\lfloor 2\gamma \rfloor$ even, and $c_{i+1} = 1$. By Lemma 1, we have that $d_i = \lfloor 2\gamma \rfloor$, so $b_i = 1$. We get a contradiction to condition (W1) of Claim 1 both when $c_{i+2} = 1$ and $c_{i+2} = 2$ (in which case also $b_{i+1} = 1$).

(S4) Assume, by way of contradiction, that $c_i = 1$ and $d_{i-1} = d_i = 2\gamma$ even. Due to (S3) and Lemma 1, it then follows that $c_{i-1} = c_{i+1} = 1$. Moreover, $a_{i-1} = a_i = \gamma - 1$. Hence, a player with utility γ could move to node θ_i and so improve his utility to (at least) $\frac{1}{3} + \frac{a_{i-1}+a_i+1}{2} + \frac{1}{3} = \gamma + \frac{1}{6}$.

(S5) We only show the first implication as the second one is symmetric. Assume, by way of contradiction, that $c_i = c_{i+1} = 1$ and $d_{i-1}+d_i = d_{i+1} = 2\gamma$ even, i.e., $b_{i-1} = b_i$ and $b_{i+1} = 1$. Due to (S3) and Lemma 1, it then follows that $c_{i-1} = c_{i+2} = 1$ and $v_i = \frac{d_{i-1}+d_i}{2} = \gamma$. Moreover, $a_{i-1} + a_i = \gamma - (b_{i-1} \vee b_i) - 1$ and $a_{i+1} = \gamma - 1$. Hence, the player with strategy θ_i could move to θ_{i+1} and so improve his utility to $\frac{1}{3} + \frac{a_{i-1}+a_i+a_{i+1}+b_{i-1}\vee b_i+1}{2} + \frac{1}{3} = \gamma + \frac{1}{6}$.

In the remainder of the proof, we establish that (S1)–(S5) are indeed sufficient ("$\Longleftarrow$"): Clearly, we have to verify all conditions of Claim 1.

(W1) Assume $c_i = 1$. If d_{i-1} and d_i are even, it holds by condition (S3) that $c_{i-1} = c_{i+1} = 1$. Now, if d_{i-1} is even and d_i odd, then (S3) implies $c_{i-1} = 1 \leq c_{i+1}$. Correspondingly, d_{i-1} odd and d_i even implies $c_{i-1} \geq c_{i+1}$.

(W2) Condition (S2) implies that if a player moves to a node that is not in the support, then his new utility is at most $\frac{2\gamma}{2} = \gamma$.

(W3) Due to Lemma 1, a player can only improve to a non-neighboring strategy θ_j if $c_j = 1$ and $d_{j-1} = d_j = 2\gamma$. Then 2γ is odd by (S4), hence $v' = \gamma$.

(W4) The same argument as for (W3) applies.

(W5) Let $i \in \mathbb{Z}_\ell$ and consider the unique player $p \in [n]$ with strategy $s_p = \theta_i$. Let v' be his new utility if he moved to θ_{i+1}. Assume for the moment that $c_{i+1} = 1$. Then, $v_i = a_{i-1} + a_i + \frac{b_{i-1}}{c_{i-1}+1} + \frac{b_i}{2} + 1$ and

$$v' = \frac{a_{i-1}+a_i}{2} + \frac{a_{i+1}}{2} + \frac{b_{i-1} \leftrightarrow b_i}{c_{i-1}+2} + \frac{b_{i-1} \vee b_i}{2} + \frac{b_{i+1}}{2+c_{i+2}} + \frac{1}{2}.$$

We now argue that it is sufficient to show the claim for $c_{i+1} = 1$. Otherwise, if $c_{i+1} = 2$, the old utility of player p would be $v_i - \frac{b_i}{2} + \frac{b_i}{3}$ and his new utility (after moving to θ_{i+1}) would be at most $v' - \frac{1}{2} + \frac{1}{3}$; hence, the gain in utility cannot be larger than in the case $c_{i+1} = 1$.

Since $c_i = 1$, we have $c_{i-1} = 1$ or $b_{i-1} = 0$ due to (S3) and Lemma 1. Now, it is sufficient to consider the case $b_{i-1} = 0$. Otherwise, if $b_{i-1} = 1$, then $c_{i-1} = 1$ and the utility of player p would remain v_i when moving him to $\theta_i - 1$ (to change parity). We have now $v_i = a_{i-1} + a_i + \frac{b_i}{2} + 1$ and

$$v' = \frac{a_{i-1}+a_i}{2} + \frac{a_{i+1}}{2} + \frac{\neg b_i}{c_{i-1}+2} + \frac{b_i}{2} + \frac{b_{i+1}}{2+c_{i+2}} + \frac{1}{2}. \tag{1}$$

Since $b_{i-1} = 0$ and $c_i = c_{i+1} = 1$, it holds that $d_{i-1} + d_i = 2v_i \geq 2\gamma$. Consequently, there are two cases:

- $d_{i-1} + d_i = 2\gamma$

  Due to Lemma 1, the move could only improve p's utility if $d_{i-1}+d_i = d_{i+1} = 2\gamma$. Then 2γ is odd due to condition (S5), so $v' = \gamma = v_i$.

– $d_{i-1} + d_i > 2\gamma$

Since $b_{i-1} = 0$ and $c_{i+1} = 1$, we have $v_i = \frac{d_{i-1}+d_i}{2} \geq \frac{2\gamma+1}{2}$, i.e., $2\gamma \leq 2v_i - 1$. Now condition (S2) implies $d_{i+1} = 2a_{i+1} + b_{i+1} + 1 \leq 2\gamma \leq 2v_i - 1 = 2(a_{i-1} + a_i + \frac{b_i}{2} + \frac{1}{2})$, so $\frac{a_{i+1}}{2} \leq \frac{a_{i-1}+a_i}{2} + \frac{b_i}{4} - \frac{b_{i+1}}{4}$.
Inserting into (1) yields

$$v' \leq a_{i-1} + a_i + \frac{b_i}{4} - \frac{b_{i+1}}{4} + \frac{\neg b_i}{c_{i-1}+2} + \frac{b_i}{2} + \frac{b_{i+1}}{2+c_{i+2}} + \frac{1}{2}$$

$$\leq a_{i-1} + a_i + \frac{3b_i}{4} + \frac{\neg b_i}{3} + \frac{7}{12}.$$

Hence $v' \leq a_{i-1} + a_i + \frac{11}{12} < v_i$ if $b_i = 0$ and $v' \leq a_{i-1} + a_i + \frac{4}{3} < v_i$ if $b_i = 1$.

Due to symmetry, we have hence shown that no player using a node alone may improve by moving to a neighboring strategy. □

3   Existence of Nash Equilibria

In this section, we give a condition for the existence of Nash equilibria in cycle graphs that is both necessary and sufficient. This condition only depends on the ratio between the number of players and the number of nodes in the cycle graph.

Theorem 2. $C(n, k)$ does not have a Nash equilibrium if $\frac{2n}{3} < k < n$.

Proof. By way of contradiction, let $\frac{2n}{3} < k < n$ and assume there is a Nash equilibrium with minimum utility $\gamma := \min_{i \in \mathbb{Z}_\ell}\{v_i\}$. Note that $n \geq 4$ and $k \geq 3$. Clearly, $1 \leq \gamma \leq \frac{n}{k} < \frac{3}{2}$, so Lemma 1 implies $\gamma = 1$. Hence, no two players may have the same strategy as otherwise, by (S3) and Lemma 1, it holds for all $i \in \mathbb{Z}_\ell$ that $c_i = 2$ and $d_i = 2$. This implies $k = n$ (and n even). A contradiction.

Consequently, we have that $\ell = k$ and for all $i \in \mathbb{Z}_\ell$ that $c_i = 1$. Since $k > \frac{2n}{3}$, there has to be some $i \in \mathbb{Z}_\ell$ with $d_{i-1} = d_i = 1$ and $d_{i+1} = 2$. This is a contradiction to (S5), as 2γ is even. Specifically, the player on node θ_i has utility $v_i = 1$, but by switching to strategy θ_{i+1} he could improve to a utility of at least $\frac{1}{3} + \frac{1}{2} + \frac{1}{3} = \frac{7}{6} > 1$. See Figure 2 (d). □

Definition 2. *A strategy profile with distances* $(d_i)_{i \in \mathbb{Z}_\ell}$ *is called* standard *if* $\ell = k$ *and* $\forall i \in \mathbb{Z}_k : d_i \in \{\lfloor \frac{n}{k} \rfloor, \lceil \frac{n}{k} \rceil\}$.

Theorem 3. *If* $k \leq \frac{2n}{3}$ *or* $k = n$, *then* $C(n, k)$ *has a standard strategy profile that is a Nash equilibrium.*

Proof. If $k = n$, then $s = (0, 1, \ldots, n-1)$, i.e., $\ell = n$, $(c_i)_{i \in [n]} = (d_i)_{i \in [n]} = (1, \ldots, 1)$ is trivially a standard Nash equilibrium.

Consider now the case $k < \frac{2n}{3}$. Define $p \in \mathbb{N}_0$, $q \in [k-1]_0$ by $n = p \cdot k + q$. Moreover, define a strategy profile by $\ell = k$, $(c_i)_{i \in \mathbb{Z}_\ell} = (1, \ldots, 1)$, and if

$q \leq \frac{k}{2}$ then $(d_i)_{i \in \mathbb{Z}_\ell} = (\underbrace{p, p+1, p, p+1, \ldots, p, p+1}_{2q \text{ elements}}, p, p, \ldots, p)$ and otherwise

$(d_i)_{i \in \mathbb{Z}_\ell} = (\underbrace{p, p+1, p, p+1, \ldots, p, p+1}_{2(k-q) \text{ elements}}, p+1, p+1, \ldots, p+1)$. Clearly, both

are valid profiles because $\sum_{i \in \mathbb{Z}_\ell} d_i = p \cdot k + q = n$. Moreover, let again $\gamma := \min_{i \in \mathbb{Z}_\ell} \{v_i\}$ be the minimum utility of any player. Then $1 \leq p \leq \gamma < p+1 \leq 2p \leq 2\gamma$, so conditions (S1)–(S3) of Theorem 1 are fulfilled. In order to verify also conditions (S4) and (S5), we show that $p+1 < 2\gamma$: If $\frac{n}{2} < k \leq \frac{2n}{3}$ then $p = 1$ and $q \geq \frac{k}{2}$; so $\gamma = \frac{3}{2}$. Hence, $p + 1 = 2 < 3 = 2\gamma$. Otherwise, if $k \leq \frac{n}{2}$, then $p \geq 2$ and so $p + 1 < 2p \leq 2\gamma$. □

4   Social Cost and the Prices of Anarchy and Stability

In this section, we first show that standard profiles are optimal; hence, if $k \leq \frac{2n}{3}$ or $k = n$, then the price of stability is 1. We then continue by proving that the price of anarchy is at most $\frac{9}{4}$. Furthermore, we give families of Voronoi games on cycle graphs where these ratios are attained exactly.

Consider the following optimization problem on a vector $\boldsymbol{\lambda} \in \mathbb{N}^n$:

$$
\begin{aligned}
\text{Minimize } & \sum_{i=1}^n i \cdot x_i &&(2)\\
\text{subject to } & \sum_{i=1}^n x_i = n \\
& 0 \leq x_i \leq \lambda_i \;\; \forall i \in [n] \\
\text{where } & x_i \in \mathbb{N}_0 \qquad \forall i \in [n]
\end{aligned}
$$

Lemma 2. *Let $\boldsymbol{\lambda} \in \mathbb{N}^n$ and $r := \min\{i \in [n] : \sum_{j=1}^i \lambda_j \geq n\}$. Then, the unique optimal solution of (2) is $\boldsymbol{x}^* := (\lambda_1, \ldots, \lambda_{r-1}, n - \sum_{i=0}^{r-1} \lambda_i, 0, \ldots, 0) \in \mathbb{N}_0^n$.*

Theorem 4. *A standard strategy profile has optimal social cost.*

Proof. Consider the Voronoi game $\mathcal{C}(n, k)$. We first observe the following relationship between the optimization problem (2) on $\boldsymbol{\lambda} := (k, 2k, 2k, \ldots, 2k) \in \mathbb{N}^n$ and profiles with optimal social cost. For any strategy profile $\boldsymbol{s} \in \mathscr{S}$ define $\boldsymbol{x}(\boldsymbol{s}) \in \mathbb{N}_0^n$ by $x_i(\boldsymbol{s}) := |\{u \in V : \min_{j \in [k]}\{\text{dist}(s_j, u)\} = i - 1\}|$. It is easy to see that, for all $\boldsymbol{s} \in \mathscr{S}$, $\boldsymbol{x}(\boldsymbol{s})$ is a feasible solution to optimization problem (2) on vector $\boldsymbol{\lambda}$ and $SC(\boldsymbol{s}) = \sum_{i=1}^n i \cdot x_i(\boldsymbol{s})$. Hence, if $\boldsymbol{x}(\boldsymbol{s})$ is an optimal solution to (2) then $\boldsymbol{s}$ is a profile with optimal social cost.

Now let $\boldsymbol{s} \in \mathscr{S}$ be a standard profile. By definition, $\ell = k$, and for all $i \in [k]$ it holds that $c_i = 1$ and $d_i \in \{\lfloor \frac{n}{k} \rfloor, \lceil \frac{n}{k} \rceil\}$. Hence, since $\frac{1}{2} \cdot \lceil \frac{n}{k} \rceil \leq \lceil \frac{n}{2k} \rceil$, we have for all $u \in V$ that $\min_{j \in [k]}\{\text{dist}(s_j, u)\} \leq \lfloor \frac{1}{2} \cdot (\lceil \frac{n}{k} \rceil + 1) \rfloor \leq \lfloor \lceil \frac{n}{2k} \rceil + \frac{1}{2} \rfloor \leq \lfloor \frac{n}{2k} \rfloor + 1$. Moreover, $x_1(\boldsymbol{s}) = k$, and for all $i \in \{2, \ldots, \lfloor \frac{n}{2k} \rfloor\}$ we have $x_i(\boldsymbol{s}) = 2k$. Hence, according to Lemma 2, $\boldsymbol{x}(\boldsymbol{s})$ is the optimal solution to (2). By the above observation, it then follows that $\boldsymbol{s}$ has optimal social cost. □

We will now determine tight upper bounds for the social cost of worst Nash equilibria. Therefore, consider the following optimization problem on a tuple (n, μ, f) where $n \in \mathbb{N}$, $\mu \in \mathbb{N}$, and $f : \mathbb{R} \to \mathbb{R}$ is a function.

$$\text{Maximize } \sum_{i=1}^{\ell} f(d_i) \qquad\qquad (3)$$
$$\text{subject to } \sum_{i=1}^{\ell} d_i = n$$
$$1 \leq d_i \leq \mu \quad \forall i \in [\ell]$$
$$\text{where } \ell, d_i \in \mathbb{N} \quad \forall i \in [\ell]$$

Recall that a function f is *superadditive* if it satisfies $f(x + y) \geq f(x) + f(y)$ for all of its domain. We need:

Lemma 3. *Let $n \in \mathbb{N}$, $\mu \in [n] \setminus \{1\}$, and f be a superadditive function. Then, $(\ell^*, \boldsymbol{d}^*)$ with $\ell^* = \lceil \frac{n}{\mu} \rceil \in \mathbb{N}$ and $\boldsymbol{d}^* = (\mu, \ldots, \mu, n - (\ell^* - 1) \cdot \mu) \in \mathbb{N}^{\ell^*}$ is an optimal solution of (3).*

In the following, let $f : \mathbb{R}_{\geq 0} \to \mathbb{R}_{\geq 0}$ be an auxiliary function that associates the distance between two strategies to the social cost corresponding to such a segment; define f by

$$f(x) := \begin{cases} \frac{x^2}{4} & \text{if } x \in \mathbb{N}_0 \text{ and } x \text{ is even} \\ \frac{x^2 - 1}{4} & \text{if } x \in \mathbb{N} \text{ and } x \text{ is odd,} \end{cases}$$

and by linear interpolation for all other points. That is, if $x \in \mathbb{R}_{>0} \setminus \mathbb{N}$, then $f(x) := (\lceil x \rceil - x) \cdot f(\lfloor x \rfloor) + (x - \lfloor x \rfloor) \cdot f(\lceil x \rceil)$. By definition, the social cost of a strategy profile is $\sum_{i=1}^{\ell} f(d_i)$. It is straightforward (and omitted here) to verify that f is superadditive. Note also that

$$f(2x) = \begin{cases} x^2 = 4f(x) & \text{if } x \in \mathbb{N} \text{ is even} \\ x^2 = 4f(x) + 1 & \text{if } x \in \mathbb{N} \text{ is odd} \\ x^2 - \frac{1}{4} = 2f(x - \frac{1}{2}) + 2f(x + \frac{1}{2}) = 4f(x) & \text{if } 2x \in \mathbb{N} \text{ is odd .} \end{cases}$$

Theorem 5. *Consider an arbitrary Voronoi game $\mathcal{C}(n, k)$ where $k \leq \frac{n}{2}$ and let $s \in \mathscr{S}$ be a Nash equilibrium. Define $\gamma := \frac{1}{2} \cdot \lfloor \frac{2n}{k} \rfloor$. The following holds:*

1. *If γ is an odd integer, then $\mathrm{SC}(s) \leq \frac{9}{4} \mathrm{OPT}$.*
2. *Otherwise, $\mathrm{SC}(s) \leq 2 \, \mathrm{OPT}$.*

Proof. Theorem 1 and Lemma 1 imply that, in any Nash equilibrium, the minimum utility of all players can be no more than γ. Hence, the maximum distance between two strategies is 2γ. Now let s be a strategy profile with $\ell = \lceil \frac{n}{2\gamma} \rceil$ and $\boldsymbol{d} = (2\gamma, \ldots, 2\gamma, n - (\ell - 1) \cdot 2\gamma)$. Due to Lemma 3 (with $\mu := 2\gamma$), no Nash equilibrium can have social cost larger than $\mathrm{SC}(s)$.

Let $p \in \mathbb{N}_0, q \in [k - 1]_0$ be defined by $n = p \cdot k + q$. Similarly, let $t \in \mathbb{N}_0, u \in [2\gamma - 1]_0$ be defined by $n = t \cdot 2\gamma + u$. Clearly, $\mathrm{SC}(s) \leq t \cdot f(2\gamma) + f(u)$.

Finally, in order to compare $\mathrm{SC}(s)$ with OPT, define $v \in \mathbb{N}_0, w \in [0, 2\gamma)$: If $\gamma \in \mathbb{N}$, then by $q = v \cdot 2\gamma + w$ and otherwise (if $2\gamma \in \mathbb{N}$ is odd) by $(q - \frac{k}{2}) = v \cdot 2\gamma + w$. Note here that $2w \in \mathbb{N}_0$ and $(2\gamma \text{ odd} \iff q \geq \frac{k}{2} \iff \gamma = p + \frac{1}{2})$.

Claim 2. $SC(s) \le \left(\frac{k}{2} + v\right) \cdot f(2\gamma) + w \cdot \frac{\gamma}{2}$.

The proof of the claim is omitted here. We now examine the optimal cost. Note first that if $\gamma \in \mathbb{N}$, then $OPT = k \cdot f(\gamma) + q \cdot \lfloor \frac{\gamma+1}{2} \rfloor$. Consider the following cases:

- γ is even

  Since $q \cdot \lfloor \frac{\gamma+1}{2} \rfloor = v \cdot \gamma^2 + w \cdot \frac{\gamma}{2} = v \cdot f(2\gamma) + w \cdot \frac{\gamma}{2}$, we have

$$OPT = k \cdot f(\gamma) + v \cdot f(2\gamma) + w \cdot \frac{\gamma}{2} = \left(\frac{k}{4} + v\right) \cdot f(2\gamma) + w \cdot \frac{\gamma}{2} \ge \frac{1}{2} \cdot SC(s).$$

- γ is odd

  Since now $q \cdot \lfloor \frac{\gamma+1}{2} \rfloor = v \cdot \gamma^2 + w \cdot \frac{\gamma}{2} + \frac{q}{2} = v \cdot f(2\gamma) + w \cdot \frac{\gamma}{2} + \frac{q}{2}$, we have

$$OPT = \left(\frac{k}{4} + v\right) \cdot f(2\gamma) - \frac{k}{4} + w \cdot \frac{\gamma}{2} + \frac{q}{2} \ge \frac{1}{2} \cdot SC(s) - \frac{k}{4}.$$

  Now, a trivial bound is always $OPT \ge n - k$. Since $n \ge \gamma k$, as otherwise $\gamma = \frac{1}{2} \cdot \lfloor \frac{2n}{k} \rfloor > \frac{n}{k}$, this implies $OPT \ge (\gamma - 1) \cdot k$. Finally, due to $k \le \frac{n}{2}$ and since γ is odd, we have $\gamma \ge 3$; so $OPT \ge 2k$ and $SC(s) \le \frac{9}{4} OPT$.
- $2\gamma \in \mathbb{N}$ is odd

  Since $f(2\gamma) = \gamma^2 - \frac{1}{4}$, it follows that

$$SC(s) \le \frac{k}{2} \cdot f(2\gamma) + v \cdot \left(\gamma^2 - \frac{1}{4}\right) + w \cdot \frac{\gamma}{2} = \frac{k}{2} \cdot f(2\gamma) + \left(q - \frac{k}{2}\right) \cdot \frac{\gamma}{2} - \frac{v}{4}.$$

  Note that $n = \gamma \cdot k + (q - \frac{k}{2})$ and $OPT = k \cdot f(\gamma) + (2q - k) \cdot (f(p+1) - f(\gamma))$.
  If $p = \lfloor \gamma \rfloor$ is even, then $2 \cdot (f(p+1) - f(\gamma)) = \frac{(p+1)^2 - 1 - p^2}{4} = \frac{p}{2} = \frac{\gamma}{2} - \frac{1}{4}$, so

$$OPT = \frac{k}{4} \cdot f(2\gamma) + \left(q - \frac{k}{2}\right) \cdot \left(\frac{\gamma}{2} - \frac{1}{4}\right).$$

  Since $\frac{\gamma}{2} \le \gamma - \frac{1}{2}$, we get $SC(s) \le 2 \, OPT$.
  If $p = \lfloor \gamma \rfloor$ is odd, then $OPT = \frac{k}{4} \cdot f(2\gamma) + \left(q - \frac{k}{2}\right) \cdot \left(\frac{\gamma}{2} + \frac{1}{4}\right)$, Clearly, we again have $SC(s) \le 2 \, OPT$. $\qquad \square$

Theorem 6. *The bounds in Theorem 5 are tight.*

Proof. Let $k \in \mathbb{N}$ even and $n = \gamma \cdot k$, where $2\gamma \in \mathbb{N}$. Consider a profile s with $\ell = \frac{k}{2}$ and $d_1 = \cdots = d_\ell = 2\gamma$. Clearly, a standard (and thus optimal) profile s' has $\ell' = k$ and $d'_1 = \cdots = d'_k = \gamma$. Then $SC(s') = OPT = k \cdot f(\gamma)$.

  If γ is even or $\gamma \notin \mathbb{N}$, then $SC(s) = \ell \cdot f(2\gamma) = \ell \cdot 4f(\gamma) = 2k \cdot f(\gamma) = 2 \, OPT$. On the other hand, if γ is odd, then $SC(s) = \ell \cdot f(2\gamma) = \ell \cdot (4f(\gamma) + 1) = 2k \cdot (f(\gamma) + \frac{1}{4}) = (2 + \frac{1}{2 \cdot f(\gamma)}) \cdot OPT$. To see the last equality, recall that $\frac{k}{2} = \frac{OPT}{2f(\gamma)}$. For the case $\gamma = 3$ this means $SC(s) = \frac{9}{4} \cdot OPT$. $\qquad \square$

Theorem 7. *Consider $\mathcal{C}(n, k)$. Up to rotation, the following holds:*

1. *If $\frac{n}{2} < k \le \frac{2}{3}n$, then the best Nash equilibrium has social cost $OPT = n - k$, whereas the worst Nash equilibrium has social cost $\lfloor \frac{2n}{3} \rfloor \le 2 \, OPT$.*
2. *If $k = n$, then the best Nash equilibrium has social cost 0. If n is even, then the only other Nash equilibrium has social cost $\frac{n}{2}$. Otherwise, there is no other Nash equilibrium.*

5   Conclusion

Hotelling's famous "Stability in Competition" [6] from 1929 has attracted an immense but also belated interest in competitive location games, from researchers in various disciplines [5]. While the Voronoi games on graphs studied here imply several idealistic assumptions, they still provide first steps for predicting sellers' positions in *discrete markets*; e.g., locations of competitive service providers in a *computer network*. In this work, we looked at Voronoi games from the stability angle by a comprehensive examination of their Nash equilibria. As a starting point, we assumed that the network is merely a cycle graph. Even for these very simple graphs, the analysis turned out to be non-trivial and much more complex than for the continuous case [4]; with much of the complexity owed to the discrete nature of graphs and parity issues. While we consider now Voronoi games on cycle graphs to be fully understood—by giving an exact characterization of all Nash equilibria, an existence criterion and exact prices of anarchy and stability—a generalization to less restrictive classes of graphs remains open.

Acknowledgment. We thank Martin Gairing, Tobias Tscheuschner, and the anonymous referees for helpful comments.

References

1. Anshelevich, E., Dasgupta, A., Kleinberg, J., Roughgarden, T., Tardos, É., Wexler, T.: The price of stability for network design with fair cost allocation. In: Proceedings of the 45th Annual Symposium on Foundations of Computer Science (FOCS 2004), pp. 295–304 (2004), doi:10.1109/FOCS.2004.68
2. Boulding, K.E.: Economic Analysis, 4th edn. Harper & Row, New York (1966)
3. Dürr, C., Thang, N.K.: Nash equilibria in Voronoi games on graphs. In: Arge, L., Hoffmann, M., Welzl, E. (eds.) ESA 2007. LNCS, vol. 4698, pp. 17–28. Springer, Heidelberg (2007), http://arxiv.org/abs/cs.IT/0702054
4. Eaton, B.C., Lipsey, R.G.: The principle of minimum differentiation reconsidered: Some new developments in the theory of spatial competition. Review of Economic Studies 42(129), 27–49 (1975), http://www.jstor.org/stable/2296817
5. Eiselt, H.A., Laporte, G., Thisse, J.-F.: Competitive location models: A framework and bibliography. Transportation Science 27(1), 44–54 (1993)
6. Hotelling, H.: Stability in competition. Computational Geometry: Theory and Applications 39(153), 41–57 (1929)
7. Koutsoupias, E., Papadimitriou, C.: Worst-case equilibria. In: Meinel, C., Tison, S. (eds.) STACS 1999. LNCS, vol. 1563, pp. 404–413. Springer, Heidelberg (1999)
8. Wendell, R.E., McKelvey, R.D.: New perspectives in competitive location theory. European Journal of Operational Research 6(2), 174–182 (1981)

Colouring Random Empire Trees

Andrew R. McGrae and Michele Zito

Department of Computer Science
University of Liverpool
Liverpool L69 3BX, UK
{andrew,michele}@liverpool.ac.uk

Abstract. We study the empire colouring problem (as defined by Percy Heawood in 1890) for maps whose dual planar graph is a tree, with empires formed by exactly r countries. We first notice that $2r$ colours are necessary and sufficient to solve the problem in the worst-case. Then we define the notion of a random r-empire tree and, applying a method for enumerating spanning trees in a particular class of graphs, we find exact and asymptotic expressions for all central moments of the number of (balanced) s-colourings of such graphs. Such result in turns enables us to prove that, for each $r \geq 1$, there exists a positive integer $s_r < r$ such that, for large n, almost all n country r-empire trees need more than s_r colours, and then to give lower bounds on the proportion of such maps that are colourable with $s > s_r$ colours.

1   Introduction

The question of whether four distinct colours suffice to colour all maps so that no two countries sharing a border receive the same colour has a long history dating back to Francis Guthrie (circa 1852). After a number of failing attempts (the interested reader may consult [6]), the first convincing solution was obtained by Appel and Haken (see for instance [2]) in the seventies and, eventually, simplified and improved in several ways by Robertson *et al.* [18] almost 20 years later.

Although these results settle the original question, several related issues remain open. For our convenience from now on we switch to the dual graph-theoretic representation. Countries will be vertices of a graph G and vertices corresponding to neighbouring countries will be adjacent in G. In this setting, even though Appel and Haken's theorem guarantees that four colours always suffice, it is not clear how many of the resulting planar graphs cannot be coloured with less than four colours. Clearly two colours are enough if and only if G is bipartite. Other important classes of planar graphs are 3-colourable including graphs not containing triangles [8] or cycles of a certain size (see [19] and references therein) and some classes of triangulated or nearly-triangulated planar graphs. However no satisfactory characterization is known and the problem of deciding whether the *chromatic number* of a planar graph is three is NP-complete [7]. Note that several issues related to the chromatic number of planar graphs are also difficult from the probabilistic point of view as models or random planar graphs are not easy to work with [4,5,15].

E. Ochmański and J. Tyszkiewicz (Eds.): MFCS 2008, LNCS 5162, pp. 515–526, 2008.

A linked (but less famous) problem concerns the colourability of planar graphs with the additional constraint that groups of vertices (or *empires*) must receive the same colour. The problem has some practical interest as several countries in the world (e.g. the United States, China, and Azerbaijan) are formed by a number of non-contiguous regions. This *empire colouring problem* (or ECP), was first defined by Heawood [9] who also proved that if at most r countries belong to any given empire then $6r$ colours are enough (viz. a simple linear time algorithm returns a $6r$-colouring of any such map) and 12 colours are necessary when $r = 2$. Later on other authors showed that $6r$ colours are also necessary when $r > 2$ [12,21]. In all cases the graphs requiring a large number of colours have a rather special structure. Hence such results do not really say much about the number of colours sufficient for the "typical" empire graph.

As a first step in the direction of settling this question, in this paper we investigate the chromatic properties of empire maps whose dual planar graph is a tree, with empires formed by exactly r countries. After obtaining some preliminary, worst case, results, our analysis is mostly probabilistic. We define the notion of a random r-empire tree and investigate the chance that a certain number of colours is sufficient to solve the ECP in such graphs. Our analysis stems from the study of all central moments of the relevant random variables. A lower bound on the number of colours needed to solve the ECP (at least with overwhelming probability) is then proved using the first moment method. Finally, as for other types of random graphs [1], the second moment method gives estimates on the proportion of r-empire trees that are colourable with a certain number of colours.

Understandably the reader may question the motivations behind our work. First the ECP is a variant of the classical graph colouring problem that has received less attention and, nevertheless, has a number of interesting features. As it will become apparent very soon the ECP reduces to classical graph colouring but the two problems are not equivalent. Also, the ECP is related to the colouring of graphs of given *thickness* [10]. Second, we contend that even in the simplified, and perhaps artificial, setting considered in this paper, the ECP has a rich combinatorial structure that is worth looking at. Apart from shedding some light on an interesting type of graph colouring, we believe that our analysis uses several techniques that may be of independent interest, notably two different ways of counting the number of spanning trees in a graph, and a way of approximating the moments of the relevant random variables by multivariate normal integrals. Last but not least our combinatorial investigation was motivated by the empirical evidence that simple list colouring strategies to solve the ECP often outperform Heawood colouring strategy on randomly generated empire graphs.

The rest of the paper is organised as follows. After describing some elementary worst-case results, we end this section defining our probabilistic model and stating our main results. In Section 2 we look at the expected number of colourings of a random empire tree and prove that, as their size grows, the proportion of such structures that can be coloured with only few (say two) colours is vanishingly small. Generalizing on this, in Section 3 we compute the higher moments of the relevant random variables. Finally, in Section 4 we derive asymptotic approximations for all moments considered and, as a consequence, lower bounds on the probability that a certain number of colours is enough to solve the ECP in a random empire tree.

Preliminaries and worst-case results. Given a planar graph $G = (V, E)$ whose vertex set is partitioned in empires $V_1, V_2, \ldots$ (a.k.a. *empire graph*), the *empire colouring problem* (or ECP) asks for a colouring of the vertices of G that never assigns the same colour to adjacent vertices in different empires and, conversely, assigns the same colour to all vertices in the same empire, disregarding adjacencies. For any integer $s \geq 1$, an *s-(empire) colouring* is one such colouring using s distinct colours. For any integer $r \geq 1$, an *r-empire graph* is an empire graph whose empires contain exactly r vertices. The *reduced graph* of an empire graph G has one vertex for each empire, and an edge between two vertices x and y if and only if at least two vertices of G (one belonging to empire x one to y) are adjacent. The reduced graph of an r-empire graph on e empires (or *r-reduced graph*) has e vertices and at most $3re - 6$ edges. Therefore its average degree is less than $6r$, and this property is true for any of its subgraphs. It is obvious that each instance of the ECP reduces to an instance of the classical colouring problem on a reduced graph. Heawood's $6r$-colouring algorithm [9] repeatedly finds a vertex v of degree at most $6r - 1$ in the reduced graph of G. Removing v leaves a graph on $e - 1$ vertices, which can be coloured recursively. Once this is done no more than $6r - 1$ colours will be in the neighbourhood of v. Hence v can be given a (spare) colour from a palette of $6r$ colours. In this paper we look at the colourability of r-empire *trees* (i.e. collections of empires whose adjacencies define a tree). Of course the simplified topology reduces the number of colours that are sufficient to solve the ECP. The following Theorem gives the best possible worst-case results in this setting.

Theorem 1. *Let $G = (V, E)$ be an r-empire tree on n vertices. Then its reduced graph can be coloured using $2r$ colours. Furthermore there is a family of r-empire trees $(T_r)_{r \geq 1}$ whose reduced graph cannot be coloured with less than $2r$ colours.*

Proof. The upper bound follows from Heawood's argument as trees have average degree less than two. We define a family of trees $(T_r)_{r \geq 1}$ such that, for all integers $r \geq 1$, the reduced graph of T_r is K_{2r}, the complete graph on $2r$ vertices. The tree T_r will have $2r^2$ vertices $r^2 - r + 2$ of which have degree one. Furthermore, if C_1 and C_2 are two special vertices called the *centres* of T_r then there will be exactly r vertices of degree one, belonging to empires $1, 3, \ldots, 2r - 1$ at distance $2(r - 1)$ from C_1 and r vertices of degree one, belonging to empires $2, 4, \ldots, 2r$ at distance $2(r - 1)$ from C_2. These sets of vertices are called Far$_1$ and Far$_2$ respectively.

The empire tree $T_1 \equiv K_2$. Assume that T_{r-1} is given consisting of empires of size $r - 1$, labelled from one to $2(r - 1)$, which satisfies all properties above. Add $r - 1$ new vertices belonging to empire $2r - 1$ and $r - 1$ vertices belonging to empire $2r$. Connect each of the new vertices in empire $2r - 1$ (resp. $2r$) with a distinct element of Far$_1$ (resp. Far$_2$). By adding one more vertex to each empire we can change this so that any two empires are adjacent. Choose one vertex from

Table 1. A.a.s. lower bounds on the chromatic number of $G_{r,n}$ for different values of r

r	2 3 4 5 6 7 8 9 10 ... 20 ... 50
s_r	2 3 3 4 4 4 5 5 6 ... 9 ... 17

empire $2r - 1$ (resp. $2r$). Attach r new leaves belonging to empires $2, 4, \ldots, 2r$ (resp. $1, 3, \ldots, 2r-1$) to such vertex. The resulting tree is an r-empire tree. It has $l(r) = l(r-1) + 2(r-1)$ vertices of degree one and $v(r) = v(r-1) + 2(r-1) + 2r$ vertices in total and reduces to K_{2r}.   □

Notice that there are 2-empire trees that need four colours but whose reduced graphs do not have K_4 as a subgraph (e.g. a wheel with five spokes). However we believe that r-empire trees requiring $2r$ colours are the exception rather than the norm. In the rest of the paper we provide partial support to our claim by showing that, for each r, at least a constant fraction of all r-empire trees can be coloured with less than $2r$ colours.

Average-case analysis, setting and results. To study the distribution of the s-colourable r-empire trees we first define the notion of a random r-empire tree. Let T_n be a random tree on n labelled vertices (see for instance [16, Chap. 7]). Formally T_n denotes a probability space obtained by associating the uniform probability measure to the set of all n^{n-2} labelled trees on n vertices, but we will often hide the distinction between the whole space and its typical element. A *random r-empire tree $G_{r,n}$* is obtained from T_n by assuming that the ith empire, for $i \in \{1, \ldots, \frac{n}{r}\}$ is formed by vertices labelled $(i-1)r+1, \ldots, ir$ (from now on we assume w.l.o.g. that r divides n). Using $G_{r,n}$ we investigate the density of s-colourable r-empire trees over the whole population. In what follows we say that a result holds *asymptotically almost surely* (a.a.s.) if it is true with probability approaching one as n tends to infinity.

The obvious starting point is the question of whether Heawood algorithm is really all that good on $G_{r,n}$. A first consequence of our probabilistic setting is that families of r-empire trees requiring few colours have a vanishing effect on the typical chromatic properties of $G_{r,n}$. In the following statement, for any fixed integer $r \geq 2$, s_r is the largest integer s such that $c_{s,r} = s^{\frac{1}{r}-1}(s-1) < 1$.

Theorem 2. *For any fixed integer $r \geq 2$ a random r-empire tree admits a.a.s. no s-colouring for any $s \in \{1, \ldots, s_r\}$. Furthermore for large r, $s_r = \lceil \frac{r}{\log r} \rceil (1 + O(\frac{1}{\log \log r}))$.*

Table 1 gives the values of s_r for the first few values of r. This result implies that, in a sense, Heawood algorithm is not too bad. Furthermore, one may be lead to believe that Theorem 2 may be strengthened to prove that, in fact, very few r-empire trees are colourable with less than $2r$ colours. However, the main contribution of this paper shows that this is not the case. Many r-empire trees can be coloured with a few as $s_r + 1$ colours. The statement is formalised.

Theorem 3. *For any fixed integers $r \geq 2$ and $s > s_r$ a random r-empire tree is s-colourable with (at least) constant positive probability.*

Table 2 shows the lower bounds on the proportions of s-colourable r-empire trees derived in the proof of Theorem 3 for various values of r and s. At least almost 50% of all 3-empire trees can be coloured with only five colours, at least almost 20% only need four colours. Thus the guarantees of Heawood's algorithm are often not very good.

The results stem from the use of tight estimates for the expected value of the random variable $Y_{s,r}$, counting (up to colour class relabelling) s-colourings of $G_{r,n}$ and precise asymptotics for all central moments of its close relative $Z_{s,r}$, counting *balanced* colourings, in the sense that each colour class has size $\frac{n}{sr}$ (from now on we assume that $\frac{n}{sr}$ is a positive integer).

2   Proof of Theorem 2: First Moment of $Y_{s,r}$

The first part of Theorem 2 is a consequence of the first moment method. By Markov's inequality the probability that $Y_{s,r} > 0$ is at most $\mathbf{E}Y_{s,r}$. Thus if we can prove that if $s \le s_r$, $\mathbf{E}Y_{s,r}$ tends to zero as n tends to infinity we have that $Y_{s,r} = 0$ a.a.s. An s-colouring of $G_{r,n}$ is a partition of $\{1, \ldots, n\}$ into s blocks such that

P1 no empire is split between two or more blocks and
P2 all edges of T_n connect vertices in different blocks or in a same empire.

By linearity of expectation, $\mathbf{E}Y_{s,r}$ can be computed as a sum of terms, one for each such partition, representing the probability that the given partition corresponds to an s-colouring of $G_{r,n}$. Such probability is simply the ratio of the number of trees that satisfy **P1** and **P2** for the given partition over n^{n-2}. Let $\mathcal{H}_{s,r}$ be a graph whose vertex set is partitioned into s blocks of size $rn_1, rn_2, \ldots, rn_s$. For $i \in \{1, \ldots, s\}$, the ith block is further partitioned in n_i groups of vertices each of size r. Each vertex of $\mathcal{H}_{s,r}$ is adjacent to all vertices in the same group or in different blocks. The blocks represent the colour classes of a possible s-colouring of $G_{r,n}$, the groups correspond to the empires, the edges describe the allowed adjacencies. The spanning trees of $\mathcal{H}_{s,r}$ define precisely those r-empire trees for which the given set of s blocks represent a valid s-colouring. We denote by $\kappa(G)$ the number of spanning trees of graph G. The next result uses classical algebraic methods (and elementary linear algebra) to find $\kappa(\mathcal{H}_{s,r})$.

Table 2. Lower bounds on the proportion of s-colourable r-empire trees for various values of r and $s \in \{s_r + 1, \ldots, 2r - 1\}$, and large n

$r \setminus s$	3	4	5	6	7	8	9	10	11	12	13
2	0.1045	—	—	—	—	—	—	—	—	—	—
3	—	0.1955	0.4947	—	—	—	—	—	—	—	—
4	—	0.011	0.2336	0.4479	0.5935	—	—	—	—	—	—
5	—	—	0.07	0.2568	0.4237	0.549	0.6407	—	—	—	—
6	—	—	0.0095	0.1194	0.2729	0.4094	0.5183	0.6029	0.6686	—	—
7	—	—	0.0002	0.0421	0.1558	0.2848	0.4003	0.4961	0.5738	0.6364	0.6870
8	—	—	—	0.0101	0.0771	0.1832	0.294	0.394	0.4794	0.5508	0.61

It is worth pointing out that in Section 3 we will generalize considerably the construction of $\mathcal{H}_{s,r}$, Lemma 1 is a special case (for $k = 1$) of Lemma 6 proved there, and, in fact, Lemma 3 can be improved to a result similar to Theorem 4. However the current presentation is simple, relatively self-contained and provides a gentle introduction to the rest of the paper.

Lemma 1. *For any fixed integers $r \geq 2$ and $s \geq 2$, let $\mathcal{H}_{s,r}$ be defined on a partition of $\{1, \ldots, n\}$ with blocks of size rn_i, for $i \in \{1, \ldots, s\}$. Then*
$$\kappa(\mathcal{H}_{s,r}) = n^{s-2} \prod_{i=1}^{s} (n - r(n_i - 1))^{(r-1)n_i} (n - rn_i)^{n_i - 1}.$$

Proof. By a result of Temperley [20] $n^2 \kappa(\mathcal{H}_{s,r}) = \det(\text{One}_n + L(\mathcal{H}_{s,r}))$ where $L(G) = D(G) - A(G)$ is the Laplacian of graph G, defined in terms of a diagonal matrix $D(G)$ whose i^{th} row entry is the degree of vertex i in G and $A(G)$ the adjacency matrix of G, and One_n is an $n \times n$ matrix whose entries are all ones. Note that $\text{One}_n + L(\mathcal{H}_{s,r}) = D(\mathcal{H}_{s,r}) + I_n + A((\mathcal{H}_{s,r})^c)$. Here I_n is the $n \times n$ identity matrix and $(\mathcal{H}_{s,r})^c$ is the complement of $\mathcal{H}_{s,r}$. From this we have that $n^2 \kappa(\mathcal{H}_{s,r}) = \det(nI_n - L((\mathcal{H}_{s,r})^c))$. The graph $(\mathcal{H}_{s,r})^c$ is formed by s disjoint connected components $(\mathcal{H}_{s,r})_1^c, \ldots, (\mathcal{H}_{s,r})_s^c$. Hence $\det(nI - L((\mathcal{H}_{s,r})^c)) = \prod_{i=1}^{s} \det(nI - L((\mathcal{H}_{s,r})_i^c))$. For each $i \in \{1, \ldots, s\}$, $(\mathcal{H}_{s,r})_i^c$ is a circulant graph on rn_i vertices and it is regular of degree $r(n_i - 1)$. Thus [3, Chap. 3] for each i, the characteristic polynomial of $A((\mathcal{H}_{s,r})_i^c)$ is $(\lambda - r(n_i - 1)) \lambda^{(r-1)n_i} (\lambda + r)^{n_i - 1}$. Putting all this together we have that
$$n^2 \kappa(\mathcal{H}_{s,r}) = (-1)^n \prod_{i=1}^{s} \det((r(n_i - 1) - n)I - A((\mathcal{H}_{s,r})_i^c))$$
$$= \prod_{i=1}^{s} n \, (n - r(n_i - 1))^{(r-1)n_i} (n - rn_i)^{n_i - 1}.$$

In the first equality we've used the fact [3, Chap. 6] that for any μ, $\det(\mu I - L(G)) = (-1)^n \det((d - \mu)I - A(G))$, for any n-vertex d-regular graph G. $\quad\square$

The next result now follows almost immediately from Lemma 1 and the earlier argument.

Lemma 2. *For each integer $r \geq 2$ and $s \geq 2$,*
$$\mathbf{E}Y_{s,r} = \frac{1}{s!} \sum_{n_1, \ldots, n_s} \binom{\frac{n}{r}}{n_1, \ldots, n_s} \prod_{i=1}^{s} (1 - \frac{r(n_i - 1)}{n})^{(r-1)n_i} (1 - \frac{rn_i}{n})^{n_i - 1},$$
where the sum is over all sequences of s positive integers adding up to $\frac{n}{r}$.

Proof. First notice that $\mathbf{E}Y_{1,r} = 0$ for any positive integer r. For $s \geq 2$, expression $n^{n-2} \mathbf{E}Y_{s,r}$ may be computed as a sum of terms $\kappa(\mathcal{H}_{s,r})$ over all possible ways to partition the $\frac{n}{r}$ empires into blocks $B_1, \ldots, B_s$. Hence, by Lemma 1
$$\mathbf{E}Y_{s,r} = \frac{n^{s-n}}{s!} \sum_{B_1, \ldots, B_s} \prod_{i=1}^{s} (n - r(|B_i| - 1))^{(r-1)|B_i|} (n - r|B_i|)^{|B_i| - 1}.$$

The result follows by noticing that $\kappa(\mathcal{H}_{s,r})$ only depends on the sizes of the blocks. $\quad\square$

Lemma 3. $\lim_{n \to +\infty} (\mathbf{E}Y_{s,r})^{\frac{1}{n}} = c_{s,r}$, *for each integer $r \geq 2$ and $s \geq 2$.*

Proof. We claim that for each integer $r \geq 2$ and $s \geq 2$

$$\frac{e^{\frac{s(r-1)}{s-1}} s^{\frac{3s}{2}}}{s! \,(s-1)^s} \left(\frac{r}{2\pi n}\right)^{\frac{s-1}{2}} (c_{s,r})^n (1 - o(1)) \leq \mathbf{E}Y_{s,r} \leq \frac{e^{\frac{s(r-1)}{s-1}} s^s}{s! \,(s-1)^s} (c_{s,r})^n (1 + o(1)).$$

Note that $\mathbf{E}Y_{s,r}$ is always at least $\left(\frac{\frac{n}{r}}{\frac{n}{rs},\dots,\frac{n}{rs}}\right) \frac{1}{s!} \prod_{i=1}^s (1 - \frac{1}{s} + \frac{r}{n})^{\frac{r-1}{r}\frac{n}{s}} (1 - \frac{1}{s})^{\frac{n}{rs}-1}$.
Rewriting $(1 - \frac{1}{s} + \frac{r}{n})$ as $(1 - \frac{1}{s})(1 + \frac{rs}{(s-1)n})$ the expression above becomes
$\left(\frac{\frac{n}{r}}{\frac{n}{rs},\dots,\frac{n}{rs}}\right) \frac{1}{s!} \left(\frac{s}{s-1}\right)^s \left(\frac{s-1}{s}\right)^n (1 + \frac{rs}{(s-1)n})^{\frac{r-1}{r}n}$. The stated lower bound on $\mathbf{E}Y_{s,r}$ follows using Stirling's approximations for the various factorials and replacing the term $(1 + \frac{rs}{(s-1)n})^{\frac{r-1}{r}n}$ with $\exp\{\frac{s(r-1)}{s-1}\}(1 - o(1))$.

For the upper bound, $s! \times \mathbf{E}Y_{s,r}$ is at most

$$\sum_{n_1,\dots,n_s} \left(\frac{\frac{n}{r}}{n_1,\dots,n_s}\right) \max_{n_1,\dots,n_s} \prod_{i=1}^s (1 - \frac{r}{n}(n_i - 1))^{(r-1)n_i} (1 - \frac{rn_i}{n})^{n_i-1},$$

and the sought maximum is achieved when $n_1 = \dots = n_s = \frac{n}{rs}$. Therefore $\mathbf{E}Y_{s,r}$ is at most $(1 + \frac{rs}{(s-1)n})^{\frac{(r-1)n}{r}} \left(\frac{s-1}{s}\right)^{n-s} \times \frac{s^{\frac{n}{r}}}{s!}$. $\qquad\square$

It follows immediately from Lemma 3 that $\mathbf{E}Y_{s,r}$ tends to zero as n tends to infinity if $s \leq s_r$. To complete the proof of Theorem 2 we need the following statement about s_r.

Lemma 4. *For large integers r, $s_r = \lceil \frac{r}{\log r} \rceil (1 + O(\frac{1}{\log \log r}))$.*

Proof. Let $s' = \frac{r}{\log r}$ and $s'' = (1 + \frac{1}{\log \log r}) \frac{r}{\log r}$. The result follows since, for large r, $c_{s',r} < 1$, and $c_{s'',r} > 1$. $\qquad\square$

3   A Detour in Enumerative Combinatorics

The approach used in Section 2 could be followed to study the higher moments of $Y_{s,r}$. However a slightly different method leads to exact expressions for all moments of $Y_{s,r}$ and $Z_{s,r}$ and to the proof of Theorem 3. As before, the idea is to reduce the computation of the moments of the random variables under investigation to the enumeration of the spanning trees of particular classes of graphs. Colourings of a graph G can be seen as homomorphisms from G to another graph whose vertices correspond to the colour classes [11]. This correspondence can be extended to k-tuples of colourings. For any integer $s \geq 2$ and $k \geq 1$, let a vertex of graph $B_{s,k}$ be labelled by a sequence $\imath \equiv (i_1,\dots,i_k)$ where $i_j \in \{1,\dots,s\}$ for each $j \in \{1,\dots,k\}$. When lists of such sequences are needed we will assume they are produced in lexicographic order and we'll denote the elements of such lists by $\imath(1), \imath(2), \dots$. If E is an expression involving $\imath(j)$ for some $j \in \{1,\dots,s^k\}$, then $\sum_{\imath} E(\imath)$ (or $\prod_{\imath} E(\imath)$) is a shorthand for $\sum_{j=1}^{s^k} E(\imath(j))$ (or $\prod_{j=1}^{s^k} E(\imath(j))$). Two vertices, labelled $\imath$ and $\imath'$, are adjacent if and only if $i_j \neq i'_j$ for all j's. Thus, $B_{s,k}$ is an $(s-1)^k$-regular graph on s^k vertices. Any k-tuple of s-colourings in G defines a homomorphism from G to $B_{s,k}$. If $v \in V(G)$ is mapped to $\imath$ the sequence $(i_1,\dots,i_k)$ gives the colour of vertex v in each of the k given colourings. Thus we call $B_{s,k}$ the *constraint graph* on the class of all k-tuples of s-colourings. The following result will be used in the asymptotic results of Section 4.

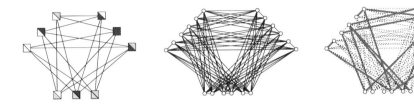

Fig. 1. The graph $B_{3,2}$ (left, with vertex labels represented by pairs of colours), one possible $\mathcal{H}_{3,2,2}$ (centre) with $n_{(i,j)} = 1$ for all $i, j \in \{1, 2, 3\}$, and a tree legally coloured by the two colourings (right, in red)

Lemma 5. $s^k \kappa(B_{s,k}) = \prod_{i=1}^{k}((s-1)^k - (-1)^i(s-1)^{k-i})^{\binom{k}{i}(s-1)^i}$, *for each integer $s \geq 2$ and $k \geq 1$.*

Proof. (Sketch) It is easy to verify that, for $i \in \{0, \ldots, k\}$, $(-1)^i(s-1)^{k-i}$ is an eigenvalue of $A(B_{s,k})$ of multiplicity $\binom{k}{i}(s-1)^i$. The result follows from the relationship between the spectrum of $A(B_{s,k})$, and that of $L(B_{s,k})$ and the fact that the number of spanning trees of $B_{s,k}$ multiplied by s^k is the product of the Laplacian's non-zero eigenvalues. □

The graph $B_{s,k}$ can be "inflated" to describe the set of admissible edges for an r-empire tree admitting k given s-colourings. First replace the node labelled 1 by n_1 *pseudo*-nodes and connect two pseudo-nodes by an edge if and only if the two original nodes were adjacent in $B_{s,k}$. Call the resulting multipartite graph $\mathcal{B} = \mathcal{B}(n_{1(1)}, n_{1(2)}, \ldots, n_{1(s^k)})$. Then replace each pseudo-node by a copy of K_r whose vertices are labelled by the elements of one of the empires of $G_{r,n}$. Two vertices in different cliques are connected if and only if the pseudo-nodes they replaced were adjacent in $\mathcal{B}$. Call the resulting graph $\mathcal{H}_{s,r,k}$ (note that $\mathcal{H}_{s,r}$ of Section 2 is $\mathcal{H}_{s,r,1}$). Obviously an r-empire tree is coloured by a given k-tuple of s-colourings if and only if it is a spanning tree of $\mathcal{H}_{s,r,k}$ (Fig. 1 gives an example). Note that the construction of both $\mathcal{B}$ and $\mathcal{H}_{s,r,k}$ follows a simple pattern. In each case each of the vertices of a given base graph ($B_{s,k}$ in the case of $\mathcal{B}$, and $\mathcal{B}$ itself in the case of $\mathcal{H}_{s,r,k}$) is replaced by another graph (an empty graph on n_1 vertices for $1 \in V(B_{s,k})$ in the case of $\mathcal{B}$, a copy of K_r in the case of $\mathcal{H}_{s,r,k}$). The adjacencies on the resulting set of vertices are defined in terms of the adjacencies in the base graph and the replacement graphs. An old result of Knuth [14], rediscovered by Pak and Postnikov [17] gives a formula for the enumeration of the spanning trees of any graph built in this way. The following lemma is a simple consequence of such result. Denote by $d(1) = \sum_{J}(A(B_{s,k}))_{1,J} n_J$ the degree, in $\mathcal{B}$, of each pseudo-node replacing $1 \in V(B_{s,k})$.

Lemma 6. *For each integer $r \geq 2$, $s \geq 2$, and $k \geq 1$,*

$$\kappa(\mathcal{H}_{s,r,k}) = r^{n-2} \prod_{l=1}^{\frac{n}{r}} (\deg_{\mathcal{B}}(l) + 1)^{r-1} \prod_{1} d(1)^{n_1-1} \sum_{T} \prod_{J} n_J^{\deg_T(J)-1},$$

where the sum is over all spanning trees of $B_{s,k}$.

Proof. Since K_r has r vertices, it follows from the main result in [17] that the number of spanning trees of $\mathcal{H}_{s,r,k}$ is $r^{\frac{n}{r}-2} \prod_{l=1}^{\frac{n}{r}} \left(\sum_{i=1}^{r} f_l(i) d(l)^{i-1} \right) \kappa(\mathcal{B})$, where the term $f_l(i)$ counts the number of spanning forests consisting of i rooted trees in each copy of K_r. Since the i roots can be chosen in $\binom{r}{i}$ ways it follows that $f_l(i) = \binom{r}{i}ir^{r-i-1}$. Also, $d(l) = r\deg_{\mathcal{B}}(l)$, for each vertex l of $\mathcal{B}$. Hence $\kappa(\mathcal{H}_{s,r,k}) = r^{\frac{n}{r}-2}\prod_{l=1}^{\frac{n}{r}}\left(\sum_{i=1}^{r}\binom{r}{i}ir^{r-i-1}r^{i-1}\deg_{\mathcal{B}}(l)^{i-1}\right)\kappa(\mathcal{B})$. Grouping together all r's gives an $r^{(1-\frac{1}{r})n-2}$ term which can be moved out of the sum. The result now follows since $\sum_{i=1}^{r}\binom{r}{i}i(\deg_{\mathcal{B}}(l))^{i-1} = r(1+\deg_{\mathcal{B}}(l))^{r-1}$ and $\kappa(\mathcal{B}) = \prod_l d(l)^{n_l-1}\sum_T \prod_{\mathsf{J}} n_{\mathsf{J}}^{\deg_T(\mathsf{J})-1}$ (the sum being over all spanning trees T of $B_{s,k}$), by Corollary 4 in [17]. □

Lemma 6 enables us to derive an exact expression for all moments of $Y_{s,r}$ and $Z_{s,r}$. Its proof is conceptually identical to that of Lemma 2, using Lemma 6 instead of Lemma 1 for the enumeration of the spanning trees in the appropriate graph.

Lemma 7. *For each integer $r \geq 2$, $s \geq 2$, and $k \geq 1$. Let $\mathcal{X}_{s,r}$ be either $Y_{s,r}$ or $Z_{s,r}$. Then $(s!)^k\mathbf{E}\mathcal{X}_{s,r}^k$ is equal to*

$$\sum_{n_{1(1)},\ldots,n_{1(s^k)}}\binom{\frac{n}{r}}{n_{1(1)},\ldots,n_{1(s^k)}}r^{n-2}\left(\prod_l\left(\frac{d(l)+1}{n}\right)^{(r-1)n_l}\left(\frac{d(l)}{n}\right)^{n_l-1}\right)\frac{\sum_T\prod_{\mathsf{J}}n_{\mathsf{J}}^{\deg_T(\mathsf{J})-1}}{n^{s^k-2}},$$

where the first sum is over all sequences of positive integers $n_{1(1)},\ldots,n_{1(s^k)}$ adding up to $\frac{n}{r}$ if $\mathcal{X}_{s,r} = Y_{s,r}$ (whereas only those sequences corresponding to balanced colourings define $\mathbf{E}Z_{s,r}^k$), and the second one over all spanning trees of $B_{s,k}$.

4   Proof of Theorem 3

For each integer $r \geq 2$, and $s \geq 2$ let $a_n = n^{-\frac{s-1}{2}}(c_{s,r})^n$. The main result of this section is the following:

Theorem 4. *For each integer $r \geq 2$, $s \geq 2$, and $k \geq 1$, there exists a positive real number $C_{s,r,k}$, independent of n, such that*

$$\mathbf{E}Z_{s,r}^k \sim C_{s,r,k} \times (a_n)^k.$$

Furthermore, for fixed integers $r \geq 2$, and $s \geq 2$,

$$C_{s,r,2} = \frac{e^{\frac{s^2(r-1)}{(s-1)^2}}s^{s^2+s+1}(s-2)^{(s-1)^2}}{(s!)^2(s-1)^2(r-2r(s-1)^2+(s-1)^4)^{\frac{(s-1)^2}{2}}}\left(\frac{r}{2\pi}\right)^{s-1}.$$

Proof. The result on $\mathbf{E}Z_{s,r}$ follows from Lemma 3 as $\mathbf{E}Z_{s,r}$ is just the central term of $\mathbf{E}Y_{s,r}$. For $k \geq 2$ we argue that the main component of the sum defining $\mathbf{E}Z_{s,r}^k$ consists of all terms close to the one having $n_{1(j)} = \frac{n}{s^k r}$ for all $j \in \{1,\ldots,s^k\}$ (see Lemma 10 below). This expression, for large n, can be approximated by a multivariate Gaussian integral (see Lemma 9). For each integer $r \geq 2$ and $s \geq 2$ let $X_{s,r} = s^2r(\frac{r-2r(s-1)^2+(s-1)^4}{(s-1)^4})$. For $k = 2$ the matrix $A_{s,r,2}$ has eigenvalues $X_{s,r}$ with multiplicity $(s-2)^2$, $sX_{s,r}$ with multiplicity $2(s-2)$ and $s^2X_{s,r}$. If A is a non-singular real symmetric matrix, for each positive m,

$$\int_{\mathbb{R}^m}e^{-\frac{1}{2}yAy^T}dy = \prod_{i=1}^{m}\frac{1}{\sqrt{\lambda_i}}\int_{\mathbb{R}^m}e^{-\frac{||z||^2}{2}}dz.$$

Hence for $k = 2$ the integral in Lemma 9 is asymptotically equal to $\prod_{i=1}^{(s-1)^2} \frac{2\pi}{\sqrt{\lambda_i}}$, where $\lambda_1, \lambda_2, \ldots$ are the eigenvalues of $A_{s,r,2}$. The result on $C_{s,r,2}$ follows, using Lemma 5. □

Theorem 4 implies Theorem 3, through the use of the following refined version of Chebyshev's inequality:

Lemma 8. *(see for instance [13, Chap. 3]) Let X be a non-negative integer valued random variable. Then* $\Pr[X > 0] \geq (\mathbf{E}X)^2/\mathbf{E}X^2$.

The remainder of this paper gives a few more details on our approximations.

Approximating $\mathbf{E}Z_{s,r}^k$. For each integer $k \geq 2$ each colouring in a k-tuple of balanced s-colourings splits each of the other colourings' colour classes into s subclasses. The sizes of the classes in the resulting partition of the $\frac{n}{r}$ empires satisfy $\sum n_{i_1,\ldots,i_k} = \frac{n}{r}$ and, furthermore, for each $j \in \{1, \ldots, k\}$ and each fixed value ι for i_j, $\sum n_{i_1,\ldots,i_{j-1},\iota,i_{j+1},\ldots,i_k} = \frac{n}{sr}$. It will be convenient, in the forthcoming treatment, to define each partition in terms of the one having all parts equal to $\frac{n}{s^k r}$. Thus the block whose vertices are coloured i_1 in the first colouring, i_2 in the second one and so on has size $n_1 = \frac{n}{s^k r} + x_1$, where the x_1 are integers adding up to zero, and the sum of $x_{i_1,\ldots,i_{j-1},\iota,i_{j+1},\ldots,i_k}$ over all possible choices of $i_1, \ldots, i_k$ with $i_j = \iota$ fixed is also zero, for each $j \in \{1, \ldots, k\}$. To prove Theorem 4 we will focus on sequences of x_1 with $|x_1| = O(\sqrt{n \log n})$. We call these sequences and the corresponding partitions *nice*. We denote by $\mathcal{H}_{s,r,k}(x_{1(1)}, \ldots, x_{1(s^k)})$ the instance of $\mathcal{H}_{s,r,k}$ associated with k s-colourings defining colour blocks of size $n_{1(1)}, \ldots, n_{1(s^k)}$. Define $f(\boldsymbol{x}')$ and $g(\boldsymbol{x}')$ by
$$((\tfrac{n}{s^k r})!)^{s^k} f(\boldsymbol{x}') = \prod_1 (\tfrac{n}{s^k r} + x_1)! \quad \text{and} \quad \kappa(\mathcal{H}_{s,r,k}(\mathbf{0}'))g(\boldsymbol{x}') = \kappa(\mathcal{H}_{s,r,k}(\boldsymbol{x}')),$$
where $\boldsymbol{x}'$ stands for the sequence of all x_1 except those of the form $x_{s,\ldots,i_j,\ldots,s}$ for each j and each value of i_j.

Lemma 9. *For fixed integer $r \geq 2$, $s \geq 2$, and $k \geq 2$, let $m = s^k - k(s-1) - 1$. Then there exists a non-singular real symmetric matrix $A_{s,r,k}$ such that*
$$\sum g(\boldsymbol{x}')f(\boldsymbol{x}') \sim n^{\frac{m}{2}} \int_{\mathrm{IR}^m} e^{-\frac{1}{2}\boldsymbol{y}A_{s,r,k}\boldsymbol{y}^T} \, d\boldsymbol{y}.$$
where the sum is over all sequences corresponding to nice partitions of $\frac{n}{r}$.

Proof. Using simple properties of exponentials and logarithms we prove that there are two $m \times m$ symmetric matrices F and G such that $f(\boldsymbol{x}') \sim e^{-\frac{1}{2}\boldsymbol{z}F\boldsymbol{z}^T}$ and $g(\boldsymbol{x}') \sim e^{-\frac{1}{2}\boldsymbol{z}G\boldsymbol{z}^T}$, where $z_1 = x_1/\sqrt{n}$. Finally, the approximation of the sum by the integral is obvious. We give all details about $f(\boldsymbol{x}')$. The proof about $g(\boldsymbol{x}')$ will be sketched. Let $N = \frac{n}{s^k r}$. First note that, for each positive integer S,
$$\frac{(N!)^S}{\prod_{i=1}^S (N+x_i)!} \sim e^{-\frac{1}{2N}\boldsymbol{x}I_S\boldsymbol{x}^T}$$

(here we assume w.l.o.g. that there exists $i_0 \in \{1, \ldots, S\}$ such that $x_i > 0$ iff $i \leq i_0$ and $\sum_{i \leq i_0} x_i = \sum_{i > i_0} |x_i|$). To believe this notice that

$$\frac{(N!)^S}{\prod_{i=1}^{S}(N+x_i)!} = \frac{N \cdot N \cdot \ldots \cdot N}{\prod_{i=1}^{i_0}\prod_{k=0}^{x_i-1}(N+x_i-k)} \times \frac{\prod_{i=i_0+1}^{S}\prod_{k=0}^{|x_i|-1}(N-k)}{N \cdot N \cdot \ldots \cdot N} =$$

$$= \prod_{i=1}^{i_0}\prod_{k=0}^{x_i-1}\left(1 - \frac{x_i-k}{N+x_i-k}\right) \times \prod_{i=i_0+1}^{S}\prod_{k=0}^{|x_i|-1}\left(1 - \frac{k}{N}\right),$$

and then that, for instance, using elementary properties of logarithms and exponentials, one can prove that

$$\prod_{i=i_0+1}^{S}\prod_{k=0}^{|x_i|-1}\left(1 - \frac{k}{N}\right)$$

is asymptotic to $\exp\left\{-\frac{1}{N}\sum_{i=i_0+1}(\sum_{j\leq i}x_j)x_i\right\}$, provided the x_i are not too large. We then get the result for $f(\boldsymbol{x}')$, by noticing that it is defined by

$$\prod_{i=1}^{s}\frac{(N!)^{s^{k-1}}}{\prod_{j=1}^{s^{k-1}}(N+x_{i,1(j)})!},$$

replacing all x_1 whose index contains at least $s-1$ values equal to s by a combination of the remaining variables. For k-tuples 1 and J let $b_{1,J}$ be the number of index values shared by 1 and J, s_1 be the number of values equal to s in 1 and $s_{1,J}$ the number of corresponding indices in 1 and J having value s. The $m \times m$ matrix F is defined by (here $\delta_{a,b} = 1$ if $a = b$, and zero otherwise)

$$F_{1,J} = (b_{1,J} - s_{1,J}) + (k-1-s_1)(k-1-s_J) + \delta_{b_{1,J},k}.$$

As to g, using $\log(1+y) = y - \frac{y^2}{2} + O(y^3)$, valid if $|y|$ is sufficiently smaller than one, one can prove that

$$g(\boldsymbol{x}') \sim \exp\left\{\frac{r}{(s-1)^k N}\sum_1 x_1\left[\sum_{\{J:1\cap J=\emptyset\}} x_J - \frac{1}{2(s-1)^k}\left(\sum_{\{J:1\cap J=\emptyset\}} x_J\right)^2\right]\right\}.$$

Substitutions identical to those used for f lead to the definition of the required $m \times m$ matrix G. $\qquad\square$

For $k \geq 2$, we can write $(s!)^k \mathbf{E}Z_{s,r}^k$ as $\sum T(x_{1(1)}, \ldots, x_{1(s^k)})$, where the sum is over all admissible tuples of x_1's and

$$T(y_1, \ldots, y_{s^k}) = \left(\frac{n}{\frac{n}{s^k r}+y_1, \ldots, \frac{n}{s^k r}+y_{s^k}}\right) r^{n-2} \times$$

$$\times \left(\prod_{l=1}^{s^k}\left(\frac{d(l)+1}{n}\right)^{(r-1)(\frac{n}{s^k r}+y_l)}\left(\frac{d(l)}{n}\right)^{\frac{n}{s^k r}+y_l-1}\right)\frac{\sum_T \prod_{i=1}^{s^k}\left(\frac{n}{s^k r}+y_i\right)^{\deg_T(i)-1}}{n^{s^k-2}}.$$

The following result states that the $T(x_{1(1)}, \ldots, x_{1(s^k)})$ corresponding to not nice partitions add up to a negligible portion of $\mathbf{E}Z_{s,r}^k$.

Lemma 10. *For fixed integers $r \geq 2$, $s \geq 2$, and $k \geq 2$*

$$(s!)^k \mathbf{E}Z_{s,r}^k = T(0, \ldots, 0) \times \sum g(\boldsymbol{x}')f(\boldsymbol{x}') + o((a_n)^k).$$

where the sum is over all sequences corresponding to nice partitions of $\frac{n}{r}$.

Finally, the next result estimates the "central" term $T(0, \ldots, 0)$.

Lemma 11. *For fixed integers $r \geq 2$, $s \geq 2$, and $k \geq 2$*

$$T(0, \ldots, 0) \sim e^{\frac{s^k(r-1)}{(s-1)^k}} s^{\frac{k s^k}{2}+2k}(s-1)^{-ks^k}\kappa(B_{s,k})\left(\frac{r}{2\pi n}\right)^{\frac{s^k-1}{2}}(c_{s,r})^{kn}.$$

Proof. (Sketch) Note that $T(0, \ldots, 0) = n^{2-n}\left(\frac{n}{\frac{n}{s^k r}, \ldots, \frac{n}{s^k r}}\right)\kappa(\mathcal{H}_{s,r,k}(0, \ldots, 0))$. Approximate the multinomial using Stirling's approximation to the factorial. Finally, since all blocks have the same size, $\sum_T \prod_J n_J^{\deg_T(J)-1} = \kappa(B_{s,k})\left(\frac{n}{s^k r}\right)^{s^k-2}$ and $d(1) = \frac{(s-1)^k n}{s^k r}$. $\qquad\square$

References

1. Achlioptas, D., Naor, A.: The two possible values of the chromatic number of a random graph. Annals of Mathematics 162 (2005)
2. Appel, K., Haken, W.: Every planar map is four colorable. American Mathematical Society Bulletin 82(5), 711–712 (1976)
3. Biggs, N.: Algebraic Graph Theory, 2nd edn. Cambridge University Press, Cambridge (1993)
4. Bodirsky, M., Gröpl, C., Kang, M.: Generating labeled planar graphs uniformly at random. In: Baeten, J.C.M., Lenstra, J.K., Parrow, J., Woeginger, G.J. (eds.) ICALP 2003. LNCS, vol. 2719, pp. 1095–1107. Springer, Heidelberg (2003)
5. Denise, A., Vasconcellos, M., Welsh, D.J.A.: The random planar graph. Congressus Numerantium 113, 61–79 (1996)
6. Fritsch, R., Fritsch, G.: The Four-Color Theorem. Springer, Heidelberg (1998)
7. Garey, M.R., Johnson, D.S.: Computer and Intractability, a Guide to the Theory of NP-Completeness. Freeman and Company, New York (1979)
8. Grötsch, H.: Ein dreifarbensatz für dreikreisfreie netze auf der kugel. Wiss. Z. Martin Luther-Univ. Halle Wittenberg, Math.-Nat. Reihe 8, 109–120 (1959)
9. Heawood, P.J.: Map colour theorem. Quarterly Journal of Pure and Applied Mathematics 24, 332–338 (1890)
10. Hutchinson, J.P.: Coloring ordinary maps, maps of empires, and maps of the moon. Mathematics Magazine 66, 211–226 (1993)
11. Hell, P., Nešetřil, J.: On the complexity of H-coloring. Journal of Combinatorial Theory, B 48(1), 92–110 (1990)
12. Jackson, B., Ringel, G.: Solution of Heawood's empire problem in the plane. Journal für die Reine und Angewandte Mathematik 347, 146–153 (1984)
13. Janson, S., Łuczak, T., Ruciński, A.: Random Graphs. J. Wiley & Sons, Chichester (2000)
14. Knuth, D.E.: Another Enumeration of Trees. Can. J of Math. 20, 1077–1086 (1968)
15. McDiarmid, C., Steger, A., Welsh, D.J.A.: Random planar graphs. Journal of Combinatorial Theory 93 B, 187–205 (2005)
16. Moon, J.W.: Counting Labelled Trees. Canadian Mathematical Monographs, vol. 1, Canadian Mathematical Congress (1970)
17. Pak, I.M., Postnikov, A.E.: Enumeration of spanning trees of certain graphs. Russian Mathematical Survey 45(3), 220 (1994)
18. Robertson, N., Sanders, D., Seymour, P., Thomas, R.: The four-colour theorem. Journal of Combinatorial Theory 70 B, 2–44 (1997)
19. Salavatipour, M.: Graph Colouring via the Discharging Method. PhD thesis, Department of Computer Science - University of Toronto (2003)
20. Temperley, H.N.V.: On the mutual cancellation of cluster integrals in Mayer's fugacity series. Proceedings of the Physical Society 83, 3–16 (1964)
21. Wessel, W.: A short solution of Heawood's empire problem in the plane. Discrete Mathematics 191, 241–245 (1998)

A Random Oracle Does Not Help Extract the Mutual Information

Andrei Muchnik[1] and Andrei Romashchenko[2]

[1] Andrei Muchnik (24.02.1958 – 18.03.2007) worked at the Institute of New Technologies in Education
[2] Laboratoire de l'Informatique du Parallélisme (Lyon) and Institute for Information Transmission Problems (Moscow)
`Andrei.Romashchenko@ens-lyon.fr`

Abstract. Assume a tuple of words $\bar{x} = \langle x_1, \ldots, x_n \rangle$ has negligible mutual information with another word y. Does this mean that properties of Kolmogorov complexity for $\bar{x}$ do not change significantly if we relativize them conditional to y ? This question becomes very nontrivial when we try to formalize it. We investigate this question for a very particular kind of properties: we show that a random (conditional to $\bar{x}$) oracle y cannot help extract the mutual information from x_i's.

1 Introduction

Kolmogorov complexity $K(x)$ of a word x is the length of a minimal description of this word for an optimal algorithmic description method (see [1,4]). Respectively, conditional Kolmogorov complexity $K(x|y)$ is the length of a minimal description of x when y is known. In other words, $K(x|y)$ is Kolmogorov complexity of x with the oracle y.

The difference between plain and conditional complexities

$$I(x : y) = K(y) - K(y|x)$$

is called *information in x on y*. The basic result of the algorithmic information theory is the fact that $I(x : y)$ is symmetric up to a small additive term:

Theorem 1 (Kolmogorov–Levin, [1])

$$I(x : y) = I(y : x) + \mathcal{O}(\log K(x, y)) = K(x) + K(y) - K(x, y) + \mathcal{O}(\log N)$$

If the value $I(x : y)$ is negligible (logarithmic in $K(x, y)$), the words x and y are often called *independent*.

Intuitively it seems that if x and y are 'independent' then 'reasonable' algorithmic properties of x (expressible in terms of Kolmogorov complexity) should not change significantly when we relativize them conditional to y.

E. Ochmański and J. Tyszkiewicz (Eds.): MFCS 2008, LNCS 5162, pp. 527–538, 2008.

Let us find a formal statement corresponding to this intuition. Let us take a tuple $\bar{x} = \langle x_1, x_2, \ldots, x_n \rangle$ instead of a single word[1] x. Suppose that the mutual information between $\bar{x}$ and some y is negligible. Then it is easy to see that *the basic* properties of Kolmogorov complexity for $\bar{x}$ do not really change when we relativize them conditional to y:

$$K(x_i) \approx K(x_i|y), \ K(x_i, x_j) \approx K(x_i, x_j|y), \ldots,$$

for all i, j, etc. (the approximative equations hold up to $I(y : \bar{x}) + \mathcal{O}(\log K(\bar{x}))$, which is negligible by the assumption).

Further we deal with less trivial properties of Kolmogorov complexity. Probably the simplest appropriate example is the property of extractability of common information. Let $\bar{x} = \langle x_1, x_2 \rangle$ be a pair of binary words. We say that α bits of the common information can be extracted from this pair for a precision threshold k if

$$\exists z \text{ such that for } i = 1, 2 \ K(z|x_i) < k \text{ and } K(z) \geq \alpha$$

Straightforward arguments imply that for such a word z

$$K(z) \leq I(x_1 : x_2) + \mathcal{O}(k + \log K(x_1, x_2))$$

This is a very natural fact: it means that for a small threshold k we cannot extract from x_1, x_2 much more than $I(x_1 : x_2)$ bits of information.

The question on extracting common information cannot be reduced to the values of complexities $K(x_1)$, $K(x_2)$, $K(x_1, x_2)$. For example, given that $K(x_1) = K(x_2) = 2n$ and $K(x_1, x_2) = 3n$ we cannot say anything nontrivial about extracting common information. On one hand, there exist pairs $\langle x_1, x_2 \rangle$ with the given complexities, such that n bits of common information can be extracted from these words for a very small threshold $k = \mathcal{O}(1)$. On the other hand, there exist pairs with the same complexities such that only negligible amount of information can be extracted for pretty large k. See detailed discussions on this topic in [2,3,6,11]. A similar property of extracting common information can be investigated not only for pairs but also for all finite tuples $\langle x_1, \ldots, x_n \rangle$. For the sake of simplicity in the sequel we restrict ourselves to the case $n = 2$ (though our technique is suitable for all n).

Once again, our intuition says that *negligible mutual information* between $\langle x_1, \ldots, x_n \rangle$ and y actually means that *the relativization conditional to y should not change properties of $x_1, \ldots, x_n$*. Let us formalize this intuitive idea for the problem of extracting common information:

Assume the mutual information between $\bar{x} = \langle x_1, x_2 \rangle$ and y is negligible. Then α bits of common information between x_1 and x_2 can be extracted for a precision

---

[1] More formally, we fix a computable bijection between the set of binary words and the set of all finite tuples of binary words. Now every tuple has a code. When we talk about Kolmogorov complexity of pairs, triples, etc., we mean Kolmogorov complexity of codes of these tuples. There is no natural canonical encoding of all tuples. However the choice of a particular code is not essential. Changing this encoding we change Kolmogorov complexity of tuples by only $\mathcal{O}(1)$ additive term.

threshold k iff the same is true given y as an oracle (for possibly a little different precision threshold).

The 'if' part of the equivalence above is trivial (if some information can be extracted without any oracle, the same can be done also given an oracle). The interesting part is the 'only if' statement. Let us formulate it in the most natural way, with logarithmic thresholds:

Conjecture 1. *For every integer $C_1 > 0$ there exists an integer $C_2 > 0$ such that for all $\bar{x} = \langle x_1, x_2 \rangle$ and y, if $I(y : \bar{x}) \leq C_1 \log N$ and*

$$\exists w: \ K(w|y) \geq \alpha, \ K(w|x_i, y) \leq C_1 \log N \ (i = 1, 2),$$

where $N = K(\bar{x}, y)$, (i.e., α bits of information can be extracted from x_1, x_2 for the precision threshold $C_1 \log N$, assuming y is given as an oracle) then

$$\exists z: \ K(z) \geq \alpha, \ K(z|x_i) \leq C_2 \log N \ (i = 1, 2),$$

i.e., the same α bits of common information can be extracted without oracles (for another threshold $C_2 \log N$).

This natural statement is surprisingly hard to prove. In [7] this conjecture was proven for $\alpha = I(x_1 : x_2)$. The general case is still an open problem.

In this paper we prove a version of this conjecture for $o(N)$ thresholds instead of logarithmic ones.

Theorem 2. *For every function $f(N)$, $f(N) = o(N)$ there exists a function $g(N)$ (also $g(N) = o(N)$) such that for every $\bar{x} = \langle x_1, x_2 \rangle$ and y if $I(y : \bar{x}) \leq f(N)$ and*

$$\exists w: \ K(w|y) \geq \alpha, \ K(w|x_i, y) \leq f(N) \ (i = 1, 2),$$

where $N = K(\bar{x}, y)$, (i.e., α bits of information can be extracted from x_1, x_2 for the precision threshold $f(N)$, assuming y is given as an oracle) then

$$\exists z: \ K(z) \geq \alpha, \ K(z|x_i) \leq g(N) \ (i = 1, 2),$$

i.e., the same α bits of common information can be extracted without oracles (for another threshold $g(N)$).

It is rather uncommon for algorithmic information theory that a natural statement is proven with $o(\cdot)$-precision but not up to logarithmic terms. Thus, the challenge is to prove Theorem 2 for $g(N) = \mathcal{O}(f(N))$, or at least to show that Conjecture 1 is true.

In the rest of the paper we prove Theorem 2, and in Conclusion discuss some variant of Conjecture 1 that is known to be true.

2 Preliminaries and Technical Tools

The main proof of this article is based on two technical tools: typization of words with a given profile, and extracting the common information from bunches of words.

2.1   Complexity Profiles

For an n-tuple of words $\bar{x} = \langle x_1, \ldots, x_n \rangle$ and a set of indexes $V = \{i_1, \ldots, i_k\} \subseteq \{1, \ldots, n\}$ ($i_1 < i_2 < \ldots < i_k$) we denote by $\bar{x}_V$ the tuple of words x_j for $j \in V$:

$$\bar{x}_V = \langle x_{i_1}, \ldots, x_{i_k} \rangle.$$

Thus, $K(\bar{x}_V) := K(x_{i_1}, \ldots, x_{i_k})$. We let $K(\bar{x}_\emptyset) := K(\lambda)$ (where λ is the empty word). We use similar notations for conditional complexities: if $V = \{i_1, \ldots, i_k\} \subseteq \{1, \ldots, n\}$ and $W = \{j_1, \ldots, j_l\} \subseteq \{1, \ldots, n\}$ we denote

$$K(\bar{x}_V | \bar{x}_W) := K(x_{i_1}, \ldots, x_{i_k} | x_{j_1}, \ldots, x_{j_l}).$$

We also let $K(\bar{x}_V | \bar{x}_\emptyset) := K(\bar{x}_V | \lambda)$ (which is equal to $K(\bar{x})$ up to an additive constant).

Definition 1. *We call by* complexity profile $\boldsymbol{K}$ *of an n-tuple $x_1, \ldots, x_n$ the vector of integers that consists of all complexity quantities $K(\bar{x}_V | \bar{x}_W)$, where $V, W \subseteq \{1, \ldots, n\}$, $V \cap W = \emptyset$ and $V \neq \emptyset$. Note that complexity profile implicitly contains unconditional complexity quantities: if $W = \emptyset$ we have $K(\bar{x}_V | \bar{x}_\emptyset) = K(\bar{x}_V) + \mathcal{O}(1)$. We need to fix somehow the order of components in the complexity profile. Let us suppose that all pairs (V, W) are arranged in the lexicographical order, i.e.,*

$$\boldsymbol{K}(x_1, \ldots, x_n) = (K(x_1), K(x_1 | x_2), \ldots, K(x_2 | x_1), K(x_2 | x_3), \ldots).$$

Similarly we define the conditional complexity profile *of $x_1, \ldots, x_n$ conditional to some y. It is the vector of all complexity quantities $K(\bar{x}_V | \bar{x}_W, y)$:*

$$\boldsymbol{K}(x_1, \ldots, x_n | y) = (K(x_1 | y), K(x_1 | x_2, y), \ldots, K(x_2 | x_1, y), K(x_2 | x_3, y), \ldots).$$

We say that a profile $\bar{\alpha}$ is not greater than another profile $\bar{\beta}$ (notation: $\bar{\alpha} \leq \bar{\beta}$) if *every* component of the first vector is not greater than the corresponding component of the second vector.

Denote by $\rho(\alpha, \beta)$ the l_∞-norm of the difference between the vectors α and β.

2.2   Typization

The method if *typization* was proposed in [8,10,9].

Definition 2. *Let $\bar{x} = \langle x_1, \ldots, x_n \rangle$ and $\bar{y} = \langle y_1, \ldots, y_m \rangle$ be tuples of words. The* typization *of $\bar{x}$ conditional to $\bar{y}$ is the following set of n-tuples:*

$$T(\bar{x} | \bar{y}) := \{\bar{x}' = \langle x_1', \ldots, x_n' \rangle \mid \boldsymbol{K}(\bar{x}', \bar{y}) \leq \boldsymbol{K}(\bar{x}, \bar{y})\}.$$

Further, the k-strong typization *of $\bar{x}$ conditional to $\bar{y}$ is the following set:*

$$ST_k(\bar{x} | \bar{y}) := T(\bar{x} | \bar{y}) \cap \{\bar{x}' = \langle x_1', \ldots, x_n' \rangle \mid \rho(\boldsymbol{K}(\bar{x}', \bar{y}), \boldsymbol{K}(\bar{x}, \bar{y})) \leq k\}.$$

Obviously there exists an algorithm that enumerates the list of all elements of $T(\bar{x}|\bar{y})$ given as an input the tuple $\bar{y}$ and the profile $\boldsymbol{K}(\bar{x}, \bar{y})$.

The following Lemmas are proven in [8,9]:

Lemma 1. *For every $\bar{x} = (x_1, \ldots, x_n)$ and $\bar{y} = (y_1, \ldots, y_m)$*

$$\log |T(\bar{x}|\bar{y})| = K(\bar{x}|\bar{y}) + \mathcal{O}(\log N),$$

where $N = K(\bar{x}, \bar{y})$. The constant in $\mathcal{O}(\cdot)$-notation depends on n and m.

Lemma 2. *There exists a computable function $C = C(n, m)$ such that for every n-tuple $\bar{x} = \langle x_1, \ldots, x_n \rangle$ and for every m-tuple $\bar{y} = \langle y_1, \ldots, y_m \rangle$ it holds*

$$|ST_{C(n,m) \log N}(\bar{x}|\bar{y})| > \frac{1}{2}|T(\bar{x}|\bar{y})|,$$

where $N = K(\bar{x}, \bar{y})$.

For brevity we denote by $ST(\bar{x}|\bar{y})$ the set $ST_{C \log N}(\bar{x}|\bar{y})$, where C is the value from Lemma 2.

2.3 Bunches

The following definition of a *bunch* was given in [12]:

Definition 3. *A set $X \subset \{0,1\}^*$ is called an (α, β, γ)-bunch if*

1. $|X| = 2^\alpha$,
2. $K(x_1|x_2) < \beta$ for every $x_1, x_2 \in X$,
3. $K(x) < \gamma$ for all $x \in X$.

The usage of this definition is based on the following combinatorial lemma:

Lemma 3 ([12]). *There exists an algorithm that takes (α, β, γ) as an input and prints a list of* standard (α, β, γ)-branches $U_0, \ldots, U_q$ *such that:*

- *for every (α, β, γ)-bunch U there exists a number $i \leq q$ such that $|U \cap U_i| \geq 2^{\beta - \epsilon}$, $\epsilon = 2(\beta - \alpha) + \mathcal{O}(1)$,*
- *$q < 2^{\beta + \gamma - 2\alpha + \mathcal{O}(1)}$.*

Here is a typical usage of Lemma 3: Assume we are given 2^n words a_i of complexity $2n$, and for every pair a_i, a_j it holds $K(a_i|a_j) \leq n$. Then the given family of words is an $(n, n, 2n)$-bunch. From the lemma it follows that some U_s from the list of 'standard bunches' (here $s < 2^n$) contains at least $\Omega(2^n)$ of the words a_i. It is not hard to show that for *all* given a_i

$$K(a_i|s) \leq n + \mathcal{O}(\log n) \text{ and } K(s|a_i) = \mathcal{O}(\log n).$$

Thus, the ordinal number s of a standard bunch U_s is an n-bit 'kernel' of the given family of a_i's; it is a materialization of the mutual information of all these words. See a more detailed discussion and corollaries of these arguments in [12].

We need to modify the definition of a bunch:

Definition 4. *A set $X \subset \{0,1\}^*$ is called an (α, β, γ)-semi-bunch if*

1. $|X| = 2^\alpha$,
2. *for every $x_1 \in X$, for the majority of all words $x_2 \in X$ it holds $K(x_1|x_2) < \beta$*
3. $K(x) < \gamma$ for all $x \in X$.

The following statement generalizes Lemma 3:

Lemma 4. *There exists an algorithm that takes (α, β, γ) as an input and prints a list of (α, β, γ)-semi-bunches $U_0, \ldots, U_q$ such that:*

- *for every (α, β, γ)-semi-bunch U there exists a number $i \leq q$ such that $|U \cap U_i| \geq 2^{\beta - \epsilon}$, where $\epsilon = 2(\beta - \alpha) + \mathcal{O}(1)$,*
- $q < 2^{\beta + \gamma - 2\alpha + \mathcal{O}(1)}$.

The proof of Lemma 4 is almost the same as the proof of Lemma 3 in [12]. We prove this lemma in Appendix. Let us call the semi-bunches $U_0, \ldots, U_q$ from Lemma 4 *standard semi-bunches* (i.e., for each triple of parameters α, β, γ we fix a canonical list of standard semi-bunches).

3   Proof of Theorem 2

Let us define some notations and make several assumptions. W.l.o.g. we may suppose that $f(N) > \log N$, and $f(N)$ does not decrease ($f(N+1) \geq f(N)$ for all N).

We chose $g(N)$ and $\delta(N)$ that are not 'too large' and not 'too small', so that the construction of the proof works. Let $\delta(N) = N/\sqrt{\log \frac{N}{f(N)}}$ and

$$g(N) = C(3^{D\sqrt{\log \frac{N}{f(N)}}} \cdot f(N) + \delta(N))$$

(we will fix the constants C and D later). For brevity we will write just δ if the value of N is clear from the context.

The main construction

Informal idea
The main trick of the proof is typization of y and w conditional to $\bar{x}$. We take the set of all 'clones' of the pair $\langle y, w \rangle$, which have approximately the same complexity profile (conditional to $\bar{x}$). The two cases are possible:

The good case: Assume this set of 'clones' is well consolidated in the sense that most clones have large enough mutual information. Then we apply Lemma 4 and extract from the class of clones some common kernel z. This word z contains about α bits of information, and it is rather simple conditional to each of x_i. Thus we extract from the words x_i about α bits of common information without any oracle, and we are done.

The bad case: Assume the set of 'clones' is not well consolidated. Then there exist pairs of different clones that have rather small mutual information. At this stage we cannot extract from x_i's their common information. Instead we change the word y to some y_1 such that conditional to y_1 at least α_1 bits of common information (where α_1 is greater than α) can be extracted from the words x_1, x_2. Thus, we come back to the assumption of the theorem, but with a greater value α_1 instead of α and a new oracle y_1 instead of y. The price for this modification is some loss of precision: instead of the term $f(N)$ we get some greater threshold $f_1(N)$.

The technical question is how to get such a word y_1. The answer is based on the fact that the set of 'clones' is not well consolidated. If we take two of them at random (denote them $\langle y', w' \rangle$ and $\langle y'', w'' \rangle$) then the pair $\langle y', y'' \rangle$ can play the role of y_1. Indeed, with the new oracle we can extract from x_i's both w' and w'', which make up α_1 bits of common information ($\alpha_1 > \alpha$; technically, we will get $\alpha_1 \geq \alpha + \delta/2$).

Then we iterate the trick above again and again, until at some stage we get a well consolidated set of clones...

The formal arguments

We are given a w such that $K(w|x_i, y) \leq f(N)$ (for $i = 1, 2$). W.l.o.g. we assume that $\alpha = K(w|y)$ (if $K(w|y) > \alpha$, we increase the value of α; this makes the statement only stronger). Denote $m = K(y)$. The aim is to construct z such that $K(z|x_i) \leq g(n)$ and $K(z) \geq \alpha - g(N)$.

We take the strong typization of $\langle y, w \rangle$ conditional to x: $A = ST(y, w|\bar{x})$. From Lemma 1 it follows $|A| = 2^{K(y,w|\bar{x}) - \mathcal{O}(f(N))}$. We have

$$K(y, w|\bar{x}) = K(y|\bar{x}) + K(w|y, \bar{x}) + O(\log N),$$

$K(y|\bar{x}) \geq K(y) - f(N)$ (the mutual information between y and $\bar{x}$ is negligible) and $K(w|y, \bar{x}) \leq f(N)$ (w can be easily extracted from any x_i given y as an oracle). Hence, $|A| = 2^{m - \mathcal{O}(f(N))}$. Note that for every $\langle y', w' \rangle \in A$ it holds

$$K(y', w') = K(y') + K(w'|y) + \mathcal{O}(\log N) = m + \alpha + \mathcal{O}(f(N)).$$

Two cases are possible:

Case 1^0: For every $\langle y', w' \rangle \in A$ for the majority of $\langle y'', w'' \rangle \in A$

$$I(y'w' : y''w'') \geq \alpha - \delta.$$

This inequality implies that

$$K(y'w'|y''w'') = K(y', w') - I(y'w' : y''w'') \leq m + \delta + \mathcal{O}(f(N)).$$

In this case the set A is a semi-bunch with the parameters

$$(m - \mathcal{O}(f(N)), m + \delta + \mathcal{O}(f(N)), m + \alpha + \mathcal{O}(f(N)).$$

We apply Lemma 4: it follows that there exists a *standard semi-bunch* U_j (with the same parameters) such that

$$|A \cap U_j| \geq 2^{m - \delta + \mathcal{O}(f(N))},$$

and j is an integer less than $2^{\alpha+\delta+\mathcal{O}(f(N))}$. So Kolmogorov complexity of j is not greater than $\alpha + \delta + \mathcal{O}(f(N))$.

Further, the words x_i ($i = 1, 2$) have two properties:

- for every pairs $\bar{v} \in A \cap U_j$ it holds $K(x_i|\bar{v}) \leq K(x_i|y, w)$ (by the definition of $A = ST(y, w|\bar{x})$);
- for every pair $\bar{v} \in A \cap U_j$ it holds $K(\bar{v}|j) \leq \log |U_j| + \mathcal{O}(\log N) \leq m$ (given the number j, the elements of a standard semi-bunch U_j can be enumerated algorithmically).

This means that x_i belong to the set

$$X(i) = \{\hat{x} \mid \text{ there exists at least } 2^{m-\delta+\mathcal{O}(f(N))} \text{ words } \bar{v}$$
$$\text{s.t. } K(\hat{x}|\bar{v}) \leq K(x_i|y, w) \leq K(x_i) - \alpha + f(N) \text{ and } K(\bar{v}|j) \leq m\}.$$

The set $X(i)$ is enumerable given j and additional $\mathcal{O}(\log N)$ bits of information (we need these additional bits to specify the parameters of the semi-bunch). Also we can bound the size of $X(i)$. Indeed, for each fixed j there exist at most 2^{m+1} different tuple $\bar{v}$ such that $K(\bar{v}|j) \leq m$; for every $\bar{v}$ there exist at most $2^{K(x_i)-\alpha+f(N)}$ different $\hat{x}$ such that $K(\hat{x}|\bar{v}) \leq K(x_i) - \alpha + f(N)$. Since for every $\hat{x} \in X(i)$ there is at least $2^{m-\delta+\mathcal{O}(f(N))}$ different $\bar{v}$, we get

$$\log |X(i)| \leq \log \frac{2^m \cdot 2^{K(x_i)-\alpha+f(N)}}{2^{m-\delta+O(f(N))}} \leq K(x_i) - \alpha + \delta + \mathcal{O}(f(N)).$$

It follows that $K(x_i|j) \leq K(x_i) - \alpha + \delta + \mathcal{O}(f(N))$ (in a word, the mutual information between j and x_i is at least $\alpha - \delta - \mathcal{O}(f(N))$). From symmetry of the mutual information we have

$$K(j|x_i) = K(x_i|j) + K(j) - K(x_i) + \mathcal{O}(\log N) \leq 2\delta + \mathcal{O}(f(N)).$$

We set $z = j$. Since $K(z) \geq I(z : x_i) \geq \alpha - \delta - \mathcal{O}(f(N))$, we get $K(z) \geq \alpha - g(N)$.

Thus for the function $g(n)$ defined above it holds $K(z) \geq \alpha - g(N)$ and $K(z|x_i) \leq g(N)$, and we are done.

Case 2^0. For some pair $\langle y', w' \rangle \in A$ and for the majority of $\langle y'', w'' \rangle \in A$ it holds

$$I(y'w' : y''w'') < \alpha - \delta.$$

This means that

$$K(y'y''w'w'') \geq 2m + \alpha + \delta - \mathcal{O}(\log N) \tag{1}$$

Since this inequality holds for the majority of pairs $\langle y'', w'' \rangle \in A$, we can choose one of them such that $\langle y', w' \rangle$ and $\langle y'', w'' \rangle$ are independent conditional to $\bar{x}$. In particular, the words y' and y'' are also independent conditional to $\bar{x}$ (i.e., $I(y' : y''|\bar{x}) = \mathcal{O}(\log N)$). Further, for all $\bar{x}, y', y''$ the following inequality holds:

$$I(y'y'' : \bar{x}) \leq I(y' : \bar{x}) + I(y'' : \bar{x}) + I(y' : y''|\bar{x}) + \mathcal{O}(\log N)$$

(in fact this inequality is equivalent to the sum of two trivial ones:

$$K(y'y'') \leq K(y') + K(y'') + \mathcal{O}(\log N),$$
$$K(y'|x) + K(y''|x) = K(y'y''|x) + I(y' : y''|x) + \mathcal{O}(\log N),$$

which follow immediately from the Kolmogorov–Levin theorem [1]). For the given words, the quantities $I(y' : \bar{x})$ and $I(y'' : \bar{x})$ are bounded by $f(N)$ ($\bar{x}$ and y are independent), and $I(y' : y''|\bar{x}) = \mathcal{O}(\log N) \ll f(N)$. Thus, we have

$$I(y'y'' : \bar{x}) \leq 3f(N) \qquad (2)$$

Also we have $K(y'y'') \leq 2K(y) + 3f(N) \leq 3N$ (a very rough bound).

From (1) and (2) it follows that for $y^1 = \langle y', y'' \rangle$ and $w^1 = \langle w', w'' \rangle$ it holds

$$K(w^1|y^1) \geq \alpha + \delta - 3f(N) - \mathcal{O}(\log N) \geq \alpha + \delta/2.$$

Thus, we have constructed a word y^1 such that $I(y^1 : \bar{x}) \leq 3f(N)$ and

$$\exists w^1 : \ K(w^1|y^1) \geq \alpha + \delta/2, \ K(w^1|x_i, y^1) \leq 3f(N) \ (i = 1, 2).$$

We have got a new pair $\langle y^1, w^1 \rangle$ instead of the original one $\langle y, w \rangle$. By the construction, the word y^1 is independent from $\bar{x}$ (though the precision of 'independence' becomes three times worse: $I(y^1 : \bar{x}) \leq 3f(N)$). Given y^1 as an oracle, the word w^1 is simple conditional to each x_i (the precision of 'simplicity' also becomes $3f(N)$). Complexity of w^1 conditional to y^1 is not less than $\alpha + \delta/2$. Thus, $\alpha + \delta/2$ bits of common information can be extracted from the words x_1, x_2 with the precision threshold $3f(N)$ given y^1 as an oracle. Note that complexities of the words w^1, y^1 are not greater than $3N$.

Further we iterate the arguments above. We repeat the same procedure with the pair w^1, y^1. Denote $\alpha^1 = \alpha + \delta/2$, $m_1 = K(y^1)$, and $f_1(N) = 3f(N)$. We take the strong typization of the pair $\langle y^1, w^1 \rangle$ conditional to $\bar{x}$:

$$A^1 = ST(y^1, w^1|\bar{x}).$$

Once again, we consider two cases.

Case 1^1. For every $\langle y', w' \rangle \in A^1$ for the majority $\langle y'', w'' \rangle \in A^1$

$$I(y'w' : y''w'') \geq \alpha_1 - \delta.$$

In this case A^1 is a semi-bunch with the following parameters:

$$(m_1 - \mathcal{O}(f_1(N)), m_1 + \delta + \mathcal{O}(f_1(N)), m_1 + \alpha_1 + \mathcal{O}(f_1(N))).$$

From Lemma 4 we get a number j such that for $i = 1, 2$

$$K(j|x_i) \leq 2\delta + \mathcal{O}(f_1(N)), \ \ I(j : x_i) \geq \alpha_1 - \delta + \mathcal{O}(f_1(N)).$$

Similarly to Case 1^0, we define $z := j$, and we are done.

Case 2^1. Assume that for some $\langle y', w' \rangle \in A^1$ and for the majority of $\langle y'', w' \rangle \in A^1$ it holds $I(y'w' : y''w'') < \alpha_1 - \delta$. Then there exists a pair $\langle y^2, w^2 \rangle$ such that

1. $K(y^2) = m_2 < 3m_1$,
2. $I(y^2 : \bar{x}) \leq f_2(N) := 3f_1(N)$,
3. $K(w^2|y^2, x_i) \leq f_2(N)$,
4. $K(w^2|y^2) = \alpha_2 \geq \alpha_1 + \delta/2$.

Iterating these arguments again and again, at stage s we get some words w^s, y^s such that

1. $K(y^s) = m_s = 3m_{s-1}$,
2. $I(y^s : \bar{x}) \leq f_s(N) := 3f_{s-1}(N) = 3^s f(N)$,
3. $K(w^s|y^s, x_i) \leq f_s(N)$,
4. $K(w^s|y^s) = \alpha_s > \alpha_{s-1} + \delta/2 = \alpha + s\delta/2$.

We are iterating the same construction for the 'bad' cases $2^1, 2^2, 2^3 \ldots, 2^j, \ldots$ until at some step s_{max} we come to the 'good' case $1^{j_{max}}$.

This iteration process cannot be too long. Indeed, after $s = D\sqrt{\log \frac{N}{f(N)}}$ steps of the iteration (for large enough D) we get a contradiction with the inequality

$$K(w^s|y^s) \leq K(w^s|x_1, y^s) + K(w^s|x_2, y^s) + I(x_1 : x_2|y^s) + \mathcal{O}(\log N)$$

(it is easy to check that this inequality holds for all words, see e.g., the proof of inequality (6) in [8]): the value on the left-hand side of the inequality is at least $DN/2$, and the right-hand side is only

$$2f_s(N) + I(x_1 : x_2|y_s) + \mathcal{O}(\log N) \ll N.$$

Remark: In all the arguments above we ignore additive terms of the order $\mathcal{O}(\log K(y^s, w^s))$ because $\log K(y^s, w^s) \ll f(N)$. This bound is valid since $K(y^s)$, $K(w^s) < N^2$ for $s \ll \log N$.

Thus, after several iterations of **Case** 2^s, for some $s_{max} < D\sqrt{\log \frac{N}{f(N)}}$ we get **Case** $1^{s_{max}}$. We obtain some word z such that

$$K(z) \geq \alpha + s_{max}\delta/2 - \mathcal{O}(f_{s_{max}}(N)) > \alpha - g(N)$$

and

$$K(z|x_i) \leq 2\delta + f_{s_{max}} < 2\delta + 3^{D\sqrt{\log \frac{N}{f(N)}}} f(N) < g(N) \; (i = 1, 2).$$

In other words, at least α bits of common information can be extracted from the words x_i for the precision threshold $g(N)$.

4   Conclusion

We cannot prove Conjecture 1 in the general case. However we know that it is true for *stochastic* pairs $\langle x_1, x_2 \rangle$.

Definition 5. *A tuple $\bar{x}$ is called (α, β)-stochastic if there exists a finite set $A \ni \bar{x}$ such that (a) complexity of the list of all elements of A (in lexicographical order) is at most α, and (b) $K(\bar{x}|[\text{list of all elements of } A]) \geq \log|A| - \beta$ (c.f. the definition of (α, β)-stochastic sequences [4]).*

In most applications of Kolmogorov complexity all tuples under consideration are (α, β)-stochastic with logarithmic α, β. For stochastic tuples Conjecture 1 is true:

Theorem 3. *For every integer $C_1 > 0$ there exists an integer $C_2 > 0$ such that for all y and all $(C_1 \log N, C_1 \log N)$-stochastic $\bar{x} = \langle x_1, x_2 \rangle$ if $I(y : \bar{x}) \leq C_1 \log N$ and*

$$\exists w : \; K(w|y) \geq \alpha, \; K(w|x_i, y) \leq C_1 \log N \; (i = 1, 2), \; \text{where } N = K(\bar{x}, y),$$

then $\exists z : \; K(z) \geq \alpha, \; K(z|x_i) \leq C_2 \log N \; (i = 1, 2)$.

(We skip the proof due to the lack of space).

Thus, Conjecture 1 is still an open problem. Also there is another interesting question: Does any counterpart of the results above hold for infinite oracles ?

References

1. Zvonkin, A.K., Levin, L.A.: The Complexity of Finite Objects and the Algorithmic Concepts of Information and Randomness. Russian Math. Surveys 25(6), 83–124 (1970)
2. Gács, P., Körner, J.: Common Information Is far Less Than Mutual Information. Problems of Control and Information Theory 2, 49–62 (1973)
3. Ahlswede, R., Körner, J.: On Common Information and Related Characteristics of Correlated Information Sources. In: Proc. 7th Prague Conf. on Inf. Th. Stat. Dec. Fct's and Rand (1974)
4. Li, M., Vitányi, P.: An introduction to Kolmogorov complexity and its applications, 2nd edn. Springer, New York (1997)
5. Zhang, Z., Yeung, R.W.: On Characterization of Entropy Functions via Information Inequalities. IEEE Trans. on Information Theory 44, 1440–1452 (1998)
6. Muchnik, A.A.: On Common Information. Theoretical Computer Science 207, 319–328 (1998)
7. Romashchenko, A.: Pairs of Words with Nonmaterializable Mutual Information. Problems of Information Transmission 36(1), 1–18 (2000)
8. Hammer, D., Romashchenko, A., Shen, A., Vereshchagin, N.: Inequalities for Shannon Entropy and Kolmogorov Complexity. Journal of Computer and System Sciences 60, 442–464 (2000)
9. Makarychev, K., Makarychev, Yu., Romashchenko, A., Vereshchagin, N.: A New Class of non Shannon Type Inequalities for Entropies. Communications in Information and Systems 2(2), 147–166 (2002)
10. Romashchenko, A., Shen, A., Vereshchagin, N.: Combinatorial Interpretation of Kolmogorov Complexity. Theoretical Computer Science 271, 111–123 (2002)
11. Chernov, A., Muchnik, A.A., Shen, A., Romashchenko, A., Vereshchagin, N.K.: Upper Semi-Lattice of Binary Strings with the Relation "x Is Simple Conditional to y". Theoretical Computer Science 271, 69–95 (2002)
12. Romashchenko, A.: Extracting the Mutual Information for a Triple of Binary Strings. In: Proc. 18th Annual IEEE Conference on Computational Complexity (2003)

Appendix

Proof of Lemma 4: First of all, let us fix an algorithm that gets integers α, β, γ as an input, and enumerates the list of *all* (α, β, γ)-semi-bunches. We call this algorithm the *complete enumerator*. Though the number of semi-bunches (for given parameters) is finite, the complete enumerator never stops. We cannot decide effectively if it has already found *all* semi-bunches. We only guarantee that each semi-bunch must be enumerated in the list, soon or late.

Now we describe another enumerator, which chooses some subsequence from the complete enumeration of all semi-bunches as follows. The complete enumerator prints semi-bunches one by one, and we need to select some of them. Assume some semi-bunches $U_0, \ldots, U_s$ are already selected, and the complete enumerator finds a new semi-bunch V. If $|V \cap U_i| < 2^{\beta-\epsilon}$ for all $i = 0, \ldots, s$, where $\epsilon = 2(\beta - \alpha + 2)$, then we *select* this semi-bunch and let $U_{s+1} = V$. Otherwise we skip V and wait for the next item from the complete enumeration.

Let $U_0, \ldots, U_q$ be the list of all selected semi-bunches for given α, β, γ. From the construction it is evident that for every semi-bunch V either $V = U_i$ or at least $|V \cap U_i| \geq 2^{\beta-\epsilon}$ for some $i \leq q$. Also it follows from the construction that $|U_i \cap U_j| < 2^{\beta-\epsilon}$ for every two different *selected* semi-bunches U_i, U_j. It remains to prove that q is not too large.

In fact it is enough to prove that every x belongs to less than $2^{\beta-\alpha+2}$ selected semi-bunches. Indeed, there are less than 2^γ words x such that $K(x) < \gamma$. If every x belongs to at most $2^{\beta-\alpha+2}$ selected semi-bunches, and every semi-bunch U_i contains 2^α words then the number of all selected semi-bunches is bounded by

$$\frac{2^\gamma \cdot 2^{\beta-\alpha+2}}{2^\alpha} = 2^{\beta+\gamma-2\alpha+2}$$

Thus, it remains to bound the number of selected semi-bunches that contain one fixed word x.

Assume that there exist $N = 2^{\beta-\alpha+2}$ different selected semi-bunches U_i that contain the same word x. Denote

$$U_i' = U_i \cap \{y \mid K(y|x) < \beta\}$$

for all these semi-bunches U_i. From the definition of a semi-bunch it follows that U_i' contains at least $2^{\alpha-1}$ elements.

On one hand, we have

$$\left\| \bigcup U_i' \right\| \leq \|\{y \mid K(y|x) < \beta\}\| < 2^\beta$$

On the other hand,

$$\left\| \bigcup U_i' \right\| \geq \sum_i \|U_i'\| - \sum_{i<j} \|U_i' \cap U_j'\|$$

As $\|U_i'\| \geq 2^{\alpha-1}$ and $\|U_i' \cap U_j'\| \leq \|U_i \cap U_j\| \leq 2^{\beta-\epsilon}$, it follows that

$$\left\| \bigcup U_i' \right\| \geq N \cdot 2^{\alpha-1} - N^2 \cdot 2^{\beta-\epsilon} = 2^\beta$$

and we get a contradiction. The lemma is proven.

Approximating Independent Set and Coloring in Random Uniform Hypergraphs

Kai Plociennik

TU Chemnitz, Straße der Nationen 62, 09107 Chemnitz, Germany
`kai.plociennik@informatik.tu-chemnitz.de`

Abstract. We consider the problems Independent Set and Coloring in uniform hypergraphs with n vertices. If $\mathcal{NP} \not\subseteq \mathcal{ZPP}$, there are no polynomial worst case running time approximation algorithms with approximation guarantee $n^{1-\varepsilon}$ for any $\varepsilon > 0$. We show that the problems are easier to approximate in polynomial expected running time for random hypergraphs. For $d \geq 2$, we use the $H_d(n,p)$ model of random d-uniform hypergraphs on n vertices, choosing the edges independently with probability p. We give deterministic algorithms with polynomial expected running time for random inputs from $H_d(n,p)$, and approximation guarantee $O(n^{1/2} \cdot p^{-(d-3)/(2d-2)}/(\ln n)^{1/(d-1)})$.

1 Introduction

A *hypergraph* $H = (V, E)$ consists of a finite set V of *vertices* (in the following, n always denotes $|V|$) and a set E of *edges*, which are subsets of V. For an integer d, H is *d-uniform* if all edges have cardinality d. Hence, a graph (without loops) is a 2-uniform hypergraph. For uniform hypergraphs $H = (V, E)$ as inputs, we consider two problems which we define next. The problem *Independent Set* (*IS*) is to compute a set $I \subseteq V$ with maximum cardinality such that I is *independent*, i. e., it spans no edges of E. The size of a largest independent set in H is its *independence number* $\alpha(H)$. A *coloring* with k colors is a partition $C = \{C_1, \ldots, C_k\}$ of V into k *color classes* such that all classes are independent. For a coloring C, we denote by $\text{big}(C)$ its largest class (break ties arbitrarily). The problem *Coloring* (*COL*) is to compute a coloring with as few as possible colors, the smallest possible number of colors being H's *chromatic number* $\chi(H)$. Notice that different generalizations of independence from graphs to hypergraphs have been studied. Ours results in the so called *weak chromatic number*, where in a coloring no edge may be monochromatic. For the *strong chromatic number*, the vertices of every edge must have pairwise different colors.

Our problems are well-studied with respect to their computational complexity. IS and COL are $\mathcal{NP}$-hard for graphs as inputs (see Karp [7]). Since for such problems it is unlikely that there are *efficient*, i. e., polynomial worst case running time algorithms which always compute optimal solutions, one considers approximation algorithms (for a survey, see e. g. Vazirani [11]). In this context, for the problem IS, we let the *approximation ratio* of an independent set I in

E. Ochmański and J. Tyszkiewicz (Eds.): MFCS 2008, LNCS 5162, pp. 539–550, 2008.
© Springer-Verlag Berlin Heidelberg 2008

a hypergraph H be $\alpha(H)/|I|$. For COL, a coloring with k colors has approximation ratio $k/\chi(H)$. Since optimal solutions have approximation ratio 1 and larger approximation ratios correspond to worse solutions, we seek for solutions with small approximation ratios. For a function $f = f(n)$, we say that an approximation algorithm has *approximation guarantee* $f(n)$ if for every uniform hypergraph with n vertices, it outputs a solution with approximation ratio at most $f(n)$. For IS (see Feige [1]) and COL (see Halldórsson [4]) in graphs, the best efficient approximation algorithms known have approximation guarantee $O(n \cdot (\log \log n)^2/(\log n)^3)$. For both problems, Hofmeister and Lefmann [6] presented efficient algorithms for d-uniform hypergraphs, $d \geq 2$ fixed, with approximation guarantee $O(n/(\log^{(d-1)} n)^2)$, where $\log^{(d-1)}$ denotes the $(d-1)$-fold iterated logarithm (later, Halldórsson [3] showed that the approximation guarantee can be shown to be $O(n/\log n)$ for all $d \geq 3$). For fixed $d \geq 3$, Krivelevich and Sudakov [9] gave an efficient algorithm for COL in d-uniform hypergraphs with approximation guarantee $O(n \cdot (\log \log n)^2/(\log n)^2)$.

Unfortunately, it seems that no considerably better approximation guarantee is achievable for our problems by efficient algorithms: For graphs as inputs, Håstad [5] showed for IS, and Feige and Kilian [2] for COL, that there is no efficient algorithm with approximation guarantee $n^{1-\varepsilon}$ for any $\varepsilon > 0$, assuming $\mathcal{NP} \not\subseteq \mathcal{ZPP}$. For fixed $d \geq 3$, Hofmeister and Lefmann [6] extended these results to IS and COL in d-uniform hypergraphs, and Krivelevich and Sudakov [9] did the same for COL. A recent result of Zuckerman [12] shows that the above inapproximability results for graphs also hold under the weaker assumption $\mathcal{P} \neq \mathcal{NP}$. Often, inapproximability results as above are due to "a small fraction of difficult inputs". Then, it may be possible to devise algorithms which perform considerably better on the average (which in this paper means for random inputs) than in the worst case. In the following, we use the $H_d(n, p)$ model of random hypergraphs. For an integer $d \geq 2$, $n \in \mathbb{N}$, and a probability $p \in [0, 1]$, the model $H_d(n, p)$ creates a random d-uniform hypergraph on n labeled vertices by inserting every possible edge of cardinality d independently of the others with probability p. This model extends the well-known $G(n, p)$ model of random graphs, and $H_2(n, p) = G(n, p)$. In [8], Krivelevich and Vu proved the following interesting lemma regarding the independence number and a certain matrix of a random graph from $G(n, p)$.

Lemma 1. *For a graph $H = (V = \{v_1, \ldots, v_n\}, E)$ and a probability $p \in (0, 1)$, define the $n \times n$-matrix $M(H, p) = (m_{ij})$ by*

$$m_{ij} := \begin{cases} 1 & \text{if } \{v_i, v_j\} \notin E \\ -(1-p)/p & \text{otherwise ,} \end{cases}$$

and let $\lambda_1(M)$ denote its largest eigenvalue. Then, for every graph H and every probability $p \in (0, 1)$, we have $\alpha(H) \leq \lambda_1(M(H, p))$. Furthermore, for $p = p(n) = \omega(n^{-1})$, $p < 1$, and a random graph H from $G(n, p)$, it holds that $\Pr[\lambda_1(M(H, p)) \geq 4 \cdot (n/p)^{1/2}] \leq 2^{-np/8}$.

Given p and a random graph H from $G(n, p)$, one can compute $\lambda_1(M(H, p))$ in polynomial time, since this can be done for real symmetric matrices in general (see e. g. Ralston [10]). Thus, one can efficiently compute an upper bound of $\lambda_1(M(H, p))$ on $\alpha(H)$, and Lemma 1 yields that this bound is less than $4 \cdot (n/p)^{1/2}$ with high probability (at least $1 - 2^{-np/8}$). Krivelevich and Vu used this tool to devise deterministic algorithms for IS and COL with the following properties: For random inputs from $G(n, p)$, with $p = p(n)$ in the range $n^{-1/2+\varepsilon} \le p \le 3/4$ with $\varepsilon > 0$ fixed, the expected running time is polynomial and the approximation guarantee is $O((np)^{1/2}/\log n)$, which is considerably better than what is possible in polynomial worst case running time due to the above inapproximability results. The general idea of the algorithms is as follows. Consider the problem IS. Given a random graph $H = (V, E)$ from $G(n, p)$, one computes an independent set I of H using a simple greedy algorithm and checks for a certain function $f_1(n, p)$, whether $|I| \ge f_1(n, p)$. If this holds, one continues and tries to prove that $\alpha(H) \le f_2(n, p)$ for another function $f_2(n, p)$. The basis of this step is the technique using Lemma 1 explained above. In case both steps succeed, I is output, achieving an approximation ratio of $\alpha(H)/|I| \le f_2(n, p)/f_1(n, p)$, where the latter is the approximation guarantee one claims for the algorithm. In case one of the steps fails, one performs an exhaustive search over all possible subsets of V, finding an optimal solution in exponential time. A polynomial expected running time is achieved since the probability that the exhaustive search with its exponential running time is performed is exponentially small. In this paper, we show how the algorithms of Krivelevich and Vu can be generalized to the case of d-uniform hypergraphs, $d \ge 2$. We present algorithms for IS and COL which for fixed $d \ge 2$ have approximation guarantee $O(n^{1/2} \cdot p^{-(d-3)/(2d-2)}/(\ln n)^{1/(d-1)})$ and polynomial expected running time for random inputs from $H_d(n, p)$. For a constant $c(d) > 0$, the edge probability $p = p(n)$ must be in the range $c(d) \cdot (\ln n)/n^{1-1/d} \le p \le 3/4$, except that for COL in case $d = 2$, it is required that $n^{-1/2+\varepsilon} \le p \le 3/4$ for fixed $\varepsilon > 0$. Throughout the rest of this paper, we implicitly assume n large enough if necessary.

2   Results

We analyze a simple greedy algorithm $Color(H = (V, E))$ for hypergraphs H. Later, it is used as a subroutine in our algorithms. Denote the vertices by $V = \{v_1, \ldots, v_n\}$. The algorithm computes a coloring C of the vertices of H with classes C_i, greedily assigning to each vertex the smallest possible color.

Algorithm Color($H = (V, E)$)
1. Set $C \leftarrow \{C_1 \leftarrow \{v_1\}\}$.
2. For $v = v_2, \ldots, v_n$: If for some class $C_i \in C$, $C_i \cup \{v\}$ is independent, set $C_i \leftarrow C_i \cup \{v\}$ for the smallest such i. Otherwise, create a new class by setting $C \leftarrow C \cup \{C_{|C|+1} \leftarrow \{v\}\}$.
3. Output C.

Lemma 2. *Fix an integer $d \ge 2$ and $0 < \varepsilon \le 1$. There is a constant $c(d, \varepsilon) > 0$ such that for probability $p = p(n)$ with $c(d, \varepsilon) \cdot (\ln n)^d/n^{d-1-\varepsilon} \le p \le 3/4$, the following holds. Let C be the coloring computed by algorithm $Color(H)$ for a* random hypergraph $H = (V, E)$ from $H_d(n, p)$. Then, $\Pr[|\mathrm{big}(C)| \leq ((d-2)! \cdot \varepsilon \cdot (\ln n)/(2 \cdot p))^{1/(d-1)}] \leq e^{-n \ln n}$.

Proof. For the constant $c(d, \varepsilon)$, we set $c(d, \varepsilon) := (d-2)! \cdot 8^{d-1} \cdot \varepsilon/2$. Furthermore, we set $s := ((d-2)! \cdot \varepsilon \cdot (\ln n)/(2 \cdot p))^{1/(d-1)}$ and $t := n/(2s)$. Let $\mathcal{E}$ denote the event "the size of $\mathrm{big}(C)$ is at most s". Then, the lemma states that $\Pr[\mathcal{E}] \leq e^{-n \ln n}$. If $\mathcal{E}$ happens, there are at least $n/s > t$ color classes in the coloring C. In this case, let $C^* := \{C_1, \ldots, C_t\}$, i.e., C^* contains the first t color classes of C. We call a set $D = \{D_1, \ldots, D_t\}$ of pairwise disjoint classes $D_i \subseteq V$ with $|D_i| \leq s$ for $i = 1, \ldots, t$ a *partial vertex coloring*. Furthermore, D is *bad* if every vertex $v \in V \setminus (D_1 \cup \ldots \cup D_t)$ is *bad*, i.e., for every class D_i, $i = 1, \ldots, t$, there is an edge $e \in E$ with $v \in e$ and $e \setminus \{v\} \subseteq D_i$.

Consider the set C^* in case that $\mathcal{E}$ happens. All classes in C^* are of size at most s. Furthermore, every vertex $v \in V \setminus (C_1 \cup \ldots \cup C_t)$ is not assigned to any of the classes in C^* by the algorithm. Thus, for every vertex v outside of C^* and every class $C_i \in C^*$, adding v to C_i violates the independence of C_i, so for every such vertex v and class C_i, there has to be an edge $e \in E$ with $v \in e$ and $e \setminus \{v\} \subseteq C_i$. Thus, we can conclude that if $\mathcal{E}$ happens, then C^* is a bad partial vertex coloring, so $\Pr[\mathcal{E}] \leq \Pr[\text{there is a bad partial vertex coloring}]$.

We estimate the probability that a given partial vertex coloring $D = \{D_1, \ldots, D_t\}$ is bad. The probability that a fixed vertex $v \in V \setminus (D_1 \cup \ldots \cup D_t)$ is bad equals $\prod_{i=1}^{t} \left(1 - (1-p)^{\binom{|D_i|}{d-1}}\right)$, as $(1-p)^{\binom{|D_i|}{d-1}}$ is the probability that there is no edge $e \in E$ with $v \in e$ and $e \setminus \{v\} \subseteq D_i$. Since for $i = 1, \ldots, t$, it holds that $|D_i| \leq s$, we obtain $|V \setminus (D_1 \cup \ldots \cup D_t)| \geq n - t \cdot s = n/2$. With $1 + x \leq e^x$ for $x \in \mathbb{R}$ and $\binom{a}{b} \leq a^b/b!$ for $a, b \in \mathbb{N}$, we infer that

$$\Pr[D \text{ is bad}] = \prod_{v \in V \setminus (D_1 \cup \ldots \cup D_t)} \Pr[v \text{ is bad}] \leq \left(\prod_{i=1}^{t} \left(1 - (1-p)^{\binom{|D_i|}{d-1}}\right)\right)^{n/2}$$

$$\leq e^{-\left(\sum_{i=1}^{t}(1-p)^{\binom{|D_i|}{d-1}}\right) \cdot n/2} \leq e^{-(n/2) \cdot \sum_{i=1}^{t}(1-p)^{|D_i|^{d-1}/((d-1)!)}}$$

$$\leq e^{-(tn/2) \cdot (1-p)^{s^{d-1}/((d-1)!)}} \leq e^{-(tn/2) \cdot n^{-\varepsilon/(d-1)}}, \tag{1}$$

again using $|D_i| \leq s$ for $i = 1, \ldots, t$. With $1 - x \geq e^{-2x}$ for $0 \leq x \leq 3/4$ and $p \leq 3/4$, (1) follows from

$$(1-p)^{s^{d-1}/((d-1)!)} \geq n^{-\varepsilon/(d-1)} \Leftarrow e^{-2ps^{d-1}/((d-1)!)} \geq n^{-\varepsilon/(d-1)}$$
$$\Leftrightarrow s \leq ((d-2)! \cdot \varepsilon \cdot (\ln n)/(2 \cdot p))^{1/(d-1)},$$

which holds by choice of s. Now, let $\widehat{D}$ be the set of all partial vertex colorings. Since there are at most $\sum_{i=1}^{s} \binom{n}{i}$ possible choices for each of the t color classes of a partial vertex coloring $D \in \widehat{D}$, (1) yields

$$\Pr[\mathcal{E}] \leq \Pr[\exists D \in \widehat{D}: D \text{ is bad}] \leq \sum_{D \in \widehat{D}} \Pr[D \text{ is bad}]$$

$$\leq \left(\sum_{i=1}^{s} \binom{n}{i}\right)^t \cdot e^{-(tn/2) \cdot n^{-\varepsilon/(d-1)}} \leq \binom{n}{s+1}^t \cdot e^{-(tn/2) \cdot n^{-\varepsilon/(d-1)}}, \tag{2}$$

where we used that $\sum_{i=1}^{s} \binom{n}{i} \le \binom{n}{s+1}$ for $s+1 \le n/3$, where the latter holds since $p \ge c(d, \varepsilon) \cdot (\ln n)^d/n^{d-1-\varepsilon}$. Now, observe that $\binom{n}{s+1}^t \le n^{(s+1)\cdot t} = n^{(s+1)\cdot n/(2s)} \le n^n = e^{n \ln n}$. Furthermore, by choice of p, we have $(tn/2) \cdot n^{-\varepsilon/(d-1)} \ge 2 \cdot n \ln n$. Thus, with (2), we finally get that $\Pr[|\text{big}(C)| \le s] = \Pr[\mathcal{E}] \le \binom{n}{s+1}^t \cdot e^{-(tn/2)\cdot n^{-\varepsilon/(d-1)}} \le e^{n \ln n - 2 \cdot n \ln n} = e^{-n \ln n}$ as claimed. $\square$

Lemma 3. *Fix an integer $d \ge 2$ and $0 < \varepsilon < 1/2$. Let $p = p(n)$ be a probability with $2^{(d-1)/2} \cdot (\ln n)^{(d+1)/2}/n^{(d-1)/2-\varepsilon/((d-2)!)} \le p \le 3/4$, and let C be the coloring computed by algorithm $\text{Color}(H)$ for a random hypergraph $H = (V, E)$ from $H_d(n, p)$. Then, for the number $|C|$ of classes in C, it holds that $\Pr[|C| \ge 2n \cdot (p/(\varepsilon \cdot \ln n))^{1/(d-1)}] < e^{-n \ln n}$.*

Proof. Let $k_0 := n \cdot (p/(\varepsilon \cdot \ln n))^{1/(d-1)}$. We prove that $\Pr[|C| \ge 2k_0] < e^{-n \ln n}$. Denote the vertices by $V = \{v_1, \ldots, v_n\}$. For $(k-1)(d-1)+1 \le j \le n$, let $\mathcal{B}_j^k$ denote the event that vertex v_j gets color k, i.e., it is assigned to class C_k (a vertex v_j with $j \le (k-1)(d-1)$ cannot get color k, since there are not enough sufficiently large color classes at the time of coloring v_j), and let $\mathcal{A}_j^k$ denote the event that for coloring the first j vertices, the algorithm uses at least k colors. Finally, let $\mathcal{A}^k$ be the event $|C| \ge k$. We estimate the conditional probability $\Pr[\mathcal{A}^{k+1}|\mathcal{A}^k]$. Since $\mathcal{A}^{k+1} = \bigcup_{j=k(d-1)+1}^{n} \mathcal{B}_j^{k+1}$, we infer that $\Pr[\mathcal{A}^{k+1}|\mathcal{A}^k] \le \sum_{j=k(d-1)+1}^{n} \Pr[\mathcal{B}_j^{k+1}|\mathcal{A}^k]$. Furthermore, since $\mathcal{B}_j^{k+1} \subseteq \mathcal{A}_{j-1}^k \subseteq \mathcal{A}^k$, it follows that $\Pr[\mathcal{B}_j^{k+1}|\mathcal{A}^k] \le \Pr[\mathcal{B}_j^{k+1}|\mathcal{A}_{j-1}^k]$, and hence,

$$\Pr[\mathcal{A}^{k+1}|\mathcal{A}^k] \le \sum_{j=k(d-1)+1}^{n} \Pr[\mathcal{B}_j^{k+1}|\mathcal{A}_{j-1}^k] \; . \tag{3}$$

We estimate $\Pr[\mathcal{B}_j^{k+1}|\mathcal{A}_{j-1}^k]$ for a fixed vertex v_j. The vertex v_j gets color $k+1$ only if for all classes C_i, $1 \le i \le k$, which exist at the time of coloring v_j since $\mathcal{A}_{j-1}^k$ occurs, there is an edge connecting v_j with $d-1$ vertices in C_i. Therefore,

$$\Pr[\mathcal{B}_j^{k+1}|\mathcal{A}_{j-1}^k] \le \prod_{i=1}^{k}(1 - (1-p)^{\binom{|C_i|}{d-1}}) \le e^{-\sum_{i=1}^{k}(1-p)^{\binom{|C_i|}{d-1}}}$$

$$\le e^{-\sum_{i=1}^{k}(1-p)^{|C_i|^{d-1}/((d-1)!)}} \le e^{-k\cdot(1-p)^{(n/k)^{d-1}/((d-1)!)}} \tag{4}$$

$$\le e^{-k\cdot e^{-2p\cdot(n/k)^{d-1}/((d-1)!)}} \; . \tag{5}$$

In (4), we used that for $s := \sum_{i=1}^{k} |C_i|$, it follows with $s \le n$ that $\sum_{i=1}^{k}(1-p)^{|C_i|^{d-1}/((d-1)!)} \ge k \cdot (1-p)^{(s/k)^{d-1}/((d-1)!)} \ge k \cdot (1-p)^{(n/k)^{d-1}/((d-1)!)}$. To get (5), we used that $1 - x \ge e^{-2x}$ for $0 \le x \le 3/4$ and $p \le 3/4$. In the following, we assume $k \ge k_0$. With $k_0 = n \cdot (p/(\varepsilon \cdot \ln n))^{1/(d-1)}$, (3), and (5) we infer

$$\Pr[\mathcal{A}^{k+1}|\mathcal{A}^k] \le \sum_{j=k(d-1)+1}^{n} \Pr[\mathcal{B}_j^{k+1}|\mathcal{A}_{j-1}^k] \le ne^{-k\cdot e^{-2p\cdot(n/k)^{d-1}/((d-1)!)}}$$

$$\le ne^{-k_0\cdot e^{-2p\cdot(n/k_0)^{d-1}/((d-1)!)}} = e^{\ln n - k_0\cdot n^{-2\varepsilon/((d-1)!)}} \le e^{-k_0\cdot n^{-2\varepsilon/((d-1)!)}/2} \; . \tag{6}$$

In (6), we used that since $p \ge 2^{(d-1)/2}\cdot(\ln n)^{(d+1)/2}/n^{(d-1)/2-\varepsilon/((d-2)!)}$, it follows that $k_0 \cdot n^{-2\varepsilon/((d-1)!)} = n \cdot (p/(\varepsilon \cdot \ln n))^{1/(d-1)} \cdot n^{-2\varepsilon/((d-1)!)} \ge (2^{1/2}/\varepsilon^{1/(d-1)})$.

$n^{1/2-\varepsilon/((d-1)!)} \cdot (\ln n)^{1/2}$. Furthermore, since $\varepsilon < 1/2$ and $d \geq 2$, it holds that $n^{1/2-\varepsilon/((d-1)!)} = n^{\Omega(1)} = \omega(\ln n)$, yielding $k_0 \cdot n^{-2\varepsilon/((d-1)!)} \geq 2\ln n$. Now, with (6) we get

$$\Pr\left[|C| \geq 2n \cdot (p/(\varepsilon \cdot \ln n))^{1/(d-1)}\right] = \Pr[\mathcal{A}^{2k_0}] \leq \prod_{k=k_0}^{2k_0-1} \Pr[\mathcal{A}^{k+1}|\mathcal{A}^k]$$
$$\leq e^{-k_0^2 \cdot n^{-2\varepsilon/((d-1)!)}/2} = e^{-n^2 \cdot (p/(\varepsilon \cdot \ln n))^{2/(d-1)} \cdot n^{-2\varepsilon/((d-1)!)}/2}$$
$$\leq e^{-n^2 \cdot (2 \cdot (\ln n)/n^{1-2\varepsilon/((d-1)!)}/\varepsilon^{2/(d-1)}) \cdot n^{-2\varepsilon/((d-1)!)}/2}$$
$$= e^{-n^{2-1+2\varepsilon/((d-1)!)-2\varepsilon/((d-1)!)} \cdot (\ln n)/\varepsilon^{2/(d-1)}} = e^{-n \cdot (\ln n)/\varepsilon^{2/(d-1)}} < e^{-n \ln n} \ . \qquad \square$$

2.1   Algorithms for Independent Set

Lemma 2 immediately yields the following approximation algorithm for IS with approximation guarantee $O(n \cdot (p/\ln n)^{1/(d-1)})$ and polynomial expected running time for random inputs from $H_d(n, p)$. Notice that in addition to the random hypergraph H from $H_d(n, p)$, it is given the edge probability p used to create H, and a parameter $\varepsilon \in (0, 1]$ which is used to adapt the algorithm to different lower bounds for the possible edge probabilities p. All following algorithms perform a sequence of steps. If a step outputs a solution, it also terminates. If it does not (but could have done so in general), we say that the step *fails*.

Algorithm IndepSet1$(H = (V, E), p, \varepsilon)$
1. $C \leftarrow Color(H)$. If $|\text{big}(C)| > ((d-2)! \cdot \varepsilon \cdot (\ln n)/(2 \cdot p))^{1/(d-1)}$, output big$(C)$.
2. Test all subsets of V for independence and output a largest independent subset found.

Theorem 1. *Fix an integer $d \geq 2$ and $0 < \varepsilon \leq 1$. There is a constant $c(d, \varepsilon) > 0$ such that for probability $p = p(n)$ with $c(d, \varepsilon) \cdot (\ln n)^d/n^{d-1-\varepsilon} \leq p \leq 3/4$, algorithm IndepSet1$(H, p, \varepsilon)$ has approximation guarantee $O(n \cdot (p/\ln n)^{1/(d-1)})$ and polynomial expected running time for random inputs $H = (V, E)$ from $H_d(n, p)$.*

Proof. We start with the expected running time of the algorithm, which is the sum of the expected running times of the algorithm's steps. A step's expected running time is the product of its *effort*, i.e., the time spent if it is executed, and the probability that it is executed. Step 1 has polynomial effort and hence also polynomial expected running time, since algorithm *Color* has polynomial worst case running time: For $n-1$ vertices v, it checks for at most n color classes C_i whether there is an edge among the $O(n^d)$ edges of H which consists of v and $d-1$ vertices in C_i. Thus, *Color* performs a total of $O(n^{d+2})$ polynomial time tests. Step 2's effort is $O(q(n) \cdot 2^n)$ for a polynomial $q(n)$ (in the following, $q(n)$ always denotes a suitably chosen polynomial), since it tests 2^n subsets of V, each of which can be tested in polynomial time. Since it is executed only if $|\text{big}(C)| \leq ((d-2)! \cdot \varepsilon \cdot (\ln n)/(2 \cdot p))^{1/(d-1)}$, Lemma 2 yields an execution probability of at most $e^{-n \ln n} \leq 2^{-n}$, and thus its expected running time is $O(q(n) \cdot 2^n \cdot 2^{-n}) = O(q(n))$, which is polynomial. Lemma 2 is applicable, since its interval of legal edge probabilities p coincides with the one in our theorem.

We turn to the approximation guarantee of the algorithm and show that each step achieves the approximation ratio claimed in the theorem. Step 2 achieves an approximation ratio of 1, since it computes an optimal solution. If Step 1 outputs $\text{big}(C)$, we have $|\text{big}(C)| > ((d-2)! \cdot \varepsilon \cdot (\ln n)/(2 \cdot p))^{1/(d-1)} = \Omega(((\ln n)/p)^{1/(d-1)})$. Trivially, $\alpha(H) \leq n$. Therefore, the approximation ratio achieved in Step 1 is $\alpha(H)/|\text{big}(C)| = n/\Omega(((\ln n)/p)^{1/(d-1)}) = O(n \cdot (p/\ln n)^{1/(d-1)})$. $\qquad\square$

In the following, we improve the approximation guarantee of algorithm *IndepSet1* by adding some steps which try to upper bound $\alpha(H)$ by $O((n/p)^{1/2})$. Then, we can use $O((n/p)^{1/2})$ instead of the trivial upper bound n on $\alpha(H)$ in the analysis. By choice of p in the following, $(n/p)^{1/2} = o(n)$, so this improves the approximation guarantee. The resulting algorithm *IndepSet2* is very similar to the corresponding one in [8]. We start with some definitions and lemmas. For an integer $d \geq 2$, a d-uniform hypergraph $H = (V, E)$, and a set $S \subseteq V$ with $|S| = d - 2$, the *projection graph* of H on S is the graph $G_\text{p}(H, S) := (V \setminus S, E')$, such that $\{u, w\} \in E'$ iff $(S \cup \{u, w\}) \in E$.

Lemma 4. *Let $d \geq 2$ be a fixed integer. For every d-uniform hypergraph $H = (V, E)$, we have $\alpha(H) \leq \max_{S \subseteq V, \, |S| = d-2} \alpha(G_\text{p}(H, S)) + d - 2$.*

Proof. First, we prove that, given H, for every subset $S \subseteq V$ with $|S| = d-2$, the size of a largest independent set $I \subseteq V$ such that $S \subseteq I$ is at most $\alpha(G_\text{p}(H, S)) + d - 2$. To prove this, fix S and let I be such a set. We show that $I \setminus S$ is independent in the projection graph $G_\text{p}(H, S)$, yielding $\alpha(G_\text{p}(H, S)) \geq |I \setminus S|$, or equivalently, $|I| \leq \alpha(G_\text{p}(H, S)) + d - 2$ as claimed. For contradiction, assume that in $G_\text{p}(H, S)$, there is an edge $\{u, w\}$ with $u, w \in I \setminus S$. Then, by definition of $G_\text{p}(H, S)$, H contains the edge $(S \cup \{u, w\}) \subseteq I$, a contradiction to the independence of I. Now, the lemma is an easy consequence of the above observation. $\qquad\square$

For a hypergraph $H = (V, E)$ and $S \subseteq V$, let the *non-neighborhood* of S be $\overline{N}(S) := \{v \in V \setminus S : \text{there is no edge } (\{v\} \cup T) \text{ in } E \text{ with } T \subseteq S\}$, i. e., the non-neighborhood of S is the set of all vertices v outside of S such that there is no edge in E connecting v and some vertices in S.

Lemma 5. *For every hypergraph $H = (V, E)$ and all $a, b \in \mathbb{N}$, the following holds: If for all sets $S \subseteq V$ with $|S| = a$, we have $|\overline{N}(S)| \leq b$, then $\alpha(H) \leq a + b$.*

Proof. Fix a hypergraph H and $a, b \in \mathbb{N}$, and assume that $\alpha(H) > a + b$. Choose $I \subseteq V$ independent with $|I| > a + b$, and $S \subseteq I$ with $|S| = a$. Clearly, for each vertex $v \in I \setminus S$, there is no edge connecting v and some vertices in S, since I is independent. Thus, $|\overline{N}(S)| \geq |I \setminus S| > (a + b) - a = b$. Therefore, not for all subsets S with $|S| = a$, we have $|\overline{N}(S)| \leq b$. $\qquad\square$

Algorithm IndepSet2($H = (V, E), p$)
1. $C \leftarrow Color(H)$. If $|\text{big}(C)| \leq ((d-2)! \cdot (\ln n)/(6 \cdot p))^{1/(d-1)}$, go to Step 5.
2. Set $m \leftarrow \max_{S \subseteq V, \, |S| = d-2} \lambda_1(M(G_\text{p}(H, S), p))$ by computing the $\binom{n}{d-2}$ necessary eigenvalues. If $m \leq 4 \cdot (n/p)^{1/2}$, output $\text{big}(C)$.
3. For $s' := (d-1) \cdot (4 \cdot (\ln n)/p)^{1/(d-1)}$, compute $|\overline{N}(S')|$ for all sets $S' \subseteq V$ with $|S'| = s'$. If $|\overline{N}(S')| \leq (n/p)^{1/2}$ for all the subsets S', output $\text{big}(C)$.
4. Check all subsets $S'' \subseteq V$ with $|S''| = 2 \cdot (n/p)^{1/2}$. If none of them is independent, output $\text{big}(C)$.

5. Test all subsets of V for independence and output a largest independent subset found.

Theorem 2. *Let $d \geq 2$ be a fixed integer. There is a constant $c(d) > 0$ such that for probability $p = p(n)$ with $c(d) \cdot (\ln n)/n^{1-1/d} \leq p \leq 3/4$, algorithm IndepSet2$(H, p)$ achieves an approximation guarantee of $O(n^{1/2} \cdot p^{-(d-3)/(2d-2)}/(\ln n)^{1/(d-1)})$ and has polynomial expected running time for random inputs $H = (V, E)$ from $H_d(n, p)$.*

Proof. We start with the approximation guarantee. Step 5 achieves approximation ratio 1. If big(C) is output in any other step, its size is $|\text{big}(C)| > ((d-2)! \cdot (\ln n)/(6 \cdot p))^{1/(d-1)} = \Omega(((\ln n)/p)^{1/(d-1)})$. In case Step 2 outputs big(C), $\max_{S \subseteq V, |S|=d-2} \lambda_1(M(G_p(H, S), p)) \leq 4 \cdot (n/p)^{1/2}$. Lemmas 1 and 4 yield that $\alpha(H) \leq 4 \cdot (n/p)^{1/2} + d - 2 = O((n/p)^{1/2})$. Thus, if the algorithm outputs big(C) in Step 2, its approximation ratio is $\alpha(H)/|\text{big}(C)| = O((n/p)^{1/2}/((\ln n)/p)^{1/(d-1)}) = O(n^{1/2} \cdot p^{-(d-3)/(2d-2)}/(\ln n)^{1/(d-1)})$. If the algorithm outputs big(C) in Step 3, all subsets $S' \subseteq V$ with $|S'| = (d-1) \cdot (4 \cdot (\ln n)/p)^{1/(d-1)}$ have a non-neighborhood with at most $(n/p)^{1/2}$ vertices. Lemma 5 yields that in this case, $\alpha(H) \leq (d-1) \cdot (4 \cdot (\ln n)/p)^{1/(d-1)} + (n/p)^{1/2} = O((n/p)^{1/2})$, the same upper bound on $\alpha(H)$ as in Step 2, so again approximation ratio $O(n^{1/2} \cdot p^{-(d-3)/(2d-2)}/(\ln n)^{1/(d-1)})$ follows. Finally, if Step 4 outputs big(C), clearly $\alpha(H) \leq (n/p)^{1/2}$. Again, the same approximation ratio follows.

It remains to prove that the expected running time of the algorithm is polynomial. Steps 1 and 2 have polynomial effort. Step 3 is executed only if among the $\binom{n}{d-2}$ sets S considered in Step 2, there is one with $\lambda_1(M(G_p(H, S), p)) > 4 \cdot (n/p)^{1/2}$. Since in H, the edges of cardinality d are chosen independently with probability p, the graphs $G_p(H, S)$ are random graphs according to $H_2(n - (d - 2), p) = G(n - (d - 2), p)$. Thus, Lemma 1 yields that $\Pr[\lambda_1(M(G_p(H, S), p)) > 4 \cdot (n/p)^{1/2}] \leq 2^{-(n-(d-2))p/8}$ for a fixed set S. Therefore, Step 3's execution probability is at most

$$\Pr[\exists S \subseteq V, \ |S| = d - 2 \colon \lambda_1(M(G_p(H, S), p)) > 4 \cdot (n/p)^{1/2}] \atop \leq \binom{n}{d-2} \cdot 2^{-(n-(d-2))p/8} \leq n^{d-2} \cdot 2^{-np/9} \ . \tag{7}$$

The effort of Step 3 is $O(q(n) \cdot \binom{n}{s'})$, since it tests $\binom{n}{s'}$ subsets, each of which is tested in time $O(q(n))$. Since $\binom{n}{s'} \leq n^{s'} = 2^{s' \cdot \log n}$, its expected running time is

$$O(q(n) \cdot \binom{n}{s'} \cdot n^{d-2} \cdot 2^{-np/9}) = O(q(n) \cdot n^{d-2} \cdot 2^{s' \cdot \log n - np/9})$$
$$= O(q(n) \cdot n^{d-2} \cdot 2^{(d-1) \cdot (4 \cdot (\ln n)/p)^{1/(d-1)} \cdot \log n - np/9}) \ . \tag{8}$$

For the constant $c(d)$ in the theorem, we set $c(d) := (9(d-1) \cdot 4^{1/(d-1)}/\ln 2)^{1-1/d}$. Then, by choice of p, $(d-1) \cdot (4 \cdot (\ln n)/p)^{1/(d-1)} \cdot \log n - np/9 \leq 0$ in (8), which yields that the expected running time of Step 3 is $O(q(n) \cdot n^{d-2})$, and hence polynomial. We turn to Step 4's execution probability. For a fixed set S' of size s' considered by Step 3, we have $\Pr[|\overline{N}(S')| > (n/p)^{1/2}] \leq \binom{n}{(n/p)^{1/2}} \cdot (1 - p)^{\binom{s'}{d-1} \cdot (n/p)^{1/2}}$, since the number of potential non-neighborhoods with a size of

$(n/p)^{1/2}$ is $\binom{n}{(n/p)^{1/2}}$, and the probability that there is no edge connecting any of the $(n/p)^{1/2}$ vertices of such a set with some $d-1$ vertices in the set S' is $(1-p)^{\binom{s'}{d-1}\cdot(n/p)^{1/2}}$. Since there are $\binom{n}{s'}$ subsets of size s', with $1-x \le e^{-x}$ for $x \in \mathbb{R}$ and $\binom{s'}{d-1} \ge (s'/(d-1))^{d-1}$, we conclude that the probability that Step 4 is executed since Step 3 fails is at most $\Pr[\exists S' \subseteq V, |S'| = s' : |\overline{N}(S')| > (n/p)^{1/2}] \le \binom{n}{s'} \cdot \binom{n}{(n/p)^{1/2}} \cdot (1-p)^{\binom{s'}{d-1}\cdot(n/p)^{1/2}}$, which is at most

$$e^{s'\cdot\ln n+(n/p)^{1/2}\cdot\ln n-p\cdot(s'/(d-1))^{d-1}\cdot(n/p)^{1/2}} . \tag{9}$$

Observe that by choice of $s' = (d-1) \cdot (4 \cdot (\ln n)/p)^{1/(d-1)}$, we have $p \cdot (s'/(d-1))^{d-1} \cdot (n/p)^{1/2} = 4 \cdot (\ln n) \cdot (n/p)^{1/2}$. Furthermore, $s' = o((n/p)^{1/2})$ (for $d \ge 3$, this is trivial, and for $d = 2$, $s' = o((n/p)^{1/2}) \Leftrightarrow p = \omega((\ln n)^2/n)$, which holds by choice of $p \ge c(d) \cdot (\ln n)/n^{1-1/d} = \Omega((\ln n)/n^{1/2}))$. Now, (9) yields that Step 4 is executed with probability at most $e^{s'\cdot\ln n+(n/p)^{1/2}\cdot\ln n-p\cdot(s'/(d-1))^{d-1}\cdot(n/p)^{1/2}} = e^{o((n/p)^{1/2})\cdot\ln n-3\cdot(n/p)^{1/2}\cdot\ln n}$, which is at most $e^{-(\ln n)\cdot 2\cdot(n/p)^{1/2}}$. With Step 4's effort of $O(q(n) \cdot \binom{n}{2\cdot(n/p)^{1/2}}) = O(q(n) \cdot e^{(\ln n)\cdot 2\cdot(n/p)^{1/2}})$, an expected running time of $O(q(n) \cdot e^{(\ln n)\cdot 2\cdot(n/p)^{1/2}} \cdot e^{-(\ln n)\cdot 2\cdot(n/p)^{1/2}}) = O(q(n))$ follows for Step 4, which is polynomial. Its failure probability is

$$\Pr[\exists S'' \subseteq V, |S''| = 2 \cdot (n/p)^{1/2} : S'' \text{ is independent}]$$
$$\le \binom{n}{2\cdot(n/p)^{1/2}} \cdot (1-p)^{\binom{2\cdot(n/p)^{1/2}}{d}} \le e^{(\ln n)\cdot 2\cdot(n/p)^{1/2}-p\cdot(2\cdot(n/p)^{1/2}/d)^d} ,$$

since the probability that a subset of size $2 \cdot (n/p)^{1/2}$ is independent is $(1-p)^{\binom{2\cdot(n/p)^{1/2}}{d}}$. It can be easily seen that for all $d \ge 2$, we have $p \cdot (2\cdot(n/p)^{1/2}/d)^d \ge n$. Furthermore, by choice of $p \ge c(d) \cdot (\ln n)/n^{1-1/d}$, it follows that $(\ln n) \cdot 2 \cdot (n/p)^{1/2} \le 2 \cdot ((\ln n)/c(d))^{1/2} \cdot n^{1-1/(2d)} = o(n)$, so $(\ln n) \cdot 2 \cdot (n/p)^{1/2} = o(p \cdot (2 \cdot (n/p)^{1/2}/d)^d)$, and hence

$$e^{(\ln n)\cdot 2\cdot(n/p)^{1/2}-p\cdot(2\cdot(n/p)^{1/2}/d)^d} = e^{-(1-o(1))\cdot p\cdot(2\cdot(n/p)^{1/2}/d)^d} \tag{10}$$
$$\le e^{-(\ln 2)\cdot n} = 2^{-n} . \tag{11}$$

Therefore, the probability that Step 4 fails, and consequently Step 5 is executed, is at most 2^{-n}. The only other way that Step 5 can be executed is that Step 1 finds that $|\mathrm{big}(C)| \le ((d-2)! \cdot (\ln n)/(6 \cdot p))^{1/(d-1)}$. Using $\varepsilon = 1/3$, Lemma 2 yields that the probability of this event is at most $e^{-n\ln n} \le 2^{-n}$. The lemma is applicable, since $p \in [c(d) \cdot (\ln n)/n^{1-1/d}, 3/4]$ and $d \ge 2$ in our theorem implies that p is in the lemma's legal range $[c(d, \varepsilon) \cdot (\ln n)^d/n^{d-1-\varepsilon}, 3/4]$ of edge probabilities. Since both events leading to the execution of Step 5 have probability at most 2^{-n}, and Step 5's effort is $O(q(n) \cdot 2^n)$, the expected running time of Step 5 is $O(q(n) \cdot 2^n \cdot 2^{-n}) = O(q(n))$. Altogether, we conclude that the expected running time of algorithm *IndepSet2* is polynomial. □

2.2   Algorithms for Coloring

Analogously to Lemma 2, Lemma 3 yields the following algorithm for COL.

Algorithm Coloring1$(H = (V, E), p, \varepsilon)$
1. $C \leftarrow Color(H)$. If $|C| < 2n \cdot (p/(\varepsilon \cdot \ln n))^{1/(d-1)}$, output C.
2. For all possible partitions C' of V into nonempty subsets, test whether C' is a coloring. Output a coloring with the smallest number of colors found.

Theorem 3. *Fix an integer $d \geq 2$ and $0 < \varepsilon < 1/2$. Then, for probability $p = p(n)$ with $2^{(d-1)/2} \cdot (\ln n)^{(d+1)/2}/n^{(d-1)/2-\varepsilon/((d-2)!)} \leq p \leq 3/4$, algorithm Coloring1$(H, p, \varepsilon)$ has approximation guarantee $O(n \cdot (p/\ln n)^{1/(d-1)})$ and polynomial expected running time for random inputs $H = (V, E)$ from $H_d(n, p)$.*

We omit the simple proof of this theorem, which is analogous to the one of Theorem 1. Again, we can improve the algorithm's approximation guarantee for the range of edge probabilities p in Theorem 2. A difference is the case $d = 2$, where the range of p is slightly smaller. The reason for this is that instead of Lemma 2, we use Lemma 3 in the analysis, which in case $d = 2$ has a larger lower bound on p of $p \geq 2^{1/2} \cdot (\ln n)^{3/2}/n^{1/2-\varepsilon}$. Hence, the lower bound $p \geq c(2) \cdot (\ln n)/n^{1/2}$ of Theorem 2 does not suffice anymore. Notice Step 4a, which is introduced for $d = 2$ to have execution probability $O(e^{-n \ln n})$ for Step 5.

Algorithm Coloring2$(H = (V, E), p, \varepsilon)$
1. $C \leftarrow Color(H)$. If $|C| \geq 2n \cdot (p/(\varepsilon \cdot \ln n))^{1/(d-1)}$, go to Step 5.
2. Set $m \leftarrow \max_{S \subseteq V, |S|=d-2} \lambda_1(M(G_p(H, S), p))$ by computing the $\binom{n}{d-2}$ necessary eigenvalues. If $m \leq 4 \cdot (n/p)^{1/2}$, output C.
3. For $s' := (d-1) \cdot (4 \cdot (\ln n)/p)^{1/(d-1)}$, compute $|\overline{N}(S')|$ for all sets $S' \subseteq V$ with $|S'| = s'$. If $|\overline{N}(S')| \leq (n/p)^{1/2}$ for all the subsets S', output C.
4. Check all subsets $S'' \subseteq V$ with $|S''| = 2 \cdot (n/p)^{1/2}$. If none of them is independent, output C.
4a. If $d = 2$: For all sets $D = \{D_1, \ldots, D_{8 \ln n}\}$ of $8 \ln n$ pairwise disjoint classes $D_i \subseteq V$ with size $|D_i| = (n/p)^{1/2}$, check whether all classes in D are independent. If for no D, all classes are independent, output C.
5. For all possible partitions C' of V into nonempty subsets, test whether C' is a coloring. Output a coloring with the smallest number of colors found.

Theorem 4. *Let $d \geq 3$ be a fixed integer. There is a constant $c(d) > 0$ such that for probability $p = p(n)$ with $c(d) \cdot (\ln n)/n^{1-1/d} \leq p \leq 3/4$, algorithm Coloring2$(H, p, \varepsilon)$ with $\varepsilon := 1/(2d)$ achieves an approximation guarantee of $O(n^{1/2} \cdot p^{-(d-3)/(2d-2)}/(\ln n)^{1/(d-1)})$ and polynomial expected running time for random inputs $H = (V, E)$ from $H_d(n, p)$.*

Proof. We start with the expected running time of the algorithm. Since Steps 1–4 are essentially the same as in algorithm *IndepSet2* and hence have the same efforts and execution probabilities, a polynomial expected running time follows for these steps from the proof of Theorem 2. Next, we upper bound the execution probability of Step 5, which is only executed if Step 4 fails or if Step 1 finds that $|C| \geq 2n \cdot (p/(\varepsilon \cdot \ln n))^{1/(d-1)}$. Equation (10) upper bounds the failure probability of Step 4 by $e^{-(1-o(1))\cdot p\cdot(2\cdot(n/p)^{1/2}/d)^d}$. Thus, for $d \geq 3$, Step 4 fails with probability at most

$$e^{-(1-o(1))\cdot p\cdot(2\cdot(n/p)^{1/2}/d)^d} = e^{-\Omega(p\cdot(n/p)^{d/2})} = e^{-\Omega(n^{d/2})} \leq e^{-n\ln n} \ . \tag{12}$$

Lemma 3 yields that the probability that Step 1 finds that $|C| \geq 2n \cdot (p/(\varepsilon \cdot \ln n))^{1/(d-1)}$ is less than $e^{-n\ln n}$. Together with (12), we conclude that Step 5 is executed with probability $O(e^{-n\ln n})$. The step has an effort of $O(q(n) \cdot e^{n\ln n})$, since it checks at most $n^n = e^{n\ln n}$ partitions, each of which is tested in time $O(q(n))$. Therefore, the expected running time of Step 5 is $O(q(n) \cdot e^{n\ln n} \cdot e^{-n\ln n}) = O(q(n))$, which is polynomial. It remains to show that p is legal with respect to Lemma 3, i.e., $p \geq 2^{(d-1)/2} \cdot (\ln n)^{(d+1)/2}/n^{(d-1)/2-\varepsilon/((d-2)!)}$. This holds, since $p \geq c(d) \cdot (\ln n)/n^{1-1/d}$, $\varepsilon = 1/(2d)$, and $d \geq 3$, and hence $p \geq c(d)\cdot(\ln n)/n^{1-1/d} \geq 1/n^{1-1/d} \geq 2^{(d-1)/2}\cdot(\ln n)^{(d+1)/2}/n^{1-1/(2d)} \geq 2^{(d-1)/2} \cdot (\ln n)^{(d+1)/2}/n^{(d-1)/2-\varepsilon/((d-2)!)}$. We turn to the approximation guarantee of the algorithm. First, observe that for every hypergraph H, it holds that $\chi(H) \geq n/\alpha(H)$, since in every coloring, all color classes are independent and therefore of size at most $\alpha(H)$. The proof of Theorem 2 shows that if one of the Steps 2–4 outputs C, it holds that $\alpha(H) = O((n/p)^{1/2})$. Thus, in this case we have $\chi(H) = \Omega(n/(n/p)^{1/2}) = \Omega((np)^{1/2})$. Furthermore, if C is output in Steps 2–4, we have $|C| < 2n \cdot (p/(\varepsilon \cdot \ln n))^{1/(d-1)} = O(n \cdot (p/\ln n)^{1/(d-1)})$. It follows that the approximation ratio $|C|/\chi(H)$ achieved by Steps 2–4 is $O(n \cdot (p/\ln n)^{1/(d-1)}/(np)^{1/2}) = O(n^{1/2} \cdot p^{-(d-3)/(2d-2)}/(\ln n)^{1/(d-1)})$. Finally, Step 5 outputs an optimal solution achieving approximation ratio 1. □

Theorem 5. *Fix $d = 2$ and $0 < \varepsilon < 1/2$. For probability $p = p(n)$ with $2^{1/2}\cdot(\ln n)^{3/2}/n^{1/2-\varepsilon} \leq p \leq 3/4$, algorithm Coloring2$(H, p, \varepsilon)$ has approximation guarantee $O((np)^{1/2}/\ln n)$ and polynomial expected running time for random inputs $H = (V, E)$ from $H_d(n, p)$.*

The proof is basically the same as for Theorem 4. We sketch additional details. To begin, we show that like in Steps 2–4, $\chi(H) = \Omega((np)^{1/2})$ if Step 4a outputs C. If the step outputs C, there is no set of $8\ln n$ pairwise disjoint and independent classes $D_i \subseteq V$ of size $(n/p)^{1/2}$. Thus, in the optimal coloring C^*, there are less than $8\ln n$ color classes of size at least $(n/p)^{1/2}$ (*large* classes), and furthermore, the largest color class has a size of less than $(8\ln n) \cdot (n/p)^{1/2}$. It follows that the large classes together contain less than $64\cdot(\ln n)^2\cdot(n/p)^{1/2}$ vertices, which is $o(n)$ since $p \geq 2^{1/2} \cdot (\ln n)^{3/2}/n^{1/2-\varepsilon}$. We conclude that $n - o(n) \geq n/2$ vertices must be contained in color classes smaller than $(n/p)^{1/2}$ in C^*, yielding more than $(n/2)/(n/p)^{1/2} = \Omega((np)^{1/2})$ such color classes in C^*, so $\chi(H) = \Omega((np)^{1/2})$. The proof of Theorem 4 yields a polynomial expected running time for Steps 1–4. Step 4a tests at most $\binom{n}{(n/p)^{1/2}}^{8\ln n}$ sets, each in time $O(q(n))$. Its execution probability is at most 2^{-n} due to (11). By choice of p, a polynomial expected running time follows for this step. If it fails, a set of $8\ln n$ pairwise disjoint, independent classes of size $(n/p)^{1/2}$ exists. With the number of tested sets given above and the probability of $(1 - p)^{\binom{(n/p)^{1/2}}{2}}$ that a class is independent, it

follows that the probability that Step 5 is reached from Step 4a is at most
$$\left(\textstyle\binom{n}{(n/p)^{1/2}}\right)^{8\ln n} \cdot (1-p)^{(8\ln n)\cdot\binom{(n/p)^{1/2}}{2}} \leq e^{-n\ln n}.$$

3   Conclusions

We have shown, how the algorithms for IS and COL of Krivelevich and Vu in [8] can be generalized to not only handle random graphs from $G(n,p)$ but random d-uniform hypergraphs from $H_d(n,p)$, $d \geq 2$, as inputs. Also, we achieved some improvements for $d = 2$: Firstly, we improved the lower bound on p of the algorithm for IS in [8] from $p \geq 1/n^{1/2-\varepsilon}$, $\varepsilon > 0$ fixed, to $p \geq c \cdot (\ln n)/n^{1/2}$, $c > 0$ constant, an improvement of factor $\Theta(n^\varepsilon / \ln n)$. This is achieved since Lemma 2's lower bound on p is smaller than the one of a corresponding Lemma in [8] by the above factor. Secondly, for $d = 2$, Lemma 3 states that $\Pr[|C| \geq 2np/(\varepsilon \cdot \ln n)] < e^{-n\ln n}$. Notice that the upper bound $e^{-n\ln n}$ improves on the larger bound $2^{-2np/\ln n}$ of a corresponding lemma in [8]. Due to this, algorithm *Coloring2* is a bit simpler than the corresponding one in [8], i.e., two steps are missing for $d \geq 3$ and one for $d = 2$.

References

1. Feige, U.: Approximating Maximum Clique by Removing Subgraphs. SIAM J. Discr. Math. 18, 219–225 (2004)
2. Feige, U., Kilian, J.: Zero Knowledge and the Chromatic Number. In: 11th IEEE Conf. Comput. Compl., pp. 278–287. IEEE Computer Society Press, Washington (1996)
3. Halldórsson, M.M.: Approximations of Weighted Independent Set and Hereditary Subset Problems. J. Graph Algor. and Appl. 4(1), 1–16 (2000)
4. Halldórsson, M.M.: A Still Better Performance Guarantee for Approximate Graph Coloring. Inf. Process. Lett. 45, 19–23 (1993)
5. Håstad, J.: Clique is hard to approximate within $n^{1-\varepsilon}$. Acta Mathematica 182, 105–142 (1999)
6. Hofmeister, T., Lefmann, H.: Approximating Maximum Independent Sets in Uniform Hypergraphs. In: Brim, L., Gruska, J., Zlatuška, J. (eds.) MFCS 1998. LNCS, vol. 1450, pp. 562–570. Springer, Heidelberg (1998)
7. Karp, R.M.: Reducibility Among Combinatorial Problems. In: Miller, R.E., Thatcher, J.W. (eds.) Complexity of Computer Computation, pp. 85–103. Plenum Press, New York (1972)
8. Krivelevich, M., Vu, V.H.: Approximating the Independence Number and the Chromatic Number in Expected Polynomial Time. J. Comb. Opt. 6(2), 143–155 (2002)
9. Krivelevich, M., Sudakov, B.: Approximate coloring of uniform hypergraphs. J. of Algorithms 49, 2–12 (2003)
10. Ralston, A.: A First Course in Numerical Analysis. McGraw-Hill, New York (1985)
11. Vazirani, V.V.: Approximation Algorithms. Springer, Heidelberg (2001)
12. Zuckerman, D.: Linear degree extractors and the inapproximability of max clique and chromatic number. In: Kleinberg, J.M. (ed.) 38th ACM Symp. Th. of Comp., pp. 681–690. ACM, New York (2006)

A New Upper Bound for Max-2-SAT: A Graph-Theoretic Approach

Daniel Raible and Henning Fernau

University of Trier, FB 4—Abteilung Informatik, 54286 Trier, Germany
{raible,fernau}@informatik.uni-trier.de

Abstract. In MaxSat, we ask for an assignment which satisfies the maximum number of clauses for a boolean formula in CNF. We present an algorithm yielding a run time upper bound of $\mathcal{O}^*(2^{\frac{K}{6.2158}})$ for Max-2-Sat (each clause contains at most 2 literals), where K is the number of clauses. The run time has been achieved by using heuristic priorities on the choice of the variable on which we branch. The implementation of these heuristic priorities is rather simple, though they have a significant effect on the run time. Also the analysis uses a non-standard measure.

1 Introduction

Our Problem. MaxSat is an optimization version of the well-known decision problem SAT: given a boolean formula in CNF, we ask for an assignment which satisfies the maximum number of clauses. The applications for MaxSat range over such fields as combinatorial optimization, artificial intelligence and database-systems as mentioned in [5]. We put our focus on Max-2-Sat, where every formula is constrained to have at most two literals per clause, to which problems as Maximum Cut and Maximum Independent Set are reducible. Therefore Max-2-Sat is $\mathcal{NP}$-complete.

Results So Far. The best published upper bound of $\mathcal{O}^*(2^{\frac{K}{5.88}})$ has been achieved by Kulikov and Kutzov in [6] consuming only polynomial space. They build up their algorithm on the one of Kojevnikov and Kulikov [5] who were the first who used a non-standard measure yielding a run time of $\mathcal{O}^*(2^{\frac{K}{5.5}})$. If we measure the complexity in the number n of variables the current fastest algorithm is the one of R. Williams [10] having run time $\mathcal{O}^*(2^{\frac{\omega}{3}n})$, where $\omega < 2.376$ is the matrix-multiplication exponent. A drawback of this algorithm is its requirement of exponential space. Scott and Sorkin [9] presented a $\mathcal{O}^*(2^{1-\frac{1}{d+1}n})$-algorithm consuming polynomial space, where d is the average degree of the variable graph. Max-2-Sat has also been studied with respect to approximation [3,7] and parameterized algorithms [1,2].

Our Results. The major result we present is an algorithm solving Max-2-Sat in time $\mathcal{O}^*(2^{\frac{K}{6.2158}})$. Basically it is a refinement of the algorithm in [5], which also in turn builds up on the results of [1]. The run time improvement is twofold. In [5] an upper bound of $\mathcal{O}^*(1.1225^n)$ is obtained if the variable graph is cubic. Here n denotes the number of variables. We could improve this to $\mathcal{O}^*(1.11199^n)$

E. Ochmański and J. Tyszkiewicz (Eds.): MFCS 2008, LNCS 5162, pp. 551–562, 2008.
© Springer-Verlag Berlin Heidelberg 2008

by a more accurate analysis. Secondly, in the case where the maximum degree of the variable graph is four, we choose a variable for branching according to some heuristic priorities. These two improvements already give a run time of $\mathcal{O}^*(2^{\frac{K}{6.1489}})$. Moreover we like to point out that these heuristic priorities can be implemented such that they only consume $\mathcal{O}(n)$ time. The authors of [6] improve the algorithm of [5] by having a new branching strategy when the variable graph has maximum degree five. Now combining our improvements with the ones from [6] gives the claimed run time.

Basic Definitions and Terminology. Let $V(F)$ be the set of variables of a given boolean formula F. For $v \in V(F)$ by $\bar{v}$ we denote the negation of v. If v is set, then it will be assigned the values true or false. By the word *literal*, we refer to a variable or its negation. A *clause* is a disjunction of literals. We consider formulas in *conjunctive normal form (CNF)*, that is a conjunction of clauses. We allow only 1- and 2-clauses, i.e., clauses with at most two literals. The weight of v, written $\#_2(v)$, refers to the number of 2-clauses in which v or $\bar{v}$ occurs. For a set $U \subseteq V(F)$ we define $\#_2(U) := \sum_{u \in U} \#_2(u)$. If v or $\bar{v}$ occurs in some clause C we write $v \in C$. A set A of literals is called *assignment* if for every $v \in A$ it holds that $\bar{v} \notin A$. Loosely speaking if $l \in A$ for a literal l, than l receives the value true. We allow the formula to contain truth-clauses of the form $\{\mathcal{T}\}$ that are always satisfied. Furthermore, we consider a MAX-2-SAT instance as multiset of clauses. A $x \in V(F)$ is a *neighbor* of v, written $x \in N(v)$, if they occur in a common 2-clause. Let $N[v] := N(v) \cup \{v\}$. The *variable graph* $G_{var}(V, E)$ is defined as follows: $V = V(F)$ and $E = \{\{u, v\} \mid u, v \in V(F), u \in N(v)\}$. Observe that G_{var} is a undirected multigraph and that it neglects clauses of size one. We will not distinguish between the words "variable" and "vertex". Every variable in a formula corresponds to a vertex in G_{var} and vice versa. By writing $F[v]$, we mean the formula which emerges from F by setting v to true the following way: First, substitute all clauses containing v by $\{\mathcal{T}\}$, then delete all occurrences of $\bar{v}$ from any clause and finally delete all empty clauses from F. $F[\bar{v}]$ is defined analogously: we set x to false.

2   Reduction Rules and Basic Observations

We state well-known reduction rules from previous work [1,5]:

RR-1. Replace any 2-clause C with $l, \bar{l} \in C$, for a literal l, with $\{\mathcal{T}\}$.

RR-2. If for two clauses C, D and a literal l we have $C \setminus \{l\} = D \setminus \{\bar{l}\}$, then substitute C and D by $C \setminus \{l\}$ and $\{\mathcal{T}\}$.

RR-3. A literal l occurring only positively (negatively, resp.) is set to true (false).

RR-4. If $\bar{l}$ does not occur in more 2-clauses than l in 1-clauses, such that l is a literal, then set l to true.

RR-5. Let x_1 and x_2 be two variables, such that x_1 appears at most once in another clause without x_2. In this case, we call x_2 the *companion* of x_1. **RR-3** or **RR-4** will set x_1 in $F[x_2]$ to α and in $F[\bar{x}_2]$ to β, where $\alpha, \beta \in \{\text{true}, \text{false}\}$. Depending on α and β, the following actions will be carried out:

If $\alpha = $ false, $\beta = $ false, set x_1 to false.
If $\alpha = $ true, $\beta = $ true, set x_1 to true.
If $\alpha = $ true, $\beta = $ false, substitute every occurrence of x_1 by x_2.
If $\alpha = $ false, $\beta = $ true, substitute every occurrence of x_1 by $\bar{x}_2$.

From now on we will only consider reduced formulas F. This means that to a given formula F we apply the following procedure: **RR-i** is always applied before **RR-i+1**, each reduction rule is carried out exhaustively and after **RR-5** we start again with **RR-1** if the formula changed. A formula for which this procedure does not apply will be called *reduced*. Concerning the reduction rules we have the following lemma [5]:

Lemma 1.   *1. If $\#_2(v) = 1$, then v will be set.*
2. For any $u \in V(F)$ in a reduced formula with $\#_2(u) = 3$ we have $|N(v)| = 3$.
3. If the variables a and x are neighbors and $\#_2(a) = 3$, then in at least one of the formulas $F[x]$ and $F[\bar{x}]$, the reduction rules set a.

We need some auxiliary notions: A sequence of distinct vertices $a_1, v_1, \ldots, v_j, a_2$ ($j \geq 0$) is called *lasso* if $\#_2(v_i) = 2$ for $1 \leq i \leq j$, $a_1 = a_2$, $\#_2(a_1) \geq 3$ and $G_{var}[a_1, v_1, \ldots, v_j, a_2]$ is a cycle. A *quasi-lasso* is a lasso with the difference that $\#_2(v_j) = 3$. A lasso is called *3-lasso* (resp. 4-lasso) if $\#_2(a_1) = 3$ ($\#_2(a_1) = 4$, resp.). *3-quasi-lasso* and *4-quasi-lasso* are defined analogously.

Lemma 2.   *1. Let $v, u, z \in V(F)$ be pairwise distinct with $\#_2(v) = 3$ such that there are clauses C_1, C_2, C_3 with $u, v \in C_1, C_2$ and $v, z \in C_3$. Then either v is set or the two common edges of u and v will be contracted in G_{var}.*
2. The reduction rules delete the variables $v_1, \ldots, v_j$ of a lasso (quasi-lasso, resp.) and the weight of a_1 drops by at least two (one, resp.).

Proof. 1. If v is not set it will be substituted by u or $\bar{u}$ due to **RR-5**. The emerging clauses C_1, C_2 will be reduced either by **RR-1** or become 1-clauses. Also we have an edge between u and z in G_{var} as now $u, z \in C_3$.
2. We give the proof by induction on j. In the lasso case for $j = 0$, there must be a 2-clause $C = \{a_1, \bar{a}_1\}$, which will be deleted by **RR-1**, so that the initial step is shown. So now $j > 0$. Then on any v_i, $1 \leq i \leq j$, we can apply **RR-5** with any neighbor as companion, so, w.l.o.g., it is applied to v_1 with a_1 as companion. **RR-5** either sets v_1, then we are done with Lemma 1.1, or v_1 will be substituted by a_1. By applying **RR-1**, this leads to the lasso $a_1, v_2, \ldots, v_j, a_2$ in G_{var} and the claim follows by induction. In the quasi-lasso case for $j = 0$, the arguments from above hold. For $j = 1$, item 1. is sufficient. For $j > 1$, the induction step from above also applies here. $\square$

3   The Algorithm

We set $d_i(F) := |\{x \in V(F) \mid \#_2(x) = i\}|$. To measure the run time, we choose a non standard measure approach with the measure γ defined as follows:

$$\gamma(F) = \sum_{i=3}^{n} \omega_i \cdot d_i(F) \text{ with } \omega_3 = 0.94165, \omega_4 = 1.80315, \omega_i = \frac{i}{2} \text{ for } i \geq 5.$$

Clearly, $\gamma(F)$ never exceeds the number of clauses K in the corresponding formula. So, by showing an upper bound of $c^{\gamma(F)}$ we can infer an upper bound c^K. We set $\Delta_3 := \omega_3$, $\Delta_i := \omega_i - \omega_{i-1}$ for $i \geq 4$. Concerning the ω_i's we have $\Delta_i \geq \Delta_{i+1}$ for $i \geq 3$ and $\omega_4 \geq 2 \cdot \Delta_4$. The algorithm presented in this paper proceeds as follows: After applying the above-mentioned reduction rules exhaustively, it will branch on a variable v. That is, we will reduce the problem to the two formulas $F[v]$ and $F[\bar{v}]$. In each of the two branches, we must determine by how much the original formula F will be reduced in terms of $\gamma(F)$. Reduction in $\gamma(F)$ can be due to branching on a variable or to the subsequent application of reduction rules. By an $(a_1, \ldots, a_\ell)$-*branch*, we mean that in the i-th branch $\gamma(F)$ is reduced by at least a_i. The *i-th component* of a branch refers to the search tree evolving from the i-th branch (i.e., a_i). By writing $(\{a_1\}^{i_1}, \ldots, \{a_\ell\}^{i_\ell})$-branch we mean a $(a_1^1, \ldots, a_1^{i_1}, \ldots, a_\ell^1, \ldots, a_\ell^{i_\ell})$-branch where $a_j^s = a_j$ with $1 \leq s \leq i_j$. A $(a_1, \ldots, a_\ell)$-branch *dominates* a $(b_1, \ldots, b_\ell)$-branch if $a_i \geq b_i$ for $1 \leq i \leq \ell$.

Heuristic Priorities. If the maximum degree of G_{var} is four, variables v with $\#_2(v) = 4$ will be called *limited* if there is another variable u appearing with v in two 2-clauses (i.e., we have two edges between v and u in G_{var}). We call such u, v a *limited pair*. Note that also u is limited and that at this point by **RR-5** no two weight 4 variables can appear in more than two clauses together. We call $u_1, \ldots u_\ell$ a *limited sequence* if $\ell \geq 3$ and u_i, u_{i+1} with $1 \leq i \leq \ell - 1$ are limited pairs. A *limited cycle* is a limited sequence with $u_1 = u_\ell$. To obtain an asymptotically fast algorithmic behavior we introduce heuristic priorities (**HP**), concerning the choice of the variable used for branching.

1. Choose any v with $\#_2(v) \geq 7$.
2. Choose any v with $\#_2(v) = 6$, preferably with $\#_2(N(v)) < 36$.
3. Choose any v with $\#_2(v) = 5$, preferably with $\#_2(N(v)) < 25$.
4. Choose any unlimited v with $\#_2(v) = 4$ and a limited neighbor.
5. Choose the vertex u_1 in a limited sequence or cycle.
6. Pick a limited pair u_1, u_2. Let $c \in N(u_1) \setminus \{u_2\}$ with $s(c) := |(N(c) \cap (N(u_1)) \setminus \{c, u_1\})|$ maximal. If $s(c) > 1$, then choose the unique vertex in $N(u_1) \setminus \{u_2, c\}$, else choose u_1.
7. From $Y := \{v \in V(F) \mid \#_2(v) = 4, \exists z \in N(v) : \#_2(z) = 3 \wedge N(z) \not\subseteq N(v)\}$ choose v, preferably such that $\#_2(N(v))$ is maximal.
8. Choose any v, with $\#_2(v) = 4$, preferably with $\#_2(N(v)) < 16$.
9. Choose any v, with $\#_2(v) = 3$, such that there is $a \in N(v)$, which forms a triangle a, b, c and $b, c \notin N[v]$ (we say v has *pending triangle* a, b, c).
10. Choose any v, such that we have a $(6\omega_3, 8\omega_3)$- or a $(4\omega_3, 10\omega_3)$-branch.

From now on v denotes the variable picked according to **HP**.

Key Ideas. The main idea is to have some priorities on the choice of a weight 4 variable such that the branching behavior is beneficial. For example limited variables tend to be unstable in the following sense: If their weight is decreased due to branching they will be reduced due to Lemma 2.1. This means we can get an amount of ω_4 instead of Δ_4. In a graph lacking limited vertices we want a variable v with a weight 3 neighbor u such that $N(u) \not\subseteq N(v)$. In the branch

---

Algorithm 1. An algorithm for solving MAX-2-SAT

Procedure: SolMax2Sat(F)

1: Apply SolMax2Sat on every component of G_{var} separately.
2: Apply the reduction rules exhaustively to F.
3: Search exhaustively on any sub-formula being a component of at most 9 variables.
4: **if** $F = \{\mathcal{T}\}\ldots\{\mathcal{T}\}$ **then**
5:     **return** $|F|$
6: **else**
7:     Choose a variable v according to **HP**.
8:     **return** $\max\{\text{SolMax2SAT}(F[v]), \text{SolMax2Sat}(F[\bar{v}])\}$.

---

on v where u is set (Lemma 1.3) we can gain some extra reduction (at least Δ_4) from $N(u) \setminus N(v)$. If we fail to find a variable according to priorities 5-7 we show that either v as four weight 4 variables and that the graph is 4-regular, or otherwise we have two distinct situations which can be handled quite efficiently. Further, the most critical branches are when we have to choose v such that all variables in $N[v]$ have weight ω_i. Then the reduction in $\gamma(F)$ is minimal (i.e., $\omega_i + i \cdot \Delta_i$). We analyze this regular case together with its immediate preceding branch. Thereby we prove a better branching behavior compared to a separate analysis. In [9] similar ideas were used for MAX-2-CSP. We are now ready to present our algorithm, see Alg. 1. Reaching step 7 we can rely on the fact that G_{var} has at least 10 vertices. We call this the *small component property (scp)* which is crucial for some cases of the analysis.

4   The Analysis

In this section we investigate the cases when we branch on vertices picked according to items 1-10 of **HP**. For each item we will derive a branching vector which upper bounds this case in terms of K. In the rest of this section we show:

Theorem 1. *Algorithm 1 has a run time of* $\mathcal{O}^*(2^{\frac{K}{6.1489}})$.

4.1 G_{var} Has Minimum Degree Four

Priority 1. If $\#_2(v) \geq 7$, we first obtain a reduction of ω_7 because v will be deleted. Secondly, we get an amount of at least $7 \cdot \Delta_7$ as the weights of v's neighbors each drops by at least one and we have $\Delta_i \geq \Delta_{i+1}$. Thus, γ is reduced by at least 7 in either of the two branches (i.e., we have a $(\{7\}^2)$-branch).

Regular Branches. We call a branch *h-regular* if we branch on a variable v such that for all $u \in N[v]$ we have $\#_2(u) = h$. We will handle those in a separate part. During our considerations a 4-regular branch will have exactly four neighbors as otherwise this situation is handled by priority 4 of **HP**. The following subsections handle *non-regular* branches, which means that we can find a $u \in N(v)$ with $\#_2(u) < \#_2(v)$. Note that we already handled h-regular branches for $h \geq 7$.

Priorities 2 and 3. Choosing $v \in V(F)$ with $\#_2(v) = 6$ there is a $u \in N(v)$ with $\#_2(u) \leq 5$ due to non-regularity. Then by deletion of v, there is a reduction by ω_6 and another of at least $5\Delta_6 + \Delta_5$, resulting from the dropping weights of the neighbors. Especially, the weight of u must drop by at least Δ_5. This leads to a $(\{6.19685\}^2)$-branch. If $\#_2(v) = 5$, the same observations as in the last choice lead to a reduction of at least $\omega_5 + 4 \cdot \Delta_5 + \Delta_4$. Thus we have a $(\{6.1489\}^2)$-branch.

Priority 4. Let $u_1 \in N(v)$ be the limited variable. u_1 forms a limited pair with some u_2. After branching on v, the variable u_1 has weight at most 3. At this point, u_1 appears only with one other variable z in a 2-clause. Then, **RR-5** is applicable to u_1 with u_2 as its companion. According to Lemma 2.1, either u_1 is set or the two edges of u_1 and u_2 will be contracted. In the first case, we receive a total reduction of at least $3\omega_4 + 2\Delta_4$, in the second of at least $2\omega_4 + 4\Delta_4$. Thus, a proper estimate is a $(\{2\omega_4 + 4\Delta_4\}^2)$-branch, i.e., a $(\{7.0523\}^2)$-branch.

Priority 5. If $u_1, \ldots, u_\ell$ is a limited cycle, then $\ell \geq 10$ due to *scp*. By **RR-5** this yields a $(10\omega_4, 10\omega_4)$-branch. If $u_1, \ldots, u_\ell$ is a limited sequence, then due to priority 4 the neighbors of u_1, u_ℓ lying outside the sequence have weight 3. By **RR-5** the branch on u_1 is a $(\{3\omega_4 + 2\omega_3\}^2)$-branch, i.e, a $(\{7.29275\}^2)$-branch.

Priority 6. At this point every limited variable u_1 has two neighboring variables y, z with weight 3 and a limited neighbor u_2 with the same properties (due to priorities 4 and 5). We now examine the local structures arising from this fact and by the values of $|N(y) \setminus N(u_1)|$ and $|N(z) \setminus N(u_1)|$.

1. We rule out $|N(y) \setminus N(u_1)| = |N(z) \setminus N(u_1)| = 0$ due to *scp*.
2. $|N(y) \setminus N(u_1)| = 0, |N(z) \setminus N(u_1)| = 1$: Then, $N(y) = \{u_2, z, u_1\}, N(u_2) = \{u_1, y, s_1\}$ and $N(z) = \{u_1, y, s_2\}$, see Figure 1(a). In this case we branch on z as $s(y) > 0$ and $s(y) > s(z)$. Then due to **RR-5** y and u_1 disappear; either by being set or replaced. Thereafter due to **RR-1** and Lemma 1.1 u_2 will be set. Additionally we get an amount of $\min\{2\Delta_4, \omega_4, \omega_3 + \Delta_4\}$ from s_1, s_2. This depends on whether $s_1 \neq s_2$ or $s_1 = s_2$ and in the second case on the weight of s_1. If $\#_2(s_1) = 3$ we get a reduction of $\omega_3 + \Delta_4$ due to setting s_1. In total we have at least a $(\{2\omega_4 + 2\omega_3 + 2\Delta_4\}^2)$-branch. Analogous is the case $|N(y) \setminus N(u_1)| = 1, |N(z) \setminus N(u_1)| = 0$.
3. $|N(y) \setminus N(u_1)| = 1, |N(z) \setminus N(u_1)| = 1$: Here two possibilities occur:
   (a) $N(y) = \{u_1, u_2, s_1\}, N(z) = \{u_1, u_2, s_2\}, N(u_2) = \{u_1, y, z\}$, see Figure 1(b): Then w.l.o.g., we branch on z. Similarly to item 2. we obtain a $(\{2\omega_4 + 2\omega_3 + 2\Delta_4\}^2)$-branch.
   (b) $N(y) = \{u_1, z, s_1\}, N(z) = \{u_1, y, s_2\}$, see Figure 1(c): W.l.o.g., we branch on z. Basically we get a total reduction of $\omega_4 + 2\omega_3 + 2\Delta_4$. That is $2\omega_3$ from y and z, ω_4 from u_1 and $2\Delta_4$ from s_2 and u_2. In the branch where y is set (Lemma 1.3) we additionally get Δ_4 from s_1 and ω_4 from u_2 as it will disappear (Lemma 1.2). This is a $(2\omega_4 + 2\omega_3 + 2\Delta_4, \omega_4 + 2\omega_3 + 2\Delta_4)$-branch.
4. $|N(y) \setminus N(u_1)| = 1, |N(z) \setminus N(u_1)| = 2$, see Figure 1(d): We branch on z yielding a $(\{2\omega_4 + 2\omega_3 + 2\Delta_4\}^2)$-branch. Analogous is the case $|N(y) \setminus N(u_1)| = 2, |N(z) \setminus N(u_1)| = 1$.

5. $|N(y)\setminus N(u_1)| = 2, |N(z)\setminus N(u_1)| = 2$: In this case we chose u_1 for branching. Essentially we get a reduction of $2w_4 + 2w_3$. In the branch setting z we receive an extra amount of $2\Delta_4$ from z's two neighbors outside $N(u_1)$. Hence we have a $(2w_4 + 2w_3 + 2\Delta_4, 2w_4 + 2w_3)$-branch.

We have at least a $(2w_4 + 2w_3 + 2\Delta_4, w_4 + 2w_3 + 2\Delta_4)$-branch, i.e., a $(7.2126, 5.40945)$-branch.

Priority 7. We need further auxiliary notions: A *3-path* (*4-path*, resp.) for an unlimited weight 4 vertex v is a sequence of vertices $u_0 u_1 \ldots u_l u_{l+1}$ ($u_0 u_1 \ldots u_l$, resp.) forming a path, such that $1 \leq l \leq 4$ ($2 \leq l \leq 4$, resp.), $u_i \in N(v)$ for $1 \leq i \leq l$, $\#_2(u_i) = 3$ for $1 \leq i \leq l$ ($\#_2(u_i) = 3$ for $1 \leq i \leq l-1, \#_2(u_l) = 4$, resp.) and $u_0, u_{l+1} \notin N(v)$ ($u_0 \notin N(v)$, resp.). Due to the absence of limited vertices, every vertex v, chosen due to priority 7, must have a 3- or 4-path.

3-path. If $u_0 \neq u_{l+1}$ we basically get a reduction of $w_4 + lw_3 + (4-l)\Delta_4$. In the branch where u_1 is set, $u_2 \ldots u_l$ will be also set due to Lemma 1.1. Therefore, we gain an extra amount of at least $2\Delta_4$ from u_0 and u_{l+1}, leading to a $(w_4 + l \cdot w_3 + (6-l)\Delta_4, w_4 + l \cdot w_3 + (4-l)\Delta_4)$-branch.
   If $u_0 = u_{l+1}$ then in $F[v]$ and in $F[\bar{v}]$, $u_0 u_1 \ldots u_l u_{l+1}$ is a lasso. So by Lemma 2.2, $u_1, \ldots, u_l$ are deleted and the weight of u_0 drops by 2. If $\#_2(u_0) = 4$ this yields a reduction of $l \cdot w_3 + w_4$. If $\#_2(u_0) = 3$ the reduction is $(l+1) \cdot w_3$ but then u_0 is set. It is not hard to see that this yields a bonus reduction of Δ_4. Thus, we have a $(\{w_4 + (l+1) \cdot w_3 + (5-l)\Delta_4\}^2)$-branch.

4-path. We get an amount of $w_4 + (l-1)w_3 + (5-l)\Delta_4$ by deleting v. In the branch where u_1 is set we get a bonus of Δ_4 from u_0. Further u_l will be deleted completely. Hence we have a $(2w_4 + (l-1)w_3 + (5-l)\Delta_4, w_4 + (l-1)w_3 + (5-l)\Delta_4)$-branch.

The first branch is worst for $l = 1$, the second and third for $l = 2$ (as $l = 1$ is impossible). Thus, we have $(\{7.2126\}^2)$-branch for the second and a $(7.0523, 5.3293)$-branch for the first and third case which is sharp.

Priority 8. If we have chosen a variable v with $\#_2(v) = 4$ according to priority 8, such that $\#_2(N(v)) < 16$, then we have two distinct situations. By branching on v, we get at least a $(\{2w_4 + 2w_3 + 2\Delta_4\}^2)$-branch.

The 4- 5- and 6-regular case. The part of the algorithm when we branch on variables of weight $h \neq 4$ will be called *h-phase*. Branching according to priorities 4-8 is the *4-phase*, according to priorities 9 and 10 the *3-phase*.

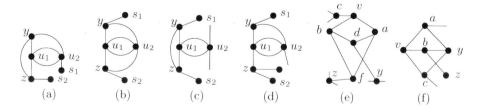

Fig. 1.

In the following we have $4 \leq h \leq 6$. Any h-regular branch which was preceded by a branch from the $(h + 1)$-phase can be neglected. This situation can only occur once on each path from the root to a leaf in the search tree. Hence, the run time is only affected by a constant multiple. We now classify h-regular branches: An *internal h-regular branch* is a h-regular branch such that another h-regular branch immediately follows in the search tree in at least one component. A *final h-regular branch* is a h-regular branch such that no h-regular branch immediately succeeds in either of the components. When we are forced to do an h-regular branch, then according to **HP** the whole graph must be h-regular at this point.

Observation 2. *If a branch is followed by a h-regular branch in one component, say in $F[v]$, then in $F[v]$ any $u \in V(F)$ with $\#_2(u) < h$ will be reduced.*

Due to Observation 2 every vertex in $N(v)$ must be completely deleted in $F[v]$.

Proposition 1. $\mathcal{O}^*(1.1088^K)$ *upper bounds any internal h-regular branch.*

Proof. By Observation 2 for $h = 4$ this yields at least a $(5\omega_4, \omega_4 + 4\Delta_4)$-branch as v must have 4 different weight 4 neighbors due to **HP**. If both components are followed by an h-regular branch we get a total reduction of $5\omega_4$ in both cases. The same way we can analyze internal 5- and 6-regular branches. This yields $(3\omega_5, \omega_5 + 5\Delta_5)$-, $(\{3\omega_5\}^2)$-, $(3\omega_6, \omega_6 + 6\Delta_6)$- and $(\{3\omega_6\}^2)$-branches as for any $v \in V(F)$ we have $|N(v)| \geq 2$. $\square$

We now analyze a final h-regular $(\{b\}^2)$-branch with its preceding (a_1, a_2)-branch. The final h-regular branch might follow in the first, the second or both components of the (a_1, a_2)-branch. So, the *combined analysis* would be a $(\{a_1 + b\}^2, a_2)$, a $(a_1, \{a_2 + b\}^2)$- and a $(\{a_1 + b\}^2, \{a_2 + b\}^2)$-branch.

Proposition 2. *Any final h-regular branch ($h \in \{5, 6\}$) considered together with its preceding branch can be upper bounded by $\mathcal{O}^*(1.1172^K)$.*

Proof. We will apply a combined analysis for both branches. Due to Observation 2 $N(v)$ will be deleted in the corresponding component of the preceding branch. The least amount we can get by deleting $N(v)$ is $\omega_5 + \omega_4$ in case $h = 5$ and $\omega_6 + \omega_4$ in case $h = 6$. Hence, we get four different branches: A $(\{3\omega_5 + \omega_4 + 5\Delta_5\}^2, \omega_5 + 5\Delta)$-, a $(\{3\omega_6 + \omega_4 + 6\Delta_6\}^2, \omega_6 + 6\Delta)$-, a $(\{3\omega_5 + \omega_4 + 5\Delta_5\}^4)$- and a $(\{3\omega_6 + \omega_4 + 6\Delta_6\}^4)$-branch, respectively. $\square$

Proposition 3. *Any final 4-regular branch considered with its preceding branch can be upper bounded by $\mathcal{O}^*(2^{\frac{K}{6.1489}}) \approx \mathcal{O}^*(1.11933^K)$.*

Proof. We must analyze a final 4-regular branch together with any possible predecessor. These are all branches derived from priorities 4-8.
Internal 4-regular branch. The two corresponding branches are a $(\{6\omega_4 + 4\Delta_4\}^2, \omega_4 + 4\Delta_4)$-branch and a $(\{6\omega_4 + 4\Delta_4\}^4)$-branch.
Priorities 4, 5 and 8 are all dominated by a $(\{2\omega_4 + 4\Delta_4\}^2)$-branch. Analyzing these cases together with a succeeding final 4-regular branch gives a $(\{3\omega_4 + 8\Delta_4\}^2, 2\omega_4 + 4\Delta_4)$-branch and a $(\{3\omega_4 + 8\Delta_4\}^4)$-branch.

Priority 7. Let o be the number of weight 4 vertices from $N(v)$ and the 3- or 4-path, respectively. If in one component a final 4-regular branch follows then the worst case is when $o = 0$ as any weight 4 vertex would be deleted completely and $w_4 > w_3$. On the other hand if there is a component without an immediate 4-regular branch succeeding then the worst case appears when o is maximal as $w_3 \geq \Delta_4$. So in the analysis we will consider for each case the particular worst case even though both together never appear.

3-path with $u_0 \neq u_{l+1}$: First if there is a weight 4 variable in $N(u)$ we have at least the following branches: a) $(\{3w_4 + 5w_3 + 4\Delta_4\}^2, w_4 + w_3 + 3\Delta_4)$, b) $(w_4 + w_3 + 5\Delta_4, \{3w_4 + 3w_3 + 4\Delta_4\}^2)$ and c) $(\{3w_4 + 5w_3 + 4\Delta_4\}^2, \{3w_4 + 3w_3 + 4\Delta_4\}^2)$. Any of those is upper bounded by $\mathcal{O}^*(2^{\frac{K}{6.1489}})$. Now suppose for all $y \in N(v)$ we have $\#_2(y) = 3$. Table 1 captures the derived branches for certain combinations. Here we will also consider the weights of u_0 and u_l. Any entry is upper bounded

Table 1.

$\#_2(u_0), \#_2(u_{l+1})$	left component	right component	both components
$\#_2(u_0) = 3$ $\#_2(u_{l+1}) = 3$	$(\{2w_4 + 6w_3 + 4\Delta_4\}^2,$ $w_4 + 4w_3)$	$(w_4 + 6w_3,$ $\{2w_4 + 6w_3 + 4\Delta_4\}^2)$	$(\{2w_4 + 6w_3 + 4\Delta_4\}^2,$ $\{2w_4 + 6w_3 + 4\Delta_4\}^2)$
$\#_2(u_0) = 3$ $\#_2(u_{l+1}) = 4$	$(\{3w_4 + 5w_3 + 4\Delta_4\}^2,$ $w_4 + 4w_3)$	$(w_4 + 5w_3 + \Delta_4,$ $\{2w_4 + 5w_3 + 4\Delta_4\}^2)$	$(\{3w_4 + 5w_3 + 4\Delta_4\}^2,$ $\{2w_4 + 5w_3 + 4\Delta_4\}^2)$
$\#_2(u_0) = 4$ $\#_2(u_{l+1}) = 4$	$(\{4w_4 + 4w_3 + 4\Delta_4\}^2,$ $, w_4 + 4w_3)$	$(w_4 + 4w_3 + 2\Delta_4,$ $\{2w_4 + 4w_3 + 4\Delta_4\}^2)$	$(\{4w_4 + 4w_3 + 4\Delta_4\}^2,$ $\{2w_4 + 4w_3 + 4\Delta_4\}^2)$

by $\mathcal{O}^*(2^{\frac{K}{6.1489}})$ except α) $(\{2w_4 + 6w_3 + 4\Delta_4\}^2, w_4 + 4w_3)$ the left upper entry and β) $(w_4 + 4w_3 + 2\Delta_4, \{2w_4 + 4w_3 + 4\Delta_4\}^2)$ the middle entry of the last row. For $U \subseteq V(F)$ we define $E_3(U) := \{\{u, v\} \mid u \in U, \#_2(u) = 3, v \notin U\}$.

Claim. 1. Suppose for all $y \in Q := N(v) \cup \{u_0, u_{l+1}\}$ we have $\#_2(y) = 3$. Then there must be some $y' \in V \setminus (N(v) \cup \{u_0, u_{l+1}\})$ with $\#_2(y') = 3$.
2. Suppose for all $y \in N(v)$ we have $\#_2(y) = 3$ and $\#_2(u_0) = \#_2(u_{l+1}) = 4$. Then there must be some $y' \in V \setminus (N(v) \cup \{u_0, u_{l+1}\})$ with $\#_2(y') = 3$.

Proof. 1. Assume the contrary. For any $1 \leq l \leq 4$ we have $|E_3(Q \cup \{v\})| \leq 10$. Due to *scp* there is a weight 4 vertex r adjacent to some vertex in Q. Observe that we must have $r \in Y$ as either there is $u \in N(v)$ with $u \in N(r)$ and $v \notin N(r)$ or w.l.o.g. $u_0 \in N(r)$ but $u_1 \notin N(r)$. Hence, r has 4 weight 3 neighbors from Q due to the choice of v. Hence we must have $|E_3(Q \cup \{v, r\})| \leq 6$. Using the same arguments again we find some $r' \in Y$ with $|E_3(Q \cup \{v, r, r'\})| \leq 2$. Again, due to *scp* we find a $r'' \in Y$ with 4 weight 3 neighbors where at most two are from Q, a contradiction.
2. Assume the contrary. Observe that $u_0, u_{l+1} \in Y$ and due to the choice of v both have 4 weight 3 neighbors which must be from $N(v)$. From $|E_3(N[v])| \leq 8$ follows that $|E_3(N[v] \cup \{u_0, u_{l+1}\})| = 0$ which contradicts *scp*. $\square$

Due to the last claim and Observation 2 we have a $(\{2w_4 + 7w_3 + 4\Delta_4\}^2, w_4 + 4w_3)$-branch for case α) and a $(w_4 + 4w_3 + 2\Delta_4, \{2w_4 + 5w_3 + 4\Delta_4\}^2)$-branch for case β). Both are upper bounded by $\mathcal{O}^*(2^{\frac{K}{6.1489}})$. $\square$

4.2   The Cubic Case

Priority 9. Observe that when we have arrived at this point, the graph G_{var} must be 3-regular and each variable has three different neighbors, due to G_{var} being reduced and due to Lemma 1.2. Also, any 3-regular graph has an even number of vertices, because we have $3n = 2m$. Thus any branching must be of the form $(2i \cdot \omega_3, 2j \cdot \omega_3)$ for some $1 \leq i, j$. Also, branching on any variable will at least result in a $(4\omega_3, 4\omega_3)$-branch (see Lemma 1.2). Note that any $u \in N(v)$ will be either set in $F[v]$ or in $F[\bar{v}]$, due to Lemma 1.3.

Lemma 3. *Let v have a pending triangle a, b, c and $N(v) = \{a, p, q\}$. Then by branching on v, we have an $(8\omega_3, 6\omega_3)$-branch.*

Proof. In $F[v]$ and $F[\bar{v}]$, the variables a, b, c form a 3-quasi-lasso. Hence, due to Lemma 2.2 w.l.o.g., only b remains in the reduced formula with $\#_2(b) = 2$ (Lemma 2 2). Also, in both branches, q and p are of weight two and therefore deleted. Note that $N(\{q, p\}) \cup \{q, p\} \subseteq \{v, a, b, c, q, p\}$, contradicts *scp*. Therefore, w.l.o.g., there is a variable $z \in N(q)$ such that $z \notin \{v, a, b, c, q, p\}$. So, in the branch where q is set, also z will be deleted. Thus, seven variables will be deleted.   □

Priority 10. From now on, due to **HP**, G_{var} is triangle-free and cubic. We show that if we are forced to choose a vertex v to which none of the priorities 1-9 fits, we can choose v such that we obtain either a $(6\omega_3, 8\omega_3)$- or a $(4\omega_3, 10\omega_3)$-branch.

Lemma 4. *Let v be a vertex in G_{var} and $N(v) = \{a, b, c\}$. Suppose that, w.l.o.g., in $F[v]$ a, b and in $F[\bar{v}]$ c will be set. Then we have a $(6\omega_3, 8\omega_3)$-branch.*

Proof. If $|(N(a) \cap N(b)) \setminus \{v\}| \leq 1$, then by setting a and b in $F[v]$, five variables will be reduced. Together with v and c, this is a total of seven. If $|(N(a) \cap N(b)) \setminus \{v\}| = 2$, then situation 1(e) must occur (note the absence of triangles). If $z = y$ then also $z \neq c$ due to *scp*. Then in $F[v]$ due to Lemma 1.1 v, a, b, c, d, f, z will be deleted. If $z \neq y$ then v, a, b, d, f, z, y will be deleted. Together with $F[\bar{v}]$ where c is set, we have a $(6\omega_3, 8\omega_3)$-branch.   □

Lemma 5. *If for any $v \in V(F)$ all its neighbors are set in one branch (say, in $F[v]$), we can perform a $(6\omega_3, 8\omega_3)$- or a $(4\omega_3, 10\omega_3)$-branch due to cubicity.*

Proof. If $|N(a, b, c) \setminus \{v\}| \geq 5$, then in $F[v]$, 9 variables are deleted, so that we have a $(4\omega_3, 10\omega_3)$-branch. Otherwise, either one of the two following situations must occur: a) There is a variable $y \neq v$, such that $N(y) = \{a, b, c\}$, see Figure 1(f). Then branch on b. In $F[\bar{b}]$, v, y, a, c, z will disappear (due to **RR-5** and Lemma 2.1). In $F[b]$, due to setting z, additionally a neighbor $f \notin \{a, b, c, v, y\}$ of z will be deleted due to *scp*. This is a total of seven variables. b) There are variables p, q, such that $|N(p) \cap \{a, b, c\}| = |N(q) \cap \{a, b, c\}| = 2$. The last part of Theorem 4.2 of [5] handles b).   □

Due to the last three lemmas, branchings according to priorities 9 and 10 are upper bounded by $\mathcal{O}^*(2^{\frac{K}{6.1489}})$. Especially, the $(4\omega_3, 10\omega_3)$-branch is sharp.

5   Combining Two Approaches

Kulikov and Kutzov [6] achieved a run time of $\mathcal{O}^*(2^{\frac{K}{5.88}})$. This was obtained by speeding up the 5-phase by a concept called 'clause learning'. As in our approach the 3- and 4-phase was improved we will show that if we use both strategies we can even beat our previous time bound. This means that in **HP** we substitute priority 3 by their strategy with one exception: we prefer variables v with a non weight 5 neighbor. Forced to violate this preference we do a simple branching of the form $F[v]$ and $F[\bar{v}]$. For the analysis we redefine the measure $\gamma(F)$: we set $\omega_3 = 0.9521$, $\omega_4 = 1.8320$, $\omega_5 = 2.488$ and keep the other weights. We call this measure $\tilde{\gamma}(F)$. We will reproduce the analysis of [6] briefly with respect to $\tilde{\gamma}(F)$ to show that their derived branches for the 5-phase are upper bounded by $\mathcal{O}^*(2^{\frac{K}{6.2158}})$. It also can be checked that this is also true for the branches derived for the other phases by measuring them in terms of $\tilde{\gamma}(F)$. Let k_{ij} denote the number of weight j variables occurring i times in a 2-clause with some $v \in V(F)$ chosen for branching. Then we must have: $k_{13}+k_{14}+k_{15}+2k_{24}+2k_{25}+3k_{35} = 5$. If F' is the the formula obtained by assigning a value to v and by applying the reduction rules afterwards we have:

$$\tilde{\gamma}(F) - \tilde{\gamma}(F') \geq 5\Delta_5 + \omega_5 + (\omega_3 - \Delta_5)k_{13} + (\Delta_4 - \Delta_5)k_{14} + (\tfrac{\omega_4}{2} - \Delta_5)2k_{24} + (\Delta_4 -$$
$$\Delta_5)k_{25} + (\tfrac{\omega_5}{2} - \tfrac{3}{2}\Delta_5)2k_{35} = 5.768 + 0.2961k_{13} + 0.2239(k_{14}+k_{25}) + 0.26\cdot 2(k_{24}+k_{35})$$

Basically we reduce $\tilde{\gamma}(F)$ by at least $\omega_5 + 5\Delta_5$. Now the coefficients of the k_{ij} in the above equation express how the reduction grows if $k_{ij} > 0$. If $k_{13} + k_{14} + 2k_{24} + k_{25} + 2k_{35} \geq 2$ we are done as $\tilde{\gamma}(F) - \tilde{\gamma}(F') \geq 6.2158$.

If $k_{13} = 1$ and $k_{15} = 4$ then [6] stated a $(5\Delta_5 + \omega_5 + (\omega_3 - \Delta_5), 5\Delta_5 + \omega_5 + (\omega_3 - \Delta_5) + 2\Delta_5)$-branch and for $k_{25} = 1$ and $k_{15} = 3$ a $(5\Delta_5 + \omega_5 + (\Delta_4 - \Delta_5), 5\Delta_5 + \omega_5 + (\Delta_4 - \Delta_5) + \omega_3)$-branch. If $k_{14} = 1$ and $k_{15} = 4$ a branching of the kind $F[v], F[\bar{v}, v_1], F[\bar{v}, \bar{v}_1, v_2, v_3, v_4, v_5]$ is applied, where $\{v_1, \ldots, v_5\} = N(v)$. From this follow a $(5\Delta_5 + \omega_5 + (\Delta_4 - \Delta_5), 4\Delta_5 + \omega_5 + \Delta_4 + \omega_4 + 3\Delta_4 + \Delta_5, 5\omega_5 + \omega_4)$- and a $(\omega_5 + 4\Delta_5 + \Delta_4, \omega_5 + 4\Delta_5 + \Delta_4 + \omega_4 + 4\Delta_5, 5\omega_5 + \omega_4 + 3\omega_3)$-branch. This depends on whether v_1 has at least three neighbors of weight less than 5 in $F[\bar{v}]$ or not. We observed that we can get a additional reduction of Δ_5 in the third component of the first branch as $N[v]$ cannot be a component in $V(F)$ after step 3 of Alg. 1 yielding a $(4\Delta_5 + \omega_5 + \Delta_4, 5\Delta_5 + \omega_5 + 4\Delta_4 + \omega_4, 5\omega_5 + \omega_4 + \Delta_5)$-branch. The analysis of the 5-regular branch (i.e. $k_{15} = 5$) proceeds the same way as in the simple version of the algorithm except that we have to take into account the newly introduced branches.

Theorem 3. MAX-2-SAT *can be solved in time* $\mathcal{O}^*(2^{\frac{K}{6.2158}}) \approx \mathcal{O}^*(1.118^K)$.

6   Conclusion

We presented an algorithm solving MAX-2-SAT in $\mathcal{O}^*(2^{\frac{K}{6.2158}})$, with K the number of clauses of the input formula. This is currently the end of a sequence of polynomial-space algorithms each improving on the run time: beginning with

$\mathcal{O}^*(2^{\frac{K}{2.88}})$ which was achieved by [8], it was subsequently improved to $\mathcal{O}^*(2^{\frac{K}{3.742}})$ by [2], to $\mathcal{O}^*(2^{\frac{K}{5}})$ by [1], to $\mathcal{O}^*(2^{\frac{K}{5.217}})$ by [4], to $\mathcal{O}^*(2^{\frac{K}{5.5}})$ by [5] and finally to the hitherto fastest upper bound of $\mathcal{O}^*(2^{\frac{K}{5.88}})$ by [6]. Our improvement has been achieved due to heuristic priorities concerning the choice of variable for branching in case of a maximum degree four variable graph. As [6] improved the case where the variable graph has maximum degree five, it seems that the only way to speed up the generic branching algorithm is to improve the maximum degree six case. Our analysis also implies that the situation when the variable graph is regular is not that harmful. The reason for this that the preceding branch must have reduced the problem size more than expected. Thus considered together these two branches balance each other. Though the analysis is to some extent sophisticated and quite detailed the algorithm has a clear structure. The implementation of the heuristic priorities for the weight 4 variables should be a straightforward task. Actually, we have already an implementation of Alg. 1. It is still in an early phase but nevertheless the performance is promising. We are looking forward to report on these results on another occasion.

References

1. Gramm, J., Hirsch, E.A., Niedermeier, R., Rossmanith, P.: Worst-case upper bounds for MAX-2-SAT with an application to MAX-CUT. Discrete Applied Mathematics 130, 139–155 (2003)
2. Gramm, J., Niedermeier, R.: Faster exact solutions for MAX2SAT. In: Bongiovanni, G., Petreschi, R., Gambosi, G. (eds.) CIAC 2000. LNCS, vol. 1767, pp. 174–186. Springer, Heidelberg (2000)
3. Hofmeister, T.: An approximation algorithm for MAX-2-SAT with cardinality constraint. In: Di Battista, G., Zwick, U. (eds.) ESA 2003. LNCS, vol. 2832, pp. 301–312. Springer, Heidelberg (2003)
4. Kneis, J., Mölle, D., Richter, S., Rossmanith, P.: Algorithms based on the treewidth of sparse graphs. In: Kratsch, D. (ed.) WG 2005. LNCS, vol. 3787, pp. 385–396. Springer, Heidelberg (2005)
5. Kojevnikov, A., Kulikov, A.S.: A new approach to proving upper bounds for MAX-2-SAT. In: SODA, pp. 11–17. ACM Press, New York (2006)
6. Kulikov, A.S., Kutzkov, K.: New bounds for max-sat by clause learning. In: Diekert, V., Volkov, M.V., Voronkov, A. (eds.) CSR 2007. LNCS, vol. 4649, pp. 194–204. Springer, Heidelberg (2007)
7. Lewin, M., Livnat, D., Zwick, U.: Improved rounding techniques for the MAX 2-SAT and MAX DI-CUT problems. In: Cook, W.J., Schulz, A.S. (eds.) IPCO 2002. LNCS, vol. 2337, pp. 67–82. Springer, Heidelberg (2002)
8. Niedermeier, R., Rossmanith, P.: New upper bounds for maximum satisfiability. Journal of Algorithms 36, 63–88 (2000)
9. Scott, A., Sorkin, G.: Linear-programming design and analysis of fast algorithms for Max 2-CSP. Discrete Optimization 4(3-4), 260–287 (2007)
10. Williams, R.: A new algorithm for optimal 2-constraint satisfaction and its implications. Theoretical Computer Science 348(2-3), 357–365 (2005)

Directed Percolation Arising in Stochastic Cellular Automata Analysis

Damien Regnault

IXXI-LIP, École Normale Supérieure de Lyon, 46 allée d'Italie
69364 Lyon Cedex 07, France
http://perso.ens-lyon.fr/damien.regnault

Abstract. Cellular automata are both seen as a model of computation and as tools to model real life systems. Historically they were studied under synchronous dynamics where all the cells of the system are updated at each time step. Meanwhile the question of probabilistic dynamics emerges: on the one hand, to develop cellular automata which are capable of reliable computation even when some random errors occur [24,14,13]; on the other hand, because synchronous dynamics is not a reasonable assumption to simulate real life systems.

Among cellular automata a specific class was largely studied in synchronous dynamics : the elementary cellular automata (ECA). These are the "simplest" cellular automata. Nevertheless they exhibit complex behaviors and even Turing universality. Several studies [20,7,8,5] have focused on this class under α-asynchronous dynamics where each cell has a probability α to be updated independently. It has been shown that some of these cellular automata exhibit interesting behavior such as phase transition when the asynchronicity rate α varies.

Due to their richness of behavior, probabilistic cellular automata are also very hard to study. Almost nothing is known of their behavior [20]. Understanding these "simple" rules is a key step to analyze more complex systems. We present here a coupling between oriented percolation and ECA 178 and confirms observations made in [5] that percolation may arise in cellular automata. As a consequence this coupling shows that there is a positive probability that the ECA 178 does not reach a stable configuration as soon as the initial configuration is not a stable configuration and $\alpha > 0.996$. Experimentally, this result seems to stay true as soon as $\alpha > \alpha_c \approx 0.5$.

1 Introduction

A cellular automaton is a process where several cells, characterized by a state, evolve according to the states of their neighboring cells. Cellular automata can both model parallel computing and real life systems [22] .

Historically they have been studied under synchronous dynamics where all the cells update at the same time. Meanwhile models of probabilistic cellular automata have emerged. Several studies focus on a model of cellular automata evolving synchronously but where some random errors can occur. In [24], Toom

E. Ochmański and J. Tyszkiewicz (Eds.): MFCS 2008, LNCS 5162, pp. 563–574, 2008.
© Springer-Verlag Berlin Heidelberg 2008

gives a 2D cellular automata capable of remembering one bit of information in presence of random error. This result is used in [14] to develop a 3D cellular automaton capable of reliable computation. Later on Gács proves the existence of a 1D cellular automaton exhibiting the same reliability [13]. In [1] the authors try to apply the mean field approach on a probabilistic model of cellular automata and show that complex behaviors cannot be explained by this method.

Several empirical studies have shown that the behavior of cellular automata changes drastically under asynchronous dynamics [2,3,6,17,23]. Only few theoretical results are known. Mainly, either they concern specific cellular automata or show that it is difficult to describe the global behavior of cellular automata under probabilistic updates [1,12,13,10,11,20,7,8,21].

Percolation theory was introduced to model the fact that a liquid or a gas can flow through a solid due to porosity. Other applications were found for this model and it has been extensively studied in the last decades on which probability theory made a lot of progress. A good introduction to percolation theory can be found in [15]. Ising models, Potts models and percolation were unified into the random cluster model [9,16]. We define here a coupling between oriented percolation and a "simple" probabilistic cellular automaton. This coupling shows that "simple" rules may embed very complex phenomena. The link between elementary cellular automata and percolation theory was already observed in experiments in [5] but it is the first time that a correlation is proved.

The probabilistic dynamics we study here is the α-*asynchronous dynamics* where each cell has an independent probability α to be updated and a probability $1 - \alpha$ to stay in its current state at each time step. A particular class of cellular automata is the 256 elementary cellular automata (ECA). This class gathers the "simplest" cellular automata : cells are placed on a line, they are characterized by a state which is 0 (white) or 1 (black) and they can communicate only with their two closest neighbors. Nevertheless, studies have shown that this class exhibits a wide range of behavior including Turing universality [4]. Even if this class seems simple, questions remain open in the deterministic synchronous case such as the intrinsic universality of ECA 110 [18]. Current works study this class under α-asynchronous dynamics [20,7,8], in particular the 64 cellular automata for which the two configurations all black and all white are stable. The understanding of these "simple" rules is a key step to understand more complex phenomena. First in [7] it was shown that when only one cell is updated at each time step (which may be under certain circumstances the limit as $\alpha \to 0$) the behavior of these cellular automata are similar to behavior of coupon collectors or random walks. In [8], the behavior of some of these cellular automata has been determined under α-asynchronous dynamics, and the authors have isolated some cellular automata exhibiting rich behavior such as phase transitions. The elementary cellular automaton studied here ECA 178, FLIP-IF-NOT-ALL-EQUAL is one of these complex automata. These automata are hard to study and only few results are known yet. Here we present a new result on the ECA 178 and show that there exists a positive probability that the process will never reach a stable configuration when $\alpha > 0.996$ even when the initial configuration contains only one black cell. This is the first result proved for this automaton. Moreover ongoing work shows that ECA 178 arises naturally in the study of 2D Minority [21] which is a model of anti-ferromagnetism in physics [19] and also one of the simplest non-monotonic gene network model in biology. Since the proof of this result is based on a coupling between the space-time diagram of ECA 178 and oriented percolation on a graph, it tends to support the fact that the behavior of this automaton is indeed complex since very little is known on directed percolation yet.

2   Asynchronous Cellular Automata

2.1   Definition

We give here a formal definition of ECA 178. The next part presents informally its behavior and the underlying difficulties of its analysis.

Definition 1 (Configuration). *We denote by $\mathbb{Z}$ the set of cells and $Q = \{0, 1\}$ the set of states (0 stands for white and 1 for black in the figures). The neighborhood of a cell i consists of the cells $i - 1$, i and $i + 1$. A configuration c is a function $c : \mathbb{Z} \to Q$; c_i is the state of the cell i in configuration c.*

Definition 2 (ECA 178: FLIP-IF-NOT-ALL-EQUAL). *The rule of a cellular automaton is a function which associates a state to a neighborhood. The rule δ of the ECA 178 is defined as follows:*

$$\delta(c_{i-1}, c_i, c_{i+1}) = \begin{cases} c_i & \text{if } c_{i-1} = c_i = c_{i+1} \\ 1 - c_i & \text{otherwise} \end{cases}$$

Time is discrete and in the classic deterministic synchronous dynamics all the cells of a configuration are updated at each time step according to the transition rule of the cellular automaton (see figure 1). Here we consider a stochastic asynchronous dynamics where only a random subset of cells is updated at each time step.

Definition 3 (Asynchronous dynamics). *Given $0 < \alpha \leqslant 1$, we call α-asynchronous dynamics the following process : time is discrete and c^t denotes the random variable for the configuration at time t. The configuration c^0 is the initial configuration. The configuration at time $t + 1$ is the random variable defined by the following process : each cell has independently a probability α to be updated according to the rule δ (we say that the cell fires at time t) and a probability $1 - \alpha$ to remain in its current state. A cell is said active if its state changes when fired.*

Figure 1 presents different space-time diagrams of ECA 178 for different values of α. The initial configuration consists in one single black cell and is displayed horizontally at the bottom of the diagram (time flows upwards).

$\alpha = 0$ $\alpha = 0.1$ $\alpha = 0.25$ $\alpha = 0.5$ $\alpha = 0.75$ $\alpha = 0.9$ synchronous dynamics

Fig. 1. ECA 178 under different dynamics ($\alpha = 0$ stands for fully asynchronous dynamics, for this diagram only one every 50 time step is displayed)

Definition 4 (Stable configuration). *A configuration c is a stable if for all* $i \in \mathbb{Z}$, $\delta(c_{i-1}, c_i, c_{i+1}) = c_i$.

ECA 178 (FLIP-IF-NOT-ALL-EQUAL) admits only two stable configurations : the configurations all white and all black.

Definition 5 (Convergence). *We say that a random sequence* $(c^t)_{t \geqslant 0}$ *defined by ECA 178 converges under α-asynchronous dynamics if there exists $t < \infty$ such that c^t is a stable configuration. We denote by $P_\alpha(c^0)$ the probability that such a t exists.*

From the definition of stable configuration, it follows that if there exists t such that c^t is a stable configuration then for all $t' \geqslant t$ the configuration $c^{t'}$ is the same stable configuration. Note that since the configuration is infinite, only specific initial configurations may converge with positive probability. Here we will consider only a particular initial configuration which is very "close" to a stable configuration and show that when α is large enough, being this close is not enough to guarantee convergence almost surely.

Definition 6 (Initial configuration). *We define c^{init} as the configuration where $c_0^{init} = 1$ and for all $i \neq 0$, $c_i^{init} = 0$.*

From now on, we will consider that the initial configuration is always c^{init}. The configuration c^{init} differs from the configuration all white by only one cell. Nevertheless we show the following result :

Theorem 1 (Main result). *If $\alpha \geqslant \sqrt[3]{80/81} \approx 0.996$ then $P_\alpha(c^{init}) < 1$.*

Section 4 is dedicated to the proof of this result. This is the first result on ECA 178. This result shows that this rule can exhibit very complex behavior and shows how simple rules may turn out to be hard to analyze. Before the proof, the following section exposes experimental results on the behavior of ECA 178.

2.2   Discussion

In [8] it was conjectured that ECA 178 admits a phase transition which occurs experimentally at $\alpha = \alpha_c \approx 0.5$. Figure 1 illustrates the changes in the space time diagrams of ECA 178 when α varies. In [7] it was proven that this automaton behaves as a non-biased random walk on a finite configuration with periodic boundary condition under the fully asynchronous dynamics (when only one random cell fires at each time step). This proof can easily be extended to infinite configurations when $c^0 = c^{init}$ to prove that its converges in polynomial time to all white almost surely under the fully asynchronous dynamics.

When $0 < \alpha < \alpha_c$, despite of the fact that some small "errors" may occur, the global behavior seems to be similar to the fully asynchronous dynamics. When $\alpha > \alpha_c$ the behavior changes drastically : an alternating background pattern (0101010) appears and extends quickly in expense of the black and white regions preventing the configuration from ever reaching a fixed point. Cells inside a big white or black region are inactive whereas cells of a 010101 region are all active. When α is very small, regions of 010101 are highly unstable and the presence of patterns 010101 is marginal. Nevertheless this pattern does not exist in the fully asynchronous case and since the study of this dynamics relies on a perfect symmetry between black and white regions, the emergence of a third kind of region, which behave drastically differently, prevents us from deriving a lower bound from the fully asynchronous dynamics. We believe that any lower bound on α_c would be a huge achievement.

We prove next that when $\alpha > 0.996$ there is a strictly positive probability that the process will not reach the stable configuration all white even if only one cell is black in the initial configuration. The global behavior here is thus no more related to a unbiased random walk. Since rule 178 acts symmetrically on black and white states the same result holds when the initial configuration has only one white cells. Big white regions and big black regions tends to disappear in favor of 0101 patterns and our result confirms the fact that the pattern 0101 ends up dominating in the configuration when $\alpha > 0.996$. Finally our proof is based on a coupling between cellular automaton and oriented percolation. This correlation was already spotted in [5] but it is the first time that it is theoretically proved and the emergence of oriented percolation in ECA 178 is also a strong evidence of the richness of its behavior.

3 Oriented Percolation on $(\mathbb{Z}^+)^2$

Definition 7 (Oriented bond percolation). *Consider a probability p and the randomly labeled graph $\mathbb{L}(p) = ((\mathbb{Z}^+)^2, \mathbb{E})$ where $(\mathbb{Z}^+)^2$ is called the set of sites and $\mathbb{E}$ the set of bonds. For all $i, j \in \mathbb{Z}^+$, there are oriented bonds between site (i, j) and sites $(i + 1, j)$ and $(i, j + 1)$. Each bond has independently a probability p to be labeled open and a probability $1 - p$ to be labeled closed.*

Figure 2 illustrates several examples of oriented percolation for different values of p. Only open bonds are shown in the figure. The main question of percolation theory is the size of the open cluster.

Definition 8 (Open cluster). *We denote by C the open cluster of site $(0,0)$: a site (i, j) is in C if and only if there is an oriented path from site $(0,0)$ to site (i, j) only made of open bonds. We call $\theta(p)$ the propability that C is infinite (i.e. that there exists an infinite open path from cell $(0,0)$).*

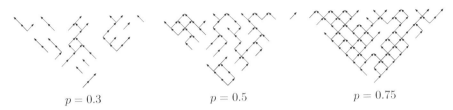

$p = 0.3$ $\qquad\qquad\qquad$ $p = 0.5$ $\qquad\qquad\qquad$ $p = 0.75$

Fig. 2. Oriented percolation with different probability of open bonds

It is easy to show that $\theta(p)$ is an increasing function of p. Moreover there exists a critical value p_c such that $\theta(p) > 0$ for $p > p_c$. Bounds on this critical value can be found in [15].

Theorem 2 (Critical value [15]). *There exists a critical value p_c such that if $p < p_c$ then $\theta(p) = 0$, and if $p > p_c$ then $\theta(p) > 0$. Moreover, $0.6298 < p_c < 2/3$.*

4   Coupling Cellular Automaton 178 with Percolation

There is no notion of time in percolation but the height will stand for it in our coupling. Indeed in order to know if a site of height $t + 1$ is in the open cluster or not, we only need information about sites of height t. In a cellular automata in order to know the state of a cell at time $t + 1$, we only need to know the states of cells at time t.

Definition 9 (Height). *The* height *of a percolation site (i, j) is the length of any path from $(0, 0)$ to (i, j) in $\mathbb{L}(p)$, that is to say $i + j$. We denote by C^t the sites of height t which are in the open cluster C. The height of a bond is the height of its origin.*

Definition 10 (Candidate). *A site is a* candidate *of height $t + 1$ if and only if at least one of its predecessors is in C^t. We denote by $\hat{C}^{t+1}$ the set of candidates of height of $t + 1$.*

Clearly for all $t > 0$ we have $C^t \subset \hat{C}^t$. Figure 4 illustrates the notion of height and candidate.

In this part, $(c^t)_{t \geq 0}$ denotes the random sequence of configurations updated according to rule 178 under α-asynchronous dynamics where $c^0 = c^{init}$.

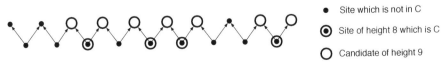

- ● Site which is not in C
- ◉ Site of height 8 which is C
- ○ Candidate of height 9

Fig. 3. An example of sites and candidates of height 8 and 9. All bonds (indifferently open or closed) are represented.

Definition 11 (Mapping). *We define the set of cells $\mathbb{T}^t$ of configuration c^t as follows : cell c_i^t is in $\mathbb{T}^t$ if and only if $-t \leqslant i \leqslant t$ and $i \equiv t \bmod 2$. Let $\mathbb{T} = \cup_{t=0}^{\infty} \mathbb{T}^t$. We define $g : (\mathbb{Z}^+)^2 \to \mathbb{T}$ as the bijection which associates the percolation site (i,j) to the cell $2i - \lceil \frac{i+j}{2} \rceil$ of configuration c^{i+j}. From these definitions, it follows that the image of the sites of height t in $\mathbb{L}(p)$ by g are the cells $\mathbb{T}^t$ in configuration c^t.*

Figure 4 shows the mapping of $\mathbb{L}(p)$ on a space-time diagram. We can notice that the cells of $\mathbb{T}$ correspond exactly to the black cells of the space-time diagram of figure 1 under synchronous dynamics. Our aim is to define a coupling such that the corresponding cells of sites in C will always be active. The following criterion formulates this property.

time / height

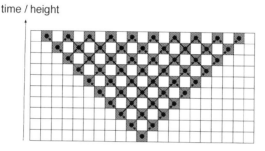

Fig. 4. Mapping of the graph $\mathbb{L}(p)$ on the space-time of a cellular automaton; the cells of $\mathbb{T}$ are colored in black

Definition 12 (Correspondence criterion). *We say that a space-time diagram $(c^t)_{t>0}$ and a labelled directed graph $\mathbb{L}(p)$ satisfy the correspondence criterion at time t if and only if the cells of $g(C^t)$ are all active. We say that they satisfy the correspondence criterion if and only if they satisfy the correspondence criterion for all $t \geqslant 0$ (i.e. the cells of $g(C)$ are all active).*

Since in c_0^{init} site $(0,0) \in C$ is active, the configuration $c^0 = c^{init}$ and any randomly labeled directed graph $\mathbb{L}(p)$ always satisfy the correspondence criterion at time 0. Moreover if they satisfy the correspondence criterion and the open cluster is infinite in $\mathbb{L}(p)$ then for all t, c^t admits at least one active cell and thus the sequence (c^t) does not converge. Figure 5 gives a configuration which satisfies the correspondence criterion with the directed graph of figure 4 at time 8.

Consider a configuration c^t such that the correspondence criterion is true at time t. We want in our coupling that the correspondence criterion stays true at time $t + 1$. Consider a site in $\hat{C}^{t+1}$, this site may be in C. If it is the case, the cell c_i^{t+1} corresponding to this site has to be active. Thus, we have to focus on cell c_i^t at time t and design a coupling such that this cell is active at time $t + 1$ if the corresponding site is in the open cluster. We say that there is a constraint on this cell.

Definition 13 (Constrained cells). *A cell c_i^t is constrained at time t if and only if $c_i^{t+1} \in \mathbb{T}$ and $g^{-1}(c_i^{t+1})$ is in $\hat{C}^{t+1}$.*

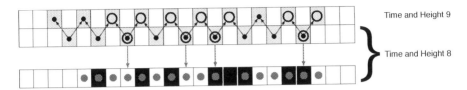

Fig. 5. Cells of $\mathbb{T}^8$ and $\mathbb{T}^9$ are gray-colored. Active cells of c^8 are marked by a dot. Arrows link the sites of C^8 to their corresponding cells in c^8. In this example, these cells are active in c^8. Thus the correspondence criterion is verified at time 8.

We may have to force a constrained cell to be active at time $t+1$ in our coupling. This turn out to be possible because a constrained cell always has an active cell in its neighborhood when the correspondence criterion is verified (Lemma 1). It follows that one can associate to each constrained cell a partner which is an active cell in its neighborhood (Definition 14). Firing these two cells makes the constrained cell active at time $t+1$ (Lemma 2) which can be coupled with the opening of the bonds when α is large enough (Theorem 3).

Lemma 1. *If $(c^t)_{t \geqslant 0}$ and $\mathbb{L}(p)$ verify the correspondence criterion at time t' each constrained cell of configuration $c^{t'}$ has an active neighbor.*

Proof. Suppose that cell c_i^t is constrained at time t. Then site $(k, l) = g^{-1}(c_i^{t+1})$ is in $\hat{C}^{t+1}$. Thus at least one of the two sites $(k, l-1)$ and $(k-1, l)$ is in C^t. Since the correspondence criterion is verified at time t, at least one of the two cells $g(k, l-1) = c_{i+1}^t$ and $g(k-1, l) = c_{i-1}^t$ is active.   □

Definition 14 (Partner). *The partner of a constrained cell c_i^t is defined as follows:*

- *if c_i^t is the pointed cell in the neighborhood* ▢▮ *or* ▢▮▮▮ *then its partner is c_{i-1}^t.*

- *if c_i^t is the pointed cell in the neighborhood* ▮▮▢ *or* ▮▮▮▮▢ *then its partner is c_{i+1}^t.*

Figure 6 illustrates the constrained cells of Figure 4 and their partners. Since two cells of $\mathbb{T}$ cannot be neighbors, two constrained cells cannot be neighbors. Moreover since the partner of a constrained cell is a neighbor of this cell then a cell cannot be at the same time constrained and the partner of an other cell. Note that however a cell may be the partner of two constrained cells.

Lemma 2. *For all $t' > 0$, if $(c^t)_{t \geqslant 0}$ and $\mathbb{L}(p)$ verify the correspondence criterion at time t', each constrained cell at time t' has a partner and if these two cells fire at time t' then the constrained cell is active at time $t' + 1$.*

Proof. Suppose that cell c_i^t is constrained at time t. Then w.l.o.g. we suppose that the state of c_i^t is 1 (the other case is symmetric). We do a case study on the

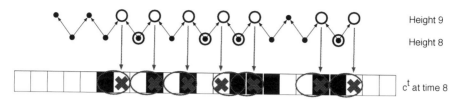

Fig. 6. Arrows map the sites of $\hat{C}^9$ to their corresponding constrained cells. These constrained cells are marked by a cross. Rounds surround a constrained cell and its partner.

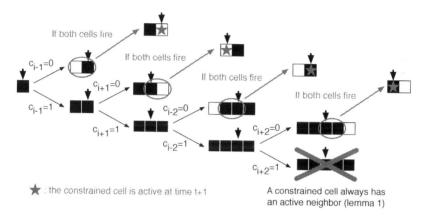

Fig. 7. Proof of lemma 2

states of its neighboring cells and show that in every cases, if c_i^t and its partner fire then c_i^{t+1} is active. This case study is done in figure 7.   □

Theorem 3. *If* $\alpha \geqslant \sqrt[3]{1 - (1 - p_c)^4}$ *then* $P_\alpha(c^{init}) < 1$.

Proof. Consider a random sequence $(c^t)_{t \geqslant 0}$ of configurations updated according to rule 178 under α-asynchronous dynamics where $c^0 = c^{init}$ and a randomly labeled graph $\mathbb{L}(p)$. Assume $\alpha \geqslant \sqrt[3]{1 - (1 - p)^4}$. We design a coupling between these two processes.

The coupling is defined recursively for all t between c^t and bonds of height t in $\mathbb{L}(p)$. At time t we have to label the edges of height t of $\mathbb{L}$ such that each edge is open with independent probability p. We have to chose which cells fire such that each cell fires with independent probability α. We also design our coupling such that if at least one of the two bonds which ends to a site of $\hat{C}^{t+1}$ is open then the corresponding cell c_i^{t+1} is active. In order to do so, we use Lemma 2 and assuming that the correspondence criterion is true at time t, we force the constrained cell c_i^t and its partner to fire. Since the correspondence criterion is true at time 0, it will recursively stay true for all t.

To achieve this, we partition the cells of c^t and the bonds of height t into buckets. Each bucket consists in:

– type 1 : a unique cell c_i^t if it is neither constrained nor the partner of any other cell.
– type 2 : a unique bond b if it does not end to a site of $\hat{C}^{t+1}$.
– type 3 : two cells c_i^t and $c_{i+\epsilon}^t$ with $\epsilon = 1$ or -1 and the two bonds of height t pointing to the site corresponding to cell $c_{i+\epsilon}^{t+1}$ if c_i^t is the partner of only one constrained cell $c_{i+\epsilon}^t$.
– type 4 : three cells c_{i-1}^t, c_i^t and c_{i+1}^t and the four bonds of height t pointing to the sites corresponding to cells c_{i-1}^{t+1} and c_{i+1}^{t+1} if c_i^t is the partner of the two constrained cells c_{i-1}^t and c_{i+1}^t.

One can easily verify that each bond of height t and each cell of c^t belongs to exactly one bucket. We assign to each bucket k an independent random variable X_k^t uniformly distributed in $[0, 1]$. Now we define the coupling within each bucket:

– type 1 : if $X_k^t < \alpha$ then c_i^t fires and if $X_k^t > \alpha$ then c_i^t does not fire.
– type 2 : if $X_k^t < p$ then b is open and if $X_k^t > p$ then b does is closed.
– type 3 : bounds are open and cells fires according to this diagram :

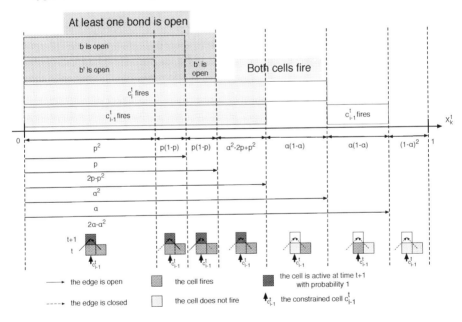

Each bond is open with probability p. Bonds b and b' are opened with probability p^2, b is open and b' is closed with probability $p(1 - p)$ and b is closed and b' is open with probability $p(1 - p)$. Thus each bond is open independently of the other. Same thing holds for the cells. Since $\alpha \geqslant \sqrt[3]{1 - (1 - p)^4}$, we have $\alpha^2 - 2p + p^2 > 0$ and the probability that at least one bond is open is less than the probability that both cells fire.

– type 4: As in the previous case we define a coupling which respect the distribution of probability and the independence of updates/labeling such that:

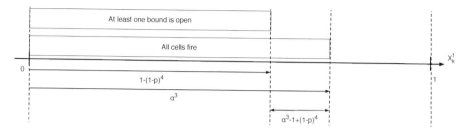

The probability that all cells fire is α^3 and the probability that at least one bound is open is $1 - (1-p)^4$. Since $\alpha \geqslant \sqrt[3]{1 - (1-p)^4}$, the three cells always fire when at least one bond is open in this coupling.

Thus as soon as one edge ending to site of $\hat{C}^{t+1}$ is open, the corresponding cell at time $t+1$ is active. Thus the correspondence criterion stays true at time $t+1$. Since the correspondence criterion is true at time 0, it will recursively stay true for all t. If $p > p_c$ then $\theta(p) > 0$ which means that for all $t \geqslant 0$ there exists a cell of height t which is in the open cluster with positive probability. Thus there exists an active cell in c^t for all $t \geqslant 0$, with positive probability. Then if $\alpha \geqslant \sqrt[3]{1 - (1 - p_c)^4}$ then $P_\alpha(c^{init}) < 1$. $\qquad\square$

Theorem 2 and theorem 3 yields the theorem 1.

Acknowledgements. Thanks to Nicolas Schabanel and Éric Thierry for their useful remarks that helped to improve the writing of the paper.

References

1. Balister, P., Bollobás, B., Kozma, R.: Large deviations for mean fields models of probabilistic cellular automata. Random Structures & Algorithms 29, 399–415 (2006)
2. Bersini, H., Detours, V.: Asynchrony induces stability in cellular automata based models. In: Proceedings of Artificial Life IV, pp. 382–387. MIT Press, Cambridge (1994)
3. Buvel, R.L., Ingerson, T.E.: Structure in asynchronous cellular automata. Physica D 1, 59–68 (1984)
4. Cook, M.: Universality in elementary cellular automata. Complex system 15, 1–40 (2004)
5. Fatés, N.: Directed percolation phenomena in asynchronous elementary cellular automata. In: El Yacoubi, S., Chopard, B., Bandini, S. (eds.) ACRI 2006. LNCS, vol. 4173, pp. 667–675. Springer, Heidelberg (2006)
6. Fatés, N., Morvan, M.: An experimental study of robustness to asynchronism for elementary cellular automata. Complex Systems 16(1), 1–27 (2005)
7. Fatés, N., Morvan, M., Schabanel, N., Thierry, É.: Asynchronous behaviour of double-quiescent elementary cellular automata. In: Jedrzejowicz, J., Szepietowski, A. (eds.) MFCS 2005. LNCS, vol. 3618, pp. 316–327. Springer, Heidelberg (2005)

8. Fatés, N., Regnault, D., Schabanel, N., Thierry, É.: Asynchronous behaviour of double-quiescent elementary cellular automata. In: Correa, J.R., Hevia, A., Kiwi, M. (eds.) LATIN 2006. LNCS, vol. 3887. Springer, Heidelberg (2006)

9. Fortuin, C.M., Kasteleyn, P.W.: On the random cluster model. i. Introduction and relation to other models. Physica 57, 536–564 (1972)

10. Fukś, H.: Non-deterministic density classification with diffusive probabilistic cellular automata. Phys. Rev. E 66(2) (2002)

11. Fukś, H.: Probabilistic cellular automata with conserved quantities. Nonlinearity 17(1), 159–173 (2004)

12. Gács, P.: Reliable computation with cellular automata. Journal of Computer and System Sciences 32(1), 15–78 (1986)

13. Gács, P.: Reliable cellular automata with self-organization. Journal of Statistical Physics 103(1/2), 45–267 (2001)

14. Gács, P., Reif, J.: A simple three-dimensional real-time reliable cellular array. Journal of Computer and System Sciences 36(2), 125–147 (1988)

15. Grimmett, G.: Percolation, 2nd edn. Grundlehren der mathematischen Wissenschaften, vol. 321. Springer, Heidelberg (1999)

16. Grimmett, G.: The Random-Cluster Model. Grundlehren der mathematischen Wissenschaften, vol. 333. Springer, Heidelberg (2006)

17. Lumer, E.D., Nicolis, G.: Synchronous versus asynchronous dynamics in spatially distributed systems. Physica D 71, 440–452 (1994)

18. Ollinger, N.: The intrinsic universality problem of one-dimensional cellular automata. In: Alt, H., Habib, M. (eds.) STACS 2003. LNCS, vol. 2607, pp. 632–641. Springer, Heidelberg (2003)

19. Ovchinnikov, A.A., Dmitriev, D.V., Krivnov, V.Y., Cheranovskii, V.O.: Antiferromagnetic ising chain in a mixed transverse and longitudinal magnetic field. Physical review B. Condensed matter and materials physics 68(21) 214406.1–214406.10 (2003)

20. Regnault, D.: Abrupt behavior changes in cellular automata under asynchronous dynamics. In: Proceedings of 2nd European Conference on Complex Systems (ECCS), Oxford, UK (to appear, 2006)

21. Regnault, D., Schabanel, N., Thierry, É.: Progresses in the analysis of stochastic 2D cellular automata: a study of asynchronous 2D Minority. In: Kučera, L., Kučera, A. (eds.) MFCS 2007. LNCS, vol. 4708, pp. 320–332. Springer, Heidelberg (2007)

22. Sarkar, P.: A brief history of cellular automata. ACM Computing Surveys 32(1), 80–107 (2000)

23. Schönfisch, B., de Roos, A.: Synchronous and asynchronous updating in cellular automata. BioSystems 51, 123–143 (1999)

24. Toom, A.: Stable and attractive trajectories in multicomponent systems. Advances in Probability 6, 549–575 (1980)

Resolution Width and Cutting Plane Rank Are Incomparable

Mark Rhodes

Durham University, Department of Computer Science,
South Road, Durham, Co. Durham, DH1 3LE, UK
m.n.c.rhodes@dur.ac.uk
http://www.dur.ac.uk/m.n.c.rhodes

Abstract. We demonstrate that the Cutting Plane (CP) rank of a polytope defined by a system of inequalities derived from a set of unsatisfiable clauses can be arbitrarily larger than the Resolution width of the
clauses, thus demonstrating the two measures are incomparable. More
specifically, we show there exists an infinite family of unsatisfiable clauses
defined over $n \in \mathbb{N}$, which have constant Resolution width, but, yield
polytopes which have CP rank $\Omega(\log_2 n)$.

Keywords: Propositional Proof Complexity, Cutting Plane Proofs, Rank
Lower Bounds, Resolution Width.

1 Introduction

The problem of satisfiability of propositional CNF formulae (SAT) is perhaps the
most well studied in theoretical computer science. Resolution provides the logical
basis for almost all the most widely used SAT solvers (cf. [16]) and is also the
most intensively studied and well understood proof system in the area of proof
complexity. One interesting complexity measure for Resolution is the required
width of a proof, which under certain conditions is known to be closely related
to the size of the Resolution proof as well as its minimum space complexity (see
[2] and [1] respectively).

Since Linear Programming (LP) has been shown to be solvable in polynomial
time, by Khachain in [11], there has been an increasing interest in using Integer
Linear Programming (ILP) algorithms as a means of solving SAT instances. One
of the oldest and most commonly used methods for solving ILP is Cutting Planes
(CP) (also known as Gomory Cuts), which was first introduced in [9] and first
considered as a proof system in [5]. An interesting measure of complexity of CP
proofs, conceived in [4] and put forward for study in proof complexity in [3], is
the rank of a bounded polyhedron (polytope), namely the minimum number of
rounds cuts required to reach its integer hull. Whilst it is clear that CP can p-
simulate Resolution with respect to the size of proofs [5], it is not clear where the
measures of Resolution width and CP rank are related. Currently the only CP
rank bounds for polytopes derived from unsatisfiable SAT instances are those
given in [3]. These show that the CP rank of a polytope can be arbitrarily smaller

E. Ochmański and J. Tyszkiewicz (Eds.): MFCS 2008, LNCS 5162, pp. 575–587, 2008.

than the Resolution width, however, it is unknown whether an upper bound for the Resolution width of an unsatisfiable SAT instance implies an upper bound on the CP rank of its corresponding polytope. We show that this is not true by demonstrating that there is an infinite family of tautologies which require constant Resolution width, yet at least logarithmic CP rank. Our proof uses a novel technique based on a simple card game, since the method used in [3] is ineffective on the tautologies we consider.

2   Preliminaries

Resolution is perhaps the most simplistic of all proof systems. A Resolution proof of a CNF formula F, can be considered to be an ordered list of clauses $C_1, \ldots, C_s$, where each C_s is the empty clause, and each C_i is either a clause in F or is derived from two other clauses $C_{i'}$ and $C_{i''}$ where $i', i'' < i$, using the following rule, known as the Resolution rule:

$$\frac{A \vee l \quad B \vee \neg l}{A \vee B}.$$

The measure of the width of a Resolution refutation was introduced in [2] and is defined to be the size of the largest clause in the refutation. The Resolution width of a given set of clauses is the minimum possible width of any Resolution refutation of these clauses. One problem with this definition is that instances with large clauses, must have high width, even though it maybe possible to reduce the width simply by rewriting the clauses as a 3-CNF formula. We consider a more robust definition of the Resolution width, introduced in [8], which is the minimum value k, such that the clauses have a width k narrow Resolution refutation. A width k narrow Resolution refutation of a CNF formula F is a sequence of clauses $C_1 \ldots C_s$ where C_s is the empty clause and each C_i is either a clause in F or is derived using one of the following three rules:

1. From B we can derive $B \vee x$, or $B \vee \neg x$ (weakening);
2. From $A \vee x$ and $B \vee \neg x$ we can derive $A \vee B$ (Resolution);
3. From $x_1 \vee \cdots \vee x_m$ (where m can be $\geq k$) in F, and $B \vee \neg x_1, \ldots, B \vee \neg x_m$ we can derive B (Resolution by cases);

crucially all clauses derived from any of the three rules must have $\leq k$ literals. In [8], they show how the narrow Resolution proof system is related to normal Resolution, later we use these relations to show that our result holds for both definitions of Resolution width.

The CP proof system, can be considered as a refutation proof system operating on linear inequalities (i.e. it derives the contradiction $1 \leq 0$) which has the axioms $x_i \leq 1$ and $x_i \geq 0$ for any variable x_i and the following inference rule, which we will call the cut rule:

$$a_{11}x_1 + \cdots + a_{1n}x_n \geq b_1$$
$$\cdots$$
$$\frac{a_{n1}x_1 + \cdots + a_{nn}x_n \geq b_n}{(\sum_{i=1}^{n} \lambda_i a_{i1})x_1 + \cdots + (\sum_{i=1}^{n} \lambda_i a_{in})x_n \geq c}$$

where c is an integer such that $\lceil \sum_{i=1}^{n} \lambda_i b_i \rceil \geq c$ and the λ_i's are non-negative rational coefficients satisfying $\sum_{i=1}^{n} \lambda_i a_{ij} \in \mathbb{Z}$ for all $1 \leq j \leq n$.

We translate the clauses of the original CNF formula into inequalities as follows, the clause

$$x_{i_1} \vee \cdots \vee x_{i_t} \vee \neg x_{j_1} \vee \cdots \vee \neg x_{j_f}$$

becomes the inequality

$$x_{i_1} + \cdots + x_{i_t} + (1 - x_{j_1}) + \cdots + (1 - x_{j_f}) \geq 1.$$

In [5], it is shown that under this translation, CP p-simulates Resolution.

The rank of a polytope is the minimum number of rounds of applications of the cut rule required to reach its integer hull. If the converted CNF is contradictory as a linear program, then its CP rank is 0, if one round of the cut rule is enough to reach a contradiction, its CP rank is 1 and so on. We refer to the polytope defined by the converted CNF as P^0 and the polytope containing only points that can not be removed form P^0 after i rounds of cuts as P^i. When studying CP rank it is often useful to consider the following definition of the polytope remaining after a round of cuts, presented in [3]. In the definition P refers to the current polytope and P' is the one remaining after a round of applications of the cut rule.

$$P' = \{x \in P' : \langle a, x \rangle \geq \lceil b \rceil \text{ whenever } a \in \mathbb{Z}, b \in \mathbb{R}, \text{ and } \langle a, y \rangle \geq b \text{ for all } y \in P\}.$$

The family of unsatisfiable CNF formulae that we shall consider we will call $RHSP2_n$, which stands for the Relativized House Sitting Principle with 2 sets. This is the family of CNF formulae generated by taking the normal House Sitting Principle (HSP), which is defined in first-order logic as the constraints: $\forall x \exists y ((y \geq x) \wedge W(x, y))$, $\neg W(0, 0)$, $\forall x, y ((x < y) \wedge W(y, y) \rightarrow \neg W(x, y))$, relativising it twice (see [6]) and converting it to a purely propositional sentence over n variables (see [14]). The reason we consider the relativised version of HSP is that the original version has CP rank 0, so would not be appropriate. The reason we consider the version with two sets instead of one is simply that the method we employ fails on the later version.

The formula $RHSP2_n$ can informally be considered to represent the contradictory scenario where there there is a street with n houses on it, and the higher the house number, the better it is. House one is a run-down shack and house n is a luxurious mansion. The residents living on the street decide to play a game in which they can go to each other's houses, under a number of conditions. Since no-one wants a bad deal there agree only to go into a house at least as good as their own. The residents of house one, obviously don't like their own house and so don't want to stay there and belong to two groups, the neighborhood watch (q) and a risky pyramid scheme (r). The residents of the street decide that if people who are in both these groups go into their house, they should join these groups as well. They also decide that if the owners of a house are in both these groups and stay their own homes, then no-one belonging to both the groups is allowed to visit their house. We represent the proposition "some of the owners of house i go to house j" with the variable $W_{i,j}$ and the proposition "the owners of house i belong to q (r)" using the variable $S_{i,q}$ $(S_{i,r})$. We consider the formula $RHSP2_n$, where $n \geq 2$, as being defined by the following inequalities (clauses):

$\sum_{j=i}^{n} W_{i,j} \geq 1$ $(\bigvee_{j=i}^{n} W_{i,j})$, for all $1 \leq i \leq n$, which we shall call the *witnessing* inequalities and can be considered to state that the residents of house i must go to a house at least as good as there own.

$2 + S_{j,t} \geq W_{i,j} + S_{i,q} + S_{i,r}$ $(S_{j,t} \vee \neg W_{i,j} \vee \neg S_{i,q} \vee \neg S_{i,r})$ for all $t \in \{q,r\}$, $i \leq n - 1$ and all $j \geq i + 1$, $j \leq n$. We refer to this set of inequalities as the *inductive* ones and they can be thought of as stating that if residents of house i are in both groups and they go to house j, then j must be in the set t.

$W_{i,j} + W_{j,j} + S_{i,q} + S_{i,r} + S_{j,q} + S_{j,r} \leq 5$ $(\neg W_{i,j} \vee \neg W_{j,j} \vee \neg S_{i,q} \vee \neg S_{i,r} \vee \neg S_{j,q} \vee \neg S_{j,r})$, for all $1 \leq i \leq n - 1$, $2 \leq j \leq n$ which we will refer to as the *fullhouse* inequalities. These can be considered as stating that if the residents of houses i and j are in both groups, and j goes into their own house then i can't go to j's house.

A set of single clause inequalities, $W_{1,1} \leq 0$ $(\neg W_{1,1})$ which states that the residents of house one don't stay in their own house, $S_{1,1} \geq 1$ $(S_{1,1})$ and $S_{1,2} \geq 1$ $(S_{1,2})$, which state that residents of house one are in both sets.

3   Results

To get our upper bound on the narrow Resolution width of $RHSP2_n$ we use the following witnessing pebbling game, introduced in [8].

Let F be a CNF formula. The witnessing pebble game in F is played between a Prover and a Delayer on a set of literals arising from the variables in F. A pebble can never appear on both the positive and negative literals of any variable. In each round, one of three things can happen.

1. The Prover lifts a pebble from the board; Delayer makes no response.
2. (Querying a Variable.) The Prover gives a pebble to the Delayer and names an empty variable x (i.e. neither x nor $\neg x$ is pebbled already). The Delayer then places the pebble on x or $\neg x$.
3. (Querying a Clause.) The Prover give the Delayer a pebble and names a clause C from F. The Delayer must then place it one of the literals of C, without contradicting a pebble already on the board. If this is impossible then Prover wins the game.

When the game is limited to a given number of pebbles k, then we call this the k-pebble witnessing game. Notice that Prover can only win if the pebbles on the board falsify some of the clauses of F and Prover has at least one pebble left. From [8], we also get the following lemma, linking the witnessing pebble game with the narrow Resolution width of a proof.

Lemma 1. *([8], Proposition 4)*
Let F be an CNF formula. If there is a winning strategy for Prover in the k-pebble witnessing game for F then there is a narrow Resolution proof of width k of F.

We can now prove our constant upper bound on the Resolution width of $RHSP2_n$, by demonstrating that the value of n does not affect the number of pebbles required for Prover to win the witnessing pebble game on the clauses of $RHSP2_n$.

Theorem 1. *For every $n \geq 3$, $RHSP2_n$ has a narrow Resolution proof of width ≤ 6.*

Proof. At the start of the game Prover queries the single literal clauses $S_{1,q}$, $S_{2,r}$ and $\neg W_{1,1}$.

If there are pebbles on $S_{i,q}$, $S_{i,r}$, $\neg W_{i,i}$, then Prover can use only 3 more pebbles to force Delayer to placed pebbles on $S_{j,q}$, $S_{j,r}$, $\neg W_{j,j}$ where $1 \leq i, j \leq n$ and $i < j$. Prover first queries the *witnessing* clause $\bigvee_{q=1}^{n} W_{i,q}$, Delayer must put a pebble on some $W_{i,j}$. Prover then queries the variables $S_{j,q}$ and $S_{j,r}$, Delayer must play on the positive literals of both of these variables or Prover could then query the one of the *inductive* clauses $S_{j,t} \vee \neg W_{i,j} \vee \neg S_{i,q} \vee \neg S_{i,r}$ for each $t \in \{q, r\}$. Prover then queries the *fullhouse* clause $\neg W_{j,j} \vee \neg S_{i,q} \vee \neg S_{i,r} \vee \neg S_{j,q} \vee \neg S_{j,r}$; in response Delayer must place a pebble on $\neg W_{j,j}$.

Note that is Prover plays in this manner, he can pick up the pebbles on $S_{i,q}$, $S_{i,r}$ and $W_{i,j}$, so never needs to put more than 6 pebbles down at any time. If Player continually uses this strategy, then eventually $j = n$, then Prover can win by querying the single literal clause $W_{n,n}$. □

To prove our lower bound for the rank of cutting planes on $RHSP2_n$, we use the following *protection* lemma presented in [3]. The name *protection* comes from the fact that it ensures a certain point is protected from being removed in the next round by the cut rule provided that a number of other points are present in the current round. Throughout the rest of this paper, $E(x)$ denotes the set of all variables in which the point x is not an integer. Recall that P' is defined to be the polytope remaining after applying a round of cuts to P.

Lemma 2. *([3], Lemma 3.1) Let P be a bounded polytope in $\mathbb{R}^n$. Let $x \in \frac{1}{2}\mathbb{Z}^n$ and let $E = E(x)$ be partitioned into sets $E_1, E_2, \ldots, E_t$. Suppose for every $j \in 1, 2, \ldots, t$ we can represent x as an average of vectors in P that are $0 - 1$ on E_j but agree with x elsewhere. Then $x \in P'$.*

The proof of our lower bound involves showing that a point, which we will from here on refer to as x, defined as having the variables set to the following values, must be present in $P^{\log_2(n)-2}$.

$W_{n,n}$, $S_{1,q}$, $S_{1,r} = 1$.
$W_{i,i+1}$, $W_{i,i+2} = \frac{1}{2}$ for all odd i, where $1 \leq i \leq n - 3$.
$W_{i,i+2}$, $W_{i,i+3} = \frac{1}{2}$ for all even i where $2 \leq i \leq n - 4$.
$S_{i,q}$, $S_{i,r} = \frac{1}{2}$ for all $2 \leq i \leq n$.
$W_{n-2,n-2}$, $W_{n-2,n-1}$, $W_{n-1,n-1}$, $W_{n-1,n} = \frac{1}{2}$.
All other variables are set to 0.

Note that by substituting the values given in x, each inequality of $RHSP2_n$ is satisfied and therefore $x \in P^0$.

Theorem 2. *The CP rank of* $RHSP2_n > \log_2(n) - 2$, *where* $n \geq 8$ *and a power of 2.*

Proof. We demonstrate the $P^{\log_2(n)-2}$ is non-empty by demonstrating that it contains x.

We consider a card game played between two players Dealer and Player, which runs over a number of rounds. In each round Player is at some point y in the space and in the first round $y = x$. It is his job to try and find a point $y \notin P^0$. At the start of the game, Dealer has a pack of unique cards each one representing a single variable in $E = E(x)$. When we talk of a card we are really talking about the variable the card represents. At the beginning of each round he deals all the cards he has out on to a table and arranges them into sets. These sets are are a partitioning of the variables of E, as in lemma 2. For each set of cards Dealer decides on two possible assignments of 0/1 values to these cards, where each card is assigned 0 in one assignment and 1 in the other. Player then picks up one of these sets and chooses one of the two possible assignments. Player updates their position y by setting these variables to their chosen values. We call a position reached in this manner a child of y. Note that if the Player is at a point p, the average of all the possible children of p is the point p. The round ends when Dealer picks up all the remaining cards and the next round starts when he deals them out again.

Since the inequalities of $RHSP2_n$ are unsatisfiable, it is clear that Player will eventually win the game (i.e. if he picked up all the cards, he would have set all the variables to 0/1 values and must then have reached a point $y \notin P^0$). The link between the card game and the CP rank is that if Dealer can play so that the game lasts until the end of round i, by lemma 2, the CP rank of the polytope is at least $i + 1$. We therefore demonstrate a strategy for Dealer that allows him to ensure the game lasts until the end of round $\log_2(n) - 2$.

Figures 1 to 6, define the possible sets of cards into which Dealer will partition E together with their associated 0/1 values which define the child points of a given current point. An edge uv in these figures represents the variable $W_{u,v}$ if v is not labeled q or r otherwise it represented the variable $S_{u,v}$. The two possible sets of 0/1 values associated with each possible set of cards are defined as having all the dashed edges set to 0, and the solid edges 1 and vice-versa.

Figures 1 and 2 show how specific elements of E are partitioned into subsets, however figures 3 to 6 give a template defining how elements of E can be partitioned according to some parameter i. We call the set matching template T_a having $i = p$, $E_{a,p}$ and we say $E_{a,p}$ has 'i' value p.

Note that E_{start} and T_{left} are constructed so that if someone goes into some house h, the residents of h are always in both q and r. By comparison, the other sets are constructed as to prevent the residents of a particular house belonging to both q and r.

Intuition. There are distinct numbers, (1) the biggest number so that the residents of house h are in both q and r and (2) the lowest number so that the residents of house are not in one of q or r. Only the segment between (1) and

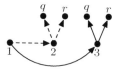

Fig. 1. The set E_{start}, and associated 0/1 values

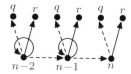

Fig. 2. The set E_{end}, and associated 0/1 values

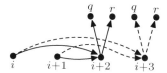

Fig. 3. The template T_{left}, and associated 0/1 values

(2) is inconsistent so Player tries to narrow the size of this segment, whilst Dealer makes sure it can only be halved in each round, hence the $\Omega(\log_2 n)$ lower bound. A valid, although ineffective move for Player is to play in the consistant region to the left of (1), or in the consistant region to the right of (2).

Before describing Dealer's strategy we need a few definitions. r is the number of the current round, and at the start $r = 1$. m is the 'i' value of the set matching T_{mid} in the current round (there is exactly one such set in each deal). $rmin$ and $lmax$ represent the smallest 'i' value of a set matching T_{right} and the largest 'i' value of a set matching T_{left} respectively, that can be chosen by Player so that Dealer plays in the same way as the previous round. Initially $rmin = n - 4$ and $lmax = 2$.

In round 1, Dealer partitions the cards into the sets E_{start}, E_{end}, $E_{mid, \frac{n}{2}}$, $E_{left,i}$ for each $2 \leq i \leq \frac{n}{2} - 2$, where i is even, and $E_{right,j}$ for each even j where $\frac{n}{2} + 2 \leq j \leq n - 4$. It is an easy exercise to see that this constitutes a valid partitioning of E (i.e. each element of E appears in exactly one of these sets). In subsequent rounds, Dealer does one of the following three things, depending on the actions of Player:

1. If Player picks any of the sets of cards E_{start}, $E_{left,i}$ for $2 \leq i \leq lmax$, $E_{right,j}$, where $rmin \leq j \leq n - 4$, E_{end} or any set matching T_{other}, Dealer simply deals in the same fashion as he did in the previous round. Intuitively this case represents the situation where Player plays in either of the consistent regions or on one the the very edges of the inconsistent region.
2. If Player picks a set $E_{left,l}$ where $l > lmax$, then Dealer moves the "mid point" to the right by setting the 'i' value of the middle set (i.e. the one

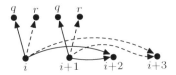

Fig. 4. The template T_{right}, and associated 0/1 values

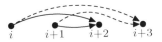

Fig. 5. The template T_{mid}, and associated 0/1 values

Fig. 6. The template T_{other}, and associated 0/1 values

matching T_{mid}) to $m' = m + \frac{n}{2^{r+1}}$. Any remaining cards "to the left of this set" are partitioned as many of E_{start} and $E_{left,j}$ for each even j where $2 \leq j \leq m'-2$ as can be made with the cards Dealer has left. The remaining cards (i.e. those representing $S_{i,q}$, $S_{i,r}$ and $W_{i,j}$ where $m' + 2 \leq i, j \leq n$), are dealt into the same sets as they appeared in the previous round. Dealer then increases $lmax$ to $m + 2$. Intuitively this case represents the situation where Player plays in the inconsistent region to the left of the "mid point".

3. If Player removes a set $E_{right,i}$, where $i < rmin$, then Dealer moves the "mid point" to the left by setting the 'i' value of the middle set to $m' = m - \frac{n}{2^{r+1}}$. He then makes as many of the sets E_{start}, $E_{left,j}$ where $2 \leq j \leq m'-2$ and j is even, as he can with the remaining cards "on the left of the new mid point". The cards between the new "mid point" and the previous one ($S_{i,q}$, $S_{i,r}$ and $W_{i,j}$ where $m'+2 \leq i \leq m$ and j is any number) are partitioned into the sets $E_{right,k}$ for each even k where $m' + 2 \leq k \leq m$. The remaining cards (those to the right of the previous "mid point") are dealt into the same sets as in the previous round. Dealer then decreases $rmin$ to $m - 2$. Intuitively this case represents the situation where Player plays in the inconsistent region to the right of the "mid point".

4. If Player picks the exact "mid point" by selecting the set $E_{mid,m}$, then Dealer moves the "mid point" to the left by setting the 'i' value of the middle set to $m' = m - \frac{n}{2^{r+1}}$. In this case Dealer lays down the set $E_{other,m}$ which deals with the "left over" cards $S_{m,q}$, $S_{m,r}$, $S_{m+1,q}$ and $S_{m+1,r}$. The cards between the new "mid point" and the previous one ($S_{i,q}$, $S_{i,r}$ and $W_{i,j}$ where $m'+2 \leq i \leq m-2$ and j is any number) are partitioned into the sets $E_{right,k}$ for each even k where $m' + 2 \leq k \leq m - 2$. The remaining cards (those to the right of the previous "mid point") are dealt into the same sets as in the previous round. Dealer then decreases $rmin$ to $m - 2$.

Observation 1. The maximum possible difference between m and m' (the value of m in the next round) decreases by $\frac{1}{2}$ each round. Since after round 1 it can be at most $\frac{n}{4}$, and there are only $\log_2(n) - 2$ rounds then the maximum range of m is $4 \leq m \leq n - 4$.

Observation 2. Dealer never deals a set $E_{left,i}$, where $i > m - 2$ on the table.

Observation 3. Dealer never deals a set matching T_{other}, T_{right} or T_{mid}, with an 'i' value $< m$ on the table.

Observation 4. Dealer always deals his remaining cards $S_{i,q}, S_{i,r}$ and $W_{j,j'}$ where $1 \leq i \leq m + 1$, $1 \leq j \leq m - 1$ and j' is any number, into either E_{start} or sets matching T_{left}. All other cards are put into sets not matching T_{left} nor being E_{start}.

To prove that this strategy allows Dealer to play up until the end of round $\log_2(n) - 2$, we will show that it satisfies two properties, namely that it always provides a valid partitioning of the cards into sets matching the templates, E_{start} or E_{end} and y remains in P^0. To demonstrate these properties hold, we use the following claim, where a refers to the highest 'i' value of a set picked up by Player matching T_{left} and b is the smallest 'i' value of a set held by Player matching any of the other templates and initially where no such sets have yet been picked, we consider $a = 0$ and $b = n$:

Claim 1. At the start of round r, where $r \leq \log_2(n) - 2$, $a + 2 \leq lmax \leq m \leq rmin \leq b - 2$.

Key to showing this claim is true is observation 1. This observation implies that if Dealer ever increases (decreases) the value of m, then no matter how Player decides to play in any future rounds, it is never the case that we can arrive at a point where m is less than (more than) what it was before Dealer increased (decreased) it plus (minus) four.

To show that $lmax \leq m$ while $r \leq \log_2(n) - 2$, we use induction on r. It is trivially true for $r = 1$, as $m = \frac{n}{2}$ and $lmax = 4$. It remains true throughout as we only increase $lmax$ when we increase m. This only occurs when Dealer plays option 2, where $lmax$ is set to the value of m in the previous round plus two, yet the new value for m, m', is $\geq m + 4$ due to observation 1.

The first part of claim 1, namely that $a + 2 \leq lmax$, is trivially true at the start of the game and it remains true throughout as if a set $E_{left,j}$, where $j > lmax$, is picked up by Player, then dealer increases the value of $lmax$ to $m + 2$, and $m \geq j$ by observation 2.

The final part of claim 1, stating that $rmin \leq b - 2$, is clearly true at the start of the game. To see that it remains true, note that if ever a set matching, T_{right} or T_{mid} with an 'i' value $< rmin$ is selected by Player, Dealer reduces the value of $rmin$ to $m - 2$. Due to observation 3, this is sufficient.

To see that the strategy always ensures that at the start of any round $m \leq rmin$, we again use induction. When $r = 1$, it trivially holds as $rmin = n$ and $m = \frac{n}{2}$. It remains true as $rmin$ only decreases when m decreases (when Dealer plays options 3 or 4). When this occurs, $rmin$ is set to the value m was in the previous round minus two, yet the new value for m, m' is at most $m - 4$ by observation 1. This completes the proof of claim 1.

Claim 2. The strategy always gives a valid partition of all the remaining cards into sets matching one of the templates, E_{start} or E_{end}, while $r \leq log_2(n) - 2$.

We will show this claim to be true by induction or r. When $r = 1$, it trivially holds since we know exactly how Dealer plays and one can check this is a valid partitioning. The inductive case has four subcases, depending on which of the plays Dealer makes. Assume claim 2 holds for round $r - 1$. The trivial case for round r is when Dealer plays using option 1, since he will clearly deal all the remaining cards into valid sets by the inductive hypothesis. Notice that even if Player was able to pick all the sets in the consistant region to the left, they could set each of them to either of the two possible 0/1 values to create a single consistant chain beginning from house one to house $lmax$ in which the residents of each house go into another house and the residents of every occupied house are in both sets.

If Dealer plays using option 2, then Player selected a set $E_{left,j}$, where $j > lmax$. By observation 4, we know all remaining cards $S_{i,q}, S_{i,r}$ and $W_{g,g'}$ where $1 \leq i \leq m + 1, 1 \leq g \leq m - 1$ and g' is any number, were partitioned into set matching T_{left} or E_{start}. By the inductive hypothesis, it is valid for Dealer to put these cards down in the same sets in round r (without the set $E_{left,j}$ removed by Player). The cards $S_{i,q}, S_{i,r}$ and $W_{g,g'}$ where $m + 2 \leq i \leq m' + 1, m \leq g \leq m' + 1$, m' is the new value of m in the next round, and g' is any value, must still be in the pack available to Dealer, because by claim 1 and the fact that b is unchanged during this round, $m' \leq b - 2$. This enables Dealer is deal out the sets $E_{left,q}$ for all $m + 2 \leq q \leq m'$, and $E_{mid,m'}$ next time, as described in option 2. The rest of the cards can be partitioned precisely as they were in the previous round, these are $S_{i,q}$, $S_{i,r}$ or $W_{i,i'}$ where $m + 2 \leq i \leq n$ that aren't in Player's hand. By observation 4 these could not have been in E_{start} or a set matching T_{left} in round $r - 1$. Therefore, none of the possible sets these cards could have been in require a card $W_{i,i'}$, where $i < m' + 2$ and i' is any number. Therefore, Dealer has not yet dealt any of the cards required to construct these sets.

When Dealer plays using option 3, we know that Player picked a set matching T_{right} with an 'i' value $< rmin$. In this case, all the cards left from $S_{i,q}$, $S_{i,r}$ and $W_{i,i'}$ where $m + 2 \leq i \leq n$ and i' is any number, are placed in the same sets they appeared in the last round. This is acceptable because from observation 4 these cards could not have been in sets matching T_{left} or E_{start} and therefore as previously mentioned can not contain any variable $S_{g,q}$, $S_{g,r}$ or $W_{g,i'}$ for any $g < m + 2$. All the cards for $S_{i,q}, S_{i,r}$ and $W_{g,g'}$ where $m' + 2 \leq i \leq m + 1$, $m' \leq g \leq m + 1$, m' represents the new value of m at the end of round r and g' is any value, must still be in the pack available to Dealer, because by claim 1, $a + 2 \leq m$. Therefore, as detailed in option 3, Dealer can partition all these cards into the sets $E_{mid,m'}$ and $E_{right,j}$, where $m' + 2 \leq j \leq m$, without causing conflict.

If however Dealer plays using option 4, (i.e. Player selected $E_{mid,m}$), by the same reasoning as in the previous case, we know that Dealer has all the cards $S_{i,q}, S_{i,r}$ and $W_{g,g'}$ where $m' + 2 \leq i \leq m + 1$, $m' \leq g \leq m + 1$, m' represents the new value of m at the end of round r and g' is any value, except the cards for $W_{m,m+2}$, $W_{m,m+3}$, $W_{m+1,m+2}$ and $W_{m+1,m+3}$. Therefore, he is able to deal all these cards into the sets $E_{other,m}$, $E_{mid,m'}$ and $E_{right,j}$, where j is even and $m' + 2 \leq j \leq m - 2$, as detailed in option 4. One should note that the template T_{other} is only used when Dealer plays option 4; it is simply a way of dealing with the left over variables $S_{m,q}$, $S_{m,r}$, $S_{m+1,q}$ and $S_{m+1,r}$. In option 4, the rest of the cards are dealt into as many of the sets E_{start} and $E_{left,j}$, where $2 \leq j \leq m' - 2$ and j is even, as is possible. By observation 4, in the previous (successful) deal these cards must have been allocated into these sets and therefore, can be partitioned likewise in this round.

Since we know the strategy of Dealer creates a valid partitioning all the cards by following the four possible options, the inductive step holds and the proof of claim 2 is complete.

Claim 3. The strategy ensures Player's position y, is in P^0 at the end of round $r = \log_2(n) - 2$.

Note that the only inconsistent situations are (1) when the residents of a house don't go anyway, (2) the residents of a house are in both sets and go to another house in which the residents are not in both sets and (3) the residents of a house in both sets occupy their own house and other people also in both sets go into that house.

From claim 1, it is clear that there are two distinct non-intersecting regions from which Player has selected sets: the left of the middle and the right of the middle. From the way in which the sets are constructed, one can see that the residents of a house on the right of the middle are never in both sets whilst the residents of any house j on the left of the middle that has people go into it (i.e. $W_{i,j} = 1$ for some i), are in both sets (i.e. $S_{j,q}, S_{j,r} = 1$), for this reason situation (2) can not arise in the position defined by y.

It is also clear that in any position defined by y, situation (3) never occurs since only residents of houses $n - 3$ to n have the option of going into their own houses and by claim 1 they must always be on the right of the middle. Finally, situation (1) can not occur in y since whenever Player sets one variable $W_{i,j}$ to zero, they set another variable $W_{i,j'}$ to one. This concludes the proof of claim 3.

Since we have shown that $y \in P^0$, at the end of round $\log_2(n) - 2$ and that the strategy always creates a valid partitioning of the set E, our proof of theorem 2 is complete.   □

From theorems 1 and 2, we get the following corollary.

Corollary 1. *Narrow Resolution width is incomparable to CP rank.*

As in the proof of theorem 2 all variables $W_{i,j} = 0$ where $j > i + 3$ in x, it is clear that we could get precisely the same CP rank lower bound for the polytope defined by all the inequalities in $RHSP2_n$, except having all these variables removed. This would give us an instance which had a maximum clause length of six, i.e. a 6-CNF. In [8], they prove the following lemma.

Lemma 3. *([8], **Proposition** 2) If a r-CNF, F, has a width k narrow Resolution refutation, the F has a width $r + k - 2$ "normal" Resolution refutation.*

Since the narrow Resolution width of a set of clauses can not increase if variables are removed, this lemma allows us to see that our altered version of $RHSP2_n$ has a "normal" Resolution refutation of width ten. This implies the following corollary.

Corollary 2. *Resolution width, defined in terms of the maximum size of any clause in the proof, is incomparable to CP rank.*

4   Further Work

Our result fills in a gap in our knowledge about whether CP rank and Resolution width are related. There are similar other gaps about the rank of proofs in other systems such as the well known Lovásv-Schrijver proof systems and the Sherali-Adams (SA) proof system (see [12] and [15] respectively). For instance it is unknown whether CP and LS rank are incomparable. For the SA operator, before this result, the same was true (see [13]), however corollary 2 combined with proposition 1 from [7], tells us that the SA rank of a polytope is also incomparable to its CP rank.

We conjecture that the narrow Resolution width is exactly the SA rank of any unsatisfiable CNF and that the SA rank maybe arbitrarily smaller than the LS rank of a polytope, although never larger. A proof of the first conjecture (linking SA rank and narrow Resolution width) would be particularly interesting since it is likely to involve developing a means of assigning values to consistent partial assignments of a arbitrary instance.

One direct open question is whether $\Omega(\log_2(n))$ is a tight lower bound on the CP rank of $RHSP2_n$, since it is possible that it could require linear CP rank.

References

1. Atserias, A., Dalmau, V.: A combinatorial characterization of Resolution width. In: Proceedings of the 18th IEEE Conference on Computational Complexity, pp. 239–247 (2003)
2. Ben-Sasson, E., Wigderson, A.: Short proofs are narrow - Resolution made simple. Journal of the ACM 48(2), 149–168 (2001)
3. Buresh-Oppenheim, J., Galesi, N., Hoory, S., Magen, A., Pitassi, T.: Rank bounds and integrality gaps for cutting planes procedures. Theory of Computing 2, 65–90 (2006)
4. Chvátal, V.: Edmonds polytopes and a hierarchy of combinatorial problems. Discrte Mathematics 4, 205–337 (1973)
5. Cook, W., Coullard, R., Turan, G.: On the complexity of cutting plane systems. Discrete Applied Maths 18, 25–38 (1987)
6. Dantchev, S.: Relativisation Provides Natural Separations for Resolution-Based Proof Systems. In: Grigoriev, D., Harrison, J., Hirsch, E.A. (eds.) CSR 2006. LNCS, vol. 3967, pp. 147–158. Springer, Heidelberg (2006)

7. Dantchev, S., Martin, B., Rhodes, M.: Tight rank bounds for the Sherali-Adams proof system (manuscript, 2007)
8. Galesi, N., Thapen, N.: Resolution and Pebbling Games. In: Bacchus, F., Walsh, T. (eds.) SAT 2005. LNCS, vol. 3569, pp. 76–90. Springer, Heidelberg (2005)
9. Gomory, R.: Outline of an algorithm for integer solutions to linear programs. Bulletin of the AMS 64, 275–278 (1958)
10. Grigoriev, D., Hirsch, E., Pasechnik, D.: Complexity of semi-algebraic proofs. Moscow Mathematical Journal 4(2), 647–679 (2002)
11. Khachain, L.G.: A polynomial time algorithm for linear programming. Doklady Akademii Nauk SSSR, n.s. 244(5), 1063–1096 (1979); English translation in Soviet Math. Dokl. 20, 191–194
12. Lovász, L., Schrijver, A.: Cones of matricies and set-functions and 0-1 optimization. SIAM J. Optimization 1(2), 166–190 (1991)
13. Rhodes, M.: Rank lower bounds for the Sherali-Adams operator. In: Cooper, S.B., Löwe, B., Sorbi, A. (eds.) CiE 2007. LNCS, vol. 4497, pp. 648–659. Springer, Heidelberg (2007)
14. Riis, S.: A complexity gap for tree-Resolution (manuscript, 1999), http://citeseer.ist.psu.edu/riis99complexity.html
15. Sherali, H.D., Adams, W.P.: A hierarchy of relaxations between the continuous and convex hull representations for zero-one programming problems. SIAM Journal of Discrete Mathematics 3, 411–430 (1990)
16. Warners, J.P.: Nonlinear approaches to satisfiability problems. PhD thesis, Eindhoven University of Technology, The Netherlands (1999)

On the Decidability of
Bounded Valuedness for Transducers
(Extended Abstract)

Jacques Sakarovitch[1] and Rodrigo de Souza[2,⋆]

[1] LTCI, ENST/CNRS, Paris (France)
sakarovitch@enst.fr
[2] ENST, 46, rue Barrault, 75634 Paris Cedex 13 (France)
rsouza@enst.fr

Abstract. We give a new and conceptually different proof for the decidability of k-valuedness of transducers (a result due to Gurari and
Ibarra), without resorting to any other kind of machines than transducers. In contrast with the previous proof, our algorithm takes into account
the structure of the analysed transducers and yields better complexity
bounds. With the same techniques, we also present a new proof, hopefully
more easily understandable, for the decidability of bounded valuedness
(a result due to Weber).

1 Introduction

This communication is part of a complete reworking of the theory of k-valued
rational relations and transducers which makes it appear as a natural generalisation of the theory of rational functions (the 1-valued ones) and functional
transducers, not only at the level of results but *also at the level of proofs*.

In one word, it is decidable whether a finite transducer is functional (Schützenberger [1]), the equivalence of functional transducers is decidable (consequence
of the previous result), and every functional transducer is equivalent to an unambiguous one (Eilenberg [2]). These results generalise in a remarkable way to
bounded valued transducers: it is decidable whether the cardinality of the image
of every word by a given transducer is bounded (Weber [3]) and whether it is
bounded by a given integer k (Gurari and Ibarra [4], by reduction to the emptiness problem for a class of multi-counter automata); the equivalence of k-valued
transducers is decidable (Culik and Karhumäki [5] in the context of the study
of Ehrenfeucht's conjecture, and Weber [6]), and every k-valued transducer is
equivalent to the sum of k functional and unambiguous ones (Weber [7]).

In [8], we have given a new and shorter proof for this last result, with a gain
of one exponential in the size of the result with respect to the original proof. It
is based on a construction that we call the *lag separation covering* (of real-time

⋆ A financial support of CAPES Foundation (Brazilian government) for doctoral studies is gratefully acknowledged by this author.

E. Ochmański and J. Tyszkiewicz (Eds.): MFCS 2008, LNCS 5162, pp. 588–600, 2008.

transducers). This construction itself uses the *Lead or Delay Action* (LDA for short) introduced in [9] to describe an efficient construction for the decidability of the functionality of transducers.

In this communication we present a new proof for the following result:

Theorem 1 (Gurari-Ibarra [4]). *Let T be a transducer and k a positive integer. It is decidable in polynomial time whether T is k-valued.*

We also present here a new proof for the decidability of the bounded valuedness, which comes very naturally together with the proof of Theorem 1:

Theorem 2 (Weber [3]). *Let T be a transducer. It is decidable in polynomial time whether there exists an integer k such that T is k-valued.*

In a third part [10], we tackle the decidability of the equivalence of k-valued transducers by using together the methods we present here and in [8].

In the original proof of Theorem 1 by Gurari and Ibarra, a nondeterministic $k(k+1)$-counter 1-turn automaton $\mathcal{A}$ (see [11] for definitions) is built. A computation in $\mathcal{A}$ corresponds to $k+1$ computations of T with the same input, and each pair of these computations of T is associated with two counters. The counters are incremented by the lengths of the outputs until a position, guessed nondeterministically, where these outputs become different, and a computation of $\mathcal{A}$ is successful iff the outputs of its $k+1$ projections are pairwise distinct. Theorem 1 follows then from the decidability, in polynomial time, of the emptiness of a finite turn r-counter automaton, another result due to Ibarra [11].

If this theoretical scheme is clear, the actual complexity of the corresponding procedure is difficult to estimate beyond the fact that "it is polynomial". This is particularly true for the procedure which decides of the emptiness of a multi-counter automaton, for it is based on general arguments of complexity theory: if the r-counter automaton accepts some input, then there exists a constant c such that it accepts an input of length bounded by $(rm)^{cr}$ (where m is the number of transitions); it is possible to test these bounded inputs with a nondeterministic Turing machine working in space proportional to $cr\log(rm)$; for each nondeterministic Turing machine working in space $f(n)$ there exists an equivalent deterministic one working in time $d^{f(n)}$, for some constant d (cf. [12]). It is not clear how these two constants c and d can be effectively computed and if their actual values have any relationship with the transducer under inspection.

Our proof of Theorem 1 (Section 3) stems from a generalisation of the characterisation of functional transducers with the Lead or Delay Action (LDA) $\mathcal{G}$ in [9]. Roughly speaking, a computation in the product $T \times T = T^2$ projects on two computations with equal inputs in T, thus T is functional iff every successful computation in T^2 outputs a pair of equal words. Differences between words are witnessed by the LDA $\mathcal{G}$, and T is functional iff the product $T^2 \times \mathcal{G}$ is isomorph to T^2 (being hence finite) and assigns the empty word to the final states of T^2.

At first we generalise the LDA to an action $\mathcal{G}_{k+1}$ which measures the differences between the outputs in the $(k+1)$-tuples of computations of T. It is not difficult to get a necessary and sufficient condition on $T^{k+1} \times \mathcal{G}_{k+1}$ for T

be k-valued (Proposition 1). The problem is that this condition *is not effective anymore* for $\mathcal{T}^{k+1} \times \mathcal{G}_{k+1}$ may be infinite for $k > 1$ even if $\mathcal{T}$ is k-valued. The core of our method – and this is of course more complicated – is the proof that it is possible to attach to every state of $\mathcal{T}^{k+1}$ a finite set of information, effectively computable from $\mathcal{T}$, which retains all the useful information from $\mathcal{T}^{k+1} \times \mathcal{G}_{k+1}$ to decide whether $\mathcal{T}$ is k-valued (Theorem 3). These sets are what we call the *Lead or Delay Valuation* of $\mathcal{T}^{k+1}$ (LDV). We explain in Section 3.4 how the LDV can be constructed in $\mathcal{O}(\ell n^{k+1} m^{k+1})$, where n and m are the numbers of states and transitions of $\mathcal{T}$ and ℓ is the maximal length of outputs of transitions. By comparison with the complexity of the procedure to decide the functionality in [9], this is probably the best that can be hoped for. On the other hand it is to be acknowledged that the constant hidden in the "big O" is handed by a function which grows exponentially fast with the valuedness k, namely, $2^{5(k+1)^4}$.

Weber's proof of Theorem 2 is somewhat similar to the classical characterisation of bounded $\mathbb{N}$-automata of Mandel and Simon [13] (made more explicit in [14]): $\mathcal{T}$ is bounded valued iff $\mathcal{T}$ does not contain certain *forbidden* computations (Theorem 4). Weber gives in [3] an algorithm to detect these computations.

We give another proof for Theorem 2, which uses a construct, the *lag separation covering*, that we have defined in [8] in order to establish the decomposition resulted quoted above. We first describe the forbidden computations in a slightly different way (Theorem 5). With the help of the lag separation covering, the proof that the absence of these computations implies the bounded valuedness is straightforward: if this holds for $\mathcal{T}$, then the covering has an equivalent subtransducer whose underlying input automaton is finitely ambiguous; in other words, every input word can be read by a bounded number of computations in $\mathcal{T}$, thus $\mathcal{T}$ is bounded valued. We explain in Section 4.2 how this characterisation can be tested in a certain subtransducer of the product of $\mathcal{T}^3$ by the LDA in complexity $\mathcal{O}(\ell n^3(n^3 + m^3))$. To some extent the complexity claimed in [3] is of the same order as it is in $\mathcal{O}(\ell^2 n^9)$ but the proof is indeed difficult to follow.

Due to space constraints most of the proofs have not been included, but they can be found in [15] and hopefully in a forthcoming paper which is in preparation. We have tried our best to give here the ideas underlying the proofs. This is anyway a highly technical matter of which it would be futile to disguise the intrinsic complexity.

2   Preliminaries

We follow the definitions and notation in [16,2,17]. The set of words over a finite alphabet A (the free monoid over A) is denoted by A^*, and the empty word by 1_{A^*}, or simply 1 in figures. The length of a word u in A^* is denoted by $|u|$.

Let M be a monoid. An *automaton* $\mathcal{A} = (Q, M, E, I, T)$ is a directed graph given by sets Q of states, $I, T \subseteq Q$ of initial and final states, respectively, and $E \subseteq Q \times M \times Q$ of transitions labelled by M. It is finite if Q and E are finite.

A *computation* in $\mathcal{A}$ is a sequence of transitions $c : p_0 \xrightarrow{m_1} p_1 \xrightarrow{m_2} \ldots \xrightarrow{m_l} p_l$, also denoted by $c : p_0 \xrightarrow{m_1 \ldots m_l} p_l$. Its *label* is the element $m_1 \ldots m_l$ of M and

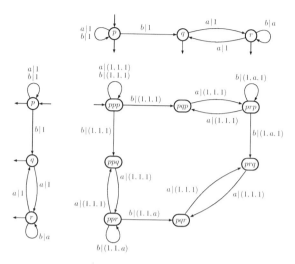

Fig. 1. A transducer T_1 (drawn on the left and above), and the part of T_1^3 accessible from (p, p, p) and co-accessible to (p, q, r). All states are final.

its length is $|c| = l$. It is *successful* if $p_0 \in I$ and $p_l \in T$. The *behaviour* of $\mathcal{A}$ is the set $\|\mathcal{A}\| \subseteq M$ of labels of successful computations. The behaviour of finite automata over M coincide with the family Rat M of *rational subsets* of M [2].

If M is a free monoid A^* and the labels of transitions are letters, then $\mathcal{A}$ is a (boolean) automaton over A. If M is a product $A^* \times B^*$, then every transition is labelled by an input word $u \in A^*$ and an output one $x \in B^*$ — this is denoted by $u|x$ — and $\mathcal{A}$ is a transducer realising a *rational relation* from A^* to B^*.

The image of a word u by a transducer T is the set of outputs of the successful computations reading u; T is called *k-valued* (for an integer $k > 0$) if the cardinalities of these images are at most k, and *bounded valued* if there exists such k.

We shall only consider *real-time* transducers: their labels are pairs $a|K$ formed by a letter a and a set $K \in \text{Rat } B^*$, and I and T are functions from Q to Rat B^*. By using classical constructions on automata, every transducer can be transformed into a real-time one. For bounded valued relations we may suppose that the transitions output a single word, and in order to avoid inessential details we can also suppose that the image of every initial or final state is the empty word.[1] In this case, the transducer is denoted rather as $T = (Q, A, B^*, E, I, T)$.

We shall make systematic use of product of automata. For real-time transducers, this operation is defined in the same way as for boolean automata, with the difference that the outputs have to be taken into account. Formally, the *square* of $T = (Q, A, B^*, E, I, T)$ is the transducer $T^2 = (Q^2, A, B^{*2}, E^{(2)}, I^2, T^2)$ where $(p, q) \xrightarrow{a|(u,v)} (p', q')$ is in $E^{(2)}$ iff both $p \xrightarrow{a|u} p'$ and $q \xrightarrow{a|v} q'$ are in E (see [9] for details). We define likewise the product of T by itself l times: a transducer T^l labelled by $A \times B^{*l}$ whose state set is Q^l, and the set of transitions is $E^{(l)}$.

---

[1] Such transducers are also called *nondeterministic generalised sequential machines*.

A last, and useful, convention: all automata (or transducers) considered or built by the various algorithms described here are implicitly assumed to be *accessible*. In particular, we write that Q^l is the set of states of $\mathcal{T}^l$, but indeed when we say that $\mathbf{q}$ is a state[2] of $\mathcal{T}^l$, we mean that $\mathbf{q}$ is accessible in $\mathcal{T}^l$.

3   Deciding k-Valuedness

Our proof for Theorem 1 consists in testing the k-valuedness of a transducer $\mathcal{T}$ in the cartesian product $\mathcal{T}^{k+1}$ in the same way as the functionality may be witnessed in the product of $\mathcal{T}^2$ by the Lead or Delay Action (LDA) $\mathcal{G}$ [9].

At first, the road to the generalisation seems easy: $(k+1)$-tuples of distinct computations in $\mathcal{T}$ with the same input are seen as computations in $\mathcal{T}^{k+1}$ and $\mathcal{T}$ is k-valued if in such computations at least two of the outputs are equal. To that end, the LDA is generalised to a *Pairwise Lead or Delay Action*, denoted $\mathcal{G}_{k+1}$, and the wanted property is expressed in $\mathcal{T}^{k+1} \times \mathcal{G}_{k+1}$ (Proposition 1).

The difficulty arises with the fact that $\mathcal{T}^{k+1} \times \mathcal{G}_{k+1}$ *may be infinite for $k > 1$, even if $\mathcal{T}$ is k-valued* (as in Figure 2). Here comes the crux of the proof: with the definition of *partially defined pairwise differences*, or PDPD (Section 3.3), we are able to attach to every state $\mathbf{q}$ of $\mathcal{T}^{k+1}$ a *finite* set $\mathsf{m}(\mathbf{q})$ of PDPDs. This $\mathsf{m}(\mathbf{q})$ subsumes the essential information contained in the states of $\mathcal{T}^{k+1} \times \mathcal{G}_{k+1}$ that map onto $\mathbf{q}$ and makes it possible to characterise the k-valuedness within a finite object, the *Lead or Delay Valuation* (LDV) of $\mathcal{T}^{k+1}$ (Theorem 3). As we explain in Section 3.4, the LDV can be built in polynomial time with a traversal of $\mathcal{T}^{k+1}$.

3.1   The Lead or Delay Action

Let B be an alphabet and $\overline{B}$ a disjoint copy of B. The underlying structure of the LDA is the free group $F(B)$ generated by B, that is, the quotient of $(B \cup \overline{B})^*$ by the relations $x\overline{x} = \overline{x}x = 1_{B^*}$, for every x in B. The inverse of an element u in $F(B)$ is denoted by $\overline{u}$ (for example, $\overline{xxy} = \overline{y}\,\overline{x}\,\overline{x}$). We write $\Delta = B^* \cup \overline{B}^* \cup \{\mathbf{0}\}$, where $\mathbf{0}$ is a new element, a zero, not in $F(B)$, and define a function $\rho : F(B) \cup \{\mathbf{0}\} \to \Delta$ by $w\rho = w$, if $w \in \Delta$, and $w\rho = \mathbf{0}$ otherwise.[3]

Definition 1 ([9,17]). *The Lead or Delay Action (LDA) of $B^* \times B^*$ on Δ, denoted by $\mathcal{G}$, is defined as follows: for every $w \in \Delta$ and $(u,v) \in B^* \times B^*$, $w \cdot (u,v) = (\overline{u}\,w\,v)\rho$ (where the product is taken with the rules $\mathbf{0}u = u\mathbf{0} = \mathbf{0}$).*

Intuitively, $1_{B^*} \cdot (u,v)$ represents the "difference" of the words u and v, being a positive word if u is a prefix of v (the *lead* of v with respect to u), a negative word if v is a prefix of u (the *delay* of v with respect to u), and $\mathbf{0}$ if u and v are not prefixes of a common word. In [9], an effective characterisation of the functionality is made with the product $\mathcal{T}^2 \times \mathcal{G}$ (cf. Definition 3), which shows the differences between pairs of computations of $\mathcal{T}$: $\mathcal{T}$ is functional iff $\mathcal{T}^2 \times \mathcal{G}$ assigns an unique value of $\Delta - \{\mathbf{0}\}$ to *every useful state* of $\mathcal{T}^2$ and 1_{B^*} to the final ones.

---

[2] We write tuples of states, or of words, with bold letters.

[3] We use a postfix notation for relations: $x\tau$ is the image of x by the relation τ.

3.2 The Pairwise Lead or Delay Action

In order to deal with the differences between the outputs of an arbitrary number l ($l > 1$) of computations in parallel, we generalise the LDA as follows. Let us write $D_l = \{(i,j) \mid 1 \leq i < j \leq l\}$. We write Δ_l for Δ^{D_l}, that is, the set of vectors of dimension D_l with entries in Δ, which we call *pairwise differences* or PD for short. The entry at the coordinate (i,j) of a PD δ is denoted by $\delta_{i,j}$. The PD with all entries equal to the empty word is denoted by η.

Definition 2. *For every integer $l > 1$, the Pairwise Lead or Delay Action of B^{*l} on Δ_l is the function $\mathcal{G}_l : \Delta_l \times B^{*l} \rightarrow \Delta_l$ which maps every $(\delta, \mathbf{u})$ in $\Delta_l \times B^{*l}$ to the PD γ in Δ_l such that, for every (i,j) in D_l, $\gamma_{i,j} = \delta_{i,j} \cdot (\mathbf{u}_i, \mathbf{u}_j)$.*

($\mathcal{G}_l$ is indeed an action for the LDA is applied independently to each coordinate.)

Definition 3. *For every integer $l > 1$, the product of $\mathcal{T}^l$ by $\mathcal{G}_l$ is the (accessible) transducer $\mathcal{T}^l \times \mathcal{G}_l = (Q^l \times \Delta_l, A, B^{*l}, F, I^l \times \{\eta\}, \mathcal{T}^l \times \Delta_l)$ where $(\mathbf{p}, \delta) \xrightarrow{a|\mathbf{u}} (\mathbf{q}, \delta')$ is a transition in F iff $\mathbf{p} \xrightarrow{a|\mathbf{u}} \mathbf{q}$ is a transition in $E^{(l)}$ and $\delta' = \delta \cdot \mathbf{u}$.*

The k-valuedness of $\mathcal{T}$ is witnessed by the final states of $\mathcal{T}^{k+1} \times \mathcal{G}_{k+1}$:

Proposition 1. *A transducer $\mathcal{T}$ is k-valued iff for every final state $(\mathbf{q}, \delta)$ of $\mathcal{T}^{k+1} \times \mathcal{G}_{k+1}$, δ has at least one entry equal to 1_{B^*}.*   □

This condition is not however effective. For every state $\mathbf{q}$ of $\mathcal{T}^{k+1}$, let us write $X(\mathbf{q})$ for the set of PDs in the states of $\mathcal{T}^{k+1} \times \mathcal{G}_{k+1}$ projecting on $\mathbf{q}$: $X(\mathbf{q}) = \{\delta \in \Delta_{k+1} \mid (\mathbf{q}, \delta) \text{ state of } \mathcal{T}^{k+1} \times \mathcal{G}_{k+1}\}$. Contrary to the characterisation of the functionality in [9], $X(\mathbf{q})$ may be infinite, even if $\mathcal{T}$ is k-valued (as in Figure 2).

3.3 A Finite Characterisation of k-Valuedness

The main concept for the definition of the Lead or Delay Valuation is that of *traverse* of a set of PDs. Intuitively, a traverse for $X \subseteq \Delta_l$ is a PD γ in Δ_l such that for every δ in X, there exists a coordinate (i,j) satisfying $\delta_{i,j} \neq \mathbf{0}$ and $\gamma_{i,j} = \delta_{i,j}$. In other words, each PD in X has a non null "intersection" with γ.

It may well exists some (i,j) in which no intersection arises. Such coordinates are not really useful. For this reason, we embed Δ_l in a larger set $H_l = [\Delta \cup \{\bot\}]^{D_l}$ of *partially defined pairwise differences* (PDPD), where $\bot$ fills undefined entries. Now, we say that a traverse for a set $X \subseteq H_l$ of PDPDs is a PDPD $\gamma \in H_l$ satisfying: for every $\delta \in X$, there exists a coordinate (i,j) such that $\delta_{i,j} \neq \mathbf{0}$, $\delta_{i,j} \neq \bot$, and $\gamma_{i,j} = \delta_{i,j}$; for every (i,j) such that $\gamma_{i,j} \neq \bot$, there exists at least one δ in X such that $\delta_{i,j} \neq \mathbf{0}$ and $\gamma_{i,j} = \delta_{i,j}$. A traverse has at least one defined entry, and has no entry equal to $\mathbf{0}$. We denote by $\operatorname{tv}(X)$ the set of traverses for X. As before, $\operatorname{tv}(X)$ may be infinite or empty.

The set H_l is naturally ordered by $\beta \sqsubseteq \gamma$ iff γ coincides with β on the defined entries of β. We denote by $\mathsf{m}(X) = \min(\operatorname{tv}(X))$ the set of minimal traverses for X, and for a state $\mathbf{q}$ of $\mathcal{T}^{k+1}$ we write $\mathsf{m}(\mathbf{q}) = \mathsf{m}(X(\mathbf{q}))$. The set $\mathsf{m}(\mathbf{q})$ is what we call the *value* of $\mathbf{q}$, the family of these sets is the *Lead or Delay Valuation* (LDV) of $\mathcal{T}^{k+1}$. It is not difficult to restate Proposition 1 in terms of this concept:

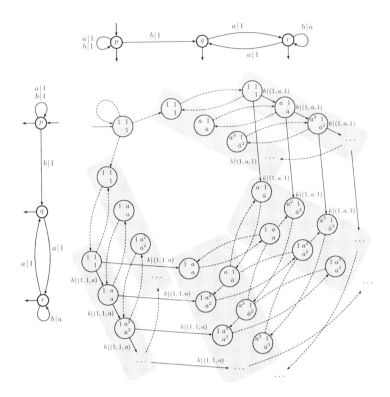

Fig. 2. The product by $\mathcal{G}_3$ of the part of T_1^3 in Figure 1. Gray regions gather states which project on a same state of $\mathcal{T}^3$. The output of the dotted transitions is $(1,1,1)$.

Theorem 3. *A transducer $\mathcal{T}$ is k-valued iff for every final state $\mathbf{q}$ of $\mathcal{T}^{k+1}$ there exists at least one γ in $\mathsf{m}(\mathbf{q})$ whose defined entries are all equal to 1_{B^*}.* □

Theorem 3 is the finite characterisation of the k-valuedness we are aiming at because the sets $\mathsf{m}(\mathbf{q})$ are finite and computable. The finiteness holds indeed for the set of minimal traverses of every set of PDs:

Proposition 2. *For every $l > 1$, for every $X \subseteq \Delta_l$, $\mathrm{card}(\mathsf{m}(X)) \leq 2^{l^4}$.* □

3.4   Making the Characterisation Effective

The effective construction of the LDV is based on two properties. The first one is a stability property in the strongly connected components (SCCs) of $\mathcal{T}^{k+1}$. The second one states that every $\mathsf{m}(\mathbf{q})$ depends uniquely on the values of the states which precede the SCC of $\mathbf{q}$.

We say that a PDPD γ is *stable* in $\mathbf{q}$ if for every circuit $\mathbf{q} \xrightarrow{f|\mathbf{u}} \mathbf{q}$, $\gamma \cdot \mathbf{u} = \gamma$.

Proposition 3. *Every $\gamma \in \mathsf{m}(\mathbf{q})$ is stable in $\mathbf{q}$. Thus, for every $\mathbf{p}$ in the same SCC as $\mathbf{q}$ and every computation $\mathbf{q} \xrightarrow{f|\mathbf{u}} \mathbf{p}$, $\mathsf{m}(\mathbf{p}) = \mathsf{m}(\mathbf{q}) \cdot \mathbf{u}$.* □

(Here $\mathcal{G}_l$ is extended to sets of PDPDs: for every $X \subseteq H_l$ and every l-tuple of words $\mathbf{u}$, $X \cdot \mathbf{u} = \{\delta \cdot \mathbf{u} \mid \delta \in X\}$, where undefined entries of δ remains undefined in $\delta \cdot \mathbf{u}$). For $X \subseteq H_{k+1}$, we denote $\mathrm{st}_{\mathbf{q}}(X) = \{\gamma \in X \mid \gamma \text{ stable in } \mathbf{q}\}$.

In order to explain the second property, we define a commutative and associative operation between sets of PDPDs. Given β and γ in (the partially ordered set) H_l, let $\beta \vee \gamma$ be their least upper bound (which exists iff β and γ are compatible on the defined coordinates). For $X, Y \subseteq H_l$, we define $X \oplus Y = \min(\{\beta \vee \gamma \mid \beta \in X, \gamma \in Y\})$. If X and Y are finite, then $X \oplus Y$ is clearly finite. Let us also fix a notation. For every SCC C of $\mathcal{T}^{k+1}$, let $I(C)$ be the set of transitions incoming in C: $I(C) = \{\mathbf{p} \xrightarrow{a|u} \mathbf{r} \mid \mathbf{p} \notin C, \mathbf{r} \in C\}$. For every $e : \mathbf{p} \xrightarrow{a|u} \mathbf{r}$ in $I(C)$ and every state $\mathbf{q}$ in C, let $\mathbf{v}_{e,\mathbf{q}}$ be the output of an arbitrary but fixed computation from $\mathbf{r}$ to $\mathbf{q}$, and $X_{e,\mathbf{q}} = \mathrm{st}_{\mathbf{q}}(m(\mathbf{p}) \cdot (\mathbf{u}\mathbf{v}_{e,\mathbf{q}}))$.

Proposition 4. *For every* $\mathbf{q}$ *(in the SCC C of $\mathcal{T}^{k+1}$), $m(\mathbf{q}) = \oplus_{e \in I(C)} X_{e,\mathbf{q}}$.*   □

Propositions 3 and 4 yield a construction of the LDV of $\mathcal{T}^{k+1}$ with a topological traversal of the SCCs of $\mathcal{T}^{k+1}$. It starts at a *hidden* initial state $\mathbf{i}$ with outgoing transitions labelled by 1_{B^*} ending in the initial states of $\mathcal{T}^{k+1}$; $m(\mathbf{i})$ is the set of PDPDs having exactly one defined entry which is equal to 1_{B^*}.

We express the complexity of the algorithm on the following parameters of $\mathcal{T}$: n (number of states), m (number of transitions) and ℓ (maximal length of the outputs of transitions)[4]. The analysis depends on the following:

Proposition 5. *For every* $\gamma \in m(\mathbf{q})$, *if* $\gamma_{i,j}$ *is defined, then*[5] $|\gamma_{i,j}| \leq \ell n^{k+1}$.   □

Testing whether a PDPD is stable in a SCC with s transitions can be made in time $\mathcal{O}(k^2 \ell n^{k+1} s)$. By Proposition 2, the cardinality of every $m(\mathbf{q})$ is finite and does not depend on the transducer $\mathcal{T}$. Therefore, the construction of each set $X_{e,\mathbf{q}}$ can be made in $\mathcal{O}(k^2 \ell n^{k+1} s)$ and each operation $\oplus$ in $\mathcal{O}(k^2 \ell n^{k+1})$. It follows that the overall complexity of our algorithm is $\mathcal{O}(\ell n^{k+1} m^{k+1})$. The multiplicative constant hidden in the "big O" comes from the bound established in Proposition 2 and is thus at most $2^{(k+1)^4}$.

Example 1. In this example and in the figures, PDs are represented as upper triangular matrices indexed by $\{p, q\} \times \{q, r\}$ (in this order).

Let $\mathbf{q} = (p, q, r)$ be a state of $\mathcal{T}_1^3$ (Figure 1). The set of PDs in $\mathcal{T}_1^3 \times \mathcal{G}_3$ attached to $\mathbf{q}$ is $X(\mathbf{q}) = \left\{ \left(\begin{smallmatrix} 1 & a^t \\ & a^t \end{smallmatrix} \right) \mid t > 0 \right\} \cup \left\{ \left(\begin{smallmatrix} a^t & 1 \\ & a^t \end{smallmatrix} \right) \mid t > 0 \right\}$ (see Figure 2). The set of traverses of $X(\mathbf{q})$ is $\mathrm{tv}(X(\mathbf{q})) = \left\{ \left(\begin{smallmatrix} 1 & 1 \\ & a^t \end{smallmatrix} \right) \mid t > 0 \right\} \cup \left\{ \left(\begin{smallmatrix} 1 & 1 \\ & \frac{1}{a^t} \end{smallmatrix} \right) \mid t > 0 \right\} \cup \left\{ \left(\begin{smallmatrix} 1 & 1 \\ & \bot \end{smallmatrix} \right) \right\}$. There is only one minimal one: $m(\mathbf{q}) = \left\{ \left(\begin{smallmatrix} 1 & 1 \\ & \bot \end{smallmatrix} \right) \right\}$. This is the value of $\mathbf{q}$, also obtained by applying the operation $\oplus$ to the PDPDs $\left(\begin{smallmatrix} 1 & 1 \\ & \bot \end{smallmatrix} \right)$ and $\left(\begin{smallmatrix} \bot & 1 \\ & \bot \end{smallmatrix} \right)$ incoming in the SCC of $\mathbf{q}$ (see Figure 3).

---

[4] Recall that the valuedness k is considered as a constant.

[5] Recall that $\gamma_{i,j}$, if defined, is a word in $B^* \cup \overline{B}^*$; $|\gamma_{i,j}|$ is as usual the length of it.

4   Deciding Finite Valuedness

Weber's proof for Theorem 2 [3] is in two steps: first, the bounded valuedness is characterised by three conditions on the computations of the transducer $\mathcal{T}$ (Theorem 4); next, it is shown that these conditions can be tested by means of a construction with the underlying graph of $\mathcal{T}$. The proof is difficult due in part to the fact that besides the decidability it gives an upper bound for the valuedness.

Our proof is akin to Weber's one, but on the other hand is different in both steps. We first describe other conditions, C1 and C2 (Theorem 5), and prove that they characterise the bounded valuedness. That stating these new conditions is useful comes from the fact that they are well-fitted with the use of a construction for transducers which we defined in [8], the *lag separation covering*. This construction together with the characterisation of bounded ambiguity for $\mathbb{N}$-automata due to Mandel and Simon (Theorem 7) yields a straightforward proof that C1 and C2 imply the bounded valuedness of $\mathcal{T}$: if $\mathcal{T}$ satisfies C1 and C2, then with the construction of a lag separation covering on $\mathcal{T}$ we obtain a transducer which is equivalent to $\mathcal{T}$, and whose underlying input automaton, say $\mathcal{A}$, satisfies the conditions S1 and S2 in Theorem 7; thus, $\mathcal{A}$ realises a series which is bounded by some integer k; as the number of outputs in $\mathcal{T}$ for every

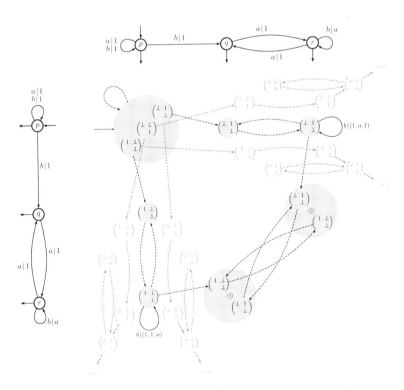

Fig. 3. The values $\mathsf{m}(\mathbf{q})$ for the states of $\mathcal{T}_1^3$ (filled regions) accessible from (p, p, p) and co-accessible to (p, q, r). Dashed transitions have output equal to $(1_{B^*}, 1_{B^*}, 1_{B^*})$.

input word u is at most the number of successful computations in $\mathcal{A}$ labelled by u, the valuedness of $\mathcal{T}$ is at most k. The condition C1 is easily testable. In order to test C2, the LDA will be useful again, and we describe in Section 4.2 a condition on $\mathcal{T}^3 \times \mathcal{G}$ equivalent to C2 which can be tested in polynomial time. It turns out that the product $\mathcal{T}^3 \times \mathcal{G}$ captures the technicalities of the constructions underlying Weber's algorithm (Section 3 of [3]).

4.1 A Characterisation of Bounded Valuedness

Weber's conditions for bounded valuedness are better described with the help of Figure 4. Let us call a computation $p \to p \to q \to q$ with $p \neq q$ a *dumbbell-computation*, and a computation such as on the right of Figure 4 a *W-computation*.

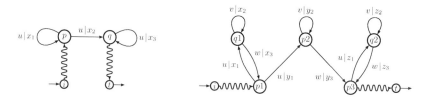

Fig. 4. A dumbbell-computation and a W-computation

Theorem 4 (Weber [3]). *A trim transducer $\mathcal{T}$ is bounded valued iff:[6] **W1)** $\mathcal{T}$ does not contain co-terminal[7] circuits with same input and distinct outputs; **W2)** $\mathcal{T}$ does not contain a dumbbell-computation $p \xrightarrow{u|x_1} p \xrightarrow{u|x_2} q \xrightarrow{u|x_3} q$ with $x_1 x_2 \neq x_2 x_3$; **W3)** $\mathcal{T}$ does not contain a W-computation with $|x_1| \neq |y_2|$.* $\square$

It is not so difficult to see that these three conditions are necessary for the bounded valuedness. The substance of the theorem is that they are sufficient.

The idea of our new conditions is to adjoin a restriction on the *lag* between the computations which allows to capture W2 and W3 on a single statement.[8]

Theorem 5. *A trim transducer $\mathcal{T}$ with n states and output lengths bounded by ℓ is bounded valued iff: **C1)** $\mathcal{T}$ does not contain a circuit which contain co-terminal transitions $p \xrightarrow{a|u} q$ and $p \xrightarrow{a|v} q$ such that $u \neq v$; **C2)** $\mathcal{T}$ does not contain a dumbbell-computation $c_1 : p \xrightarrow{u|x_1} p, c_2 : p \xrightarrow{u|x_2} q, c_3 : q \xrightarrow{u|x_3} q$ where either $x_1 x_2 \neq x_2 x_3$ or $x_1 x_2 = x_2 x_3$ and $\langle c_1 c_2, c_2 c_3 \rangle > \ell n^3$.*

---

[6] Weber's conditions are slightly different (but equivalent), for W1 and W2 are stated together. We chose other presentation in order to make clear the comparison with the statements to come.

[7] With same origin and same end.

[8] The notation $\langle c, d \rangle$ in this statement stands for the *lag* between two computations c and d and is defined in [8].

If the valuedness of $\mathcal{T}$ is bounded, clearly the condition C1 must hold, and every $c_1c_2c_3$ in C2 must satisfy $x_1x_2 = x_2x_3$. The proof that every dumbbell-computation must satisfy $\langle c_1c_2, c_2c_3 \rangle \leq \ell n^3$ is a pumping argument showing that such a lag would imply a W-computation which does not satisfies W3.

The proof of the sufficiency of the conditions C1 and C2 is straightforward with the use of two tools. The first one is the lag separation covering of $\mathcal{T}$, a construction parametrised by an integer $N > 0$ of a new and larger transducer $\mathcal{U}_N$ with a morphism from $\mathcal{U}_N$ to $\mathcal{T}$ inducing a bijection between their successful computations.[9] This covering allows to avoid pairs of computations such that the differences of lengths of outputs along them (their "lag") are bounded by N:

Theorem 6 ([8]). *For every $N > 0$, the lag separation covering $\mathcal{U}_N$ contains a subtransducer $\mathcal{V}_N$ equivalent to $\mathcal{T}$ where distinct successful computations with same label have lag larger than N.* $\quad\square$

The second tool is a classical characterisation of boundedness of $\mathbb{N}$-automata:

Theorem 7 (Mandel-Simon [13], Seidl-Weber [14]). *A trim $\mathbb{N}$-automaton $\mathcal{A}$ realises a bounded $\mathbb{N}$-series iff:[10] **S1)** $\mathcal{A}$ does not contain a circuit which contains a transition with multiplicity greater than 1; **S2)** $\mathcal{A}$ does not contain a dumbbell-computation.* $\quad\square$

The idea is to show, starting from Theorem 6, that if $\mathcal{T}$ satisfies C1 and C2, then the underlying input automaton $\mathcal{A}$ of $\mathcal{V}_N$ (with $N = \ell n^3$) satisfies S1 and S2. Now, by Theorem 7, $\mathcal{A}$ is of bounded ambiguity, and thus the valuedness of $\mathcal{V}_N$ (and that of $\mathcal{T}$, for $\mathcal{V}_N$ is equivalent to it) is bounded.

4.2   Testing the Characterisation in $\mathcal{T}^3 \times \mathcal{G}$

The condition C1 in Theorem 5 is easily testable. The substance of our algorithm is a characterisation of C2 within the product of $\mathcal{T}^3$ by the LDA $\mathcal{G}$. In Section 3.2, we defined $\mathcal{T}^3 \times \mathcal{G}_3$, the LDA applied to every pair of projections of $\mathcal{T}^3$. The product $\mathcal{T}^3 \times \mathcal{G}$ is defined likewise but in this case $\mathcal{G}$ acts on a single pair of projections, the first and the second one. Let $\mathcal{W}$ be the part of $\mathcal{T}^3 \times \mathcal{G}$ consisting of states accessible from some state of form $((p,p,q), 1_{B^*})$ and co-accessible to some state of form $((p,q,q), v)$ (where p and q are distinct states of $\mathcal{T}$).

Lemma 1. *The transducer $\mathcal{T}$ (with n states and lengths of outputs bounded by ℓ) satisfies C2 iff for every state $((r,s,t), w)$ of $\mathcal{W}$, $w \neq \mathbf{0}$ and $|w| \leq \ell n^3$.* $\quad\square$

Thus, $\mathcal{W}$ allows to test C2. But this subtransducer seems to be very large, for there are exponentially many words of length at most ℓn^3. In order to obtain a polynomial time complexity, the idea is to consider only the *harder* part of $\mathcal{W}$, the subtransducer $\mathcal{W}'$ consisting of states which are co-accessible to some circuit whose output (x, y, z) is such that either $x \neq 1_{B^*}$ or $y \neq 1_{B^*}$:

---

[9] For the definition of *covering of automata* see [18].

[10] The original statement in [13] reads instead of S1 that S1': $\mathcal{A}$ contains neither a circuit with multiplicity greater than 1 nor distinct co-terminal circuits with the same input. In the presence of S2 both formulations are equivalent.

Lemma 2. *If all the states of $\mathcal{W}'$ satisfy the conditions in Lemma 1, then the same is true for the states of $\mathcal{W}$.* □

It turns out that this part is not so large due to the following lemma[11]:

Lemma 3. *In states of $\mathcal{W}'$ projecting on a same state of T^3 the words in the second component must be prefix of a common word or C2 is not satisfied.* □

Thus, in order to construct $\mathcal{W}'$ we can maintain for every state (r, s, t) of T^3 only two words of length at most ℓn^3, a positive and a negative one, whose prefixes represent the states of $\mathcal{W}'$ already constructed. Each prefix implies a traversal of T^3, thus the complexity of our algorithm to test the bounded valuedness of T is $\mathcal{O}(\ell n^3(n^3 + m^3))$ (where m is the number of transitions of T).

References

1. Schützenberger, M.P.: Sur les relations rationnelles. In: Brakhage, H. (ed.) GI-Fachtagung 1975. LNCS, vol. 33, pp. 209–213. Springer, Heidelberg (1975)
2. Eilenberg, S.: Automata, Languages, and Machines, vol. A. Academic Press, London (1974)
3. Weber, A.: On the valuedness of finite transducers. Acta Informatica 27(8), 749–780 (1989)
4. Gurari, E., Ibarra, O.: A note on finite-valued and finitely ambiguous transducers. Mathematical Systems Theory 16, 61–66 (1983)
5. Culik, K., Karhumäki, J.: The equivalence of finite valued transducers (on HDT0L languages) is decidable. Theoretical Computer Science 47(1), 71–84 (1986)
6. Weber, A.: Decomposing finite-valued transducers and deciding their equivalence. SIAM Journal on Computing 22(1), 175–202 (1993)
7. Weber, A.: Decomposing a k-valued transducer into k unambiguous ones. RAIRO Informatique Théorique et Applications 30(5), 379–413 (1996)
8. Sakarovitch, J., de Souza, R.: On the decomposition of k-valued rational relations. In: Albers, S., Weil, P. (eds.) Proceedings of STACS 2008, pp. 621–632 (2008) arXiv:0802.2823v1, http://stacs-conf.org (to appear in Theory of Computing Systems)
9. Béal, M.P., Carton, O., Prieur, C., Sakarovitch, J.: Squaring transducers: an efficient procedure for deciding functionality and sequentiality. Theoretical Computer Science 292, 45–63 (2003)
10. de Souza, R.: On the decidability of the equivalence of k-valued transducers (submitted)
11. Gurari, E., Ibarra, O.: The complexity of decision problems for finite-turn multi-counter machines. Journal of Computer and System Sciences 22(2), 220–229 (1981)
12. Hopcroft, J.E., Motwani, R., Ullman, J.D.: Introduction to Automata Theory, Languages and Computation. Addison-Wesley, Reading (2001)
13. Mandel, A., Simon, I.: On finite semigroups of matrices. Theoretical Computer Science 5(2), 101–111 (1977)

---

[11] Let us note the similarity of this lemma with a critical property to establish the polynomial complexity of the procedure given in [9] to test the sequentiality of transducers.

14. Weber, A., Seidl, H.: On the degree of ambiguity of finite automata. Theoretical Computer Science 88, 325–349 (1991)
15. Sakarovitch, J., de Souza, R.: On the decidability of finite valuedness of transducers (manuscript), `http://www.infres.enst.fr/~rsouza/DFV.pdf`
16. Berstel, J.: Transductions and Context-Free Languages. B. G. Teubner (1979)
17. Sakarovitch, J.: Éléments de théorie des automates. Vuibert, Paris (2003); English translation: Elements of Automata Theory. Cambridge University Press, Cambridge (to appear)
18. Sakarovitch, J.: A construction on finite automata that has remained hidden. Theoretical Computer Science 204(1–2), 205–231 (1998)

Monadic Second Order Logic on Graphs with Local Cardinality Constraints

Stefan Szeider

Department of Computer Science
Durham University, UK
stefan.szeider@durham.ac.uk

Abstract. We show that all problems of the following form can be solved in polynomial time for graphs of bounded treewidth: Given a graph G and for each vertex v of G a set $\alpha(v)$ of non-negative integers. Is there a set S of vertices or edges of G such that S satisfies a fixed property expressible in monadic second order logic, and for each vertex v of G the number of vertices/edges in S adjacent/incident with v belongs to the set $\alpha(v)$? A wide range of problems can be formulated in this way, for example Lovász's General Factor Problem.

1   Introduction

Treewidth is a graph invariant that indicates the tree-likeness of a graph. It was introduced independently by several authors in different context and terminology (see, e.g., [4]). Many combinatorial graph problems that are hard in general become easy for graphs of small treewidth, for example 3-COLORABILITY and HAMILTONICITY can be decided in linear time for graphs of treewidth bounded by a constant k (albeit with a running time containing a constant factor that is exponential in k). Such algorithms are usually obtained by means of dynamic programming applied via a bottom-up traversal of a decomposition tree associated with the given graph. Courcelle's famous theorem [7,8] provides a unified framework for establishing linear-time algorithms for problems on graphs of bounded treewidth. It covers all decision problems that can be expressed in the formalism of Monadic Second Order (MSO) logic (3-COLORABILITY and HAMILTONICITY are such problems). Arnborg, Lagergren, and Seese [1] establish an extension of Courcelle's Theorem that allows to solve MSO-expressible optimization problems in linear time for graphs of bounded treewidth. This covers problems that ask for a set S of vertices or edges of a given graph that satisfies an MSO-expressible property and minimizes a linear cost function associated with vertices and edges of the given graph. Dozens of classical NP-hard problems such as VERTEX COVER and DOMINATING SET are of this type [1]. The mentioned theorems provide a powerful toolkit for a quick complexity classification of problems. However, the provided algorithms are usually not feasible in practice. Once a problem is classified as being easy in principle, one can develop more practical problem-specific algorithms.

E. Ochmański and J. Tyszkiewicz (Eds.): MFCS 2008, LNCS 5162, pp. 601–612, 2008.

In this paper we prove an extension to the known MSO theorems. We consider problems of the following form: Given a graph G and for each vertex v of G a set $\alpha(v)$ of non-negative integers. The question is whether there exists a set S of vertices or edges of G that (i) satisfies a fixed MSO-expressible property and (ii) for each vertex v of G, the number of vertices in S adjacent to v plus the number of edges in S incident with v belongs to the set $\alpha(v)$. We call such a problem an *MSO problem for graphs with local cardinality constraints*, or *MSO-LCC problem*, for short.

For example, the GENERAL FACTOR problem introduced by Lovász [14,15] is an MSO-LCC problem where the MSO property simply states that S is a set of edges. In Section 4 we discuss further examples of MSO-LCC problems.

Our main result is the following:

Theorem 1. *Every MSO-LCC problem can be solved in polynomial time for graphs of bounded treewidth.*

We establish Theorem 1 in two steps. Consider an MSO formula $\varphi(X)$ with free set variable X and a constant $k \geq 1$. Assume we are given a graph G with n vertices and treewidth k, and local cardinality constraints α. In the first step we use Bodlaender's Theorem [3] and the tree automata approach of Arnborg et al. [1] to construct in time $O(n)$ a certain auxiliary structure, a *solution tree*, which is a tree decomposition of G equipped with additional labels. The solution tree is a succinct representation of all solutions, that is, of all sets S of vertices and edges for which $\varphi(S)$ is true in G. We think that the concept of solution trees can be useful for other extensions or generalizations of the known MSO theorems. In a second step we perform dynamic programming on the solution tree. This second step is carried out in time $O(n^{2k+3})$. It is easy to modify the algorithm so that it actually produces a solution (if one exists) within the same asymptotic running time.

The running time of the algorithm is polynomial since the treewidth bound k is considered constant; however, the order of the polynomial depends on k. One might wonder if this dependency is necessary. In Section 5 we explain that, subject to a complexity theoretic assumption, this dependency cannot be eliminated.

2   Preliminaries

2.1   Graphs and Local Cardinality Constraints

All considered graphs are finite, simple, and undirected. $V(G)$ and $E(G)$ denote the vertex set and the edge set of a graph G, respectively. The *order* of a graph is the number of its vertices. We denote an edge between vertices u and v by uv (or equivalently by vu). The *degree* of a vertex v in G is denoted by $d_G(v)$. In addition we assume each vertex or edge x is labeled with an element of some fixed finite set. The labels allow us to consider graphs with different "sorts" of vertices and edges. For a vertex $v \in V(G)$ we denote by $N_G(v)$ the set of neighbors of v in G (that is, the set of vertices $u \in V(G)$ such that $uv \in E(G)$) and by $I_G(v)$

the set of edges of G incident with v (that is, the set of edges $uv \in E(G)$). We denote the subgraph of a graph G induced by a set $S \subseteq V(G)$ by $G[S]$ (that is, $V(G[S]) = S$, $E(G[S]) = \{\, uv \in E(G) \mid u, v \in S \,\}$).

A *graph with local cardinality constraints* is a pair (G, α) where G is a graph and α is a mapping that assigns to each vertex $v \in V(G)$ a set $\alpha(v)$ of non-negative integers. For a set $S \subseteq V(G) \cup E(G)$, a subgraph H of G, and a vertex $v \in V(H)$, we write

$$\operatorname{touch}(H, S, v) = |N_H(v) \cap S| + |I_H(v) \cap S|.$$

We say that a set $S \subseteq V(G) \cup E(G)$ *satisfies* α if $\operatorname{touch}(G, S, v) \in \alpha(v)$ holds for all $v \in V(G)$. Since $\operatorname{touch}(G, S, v) \leq |V(G)| - 1 + |E(G)|$, we can always assume that $\alpha(v) \subseteq \{0, \ldots, |V(G)| + |E(G)| - 1\}$.

2.2 Tree Decompositions

A *tree decomposition* of a graph G is a pair (T, χ) where T is a tree and χ is a mapping that assigns to each node $t \in V(T)$ a set $\chi(t) \subseteq V(G)$ such that the following conditions hold (we refer to the vertices of T as "nodes" to make the distinction between T and G clearer).

1. $V(G) = \bigcup_{t \in V(T)} \chi(t)$ and $E(G) \subseteq \bigcup_{t \in V(T)} \{\, uv \mid u, v \in \chi(t) \,\}$.
2. The sets $\chi(t_1) \setminus \chi(t)$ and $\chi(t_2) \setminus \chi(t)$ are disjoint for any three nodes $t, t_1, t_2 \in V(T)$ such that t lies on a path from t_1 to t_2 in T.

The *width* of (T, χ) is $\max_{t \in V(T)} |\chi(t)| - 1$. The *treewidth* $\operatorname{tw}(G)$ of a graph G is the smallest integer k such that G has a tree decomposition of width k. For a node $t \in V(T)$ we write $\chi^*(t) = \chi(t) \cup E(G[\chi(t)])$.

For fixed $k \geq 1$, one can decide for a given graph G of order n in time $O(n)$ whether $\operatorname{tw}(G) \leq k$, and if so, compute a tree decomposition of G of width $\leq k$ and with $O(n)$ nodes (Bodlaender's Theorem [3]).

A class $\mathcal{C}$ of graphs is of *bounded treewidth* if there exists a constant k such that all graphs in $\mathcal{C}$ have treewidth k or less. A graph of treewidth k and order n has at most $kn - k(k+1)/2$ edges [18]. Hence for graphs of bounded treewidth, the number of edges is linear in the number of vertices. Thus linear running times for algorithms on graphs of bounded treewidth can be expressed as $O(n)$.

For our purposes it is convenient to consider tree decompositions in a certain normal form. A triple (T, r, χ) is a *nice tree decomposition* if (T, χ) is a tree decomposition, and considering T as a rooted tree with root r, each node $t \in V(T)$ is of one of the following four types: (1) t is a *leaf.* (2) t has exactly one child t' and $|\chi(t) \setminus \chi(t')| = 1$ (we call t an *introduce node*). (3) t has exactly one child t' and $|\chi(t') \setminus \chi(t)| = 1$ (we call t a *forget node*). (4) t has exactly two children t', t'' and $\chi(t) = \chi(t') = \chi(t'')$ (we call t a *join node*). We assume an arbitrary but fixed order of the two children of a join node, so that we can speak of the right child and the left child.

It is easy to see that one can transform efficiently a tree decomposition into a nice tree decomposition of the same width. In fact, if the width is bounded by a constant, then this transformation can be carried out in linear time [11]. Thus Bodlaender's Theorem can be strengthened to provide nice tree decompositions.

2.3   Monadic Second Order Logic

We consider *Monadic Second Order* (MSO) logic on (labeled) graphs in terms of their incidence structure whose universe contains vertices and edges. We assume an infinite supply of *individual variables* $x, x_1, x_2, \ldots$ and of *set variables* $X, X_1, X_2, \ldots$ The *atomic formulas* are $E(x)$ ("x is an edge"), $V(x)$ ("x is a vertex"), $I(x, y)$ ("vertex x is incident with edge y"), $E(x, y)$ ("vertices x and y are adjacent"; actually this predicate can be defined in terms of predicate I), $x = y$ (equality), $P_a(x)$ ("vertex x has label a"), and $X(y)$ ("vertex y is element of set X). *MSO formulas* are built up from atomic formulas using the usual Boolean connectives ($\neg, \wedge, \vee, \rightarrow, \leftrightarrow$), quantification over individual variables ($\forall x, \exists x$), and quantification over set variables ($\forall X, \exists X$).

We write $\varphi(X)$ to denote an MSO formula with free set variable X. Let G be a graph, $S \subseteq V(G) \cup E(G)$, and let $\varphi(X)$ be an MSO formula. We write $G \models \varphi(S)$ to indicate that φ is true for G if X is interpreted as S. For a graph with local cardinality constraints (G, α) we write $(G, \alpha) \models \varphi(S)$ if $G \models \varphi(S)$ and S satisfies α.

2.4   Tree Automata

Let Σ be finite set. A Σ-*tree* T is a rooted binary tree whose nodes are labeled with elements of Σ. We assume that the tree is *ordered* in the sense that the children of a node are given in some arbitrary but fixed order; that allows us to speak of a left or right child of a node. We shall indicate the ordering by using the subscripts l and r or 1 and 2, respectively.

A (deterministic, bottom-up) *tree automaton* for Σ-trees is a tuple $M = (Q, \Sigma, \delta, q_0, Q_{\mathrm{acc}})$, consisting of a finite set Q of *states* disjoint from Σ, a *transition function* $\delta : Q \times Q \times \Sigma \to Q$, an *initial state* $q_0 \in Q$, and a set $Q_{\mathrm{acc}} \subseteq Q$ of *accepting states*. Given a Σ-tree T, the automaton assigns deterministically states to the nodes of T, traversing the tree in a bottom-up ordering: leaves with label $a \in \Sigma$ are assigned state $\delta(q_0, q_0, a)$; if the two children of a node v with label a have been assigned states q_l and q_r, respectively, then v is assigned state $\delta(q_l, q_r, a)$. M *accepts* T if it assigns the root an accepting state $q \in Q_{\mathrm{acc}}$.

MSO logic on Σ-trees can be defined in the obvious way, using unary predicates $P_a(v)$ ("node v has label $a \in \Sigma$") and binary predicates $L(u, v_l)$ ("v_l is the left child of u") and $R(u, v_r)$ ("v_r is the right child of u").

By a classical result of Thatcher and Wright [20], one can, given a closed MSO formula φ for Σ-trees, effectively construct a Σ-tree automaton M such that M accepts a Σ-tree T if and only if $T \models \varphi$. This result carries over to open formulas by the following considerations. For a Σ-tree T and $S \subseteq V(T)$, let T_S denote the $\Sigma \times \{0, 1\}$-tree obtained from T by extending the labels of T with one bit that indicates whether the labeled node belongs to S or not. Now, it is easy to generalize the above result as follows (see, e.g., [1]): Given an MSO formula $\varphi(X)$ for Σ-trees, one can effectively construct a $\Sigma \times \{0, 1\}$-tree automaton $M_{\varphi(X)} = (Q, \Sigma \times \{0, 1\}, \delta, q_0, Q_{\mathrm{acc}})$ such that for each Σ-tree T and $S \subseteq V(T)$, $T \models \varphi(S)$ if and only if $M_{\varphi(X)}$ accepts T_S.

2.5   Tree Interpretations

Let $\varphi(X)$ be an MSO formula on (labeled) graphs and k a positive integer. Arnborg et al. [1] show that, given a graph G and a tree decomposition of G of width k, one can construct in linear time

- a Σ-tree T^I where $\Sigma = \{0,1\}^{k'}$ for $k' \in O(k^2)$,
- an MSO formula $\varphi^I(X)$ for Σ-trees, and
- a surjective mapping π from the set $L(T^I)$ of leaves of T^I to $V(G) \cup E(G)$

such that for each set $S^I \subseteq V(T^I)$, $T^I \models \varphi^I(S^I)$ if and only if

1. $S^I \subseteq L(T^I)$,
2. for any $t, t' \in L(T^I)$ with $\pi(t) = \pi(t')$, $t \in S^I$ if and only if $t' \in S^I$, and
3. $G \models \varphi(\pi(S^I))$.

We call the tuple $(\Sigma, T^I, \varphi^I(X), \pi)$ a *tree interpretation* of $(G, \varphi(X))$.

The tree T^I and the mapping π are obtained from a rooted tree decomposition (T, r, χ) by adding to each vertex $t \in V(T)$ new children, each representing a vertex $v \in \chi(t)$ (indicated by $\pi(t) = v$) or an edge uv with $u, v \in \chi(t)$ (indicated by $\pi(t) = uv$). Thereafter additional nodes are inserted to make the tree binary. For our purposes it is convenient to start this construction with a *nice* tree decomposition. After the addition of new leaves each non-leaf has at least two children. As long as there is a node t with more than two children $t_1, \ldots, t_j$, $j > 2$, we delete the edges $tt_2, \ldots, tt_j$, add a new node t' and edges $tt_1, tt', t't_2, \ldots, t't_j$. We distinguish in T^I between *old nodes* (nodes that also belong to T) and *new nodes* (nodes newly introduced). It is easy to see that one can always carry out the above construction such that each old node t has in T^I exactly two children t_l and t_r and one of the following three cases prevails.

1. t_l and t_r are both new nodes (this happens exactly when t is a leaf of T);
2. t_l is a new node and t_r is an old node, all nodes below t_l are new nodes (t has one child in T);
3. t_l is a new node and t_r is an old node, the left child of t_l is a new node, the right child of t_l is an old node (t has two children in T).

3   Proof of Main Result

Let $\varphi(X)$ be an MSO formula on (labeled) graphs, let Q be a finite set, and let G be a graph. The tuple $\mathcal{S} = (\mathcal{T}, Q, \sigma, \lambda)$ is a *solution tree for G* if the following conditions hold.

1. $\mathcal{T} = (T, r, \chi)$ is a nice tree decomposition of G.
2. σ is a labeling that assigns to each node $t \in V(T)$ a set $\sigma(t)$ of pairs $P = (q, U)$ where $q \in Q$ and $U \subseteq \chi^*(t)$ (recall the definition of $\chi^*(t)$ in Section 2.2).

   The purpose of a pair (q, U) is to indicate that there is a solution S for $\varphi(X)$ that, projected to $G[\chi(t)]$ yields the set U; the element q represents properties of S regarding vertices and edges that appeared in sets $\chi(t')$ below t.

3. λ is a mapping that assigns to each node $t \in V(T)$ and each element $P \in \sigma(t)$ a set $\lambda(t, P)$ with the following properties.
   (a) If t is a leaf then $\lambda(t, P) = \emptyset$.
   (b) If t has exactly one child t' then $\lambda(t_1, P) \subseteq \sigma(t_1)$.
   (c) If t has exactly two children t_1 and t_2, then $\lambda(t, P) \subseteq \sigma(t_1) \times \sigma(t_2)$.

   The purpose of the mapping λ is to indicate direct dependencies of elements $P \in \sigma(t)$ from elements $P' \in \sigma(t')$ of children t' of t.

The *width* of $\mathcal{S}$ is the width of the underlying tree decomposition $\mathcal{T}$. We say that $\mathcal{S}$ accepts a set $S \subseteq V(G) \cup E(G)$ if for each tree node $t \in V(T)$ we can pick a pair $P_t = (q_t, U_t) \in \sigma(t)$ such that $U_t = S \cap \chi^*(t)$ and the following conditions hold:

1. If t has exactly one child t_1 then $P_{t_1} \in \lambda(t, P_t)$.
2. If t has exactly two children t_1 and t_2 then $(P_{t_1}, P_{t_2}) \in \lambda(t, P_t)$.

Furthermore we say that $\mathcal{S}$ *accepts* $S \subseteq V(G) \cup E(G)$ *with* (q, U) *at* t if the above holds such that $(q, U) = P_t$. A solution tree $\mathcal{S}$ for G *characterizes* $\varphi(X)$ if for each $S \subseteq V(G) \cup E(G)$, $\mathcal{S}$ accepts S if and only if $G \models \varphi(S)$.

Lemma 1. *Let* $\varphi(X)$ *be a fixed MSO formula on (labeled) graphs and let* k *be a fixed positive integer. Given a graph* G *of order* n *and treewidth* k, *we can compute in time* $O(n)$ *a solution tree* $\mathcal{S} = (\mathcal{T}, Q, \sigma, \lambda)$ *for* G *of width* k *that characterizes* $\varphi(X)$ *where* $|V(T)| = O(n)$ *and the set* Q *depends on* $\varphi(X)$ *and* k *only.*

Proof. *Step 1.* We compute a nice tree decomposition $\mathcal{T} = (T, r, \chi)$ of G with $O(n)$ nodes and of width k. This can be accomplished in time $O(n)$ (see Section 2.2).

  Step 2. We compute a tree interpretation $(\Sigma, T^I, \varphi^I(X), \pi)$ of $(G, \varphi(X))$ in time $O(n)$ (see Section 2.5). Thus, T^I is a Σ-tree, $\Sigma = \{0, 1\}^{k'}$, $k' \in O(k^2)$, and $\varphi^I(X)$ is an MSO formula on Σ-trees.

  Step 3. We compute a tree automaton $M_{\varphi^I(X)} = (Q, \Sigma \times \{0, 1\}, \delta, q_0, Q_{acc})$ for $\Sigma \times \{0, 1\}$-trees that accepts $\varphi^I(X)$ (see Section 2.4). This step depends only on $\varphi(X)$ and k and can therefore be carried out in constant time.

  Extended States. Before proceeding with the algorithm we make some considerations. Let $S^I \subseteq V(T^I)$, $S = \pi(S^I) \subseteq E(G) \cup V(G)$ and let T_S^I be the $\Sigma \times \{0, 1\}$-tree that corresponds to T^I and S^I (see the end of Section 2.4). Assume $M_{\varphi^I(X)}$ accepts $T_{S^I}^I$. By definition of tree interpretations it follows that non-leaf nodes are labeled with pairs $(a, 0)$ since S^I contains leaves only. Also, any two leaves t_1, t_2 with $\pi(t_1) = \pi(t_2)$ have the same label. When executing $M_{\varphi^I(X)}$ on $T_{S^I}^I$ we can maintain additional information on the status of edges and vertices of G regarding membership in S. For that purpose we use *extended states* which are pairs (q, U) where $q \in Q$ and $U \subseteq V(G) \cup E(G)$. Let t be a node with label (a, b). If t is a leaf then we assign it the extended state $(\delta(q_0, q_0, (a, b)), U)$ where $U = \{\pi(t)\}$ if $b = 1$ and $U = \emptyset$ otherwise. If t is a non-leaf whose children are assigned the extended states $P_l = (q_l, U_l)$ and $P_r = (q_r, U_r)$, respectively, we assign t the extended state $P = (\delta(q_l, q_r, (a, 0)), U)$ as follows (recall the definition of old and new nodes from Section 2.5). If t is a new node then we put $U = U_l \cup U_r$; if t is an old node then we put $U = (U_l \cup U_r) \cap \chi^*(t)$.

We write $\delta_{\mathrm{ext}}(t, P_l, P_r) = P$ to indicate that the extended state P is computed from P_l and P_r at node t. Observe that each old node t is assigned an extended state (q, U) with $U = S \cap \chi^*(t)$.

Step 4. We simulate in parallel the execution of $M_{\varphi'(X)}$ on trees $T_{S^I}^I$ for all possible sets $S^I \subseteq V(T^I)$ by guessing the missing bit in the labels of leaves, and by applying the standard power set construction for converting a nondeterministic automaton into a deterministic one (see [1]). The standard conversion assigns each node a set of states. However, we apply the transformation with respect to extended states as considered above and assign each node t of T^I a set $Z(t)$ of extended states.

Step 5. We traverse the nodes of T^I in a top-down ordering and compute sets $Z'(t) \subseteq Z(t)$ by removing "useless" elements from the sets $Z(t)$. That is, for the root r, we remove from $Z(r)$ those extended states (q, U) where $q \notin Q_{\mathrm{acc}}$; for a node t with parent t', we remove from $Z(t)$ those extended states that have not been used in Step 4 to produce an extended state in $Z'(t')$.

Observe that each set $Z(t)$ (and so each set $Z'(t)$) is of constant size. Since $|V(T)| = O(n)$, all the Z' sets can be computed in time $O(n)$.

Step 6. Finally we compute the labelings σ and λ. For each node t of T we simply put $\sigma(t) = Z'(t)$. Let t be a node of T with two children t_1 and t_2 in T (the case where t has only one child is similar). In view of the case distinction we made at the end of Section 2.5, we may assume that t, t_1, t_2 are embedded as old nodes in T^I as follows: t has as left child a new node t' and as right child t_2; t' has as left child a new node t'' and as right child t_1. Let $P \in Z'(t)$, $P_1 \in Z'(t_1)$ and $P_2 \in Z'(t_2)$. We add (P_1, P_2) to $\lambda(t, P)$ if and only if there exist $P'' \in Z'(t'')$ and $P' \in Z'(t')$ such that $\delta_{\mathrm{ext}}(t', P'', P_1) = P'$ and $\delta_{\mathrm{ext}}(t, P', P_2) = P$.

Note that for computing $\lambda(t, P)$ we only need to consider the elements of $Z'(t')$ for a constant number of nodes t'. Since the size of each set $Z'(t)$ is constant as well, Step 6 takes time $O(n)$.   □

Proof of Theorem 1. Let $\varphi(X)$ be a fixed MSO formula on (labeled) graphs and let k be a fixed integer. We show that, given a graph (G, α) with local cardinality constraints, where G is of order n and treewidth k, we can decide in time $O(n^{2k+3})$ whether there is some $S \subseteq V(G)$ such that $(G, \alpha) \models \varphi(S)$.

First we use Lemma 1 to compute a solution tree $\mathcal{S} = (T, Q, \sigma, \lambda)$, $\mathcal{T} = (T, r, \chi)$, of width k that characterizes $\varphi(X)$. We may assume that $\chi(r)$ is a singleton (if it is not then we can simply extend the tree putting at most $k - 2$ forget nodes on top of the root). We assume an arbitrary total ordering of the vertices of G that allows us to associate with each set $\chi(t)$ a vector $\bar{\chi}(t)$ that lists the elements of $\chi(t)$ strictly increasing according to the chosen ordering.

For a node $t \in V(T)$ let $F(t)$ denote the set of vertices of G that are already "forgotten" at t; that is, $F(t) = \bigcup_{t' \leq t} \chi(t') \setminus \chi(t)$ where $t' \leq t$ means that t' belongs to the subtree of T that is rooted at t.

Let $N = \max \bigcup_{v \in V(G)} \alpha(v)$; observe that $N \in O(n)$ since $|E(G)| \in O(n)$.

For each $t \in V(T)$ and each pair $P \in \sigma(t)$ we define a set $W(t, P) \subseteq \{0, 1 \ldots, N\}^{|\chi(t)|}$ of vectors. Let $\bar{\chi}(t) = (x_1, \ldots, x_j)$. Then $(n_1, \ldots, n_j) \in W(t, P)$ if and only if there exists a set $S \subseteq V(G) \cup E(G)$ such that

1. $G \models \varphi(S)$;
2. S accepts S with P at t;
3. $\text{touch}(G, S, v) \in \alpha(v)$ holds for all $v \in F(t)$;
4. $\text{touch}(G[F(t) \cup \{x_i\}], S, x_i) = n_i$ for $1 \le i \le j$.

If we know the sets $W(r, P)$ for $P \in \sigma(r)$, then we can decide immediately whether $(G, \alpha) \models \varphi(S)$ for some $S \subseteq V(G)$: Let $\chi(r) = \{x_1\}$; the answer is *yes* if and only if there is some $(n_1) \in \bigcup_{P \in \sigma(r)} W(r, P)$ such that $n_1 \in \alpha(x_1)$.

The W sets can be computed by a bottom-up traversal of T according to the following four cases. Let t be a node of T with $\bar{\chi}(t) = (x_1, \ldots, x_j)$ and let $P \in \sigma(t)$.

Case 1: t is a leaf node. Then $W(t, P) = \{(0, \ldots, 0)\}$ since $F(t) = \emptyset$.

Case 2: t is an introduce node with child t'; w.l.o.g., assume $\chi(t) \setminus \chi(t') = \{x_j\}$. Now $W(t, P)$ contains all vectors $(n_1, \ldots, n_{j-1}, 0)$ with $(n_1, \ldots, n_{j-1}) \in W(t', P')$ for $P' \in \lambda(t, P)$, since $F(t) = F(t')$ and x_j is not adjacent with a vertex in $F(t)$ by the first property in the definition of a tree decomposition.

Case 3: t is a forget node with child t'; w.l.o.g., assume $\chi(t') \setminus \chi(t) = \{x_j\}$. $W(t, P)$ is the set of all vectors $(n_1, \ldots, n_{j-1})$ with $(n'_1, \ldots, n'_{j-1}, n'_j) \in W(t', P')$ for $P' = (q', U') \in \lambda(t, P)$ where,

$$n'_j + \text{touch}(G[\chi(t')], U', x_j) \in \alpha(x_j),$$

and

$$n_i = n'_i + \text{touch}(G[\{x_i, x_j\}], U', x_i) \quad (1 \le i \le j - 1).$$

Case 4: t is a join node with children t' and t''. Now $W(t, P_t)$ contains all vectors $(n'_1 + n''_1, \ldots, n'_j + n''_j)$ with $(n'_1, \ldots, n'_j) \in W(t', P')$ and $(n''_1, \ldots, n''_j) \in W(t'', P'')$ for $(P', P'') \in \lambda(t, P)$. The correctness of the computation of $W(t, P)$ follows from the disjointness of $F(t')$ and $F(t'')$, a consequence of the second property in the definition of a tree decomposition.

Let us review the running time. By Lemma 1, the solution tree S can be computed in time $O(n)$. It has $O(n)$ nodes, and for each node t the set $\sigma(t)$ is of constant size. Also the sets $\lambda(t, P)$ for $P \in \sigma(t)$ are of constant size. Hence we need to compute $O(n)$ many sets $W(t, P)$. For each one we have to process, in the worst case (that is, when t is a join node), $(N + 1)^{k+1} \cdot (N + 1)^{k+1}$ many pairs of vectors. Since $N = O(n)$ and T has $O(n)$ nodes, this gives in total a running time of $O(n^{2k+3})$. $\square$

4   Applications

4.1   General Factors

Lovász [14,15] introduced the following problem.

GENERAL FACTOR
Instance: A graph G and a mapping α that assigns to each vertex $v \in V(G)$ a set $\alpha(v) \subseteq \{0, \ldots, d_G(v)\}$.
Question: Is there a subset $F \subseteq E(G)$ such that for each vertex $v \in V(G)$ the number of edges in F incident with v is an element of $\alpha(v)$?

This problem clearly generalizes the polynomial-time solvable r-FACTOR problem where $\alpha(v) = \{r\}$ for all $v \in V(G)$. However, GENERAL FACTOR is easily seen to be NP-hard (say, by reduction from 3-DIMENSIONAL MATCHING). Cornuéjols [6] gives a full classification of the complexity of GENERAL FACTOR when the assigned sets are restricted to some fixed class. If the number $m = \max \bigcup_{v \in V(G)} \alpha(v)$ is bounded by a constant, then one can use Courcelle's Theorem to show that the problem is linear-time decidable for graphs of bounded treewidth [19]. For unbounded m, however, we can apply Theorem 1 with the formula $\varphi(X) = \forall x(X(x) \to E(x))$ that just states that X is a set of edges. Hence we have the following result.

Corollary 1. GENERAL FACTOR *can be solved in polynomial time for graphs of bounded treewidth.*

Theorem 1 applies also to problems that are already NP-hard without cardinality constraints. For example $\varphi(X)$ could express that X is a color class of a proper 3-coloring of G; with cardinality constraints $\alpha(v)$ we can require for each vertex v certain numbers of neighbors to be colored with color X. In addition, we can restrict the size of X to certain numbers: we add an additional vertex v_0 labeled a to G and connect it to all other vertices (the treewidth increases at most by 1). To $\varphi(X)$ we add the clause that X does not contain vertices labeled a (thus $v_0 \notin X$). With $\alpha(v_0)$ we can now apply restrictions on the size of X.

4.2   Equitable Problems

The following problem was introduced by Meyer [16] and has received a lot of attention, see Lih's survey [13].

> EQUITABLE r-COLORING
> *Instance:* A graph G.
> *Question:* Is there a proper coloring of G using colors from $\{1, \dots, r\}$ such that the sizes of any two color classes differ at most by one?

Let G be the given graph. We construct a new graph H from G by adding r new vertices $c_1, \dots, c_r$ and all edges $c_i v$ for $i \in \{1, \dots, r\}$ and $v \in V(G)$. Note that $\mathrm{tw}(H) \le \mathrm{tw}(G) + r$. Let $|V(G)| = n$. We put $\alpha(c_i) = \{\lfloor n/r \rfloor, \lceil n/r \rceil\}$, $1 \le i \le r$; for all other vertices v we put $\alpha(v) = \{1\}$. It is easy to construct an MSO formula $\varphi(X)$ that states: "X is a set of edges such that (1) each edge in X is incident with some vertex in $c_1, \dots, c_r$, and (2) any two adjacent vertices $u, v \notin \{c_1, \dots, c_r\}$ are not adjacent to the same vertex c_i, $1 \le i \le r$, via edges in X." Hence, from Theorem 1 we get the following result.

Theorem 2. EQUITABLE r-COLORING *can be solved in polynomial time for graphs of bounded treewidth.*

One can generalize this construction to (appropriately defined) "equitable MSO problems" where, instead of a proper r-coloring, one asks for other MSO-expressible properties of sets of vertices.

Bodlaender and Fomin [5] show a stronger result than Theorem 2. They show that EQUITABLE r-COLORING is polynomially solvable for graphs of bounded treewidth when the number r of colors is part of the input and not constant. Their algorithm uses a combinatorial result of Kostochka, Nakprasit, and Pemmaraju [12].

4.3 Problems with Edge Weights

An *edge weighting* of a graph G is a mapping w that assigns each edge a positive integer. An *orientation* of G is a mapping $\Lambda : E(G) \to V(G) \times V(G)$ with $\Lambda(uv) \in \{(u, v), (v, u)\}$ for each $uv \in E(G)$. The *weighted outdegree* of a vertex $v \in V(G)$ with respect to an edge weighting w and an orientation Λ is defined as

$$d^+_{G,w,\Lambda}(v) = \sum_{vu \in E(G) \text{ such that } \Lambda(vu) = (v,u)} w(vu).$$

Asahiro, Miyano, and Ono [2] consider the following problem and discuss applications and related problems.

MINIMUM MAXIMUM OUTDEGREE
Instance: A graph G, an edge weighting w of G given in unary, and a positive integer r.
Question: Is there an orientation Λ of G such that $d^+_{G,w,\Lambda}(v) \leq r$ for all $v \in V(G)$?

We assume that the edge weighting w is given in unary since otherwise the problem is already NP-complete for graphs of treewidth 2, by a simple reduction from PARTITION [2]. Asahiro et al. [2] show that MINIMUM MAXIMUM OUTDEGREE can be solved in polynomial time for graphs of treewidth 2. As an application of Theorem 1 we extend this result to arbitrary treewidth bounds.

Given a graph G with edge weighting w we construct a new graph H with three sorts of vertices (say, red, green, and blue vertices). The vertices of G correspond to the red vertices of H. For each edge $e = uv \in E(G)$ we introduce in H green vertices $g_{e,i}, g'_{e,i}$ for $1 \leq i \leq w(e)$, and blue vertices b_e, b'_e; we add the edges $ug_{e,i}, b_eg_{e,i}, vg'_{e,i}, b'_eg'_{e,i}$ and $g'_{e,i}g_{e,i}$ for $i = 1, \ldots, w(e)$. Clearly H can be constructed in polynomial time as we assume the edge weights are given in unary. Also it is easy to see that $\mathrm{tw}(H)$ is bounded in terms of $\mathrm{tw}(G)$ since we can form a tree decomposition of H by patching together tree decompositions of G and of the graphs $H[\{u, v, b_{uv}, b'_{uv}\} \cup \bigcup_{i=1}^{w(uv)} \{g_{uv,i}, g'_{uv,i}\}]$ for $uv \in E(G)$. The MINIMUM MAXIMUM OUTDEGREE problem can now be formulated as an MSO-LCC problem for H. For each red vertex $v \in V(H)$ we define $\alpha(v) = \{0, \ldots, r\}$, and for each green vertex $v \in V(H)$ we define $\alpha(v) = \{0\}$, and for each blue vertex $v \in V(H)$ we define $\alpha(v) = \{0, d_H(v)\}$. The MSO formula $\varphi(X)$ expresses the property "X is a set of green vertices containing one from any two adjacent green vertices." Hence, from Theorem 1 we get the following result.

Theorem 3. MINIMUM MAXIMUM OUTDEGREE *is solvable in polynomial time for graphs of bounded treewidth.*

5   Fixed-Parameter Intractability

The polynomial-time algorithm developed in the proof of Theorem 1 runs in time $O(n^{2k+3})$ for graphs of order n and constant treewidth k. Thus, the order of the polynomial depends on k. The question rises whether there is a better algorithm with a running time of, say, $O(n^c)$ where c is a constant independent of k (constant factors suppressed by the O-notation can be exponential in k). We give a negative answer subject to the complexity theoretic assumption $W[1] \neq FPT$ from the area of Parameterized Complexity. Let us first review some basic concepts of Parameterized Complexity; for more information see [9,10,17]. An instance of a parameterized problem is a pair (x, k), where x is the *main part* and k (usually a non-negative integer) is the *parameter*. A parameterized problem is *fixed-parameter tractable* if it can be solved in time $O(f(k)|x|^c)$ where f is a computable function and c is a constant independent of k. FPT denotes the class of all fixed-parameter tractable decision problems. Parameterized Complexity offers a completeness theory similar to the theory of NP-completeness for non-parameterized problems. A parameterized problem P *fpt-reduces* to a parameterized problem Q if we can transform an instance (x, k) of P into an instance $(x', g(k))$ of Q in time $O(f(k)|x|^c)$ (f, g are arbitrary computable functions, c is a constant) such that (x, k) is a yes-instance of P if and only if $(x', g(k))$ is a yes-instance of Q. A parameterized complexity class is the class of parameterized decision problems fpt-reducible to a certain parameterized decision problem Q (the notions of Q-hardness and Q-completeness are defined in the obvious way). Of particular interest is the class $W[1]$ that is considered as the parameterized analog to NP. It is believed that $FPT \neq W[1]$, and there is strong theoretical evidence that supports this belief, for example, $FPT = W[1]$ would imply that the Exponential Time Hypothesis fails (cf. [10]). However, as recently shown, GENERAL FACTOR is $W[1]$-hard when parameterized by the treewidth of the instance graph [19] (this even holds if the instance graph is bipartite and all vertices of one side are assigned the set $\{1\}$). Since GENERAL FACTOR can be expressed as an MSO-LCC problem, we can answer the above question negatively, subject to the assumption $FPT \neq W[1]$.

References

1. Arnborg, S., Lagergren, J., Seese, D.: Easy problems for tree-decomposable graphs. J. Algorithms 12(2), 308–340 (1991)
2. Asahiro, Y., Miyano, E., Ono, H.: Graph classes and the complexity of the graph orientation minimizing the maximum weighted outdegree. In: Proceedings of CATS 2008, Computing: The Australasian Theory Symposium. Conferences in Research and Practice in Information Technology, vol. 77, pp. 97–106. Australian Computer Society (2008)
3. Bodlaender, H.L.: A linear-time algorithm for finding tree-decompositions of small treewidth. SIAM J. Comput. 25(6), 1305–1317 (1996)
4. Bodlaender, H.L.: A partial k-arboretum of graphs with bounded treewidth. Theoret. Comput. Sci. 209(1-2), 1–45 (1998)

5. Bodlaender, H.L., Fomin, F.V.: Equitable colorings of bounded treewidth graphs. In: Fiala, J., Koubek, V., Kratochvíl, J. (eds.) MFCS 2004. LNCS, vol. 3153, pp. 180–190. Springer, Heidelberg (2004); Theoret. Comput. Sci. 349(1), pp. 22–30, (Full version published, 2005)

6. Cornuéjols, G.: General factors of graphs. J. Combin. Theory Ser. B 45(2), 185–198 (1988)

7. Courcelle, B.: Recognizability and second-order definability for sets of finite graphs. Technical Report I-8634, Université de Bordeaux (1987)

8. Courcelle, B.: Graph rewriting: an algebraic and logic approach. In: Handbook of theoretical computer science, vol. B, pp. 193–242. Elsevier Science Publishers, North-Holland, Amsterdam (1990)

9. Downey, R.G., Fellows, M.R.: Parameterized Complexity. Springer, Heidelberg (1999)

10. Flum, J., Grohe, M.: Parameterized Complexity Theory. Springer, Heidelberg (2006)

11. Kloks, T.: Treewidth: Computations and Approximations. Springer, Heidelberg (1994)

12. Kostochka, A.V., Nakprasit, K., Pemmaraju, S.V.: On equitable coloring of d-degenerate graphs. SIAM J. Discrete Math. 19(1), 83–95 (2005)

13. Lih, K.-W.: The equitable coloring of graphs. In: Handbook of combinatorial optimization, vol. 3, pp. 543–566. Kluwer Academic Publishers, Dordrecht (1998)

14. Lovász, L.: The factorization of graphs. In: Combinatorial Structures and their Applications (Proc. Calgary Internat. Conf., Calgary, Alta., 1969), pp. 243–246. Gordon and Breach, New York (1970)

15. Lovász, L.: The factorization of graphs. II. Acta Math. Acad. Sci. Hungar. 23, 223–246 (1972)

16. Meyer, W.: Equitable coloring. Amer. Math. Monthly 80, 920–922 (1973)

17. Niedermeier, R.: Invitation to Fixed-Parameter Algorithms. Oxford University Press, Oxford (2006)

18. Rose, D.J.: On simple characterizations of k-trees. Discrete Math. 7, 317–322 (1974)

19. Samer, M., Szeider, S.: Tractable cases of the extended global cardinality constraint. In: Proceedings of CATS 2008, Computing: The Australasian Theory Symposium. Conferences in Research and Practice in Information Technology, vol. 77, pp. 67–74. Australian Computer Society (2008)

20. Thatcher, J.W., Wright, J.B.: Generalized finite automata theory with an application to a decision problem of second-order logic. Math. Systems Theory 2, 57–81 (1968)

Short Proofs of Strong Normalization

Aleksander Wojdyga

[1] Faculty of Mathematics and Computer Science
Nicolaus Copernicus University
Toruń
awojdyga@mat.uni.torun.pl
[2] Institute of Computer Science,
Faculty of Electrical Engineering and Computer Science,
Lublin University of Technology,
Lublin

Abstract. This paper presents simple, syntactic strong normalization proofs for the simply-typed λ-calculus and the polymorphic λ-calculus (system $\mathbf{F}$) with the full set of logical connectives, and all the permutative reductions. The normalization proofs use translations of terms and types of $\lambda_{\rightarrow,\wedge,\vee,\bot}$ to terms and types of $\lambda_{\rightarrow}$ and from $\mathbf{F}_{\forall,\exists,\rightarrow,\wedge,\vee,\bot}$ to $\mathbf{F}_{\forall,\rightarrow}$.

Keywords: strong normalization, CPS-translation, permutative reductions, lambda calculus, system F.

1 Introduction

In this paper we consider the simply-typed and polymorphic lambda-calculus extended by type constructors corresponding to the usual logical connectives, namely conjunction, disjunction, absurdity and implication. In the polymorphic case we include both universal and existential quantification. In addition, we assume all the permutative conversions.

Different proofs of strong normalization of several variants of these calculi occur in the literature cf. [1,5,6,8,9,11]. It is however surprising that it is quite hard to find one covering the full set of connectives, applying to all the permutative conversions (in the polymorphic case none of the cited works does so) and given by a simple and straightforward argument. We can only repeat after J.Y. Girard: *I didn't find a proof really nice, and taking little space* [4, p. 130]. For instance, many proofs, like these in [6,8,9,11] are based on the computability method, or (in the polymorphic case) candidates of reducibility. This requires re-doing each time the same argument, but in a more complex way, due to the increased complexity of the language. There are sometimes difficulties with particular connective or quantifier, see the deliberation on $\exists$ in [11].

An alternative approach is to reduce the question of SN for a larger system to a known SN result for a weaker system. There are two advantages of this method. Firstly, it is simpler and thus methodologically more adequate than just reproving the whole result. Secondly, it can give some additional information on the relative complexity of normalization (the number of β-reduction steps

E. Ochmański and J. Tyszkiewicz (Eds.): MFCS 2008, LNCS 5162, pp. 613–623, 2008.
© Springer-Verlag Berlin Heidelberg 2008

and permutative conversions) in the two systems. For instance, our reduction from $\mathbf{F}_{\forall,\exists,\to,\land,\lor,\bot}$ to $\mathbf{F}_{\forall,\to}$ is arithmetical, i.e., provable in $\mathbf{PA}$. Therefore, the difference betweeen the complexity of normalization in $\mathbf{F}_{\forall,\exists,\to,\land,\lor,\bot}$ and in $\mathbf{F}_{\forall,\to}$ is bounded by a function provably total in $\mathbf{PA}$. Given that the two compared measures are not provably total in $\mathbf{PA2}$, this is a tight connection.

The first proof reduces the calculus $\lambda_{\to,\land,\lor,\bot}$ with connectives $\land,\lor,\to,\bot$ to the calculus $\lambda_{\to}$. Here we use the strong normalization of $\lambda_{\to}$ with beta-eta-reductions. The proof is based on composing the ordinary reduction of classical connectives to implication and absurdity with Ong's translation of the $\lambda\mu$-calculus to the ordinary $\lambda\eta$-calculus, as described e.g. in [7, Chapter 6]. To our knowledge this is the most direct way of showing SN for system $\lambda_{\to,\land,\lor,\bot}$.

The above method does not however extend to the polymorphic case. Indeed, the translation is strictly type-driven and requires an *a priori* knowledge of all types a given expression can obtain by polymorphic instantiation. Also the well known definition of logical connectives in system $\mathbf{F}$:

$$\land\tau \equiv \forall t.(\sigma \to \tau \to t) \to t \qquad \sigma \lor \tau \equiv \forall t.(\sigma \to t) \to (\tau \to t) \to t$$

is not adequate. The translation preserves beta-conversion, but not the permutations. The solution, first used by de Groote ([2], [3]), for first-order logic, is a CPS-translation. Our proof is similar to de Groote's but the version of CPS we use is based on Nakazawa and Tatsuta [10].

1.1 Definitions

We consider the calculi $\lambda_{\to,\land,\lor,\bot}$ and $\mathbf{F}_{\forall,\exists,\to,\land,\lor,\bot}$ in Church's style. The type τ of a term M is written informally in upper index as M^τ. However, if it is clear from the context, types will be omitted for the sake of brevity and readability–most right-hand sides of equations and reduction rules are written without types.

The Full Simply-Typed λ-Calculus. Types of $\lambda_{\to,\land,\lor,\bot}$ are built from multiple type constants; lowercase Greek letters are used to denote types.

Definition 1. Types of $\lambda_{\to,\land,\lor,\bot}$

$$\sigma,\tau,\ldots ::= p,q,\ldots,\sigma \to \tau, \sigma \land \tau, \sigma \lor \tau, \bot$$

Syntax of terms of $\lambda_{\to,\land,\lor,\bot}$ can be divided in two groups: constructor terms and eliminator terms. Lowercase Latin letters denote variables, uppercase – terms.

Definition 2. Terms of $\lambda_{\to,\land,\lor,\bot}$

$M,N,\ldots ::= $ *Variables*
$x^\sigma, y^\tau, \ldots,$
Introduction
$(\lambda x^\sigma.N^\tau)^{\sigma\to\tau}, \langle M^\sigma, N^\tau\rangle^{\sigma\land\tau}, (\mathtt{in}_1 A^\sigma)^{\sigma\lor\tau}, (\mathtt{in}_2 B^\tau)^{\sigma\lor\tau}$
Elimination
$(M^{\sigma\to\tau} N^\sigma)^\tau, (P^{\sigma\land\tau}\pi_1)^\sigma, (P^{\sigma\land\tau}\pi_2)^\tau, (W^{\sigma\lor\tau}[x^\sigma.S^\delta, y^\tau.T^\delta])^\delta,$
$(A^\bot \epsilon_\tau)^\tau$

In the above, the notation $\text{in}_1 A$ and $\text{in}_2 A$ represents the left and right injection for the sum type, π_1 and π_2 are projections and $W^{\sigma \vee \tau}[x.S^\delta, y.T^\delta]$ stands for a case statement. The epsilon represents the *ex falso*.

Reductions. The beta-reductions are written as $\to_\beta$ and commutative reductions are denoted by $\rightsquigarrow$. For any reduction $\to$ transitive closure of this relation will be denoted as $\to^+$ and transitive, reflexive closure as $\twoheadrightarrow$.

Definition 3. β-reductions in $\lambda_{\to,\wedge,\vee,\perp}$

$$(\lambda x^\tau . M^\delta) A^\tau \to_\beta M[x := A]^\delta$$
$$\langle M^\sigma, N^\tau \rangle \pi_1 \to_\beta M^\sigma$$
$$\langle M^\sigma, N^\tau \rangle \pi_2 \to_\beta N^\tau$$
$$(\text{in}_1 A)^{\sigma \vee \tau}[x^\sigma . S^\delta, y^\tau . T^\delta] \to_\beta S[x^\sigma := A^\sigma]^\delta$$
$$(\text{in}_2 B)^{\sigma \vee \tau}[x^\sigma . S^\delta, y^\tau . T^\delta] \to_\beta S[y^\tau := B^\tau]^\delta$$

Definition 4. Commutative reductions in $\lambda_{\to,\wedge,\vee,\perp}$

$$(A^\perp \epsilon_{\sigma \to \tau}) N^\sigma \rightsquigarrow A^\perp \epsilon_\tau$$
$$(A^\perp \epsilon_{\sigma \wedge \tau}) \pi_1 \rightsquigarrow A^\perp \epsilon_\sigma$$
$$(A^\perp \epsilon_{\sigma \wedge \tau}) \pi_2 \rightsquigarrow A^\perp \epsilon_\tau$$
$$(A^\perp \epsilon_{\sigma \vee \tau})[x^\sigma . S^\delta, y^\tau . T^\delta] \rightsquigarrow A^\perp \epsilon_\delta$$
$$(A^\perp \epsilon_\perp) \epsilon_\sigma \rightsquigarrow A^\perp \epsilon_\sigma$$
$$((W^{\sigma \vee \tau}[x.S^{\alpha \to \beta}, y.T^{\alpha \to \beta}]) N^\alpha)^\beta \rightsquigarrow W^{\sigma \vee \tau}[x.(SN)^\beta, y.(TN)^\beta]$$
$$((W^{\sigma \vee \tau}[x.S^{\alpha \wedge \beta}, y.T^{\alpha \wedge \beta}]) \pi_1)^\alpha \rightsquigarrow W^{\sigma \vee \tau}[x.(S\pi_1)^\alpha, y.(T\pi_1)^\alpha]$$
$$((W^{\sigma \vee \tau}[x.S^{\alpha \wedge \beta}, y.T^{\alpha \wedge \beta}]) \pi_2)^\beta \rightsquigarrow W^{\sigma \vee \tau}[x.(S\pi_2)^\beta, y.(T\pi_2)^\beta]$$
$$(W^{\sigma \vee \tau}[x.S^{\alpha \vee \beta}, y.T^{\alpha \vee \beta}])[a^\alpha . A^\delta, b^\beta . B^\delta] \rightsquigarrow$$
$$W^{\sigma \vee \tau}[x.S[a.A^\delta, b.B^\delta], y.T[a.A^\delta, b.B^\delta]]$$
$$(W^{\sigma \vee \tau}[x.S^\perp, y.T^\perp]) \epsilon_\alpha \rightsquigarrow W^{\sigma \vee \tau}[x.S\epsilon_\alpha, y.T\epsilon_\alpha]$$

Note that the above commutative reductions follow these two patterns:

$$(W[x.S, y.T]) E \rightsquigarrow W[x.SE, y.TE], \tag{1}$$
$$(A\epsilon) E \rightsquigarrow A\epsilon, \tag{2}$$

where E is an arbitrary eliminator. That is, E is either a term N or a projection, or epsilon, or it has the form $[x.S, y.T]$.

The Full Polymorphic λ-Calculus. The full polymorphic λ-calculus extends the system of the previous section by existential and universal polymorphism. Terms of the calculus are all the terms of simply-typed λ calculus plus universal and existential introduction and elimination.

Definition 5. Types of $\mathbf{F}_{\forall,\exists,\to,\wedge,\vee,\perp}$

$$\sigma, \tau, \ldots ::= p, q, \ldots, \sigma \to \tau, \sigma \wedge \tau, \sigma \vee \tau, \forall p\,\tau, \exists p\,\tau, \perp$$

In the definition below, notation $[M^{\tau[p:=\sigma]}, \sigma]$ stands for introduction of type $\exists p\,\tau$ and $[x^\tau.N^\delta]$ is a eliminator for that type.

Definition 6. Terms of $\mathbf{F}_{\forall,\exists,\to,\wedge,\vee,\perp}$

$M, N, \ldots ::= Variables$

$\quad\quad x^\sigma, y^\tau, \ldots$

$\quad\quad Introductions$

$\quad\quad (\lambda x^\sigma.N^\tau)^{\sigma\to\tau}, \langle M^\sigma, N^\tau\rangle^{\sigma\wedge\tau}, (\mathtt{in}_1 A^\sigma)^{\sigma\vee\tau}, (\mathtt{in}_2 B^\tau)^{\sigma\vee\tau},$

$\quad\quad [M^{\tau[p:=\sigma]}, \sigma]^{\exists p\,\tau}, (\Lambda p M^\tau)^{\forall p\,\tau}$

$\quad\quad Eliminations$

$\quad\quad (M^{\sigma\to\tau} N^\sigma)^\tau, (P^{\sigma\wedge\tau}\pi_1)^\sigma, (P^{\sigma\wedge\tau}\pi_2)^\tau, (W^{\sigma\vee\tau}[x^\sigma.S^\delta, y^\tau.T^\delta])^\delta,$

$\quad\quad (M^{\exists p\,\tau}[x^\tau.N^\delta])^\delta, (M^{\forall p\,\tau}\sigma)^{\tau[p:=\sigma]}$

$\quad\quad (A^\perp \epsilon_\tau)^\tau$

The β-reductions and commutative reductions in this system are as follows.

Definition 7. The β-reductions in $\mathbf{F}_{\forall,\exists,\to,\wedge,\vee,\perp}$ are as in Definition 3 and in addition

$$[M^{\tau[p:=\sigma]}, \sigma][x^\tau.N^\delta] \to_\beta (N[p:=\sigma][x:=M])^\delta \tag{3}$$

$$(\Lambda p M^\tau)\sigma \to_\beta M[p:=\sigma] \tag{4}$$

The total number of commutative reductions reaches 21. The patterns mentioned in Rules (1) and (2) are extended by the additional one:

$$(M[x.P])E \rightsquigarrow M[x.PE], \tag{5}$$

where E can also be of the form of existential ($[y.R]$) or universal (σ) eliminator.

Definition 8. Additional commutative reductions in $\mathbf{F}_{\forall,\exists,\to,\wedge,\vee,\perp}$.
Let δ abbreviate $\forall p\,\alpha$ in rules below.

$$(W^{\sigma\vee\tau}[x^\sigma.S^\delta, y^\tau.T^\delta])\gamma \rightsquigarrow W[x.(S\gamma)^{\alpha[p:=\gamma]}, y.(T\gamma)^{\alpha[p:=\gamma]}] \tag{6}$$

$$(A^\perp \epsilon_\delta)\gamma \rightsquigarrow A^\perp \epsilon_{\alpha[p:=\gamma]} \tag{7}$$

$$(M^{\exists p\,\tau}[x^\tau.P^\delta])\gamma \rightsquigarrow M^{\exists p\,\tau}[x.(P\gamma)^{\alpha[p:=\gamma]}] \tag{8}$$

In the following rules, δ abbreviates $\exists p\,\alpha$.

$$(W^{\sigma\vee\tau}[x^\sigma.S^\delta, y^\tau.T^\delta])[a^\alpha.N^\xi] \rightsquigarrow W^{\sigma\vee\tau}[x.(S[a.N])^\xi, y.(T[a.N])^\xi] \tag{9}$$

$$(A^\perp \epsilon_\delta)[a^\alpha.N^\xi] \rightsquigarrow A^\perp \epsilon_\xi \tag{10}$$

$$(M^{\exists p\,\tau}[y^\tau.P^\delta])[a^\alpha.N^\xi] \rightsquigarrow M^{\exists p\,\tau}[y.(P[a.N])^\xi] \tag{11}$$

$$A^\delta[x^\alpha.N^{\sigma\to\tau}]P^\sigma \rightsquigarrow A[x.(NP)^\tau] \tag{12}$$

$$A^\delta[x^\alpha.N^{\sigma\wedge\tau}]\pi_1 \rightsquigarrow A[x.(N\pi_1)^\sigma] \tag{13}$$

$$A^\delta[x^\alpha.N^{\sigma\wedge\tau}]\pi_2 \rightsquigarrow A[x.(N\pi_2)^\tau] \tag{14}$$

$$A^\delta[x^\alpha.N^{\sigma\vee\tau}][y^\sigma.S^\delta, z^\tau.T^\delta] \rightsquigarrow A[x.(N[y.S, z.T])^\delta] \tag{15}$$

$$A^\delta[x^\alpha.N^\perp]\epsilon_\sigma \rightsquigarrow A[x.(N\epsilon_\sigma)^\sigma] \tag{16}$$

2   The Translation for Simple Types

A type τ of the $\lambda_{\to,\wedge,\vee,\perp}$ calculus is translated to a type $|\tau|$ of $\lambda_\to$ calculus, a term M is translated to a term $|M|$.

Definition 9. Translation of types.

$$|\alpha| = \perp, \text{ for all type constants } \alpha = \perp, p, q, \dots$$
$$|\sigma \to \tau| = |\sigma| \to |\tau|$$
$$|\sigma \wedge \tau| = (|\sigma| \to |\tau| \to \perp) \to \perp$$
$$|\sigma \vee \tau| = (|\sigma| \to \perp) \to (|\tau| \to \perp) \to \perp$$

Example 1. Let $\tau = p \to q \to (p \wedge q)$. Then
$$|\tau| = \perp \to \perp \to (\perp \to \perp \to \perp) \to \perp.$$

Definition 10. (Translation of terms) It is assumed below that types $|\sigma|, |\tau|$ and $|\delta|$ are as follows: $|\sigma| = \sigma_1 \to \cdots \to \sigma_n \to \perp$, $|\tau| = \tau_1 \to \cdots \to \tau_m \to \perp$ and $|\delta| = \delta_1 \to \cdots \to \delta_k \to \perp$.

$$|x^\sigma| = x^{|\sigma|} \tag{17}$$

$$|\lambda x^\tau.M^\sigma| = \lambda x^{|\tau|}.|M|^{|\sigma|} \tag{18}$$

$$|\langle M, N\rangle^{\sigma\wedge\tau}| = \lambda z^{|\sigma|\to|\tau|\to\perp}.z|M|^{|\sigma|}|N|^{|\tau|} \tag{19}$$

$$|(\mathbf{in}_1 A)^{\sigma\vee\tau}| = \lambda x^{|\sigma|\to\perp}.\lambda y^{|\tau|\to\perp}.x|A|^{|\sigma|} \tag{20}$$

$$|(\mathbf{in}_2 B)^{\sigma\vee\tau}| = \lambda x^{|\sigma|\to\perp}.\lambda y^{|\tau|\to\perp}.y|B|^{|\tau|} \tag{21}$$

$$|(M^{\sigma\to\tau} N^\sigma)| = (|M|^{|\sigma|\to|\tau|}|N|^{|\sigma|}) \tag{22}$$

$$|(P^{\sigma\wedge\tau})\pi_1| = \lambda x_1^{\sigma_1}\dots\lambda x_n^{\sigma_n}.|P|^{|\sigma\wedge\tau|}$$
$$(\lambda x^{|\sigma|}.\lambda y^{|\tau|}.(xx_1\dots x_n)^\perp) \tag{23}$$

$$|(P^{\sigma\wedge\tau})\pi_2| = \lambda x_1^{\tau_1}\dots\lambda x_m^{\tau_m}.|P|^{|\sigma\wedge\tau|}$$
$$(\lambda x^{|\sigma|}.\lambda y^{|\tau|}.(yx_1\dots x_m)^\perp) \tag{24}$$

$$\left|A^{\sigma\vee\tau}[x.S^\delta, y.T^\delta]\right| = \lambda x_1^{\delta_1}\dots\lambda x_k^{\delta_k}.|A|^{(|\sigma|\to\perp)\to(|\tau|\to\perp)\to\perp}$$
$$(\lambda x^{|\sigma|}.|S|^{|\delta|}x_1\dots x_k)(\lambda y^{|\tau|}.|T|^{|\delta|}x_1\dots x_k) \tag{25}$$

$$\left|M^\perp\epsilon_\sigma\right| = \lambda x_1^{\sigma_1}\dots\lambda x_n^{\sigma_n}.|M|^\perp \tag{26}$$

Lemma 1 (*Soundness*). *If a term M has type δ, then $|M|$ has type $|\delta|$.*

Proof. Obvious.   □

Lemma 2. *If $R \to R'$, then $|R| \to^{+}_{\beta\eta} |R'|$.*

Proof. The proof proceeds by cases on the definition of $\to_\beta$ and $\leadsto$. Two example reductions will be elaborated here.

(23) Let $R = \langle M^\sigma, N^\tau \rangle \pi_1$ and $R \to_\beta R' = M$, where $|\sigma| = \sigma_1 \to \cdots \to \sigma_n \to \bot$.

$$
\begin{aligned}
|R| &= |\langle M, N \rangle^{\sigma \wedge \tau} \pi_1| \\
&= \lambda a_1^{\sigma_1} \ldots \lambda a_n^{\sigma_n}.|\langle M, N \rangle|^{|\sigma \wedge \tau|}(\lambda x^{|\sigma|}.\lambda y^{|\tau|}.(x a_1 \ldots a_n)^\bot) \\
&= \lambda \boldsymbol{a}.(\lambda z^{|\sigma| \to |\tau| \to \bot}.z|M||N|)(\lambda x^{|\sigma|} \lambda y^{|\tau|}.(x\boldsymbol{a})^\bot) \\
&\to_\beta \lambda \boldsymbol{a}.((\lambda x^{|\sigma|} \lambda y^{|\tau|}.(x\boldsymbol{a})^\bot)|M||N|) \\
&\to_\beta \lambda \boldsymbol{a}.(\lambda y^{|\tau|}.|M|\boldsymbol{a})|N| \to_\beta \lambda \boldsymbol{a}.|M|\boldsymbol{a} \to^{+}_\eta |M| \\
&= |R'|
\end{aligned}
$$

(25) Let $R = (W^{\sigma \vee \tau}[x.S^{\alpha \to \beta}, y.T^{\alpha \to \beta}])N^\alpha$ and let $R' = W^{\sigma \vee \tau}[x.(SN)^\beta, y.(TN)^\beta]$. Then $R \leadsto R'$, according to (25). Assuming $|\beta| = \beta_1 \to \cdots \to \beta_n \to \bot$, we have

$$
\begin{aligned}
|R| &= (\lambda a^{|\alpha|} b_1^{\beta_1} \ldots b_n^{\beta_n}.|W|(\lambda x^{|\sigma|}.|S|^{|\alpha| \to |\beta|} a \boldsymbol{b})(\lambda y^{|\tau|}.|T|^{|\alpha| \to |\beta|} a \boldsymbol{b}))|N|^{|\alpha|} \\
&\to_\beta \lambda b_1 \ldots b_n.|W|(\lambda x^{|\sigma|}.|S||N|\boldsymbol{b})(\lambda y^{|\tau|}.|T||N|\boldsymbol{b}) \\
&= |R'|
\end{aligned}
$$

Other cases are similar.   □

Theorem 1. *The calculus $\lambda_{\to,\wedge,\vee,\bot}$ is strongly normalizing.*

Proof. Suppose, by contradiction, that M^τ admits an infinite β-reduction

$$ M^\tau = M_0^\tau \to_\beta M_1^\tau \to_\beta M_2^\tau \to_\beta \cdots $$

By Theorem 2 we have an infinite reduction in $\lambda_\to$

$$ |M^\tau| = |M_0| \to^{+}_{\beta\eta} |M_1| \to^{+}_{\beta\eta} |M_2| \to^{+}_{\beta\eta} \cdots $$

This contradicts the SN property of $\lambda_\to$ □

3   Translation for Polymorphic Types

As we mentioned in the introduction, the translations in Section 3 are not adequate for the polymorphic case and therefore we apply a call-by-name CPS translation. In general, a type τ is translated to $\underline{\tau} = (\tau^* \to \bot) \to \bot$. This translation, unlike the one for simple types, does not unify type constants. The helper translation * is given below.

Definition 11. Helper translation $*$.

$$\alpha^* = \alpha, \text{ for all type constants } \alpha = \bot, p, q, \ldots$$
$$(\alpha \to \beta)^* = \underline{\alpha} \to \underline{\beta}$$
$$(\alpha \wedge \beta)^* = (\underline{\alpha} \to \underline{\beta} \to \bot) \to \bot$$
$$(\alpha \vee \beta)^* = (\underline{\alpha} \to \bot) \to (\underline{\beta} \to \bot) \to \bot$$
$$(\forall p\,\tau)^* = \forall p\,\underline{\tau}$$
$$(\exists p\,\tau)^* = (\forall p(\underline{\tau} \to \bot)) \to \bot$$

A term M^τ is translated to the term $\underline{M} = \lambda k^{\tau^* \to \bot}.(M \diamond k)$. To achieve that, two helper translations are needed: $\diamond$ and $@$. The term K in the definition below is of type $\tau^* \to \bot$. The term $M \diamond K$ is always of type $\bot$.

Definition 12. Helper translation $\diamond$

$$x^\tau \diamond K = xK \tag{27}$$
$$\lambda x^\sigma.N^\rho \diamond K = K(\lambda x^{\underline{\sigma}}.\underline{N}) \tag{28}$$
$$\langle N_1^{\tau_1}, N_2^{\tau_2} \rangle \diamond K = K(\lambda p^{\underline{\tau_1} \to \underline{\tau_2} \to \bot}.p\underline{N_1}\,\underline{N_2}) \tag{29}$$
$$(\text{in}_1 A)^{\tau_1 \vee \tau_2} \diamond K = K(\lambda a^{\underline{\tau_1} \to \bot} b^{\underline{\tau_2} \to \bot}.a\underline{A}) \tag{30}$$
$$(\text{in}_2 B)^{\tau_1 \vee \tau_2} \diamond K = K(\lambda a^{\underline{\tau_1} \to \bot} b^{\underline{\tau_2} \to \bot}.b\underline{B}) \tag{31}$$
$$\Lambda p\, N^\rho \diamond K = K(\Lambda p.\underline{N}) \tag{32}$$
$$[N^{\rho[p:=\sigma]}, \sigma] \diamond K = K(\lambda u^{\forall p(\underline{\rho} \to \bot)}.u\,\underline{\sigma}\,\underline{N}) \tag{33}$$
$$NE \diamond K = N \diamond (E @ K) \tag{34}$$

In (34) the symbol E stands for an arbitrary eliminator. That is, E is one of the expressions $\{R^\sigma, \pi_1, \pi_2, [x^{\tau_1}.S^\delta, y^{\tau_2}.T^\delta], \sigma, [x^\rho.S^\delta], \epsilon_\alpha\}$ and the omitted type of term N is appropriate for every eliminator E.

Definition 13. Helper translation $@$

$$R @ K = \lambda m^{\underline{\sigma} \to \underline{\rho}}.m\underline{R}K$$
$$\pi_1 @ K = \lambda m^{(\underline{\tau_1} \to \underline{\tau_2} \to \bot) \to \bot}.m(\lambda a^{\underline{\tau_1}} b^{\underline{\tau_2}}.aK)$$
$$\pi_2 @ K = \lambda m^{(\underline{\tau_1} \to \underline{\tau_2} \to \bot) \to \bot}.m(\lambda a^{\underline{\tau_1}} b^{\underline{\tau_2}}.bK)$$
$$[x^{\tau_1}.S^\delta, y^{\tau_2}.T^\delta] @ K = \lambda m^{(\underline{\tau_1} \to \bot) \to (\underline{\tau_2} \to \bot) \to \bot}.$$
$$m(\lambda x^{\underline{\tau_1}}.(S \diamond K))(\lambda y^{\underline{\tau_2}}.(T \diamond K))$$
$$\sigma @ K = \lambda m^{\forall p\underline{\rho}}.m\underline{\sigma}K$$
$$[x^\rho.S^\delta] @ K = \lambda m^{(\forall p(\underline{\rho} \to \bot)) \to \bot}.m(\Lambda p\lambda x^{\underline{\rho}}.(S \diamond K))$$
$$\epsilon_\alpha @ K = \lambda m^\bot.m$$

Lemma 3. [Soundness] *If a term M has type δ, then $\underline{M}$ has type $\underline{\delta}$.*

Proof. Easy. □

Lemma 4. [Properties of substitution] *For a term R and any term K and for any types τ and ρ the following holds:*

$$\underline{R[x^\delta := N^\delta]} =_\alpha \underline{R}[x := \underline{N}]; \tag{35}$$

$$(R \diamond K)[x^\delta := N^\delta] =_\alpha R[x := N] \diamond K[x := N]; \tag{36}$$

$$(R @ K)[x^\delta := N^\delta] =_\alpha R[x := N] @ K[x := N] \text{ if } R \text{ is an eliminator}; \tag{37}$$

$$\underline{\tau}[p := \rho] =_\alpha \underline{\tau[p := \rho]}; \tag{38}$$

$$(R \diamond K)[p := \rho] =_\alpha R[p := \rho] \diamond K[p := \rho]; \tag{39}$$

$$(R @ K)[p := \rho] =_\alpha R[p := \rho] @ K[p := \rho] \text{ if } R \text{ is an eliminator}. \tag{40}$$

Proof. This lemma is proved by simultaneous induction on the definition of substitution. $\qquad\square$

Lemma 5. *If $R \to_\beta R'$, then $\underline{R} \to_\beta^+ \underline{R'}$.*

Proof. Using induction on the definition of $\to_\beta$ we have 7 cases. For example, consider (3), where $R = [M^{\tau[p:=\sigma]}, \sigma][x^\tau.N^\delta]$ and $R' = (N[p := \sigma][x := M])^\delta$.

$$
\begin{aligned}
(3)\quad \underline{R} &= \lambda k.(\lambda m^{(\exists p\tau)^*}.m(\Lambda p \lambda x^\tau.(N \diamond k)))(\lambda u^{\forall p(\tau \to \bot)}.u\underline{\sigma}\,\underline{M}) \\
&\to_\beta \lambda k.(\lambda u.u\underline{\sigma}\,\underline{M})(\Lambda p \lambda x.(N \diamond k)) \\
&\to_\beta \lambda k.(\Lambda p \lambda x.(N \diamond k))\underline{\sigma}\,\underline{M} \\
&\to_\beta \lambda k.(\lambda x.(N \diamond k))[p := \underline{\sigma}]\,\underline{M} \\
&\to_\beta \lambda k.(\lambda x.(N[p := \sigma] \diamond k))\underline{M} \quad \text{(from (39))} \\
&\to_\beta \lambda k.(N[p := \sigma] \diamond k)[x := \underline{M}] \\
&=_\alpha \lambda k.(N[p := \sigma][x := M] \diamond k) \quad \text{(from (36))} \\
&= \underline{R'} \qquad\square
\end{aligned}
$$

Lemma 6. *If $R \rightsquigarrow R'$, then $\underline{R} =_\alpha \underline{R'}$.*

Proof. The complete proof consists of 21 cases. Here, two interesting commutations will be elaborated. The other cases are similar and left to the reader.

From (11) we get

$$
\begin{aligned}
\underline{\text{LHS}} &= \lambda k.(M[y.P] \diamond ([x.N] @ k)) = \lambda k.(M \diamond ([y.P] @ ([x.N] @ k))) \\
&= \lambda k.(M \diamond (\lambda m.m(\Lambda p \lambda y.(P \diamond [x.N] @ k)))) \\
\underline{\text{RHS}} &= \lambda k.(M \diamond ([y.P[x.N]] @ k)) = \lambda k.(M \diamond (\lambda m.m(\Lambda p \lambda y.(P[x.N] \diamond k)))) \\
&= \lambda k.(M \diamond (\lambda m.m(\Lambda p \lambda y.(P \diamond [x.N] @ k))))
\end{aligned}
$$

From (16) we get

$$
\begin{aligned}
\underline{\text{LHS}} &= \lambda k.(A[x.N] \diamond (\epsilon_\sigma @ k)) = \lambda k.(A[x.N] \diamond (\epsilon_\sigma @ k)) \\
&= \lambda k.(A \diamond ([x.N] @ (\epsilon_\sigma @ k))) \\
&= \lambda k.(A \diamond (\lambda m.m(\Lambda p \lambda x.(N \diamond (\epsilon_\sigma @ k))))) \\
\underline{\text{RHS}} &= \lambda k.(A \diamond ([x.N\epsilon_\sigma] @ k)) = \lambda k.(A \diamond (\lambda m.m(\Lambda p \lambda x.(N\epsilon_\sigma \diamond k)))) \\
&= \lambda k.(A \diamond (\lambda m.m(\Lambda p \lambda x.(N \diamond (\epsilon_\sigma @ k))))) \qquad\square
\end{aligned}
$$

Lemma 7. *Every sequence of commutative reductions in $\mathbf{F}_{\forall,\exists,\to,\land,\lor,\bot}$ must terminate.*

Proof. To prove this lemma we define such a measure $\chi(M) > 0$, that for any commutation $M \rightsquigarrow M'$, we have $\chi(M) > \chi(M')$. Please note, that we have 3 patterns of commutative reductions in Rules (1), (2) and (5). We use those patters to define appropriate conditions for measure χ:

$$\chi\left((W[x.S, y.T])E\right) > \chi\left(W[x.SE, y.TE]\right) \tag{41}$$
$$\chi\left((A\epsilon)E\right) > \chi\left(A\epsilon\right) \tag{42}$$
$$\chi\left((N[x.P])E\right) > \chi\left(N[x.PE]\right) \tag{43}$$
$$\chi(M) \geq 1$$

Now we give the definition of the function $\chi(M)$; it is similar to de Groote's norm $|\cdot|$ from [2] but simpler:

$\chi(x) = 1$

$\chi(\lambda x.N) = \chi(\text{in}_1 N) = \chi(\text{in}_2 N) = \chi(N), \quad \chi(\langle M_1, M_2 \rangle) = \chi(M_1) + \chi(M_2)$

$\chi(FA) = \chi(F)^2\chi(A), \quad \chi(P\pi_1) = \chi(P\pi_2) = \chi(P)^2, \quad \chi(N\sigma) = \chi(N)^2$

$\chi(W[x.S, y.T]) = \chi(W)^2(\chi(S) + \chi(T)) + 1 \quad \chi(N[x.P]) = \chi(N)^2\chi(P) + 1$

$\chi(A\epsilon) = \chi(A)^2 + 1$

There are 21 easy cases, one for each permutation from Definitions 4 and 8. We will show here one example case for each pattern mentioned above.

(41) Let $l = \chi((W[x.S, y.T])[a.A, b.B])$ and $r = \chi(W[x.S[a.A, b.B], y.T[a.A, b.B]])$.

$l = \chi(W[x.S, y.T])^2(\chi(A) + \chi(B)) + 1$

$= \left(\chi(W)^2(\chi(S) + \chi(T)) + 1\right)^2 (\chi(A) + \chi(B)) + 1$

$> \left(\chi(W)^2(\chi(S) + \chi(T))\right)^2 (\chi(A) + \chi(B)) + 1$

$= \chi(W)^4 \left((\chi(S)^2 + \chi(T)^2)(\chi(A) + \chi(B)) + 2(\chi(S)\chi(T))(\chi(A) + \chi(B))\right) + 1$

$> \chi(W)^4((\chi(S)^2 + \chi(T)^2)(\chi(A) + \chi(B)) + 2) + 1$

$r = \chi(W)^2(\chi(S[a.A, b.B]) + \chi(T[a.A, b.B])) + 1$

$= \chi(W)^2(\chi(S)^2(\chi(A) + \chi(B)) + 1 + \chi(T)^2(\chi(A) + \chi(B)) + 1) + 1$

$= \chi(W)^2((\chi(S)^2 + \chi(T)^2)(\chi(A) + \chi(B)) + 2) + 1$

$l > r$

(42) Let $l = \chi((A\epsilon_\bot)\epsilon_\sigma)$ and $r = \chi(A\epsilon_\sigma)$.

$$l = \chi(A\epsilon_\bot)^2 + 1 = (\chi(A)^2 + 1)^2 + 1 = \chi(A)^4 + 2\chi(A)^2 + 2$$
$$r = \chi(A)^2 + 1$$
$$l > r$$

(43) Let $l = \chi((N[x.P])[a.A, b.B])$ and $r = \chi(N[x.P[a.A, b.B]])$.

$$\begin{aligned}
l &= \chi(N[x.P])^2(\chi(A) + \chi(B)) + 1 \\
&= \left(\chi(N)^2\chi(P) + 1\right)^2 (\chi(A) + \chi(B)) + 1 \\
&= (\chi(N)^4\chi(P)^2 + 2\chi(N)^2\chi(P) + 1)(\chi(A) + \chi(B)) + 1 \\
&= \chi(N)^4\chi(P)^2(\chi(A) + \chi(B)) + \chi(N)^2(2\chi(P)(\chi(A) + \chi(B))) \\
&\quad + \chi(A) + \chi(B) + 1 \\
r &= \chi(N)^2\chi(P[a.A, b.B]) + 1 \\
&= \chi(N)^2(\chi(P)^2(\chi(A) + \chi(B)) + 1) + 1 \\
&= \chi(N)^2\chi(P)^2(\chi(A) + \chi(B)) + \chi(N)^2 + 1 \\
l &> r \qquad\qquad\qquad\qquad\qquad\qquad\qquad\qquad\qquad\qquad\qquad \square
\end{aligned}$$

Theorem 2. *The calculus* $\mathbf{F}_{\forall,\exists,\rightarrow,\wedge,\vee,\perp}$ *is strongly normalizing.*

Proof. Suppose that

$$M^\tau = M_0^\tau \rightarrow M_1^\tau \rightarrow M_2^\tau \rightarrow \cdots$$

If there is infinitely many β-reductions in the sequence above then we have an infinite reduction in $\mathbf{F}_{\forall,\rightarrow}$. If almost all reduction steps are of type $\rightsquigarrow$ then we use Lemma 7. In both cases we reach contradiction. $\qquad\qquad\square$

4   Summary

We have presented short proofs of strong normalization for simply-typed and polymorphic λ-calculus with all connectives. Syntax-driven translations used in those proofs allow to reduce the SN property problem to calculi with smaller number of connectives.

The CPS-translation used here for polymorphic lambda calculus may be helpful dealing with higher level λ-calculus such as $\mathbf{F}_\omega$. This is our next research problem.

References

1. David, R., Nour, K.: A Short Proof of the Strong Normalization of Classical Natural Deduction with Disjunction. Journal of Symbolic Logic 68(4), 1277–1288 (2003)
2. de Groote, Ph.: On the Strong Normalisation of Natural Deduction with Permutation-Conversions. In: Narendran, P., Rusinowitch, M. (eds.) RTA 1999. LNCS, vol. 1631, pp. 45–59. Springer, Heidelberg (1999)
3. de Groote, Ph.: On the Strong Normalisation of Intuitionistic Natural Deduction with Permutation-Conversions. Information and Computation 178(2), 441–464 (2002)
4. Girard, J.-Y.: The Blind Spot. Lectures on Logic. Rome, Autumn (2004), http://iml.univ-mrs.fr/~girard/coursang/coursang.html

5. Joachimski, F., Matthes, R.: Short Proofs of Normalization for the simply-typed λ-calculus, Permutative Conversions and Gödel's T. Archive for Mathematical Logic 42(1), 59–87 (2003)
6. Schwichtenberg, H.: Minimal Logic for Computable Functionals. In: Logic Colloquium 2005, A.K. Peters (to appear, 2005)
7. Sørensen, M.H., Urzyczyn, P.: Lectures on the Curry-Howard Isomorphism. Studies in Logic and the Foundations of Mathematics, vol. 149. Elsevier, Amsterdam (2006)
8. Tatsuta, M.: Second-order Permutative Conversions with Prawitz's Strong validity. Progress in Informatics 2, 41–56 (2005)
9. Tatsuta, M.: Simple Saturated Sets for Disjunction and Second-order existential Quantification. In: Della Rocca, S.R. (ed.) TLCA 2007. LNCS, vol. 4583, pp. 366–380. Springer, Heidelberg (2007)
10. Tatsuta, M., Nakazawa, K.: Strong Normalization of Classical Natural Deduction with Disjunctions. Annals of Pure and Applied Logic 153(1), 21–37 (2008)
11. Tatsuta, M., Mints, G.: A Simple Proof of Second-Order Strong Normalization with Permutative Conversions. Annals of Pure and Applied Logic 136(1-2), 134–155 (2005)

Author Index

Lecture Notes in Computer Science

Sublibrary 1: Theoretical Computer Science and General Issues

For information about Vols. 1– 4910
please contact your bookseller or Springer